8

ROUTLEDGE HANDBOOK OF ECOSYSTEM SERVICES

The idea that nature provides services to people is one of the most powerful concepts to have emerged over the last two decades. It is shaping our understanding of the role that biodiverse ecosystems play in the environment and their benefits for humankind. As a result, there is a growing interest in operational and methodological issues surrounding ecosystem services amongst environmental managers, and many institutions are now developing teaching programmes to equip the next generation with the skills needed to apply the concepts more effectively.

This handbook provides a comprehensive reference text on ecosystem services, integrating natural and social science (including economic). Collectively the chapters, written by the world's leading authorities, demonstrate the importance of biodiversity for people, policy and practice. They also show how the value of ecosystems to society can be expressed in monetary and non-monetary terms, so that the environment can be better taken into account in decision making. The significance of the ecosystem service paradigm is that it helps us redefine and better communicate the relationships between people and nature. It is shown how these are essential to resolving challenges such as sustainable development and poverty reduction, and the creation of a green economy in developing and developed world contexts.

Marion Potschin is a Principal Research Fellow and Director of the Centre for Environmental Management at the University of Nottingham, UK.

Roy Haines-Young is Emertus Professor, Centre for Environmental Management, School of Geography at the University of Nottingham, UK.

Robert Fish is Reader in Human Ecology in the School of Anthropology and Conservation at the University of Kent, UK.

R. Kerry Turner is a Professorial Fellow in the School of Environmental Sciences and former Professor of Environmental Economics and Management at the University of East Anglia, UK.

ROUTLEDGE HANDBOOK OF ECOSYSTEM SERVICES

Edited by
Marion Potschin,
Roy Haines-Young,
Robert Fish and
R. Kerry Turner

Routledge
Taylor & Francis Group
LONDON AND NEW YORK

earthscan
from Routledge

First published 2016
by Routledge
2 Park Square, Milton Park, Abingdon, Oxon OX14 4RN

and by Routledge
711 Third Avenue, New York, NY 10017

Routledge is an imprint of the Taylor & Francis Group, an informa business

British Library Cataloguing-in-Publication Data
A catalogue record for this book is available from the British Library

Library of Congress Cataloging-in-Publication Data
Names: Potschin, Marion, editor. | Haines-Young, R. H. (Roy H.), editor. | Fish, Robert, editor. | Turner, R. Kerry editor.
Title: Routledge handbook of ecosystem services / edited by Marion Potschin, Roy Haines-Young, Robert Fish and R. Kerry Turner.
Other titles: Handbook of ecosystem services
Description: New York, NY : Routledge, 2016. | Includes bibliographical references and index.
Identifiers: LCCN 2015023809 | ISBN 9781138025080 (hbk) | ISBN 9781315775302 (ebk)
Subjects: LCSH: Ecosystem management.
Classification: LCC QH75 .R75 2016 | DDC 333.72—dc23
LC record available at http://lccn.loc.gov/2015023809

ISBN: 978-1-138-02508-0 (hbk)
ISBN: 978-1-315-77530-2 (ebk)

Typeset in Bembo
by Apex CoVantage, LLC

Cover image: Man walking in Malham Tarn, Yorkshire Dales, England in July 2014. Photograph by James Haines-Young.

CONTENTS

ACKNOWLEDGEMENTS

Many people have contributed to this book. Our particular gratitude goes to the contributors for the stimulating material and productive collaboration. The editors would like to express their thanks to Tim Hardwick in commissioning this Handbook and for guidance in its design and technical delivery.

Marion Potschin would like to acknowledge the many valuable discussions she has had during the preparation of this book, and through the EU-funded projects of OpenNESS "Operationalising of Natural Capital and Ecosystem Services: from Concept to real-world applications" and ESMERALDA "Enhancing ecosystem Services Mapping for Policy and Decision Making". Such networking has been vital.

Roy Haines-Young would like to acknowledge his students at Nottingham, who over many years have inspired him to find ways of discussing the issues around ecosystem services in critical but constructive ways. The insights he has gained from the policy community through projects such as the UK NEA, and other work at the European scale, have also been essential for this work.

Robert Fish would like to extend his thanks to colleagues and students at the Universities of Exeter and Nottingham for the many enriching discussions he has had in the area of ecosystem services and sustainability, and to Roy Haines-Young in particular for his inspiring role as both mentor and muse. He would also like to acknowledge the University of Exeter's *Sustainable Rural Futures Research Programme*, and Professor Michael Winter for supporting his time in preparing the Handbook.

Kerry Turner would like to thank all of the contributors to the UK National Ecosystem Assessment Programme who have helped to influence his own ideas on ecosystem services and are endeavouring to mainstream the ecosystem services approach.

CONTRIBUTORS

Matthew Agarwala, Senior Post Doctoral Research Associate, Centre for Social and Economic Research on the Global Environment (CSERGE), School of Environmental Sciences, University of East Anglia, Norwich, UK

Mahbubul Alam, PhD, Director, Economics and Planning, Betty and Gordon Moore Center for Science and Oceans, Conservation International, Arlington, VA, USA

Christian Albert, PhD, Researcher, Leibniz Universität Hannover, Institute of Environmental Planning, and Helmholtz Centre for Environmental Research (UFZ), Department Environmental Politics, Germany

Tomas Badura, PhD researcher, Centre for Social and Economic Research on the Global Environment (CSERGE), School of Environmental Sciences, University of East Anglia, Norwich, UK

Patricia Balvanera, PhD, Professor, Instituto de Investigaciones en Ecosistemas y Sustentabilidad, Universidad Nacional Autónoma de México, Morelia, Michoacán, Mexico

Mario V. Balzan, PhD, Senior Lecturer, Laboratory of Terrestrial Ecology, Institute of Applied Sciences, Malta College of Arts, Science and Technology, Malta

Edmundo Barrios, PhD, Senior Scientific, World Agroforestry Centre (ICRAF), Nairobi, Kenya

David N. Barton, PhD, Norwegian Institute for Nature Research (NINA), Norway

Ian Bateman, OBE, FRSA, FSB, Professor of Environmental Economics, Centre for Social and Economic Research on the Global Environment (CSERGE), University of Exeter, UK

Nicola Beaumont, PhD, Senior Scientist, Plymouth Marine Laboratory, Prospect Place, Plymouth, Devon, UK

Janne Bengtsson, Professor, Department of Ecology, SLU (Swedish University of Agricultural Sciences), Uppsala, Sweden and STIAS (Stellenbosch Institute for Advanced Studies), Matieland, South Africa

Pam Berry, PhD, Senior Research Fellow, Environmental Change Institute (ECI), University of Oxford, UK

Amy Binner, PhD, Senior Lecturer in Environmental Economics, Centre for Social and Economic Research on the Global Environment (CSERGE), School of Environmental Sciences, University of Exeter, UK

Ryan Blanchard, PhD, Council for Scientific and Industrial Research (CSIR), Stellenbosch, South Africa

Torsten Bondo, PhD, Chief Science Officer, Sensonomic, Copenhagen, Denmark

Kate A. Brauman, PhD, Lead Scientist, Global Water Initiative, Institute on the Environment, University of Minnesota, St Paul, MN, USA

Roy Brouwer, Professor, Department of Environmental Economics, Institute for Environmental Studies, Vrije Universiteit Amsterdam, The Netherlands, and Department of Economics, University of Waterloo, Canada.

James M. Bullock, Professor, Individual Merit Scientist, NERC Centre for Ecology and Hydrology, Wallingford, Oxfordshire, UK

Benjamin Burkhard, PD, PhD, Senior Researcher, Institute for Natural Resource Conservation, Dept. of Ecosystem Management, Kiel, Germany

Claudia Carvalho-Santos, PhD, Department of Biology at Faculty of Sciences and CIBIO/InBio Associate Laboratory, University of Porto, Portugal

Kai M.A. Chan, PhD, Associate Professor and Canada Research Chair (tier 2), Institute for Resources, Environment and Sustainability (IRES), University of British Columbia (UBC), Canada

Mahé Charles, Environmental Economist, French Agency for Marine Protected Areas/Agence des aires marines protégées, Brest, France

Emmanuelle Cohen-Shacham, PhD, Department of Zoology, Tel Aviv University, Tel Aviv, Israel. Ecosystem Services Thematic Group Lead, Commission on Ecosystem Management, IUCN

Lawrence Connell, Director, Multilateral Funding, Center for Environment and Peace, Conservation International, Arlington, VA, USA

Nigel Cooper, Rev'd Canon, University Chaplain, Global Sustainability Institute, Anglia Ruskin University, Cambridge and the Diocese of Ely, UK

Steven Cork, PhD, Adjunct Professor, Crawford School of Public Policy, Australian National University and Principal, EcoInsights, Canberra, Australia

Robert Costanza, Professor, Chair in Public Policy, Crawford School of Public Policy, The Australian National University, Canberra, Australia

Wolfgang Cramer, Professor, Scientific Director, Institut Méditerranéen de Biodiversité et d'Ecologie marine et continentale (IMBE), Aix Marseille Université, CNRS, IRD, Avignon Université, France

Neville D. Crossman, PhD, Principal Research Scientist, CSIRO Land and Water Flagship, Adelaide, Australia

Gretchen C. Daily, Bing Professor of Environmental Science, Stanford University, Stanford, California; and Visiting Professor, Royal Swedish Academy of Sciences, Stockholm, Sweden

Laura Dee, Bren School of Environmental Science & Management, University of California, Santa Barbara, CA, USA

Rudolf de Groot, Associate Professor, Wageningen University, Environmental Systems Analysis Group, Wageningen, The Netherlands

Helen Ding, PhD, Senior Economist, BIO Intelligence Service (member of Deloitte Touche Tohmatsu Limited), France

Adam P. Dixon, Conservation Geographer, World Wildlife Fund – United States

Andrew J. Dougill, Dean of Faculty of Environment, Sustainability Research Institute, School of Earth and Environment, University of Leeds, UK

Guy Duke, Independent Consultant and Senior Visiting Research Associate, Oxford Environmental Change Institute, UK

Robert Dunford, PhD, Environmental Change Institute (ECI), University of Oxford, UK

Isabelle Durance, PhD, Senior Research Fellow, School of Biosciences, Cardiff University, Wales, UK

Bram Edens, Statistical Researcher, National Accounts Department, Statistics Netherlands, The Hague, The Netherlands

Benis N. Egoh, PhD, Senior Researcher, Council for Scientific and Industrial Research, Stellenbosch, South Africa. SAEE, University of KwaZulu-Natal, South Africa

Thomas Elmqvist, Professor, Stockholm Resilience Centre, Stockholm University, Stockholm, Sweden

John Finisdore, Founding Partner, Sustainable Flows, Washington, DC, USA

Robert Fish, PhD, Reader in Human Ecology in the School of Anthropology and Conservation at the University of Kent, UK

Brendan Fisher, PhD, Research Associate Professor, Rubenstein School of Environment and Natural Resources, University of Vermont, USA; World Wildlife Fund, Washington DC, USA

Janet Fisher, PhD, Chancellor's Fellow, School of GeoSciences, University of Edinburgh, UK

Giles M. Foody, Professor, School of Geography, University of Nottingham, Nottingham, UK

Susanne Frank, PhD, Senior Researcher, Center for Development Research, Dept. Ecology and Natural Resources Management, University of Bonn, Germany

Alan J. Franzluebbers, Research Ecologist, USDA – Agricultural Research Service, Plant Science Research Unit, Raleigh, North Carolina, USA

Christine Fürst, PD Dr, Senior Researcher, Institute of Meteorology and Climate Research, Atmospheric Environmental Research (IMK-IFU), Karlsruhe Institute of Technology (KIT), Germany

Gillian Galford, Dr, Research Assistant Professor, Gund Institute for Ecology Economics, University of Vermont, USA

Carolin Galler, Researcher, Leibniz Universität Hannover, Institute of Environmental Planning, Germany

Marina García-Llorente, PhD, Researcher, Madrid Institute for Rural, Agricultural and Food Research and Development-IMIDRA & Social-ecological systems Lab, Universidad Autónoma de Madrid, Spain

Ilse Geijzendorffer, PhD, Research scientist, Institut Méditerranéen de Biodiversité et d'Ecologie marine et continentale (IMBE), Aix Marseille Université, CNRS, IRD, Avignon Université, France

Davide Geneletti, PhD, Associate Professor, Planning and Design for Sustainable Places Lab, Department of Civil, Environmental and Mechanical Engineering, University of Trento, Italy

Susannah Gill, PhD, Green Infrastructure & Climate Change, The Mersey Forest, and Honorary Senior Research Fellow at the University of Manchester, UK

Erik Gómez-Baggethun, PhD, Norwegian Institute for Nature Research (NINA), Oslo, Norway

José A. González, PhD, Associate Professor, Social-ecological systems Laboratory, Department of Ecology, Universidad Autónoma de Madrid, Madrid, Spain

Geoffrey Gooch, PhD, Director DelPar Environment Consultants, Linköping, Sweden

Hedley S. Grantham, PhD, Senior Technical Director, Integrated Assessment and Planning, Betty and Gordon Moore Center for Science and Oceans, Conservation International

Eva Haas, PhD, Senior Researcher, GeoVille Information Systems GmbH, Innsbruck, Austria

Roy Haines-Young, PhD, Emeritus Professor, Centre for Environmental Management, School of Geography at the University of Nottingham, UK

Paula A. Harrison, PhD, Senior Research Fellow, Environmental Change Institute (ECI), University of Oxford, UK

Tobias Hartmann, Programme Officer, Global Nature Fund, Germany

Jennifer Hauck, PhD, Research Associate at the Department of Ecosystem Services, Helmholtz Centre for Environmental Research (UFZ), and German Centre for Integrative Biodiversity Research (iDiv) Germany

Jun He, PhD, Lecturer at College of Economics and Management, Yunnan Agricultural University, China

Lars Hein, Professor Ecosystem Services and Environmental Change, Environmental Systems Analysis Group, Wageningen University, Wageningen, The Netherlands

Julian Helfenstein, PhD Student, Institute of Agricultural Sciences, Department of Environmental Systems Science, Swiss Federal Institute of Technology (ETH) Zurich, Switzerland

Sarah Hendry, PhD, Lecturer in Law, Centre for Water Law, Policy and Science, under the auspices of UNESCO, University of Dundee, UK

Christina Hirzinger, BSc, Junior Researcher, GeoVille Information Systems GmbH, Innsbruck, Austria

Tara Hooper, PhD, Environmental Economist, Plymouth Marine Laboratory, Prospect Place, Plymouth, Devon, UK

Ying Hou, PhD, Researcher, Research Center for Eco-Environmental Sciences, Chinese Academy of Sciences, Beijing, China

Joël R.A. Houdet, PhD, Senior Research Fellow, African Centre for Technology Studies; Albert Luthuli Centre for Responsible Leadership – University of Pretoria; Synergiz, South Africa

Forest Isbell, PhD, Associate Director, Cedar Creek Ecosystem Science Reserve and Adjunct Assistant Professor, Department of Ecology, Evolution & Behaviour, University of Minnesota, USA

Louis Iverson, PhD, Research Landscape Ecologist. US Forest Service, Delaware, OH, USA

Thomas Jahn, Dr, Scientific Director, Institute for Social-Ecological Research; affiliated to Senckenberg Biodiversity and Climate Research Centre (BiK-F), Frankfurt/M, Germany

Kurt Jax, Prof. Dr., Helmholtz-Centre for Environmental Research, Dept. of Conservation Biology – UFZ, Germany and Technische Universität München, Chair for Restoration Ecology, Germany

Andrew Jordan, PhD, Professor of Environmental Science, Tyndall Centre (University of East Anglia), Norwich, UK

Daniel Juhn, Senior Director, Integrated Assessment and Planning, Betty and Gordon Moore Center for Science and Oceans, Conservation International, Arlington, VA, USA

Jasper O. Kenter, PhD, Principal Investigator in Ecological Economics & Head of Laurence Mee Centre for Society and the Sea, Scottish Association for Marine Science, University of the Highlands and Islands, Scotland

Marianne Kettunen, Principal Policy Analyst and Co-lead of Global Challenges Work Stream, IEEP; lead editor of *Social and Economic Benefits of Protected Areas – An Assessment Guide* (2013)

Hans Keune, PhD, Senior scientist, Research Institute for Nature and Forest and Belgian Biodiversity Platform, Belgium

Felix Kienast, PhD, Adjunct Professor ETH Zurich, Swiss Federal Research Institute WSL, 8903 Birmensdorf, Switzerland

Keren Klass, Israel National Ecosystem Assessment Project Manager, Hamaarag – Israel's National Nature Assessment Program, Jerusalem, Israel

Conor E. Kretsch, Executive Director, COHAB Initiative Secretariat, Ireland, and Research Fellow (Ecosystems & Health), Centre for Environmental Management, University of Nottingham, UK

Marion Kruse, PhD, Researcher, Institute for Natural Resource Conservation, Dept. of Ecosystem Management, Kiel, Germany

Dixon H. Landers, Research Environmental Scientist, US Environmental Protection Agency, ORD, National Health and Environmental Effects Research Laboratory, Western Ecology Division, Corvallis, Oregon, USA

Johannes Langemeyer, PhD, Institute of Environmental Science and Technology, Universitat Autònoma de Barcelona, Spain, and Stockholm Resilience Centre, Stockholm University, Sweden

Michael Ledwith, Remote Sensing Expert, Metria AB, Stockholm, Sweden

Harold Levrel, Professor of Economics at the Paris Institute of Technology for Life, Food and Environmental Sciences (AgroParisTech), UMR CIRED, Campus du Jardin Tropical, Nogent-sur-Marne Cedex, France

Regina Lindborg, Professor, Department of Physical Geography, Stockholm University, Stockholm, Sweden and STIAS (Stellenbosch Institute for Advanced Studies), Stellenbosch, South Africa

Camino Liquete, PhD, Researcher, European Commission, Joint Research Centre (JRC), Institute for Environment and Sustainability (IES), Ispra, Italy

Bruno Locatelli, PhD, Agricultural Research for Development (CIRAD), Montpellier, France and Center for International Forestry Research (CIFOR), Lima, Peru

Lasse Loft, Dr, Scientist, Senckenberg Biodiversity and Climate Research Centre (BiK-F), Frankfurt/M, Germany

Gary Luck, Professor, Ecology and Interdisciplinary Science, Institute for Land, Water and Society, Charles Sturt University, Albury NSW, Australia

Sandra Luque, PhD, Research Director, Irstea, National Research Institute of Science and Technology for Environment and Agriculture, France and Centre for Biological Diversity (CBD), School of Biology, University of St Andrews, Scotland, UK

Alexandra Lux, Dr, Scientist, Institute for Social-Ecological Research; affiliated to Senckenberg Biodiversity and Climate Research Centre (BiK-F), Frankfurt/M, Germany

David Lyon, MSc, MA, Principal, Irbaris LLP, London, UK and Director, ImpactAgri, London, UK

Liwei Ma, MS, PhD student, Institute for Natural Resource Conservation, Dept. of Ecosystem Management, Kiel, Germany

Georgina Mace, DPhil, Professor of Biodiversity and Ecosystems, Centre for Biodiversity and Environment Research, Department of Genetics, Evolution and Environment, University College London, UK

Neil MacLeod, Principal Research Scientist. CSIRO Agriculture, Brisbane Qld, Australia

Joachim Maes, PhD, Scientific Officer, European Commission – Joint Research Centre, Ispra, Italy

John McIvor, PhD, Research Fellow, CSIRO Agriculture, Brisbane Qld, Australia

John K. Maleganos, Environmental Economist MSc, Business Development Manager, European Sustainability Academy (ESA), Crete, Greece

Adrian Martin, Dr, Senior Lecturer in Environment and Development, School of International Development, University of East Anglia, UK

Berta Martín-López, PhD, Social-Ecological Systems Laboratory, Universidad Autónoma de Madrid, Madrid (Spain) and Leuphana University Lüneburg, Faculty of Sustainability, Lüneburg, Germany

Julia Martin-Ortega, PhD, Senior Environmental Economist, James Hutton Institute, UK

Rémi Mongruel, PhD, Researcher, Ifremer, Marine Economics Unit, Centre de Brest, UMR Amure, Plouzané, France

Carlos Montes, PhD, Social-ecological systems Laboratory, Department of Ecology, Universidad Autónoma de Madrid, Madrid, Spain

Felix Müller, Prof Dr, Head of Department, Institute for Natural Resource Conservation, Department of Ecosystem Management, Kiel, Germany

Amanda M. Nahlik, PhD, Kenyon College, Department of Biology, Gambier, Ohio, USA

Justice Nana Inkoom, Junior Researcher, Center for Development Research, Dept. Ecology and Natural Resources Management, University of Bonn, Germany

Marta Nieto-Romero, Research Centre for the Management of Agricultural and Environmental Risks (CEIGRAM). Universidad Politécnica de Madrid, Spain

Carl Obst, Fellow, Melbourne Sustainable Society Institute; University of Melbourne, Parkville, Australia

Paul Opdam, PhD, Emeritus Professor Landscape in Spatial Planning, Alterra Wageningen UR Nature and Society Group, and Wageningen University Land Use Planning Group, Wageningen, The Netherlands

Tim O'Riordan, Emeritus Professor of Environmental Sciences at the University of East Anglia, UK

Elisa Oteros-Rozas, PhD, Post Doc, Department of Geosciences and Natural Resource Management, University of Copenhagen, Denmark, and Social and Participatory Action Research Group, Department of Anthropology, Universidad Pablo de Olavide, Spain

Linwood Pendleton, International Chair, AMURE, University of West Brittany, IUEM, Plouzané, France and Senior Scholar Nicholas Institute for Environmental Policy Solutions, Durham, NC, USA

Tobias Plieninger, Associate Professor, Department of Geosciences and Natural Resource Management, University of Copenhagen, Denmark

Rosimeiry Portela, PhD, Senior Director, Economics and Planning, Betty and Gordon Moore Center for Science and Oceans, Conservation International, Arlington, VA, USA

Marion Potschin, Dr (habil), Principal Research Fellow and Director of the Centre for Environmental Management at the University of Nottingham, UK

Alison G. Power, PhD, Professor, Department of Ecology & Evolutionary Biology, Cornell University, Ithaca, NY, USA

Eeva Primmer, PhD, Head of Environmental Governance Unit, Environmental Policy Centre, Finnish Environment Institute, Finland

Sandra Quijas, PhD, Postdoctoral associate, Instituto de Investigaciones en Ecosistemas y Sustentabilidad, UNAM, Michoacán, Mexico and Centro Universitario de la Costa, Universidad de Guadalajara, Jalisco, Mexico

Dave Raffaelli, Professor, Director of Biodiversity and Ecosystem Service Sustainability, Environment, University of York, York, UK

Christopher M. Raymond, Senior Lecturer, Environmental and Sustainability Institute, University of Exeter, Penryn Campus, Cornwall, UK

Mark Reed, PhD, Professor in Interdisciplinary Environmental Research, Faculty of Computing, Engineering and the Built Environment, Birmingham City University, Birmingham, UK

Dominik Reisinger, BSc, Junior Researcher, GeoVille Information Systems GmbH, Innsbruck, Austria

Roy Remme, Post Doc, Environmental Systems Analysis Group, Wageningen University, Wageningen, The Netherlands

Charles R. Rhodes, ORISE post-doctoral fellow (supported by an interagency agreement between US EPA and DOE), working with the US EPA's Offices of Water and of Research and Development, Washington, D.C., USA

Taylor Ricketts, Professor and Director, Gund Institute for Ecology Economics, University of Vermont, USA

Mathieu Rouget, School of Agricultural Earth and Environmental, Sciences, University of KwaZulu-Natal, Scottsville, South Africa

Mark D. A. Rounsevell, PhD, Professor, David Kinloch Michie Chair of Rural Economy & Environmental Sustainability, School of GeoSciences, University of Edinburgh, UK

Duncan Russel, PhD, Associate Professor in Environmental Policy, Department of Politics, University of Exeter, UK

Mark Sagoff, Professor of Philosophy and Senior Fellow, Institute for Philosophy and Public Policy, George Mason University, Fairfax, Virginia, USA

Fernando Santos-Martín, PhD, Social-ecological systems Laboratory, Department of Ecology, Universidad Autónoma de Madrid, Madrid, Spain

Eirini Saratsi, PhD, Research Fellow, College of Social Science and International Studies, University of Exeter, UK

Terre Satterfield, PhD, Professor and Director, Institute for Resources, Environment and Sustainability (IRES), University of British Columbia (UBC), Canada

Marije Schaafsma, PhD, Senior Research Associate, Geography and Environment, and Institute for Life Sciences, University of Southampton, UK

Melanie Schnelle, BSc, Junior Researcher, GeoVille Information Systems GmbH, Innsbruck, Austria

Matthias Schröter, Postdoc, Research group Ecosystem Services, Helmholtz Centre for Environmental Research – UFZ, German Centre for Integrative Biodiversity Research (iDiv) Halle-Jena-Leipzig Germany

Thomas Sikor, Professor of Environment and Development, School of International Development, University of East Anglia, UK

James Spurgeon, Director, Sustain Value, UK

Jean L. Steiner, Supervisory Soil Scientist and Laboratory Director, USDA – Agricultural Research Service, Grazinglands Research Laboratory, El Reno, Oklahoma, USA

David Steuerman, Programme Officer, Business and Biodiversity, United Nations Environment Programme, Secretariat of the Convention on Biological Diversity, Canada

Lindsay C. Stringer, Professor of Environment and Development, Sustainability Research Institute, School of Earth and Environment, University of Leeds, UK

Elham Sumarga, Researcher, Environmental Systems Analysis Group, Wageningen University, Wageningen, The Netherlands and School of Life Sciences and Technology, Institut Teknologi Bandung, Indonesia

Patrick ten Brink, Head of Brussels office and Head of the Green Economy programme of IEEP and editor of *The Economics of Ecosystems and Biodiversity in National and International Policy Making* (2011)

Eric Thiébaut, Professor, Sorbonne Universités, UPMC Univ. Paris 06, Station Biologique de Roscoff, Roscoff, France

R. Kerry Turner, DSc, Professorial Fellow, School of Environmental Sciences and former Professor of Environmental Economics and Management at the University of East Anglia, UK

John Turnpenny, PhD, Senior Lecturer, School of Politics, Philosophy, Language and Communication Studies, University of East Anglia, Norwich, UK

Wim H. van der Putten, Head of the Department of Terrestrial Ecology, Netherlands Institute of Ecology, and Professor of Functional Biodiversity, Laboratory of Nematology, Wageningen University, NL

Christina von Haaren, Professor of Landscape Planning and Nature Conservation, Institute for Environmental Planning, Leibniz University Hannover/Germany

Diana H. Wall, University Distinguished Professor and Director, School of Global Environmental Sustainability, Colorado State University, Fort Collins, CO, USA

Peter Wangai, MS, PhD student, Institute for Natural Resource Conservation, Dept. of Ecosystem Management, Kiel, Germany

Robert Watson, Professor of Environmental Sciences, UEA, Norwich, UK

Piran White, PhD, Professor, Environment Department, University of York, York, UK

ACRONYMS AND ABBREVIATIONS

AM	adaptive management
ARIES	artificial intelligence for ecosystem services
BAP	biodiversity action plan
BBII	business and biodiversity interdependency indicator
BBN	Bayesian belief network
BESS	biodiversity and ecosystem service sustainability
BEWAS	boreal ecosystem wealth accounting system
BSA	balance sheet approach
CAFO	confined animal feed operation
CAMPFIRE	communal area management programme for indigenous resources
CAP	common agricultural policy
CBA	cost benefit analysis
CBD	convention on biological diversity
CDM	clean development mechanism
CES	cultural ecosystem services
CICES	common international classification of ecosystem services
CLA	causal layered analysis
CM	choice modelling
CVM	contingent valuation method
DPSIR	driver pressure state impact response
DSS	decision support system
DVM	deliberative monetary valuation
EA	energy analysis
EcoAIM	ecological asset information management
ECOSERVE	earth observation services for ecosystem valuation
EFCA	ecosystem function conservation area
EHS	environmentally harmful subsidies
EIA	environmental impact assessments
EO	earth observation
EROVA	environmental risk, opportunity and valuation assessment

ES	ecosystem services
ESA	ecosystem services assessments
ESB	ecosystem services benchmark
ESF	ecosystem services framework
ESP	ecosystem service provider and ecosystem service partnership
ESP-VT	ecosystem services partnership visualization tool
ESR	ecosystem services review
EUNIS	European nature information system
EVRI	environmental valuation reference inventory
FEGS	final ecosystem goods and services
FSC	forest stewardship council
G-ECO-MON	geographic ecosystem monitoring and assessment service
GPI	genuine progress indicator
GUMBO	global unified metamodel of the biosphere
GVA	gross value added
HNV	high nature value
IA	impact assessment
ILTER	international long term ecological research
InVEST	integrated valuation of ecosystem services and tradeoffs
IPBES	intergovernmental science-policy platform on biodiversity and ecosystem services
ISEW	index of sustainable economic welfare
IWRM	integrated water resources management
JFM	joint forest management
LCA	life-cycle assessment
MA	millennium ecosystem assessment
MAES	mapping and assessment of ecosystems and their services
MCA	multi-criteria analysis
MCESA	marine and coastal ecosystem services assessment
MDG	millennium development goals
MPA	marine protected area
MPI	multidimensional poverty index
MSFM	multifunctional sustainable forest management
NBSAP	national biodiversity strategy and action plan
NAP	national action programme
NAPA	national adaptation programme of action
NEA-D	German national ecosystem assessment
NESCS	national ecosystem service classification system
NDVI	normalized difference vegetation index
OpenNESS	operationalising natural capital and ecosystem services
PAR	participatory action research
PES	payment for ecosystem services
PPP	polluter pays principle
PRESET	practice-oriented ecosystem evaluation model
PWS	payment for watershed services
REDD+	reducing emissions from deforestation and forest degradation
SBA	service benefitting areas
SCEP	study of critical environmental problems

SDG	sustainable development goals
SEEA	system of environmental and economic accounting
SEF	special effect function
SES	socio-ecological systems
SHW	sustainable human well-being
SMS	safe minimum standards
SNA	system of natural accounting
Solves	social values for ecosystem services
SPU	service providing unit
SRF	special response function
TEEB	the economics of ecosystems and biodiversity
TEK	traditional ecological knowledge
TEV	total economic value
UK NEA	United Kingdom National Ecosystem Assessment
UK NEAFO	United Kingdom National Ecosystem Assessment, follow-on
UNCCD	United Nations Convention to Combat Desertification
UNFCCC	United Nations framework convention on climate change
UNSD	United Nations statistical division
WTP	willingness to pay

FOREWORD

Robert Watson

The natural world, its biodiversity and its constituent ecosystems and their services are now widely recognized to be critically important to human well-being and economic prosperity. However, even though ecosystem services have immense economic and social value, they are constantly undervalued in conventional economic analyses and rarely taken into account in decision-making by governments, the private sector or civil society.

This has resulted in biodiversity – the variety of genes, populations, species, communities, ecosystems, and ecological processes that make up life on Earth – being lost, at a rate unprecedented in the last 65 million years, jeopardizing the provision of ecosystem services that underpin human well-being, and are integral to the fabric of all the world's cultures.

The wide range of ecosystem services, from food fuel and fibre to spiritual and recreational needs, benefit people in a variety of socioeconomic conditions, across virtually every economic sector, and over a range of spatial scales, now and in the future. These ecosystem services, which contribute to our livelihoods, security, health, and social relations are fragile and being diminished across the globe. Threats to terrestrial and aquatic biodiversity are diverse, persistent, and, in some cases, increasing. The Millennium Ecosystem Assessment concluded that of the 24 ecosystem services evaluated, 15 were in decline, 4 were improving, and 5 were improving in some regions of the world and declining in other regions. The main drivers were fragmentation and conversion of natural ecosystems, over-exploitation of natural assets, introduction of alien invasive species, air, land and water pollution, and climate change.

We are at risk of losing much of biodiversity and the benefits it provides humanity. Action is critical: without it, current high rates of species loss are projected to continue what is becoming the 6th mass extinction event in Earth's history. Measures to conserve biodiversity and make a sustainable society possible need to be greatly enhanced and integrated with social, political and economic concerns.

Many of these benefits have historically been provided free of charge, and demand for them is escalating. The value of these services is being increasingly appreciated by a very large sector of society – extending from local stakeholders, the business community, farmers, environmentalists, and governmental policy makers, including development agencies. Although the global economic value of ecosystem services may be difficult to measure, it is enormous and almost certainly rivals or exceeds aggregate global gross domestic product. But while ecosystem benefits frequently outweigh costs of their conservation, the costs and benefits of conserving biodiversity often don't accrue to the same community, or at the same time or place.

Biodiversity is the most fundamental element of green economic development. However, we are squandering our natural capital for short-term gains. As noted earlier, over 60% of ecosystem services are currently being lost and will soon amount to an estimated $500 billion annually in forgone benefits. In order to move forward on the path of green economic development, technology transfers and new innovations are required to raise the value added of biological resources, especially in developing countries. In this way we can begin to shift away from the resource exploitative method of conventional development to the resource enrichment method of sustainable development.

The importance of conserving and protecting biodiversity and ecosystem services is high on the international policy agenda. There are several United Nations Biodiversity-related Conventions designed to conserve and protect biodiversity, including (i) the Convention on Biological Diversity; (ii) the Convention to Combat Desertification; (iii) the Ramsar Wetlands Convention; (iv) the Conservation of Migratory Species; and (v) the Convention on International Trade in Endangered Species. In addition, two of the goals of the UN Open Working Group on Sustainable Development Goals are directly related to biodiversity and ecosystem services: (i) conserve and sustainably use the oceans, seas and marine resources for sustainable development; and (ii) protect, restore and promote sustainable use of terrestrial ecosystems, sustainably manage forests, combat desertification, and halt and reverse land degradation and halt biodiversity loss.

Two international activities will greatly assist in improving the science-policy interface by generating and assessing the knowledge needed to protect and conserve biodiversity and ecosystem services: (i) the international transdisciplinary research program Future Earth, which builds on previous international research programs including Diversitas, the International Human Dimensions Program (IHDP) and the International Geosphere Biosphere Program (IGBP); and (ii) the Intergovernmental Panel on Biodiversity and Ecosystem Services (IPBES), which builds on international assessments such as the Global Biodiversity Assessment, the Millennium Ecosystem Assessment and the Global Biodiversity Outlook, and national assessments such as the UK National Ecosystem Assessment and its Follow-on.

The challenge is how to conserve biodiversity, halt the degradation of ecosystem services and integrate these issues into the development agenda. This *Handbook*, written by world class academic and policy experts, is long overdue, and provides a much-needed guide to address this challenge. It is an authoritative reference text written in easy to read sections that is essential reading for academics, decision-makers and civil society. The first section discusses ecosystem services concepts and frameworks, defining ecosystem services and functions and the link between biodiversity and ecosystem services, how to manage provisioning, regulating and cultural services for sustainability, as well as drivers of change. It then demonstrates the economic and social importance of these services for human well-being and sustainability. The second section discusses different methods and techniques that can be used in decision-making, ranging from global observations, modelling and scenarios, ecosystem services and accounting, diverse approaches to valuing ecosystem services, both economic and non-monetary, and issues of justice and ethics. The third section discusses ecosystem services from a range of different ecosystems, e.g., grasslands, forests, freshwater systems, cultivated lands and urban and marine ecosystems. The fourth section explores the connections of biodiversity and ecosystem services with different policy agendas, e.g., climate change, planning, poverty reduction, food and water security, human health, and the conservation of biodiversity. The final section discusses potential future directions for ecosystem services.

1

ECOSYSTEM SERVICES IN THE TWENTY-FIRST CENTURY

Marion Potschin, Roy Haines-Young, Robert Fish and R. Kerry Turner

Your true modern is separate from the land by many middlemen, and by innumerable physical gadgets. He [sic.] has no vital relation to it; to him it is the space between cities on which crops grow. Turn him loose for a day on the land, and if the spot does not happen to be a golf links or a 'scenic' area, he is bored stiff. If crops could be raised by hydroponics instead of farming, it would suit him very well. Synthetic substitutes for wood, leather, wool, and other natural land products suit him better than the originals. In short, land is something he has 'outgrown.'

The Land Ethic from 'A Sand County Almanac', Aldo Leopold, 1948

Introduction

Around 2008, the human population passed something of a milestone. For the first time there were more people living in cities than in rural areas. Looking forward, it is estimated that by 2050 roughly two-thirds of the nine or so billion people that inhabit the Earth will be urban dwellers (UNFPA, 2007). It is also projected that the absolute numbers of people living in the countryside will decline, compared to the present.

That we are becoming a predominantly urban species, and that *all* future population growth will be in built-up areas, will have many consequences. The outlook for human well-being is likely to be positive in many important respects, because there are generally better employment prospects in cities, and better access to education and health services. The United Nations Population Fund argues that social mobility in cities is greater and the chances that women can take control of their lives are greater. As a result, they suggest, fertility rates in urban areas are likely to reduce, and this will change trajectory of overall population growth.[1] But there will be other consequences, too. The fact that societies depend fundamentally on natural systems may be readily masked and obscured by the experience of city living. Many critical traditions have argued in the vein of Leopold that, with urbanity, nature is often remade as a distant 'other'; as merely background scenery for cultural processes; and as a set of commodities that conceal, or at best stylise, their origins in natural processes. While we may say that this model of spatial organisation makes society less vulnerable to environmental hazards, an appreciation of the ties that bind people and nature together will arguably be more difficult to sustain. Our planet will become no less finite just because most people will be living in cities.

The problem of the changing connection between people with nature is the one we want to address in this chapter. In thinking about this, and in particular what it means for the ecosystem services debate in the 21st century, we were reminded of Leopold's thoughts about the 'true modern'. From our current perspective, we can imagine that not only will the future descendants of the 'true modern' be less able to see the 'vital relation' that he or she has with ecosystem function, but also, since the 'golf links' and 'scenic area' are now likely to be regarded as 'valuable cultural ecosystem services', the prospect of changing his or her mind may be even more remote.

Of course, it is not inevitable that Leopold's description of the 'true modern' will apply in the future. The prospect is very much dependent on the kinds of narratives that people develop to describe their lives and the societies that they inhabit – which is why the idea of ecosystem services and the importance of natural capital may have particular significance for *Homo urbanus*. In this chapter we will examine the history of the idea of ecosystem services, and argue that while it is an idea shaped by thinking at the end of the last century, it can continue to be relevant – providing we can connect it into wider debates about what people care about.

Histories

In their 'fragmentary history' of ecosystem services, Mooney and Ehrlich (1997) attribute the first use of the term 'ecosystem services' to *Extinction: The Causes and Consequences of the Disappearance of Species* (Ehrlich and Ehrlich, 1981). However, they also recognise the idea that ecosystems can be thought of as providing 'services' to people can be found in the literature long before the particular phrase was used. They note, for example, the account of 'environmental services' a decade earlier in the *Study of Critical Environmental Problems* (SCEP, 1970), and ideas of the 'public-services' that can be provided by the global environment as described in Ehrlich and Holdren (1974) and other papers published around that time – not to mention the account of 'nature's services' provided by Westman (1977).

In their history of the concept and its link with economic valuation of the environment, Gómez-Baggethun et al. (2010) agree with Mooney and Ehrlich (1997) that while the impact of human actions on the way nature can benefit people was discussed by writers even in ancient civilisations, it was probably the publication of *Man and Nature* in 1864 by George Perkins Marsh that stimulated interest in modern times. It is from Marsh's work that we can see some of the central themes of contemporary debates about ecosystem services being rehearsed, namely the finite capacity of the earth, its limitations in providing benefits to people, and its vulnerability to human action.

As Lowenthal (2000, p.3) argues, a noteworthy feature of *Man and Nature* 'was Marsh's stress on the unforeseen and unintended consequences, as well as the heedless greed, of technological enterprise'. Marsh contended that, to sustain global resources, society needed to become aware of how it affected them (Lowenthal, 2000). While the term 'ecosystem services' did not enter scientific discourse until 1981, the idea that people directly benefit from nature, and that nature's capacity to support these benefits is limited, was thus already common currency. Indeed, the idea was shaping not only conservation debates but also institutional responses.

For example, in a 'Worldwatch' paper of 1978, we find Eckholm discussing the significance of the loss of species 'whose ecological functions are especially important to society'. He cautions: 'At the broadest level, extinctions serve as markers of the general reduction in the capacity of the earth's biological systems to provide goods and crucial, if subtle, *ecological services*' (Eckholm, 1978, p.18, author's emphasis). In 1980 the IUCN *World Conservation Strategy* explicitly used the notion of goods and services provided by ecosystems in the section on 'policy making and the integration of conservation and development', where it is used in connection with sustainable

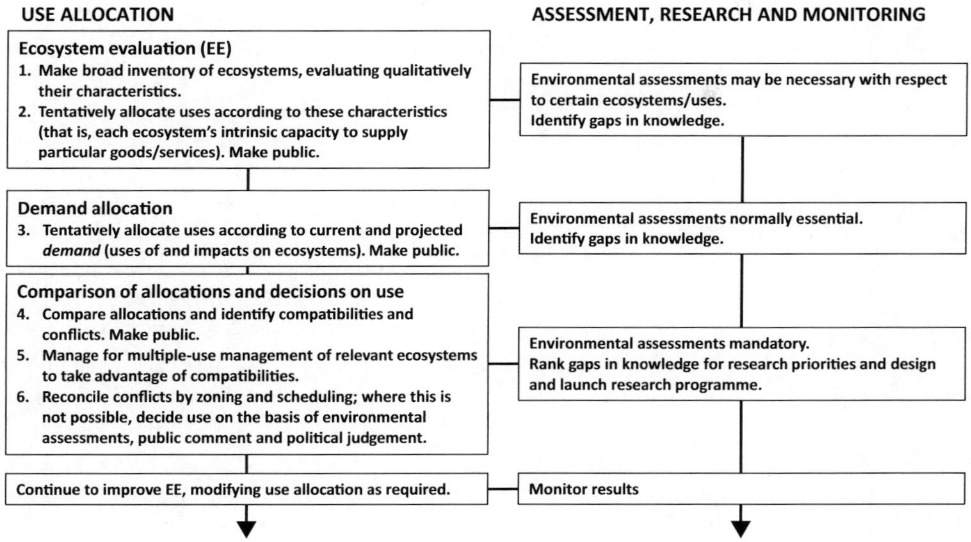

Figure 1.1 Framework for ecosystem assessment developed in the 1980 World Conservation Strategy (after IUCN, 1980). Original title: 'The relationship between the allocation of land and water uses and assessment, research and monitoring'.

forest management, and in the section on 'environmental planning and rational use of resources', where it is discussed more generally in the context of how to use ecosystem assessments to help allocate resource use (IUCN, 1980). Despite the passage of nearly four decades, the assessment framework that they proposed (Figure 1.1) remains highly applicable and consistent with current thinking.

What is perhaps surprising when we look back is that despite this early interest the idea of ecosystem services was given little attention in the 'Brundtland Report' of 1987. Reference is made to the idea only once, when the Report deals with species and ecosystems in the context of resources for development, and where it is noted that 'species and natural ecosystems make many important contributions to human welfare' (World Commission for Environment and Development, 1987, p. 125). The report observes that these resources are often not used in ways that will be able to meet the demands for the 'goods and services that depend upon these natural resources'. Background reports to the Commission did, however, emphasise the importance of natural resources and natural processes which affect human well-being (Turner, 1989). Nevertheless, the point was taken up more fully in *Agenda 21*. This was a key and influential output from the Earth Summit in 1992 and which set out the United Nation's action plan for delivering on sustainable development. In the discussion on 'combatting deforestation', the document echoes the points made in the *World Conservation Strategy* on the role of wood and non-wood goods and services as a component of sustainable forest management. Perhaps more importantly, when addressing the conservation of biodiversity, *Agenda 21* charges us to 'take measures to encourage a greater understanding and appreciation of the value of biological diversity, as manifested both in its component parts and in the ecosystem services provided' (United Nations, 1992, sect. 15.5.m). The development of integrated environmental and economic accounting methods was seen as one necessary step. The aim was to expand national accounting systems so that they better measured the '. . . crucial role of the environment as a source of natural capital . . .' (United Nations, 1992, sect. 8.41). Another key action identified in *Agenda 21* was the development of a 'science for sustainable development', in which scientific knowledge is applied 'through scientific assessments of current conditions and future prospects for the Earth system' (United Nations, 1992, sect. 35.3).

Mooney and Ehrlich (1997) describe how biodiversity assessment approaches were developed in the 1990s, eventually as part of wider related initiatives such as the Global Biodiversity Assessment of UNEP, which integrated economic and ethical issues with biodiversity science (UNEP, 1995; Perrings et al., 1995). However, despite such advances, there was a wider recognition that the new findings emerging from ecology and related fields were poorly reflected in policy discussions. A key part of the debate was the publication of *Protecting our Planet, Securing our Future: Linkages among Global Environmental Issues and Human Needs*, in 1998. This was the output of an international study sponsored by UNEP, NASA, and the World Bank.

Importantly, *Protecting our Planet* called for more integrative assessments 'that can highlight the linkages between questions relevant to climate, biodiversity, desertification, and forest issues' (Watson et al., 1998, p.56). It provided some of the impetus[2] for the 2005 Millennium Ecosystem Assessment (MA), an examination of the consequences of ecosystem change for human well-being, and a key influence on the early mainstreaming of the ecosystem services agenda (Daily et al., 2011, p.3). Significant institutional factors were, however, perhaps more decisive: the work associated with a number of international agreements such as the Convention on Biological Diversity (CBD), the Convention to Combat Desertification, the Convention on Migratory Species, and the Ramsar Convention, had shown that the needs for scientific assessments within the conventions were not being met. And so the foundations of the MA were laid. It became one of the key initiatives to help achieve the United Nations Millennium Development Goals and to carry out the Plan of Implementation for the 2002 World Summit on Sustainable Development.

As the result of the stimulus of the MA, national assessments have been made in a number of countries, including New Zealand, France, Spain, Portugal, and Israel.[3] The UK Government has also funded two national ecosystem assessments and set up a formal Natural Capital Committee[4] to audit society's use of ecosystem services (UK, NEA, 2011; UKNEA FO, 2014). Most significantly, the Inter-Governmental Platform on Biodiversity and Ecosystem Services has now been established[5] to continue to review, assess, and evaluate the growing knowledge base that has developed around the topic, and crucially to improve the capacity for using that knowledge effectively in decision-making. With its formal endorsement by the science and policy communities, the aim is to give as strong and credible a voice to issues surrounding biodiversity and ecosystem services as has been done for climate change.

Reflections on the future: people and nature?

A system of conservation based solely on economic self-interest is hopelessly lopsided. It tends to ignore, and thus eventually to eliminate, many elements in the land community that lack commercial value, but that are (as far as we know) essential to its healthy functioning. It assumes falsely, that the economic parts of the biological clock will function without the uneconomic parts.

The Land Ethic from 'A Sand County Almanac', Aldo Leopold, 1948

What our short history shows is that, despite the attention that the idea of ecosystem services is currently receiving, it is very much *a 20th-century* concept. Looked at in one way, the terminology could be seen as merely representing another way of describing a set of problems that have preoccupied people for at least 50 years, if not more, namely: the failure of 'developed' societies to take account of their dependency on nature and the prospect that the capacity of nature to support future prosperity appears to be limited – not least through the impacts of human activities. In this guise, the idea of ecosystem services is therefore no more than a 'refresh' – as in the case of the updating of the CBD Biodiversity Targets from those set in 2010 to those for 2020.

Looked at in another way there is, nevertheless, something that is fundamentally novel about the current discussions around the idea of ecosystem services. It is, we suggest, the context in which those 'old problems' must now be viewed. Part of the changed context is the prospect of what in this volume Costanza (2016) describes as a 'full world' in which, as a result of the growth of population and resource use, we are approaching some key planetary boundaries. Another aspect is, as we have described in this chapter, the prospect that we will become a predominantly urban species, with the increasing detachment from the natural world that this may involve. While the environmental problems of living on a finite planet and halting biodiversity loss may not have changed since ideas about ecosystem services were first articulated, the 'solution space' that we now have to explore has been fundamentally transformed. This new context will force us to look at the idea of ecosystem services in a new way. What kinds of change are required?

Some aspects of the development of the science of ecosystem services fit with the description of science offered by Kuhn (1962). It could be claimed, for example, that there has been a revolution in our thinking about nature with the appearance of the concept, and we are now, in small steps, working our way through the implications of the new paradigm, in a period of 'puzzle-solving'. However, there is, we suggest, an important characteristic of a 'post-normal' science like ecosystem services that makes it different from more traditional ones. Although there is indeed further basic natural and social science to be done, if the science of ecosystem services is truly 'transdisciplinary', then any puzzle-solving has to take place in parallel with a process of embedding the knowledge gained in wider societal discourses. The history of ecosystem services, and much of the contemporary work that now surrounds the concept, is largely presented from the perspective of 'nature for people' (Mace, 2016). This theme underpinned the MA, with its emphasis on the importance of ecosystem services *for* human well-being. It also characterised *The Economics of Ecosystems and Biodiversity* (TEEB), with its arguments about the value of nature *to* society. These are fundamentally 20th-century perspectives and can only take us so far. If we are to embed the idea of ecosystem services and natural capital in wider decision-making, then the future, we suggest, lies with what Mace (2016) has identified as an alternative view – namely that based on the theme of 'people and nature'.

With the development of the ecosystem service concept we have got better at making the case about nature *for* people. Many of the contributions to this *Handbook* demonstrate the richness of the basic research that has been done to underpin these kinds of arguments. We have even made 'progress' with this kind of thinking in that the sentiment expressed by Leopold at the start of this section does not seem so much at odds with contemporary views as it would have done a few years ago, when the valuation debate first took hold. The proposition that monetary values can be assigned to some components but not all of nature has captured the attention of many outside the core disciplines represented here (Turner, 2016). Many now accept that while monetary valuation of ecosystem services can be an important tool for decision-making, it is not the only one; as Gómez-Baggethun et al. (2016, p. 100) argue, 'there are various ways in which people ascribe meaning to nature', including 'ecological values'. To see what can be added by taking up the alternative 'people and nature' theme, it is useful to return to the second passage by Leopold.

What is interesting about Leopold's proposition about the limits of economic valuation is not so much its prescience but rather what he writes next. For while at first glance the quotation seems to imply that ecological values count too, the passage continues by suggesting that an approach to conservation based on 'self-interest':

> . . . tends to relegate to government many functions eventually too large, too complex, or too widely dispersed to be performed by government. An ethical obligation on the part of the private owner is the only visible remedy for these situations.

While we do not entirely endorse Leopold's prescription, the challenge he throws down in this proposition elegantly opens up the 'people and nature' debate, with its focus on issues of governance and ethics rather than just ecology. These, we suggest, are themes that need to figure more strongly in future debates about our relationships to nature, and are key to realising what many commentators mean when they write of 'operationalising' and 'mainstreaming' ecosystem services. We conclude by exploring how a focus on people *and* nature could change how the ideas of ecosystem services and natural capital are seen and used.

Mainstreaming and operationalisation

Daily (2016) argues that the 'urgent challenge today is to move from ideas to action on a broad scale'. The need for action has, in fact, long been recognised and strongly argued for as part of the 'nature for people' debate, with the development of such frameworks as the Ecosystem Approach,[6] and the call to engage with 'all relevant actors' as part of the participatory process that we now see as essential to the successful delivery of environmental policy and management. But, following Daily, it seems that more needs to be done; she argues that 'mainstreaming natural capital into decisions is a long-term game plan, requiring co-evolving advances in knowledge, social institutions, and culture' (Daily, 2016, p. 308).

One way of looking at the differences between the 'nature for people' vs the 'nature and people' perspectives is in terms of whether we are aiming ultimately to achieve action *for* nature or *through* nature. Similar kinds of debate about the focus of action have, in fact, been had in the landscape community as part of their search for ways of making a more socially relevant case for what they seek to do (Matthews and Selman, 2006). In the context of ecosystem services, the important point to note here is that rather than environment or biodiversity being seen as the target of our action, the argument to be made is that 'nature-based solutions' to a wide range of social and economic problems can be found by thinking about them in different ways. This, we suggest, is key to understanding what is being implied by the people and nature perspective. It is also key to understanding what mainstreaming, or what others have called 'operationalisation', represents. O'Riordan (2016, p. 599) sets out the task succinctly: 'To shift the default position to sustainability will require global recognition of the essentialness of ecosystem services for viable continuation of wealth creation'. Similarly, if biodiversity is indeed the most fundamental element of green economic development (Watson, 2016) then we need to *demonstrate* that it is so, and act accordingly.

To mainstream or to operationalise ideas about ecosystem services around the theme of people and nature will mean finding ways of embedding these perspectives in other debates – perhaps far removed from the core concerns of the current research and environmental policy communities. Only then will a more balanced view of the relationship between people and nature be achieved. Where we disagree with Leopold is that it seems unlikely that it can be done by focussing actions only at the individual, and specifically private, level. Despite the complexity of the problems we face, we would argue that governments and institutions have an enabling role to play too. Institutional change can alter rights and responsibilities (cf. Primmer, 2016); it can also influence and modify behaviour through practices of regulation and reward. Further still, we need to be careful of viewing individuals as economic agents alone. The socialisation of ecosystem services is by implication a wider question of environmental citizenship: how we mobilise individuals to work, think, and connect to environmental concerns as members of communities as much as property owners and consumers of resources.

The shortcomings of the 'nature for people' perspective can be seen in debates surrounding the IPBES conceptual framework (Díaz et al., 2015; Potschin and Haines-Young, 2016), where different cultural perspectives on how people and nature were connected emerged. Differences were not so much to do with recognition of the dependency of people on nature, but rather how the idea of ecosystem services could be interpreted as a one-way, utilitarian type of relationship. This is clearly one of the dangers of the 'nature for people' viewpoint, in that it could lead us to underestimate the complexity of the environment and its many facets. It could also obscure an appreciation of the deeper and more reciprocal relationships and duties of care that people have in relation to the non-human world. To shift the default position, Leopold's 'true modern' will need some kind of narrative to explain how contemporary society is as much connected to nature as more traditional ones and that this implies certain obligations. The ideas of ecosystem services and natural capital, told through the lens of 'people and nature', could be an important part of that story.

As we look to the future, there is an urgent need to build more elaborate narratives of ecosystem services and natural capital to emphasise people's rights and responsibilities in relation to nature, as well as the benefits that people derive from it. A test of whether a more balanced 'people and nature' has been achieved will be in the kinds of institutions that we construct to help us cope with a rapidly urbanising, full world. One of the ways in which we have moved away from Leopold's argument that solutions lie with private owners is that we now accept that ecosystem services can have quite different characteristics in terms of rivalry and excludability, because the systems generating them can be multi-functional. Thus, we have to manage and regulate different kinds of interactions between providers and beneficiaries and hence think of different kinds of governance arrangement for these *socio-ecological systems* (Barnaud and Antona, 2014). Moreover, since private ownership can now involve multi-national organisations of the kind not envisaged in the early part of the last century, the task of managing the mix of public and private benefits that most ecosystems can provide in 'sustainable ways' has become much more complex.

If the 'people and nature' argument is to be made effectively, it needs to start from the position that many problems previously thought to be independent of the environment are intimately connected to it. In the policy arena these include, for example, debates around human health, the economy, social justice, and national security (cf. Lubchenco, 1998). In the scientific realm the ideas of ecosystem services and natural capital have to be seen as essential ways of exploring key 'problem nexuses', linking such things as climate, energy, food, and water (cf. Liu et al., 2015). These kinds of debates in the policy and science communities will be fundamental, because they will shape the kinds of governance approaches that will be needed in the 21st century. Green economies will be as much about the institutions we design as the technologies we use. In a full, connected world, problems of 'spill-overs' from one problem area to another, and the unexpected interactions, activities, and impacts at global scales, in the form of 'telecoupling', will require ever more careful design of incentive and regulatory frameworks to cope with unwelcome effects. They will also require critical scrutiny of existing institutional and legal structures to ensure that they are fit for this purpose, or can be made so.

In a world where city living becomes the norm, it will be more difficult to make the argument for the importance of nature. In such a world, there are also dangers in presenting our dependency on nature in narrow, utilitarian terms. The countryside and wilder landscapes could be seen simply as a means to an end, where only the parts that are useful to people are retained or conserved. The challenges are therefore considerable. If we are to address them, then the story of ecosystem services perhaps has to be retold as a partnership between people *and* nature. As our history has shown, the development of such ideas can take a long time. Getting them into the mainstream will also not be easy. This *Handbook* has been written in the belief that it can be done. The descendent of Leopold's 'true modern' is already living in a city near you.

Notes

1 http://www.unfpa.org/urbanization
2 http://www.millenniumassessment.org/en/History.html
3 See IPBES database at: http://catalog.ipbes.net/
4 See: http://www.naturalcapitalcommittee.org/
5 http://www.ipbes.net/
6 https://www.cbd.int/ecosystem/

References

Barnaud, C., and Antona, M. (2014). Deconstructing ecosystem services: uncertainties and controversies around a socially constructed concept. *Geoforum*, vol 56, pp. 113–123.

Costanza, R. (2016). Ecosystem services in theory and practice. In Potschin, M., Haines-Young, R., Fish, R., and Turner, R. K. (eds) *Routledge Handbook of Ecosystem Services*. Routledge, London and New York, pp. 15–24.

Daily, G. C. (2016). Securing nature and people: can we replicate and scale success? In Potschin, M., Haines-Young, R., Fish, R., and Turner, R. K. (eds) *Routledge Handbook of Ecosystem Services*. Routledge, London and New York, pp. 304–310.

Daily, G. C., Kareiva, P. M., Polasky, S., Ricketts, T. H., and Tallis, H. (2011). Mainstreaming natural capital into decisions. In Kareiva, P., Tallis, H., Ricketts, T. H., Daily, G. C., and Polasky, S. (eds) *Natural Capital: Theory and Practice of Mapping Ecosystem Services*. Oxford University Press, Oxford.

Díaz, S., Demissew, S., Carabias, J., et al. (29 authors) (2015). The IPBES Conceptual Framework – connecting nature and people. *Current Opinion in Environmental Sustainability*, vol 14, pp. 1–16.

Eckholm, E. P. (1978). *Disappearing Species: The Social Challenge, Volume 45, Number 5*. Worldwatch Institute, Washington DC.

Ehrlich, P., and Ehrlich, A. (1981). *Extinction: The Causes and Consequences of the Disappearance of Species*. Random House, New York.

Ehrlich, P., and Holdren, J. P. (1973). Human population and the global environment. In *UN Symposium on Population, Resources, and Environment, Stockholm, Volume 26*.

Gómez-Baggethun, E., Barton, D. N., Dunford, R., and Harrison, P. (2016). Concepts and methods in ecosystem services valuation. In Potschin, M., Haines-Young, R., Fish, R., and Turner, R. K. (eds) *Routledge Handbook of Ecosystem Services*. Routledge, London and New York, pp. 99–111.

Gómez-Baggethun, E., De Groot, R., Lomas, P. L., and Montes, C. (2010). The history of ecosystem services in economic theory and practice: from early notions to markets and payment schemes. *Ecological Economics*, vol 69, no 6, pp. 1209–1218.

IUCN, UNEP, WWF. (1980). *World Conservation Strategy*. World Conservation Union, United Nations Environment Programme, World Wide Fund for Nature, Gland.

Kuhn. T. S. (1962). *The Structure of Scientific Revolutions*. University of Chicago Press, Chicago.

Leopold, A. (1948). The land ethic. In *A Sand County Almanac, and Sketches Here and There*. Oxford University Press, New York.

Liu, J., Mooney, H., Hull, V., et al. (11 authors) (2015). Systems integration for global sustainability. *Science*, vol 347, no 6225: 1258832.

Lowenthal, D. (2000). Nature and morality from George Perkins Marsh to the millennium. *Journal of Historical Geography*, vol 26, no 1, pp. 3–23.

Lubchenco, J. (1998). Entering the century of the environment: a new social contract for science. *Science*, vol 279, no 5350, pp. 491–497.

Mace, G. (2016). Ecosystem services: where is the discipline heading? In Potschin, M., Haines-Young, R., Fish, R., and Turner, R. K. (eds) *Routledge Handbook of Ecosystem Services*. Routledge, London and New York, pp. 602–605.

Matthews, R., and Selman, P. (2006). Landscape as a focus for integrating human and environmental processes. *Journal of Agricultural Economics*, vol 57, no 2, pp. 199–212.

Mooney, H. A., and Ehrlich, P. R. (1997). Ecosystem services: a fragmentary history. In Daily, D. C. (ed.) *Nature's Services: Societal Dependence on Natural Ecosystems*, pp. 11–19. Island Press, Washington DC.

O'Riordan, T. (2016). On the changing relationship between ecosystem services continuance and sustainability. In Potschin, M., Haines-Young, R., Fish, R., and Turner, R. K. (eds) *Routledge Handbook of Ecosystem Services*. Routledge, London and New York, pp. 599–601.

Perrings, C., Barbier, E. B., Brown, S., et al. (12 authors) (1995). The economic value of biodiversity. In *UNEP, Global Biodiversity Assessment*, Cambridge University Press, Cambridge.

Potschin, M., and Haines-Young, R. (2016). Frameworks for ecosystem assessments. In Potschin, M., Haines-Young, R., Fish, R., and Turner, R. K. (eds) *Routledge Handbook of Ecosystem Services*. Routledge, London and New York, pp. 125–137.

Primmer, E. (2016). An institutional perspective. In Potschin, M., Haines-Young, R., Fish, R., and Turner, R. K. (eds) *Routledge Handbook of Ecosystem Services*. Routledge, London and New York, pp. 582–585.

SCEP (1970). *Man's Impact on the Global Environment*. MIT Press, Cambridge MA.

Turner, R. K. (1989). Economics and environmentally sensitive aid. *International Journal of Environmental Studies*, vol 35, pp. 39–50.

Turner, R. K. (2016). Ecological economics and ecosystem services. In Potschin, M., Haines-Young, R., Fish, R., and Turner R. K. (eds) *Routledge Handbook of Ecosystem Services*. Routledge, London and New York, pp. 243–255.

UK National Ecosystem Assessment (2011). *The UK National Ecosystem Assessment Technical Report*. UNEP-WCMC, Cambridge.

UK National Ecosystem Assessment Follow On (2014). *The UK National Ecosystem Assessment: Synthesis of the Key Findings*. UNEP-WCMC, Cambridge.

UNEP (1995). *Global Biodiversity Assessment*. Cambridge University Press, Cambridge.

UNFPA (2007). *State of World Population 2007: Unleashing the Potential of Urban Growth*. United Nations Population Fund. ISBN 978–0–89714–807–8.

United Nations (1992). Results of the World Conference on Environment and Development: Agenda 21. UNCED United Nations Conference on Environment and Development, Rio de Janeiro, United Nations, New York.

Watson, R. (2016). Preface to the Routledge Handbook of Ecosystem Service. In Potschin, M., Haines-Young, R., Fish, R., and Turner, R. K. (eds) *Routledge Handbook of Ecosystem Services*. Routledge, London and New York, pp. xxvi–xxvii.

Watson, R. T., Dixon, J. A., and Hamburg, S. P. (1998). Protecting our planet securing our future: linkages among global environmental issues and human needs. In *Protecting Our Planet Securing Our Future: Linkages among Global Environmental Issues and Human Needs*. The World Bank, New York.

Westman, W. E. (1977). How much are nature's services worth? *Science*, vol 197, pp 960–964.

World Commission on Environment and Development (1987). *Our Common Future*. Oxford: Oxford University Press.

PART I

Ecosystem services concepts and frameworks – introduction

The concepts and approaches driving innovation and understanding in the field of ecosystem services are diverse and reflect a range of disciplinary starting points and pre-occupations. As a result, though we can recognise a distinct body of scholarship around the idea of ecosystem services, there is much room for difference and even misunderstanding. Moreover, ideas change as new research emerges and new problems are identified, and so the goal of achieving a common, accepted conceptual framework is difficult to accomplish given the rapid expansion of the field.

The aim of this first part of the *Handbook* is therefore to provide an overview of the current 'state-of-the-art' in conceptual thinking. In doing so we have not sought to impose a single perspective, but rather to highlight and explain some of the key differences between several commentators so that people, especially those unfamiliar or new to the topic, can better negotiate the conceptual landscape of 'ecosystem services'. But understanding concepts is not an end in itself. Ultimately, the value of the idea of ecosystem services lies in the problems that it helps us solve. And so, in addition to looking at the basic ingredients of the science of ecosystem services, in this part of the *Handbook* we also consider its application and the ambitions of its proponents.

The scope and significance of the ecosystem service paradigm is described at the outset in a chapter by one of the key figures in the field, Costanza. He argues that ideas about ecosystem services and the benefits nature provides to humanity are fundamental in shaping our thinking about what it means to live in a 'full world'. With a human population already in excess of seven billion, and current and rising demands in terms of resource and energy use, we have reached a stage of human development where people are altering our ecological life support systems. So profound has been the change, compared to the situation faced by our pre-industrial ancestors, that he argues that we have now entered a new geological period – the *Anthropocene*. In our full world, he concludes, 'we have to think differently about what the economy is and what it is for if we are to create sustainable prosperity'.

Recognition of the importance of ecosystem services and natural capital, and acknowledging their value in formal ways, is part of that 'different thinking' about our economy. Costanza suggests that as long as we are forced to make choices we are 'doing valuation' whether we acknowledge it or not – and so need to be smarter in our approaches and indeed the institutions we create to manage ecosystems and their services. The chapter by Gómez-Baggethun and colleagues reinforces the message that value systems play a critical role in the way societies relate to, and manage, ecosystems, and so rightly deserves the attention that they have gained in science and policy in recent years. They suggest that if valuation is ultimately to serve goals of environmental sustainability, social justice and long-term economic viability, we must as a community make good use of the available concepts and methods to capture nature's many ecological, social and economic values to people. These authors argue for 'value pluralism' rather than a narrow focus on economic values alone, which is sometimes how the ecosystem service paradigm is characterised.

Setting valuation approaches aside, however, a key theme to emerge from all the chapters in this section is that if ecosystem service thinking is to lead to better decision-making, then it has to take account of the complexity of ecosystems, the scientific uncertainties surrounding our knowledge of them, and the existence of environmental limits and threshold effects. One way forward, according to Costanza, is to engage in the 'collaborative construction of dynamic, evolutionary models of linked ecological economic systems'. In short, we need a robust understanding of the biophysical underpinnings of ecosystem services.

In addition to questions about values and valuation, the coupling of ecological and social systems is therefore a second theme to be covered in this part of the *Handbook*. The chapter by Potschin and Haines-Young traces how we can represent the links between ecosystems and human well-being and start to build a typology of ecosystem services that resolves some of the ambiguities surrounding their measurement. The chapter uses the so-called cascade model, which links ecosystem structures, processes and functions, to services, benefits and values, as a way of describing the current 'ecosystem services paradigm'. However, as the chapter by Raffaelli shows, there are likely to remain a number of underlying complexities in the relationships that we observe. For example, while it is easy to characterise some services, such as the provisioning ones, in terms of some stock of natural capital and the flows that arise from them, regulating and cultural services may need to be thought about and measured in a different way. Since regulating and cultural services do not depend on stocks that can be 'used up', he suggests that they are best thought of as 'fund-services'. Raffaelli suggests that to understand the stock-flow situation, we generally need to consider the structural properties of an ecosystem, whereas understanding fund-service relationships usually depends on knowledge about the 'functional' properties of ecosystems.

Characterisation of the functional properties of ecosystems and how they relate to ecosystem services is a critical area of concern to emerge from a reading of the chapters in this section of the *Handbook*. However, as the contributions by Potschin and Haines-Young, and Jax show, there are a number of different ways of framing the notion of 'ecosystem function' ranging from those which treat ecosystem function as synonymous with ecosystem processes to those that think of function as describing the particular capacities or properties of ecosystems that give rise to services. The chapters in this section suggest that, whatever terminology we choose to adopt, there is, nevertheless, much merit in looking at the particular characteristics of living systems that help us understand how ecosystem services are generated. Nowhere is this better illustrated than in the chapter by Balvanera and colleagues.

Although ideas about ecosystem services have been developed to make the case for biodiversity, the links between them are, according to Balvanera and her co-authors, 'complex and variable'. This chapter goes on to describe some of the recent theoretical, experimental and observational work advancing our understanding of these inter-linkages, and concludes that despite uncertainties, the conservation and management of biodiversity is critical for the long-term maintenance of social-ecological systems and the flow of benefits from nature to people at a range of spatial scales. The authors suggest that the recent literature shows that an understanding of the number, identity, functional characteristics and evenness of species is important to ecosystem functioning and consequently to the supply of different types of services, and that approaches based on the analysis of functional traits of species or species groups will be an important way forward.

As the contribution of Luck also shows, we need to find ways of quantifying the ecological units that provide ecosystem services, and unifying thinking about Service Providing Units (SPUs) based on the characteristics of species populations needs to be extended to include systems made up of multi-species functional groups, entire ecological communities, habitat types or landscapes (i.e. as 'Ecosystem Service Providers' (ESPs)). Such conceptual frameworks can be useful not only for understanding the links between biodiversity and ecosystem services, but also the drivers of change in socio-ecological systems (SES) more generally.

Although it is important to understand the biophysical underpinnings of ecosystem services, as the chapter by Loft and colleagues argues, frameworks such as the cascade model tend to emphasise the factors affecting the supply of ecosystem services rather than the demand for them. They argue for a more balanced approach that explicitly recognises the feedbacks between people and nature. To achieve this they conclude that we need to make the study of SES more central to our work. They claim that the SES framework can be used to 'structure key factors, concepts, or variables, and the assumed relationships between them, such as the spatial boundaries of systems, units of analysis, time horizons, inputs and drivers'. In this part of the *Handbook*, the contributions of Rounsevell and Harrison go on to describe how we can look at the way different drivers of change shape service supply and demand in a SES, while Hauck takes up the arguments made by Loft and colleagues in favour of a trans-disciplinary approach to the study of ecosystem services.

The need to counter-balance the biophysical perspective on ecosystem services with one that places greater emphasis on who benefits from ecosystem services is a theme taken up in the chapter by Landers and his co-workers. They argue specifically for a 'beneficiary approach', in contrast to a 'benefits approach', as a practical way forward for measuring, mapping and valuing ecosystem services. They conclude that only by matching the uses of nature's services, the benefits, to known users, or beneficiaries, are we likely to find ways of measuring and accounting for how much of which of nature's services are used or appreciated, and by whom. Their work provides a valuable contrast to the approach to classifying ecosystem services described by Potschin and Haines-Young; there is a clear potential for developing new integrated, hybrid typologies for ecosystem functions, services, benefits and beneficiaries.

Overall, the authors in this part of the *Handbook* demonstrate that considerable methodological and conceptual progress has been achieved since the publication of the Millennium Ecosystem Assessment in 2005. Indeed, many now feel that we can start to move ecosystem services 'into the mainstream' of decision-making. In concluding, however, it is important to note that this optimistic perspective is not without its critics. Sagoff succinctly sets out some of the challenges. He doubts, for example, that we can develop an 'objective' way of assessing the benefits of ecosystem services, not least because of the difficulties of disentangling the contributions made by natural and human capital in the production of many services. And even if we could, how can we convince people to pay for them? He suggests that such a persuasive case cannot easily be made. We leave the reader of the chapters in this section and the *Handbook* as a whole to decide.

Key discussion and debating points

- With its focus on the regulation and sustainable transformation of the complex human-nature interactions, the concept of social-ecological systems has become central to contemporary debates as a way of bridging concepts that describe human and natural systems as independent entities. What are the strengths and weaknesses of current conceptual frameworks for representing the links between people and nature, and what are problems and prospects of aligning these within research and policy?

- Value pluralism is the idea that there are multiple values which in principle may be equally correct and fundamental, and yet conflict with each other. What are the implications for decision-making in the context of ecosystem services? How can the problem of multiple and conflicting values be resolved?
- The knowledge of ecosystem function is essential for understanding how ecosystem services are generated. How helpful is it to distinguish the capacities and characteristics of ecosystems to generate services from a broader set of ecological structures and processes? Does the concept of 'functional traits' provide a framework for doing this?
- It has been argued that we are now living in the *Anthropocene* and that we now have to think differently about what the economy is, and what it is for, if we are to create sustainable prosperity. What might this 'different thinking' involve? What role might ecosystem services play in building a different economy?

2

ECOSYSTEM SERVICES IN THEORY AND PRACTICE

Robert Costanza

The world has changed dramatically. We no longer live in a world relatively empty of humans and their artifacts. We now live in what some are even calling a new geologic era – the "Anthropocene" (Crutzen, 2002) – a full world where humans are dramatically altering our ecological life support system (Daly, 2005).

Our traditional concepts and models of the economy were developed in the empty world. The conventional view of the "economy" is based on a number of assumptions about the way the world works, what the economy is, and what the economy is for (Table 2.1). In this "empty world" context, built capital – the houses, cars, roads, and factories of the market economy – was the limiting factor. Natural capital – our ecological life support system – and social capital – our myriad relationships with each other – were abundant. It made sense, in that context, not to worry too much about environmental and social "externalities" – effects that occurred outside the market – since they could be assumed to be relatively small and ultimately solvable. It made sense to focus on the growth of the market economy, as measured by Gross Domestic Product (GDP), as a primary means to improve human welfare. It made sense, in that context, to think of the economy as only marketed goods and services and to think of the goal as increasing the amount of the goods and services produced and consumed.

But in the new full world context, we have to think differently about what the economy is and what it is for if we are to create sustainable prosperity. If we seek "improved human well-being and social equity, while significantly reducing environmental risks and ecological scarcities", as the UN has recently proclaimed as the primary global goal (UNEP, 2011, p.2), we are going to need a new vision of the economy and its relationship to the rest of the world, one that is better adapted to the new conditions we face.

We have to first remember that the goal of the economy is to sustainably improve human well-being and quality of life. We have to remember that material consumption and GDP are merely means to that end, not ends in themselves. We have to recognize, as both ancient wisdom and new psychological research tell us, that material consumption beyond real need can actually reduce our well-being. We have to better understand what really does contribute to sustainable human well-being (SHW), and recognize the substantial contributions of natural and social capital, which are now the limiting factors to improving SHW in many countries. We have to be able to distinguish between real poverty in terms of low SHW and merely low monetary income.

Table 2.1 Basic characteristics of the current economic model and the 'ecological economics' model.

	Current Economic Model: the "Washington Consensus"	*Ecological Economics Model*
Primary policy goal	**More:** economic growth in the conventional sense, as measured by GDP. The assumption is that growth will ultimately allow the solution of all other problems. More is always better.	**Better**: focus must shift from merely growth to "development" in the real sense of improvement in quality of life, recognizing that growth has negative by-products and more is not always better.
Primary measure of progress	GDP	ISEW/GPI (or similar)
Scale/carrying capacity	Not an issue since markets are assumed to be able to overcome any resource limits via new technology and substitutes for resources are always available	A primary concern as a determinant of ecological sustainability. Natural capital and ecosystem services are not infinitely substitutable and real limits exist
Distribution/ poverty	Lip service, but relegated to "politics" and a "trickle down" policy: a rising tide lifts all boats	A primary concern since it directly affects quality of life and social capital and in some very real senses is often exacerbated by growth: a too rapidly rising tide only lifts yachts, while swamping small boats
Economic efficiency/ allocation	The primary concern, but generally including only marketed goods and services (GDP) and institutions	A primary concern, but including both market and non-market goods and services and effects. Emphasizes the need to incorporate the value of natural and social capital to achieve true allocative efficiency
Property rights	Emphasis on private property and conventional markets	Emphasis on a balance of property rights regimes appropriate to the nature and scale of the system, and a linking of rights with responsibilities. A larger role for common property institutions in addition to private and state property
Role of Government	To be minimized and replaced with private and market institutions	A central role, including new functions as referee, facilitator and broker in a new suite of common asset institutions
Principles of Governance	*Laissez faire* market capitalism	Lisbon principles of sustainable governance

Ultimately, we have to create a new vision of what the economy is and what it is for, and a new model of development that acknowledges this new full world context and vision (Table 2.1).

Planetary boundaries

Our planet's ability to provide an accommodating environment for humanity is being challenged by our own activities. The environment – our life-support system – is changing rapidly from the stable Holocene state of the last 12,000 years, during which we developed agriculture, villages, cities, and contemporary civilizations, to an unknown future state of significantly different conditions – the Anthropocene.

One way to address this challenge is to determine "safe planetary boundaries" based on fundamental characteristics of our planet and to operate within them. "Boundaries" here mean specific points related to a global-scale environmental process beyond which humanity should not go. Identifying our planet's intrinsic, nonnegotiable limits is not easy, but recently a team of scientists have specified nine areas that are most in need of well-defined planetary boundaries (Rockström *et al.*, 2009). The nine areas they identified as most in need of planetary boundaries are climate change, biodiversity loss, excess nitrogen and phosphorus production, stratospheric ozone depletion, ocean acidification, global consumption of fresh water, change in land use for agriculture, air pollution, and chemical pollution. They estimate that humanity has already transgressed three of these boundaries: climate change, biodiversity loss, and nitrogen production, with several others rapidly approaching the safe boundary.

Clearly, remedial policy responses to date have been local, partial, and inadequate. Early policy discussions and the resulting responses tended to focus on symptoms of environmental damage rather than basic causes, and policy instruments tended to be ad hoc rather than carefully designed for efficiency, fairness, and sustainability. For example, in the 1970s emphasis centered on end-of-pipe pollution control which, while a serious problem, was actually a symptom of expanding populations and inefficient technologies that fueled exponential growth of material and energy throughput while threatening the recuperative powers of the planet's life-support systems.

These problems are all evidence that the material scale of human activity is rapidly approaching, or already exceeds, the safe operating space for humanity on the earth. We are degrading our life-support systems – the ecosystem services provided by our natural capital assets.

Ecosystem services

Ecosystem services are defined as "the benefits people obtain from ecosystems" (Costanza *et al.*, 1997, MA, 2005). These include provisioning services such as food and water; regulating services such as regulation of floods, drought, and disease; supporting services such as soil formation and nutrient cycling; and cultural services such as recreational, scientific, spiritual, and other nonmaterial benefits (Costanza *et al.*, 1997, Daily, 1997, de Groot *et al.* 2002).

This is an appropriately broad and an appropriately vague definition. It includes both the benefits people perceive, and those they do not. The conventional economic approach to "benefits" is far too narrow in this regard, and tends to limit benefits to those that people both perceive and are "willing to pay" for in some real or contingent sense. But the general population's information about the world, especially when it comes to ecosystem services, is extremely limited. We can expect many ecosystem services to go almost unnoticed by the vast majority of people, especially when they are public, non-excludable services that never enter the private, excludable market. Think of the storm regulation value of wetlands (Costanza *et al.*, 2008). How can we expect the average citizen to understand the complex linkages between landscape patterns, precipitation patterns, wetlands, and flood attenuation, when even the best landscape scientists find this an extremely challenging task? We need to remember the definition of ecosystem services (the benefits provided by ecosystems), and acknowledge that the degree to which the public perceives and understands them is a separate (and very important) question. Conventional economic valuation presumes that people have well-formed preferences and enough information about trade-offs that they can adequately judge their "willingness-to-pay." Since these assumptions do not hold for many ecosystem services we must: (1) inform people's preferences by showing the underlying dynamics of the ecosystems in question using models; (2) allow groups to discuss the issues and "construct" their preferences within a framework to inform the discussions; or (3) use other techniques that do not rely directly on preferences to estimate the contribution to human well-being of ecosystem services (i.e. to directly infer marginal contributions to well-being), such as the use of computer models.

In addition, the benefits one receives from functioning ecosystems do not necessarily depend on one's ability to pay for them in monetary units. For example, indigenous populations with no money economy at all derive most of the essentials for life from ecosystem services but have zero ability to pay for them in monetary terms. To understand the value of these ecosystem services we need to understand the trade-offs involved, and these may be best expressed in units of time, energy, land, or other units not necessarily monetary, remembering that the local population may or may not understand or be able to quantify these trade-offs. Finally, if one can express the tradeoffs (value) in one set of units (numerator) and can express the trade-offs between that numerator and another, then one can convert the trade-offs into the other numerator. For example, if we can express trade-offs in units of time and can estimate the time/money trade-off, we can express the time units in monetary terms.

A second issue is that ecosystem services are, by definition, not ends or goals, but means to the end or goal of sustainable human well-being. This does not imply that ecosystems are not also valuable for other reasons, but that ecosystem services are defined as the instrumental values of ecosystems as means to the end of human well-being. An important, but different, distinction some authors have made is one between intermediate services and final services (Boyd and Banzhaf, 2007). It is certainly true that for the purposes of certain aggregation exercises adding intermediate and final services would be double counting. But that does not imply that intermediate services are not services. Think of the production of tires in an economy. Some tires are sold directly to consumers and are part of final demand, while others are sold to car companies and are intermediate products, sold to consumers as parts of cars. The tires themselves are indistinguishable from each other, the only difference being who buys them. When calculating GDP (which is the aggregate of sales to final demand) it would not be appropriate to count both the tires sold to final demand and the tires sold to car companies, since those tires are already counted as parts of the cars sold to final demand. But tires in both cases, whether intermediate or final products, are means to the end of human well-being and are not ends in themselves. Likewise, ecosystem goods and services, whether intermediate (or "supporting" in the MA typology) services or final services, are all contributors to the end of human well-being. Also, ecosystem processes (or functions) and services are not mutually exclusive categories. Some processes or functions are also services, others are not. Some services are intermediate, some are final, and some are partly both.

Ecosystems with embedded humans are complex, dynamic, adaptive systems with non-linear feedbacks, thresholds, hysteresis effects, etc. (Costanza *et al.* 1993). Ecosystem services are therefore not the product of a linear chain from production (means) to direct benefits by people (ends) with no feedbacks or any of the other complexities of the real world. All ecosystem services are, by definition, means to the end of human well-being. Ecosystem processes or functions can also be services (they are not mutually exclusive categories), and the same services can be both intermediate and final. The real world is complex and messy and our systems of classification and definition of ecosystem services should recognize that and work with it, not ignore it in a misguided attempt to impose unrealistic order and consistency.

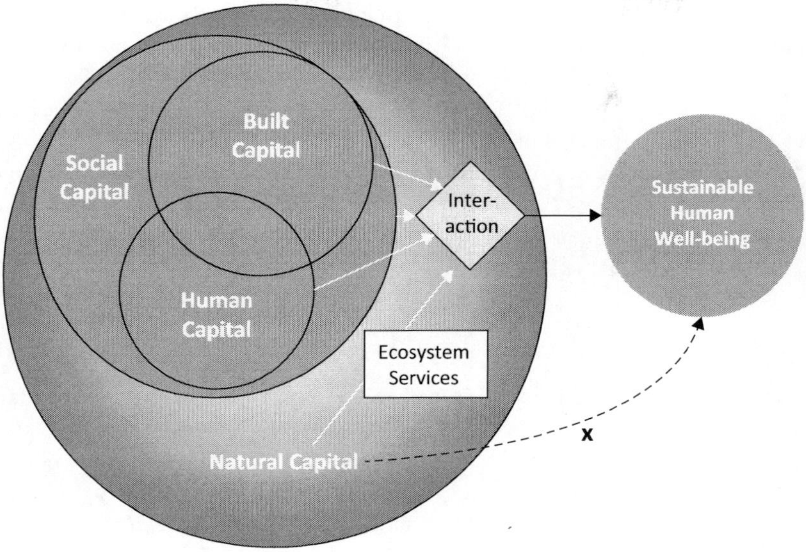

Figure 2.1 Interaction between built, social, human and natural capital required to produce human well-being. Built and human capital (the economy) are embedded in society, which is embedded in the rest of nature. Ecosystem services are the relative contribution of natural capital to human well-being; they do not flow directly. It is therefore essential to adopt a broad, transdisciplinary perspective in order to address ecosystem services.

Source: adapted from Costanza et al., 2014

Natural capital and ecosystem services

The ecosystems that provide the services are referred to as natural capital, using the general definition of capital as a stock that yields a flow of services over time (Costanza and Daly, 1992).In order for these benefits to be realized, natural capital (which does not require human activity to build or maintain) must be combined with other forms of capital that *do* require human agency to build and maintain. These include: (1) built or manufactured capital; (2) human capital; and (3) social or cultural capital (Costanza *et al.*, 1997, 2014, see Figure 2.1).

These four general types of capital are all required in complex combinations to produce any and all human benefits. *Ecosystem services thus refer to the relative contribution of natural capital to the production of various human benefits, in combination with the three other forms of capital.* These benefits can involve the use, non-use, option to use, or mere appreciation of the existence of natural capital.

This categorization of services in the MA is very broad, limited only by the requirement of a contribution to human well-being. Even without any subsequent valuation, explicitly listing the services derived from an ecosystem can help ensure appropriate recognition of the full range of potential impacts of a given policy option. This can help make the analysis of ecological systems more transparent and can help inform decision-makers of the relative merits of different options before them.

To achieve sustainability, we must incorporate natural capital, and the ecosystem goods and services that it provides, into our economic and social accounting and our systems of social choice. In estimating these values we must consider how much of our ecological life-support systems we can afford to lose. To what extent can we substitute manufactured for natural capital, and how much of our natural capital is irreplaceable? For example, could we replace the radiation screening services of the ozone layer if it were destroyed? Because natural capital is not captured in existing markets, special methods must be used to estimate its value. These range from attempts to mimic market behavior using surveys and questionnaires to elicit the preferences of current resource users (i.e. willingness-to-pay, WTP), to methods based on energy analysis (EA) of flows in natural ecosystems (which do not depend on current human preferences at all) (Farber and Costanza, 1987, Costanza *et al.*, 1989, Costanza, 2004). Because of the inherent difficulties and uncertainties in determining these values we are better off with an intelligently pluralistic approach that acknowledges and utilizes these different, independent approaches.

Valuation of ecological systems and services

The issue of valuation is inseparable from the choices and decisions we have to make about ecological systems. Some argue that valuation of ecosystems is either impossible or unwise. For example, some argue that we cannot place a value on such "intangibles" as human life, environmental aesthetics, or long-term ecological benefits. But, in fact, we do so every day. When we set construction standards for highways, bridges and the like, we value human life – acknowledged or not – because spending more money on construction would save lives. These are statistical lives, however, not particular lives, and one should not confuse the two.

People also often talk about "economic value", "ecological value", and "social value" as if they were separate things. Nothing could be further from the truth. As the discussion above makes clear, the "value" or "benefit" we are talking about here is the contribution to sustainable human well-being. None of these elements (ecological, cultural, economic) can make a contribution to that goal without interacting with the others. What we can ask is: what is the relative contribution of, for example, natural capital to sustainable human well-being, in combination with other forms of capital (built, human, social) in a particular context? We have to look at these things in context and as part of an integrated, whole system of humans embedded in cultures embedded in the rest of nature.

Another often-made argument is that we should protect ecosystems for purely moral or aesthetic reasons, and we do not need valuations of ecosystems for this purpose. But there are equally compelling moral arguments that may be in direct conflict with the moral argument to protect ecosystems. For example, the moral argument that no one should go hungry. All we have done is to translate the valuation and decision problem into a new set of dimensions and a new language of discourse. So, while ecosystem valuation is certainly difficult, one choice we do not have is whether or not to do it. Rather, the decisions we make, as a society, about ecosystems imply trade-offs and therefore valuations. We can choose to make these valuations explicit or not; we can undertake them using the best available ecological science and understanding or not; we can do them with an explicit acknowledgment of the huge uncertainties involved or not; but as long as we are forced to make choices we are doing valuation. The valuations are simply the relative weights we give to the various aspects of the decision problem. Society can make better choices about ecosystems if the valuation issue is made as explicit as possible. This means taking advantage of the best information and models we can muster and making uncertainties about valuations explicit too. It also means developing new and better ways to make good decisions in the face of these uncertainties. Ultimately, it means being explicit about our goals as a society, both in the short-term and in the long-term.

The point to stress is that the economic value of ecosystems is connected to their physical, chemical, and biological role in the long-term, global system – whether the present generation of individuals fully recognizes that role or not. If it is accepted that each species, no matter how seemingly uninteresting or lacking in immediate utility, has a role in natural ecosystems (which do provide many direct benefits to humans), it is possible to shift the focus away from our imperfect short-term perceptions and toward the goal of developing more accurate values for long-term ecosystem services. Ultimately, this will involve the collaborative construction of dynamic, evolutionary models of linked ecological economic systems that adequately address long-term responses and uncertainties, like those mentioned above.

Institutions to manage ecosystems and their services

One hears a lot of talk these days about "ecosystem service markets". The problem is, conventional markets are not the right institution for managing many ecosystem services. These services (other than provisioning services) are often "non-rival" and not easily excludable and are therefore best thought of as "public goods" or, more generally, a part of "the commons" (Farley and Costanza, 2010). While we can and should use economic incentives (fees and payments) when appropriate to manage the commons, we need a different institutional form than "markets" within which to do this – something more akin to an "ecosystem trust".

Ruhl *et al.* (2007) document the "anti-ecosystem services bias" prevalent in American property law, regulation, and social norms. One particularly interesting counter-trend to this bias emerges in the "public trust doctrine", an idea that law professor Joseph Sax identified in the 1970s as the only legal doctrine with the breadth and substance to be useful as a comprehensive approach to natural resource (and ecosystem service) management. However, so far the U.S. Supreme Court has declined to take it there. Recent proposals to expand the "commons sector" of the U.S. and global economy by creating "common asset trusts" to manage the atmosphere, water, and other natural capital assets (structured like the Alaska Permanent Fund or the many existing Land Trusts) may be one way of implementing this doctrine (Barnes, 2006, Barnes *et al.*, 2008). For example, a bill has been introduced in the Vermont Senate to create a "Vermont Common Asset Trust", based on the public trust doctrine, to "propertize" (but not privatize) the state's natural and social capital assets in order to better manage them on behalf of their common stakeholders (both living and future). Trusts are widely used and well-developed legal mechanisms designed to protect and manage assets on behalf of specific beneficiaries (Souder and Fairfax, 1996). Extending this idea to the management and protection of whole ecosystems and the services they provide is a new but straightforward extension of this idea. Trusts would define whole ecosystems as common property assets, managed by trustees on behalf of all current and future beneficiaries. Once these common assets are assigned property rights, we can use all the existing property law to manage them more effectively. For example, we can charge fees for damages and make payments for enhancement. This gives Payment for Ecosystem Services (PES) schemes a broader institutional framework within which to operate and can help to drastically reduce transaction costs (see Brouwer, 2016).

While trusts may not be the only or the best institution for managing ecosystem services, they seem to be a move in the right direction. We need to think much more creatively about the design of institutions that are better suited to the common asset nature of ecosystem services.

The promise of ecosystem services: toward a sustainable and desirable future

A new model of the economy and prosperity consistent with our new full world context (Table 2.1) would be based clearly on the goal of sustainable human well-being. It would use measures of progress that clearly acknowledge this goal (i.e. GPI instead of GDP). It would acknowledge the importance of ecological sustainability, social fairness, and real economic efficiency.

Ecological sustainability implies recognizing that natural and social capital are not infinitely substitutable for built and human capital, and that real biophysical limits – planetary boundaries – exist to the expansion of the market economy. Climate change is perhaps the most obvious and compelling of these limits.

Social fairness implies recognizing that the distribution of wealth is an important determinant of social capital and quality of life. The conventional development model, while explicitly aimed at reducing poverty, has bought into the assumption that the best way to do this is through growth in GDP. This has not proved to be the case, and explicit attention to distribution issues is sorely needed. As Frank (2007) has argued, economic growth beyond a certain point sets up a "positional arms race" that changes the consumption context and forces everyone to consume too many positional goods (like houses and cars) at the expense of non-marketed, non-positional goods and services from natural and social capital. Increasing inequality of income actually reduces overall societal well-being, not just for the poor, but across the income spectrum (Wilkinson and Pickett, 2009).

Real economic efficiency implies including all resources that affect sustainable human well-being in the allocation system, not just marketed goods and services. Our current market allocation system excludes most non-marketed natural and social capital assets and services, which are huge contributors to human well-being. The current development model ignores this and therefore does not achieve real economic efficiency. A new, sustainable ecological economic model would measure and include the contributions of natural and social capital and could better approximate real economic efficiency.

The new economic model would also acknowledge that a complex range of property rights regimes are necessary to adequately manage the full range of resources that contribute to human well-being. For example, most natural and social capital assets are public goods. Making them private property does not work well. On the other hand, leaving them as open access resources (with no property rights) does not work well either. What is needed is a third way to propertize these resources without privatizing them. Several new (and old) common property rights systems have been proposed to achieve this goal, including various forms of common property trusts.

The role of government also needs to be reinvented. In addition to government's role in regulating and policing the private market economy, it has a significant role to play in expanding the "commons sector", which can propertize and manage non-marketed natural and social capital assets. It also has a major role to play as facilitator of societal development of a shared vision of what a sustainable and desirable future would look like. As Prugh *et al.* (2000) have argued, strong democracy based on developing a shared vision is an essential prerequisite to building a sustainable and desirable future.

The conventional economic model is not working for either the developed or the developing world. It is not sustainable and it is also not desirable. It is based on a now-obsolete empty world vision and it is leading us to disaster.

We need to accept that we now live in a full world context where natural and social capital are the limiting factors. We could achieve a much higher quality of life, and one that would be ecologically sustainable, socially fair, and economically efficient, if we shifted to a new sustainable development paradigm that incorporates these principles.

The problem is that our entire modern global civilization is, as even former President Bush has acknowledged, "addicted to oil", and addicted to consumption and the conventional development model in general. An addictive substance is something one has developed a dependence on, which is either not necessary or harmful to one's longer-term well-being. Fossil fuels (and excessive material consumption in general) fit the bill. We can power our economies with renewable energy, and we can be happier with lower levels of consumption, but we must first break our addiction to fossil fuels, consumption, and the conventional development model, and as any addict can tell you: "that ain't easy". But in order to break an addiction of any kind, one must first clearly see the benefits of breaking it, and the costs of remaining addicted, facts that accumulating studies like the IPCC reports, the Stern Review (2007), the MA (2005), and many others are making more apparent every day.

What else can we do to help break this addiction? Here are just a few suggestions:

- Create and share a vision of a future with zero fossil fuel use and a quality of life higher than today. That will involve understanding that GDP is a means to an end, not the end itself, and that in some countries today more GDP actually results in less human well-being (while in others the reverse is still true). It will require a focus on sustainable scale and just distribution. It will require an entirely new and broader vision of what the economy is, what it's for, and how it functions.
- Convene a "new Bretton Woods" conference to establish the new measures and institutions needed to replace GDP, the World Bank, the IMF, and the WTO. These new institutions would promote:
  - Shifting primary national policy goals from increasing marketed economic activity (GDP) to maximizing national well-being (GPI or something similar). This would allow us to see the interconnections between built, human, social, and natural capital, and build real well-being in a balanced and sustainable way.
  - Reforming tax systems to send the right incentives by taxing negatives (pollution, depletion of natural capital, overconsumption) rather than positives (labor, savings, investment).
  - Expanding the commons sector by developing new institutions that can propertize the commons without privatizing them. Examples include various forms of common asset trusts, like the atmospheric (or sky) trust (Barnes *et al.*, 2008), payments for depletion of natural and social capital, and rewards for protection of these assets.
  - Reforming international trade to promote well-being over mere GDP growth. This implies protecting natural capital, labor rights, and democratic self-determination first and then allowing trade, rather than promoting the current trade rules that ride roughshod over all other societal values and ignore non-market contributions to well-being.

We can break our addiction to fossil fuels, overconsumption, and the current development model and create a more sustainable and desirable future. It will not be easy; it will require a new vision, new measures, and new institutions. It will require a directed evolution of our entire society (Beddoe *et al.*, 2009). But it is not a sacrifice of quality of life to break this addiction. Quite the contrary, it is a sacrifice not to.

References

Barnes, P. (2006). *Capitalism 3.0: a Guide to Reclaiming the Commons.* Berrett-Koehler, New York.

Barnes, P., Costanza, R., Hawken, P., *et al.* (7 authors) (2008). Creating an earth atmospheric trust. *Science*, vol 319, pp 724.

Beddoe, R., Costanza, R., Farley, J., *et al.* (13 authors) (2009). Overcoming Systemic Roadblocks to Sustainability: the evolutionary redesign of worldviews, institutions and technologies. *Proceedings of the National Academy of Sciences*, vol 106, pp 2483–2489.

Boyd, J., and Banzhaf, S. (2007). What are ecosystem services? The need for standardized environmental accounting units. *Ecological Economics*, vol 63, pp 616–26.

Brouwer, R. (2016). Payments for Ecosystem Services. In Potschin, M., Haines-Young, R., Fish, R., and Turner, R. K. (eds) *Routledge Handbook of Ecosystem Services*. Routledge, London and New York, pp 548–553.

Costanza, R. (2004). Value theory and energy, in: C. Cleveland (ed.) *Encyclopedia of Energy*. Elsevier, Amsterdam, vol 6, pp 337–346.

Costanza, R., Andrade, F., Antunes, P., *et al.* (16 authors) (1998). Principles for sustainable governance of the oceans. *Science*, vol 281, pp 198–199.

Costanza, R. and Daly, H. E. (1992). Natural capital and sustainable development. *Conservation Biology*, vol 6, pp 37–46.

Costanza, R., d'Arge, R., de Groot, R., *et al.* (13 authors) (1997). The value of the world's ecosystem services and natural capital. *Nature*, vol 387, pp 253–260.

Costanza, R., de Groot, R., Sutton, P., *et al.* (8 authors) (2014). Changes in the global value of ecosystem services. *Global Environmental Chang*, vol 26, pp 152–158. Available at: https://sites.google.com/a/idakub.com/www/CV/publications/2014_Costanza_GlobalValueUpdate.pdf

Costanza, R., Farber, S.C., and Maxwell, J. (1989). The valuation and management of wetland ecosystems. *Ecological Economics*, vol 1, pp. 335–361.

Costanza, R., Pérez-Maqueo, O., Martinez, M.L., *et al.* (6 authors) (2008). The value of coastal wetlands for hurricane protection. *Ambio*, vol 37, pp 241–248.

Costanza, R., Wainger, L., Folke, C., and Mäler, K-G. (1993). Modeling complex ecological economic systems: toward an evolutionary, dynamic understanding of people and nature. *BioScience*, vol 43, pp 545–555.

Crutzen, P. (2002). The effects of industrial and agricultural practices on atmospheric chemistry and climate during the Anthropocene. *Journal of Environmental Science and Health*, vol 37, pp 423–424.

Daily, G.C. (1997). *Nature's Services*. Island Press, Washington DC.

Daly, H. E. (2005). Economics in a full world. *Scientific American*, vol 293, pp 100–107.

de Groot, R. S., Wilson, M. A., and Boumans, R.M.J. (2002). A typology for the classification, description and valuation of ecosystem functions, goods and services. *Ecological Economics*, vol 41, pp 393–408.

Farber, S. and Costanza, R. (1987). The economic value of wetlands systems. *Journal of Environmental Management*, vol 24, pp 41–51.

Farley, J. and Costanza, R. (2010). Payments for ecosystem services: from local to global. *Ecological Economics*, vol 69, pp 2060–2068.

Frank, R. (2007). *Falling Behind: How Rising Inequality Harms the Middle Class*. University of California Press, Berkeley CA.

MA (2005). *Ecosystems and Human Well-being: Synthesis*. Island Press, Washington DC and Covelo CA.

Prugh, T., Costanza, R., and Daly, H. (2000). *The Local Politics of Global Sustainability*. Island Press, Washington DC and Covelo CA.

Rockström, J., Steffen, W., Noone, K. *et al.* (29 authors) (2009). A safe operating space for humanity. *Nature*, vol 461, pp 472–475.

Ruhl, J. B., Kraft, S. E. and Lant, C. L. (2007). *The law and policy of ecosystem services*. Island Press, Washington, DC.

Souder, J.A. and Fairfax, S. K. (1996). *State Trust Lands: History, Management and Sustainable Use*. University Press of Kansas, Lawrence KS.

UNEP (2011). *Towards a Green Economy: Pathways to Sustainable Development and Poverty Eradication – A Synthesis for Policy Makers*. Available at: www.unep.org/greeneconomy

Wilkinson, R. and Pickett, K. (2009). *The Spirit Level: Why Greater Equality Makes Societies Stronger*. Bloomsbury Press, New York.

3

DEFINING AND MEASURING ECOSYSTEM SERVICES

Marion Potschin and Roy Haines-Young

Introduction

The term 'ecosystem services' can mean different things to different people. On the one hand this is an advantage, because it can engage people in new conversations about the importance of biodiversity and the environment. In this sense 'ecosystem services' might be thought of as a *boundary object*, that is, an idea that can be adapted to represent different perspectives while retaining some sense of continuity across these different viewpoints (Abson et al., 2014; see also Briefing Note 7.1). On the other, that multi-faceted characteristic is a disadvantage once we come to measure and monitor these things called services: if we cannot agree what they are then people will not believe what is said about them or act on the evidence we collect. These problems of definition are amplified once we start to make a case for valuing or managing ecosystem services (see for example, Ojea et al., 2012) – that is, to apply the concept in a *normative* way.

This *Handbook* demonstrates the different ways that people think about ecosystem services; it is, in fact, a microcosm of the wider literature on the topic. Many authors start, quite legitimately, with the definition provided by the Millennium Ecosystem Assessment (MA, 2005) which describes them simply as *the benefits that ecosystems provide to people*. In contrast, others follow the guide of TEEB (The Economics of Ecosystems and Biodiversity), which views them as *the direct and indirect contributions of ecosystems to human well-being* (De Groot et al. 2010). Services, in other words, give rise to benefits; they are not the same thing. Despite these differences, however, both regard ecosystem services and goods as being synonymous. To add complexity to the debate, it is apparent that not all frame the idea in this way. The UK National Ecosystem Assessment (Mace et al., 2011), for example, suggests that it is 'goods' and 'benefits' that are one and the same, and that it is 'services' that are quite distinct (see also Mace et al., 2012; Mace, 2016). So what's the problem with all these different perspectives? In a sense, we all know what people are 'getting at', namely the importance that nature has for people. The difficulty lies in the fact that if we want to understand how ecosystems provide benefits to people, we need a way of characterising the ecological structures and processes and ecosystem characteristics that underpin them in ways that can be analysed. The aim of this chapter is to take the reader on a journey through the terminology surrounding the idea of ecosystem services, not to convince that there is a single, consistent way of thinking about them, but to provide a guide through a complex, and at times, puzzling terrain.

The ecosystem service cascade

A number of commentators have noted the problems of defining exactly what an ecosystem service is (see, for example, Boyd and Banzhaf, 2007; Wallace, 2007; Fisher and Turner, 2008; Fisher et al., 2009). Despite their differences, all agree that there is some kind of 'pathway' for delivering ecosystem services, which goes from ecological structures and processes at one end through to the well-being of people at the other. We have also represented this 'production line', describing it as a 'cascade' (Figure 3.1). Its purpose is to tease out more clearly the differences between these end-points and the steps between (Potschin and Haines-Young, 2011).

Conceptual frameworks such as the cascade serve a number of purposes. They can be used, for example, as a communication tool, a jumping-off point for discussion between experts and laypeople. Additionally or alternatively they can be used as a way of mapping out basic concepts so that they can be applied to solve problems; they identify the types of evidence that are considered relevant and so help place work on a stronger analytical footing. It is mainly for this last purpose that we use it here.

Thus, we suggest, the cascade model can help us think about the relationships between five key sets of ideas that define the ecosystem services 'paradigm'; that is, a way of looking at the world. We are clearly interested in ecosystems, and these are represented in the cascade model as the set of *ecological structures* and *processes* that we find in an area. Often we simply use some label for a habitat type, such as woodland or saltmarsh, as a catch-all to denote this box, but there is no reason why we cannot also refer to things such as 'the nitrogen cycle', with its various stores and transfers, as something that can also occupy the left-hand side of the diagram. In either case, given the complexity of most ecosystems, if we want to start to understand just how they benefit

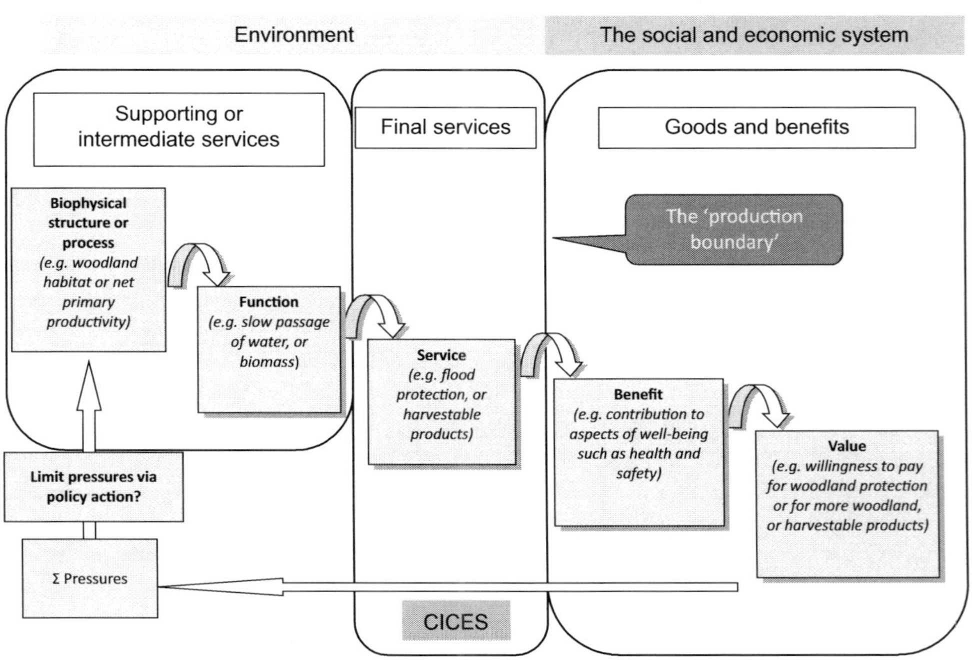

Figure 3.1 The cascade model.

Source: original, Haines-Young and Potschin, 2010; adapted from Potschin and Haines-Young, 2011

people, then it is useful to start to identify those properties and characteristics of the system that are potentially useful to people. This is where the idea of *functions* enter into the discussion. In terms of the cascade model, these are taken to be the subset of characteristics or behaviours that an ecosystem has that determine or 'underpin' its usefulness for people.

As Jax (2016) notes, the term 'function' is problematic for ecologists. For some (including those who prepared the MA) it is often just used as another way of referring to *ecological process*. Indeed, Wallace (2007) has gone so far as arguing that if ecosystem services, processes, structure and composition are adequately defined, the term 'function' is actually not required; Jax also suggests that we even might want to avoid it. However, there are, some advantages in thinking about what it is about an ecosystem that enables it to provide a service, and setting these characteristics or behaviours apart as a 'subset' can be helpful. We would suggest that it is especially helpful if we want to manage these properties in some way. In the case of woodlands, for example, their capacity to mediate runoff can be controlled by their canopy characteristics, and these are not solely determined by woodland type. Similarly, while the structural characteristics of wood that make it more or less useful for timber are determined by growth processes, these can be manipulated within the same woodland type to improve its ability to deliver a harvestable crop. Given the complex nature of ecosystem structures and processes, and that a single ecosystem may deliver a number of benefits, we need to try to be clear about just what capacities (properties, behaviours) make it useful to people; identifying these as 'functional' characteristics is, we suggest, therefore a necessary stage in understanding how ecosystems and people are linked.

Ecosystem services play a pivotal role in the cascade, which constitutes them as distinct from the functional characteristics of the ecosystems that make them useful, and the benefits that people ultimately enjoy. A defining feature of services is that they are, in some sense, the final outputs from an ecosystem. They are 'final' in that they are still connected to the structures and processes that gave rise to them; they are also final in the sense that they contribute directly to some product or condition that can be valued by people. Thus, to return to the woodland example, the standing crop of trees with particular structural characteristics is the service and the harvested, worked timber is the good or benefit. Following the logic used in the UK National Ecosystem Assessment, in the cascade *goods* and *benefits* are the things that have *value*, whether that value is expressed in monetary or non-monetary terms. 'Product' is another term that is sometimes used interchangeably with 'good'.

The distinction between functions, services, and goods and benefits can be clarified still further by recognising that a service may depend on a number of functional characteristics. For example, the utility of a standing crop of trees is dependent on a range of properties other than the characteristics of the woody material, such as the branch and stem characteristics of the stand, stem density and stand age. Similarly, a stand of trees can give rise to several different types of goods and benefits. In addition to its capacity to slow the passage of runoff, for example, those same trees can offer benefits in terms of shelter against winds, dust or noise, as well as a range of recreational activities.

A benefit is basically seen as something that can 'change people's well-being', which is understood to be things like people's health and security, or their social relations, or the kinds of choices they can make. These benefits are thus important to people, and that importance is therefore expressed by the values they assign to them. 'Value' is therefore the final box in the cascade model, on the right-hand side, and, as suggested, these values can be expressed in a number of different ways. Alongside monetary values, people can express the importance they attach to benefits using moral, aesthetic or spiritual criteria. And it is by reference to these values that people and societies chose to act (or not) to modify or manage the pressures on ecosystems and ultimately the benefits they deliver to society. This feedback is highlighted in the arrow running from values back to the left hand side of the cascade model.

Clearly, a limitation of the cascade model is that it seems to suggest a rather linear relationship between ecological structures and processes on the one hand, and benefits and values at the other. In the 'real world', of course, things are more complex and cannot easily be captured in a simple model such as this. Even for a single ecosystem, we can usually identify a network of linkages between a number of different ecological structures and processes, the different functions they support and the suite of benefits that ultimately arises. Nevertheless, the elements of the cascade do give us some of the vocabulary we need to represent and understand the richness of these relationships.

Using the cascade

The novelty of the ecosystem service paradigm stems from the willingness of researchers and practitioners to connect the study of biophysical and social systems. The cascade model can be seen as an entry-point into the discussion – as a tool for representing important elements in the production chain linking nature and people. In any real problem situation there will be many difficult judgements to be made about, for example, what counts as a function or a service or a benefit, etc., because how we interpret these ideas will change with the application context. As Boyd and Banzhaf (2007) memorably point out in their discussion of final ecosystem services: if we take potable water from a lake, the water is a final service, but if instead we eat the fish that live in the lake, then it is the fish and not the water that are the final ecosystem service. What the cascade model brings to these situations is, therefore, a framework that can structure our thinking. The ideas represented by the 'boxes' in the model are rather like words in a sentence which we can use to tell the ecosystem service story; each has meaning by virtue of the way we arrange the other 'words' (ideas) around them.

The point of the cascade model is not to put the world into tightly prescribed boxes, but to help more clearly understand the ways in which nature can influence people's well-being. Whatever terms we choose, the distinction between the contributions that an ecosystem makes and the way that well-being is changed is critical – hence our particular preoccupation with the service-benefit issue. The language used in the MA has helped all of us to make a start, but the basic concepts still need probing more deeply (cf. Lamarque et al., 2011; Portman, 2013).

We can see how the cascade model has helped people work through the logic of the ecosystem service paradigm by reference to some of the published literature. One of the key areas of analysis that it has encouraged people to think about concerns the patterns of supply and demand for ecosystem services. For example, Hansen and Pauleit (2014) have developed and modified the cascade to look at demand and supply relationships in relation to green infrastructure in urban settings, while Bürgi et al. (2015) have looked at the evolution of supply and demand interactions for ecosystem services over time in a Swiss landscape, to gain a deeper understanding of landscape history. Elsewhere, Martín-López et al. (2014) have used the cascade to undertake an empirical study of patterns of supply and demand in the Dõnana social-ecological system, in south-west Spain, while Geijzendorffer et al. (2015) have more generally reviewed some of the literature on the mismatch between the demand and supply. The latter concluded that to properly account for such mismatches studies should include multiple stakeholder groups with their different requirements, recognition that supply is not only determined by biophysical factors but also the services needed by people and hence management inputs, and temporal and spatial scale sensitivities.

Studies using the cascade to assist in understanding the services associated with particular ecosystems include those of Large and Gilvear (2014), who applied it to the analysis of 'riverscapes', and Liquete et al. (2011) in their work on mapping and assessing ecosystem services associated with freshwaters in Europe. In other applications, Plant and Prior (2014) used the cascade to develop a framework for statutory water allocation planning in Australia, while Zhang et al. (2015) applied the framework to help identify the components of plant diversity that are most correlated with ecosystem properties in a restored wetland in northern China, and Ratamäki et al. (2015) used cascade to explore pollination from a multi-level policy perspective.

There is considerable interest in the scientific policy communities in devising appropriate indicators of ecosystem services to 'mainstream' the concept (see Müller et al., 2016). Examples of the way the cascade has stimulated debate include that of Liquete et al. (2013a), who used it to propose three novel coastal protection indicators for European coastlines that cover the main anthropogenic pressures on the coastal zone. Maes et al. (2012b, 2013) have also used cascade to develop spatial indicators of potential and supply, with a view to identifying synergies and trade-offs between ecosystem service supply, biodiversity and habitat conservation, while van Oudenhoven et al. (2012) have used the cascade to develop a framework for indicator selection to assess effects of land management on ecosystem services in the southern part of the Netherlands. In the context of the work on indicators, a number of authors have attempted to make a link between the cascade and the DIPSIR framework (see Müller et al., 2016), including Hering et al. (2015) in their work on indicators for the management of Europe's water resources, and Honrado et al. (2013), who have identified a set of indicators that can sit within the DPSIR framework by looking at the relationships between the cascade concept and the environment factors assessed in Environmental Impact and Strategic Impact Analyses. As Maes et al. (2012a) argue, the policy relevance of work on ecosystem services, indicators and mapping is especially important, and have proposed a stepwise framework to support EU policies in a more effective way.

From such work it is clear that, despite its simplicity, the cascade can provide a foundation for building a number of different assessment approaches. Thus Chapman (2014) has proposed a modified cascade to support monitoring and assessment work linked to an adaptive co-management program in western Kenya; the suggested framework helped decision makers identify programme need, program activities, pathway process variables, moderating process variables, outcomes, and programme value.

In other published work, van Zanten et al. (2014) have used the cascade to explore the impact of the Common Agricultural Policy (CAP) on European agricultural landscapes and ecosystem services. These authors have adapted the cascade to help analyse the influence of commodity markets and policies on the behaviour of land managers and the influence of consumer demand on flows and values of the ecosystem services that originate from the agricultural landscape. In other developments, Cordier et al. (2014) have used the cascade to design a framework for ecosystem services monetisation in ecological-economic modelling. Their aim was to ensure that monetary valuation techniques are better able to contribute to the understanding of the impact of economic activities on changes in ecosystems services and the impact of these changes on economic activities. Applications of the cascade in a broader modelling arena include the work of McVittie et al. (2015), who use the cascade in operationalising an ecosystem services-based approach using Bayesian Belief Networks (BBNs) in the context of modelling the dynamics of riparian buffer strips, and Landuyt et al. (2013), who consider the relevance of the cascade to BBNs more generally.

In contrast to its use as an empirical, analytical framework, the cascade has been used to develop broader theoretical understandings. Pagella and Sinclair (2014), for example, have used an earlier version of the cascade to build a typology for understanding the different types of mapping tools. From their review of over 40 published studies they concluded that the major gaps in relation to our understanding of ecosystem services were the lack of analyses at scales relevant to management interventions; understandings of the pathways linking supply and use of services; synergies and trade-offs between services and the inclusion of stakeholder knowledge and uncertainty. Elsewhere, Vihervaara et al. (2013) used the cascade to categorise ecosystem service research in relation to the themes of the International Long Term Ecological Research (ILTER) network, and Kronenberg (2014) has used the cascade to look at what the current debate on ecosystem services can learn from the past in the literature dealing with 'economic ornithology'. Huang et al. (2015) have also considered agro-ecosystems, but this time from a multi-functional perspective. They observe that these systems have been studied by two scientific communities, and that while they have the same interest in understanding these landscapes from a holistic perspective an analysis of the literature suggests there has been limited interaction and exchange. Each group faces research challenges according to independently operating paradigms. These authors propose a conceptual framework based on the cascade that they suggest could stimulate a dialouge about how to analyse bundles of ecosystem services and the nature of multi-functional agriculture, and provide insights into strategies such as land sharing and land sparing.

Finally, the cascade has stimulated other theoretical approaches or readings of the links between ecosystem services and human well-being. Buchel and Frantzeskaki (2015) used the cascade as a starting point to develop a method that can be used to 'translate' ecosystem services for people using an urban park in Rotterdam. They suggested a modification to the cascade to distinguish cultural ecosystem services from other types of service, because, they argued, fundamentally they arise from the perception of nature, rather than from nature itself.

A modification to the cascade has also been suggested by Spangenberg et al. (2014, 2015) so as to include the notion of the potential of a system to generate ecosystem services. They argued for a 'reverse application' of the underlying cascade logic, so as to understand the 'full cycle of ecosystem services generation and management'. This, they suggest, is particularly helpful in a planning context, where we need to identify uncertainties, the legal and participative foundations of decision-making, and the potentially conflicting private and public interests. Von Haaren et al. (2014) go on to describe a 'practice-oriented' ES evaluation model (PRESET) as a reaction to the cascade, again specifically adapted to the requirements of local and regional planning.

The studies that have used the cascade illustrate the motivation for proposing it, namely to help focus thinking and stimulate discussion. While the cascade can be criticised because there are 'missing links', it never was proposed as a complete picture of the world. Rather, as we have suggested here, it is intended as a heuristic, a way of starting the kinds of conversation that people with different perspectives need to have in relation to the idea of ecosystem services. For particular applications, frameworks for showing the links between people and nature will need to be more fully specified – but in the case of the general use of the cascade, simplicity is perhaps a virtue.

Classifying ecosystem services

The fluid nature of the concept of ecosystem services can be an advantage in stimulating discussion, but it poses problems when we try to measure them, or design a system for classifying them so that we can report results clearly. An illustration of some of the difficulties that can arise is provided by Ojea et al. (2012), who looked at the problems that arise in the context of valuing the water services associated with forests from overlapping and ambiguous definitions in the MA classification. Elsewhere, Wong et al. (2015) have highlighted the difficulty of operationalising ecosystem services unless measurable 'endpoints' that unambiguously represent final ecosystem services can be identified. These kinds of difficulty are compounded by the fact that, even assuming that such endpoints can be agreed upon, the naming of services is often different between initiatives and service categories that appear in one system are not always included elsewhere.

In an attempt to provide a framework that could at least be used to translate between the systems, work was undertaken as part of the development of the revision of the System of Environmental and Economic Accounting (SEEA), led by the United Nations Statistical Division (UNSD). It resulted in the so-called Common International Classification of Ecosystem Services (CICES, Haines-Young and Potschin, 2013), which aimed to help people identify what constituted a final ecosystem service and navigate between the different typologies that have evolved around the ecosystem service concept, and especially to report in a standardised way (e.g. La Notte et al., 2012). While developed initially in an accounting context, it has been taken up more widely by the ecosystem services community, and is, for example, the framework being used in the EU *MAES* Process, which aims to map ecosystem services at the European scale in order to meet the commitments made under Action 5 of the EU's Biodiversity Strategy to 2020 (Maes et al., 2014). In other work, Crossman et al. (2013) suggest that such a classification might be seen as part of a more general systematic approach or 'blue print' for mapping and modelling ecosystem services. In looking to develop these more standardised approaches, Busch et al. (2012) have argued that it is especially important to develop classification systems, such as CICES, that are 'geographically and hierarchically consistent' so that we can make comparisons between regions, and integrate detailed local studies into a broader geographical understanding.

The CICES framework

The evolving nature of the science of ecosystem services and the way it is practiced, together with a field that brings together a range of disciplines, each with their own terminology, means that the design of a classification system that meets all needs is a major challenge. The development of CICES illustrates many of the issues involved, and the fact that we must probably think of the creation of a classification system as a process rather than a design problem that can be solved in a single step.

CICES was created through a consultative process, initially as part of the efforts to design integrated environmental and economic accounting systems, but latterly by involvement of the wider ecosystem service community. A key initial consideration in 2009, when the process began, was that wherever possible the system should have resonance with other widely used classifications, especially in relation to terminology. Thus CICES took as its starting point the typology of ecosystem services suggested in the Millennium Ecosystem Assessment (MA, 2005), and refined it to reflect some of the key issues identified in the wider research literature. For example, it explicitly attempted to identify what are considered to be 'final services', and was designed around the idea of a hierarchy, to accommodate the fact that people worked at different thematic as well as spatial scales.

The version of CICES that is now widely used was published in 2013, and is known as 'version 4.3' (Table 3.1).[1] At the highest or most general level are the three familiar categories used in the MA: provisioning, regulating and maintenance, and cultural. Below these major 'Sections' in the classification are a series of 'Divisions', 'Groups' and 'Classes'. Figure 3.2 shows the way in which the hierarchical structure works for Provisioning Services.

Ecosystem accounts, like more general ecosystem assessments, have to be based on a well-defined and credible metrics which are often specific to particular geographical situations or ecosystem types. For the purposes of reporting or comparison these may need to be aggregated and generalised. The hierarchical structure illustrated in Figure 3.2 allows users to go down to the most appropriate level of detail required by their application, but then group or combine results when making comparisons or more generalised reports. Thus moving down from Section, through Division, Group and Class the 'service' is increasingly more specific, and these detailed service types are nested within the broader categories that sit above them. In the classification

Table 3.1 The Common International Classification of Ecosystem services (CICES,V4.3).

Section	*Division*	*Group*	*Class*	*MA*	*TEEB*
Provisioning	Nutrition	Biomass	Cultivated crops	Food	Food
			Reared animals and their outputs		
			Wild plants, algae and their outputs		
			Wild animals and their outputs		
			Plants and algae from in-situ aquaculture		
			Animals from in-situ aquaculture		
		Water	Surface water for drinking	Water	Water
			Ground water for drinking		
	Materials	Biomass	Fibres and other materials from plants, algae and animals for direct use or processing	Fibre, timber, ornamental, biochemical	Raw materials, medicinal resources
			Materials from plants, algae and animals for agricultural use		
			Genetic materials from all biota	Genetic materials	Genetic materials
		Water	Surface water for non-drinking purposes	Water	Water
			Ground water for non-drinking purposes		
	Energy	Biomass-based energy sources	Plant-based resources	Fibre	Fuels and fibres
			Animal-based resources		
		Mechanical energy	Animal-based energy		
Regulation & Maintenance	Mediation of waste, toxics and other nuisances	Mediation by biota	Bio-remediation by micro-organisms, algae, plants, and animals	Water purification and water treatment, air quality regulation	Waste treatment (water purification), air quality regulation
			Filtration/sequestration/storage/ accumulation by micro-organisms, algae, plants, and animals		

Section	Division	Group	Class	MA	TEEB
		Mediation by ecosystems	Filtration/sequestration/storage/accumulation by ecosystems		
			Dilution by atmosphere, freshwater and marine ecosystems		
			Mediation of smell/noise/visual impacts		
	Mediation of flows	Mass flows	Mass stabilisation and control of erosion rates	Erosion regulation	Erosion prevention
			Buffering and attenuation of mass flows		
		Liquid flows	Hydrological cycle and water flow maintenance	Water regulation	Regulation of water flows, regulation of extreme events
			Flood protection	Natural hazard regulation	
		Gaseous / air flows	Storm protection		
			Ventilation and transpiration	Air quality	Air quality
	Maintenance of physical, chemical, biological conditions	Lifecycle maintenance, habitat and gene pool protection	Pollination and seed dispersal	Pollination	Pollination
			Maintaining nursery populations and habitats		
		Pest and disease control	Pest control	Pest regulation	Biological control
			Disease control	Disease regulation	
		Soil formation and composition	Weathering processes	Soil formation (supporting services)	Maintenance of soil fertility
			Decomposition and fixing processes		
		Water conditions	Chemical condition of freshwaters	Water regulation	Water
			Chemical condition of salt waters		
		Atmospheric composition and climate regulation	Global climate regulation by reduction of greenhouse gas concentrations	Atmospheric regulation	Climate regulation
			Micro and regional climate regulation	Air quality regulation	Air quality regulation

(*Continued*)

Table 3.1 (Continued)

Section	*Division*	*Group*	*Class*	*MA*	*TEEB*
Cultural	Physical and intellectual interactions with biota, ecosystems, and land-/seascapes [environmental settings]	Physical and experiential interactions	Experiential use of plants, animals and land-/seascapes in different environmental settings	Recreation and ecotourism	Recreation and tourism
			Physical use of land-/seascapes in different environmental settings		
		Intellectual and representative interactions	Scientific	Knowledge systems and educational values, cultural diversity, aesthetic values	Inspiration for culture, art and design, aesthetic information
			Educational		
			Heritage, cultural		
			Entertainment		
			Aesthetic		
	Spiritual, symbolic and other interactions with biota, ecosystems, and land-/seascapes [environmental settings]	Spiritual and/or emblematic	Symbolic	Spiritual and religious values	Information and cognitive development
			Sacred and/or religious		
		Other cultural outputs	Existence		
			Bequest		

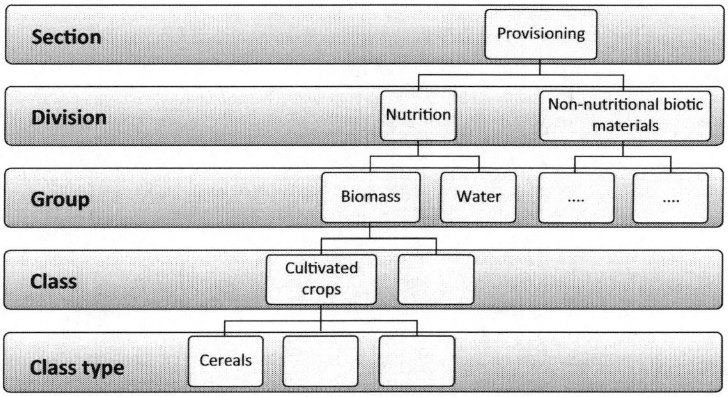

Figure 3.2 The hierarchical structure of CICES.

system there is therefore 'dependency', in the sense that the characteristics used to define services at the lower levels are inherited from the Sections, Divisions and Groups above them. There is also a sense of 'taxonomy' in that elements within the same Group or Class are conceptually more similar to each other, in terms of the ways they are used by people, than they are to services elsewhere in the classification; Table 3.2 sets out the basic definitions at the Section level. At any level in the hierarchy the categories are intended to be exclusive and non-overlapping, so that CICES can be regarded as a classification system rather than an arbitrary nomenclature.

Table 3.1 sets out the basic structure of CICES and also shows the equivalences with the categories used in the typologies of the MA and TEEB.[2] In many cases there is a fairly simple read-across at the group level, but there are categories included in CICES, such as bioenergy, that are not explicitly covered by the others.

The problem of abiotic ecosystem outputs

A key problem with any classification system is to set its boundaries – what should CICES cover and what should it exclude? A key difference that emerged during the consultation was the extent to which the notion of ecosystem services included abiotic outputs such as hydro or wind power, minerals such as salt and so on. On the one hand people argued that although such things were produced by 'natural processes', the fact they were not dependent on *living* processes, meant that the classification would 'water down' the importance that 'biodiversity' had in any future assessments. The position was reinforced by the argument that if abiotic output from nature were included where would we stop – should fossil fuels, for example, also be included? The danger here, people felt, was that if these were included their 'values' as 'ecosystem services' would outweigh many of the others. The counterargument was that many people, especially the 'public' who might be consulted during an ecosystem assessment, would not see the distinction between the biotic and abiotic outputs of ecosystems so clearly. By excluding renewable energy sources such as wind and wave power, for example, would we not tend to exclude a whole category of things that 'nature can do for us'?

The argument about whether abiotic ecosystem service outputs should be included in CICES or any other classification system is a complex one, which is not made easier by the fact that, in all the systems currently used, 'water' is generally included as a provisioning service. Although living processes certainly affect both quantity and quality issues in both the MA and TEEB, water is regarded as a provisioning service, notwithstanding the fact that the 'material' output is largely generated by abiotic, hydro-physical processes.

Table 3.2 Definition of the major categories of ecosystem services used in the CICES V4.3 Classification (after Haines, Young, and Potschin, 2013).

CICES Section	*Definition*
Provisioning	All nutritional, material and energetic outputs from living systems. In the proposed structure a distinction is made between provisioning and material outputs arising from biological or organic materials (biomass) and water. Materials can include genetic structures. The Division for energy makes a distinction between biomass based energy sources, where the organic material is consumed (e.g. fuel wood) and power provided to people by animals.
Regulating and Maintenance	All the ways in which living organisms can mediate or moderate the ambient environment that affects human performance. It therefore covers the degradation of wastes and toxic substances by exploiting living processes. Regulation and maintenance also covers the mediation of flows in solids, liquids and gases that affect people's performance. as well as the ways living organisms can regulate the physico-chemical and biological environment of people.
Cultural	All the non-material, and normally non-consumptive, outputs of ecosystems that affect physical and mental states of people. Cultural services are primarily regarded as the physical settings, locations or situations that give rise to changes in the physical or mental states of people, and whose character are fundamentally dependent on living processes; they can involve individual species, habitats and whole ecosystems. The settings can be semi-natural as well as natural settings (i.e. can include cultural landscapes) providing they are dependent on in situ living processes. In the classification we make the distinction between settings that support interactions that are used for physical activities such as hiking and angling, and intellectual or mental interactions involving analytical, symbolic and representational activities. Spiritual and religious settings are also recognised. The classification also covers the 'existence' and 'bequest' constructs that may arise from people's beliefs or understandings.

In CICES V4.3, abiotic ecosystem outputs were, in the end, excluded from the classification, although a parallel table covering these elements was provided; it applied the same classification logic to define provisioning, regulating and cultural outputs as for the services dependent on living processes. There was no attempt to restrict this list to only those abiotic outputs that were 'renewable' within the human time-frame, or to exclude 'sub-soil' assets. At this stage, however, the provisional classification of abiotic outputs is merely intended as a 'marker', to highlight the fact that we still probably need to develop a more all-encompassing vocabulary for discussing the trade-offs and synergies between the different types of output that ecosystems can provide.

Supporting services

A key difference between CICES and the typology used by the MA, for example, it that it does not include 'supporting services'. This is not because those developing CICES felt that they were unimportant, but because for them the problem was to identify the 'final' outputs from ecosystems that might form the basis of valuation and assessment. As Figure 3.1 suggests, CICES attempts to classify services which sit at the interface between the biophysical and socio-economic components of an integrated 'socio-ecological system'.

In any ecosystem assessment, once the important final services had been agreed upon or identified, then discussions about sustainability and appropriate management strategies would have to focus on the underlying ecosystem structures and processes, and the functional characteristics that give rise to them. Thus the final services are seen as the entry-points for these kinds of discussion, and it was felt that broad labels like 'nutrient cycling' or 'primary production' were not particularly helpful in this respect; for most final services there are probably multiple structures, processes and functions that 'support' them. This is not to say that some kind of agreement about how we describe these processes and functions is unnecessary – only that it is probably part of another 'conversation'.

There is, however, one aspect of the debate about supporting and intermediate services that is relevant to the structure of CICES, and it relates to the difficulty of specifying what a final service is in a particular situation. The difficulty was illustrated above in the discussion about if and when a lake's water or its fish were the final service. Thus there are other services listed in CICES that could be regarded in some situations as having an underpinning role, such as soil formation or pollination. The point here is that while the classification makes space for them, largely on the basis of customary practice, in any particular assessment a judgement has to be made about their status. Pollination might indeed be regarded as a final service if, for example, its value or importance were being compared with some alternative that depended on human intervention. Alternatively, it might be regarded as a 'supporting service' or ecological function delivered by a particular ecosystem if the harvestable fruit crop was being used to sum up the value of all the relevant ecosystem outputs (including pollination) necessary for its production. The responsibility of avoiding 'double counting' is down to the user of the classification and the purpose to which it is put – not only the designer.

Distinguishing services, goods and benefits

A particularly difficult problem that the design of CICES illustrates for those interested in classification systems for ecosystem services is the distinction between services and benefits. For those who regard services *as* benefits there is of course no problem. For those who argue that that there is a difference between them there is a problem of terminology, because services are defined as the 'activity or function of an ecosystem that *provides* benefit' while benefits are taken to be 'the many ways that human wellbeing is enhanced through the processes and functions of ecosystems via ecosystem services' (cf. Mace et al., 2012, italics ours). The problem with a consultative process such as that which led to CICES is that different people mixed the approaches and in some areas there is a blurring of categories.

In the discussion of the cascade model (see above), we suggested that final services were at the 'production boundary' where the link to ecological structures, processes and functions was broken. This is easy to visualise in the case where a crop is harvested. Thus the wheat growing in a field is the 'service' (in the sense that it is the result of all the activities or functions in the biophysical part of the socio-ecological system), while the grain in the combine harvester is the good (or benefit) – the thing that can be valued. The 'production boundary' is also easy to imagine when waste streams are reconnected to ecological processes to take advantage of 'bio-remediation services'. It is more difficult to visualise in the case of some other regulating services, especially cultural ones.

It is in the area of cultural services where many of the issues surrounding the problem of distinguishing services and benefits can be illustrated. In order to resolve the different positions in the consultative process, the design of CICES took a mixed approach by using the notion of 'environmental settings' to frame cultural services at the higher levels in the classification, and the more familiar terms used to refer to cultural services, such as 'recreation' or 'education', at the class level. As Chan et al. (2012) have noted, the classification of cultural services is particularly challenging, and these authors suggested that they might be regarded as the 'ecosystems' contributions to the non-material benefits (e.g. capabilities and experiences) that arise from 'human–ecosystem relationships' (Chan et al., 2012, p.9). This is also the approach taken in the UK National Ecosystem Assessment, where these 'contributions' are attributed to the locations (settings) or ecological features (e.g. species) that generate some benefit by virtue of some set of cultural practices (see also Church et al., 2014 and Tratalos et al., 2015). Thus 'walking' might generate the benefit of 'recreation' or 'spiritual fulfilment' in a woodland or coastal setting; the cultural practice of 'bird watching' might similarly generate a number of cultural benefits. These examples illustrate that for the non-material ecosystem outputs the 'production boundary' is crossed when the output is linked to some kind of relationship that people have with an ecosystem which then changes their well-being in some way. As Chan et al. (2012) argue, these non-material cultural benefits can include capabilities and experiences; by extension, to the non-material regulating services equivalent regulatory benefits would include such things as protection from storms or mediation of the ambient environment in which people live. An attempt to use a previous version of CICES in this way, to look at the interface between services and benefits, is provided by Staub et al. (2011) in an insightful study undertaken by the Swiss Federal Office for the Environment.

Developing our classification systems

Costanza (2008) has argued that multiple ways of classifying ecosystem services are needed, and usefully pointed to how they might be described in terms of spatial scale, or according to characteristics such as excludability and rivalness. It is indeed the case that we need to develop much richer vocabularies for describing the ways people and nature are linked. As for CICES, the purpose of stabilising the framework in 2013 as 'version 4.3' was to encourage people to test it in a practical way, so that future refinements could be informed by evidence rather than just opinion. Coming from an initiative that saw ecosystem accounting as 'experimental' meant that it was accepted that ideas need to be tested and refined.

In terms of its application, CICES has been used as the basis of the German TEEB study (Naturkapital Deutschland – TEEB DE, 2014) as well as the German National Ecosystem Assessment, NEA-D (Albert et al., 2014). Elsewhere it has been refined at the most detailed class level to meet the requirements of ecosystem assessment in Belgium (Turkelboom et al., 2013). Saastamoinen et al. (2014) have used it to classify ecosystem services associated with the boreal forests of Finland. Accounting applications include those of Schröter et al. (2014). Elsewhere, CICES has been used to look at the basis for developing or comparing indicators of ecosystem service supply and demand; examples include the work of Castro et al. (2014), Kosenius et al. (2013) and von Haaren et al. (2014). And, in other work, Bürgi et al. (2015) have used CICES to examine how ecosystem service output has changed for a Swiss landscape since about 1900; the classification framework was used to code the reports from achieve sources about whether things that we would now regard as ecosystem services were documented as important in past periods. However, while these applications of CICES suggest that the current framework is appropriate for many uses, it is clear from the work of Armstrong et al. (2012), and Liquete et al. (2013b), for example, that it may need to be adapted to ensure that it is suitable for the assessment of marine and coastal ecosystems, or integrated more closely with typologies for describing underlying ecosystem function. The recent development of the FEGS system by the US-EPA (see Landers et al., 2016) also suggests that there may be some scope to look at the way services, benefits and beneficiaries are aligned in different classification systems so that a more complete picture of the service cascade can be established.

Conclusion

Although the idea of ecosystem services is simple in concept, its application in management and policy is complex. If we are to deliver the practical benefits of managing natural capital in ways that can help sustain human well-being, it is clear that to overcome some of these challenges we need to find a means of describing and measuring ecosystems and their services consistently. Thus a discussion of how to define and classify services is not simply an academic matter, but rather central to any efforts to operationalise the ecosystem service paradigm. A critical discussion of the cascade model and the attempts to develop a Common International Classification for Ecosystem Services (CICES) is, we suggest, an important part of this evolutionary process.

Acknowledgements

The work done in this chapter was partly supported by the EU project 'Operationalising Natural Capital and Ecosystem Services (OpenNESS)' (EC grant agreement no 308428) and the European Environment Agency (EEA) Framework Service Contract No. EEA/BSS/07/007 for 'Support to the implementation of ecosystem accounting' as well as Framework contract No3421/B2014/EEA.55703/.

Notes

1 www.cices.eu
2 A simple tool for helping people make the translation is available at: http://openness.hugin.com/example/cices

References

Abson, D.J., von Wehrden, H., Baumgärtner, S., et al. (11 authors) (2014). Ecosystem services as a boundary object for sustainability. *Ecological Economics*, vol 103, pp 29–37.

Albert, C., Neßhöver, C., Wittmer, H., Hinzmann, M., and Görg, C. (2014). Sondierungsstudie für ein Nationales Assessment von Ökosystemen und ihren Leistungen für Wirtschaft und Gesellschaft in Deutschland. Helmholtz-Zentrum für Umweltforschung – UFZ, unter Mitarbeit von K. Grunewald und O. Bastian (IÖR), Leipzig.

Armstrong, C.W., Foley, N.S., Tinch, R., and van den Hove, S. (2012). Services from the deep: steps towards valuation of deep sea goods and services. *Ecosystem Services*, vol 2, pp 2–13.

Boyd, J., and Banzhaf, S. (2007). What are ecosystem services? The need for standardized environmental accounting units. *Ecological Economics*, vol 63, no 2, pp 616–626.

Buchel, S., and Frantzeskaki, N. (2015). Citizens' voice: a case study about perceived ecosystem services by urban park users in Rotterdam, the Netherlands. *Ecosystem Services*. Available online since January 2015, doi:10.1016/j.ecoser.2014.11.014

Bürgi, M., Silbernagel, J., Wu, J., and Kienast, F. (2015). Linking ecosystem services with landscape history. *Landscape Ecology*, vol 30, no 1, pp 11–20.

Busch, M., La Notte, A., Laporte, V., and Erhard, M. (2012). Potentials of quantitative and qualitative approaches to assessing ecosystem services. *Ecological Indicators*, vol 21, pp 89–103.

Castro, A.J., Verburg, P.H., Martín-López, B., et al. (7 authors) (2014). Ecosystem service trade-offs from supply to social demand: a landscape-scale spatial analysis. *Landscape and Urban Planning*, vol 132, pp 102–110.

Chan, K.M., Satterfield, T., and Goldstein, J. (2012). Rethinking ecosystem services to better address and navigate cultural values. *Ecological Economics*, vol 74, pp 8–18.

Chapman, S. (2014). A framework for monitoring social process and outcomes in environmental programs. *Evaluation and Program Planning*, vol 47, pp 45–53.

Church, A., Fish, R., Haines-Young, R., et al. (15 authors) (2014). *UK National Ecosystem Assessment Follow-on. Work Package Report 5: Cultural Ecosystem Services and Indicators*. UNEP-WCMC, LWEC. Available at: http://uknea.unep-wcmc.org/Resources/

Cordier, M., Agúndez, J.A.P., Hecq, W., and Hamaide, B. (2014). A guiding framework for ecosystem services monetization in ecological–economic modeling. *Ecosystem Services*, vol 8, pp 86–96.

Costanza, R. (2008). Ecosystem services: multiple classification systems are needed. *Biological Conservation*, vol 141, no 2, pp 350–352.

Crossman, N.D., Burkhard, B., Nedkov, S., et al. (14 authors) (2013). A blueprint for mapping and modelling ecosystem services. *Ecosystem Services*, vol 4, pp 4–14.

de Groot, R.D., Fisher, B., Christie, M., Aronson, J., Braat, L., Haines-Young, R., et al. (12 authors). (2010). Integrating the ecological and economic dimensions in biodiversity and ecosystem service valuation. In Kumar, P. (ed.) *The Economics of Ecosystems and Biodiversity (TEEB): Ecological and Economic Foundations*. Earthscan, London and Washington DC.

Fisher, B., and Turner, R.K. (2008). Ecosystem services: classification for valuation. *Biological Conservation*, vol 141, no 5, pp 1167–1169.

Fisher, B., Turner, R. K., and Morling, P. (2009). Defining and classifying ecosystem services for decision making. *Ecological Economics*, vol 68, no 3, pp 643–653.

Geijzendorffer, I.R., Martín-López, B., and Roche, P.K. (2015). Improving the identification of mismatches in ecosystem services assessments. *Ecological Indicators*, vol 52, pp 320–331.

Haines-Young, R. and Potschin, M. (2010). The links between biodiversity, ecosystem services and human well-being. In: Raffaelli, D.G and Frid, C.L.J. (eds) *Ecosystem Ecology: A New Synthesis*. Cambridge University Press, Cambridge, pp 110–139.

Haines-Young, R., and M. Potschin (2013). *Common International Classification of Ecosystem Services (CICES)*, Version 4.3. Report to the European Environment Agency. Available at: www.cices.eu

Hansen, R., and Pauleit, S. (2014). From multifunctionality to multiple ecosystem services? A conceptual framework for multifunctionality in green infrastructure planning for urban areas. *Ambio*, vol 43, no 4, pp 516–529.

Hering, D., Carvalho, L., Argillier, C., et al. (20 authors) (2015). Managing aquatic ecosystems and water resources under multiple stress – an introduction to the MARS project. *Science of The Total Environment*, vol 503, pp 10–21.

Honrado, J.P., Vieira, C., Soares, C., et al. (7 authors) (2013). Can we infer about ecosystem services from EIA and SEA practice? A framework for analysis and examples from Portugal. *Environmental Impact Assessment Review*, vol 40, pp 14–24.

Huang, J., Tichit, M., Poulot, M., et al. (6 authors) (2015). Comparative review of multifunctionality and ecosystem services in sustainable agriculture. *Journal of Environmental Management*, vol 149, pp 138–147.

Jax, K. (2016). Ecosystem functions: a critical perspective. In Potschin, M., Haines-Young, R., Fish, R., and Turner, R. K. (eds) *Routledge Handbook of Ecosystem Services*. Routledge, London and New York, pp 42–44.

Kosenius, A.K., Haltia, E., Horne, P., Kniivilä, M., and Saastamoinen, O. (2013). *Value of ecosystem services? Examples and experiences on forests, peatlands, agricultural lands, and freshwaters in Finland*. PTT Working Papers 244. Pellervo Economic Research, Helsinki.

Kronenberg, J. (2014). What can the current debate on ecosystem services learn from the past? Lessons from economic ornithology. *Geoforum*, vol 55, pp 164–177.

La Notte, A., Maes, J., Thieu, V., Bouraoui, F., and Masi, F. (2012). Biophysical Assessment and Monetary Valuation of Ecosystem Services. JRC Science and Policy Reports, Publications Office of the European Union.

Lamarque, P., Quétier, F., and Lavorel, S. (2011). The diversity of the ecosystem services concept and its implications for their assessment and management. *Comptes Rendus Biologies*, vol 334, no 5, pp 441–449.

Landers, D., Nahil, M., and Rhodes, C. R. (2016). The beneficiary perspective – benefits and beyond. In Potschin, M., Haines-Young, R., Fish, R., and Turner, R. K. (eds) *Routledge Handbook of Ecosystem Services*. Routledge, London and New York, pp 74–87.

Landuyt, D., Broekx, S., D'hondt, R., et al. (6 authors) (2013). A review of Bayesian belief networks in ecosystem service modelling. *Environmental Modelling & Software*, vol 46, pp 1–11.

Large, A.R.G., and Gilvear, D.J. (2014). Using Google Earth, a virtual-globe imaging platform, for ecosystem services-based river assessment. *River Research and Applications*. doi: 10.1002/rra.2798

Liquete, C., Maes, J., La Notte, A., and Bidoglio, G. (2011). Securing water as a resource for society: an ecosystem services perspective. *Ecohydrology & Hydrobiology*, vol 11, no 3, pp 247–259.

Liquete, C., Zulian, G., Delgado, I., Stips, A., and Maes, J. (2013a). Assessment of coastal protection as an ecosystem service in Europe. *Ecological Indicators*, vol 30, pp 205–217.

Liquete, C., Piroddi, C., Drakou, E.G., et al. (7 authors) (2013b). Current status and future prospects for the assessment of marine and coastal ecosystem services: a systematic review. *PLoS One*, vol 8, no 7, e67737.

MA (2005). *Ecosystems and Human Well-being: Current State and Trends*. Island Press, Washington, DC.

Mace, G. (2016). Ecosystem services: where is the discipline heading? In Potschin, M., Haines-Young, R., Fish, R. and Turner, R. K. (eds) *Routledge Handbook of Ecosystem Services*. Routledge, London and New York, pp 602–605.

Mace, G., Bateman, I., Albon, S., et al. (11 authors) (2011). Conceptual framework and methodology. In *The UK National Ecosystem Assessment Technical Report*. UK National Ecosystem Assessment, UNEP-WCMC, Cambridge.

Mace, G.M., Norris, K., and Fitter, A.H. (2012). Biodiversity and ecosystem services: a multilayered relationship. *Trends in Ecology & Evolution*, vol 27, no 1, pp 19–26.

Maes, J., Egoh, B., Willemen, L., et al. (14 authors) (2012a). Mapping ecosystem services for policy support and decision making in the European Union. *Ecosystem Services*, vol 1, no 1, pp 31–39.

Maes, J., Paracchini, M.L., Zulian, G., Dunbar, M.B., and Alkemade, R. (2012b). Synergies and trade-offs between ecosystem service supply, biodiversity, and habitat conservation status in Europe. *Biological Conservation*, vol 155, pp 1–12.

Maes, J., Hauck, J., Paracchini, M.L., et al. (10 authors) (2013). Mainstreaming ecosystem services into EU policy. *Current Opinion in Environmental Sustainability*, vol 5, no 1, pp 128–134.

Maes, J., Teller, A., Erhard, M., et al. (45 authors) (2014). *Mapping and Assessment of Ecosystems and their Services. Indicators for ecosystem assessments under Action 5 of the EU Biodiversity Strategy 2020*. 2nd final report, European Union, February 2014.

Martín-López, B., Gómez-Baggethun, E., García-Llorente, M., and Montes, C. (2014). Trade-offs across value-domains in ecosystem services assessment. *Ecological Indicators*, vol 37, pp 220–228.

McVittie, A., Norton, L., Martin-Ortega, J., et al. (6 authors) (2015). Operationalizing an ecosystem services-based approach using Bayesian Belief Networks: an application to riparian buffer strips. *Ecological Economics*, vol 110, pp 15–27.

Müller, M., Burkhard, B., Hou, Y., et al. (6 authors) (2016). Indicators for ecosystem services. In Potschin, M., Haines-Young, R., Fish, R., and Turner, R. K. (eds) *Routledge Handbook of Ecosystem Services*. Routledge, London and New York, pp 157–169.

Naturkapital Deutschland – TEEB DE (2014). *Naturkapital und Klimapolitik – Synergien and Konflikte. Langfassung. Technische Universität Berlin, Helmholtz-Zentrum für Umweltforschung – UFZ, Leipzig*. Available at: www.naturkapital-teeb.de

Ojea, E., Martin-Ortega, J., and Chiabai, A. (2012). Defining and classifying ecosystem services for economic valuation: the case of forest water services. *Environmental Science & Policy*, vol 19, pp 1–15.

Pagella, T.F., and Sinclair, F.L. (2014). Development and use of a typology of mapping tools to assess their fitness for supporting management of ecosystem service provision. *Landscape Ecology*, vol 29, no 3, pp 383–399.

Plant, R., and Prior, T. (2014). An ecosystem services framework to support statutory water allocation planning in Australia. *International Journal of River Basin Management*, vol 12, no 3, pp 219–230.

Portman, M.E. (2013). Ecosystem services in practice: challenges to real world implementation of ecosystem services across multiple landscapes – a critical review. *Applied Geography*, no 45, pp 185–192.

Potschin, M., and Haines-Young, R. (2011). Ecosystem services: exploring a geographical perspective. *Progress in Physical Geography*, vol 35, no 5, pp 575–594.

Ratamäki, O., Jokinen, P., Sørensen, P.B., Breeze, T., and Potts, S. (2015). A multilevel analysis on pollination-related policies. *Ecosystem Services*, online.

Saastamoinen, O., Matero, J., Horne, P. et al. (7 authors) (2014). Classification of boreal forest ecosystem goods and services in Finland. Publications of the University of Eastern Finland. *Reports and Studies in Forestry and Natural Sciences* Number 11, University of Eastern Finland, Faculty of Science and Forestry, School of Forest Sciences.

Schröter, M., Barton, D.N., Remme, R.P., and Hein, L. (2014). Accounting for capacity and flow of ecosystem services: a conceptual model and a case study for Telemark, Norway. *Ecological Indicators*, vol 36, pp 539–551.

Spangenberg, J.H., von Haaren, C., and Settele, J. (2014). The ecosystem service cascade: further developing the metaphor. Integrating societal processes to accommodate social processes and planning, and the case of bioenergy. *Ecological Economics*, vol 104, pp 22–32.

Spangenberg, J.H., Görg, C., Truong, D.T., et al. (6 authors) (2015). Provision of ecosystem services is determined by human agency, not ecosystem functions. Four case studies. *International Journal of Biodiversity Science, Ecosystem Services & Management*, vol 10, no 1, pp 40–53.

Staub, C., Ott, W., Heusi, F. et al. (12 authors) (2011). *Indicators for Ecosystem Goods and Services: Framework, methodology and recommendations for a welfare-related environmental reporting*. Federal Office for the Environment, Bern. Environmental studies no. 1102: 17 S.

Tratalos, J. A., Haines-Young, R., Potschin, M., Fish, R., and Church, A. (2015). Cultural ecosystem services in the UK: lessons on designing indicators to inform management and policy. *Ecological Indicators* (in press).

Turkelboom, F., Raquez, P., Dufrêne, M., et al. (19 authors) (2013). CICES going local: ecosystem services classification adapted for a highly populated country. In Jacobs, S., Dendoonker, N., and Keune, H. (eds) *Ecosystem Services*. Chicago.

van Oudenhoven, A.P.E., Petz, K., Alkemade, R., Hein, L., and de Groot, R. S. (2012). Framework for systematic indicator selection to assess effects of land management on ecosystem services. *Ecological Indicators*, vol 21, pp 110–122.

van Zanten, B.T., Verburg, P.H., Espinosa, M., et al. (15 authors) (2014). European agricultural landscapes, common agricultural policy and ecosystem services: a review. *Agronomy for Sustainable Development*, vol 34, no 2, pp 309–325.

Vihervaara, P., D'Amato, D., Forsius, M., et al. (18 authors) (2013). Using long-term ecosystem service and biodiversity data to study the impacts and adaptation options in response to climate change: insights from the global ILTER sites network. *Current Opinion in Environmental Sustainability*, vol 5, no 1, pp 53–66.

von Haaren, C., Albert, C., Barkmann, J., de Groot, R.S., Spangenberg, J.H., Schröter-Schlaack, C., and Hansjürgens, B. (2014). From explanation to application: introducing a practice-oriented ecosystem services evaluation (PRESET) model adapted to the context of landscape planning and management. *Landscape Ecology*, vol 29, no 8, pp 1335–1346.

Wallace, K.J. (2007). Classification of ecosystem services: problems and solutions. *Biological Conservation*, vol 139, no 3, pp 235–246.

Wong, C.P., Jiang, B., Kinzig, A.P., Lee, K.N., and Ouyang, Z. (2015). Linking ecosystem characteristics to final ecosystem services for public policy. *Ecology Letters*, vol 18, no 1, pp 108–118.

Zhang, Y., Wang, R., Kaplan, D., and Liu, J. (2015). Which components of plant diversity are most correlated with ecosystem properties? A case study in a restored wetland in northern China. *Ecological Indicators*, vol 49, pp 228–236.

Briefing Note 3.1 Ecosystem functions: a critical perspective

Kurt Jax

Different meanings of "function" in an ecosystem services context

The term "function" is used in different ways within the environmental sciences literature (see Jax 2005 for details). In an ecosystem services context, "ecosystem function" is frequently part of the frameworks used to describe the relationships between ecosystems, human benefits and well-being, e.g. the "cascade model" (Figure 3.1). Here it is usually situated on the "supply side" of the scheme and forms part of the biophysical part of the framework. In these frameworks the notion of function is used in two major meanings. First, as denoting ecosystem processes that give rise to specific services, and second, the capacity (or potential) of ecosystems to provide services to humans.

The first meaning corresponds to that used in the Millennium Ecosystem Assessment (MA, 2005, p. 895) and e.g. by Wallace (2007), or Luck et al. (2009). Luck et al. (2009, p. 249), for example, state: "Ecosystem process: Synonymous with ecosystem function. The interactions among biotic and abiotic elements of ecosystems that lead to a certain result." Other authors, however, understand ecosystem function in terms of a capability: "[W]e explicitly define ecosystem functions as 'the capacity of natural processes and components to provide goods and services that satisfy human needs [. . .]'" (de Groot et al., 2002, p. 394). In the TEEB report, "ecosystem function" is characterised as "the subset of the interactions between biophysical structures, biodiversity and ecosystem processes that underpin the capacity of an ecosystem to provide ecosystem services" (Kumar, 2010, p. xxxiii), i.e. not as the capacity itself but as something that underpins the capacity. In the graph within the same report (de Groot et al., 2010, p. 17), it even refers to "the subset of biophysical structure or process providing the service", thus not referring to a capacity any more at all. Potschin and Haines Young (2011, p. 578) take up the original definition by de Groot and state: "'[F]unction' is being used [. . .] to indicate some capacity or capability of the ecosystem to do something that is potentially useful to people." In all these latter definitions, "functions" already include a normative dimension, i.e. ecosystem functions have a specific purpose, involving the identification of nature's benefits for humans, and ultimately human well-being. Likewise, Bastian et al. (2012, p. 9), emphasise a traditional meaning of "functions" as "tasks which an area [or ecosystem] is to fulfil" for humans, i.e. as "purposes".

A critical evaluation of the concept in an ecosystem services context

As above described, different ideas on what "ecosystem function" means exist. Based on the analysis above, the main meanings of "ecosystem function(s)" found in the literature are:

- (any) ecosystem processes (and other ecosystem properties) (merely descriptive);
- selected processes (and other ecosystem properties) underpin ecosystem services;
- selected processes (and other ecosystem properties) that are ecosystem services;
- capacities/potentials of an ecosystem to provide ecosystem services;
- selected processes (and other ecosystem properties) that underpin the capacities/potentials to provide ecosystem services;
- the "tasks" of ecosystems for the benefits of humans.

One problem that obviously arises from this diversity of meanings is the danger of terminological confusion. The same word ("ecosystem functions") signifies different things. If the specific meaning is not explained or becomes clear from the context, this can (and does) lead to confusion and communication problems.

A second problem for some definitions of "ecosystem function(s)" is how to operationalise the concept, especially when understanding ecosystem function as denoting capacities to provide ecosystem services. While the idea of separating the potential of delivering ecosystem services from its actual delivery is very useful and practically relevant, hardly any of the papers referring to ecosystem functions as a capacity explain in some more detail what is meant by "capacity" or "potential" and how it might be operationalised (but see e.g. Schröter et al., 2014).

Describing ecosystem functions in the sense of ecosystem processes (and sometimes other ecosystem attributes) is, in contrast, quite straightforward and close to the everyday work of ecologists. The degree to which the concepts used must be operationisable depends on the specific task at hand. It might not be necessary when the ecosystem services concept is just used for didactic purposes (e.g. as arguing for the usefulness of nature), but will certainly be needed for quantitative ecosystem services assessments.

Conclusions

Given the high ambiguity of the term "ecosystem function(s)" in an ecosystem services context, the term should be either defined explicitly and then used consistently (!) or be avoided entirely. Examples of ecosystem services frameworks which, for the sake of clarity, explicitly avoid the term as denoting one of their basic concepts are those of Wallace (2007), Bastian et al. (2012), and Schröter et al. (2014).

Acknowledgements

The research leading to these thoughts was supported in part by funding from the European Commission's Seventh Framework Programme of the project "OpenNESS" (Grant agreement no. 308428). Thanks to my colleagues from this project for several interesting discussions on the issue.

References

Bastian, O., Haase, D., and Grunewald, K. (2012). Ecosystem properties, potentials and services – The EPPS conceptual framework and an urban application example. *Ecological Indicators*, vol 21, pp 7–16.

de Groot, R.S., Wilson, M.A., and Boumans, R.M.J. (2002). A typology for the classification, description and valuation of ecosystem functions, goods and services. *Ecological Economics*, vol 41, pp 393–408.

de Groot, R.S., Alkemade, R., Braat, L., Hein, L., and Willemen, L. (2010). Challenges in integrating the concept of ecosystem services and values in landscape planning, management and decision making. *Ecological Complexity*, vol 7, pp 260–272.

Jax, K. (2005). Function and "functioning" in ecology? What does it mean? *Oikos*, vol 111, pp 641–648.

Kumar, P. (ed.) (2010). *The Economics of Ecosystems and Biodiversity: Ecological and Economic Foundations*. Earthscan, London and Washington DC.

Luck, G.W., Harrington, R., Harrison, P., Kremen, C., Berry, P. M., Bugter, R., Dawson, T.P., de Bello, F., Diaz, S., Feld, C.K., Haslett, J.R., Hering, D., Kontogianni, A., Lavorel, S., Rounsevell, M., Samways, M.J., Sandin, L., Settele, J., Sykes, M.T., van den Hove, S., Vandewalle, M., and Zobel, M. (2009). Quantifying the contribution of organisms to the provision of ecosystem Services. *Bioscience*, vol 59, pp 223–235.

MA (2005). *Ecosystems and Human Well-being: Synthesis*, Island Press, Washington DC.

Potschin, M.B., and Haines-Young, R.H. (2011). Ecosystem services: exploring a geographical perspective. *Progress in Physical Geography*, vol 35, pp 575–594.

Schröter, M., Barton, D.N., Remme, R.P., and Hein, L. (2014). Accounting for capacity and flow of ecosystem services: a conceptual model and a case study for Telemark, Norway. *Ecological Indicators*, vol 36, pp 539–551.

Wallace, K.J. (2007). Classification of ecosystem services: problems and solutions. *Biological Conservation*, vol 139, pp 235–246.

4

THE LINKS BETWEEN BIODIVERSITY AND ECOSYSTEM SERVICES

Patricia Balvanera, Sandra Quijas, Berta Martín-López, Edmundo Barrios, Laura Dee, Forest Isbell, Isabelle Durance, Piran White, Ryan Blanchard and Rudolf de Groot

Introduction

The well-being of society depends to a large extent on the benefits derived from the functions and processes that take place within ecosystems (i.e., ecosystem services). Biodiversity plays an important role in the delivery of many of these benefits. However, human activities that derive services from ecosystems may also have adverse impacts on ecosystems and their biodiversity. This can negatively impact societal well-being, if degradation of biodiversity results in a decline in the quantity, quality, or resilience of ecosystem service provision. Understanding how biodiversity is linked to ecosystem services is critical for designing more sustainable environmental policies and landscape planning.

The significance of declines in biodiversity and the consequences for ecosystem services are increasingly being recognized. For instance, the over-exploitation of fish stocks has led to declines in marine biodiversity via by-catch and fisheries collapse (Worm *et al.*, 2006). Declines in numbers and diversity of wild insect pollinators have been linked to changes in fruit set for many highly valuable crops (Luck *et al.*, 2009). In the absence of effective management, the effects of declining biodiversity and ecosystem degradation will be exacerbated by climate change, with consequences especially for the well-being of future generations.

In this chapter, we first examine key concepts that are relevant to understanding the links between biodiversity and ecosystem services. We then review the relationship between biodiversity and ecosystem services, as well as the complexities arising from such linkages. We then provide an in-depth description of the links between biodiversity and several critical services, provided through different mechanisms and at different scales. Finally, we conclude by offering some new perspectives on addressing the links between biodiversity and ecosystem services and outline the challenges that lie ahead in this area.

Key concepts relevant to understanding the links between biodiversity and ecosystem services

Different facets of the biodiversity link to ecosystem services

Biodiversity broadly encompasses the number, abundances, functional variety, spatial distribution, and interactions of genotypes, species, populations, communities, and ecosystems. What levels of organization or components of biodiversity are likely to be most strongly linked with ecosystem services? For plant-dependent services, the local number of functional groups and total number of species (richness) can offer a crude first-order prediction for several ecosystem processes, such as productivity, and services, such as forage production. For animal-dependent services, species number and composition in mammalian communities are associated with regulation of infectious disease, although the direction of this effect (amplification or dilution of disease) depends on the types and relative abundance of different vector species in the community (Ostfeld and Keesing, 2012). The equity of the abundances (evenness) of individual species is also important in relation to biological invasions. For example, reducing evenness in plant species communities can decrease resistance to invasion by exotic plants and insect herbivores (Wilsey and Polley, 2002).

Theory predicts that increasing horizontal diversity (numbers of species within trophic levels) tends to promote several ecosystem functions that feed into ecosystem services; however, increasing vertical diversity (numbers of trophic levels) does not necessarily do so (Loreau, 2010). For plant-dependent ecosystem services, the level of service delivery probably depends most on local plant diversity because plant species interact at a local spatial scale, but there is some evidence that ecosystem services could depend on plant diversity at larger spatial scales (beta or gamma diversity; Isbell *et al.*, 2011).

The different components of ecosystem services

A critical issue in ecosystem service assessments is the scant knowledge on how ecosystem services and their components (i.e. supply, delivery, use demand, value and benefits) are produced and maintained, how they are affected by system changes, such as land use change, and how they depend on different levels of biodiversity. To improve this knowledge, we distinguish between ecological processes (called 'supporting services' in the Millennium Ecosystem Assessment; MA, 2005) and 'functions' that produce ecosystem services. Functions are intermediate products; they are necessary to the production of services but are not services themselves, i.e., not used or acknowledged directly by a beneficiary. These intermediate products or processes often underpin or determine the potential service production or supply, which can benefit society. The delivery of a service arises from the interaction between its supply and the demand from stakeholders who benefit from it (Tallis *et al.*, 2012). The benefit and value of a service reflect how people assign importance to the service, which can be evaluated in terms of market value or from a cultural perspective. For example, primary production (an ecosystem process) is needed to maintain abundance of fish population (the service supply), which can be harvested to provide food (delivery) and high nutritional value (benefit). As another example, nutrient cycling (process) is needed for water purification (supply) to provide clean water (delivery) for domestic use (benefit) (Raffaelli, 2016; Jax, 2016a).

Useful concepts for assessing the links between biodiversity and ecosystem services

To analyse the influence of biodiversity on ecosystem functions and service delivery in a structured manner, the concept of ecosystem service provider (ESP) can be useful. An 'ESP' only relates to specific organisms (from vascular plants, vertebrates to microbes) and variation of their attributes (e.g., genetic diversity, species diversity, species richness, functional diversity and vertical diversity) (Luck *et al.*, 2009, see also Luck, 2016). These ESPs can then be linked to service supply and to the needs of beneficiaries (Jax, 2016b).

The concept of functional trait can also help elucidate links between biodiversity and services. Trait-based approaches can be used to understand how species contribute to multiple ecosystem functions and also to investigate the mechanisms that contribute to trade-offs among ecosystem services.

Recently, two other concepts have been suggested in the context of the trait-based ecosystem service approach (Diaz *et al.*, 2013): the Specific Effect Function (SEF), which is the per unit capacity of a species to influence an ecosystem property or service; and the Specific Response Function (SRF), which is the ability of a species to maintain or enhance ecosystem services quantity in response to a specified change in the abiotic or biotic environment or to invade environment afresh.

Linking biodiversity to different types of services

Although the classification of ecosystem services proposed in the Millennium Ecosystem Assessment (MA, 2005) is widely adopted, there have been many other frameworks that aim to make the classification more relevant to decision-makers, economists, and ecologists. In fact, different classifications may be useful for different purposes.

The main challenge is that the same component of biodiversity may be linked in very different ways to multiple ecosystem services. Fish provide a good example. Fish are widely recognized as food, a 'provisioning service' with a direct market value. However, fish also provide indirect benefits or regulating services, like biological control. Specifically, experiments have highlighted the importance of fish predation in regulating invasive species like zebra mussels. Concurrently, the presence of fish, such as in crowded streams during spawning, can also be a highly valued cultural service.

Biodiversity and magnitude of provisioning and regulating services

The evidence to date

Our evidence that provisioning and regulating services depend directly on biodiversity remains limited. We suspect that this lack of evidence is likely due to a lack of adequate testing, rather than a lack of dependence. For ecosystem functions, hundreds of field experiments (reviewed by Cardinale *et al.*, 2011) and dozens of theoretical studies (reviewed by Loreau, 2010) have established that decreasing plant diversity can alter ecosystem functioning in directions that would likely reduce ecosystem services. They have rarely, however, tested the direct dependence of final ecosystem services, such as fodder production, on biodiversity (Balvanera *et al.*, 2014).

Possible mechanisms

Why might provisioning and regulating ecosystem services depend on biodiversity? There are three main ways in which increased biodiversity may result in increased ecosystem service provision and hence explain why decreasing biodiversity could lead to a decrease in ecosystem services.

First, complementary differences between species, combined with spatial heterogeneity, could lead to the whole community providing services at rates greater than the sum provided by its component species. This is currently referred to as the *complementarity effect* (Loreau and Hector, 2001).

Second, dominance by species that provide particularly high rates of ecosystem services could lead to a positive diversity effect on the provision of ecosystem services, on average. However, in this case, the community would never outperform the single best species that it contained. This is currently referred to as the *selection effect* (Loreau and Hector, 2001).

Third, asynchronous responses of species to environmental fluctuations could lead to greater and more stable provision of ecosystem services in mixtures than in monocultures. This is currently referred to as the *insurance hypothesis* (Yachi and Loreau, 1999).

The shape of the relationship between biodiversity and services

It is becoming increasingly clear that even the loss of a few species from a diverse community could have an adverse impact on ecosystem functioning and services. Before biodiversity experiments were conducted, most investigators predicted that ecosystem functioning would saturate at relatively low levels of biodiversity. This redundancy hypothesis posited that most species are functionally redundant, and thus their loss would not impact ecosystem functioning. Dozens of field experiments seemed to confirm this prediction (Cardinale *et al.*, 2011).

However, diversity effects tend to increase over time in long-term experiments. More importantly, the relationship between biodiversity and ecosystem functioning becomes increasingly linear over time. Thus, long-term studies tend to indicate that even the loss of a few species from diverse communities could have large impacts on ecosystem services, while results from short-term studies tend to suggest that most species are redundant.

Biodiversity and multiple services

Ecosystem services studies have often found trade-offs between ecosystem services as well as bundles, i.e., sets of different services that interact synergistically and occur simultaneously across landscapes provided by different land uses. Changes in biodiversity will likely lead to trade-offs in ecosystem service provision. For example, converting diverse grassland to cropland tends to provide high levels of crop production but low levels of many other ecosystem services. There is now considerable evidence that different ecosystem processes depend on different sets of plant species (Isbell *et al.*, 2011). Furthermore, more diverse plant communities can provide higher levels of multifunctionality and higher levels of multiple ecosystem services.

The gaps

Uncertainty about the links between biodiversity and ecosystem services remains considerable (Balvanera *et al.*, 2014). These uncertainties arise from: i) mismatches between ecosystem functions measured and final ecosystem services; ii) mismatches between study conditions and management conditions; iii) insufficient consideration of all functions upon which a service depends; iv) insufficient integration of multiple potentially critical components of biodiversity; v) confounding environmental factors; vi) trade-offs between the positive and negative effects of changes in biodiversity on various ecosystem functions that underlie ecosystem services; vii) context-dependent patterns; and viii) different scales between studies linking biodiversity and both the management and delivery of services.

Biodiversity and cultural services

The evidence to date

Cultural services supply a broad spectrum of non-tangible and non-market benefits to human well-being (i.e., psychological health, social relationships, and cohesion). These non-material benefits obtained from biodiversity can be related to different types of values (e.g., moral, spiritual, or aesthetic values), which have different importance depending on social and institutional contexts or stakeholders' groups (Chan *et al.*, 2011; Chan and Satterfield, 2016; see also Landers *et al.*, 2016). It follows that cultural services should be analysed considering the range of services (i.e., recreational activities and tourism, aesthetic values, spiritual values, local identity, etc.) and the range of values given to each service by individuals. These values are also at the very core of any decision relating to managing provisioning, or regulating services.

The different ways in which biodiversity and cultural services are linked

Differences in vegetation colour, often related to leaf nitrogen content, can be associated with the aesthetic value of landscapes. At the species level, functional traits of vegetation are significant for the supply of specific cultural services, such as recreation and aesthetics. For example, the variety of colours and green tones strongly relate to the landscape enjoyment of some ecosystems. In animals, physical traits such as large size and neotenic features (the retention of juvenile traits by adults) are important determinants of their aesthetic value.

At the level of ecosystems, recreational and aesthetic values have been attributed to those multifunctional landscapes with intermediate levels of biodiversity and some accessibility level. Multifunctional landscapes that are extensively managed have been shown to have high aesthetic and existence values (Garcia-Llorente *et al.*, 2012). People prefer them, not only for aesthetic reasons, but also because they can provide a larger set of ecosystem services than landscapes that are intensively managed or abandoned.

Social preferences towards landscapes have been also linked to the presence of visible water and vegetation, specifically referred to as hydrophilia and phytophilia, respectively. At landscape scales, expert assessments have highlighted that vegetation diversity at the species and community levels are very important for the delivery of cultural services (Quijas *et al.*, 2012).

The gaps

The analysis of the link between biodiversity and cultural services faces many scientific challenges. As has been reported for the UK (UK NEA, 2011), there is a 'cultural divide' in our knowledge because for those culturally important taxonomic groups (i.e., butterflies, fish, birds, mammals), there is high-quality information about their status and trends, but a knowledge gap on their associated cultural services (i.e., recreation, aesthetic values, existence values). In addition, for certain cultural services, such as local ecological knowledge, cultural heritage, or sense of place, there is limited knowledge about the biodiversity components that act as ecosystem service providers, such as crop or livestock varieties, native plants, and autochthonous animal breeds. Finally, there is scant information about which biodiversity components are key to religious and spiritual values.

Selected examples on the links between biodiversity and ecosystem services

Forage

Grasslands provide forage for livestock, from which we gain meat and milk products. There is considerable evidence that more aboveground plant biomass can be produced by diverse than by depauperate grassland plant communities. It remains unclear, however, whether this will necessarily lead to greater production of meat and milk products. Early trials on forage diversity were inadequately designed to test for plant diversity effects on forage production. Furthermore, most modern grassland biodiversity experiments have not: included livestock grazing; focused on species commonly used as forage; considered the impacts of changes in plant diversity on the quality of forage, including its nutritional value (crude protein content) and digestibility (lignin and cellulose content); or considered the health and growth of livestock.

A few recent studies have begun to address these limitations (e.g., Isbell and Wisley, 2011). Results show that increasing plant diversity can: i) increase biomass production under intense livestock grazing; ii) increase biomass production for common forage species; iii) increase biomass production without decreasing nutritional value or digestibility; iv) decrease the risk of mineral deficiencies or toxicities for beef cattle; and v) increase the biomass and nutrient intake by sheep. Given that managed grazing is the most extensive land use worldwide, there may be considerable value in conserving or restoring plant diversity in grasslands worldwide.

Marine fisheries

Fisheries provide a critically important source of protein, income, and jobs. Fisheries services are produced by complex interactions between environmental conditions, management, and biology. The contribution of diversity to fisheries is not yet well-resolved. The quality and quantity of evidence varies with scale and the component of diversity considered.

Theory and examples from other ecosystems suggest that biodiversity could increase or stabilize long-term yields or enhance their resistance and resilience to environmental fluctuations. Yet limited evidence is available to support these theories for fisheries, especially for resistance and resilience. On one hand, there is strong evidence that within species population diversity reduces variability in annual yields and fisheries closures for Alaskan sockeye salmon (*Oncorhynchus nerka*) via a portfolio effect (Schindler *et al.*, 2010). In contrast, future studies are needed to explore the contributions of other forms of diversity, including functional diversity, functional composition, and relative abundance of different species.

The relationship between multiple-species marine fisheries and other forms of diversity remains muddled. At a global scale, Worm *et al.*, (2006) suggest that diversifying targeted species can increase fisheries' productivity and stability while decreasing their probability of collapse. This study, however, did not control for other factors known to influence fisheries yields: environmental factors (e.g., primary productivity) and management. Others suggest that primary productivity – rather than species richness of ecosystem – drives fisheries production (Chassot *et al.*, 2010) or that this effect is instead driven by body size (Fisher *et al.*, 2010). One reason for these inconsistencies may be that these relationships were explored using broad-scale correlations at a global scale. Testing more mechanistic hypotheses about why and how different forms of biodiversity could influence fisheries could progress our understanding of the pathways through which biodiversity affects fisheries.

Soil fertility

Soil fertility can be defined as the physical, chemical, and biological condition of the soil required for plant growth and development that is inherently linked to the capacity to produce biomass in natural and agricultural ecosystems (Barrios, 2007). Complex feedbacks between plant and soil biodiversity play fundamental roles in ecosystem functions related to soil fertility generation and renewal. The soil biota contributing to soil fertility is extremely diverse, ranging from invisible microorganisms to macro-fauna such as earthworms and termites. However, the soil biodiversity-ecosystem service relationship appears to be less related to species richness and more dependent on certain key species or species with particular traits. Based on their contribution to aggregate ecosystem functions underpinning soil-based ecosystem services, soil organisms can be grouped into four keystone functional assemblages (Kibblewhite *et al.*, 2008): i) decomposers; ii) nutrient transformers; iii) ecosystem engineers; and iv) bio-controllers.

While soil organisms are largely not visible, this 'hidden' biodiversity contributes to soil fertility in many ways (Barrios, 2007). They are the principal driving agents of nutrient cycling. They regulate the dynamics of soil organic matter formation and breakdown which directly impacts soil carbon storage and greenhouse gas emissions. They modify the physical structure of soil through their effect on the aggregation of soil particles, which directly affects water regimes. They enhance the amount and efficiency of nutrient acquisition by crops and native vegetation through symbiosis with mycorrhizal fungi and nitrogen-fixing bacteria. Further, they influence plant health through the biological control of pests and diseases by their competitors, natural predators and parasites. Until recently, linkages between soil biodiversity and ecosystem services received little attention. It is now increasingly evident that soil biodiversity is critical for a sustainable human existence.

Water quality

Water quality is often determined by levels of chemical (e.g., nitrates), microbiological (e.g., fecal bacteria), or physical (e.g., soil particles) pollutants. The amount of pollutants that is acceptable varies among different types of uses (e.g., irrigation vs. drinking water) and among contexts (e.g., different countries). The avoidance, removal, and storage of these pollutants are key ecosystem services.

The chemical, microbiological, and physical quality of water at the point of human use can depend on many factors. For example, chemical pollutants can be regulated by river organisms through the processing of nutrients or toxic substances during metabolic breakdown. Transiting chemical pollutants are exported through the food web, and can find their way either downstream or to the top of the food chain. Microbiological quality is often linked to catchment management (Wilkes *et al.*, 2013). Finally, river physical quality, for example temperature or flow, is often dependent on the character of riparian vegetation.

However, we lack field evidence on how much waste can be processed by river biota before the maintenance of biodiversity, its ability to regulate water quality, or fish production are impaired. There is a clear need to quantify in situ regulation services provided by river biota for water quality and to identify resilience thresholds, in particular under different land use and climate scenarios. As pressure on quality water rises, water companies and governments are likely to increasingly look to these natural solutions to provide cheaper and more sustainable water provision.

Regulation of human disease vectors

Biodiversity is important in contributing to ecosystem services that promote livelihoods and well-being. But in some cases, biodiversity itself is much more closely linked to human health, and not necessarily in a positive way. Biodiversity represents a disease threat to humans in many circumstances. For example, many wildlife species act as hosts of disease that can have significant negative economic and well-being impacts on humans (Daszak *et al.*, 2000). Wildlife has been the origin of many emerging infectious diseases in recent years.

However, biodiversity may also have benefits to human health through protecting against or reducing disease, through a 'dilution effect'. Disease risk decreases as the diversity of an animal community of potential hosts increases and thus the relative abundance of the key host for that disease decreases. Such a dilution effect is enhanced where hosts vary in their quality for pathogens or vectors, and where an abundance of low-quality hosts or vectors reduce the encounter rates of high-quality hosts with pathogens and vectors (Ostfeld and Keesing, 2012). Empirical evidence for this link between species richness and disease regulation is limited but growing.

Moreover, empirical evidence shows that, more often than not, biodiversity loss actually increases disease transmission (Ostfeld and Keesing, 2012). In a review of the statistical relationships between biodiversity and disease transmission, it was found that 80% showed a significant negative association (dilution effect), whereas only 12% showed a significant positive association (amplification effect) (Cardinale *et al.*, 2012).

Existence value of species diversity

The existence value of species diversity arises from the satisfaction people derive from the knowledge that a species exists. This value is ultimately emotional and derived from feelings and preferences about non-human living species. The existence value of species is thus mediated by two main factors: i) people's affective and emotional responses to species; and ii) people's self-interest related to the utility derived from species. This dichotomy has been referred to as anthropomorphism vs. anthropocentric, affection vs. economic self-interest, empathy/identification vs. instrumental self-interest, affect vs. utility.

One of the most common indicators to measure the existence value of species is the willingness to pay. Specific morphological traits (e.g., size, round forms or eyes sizes) and the phylogenetic closeness of the species to humans both contribute to the existence value of species, besides various socioeconomic factors. Mammals, birds, and species with neotenic characteristics are among the most widely preferred.

Measuring the existence value of species in contexts where people cannot meet their most basic needs and where the conservation or enhancement of biodiversity may necessitate a trade-off with livelihoods or food is much more challenging on both methodological and ethical grounds.

Linking biodiversity, ecosystem services and the different components of well-being

Ecosystem services are a social construct, and they only exist where humans gain some degree of well-being from their interactions with ecosystems. Human well-being is a multifaceted concept which includes the presence of positive emotions and moods (e.g., contentment, happiness), the absence of negative emotions (e.g., depression, anxiety), satisfaction with life, fulfilment, resilience and positive functioning (Centers for Disease Control and Prevention, no date).

Wellbeing components depend differentially on financial, social, infrastructure and natural capital. Adequate nourishment, access to clean drinking water, or access to traditional medicine are examples of needs that depend largely on ecosystem services. Plant species richness can contribute to higher forage production, sufficient meat production, and meeting basic food security needs (Figure 4.1). Fish population diversity can contribute to yield stability and thus food and job security (Schindler *et al.*, 2010). Appreciated plants and animals can contribute to aesthetic enjoyment, spiritual and mental health and thus to overall health and good social relations. Yet the relative contribution of biodiversity itself to these different components of wellbeing is much harder to assess (Landers *et al.*, 2016).

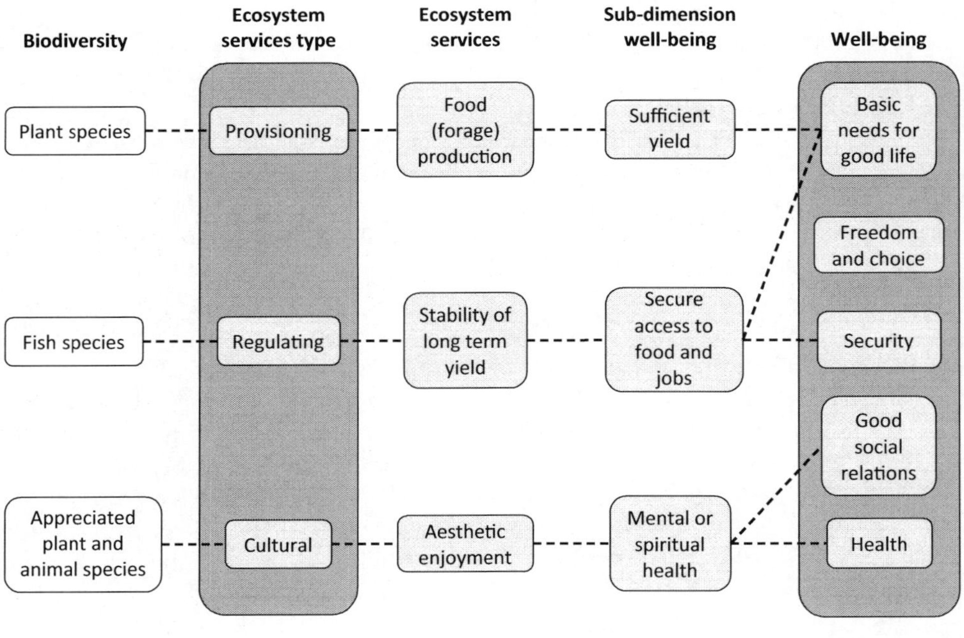

Figure 4.1 Biodiversity is linked to ecosystem services and the different components of wellbeing. Selected examples are used to illustrate these linkages.

Source: modified from MA, 2005 and Mace *et al.*, 2012

Complex interlinkages between biodiversity and services: resilience, trade-offs, scaling, substitutability, and sustainability

Global change brought by increasing population needs and altered climatic patterns is expected to affect biodiversity, ecosystems, and ecosystem services, with significant impacts on social and economic well-being. But despite mounting concern about the severity of these issues, knowledge about the complex interlinkages between biodiversity and services, involving multiple scales, non-linear dynamics, and trade-offs, is limited and disconnected.

Ecosystem services are ultimately supplied by the interactions between societies and ecosystems through social-ecological systems. Thus, the resilience of service supply will depend not only on biodiversity but rather on the whole set of biophysical and societal variables that underpin services provision. For instance, soil fertility maintenance will depend on the functional resilience of the different components of soil biodiversity contributing to the physical, chemical, and biological soil attributes that sustain land productivity (Barrios, 2007). Yet soil fertility is also influenced by soil type (e.g., parent material, texture, type of clay) and soil management: the types of crops (e.g., annuals vs. perennials), land cover (e.g., dense vs. sparse stands), diversification (e.g., monocropping vs. intercropping), management practices (e.g., tillage vs. no-tillage), and the utilization of inputs (e.g., fertilizers, irrigation).

Bundles of services rather than individual ones result from a particular configuration of biophysical, management, and societal drivers. Biodiversity may have positive effects on some services but negative or no effects on others. Agroecosystems can be managed to maximize yields or to maximize yield security through the regulation of agricultural pests and diseases, or to maintain soil fertility through time. They can also be directly or indirectly managed to maximize water infiltration in soil or soil carbon stocks. Thus assessing the contribution of plant and soil biodiversity to the services from agroecosystems and their resilience will depend on understanding how the changes to biodiversity affect individual services, as well as synergies and trade-offs among services (Balvanera *et al.*, 2014).

At larger spatial scales, maintaining habitat heterogeneity and connectivity in landscapes and seascapes allows for the maintenance of biodiversity while at the same time allowing for the supply of a larger portfolio of ecosystem services. Such heterogeneity is likely to contribute to that of key provisioning and regulating services. For example, designing marine reserve networks can simultaneously enhance conservation of biodiversity, reduce fishery costs, and even increase fishery yields and profit.

Much work is still needed. For instance, we need to explore whether species can easily be substituted and still provide an equivalent ecosystem service, although it has been shown that this is unlikely if we are interested in multiple services at large spatiotemporal scales in a changing world (Isbell *et al.*, 2011). Also, we need to assess how sustaining biodiversity meets the needs of multiple stakeholders that differentially benefit from services and trade-offs.

Also, changes in human well-being mediated by biodiversity and its interactions on ecosystem services will in turn influence decisions on biodiversity and ecosystem management. While a subject of debate, demonstrating the role of biodiversity may be beneficial for biodiversity conservation, although some species risk being at a disadvantage if their 'service record' cannot be proven. Alternatively, ecosystem services should be one of several factors considered in such decisions. However, one of the biggest remaining challenges is understanding and predicting how the perceived values of services will have feedbacks on biodiversity and ecosystems.

Increased understanding of the role of biodiversity on sustaining ecosystem services is likely to increase the relative value of services that are still currently taken for granted, such as water quality regulation. This is particularly true in the context of short-term local decision-making in which declines in ecosystem services are not apparent and thus not taken into account.

New perspectives for addressing the links between biodiversity and ecosystem services at different spatial scales

Local and landscape scale research can provide new insights if designed with a global vision to contribute to current global challenges of the Convention on Biological Diversity (i.e., Aichi targets) as well as the newly constituted Intergovernmental Science-Policy Platform on Biodiversity and Ecosystem Services (IPBES). We need to understand how and when the links between biodiversity and ecosystem services are dependent on the biophysical, management, and societal context. Integrated interdisciplinary research is needed to understand the consequences of different types of management for the key components of biodiversity that underpin different bundles of ecosystem services and for the benefits derived by stakeholders (Nagendra *et al.*, 2013).

A new network of long-term, interdisciplinary, adaptive, and participatory studies can be used to fully assess the contributions of biodiversity to ecosystem services and people's well-being (Figure 4.2). For instance, a network of sites with contrasting social and ecological conditions and a common experimental design could be used to monitor biodiversity, different bundles of services and the flow of benefits to societies (Balvanera *et al.*, 2014). Designing and monitoring management experiments with stakeholders would ensure that the experiments/monitored variables were the most relevant to society. With this design, the marginal

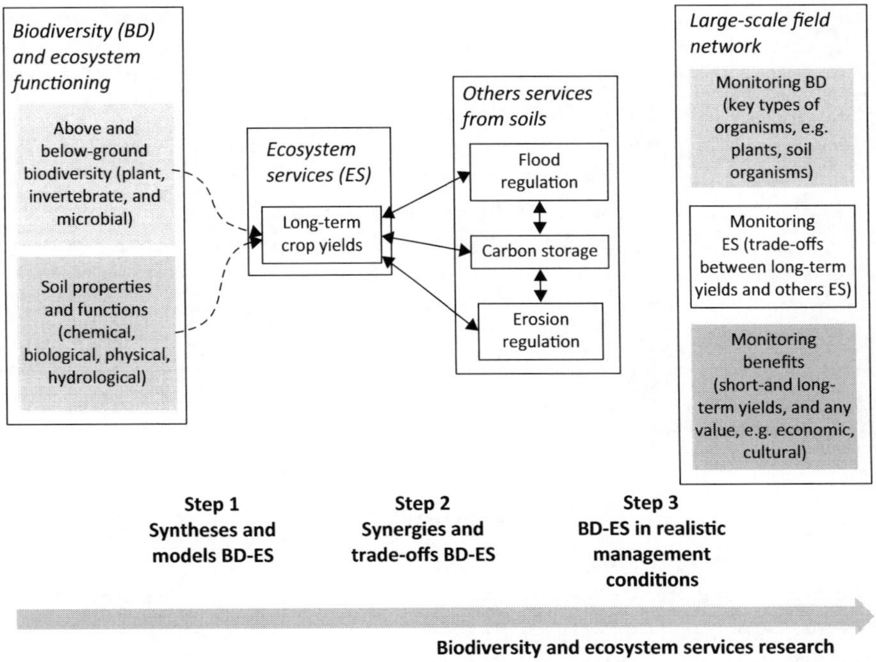

Figure 4.2 Steps leading to a new generation of biodiversity and ecosystem services research, using the effects of biodiversity through soil functions on long-term crop yields as an example.

Source: from Balvanera *et al.*, 2014

contribution of biodiversity to not only ecosystem services but also the satisfaction of stakeholder's needs could be assessed.

Studies focused on biodiversity management across gradients of intensification in tropical agricultural landscapes, currently under greatest threat of biodiversity loss, would constitute a useful starting point (Jackson *et al.*, 2012). Emphasis on agricultural practices that embrace ecological intensification through relatively high, but manageable, levels of aboveground and belowground biodiversity (e.g., agroforestry) would be key to addressing fundamental questions on the role of plant-soil feedbacks in the provision of ecosystem services required for sustainable agriculture. The strong linkage between plant functional traits and soil biodiversity highlights the potential to strategically utilize agricultural management systems to influence the provision of soil-mediated ecosystem services in agricultural landscapes. Furthermore, integrating local and technical spatial variability information on 'hot spots' of soil biological activity and focusing on functions that are relatively specific, such as the roles of ecosystem engineers or specific nutrient transformations, would help to achieve more reliable assessments of the links between biodiversity and ecosystem services (Barrios, 2007). The focused approach mentioned earlier, however, should be embedded in an interdisciplinary framework where land use decisions and the provision of ecosystem services represent the outcome of the continuous interaction between functional diversity components and the priorities of social actors.

A multi-scale approach would also be needed, given that ecosystem services are supplied, used, and managed at different spatial scales. The interlinkages between biodiversity and ecosystem services could then be assessed at different scales, ranging from local to landscape, regional, and global scales. Observations and experiments could be combined with the analysis of models for biodiversity, ecosystem services, and impacts on human well-being at different spatial scales.

Such multi-scale assessments need to be undertaken by large interdisciplinary teams of researchers and other relevant stakeholders. The good news is that such networks are currently under construction and in some cases already in place. For example, the BESS (see http://www.nerc-bess.net) programme funded by the UK Research Councils is already bringing more than 120 researchers and 50 stakeholder organizations together to investigate the role of biodiversity in sustaining ecosystem services across all UK landscapes. At a more global scale, Future Earth (www.icsu.org/future-earth), the new encompassing global environmental program and its associated projects (e.g., on Ecosystem Change and Society, www.pecs-science.org), is exploring ways to tackle some of the questions posed above. Research and communications networks such as the Ecosystem Services Partnership (www.es-partnership.org) provide opportunities to form a network of researchers and stakeholders across ecosystems and regions, and to respond to some of the pending issues. Overall, such communities can inform decision-makers at different spatial scales and across countries and to contribute to the endeavour faced by IPBES.

Conclusions

Biodiversity is closely linked to most ecosystem services but generally not in a simple way. The nature of the relationship between biodiversity and ecosystem service delivery still remains unknown for most ecosystem services. For those that are known, relationships are highly variable and may be positive, negative, or non-linear (Table 4.1). The number, identity, functional characteristics, and evenness of species are important to ecosystem functioning and consequently to the supply of different types of services. Useful concepts and approaches, such as

Table 4.1 Summary of different general types of known relationships between biodiversity (BD) and ecosystem services (ES).

Type of BD-ES relationship	*Impact of BD loss on ES*	*Basis for relationship*	*Types of services*	*Relevant examples*
Positive	Negative	Theory – complementarity effect, selection effect, insurance hypothesis Empirical – fisheries	Provisioning	Biomass production of forage for livestock (Finn *et al.*, 2013) Within-species population diversity reduces variability of annual fisheries yields (Schindler *et al.*, 2010) Globally, greater species richness of targeted species can increase fisheries productivity and stability (Worm *et al.*, 2006)
Negative	Positive	Empirical evidence	Provisioning	Some highly intensive horticulture and crop production, where much biodiversity perceived as pests
Flat	None	Empirical evidence Theory – redundancy hypothesis	Cultural	Landscape-based recreation which is not biodiversity-dependent (Quijas *et al.*, 2012)
Hump-shaped	Variable	Empirical evidence	Cultural	Landscape aesthetics – preference for extensively managed landscapes over intensively managed or unmanaged ones (García-Llorente *et al.*, 2012)
Non-linear or variable	Variable – determined by species composition and functional traits rather than biodiversity per se	Empirical evidence Theory – amplification and dilution effects for disease	Provisioning Regulating	Harvesting of individual species from extensive or non-managed systems (Chassot *et al.* 2010) Regulation of soil fertility in many ways (Barrios, 2007, Kibblewhite *et al.*, 2008) Regulation of infectious disease (Ostfeld and Keesing, 2012) Regulation of pests and invasive species (Wilsey and Polley, 2002)

ecosystem service providers and trait-based approaches, and an understanding of the mechanisms through which diversity regulates ecosystem function, have advanced our understanding of these interlinkages.

Alas, the large variety of services, the multiple scales at which they operate, and the multiple ways in which different processes and types of services link to biodiversity hinder our integrated understanding of entire systems dynamics. Evidence for most services and contexts is still in construction, largely due to lack of adequate documentation and testing, even for well-known examples such as forage, fisheries, soil fertility, water quality, and the regulation of human disease vectors.

Biodiversity likely influences the long-term maintenance of functioning social-ecological systems and the flow of benefits from nature to societies at multiple spatial scales. To understand these longer and larger-scale dynamics, multi-scale, interdisciplinary research in collaboration with stakeholders is needed. Long-term research and the assessment of multiple ecosystem services emphasize that losing a few species can be detrimental to service supply and thus human well-being. Improving our understanding of the consequences of biodiversity change can inform the design of new management interventions and policies, via platforms on Biodiversity and Ecosystem Services such as IPBES (www.ipbes.net), Future Earth's ecoSERVICES (www.futureearth.org/projects/ecoservices) and the Ecosystem Services Partnership (www.es-partnership.org).

Acknowledgements

This chapter is the result of the Working Group on Biodiversity and Ecosystem Services of the Ecosystem Services Partnership. Funding to Edmundo Barrios to contribute to this chapter was provided by the CGIAR research programs on Forests, Trees and Agroforestry, and Humid Tropics. Isabelle Durance was funded through the DURESS project NE/J014818/1, part of the BESS programme funded by the NERC.

References

Balvanera, P., Siddique, I., Dee, L., et al. (10 authors) (2014). Linking biodiversity and ecosystem services: current uncertainties and the necessary next steps. *BioScience*, vol 64, no 1, pp 49–57.

Barrios, E. (2007). Soil biota, ecosystem services and land productivity. *Ecological Economics*, vol 64, no 2, pp 269–285.

Cardinale, B.J., Matulich, K.L., Hooper, D.U., et al. (9 authors) (2011). The functional role of producer diversity in ecosystems. *American Journal of Botany*, vol 98, no 3, pp 572–592.

Cardinale, B.J., Duffy, J.E., Gonzalez, et al. (17 authors) (2012). Biodiversity loss and its impact on humanity. *Nature*, vol 486, no 7401, pp 59–67.

Centers for Disease Control and Prevention (no date). Health-Related Quality of Life (HRQOL): Wellbeing concepts. Available at: http://www.cdc.gov/hrqol/wellbeing.htm (accessed 19 February 2014).

Chan, K.M.A. and Satterfield, T. (2016). Managing cultural ecosystem services for sustainability. In Potschin, M., Haines-Young, R., Fish, R. and Turner, R.K. (eds) *Routledge Handbook of Ecosystem Services*. Routledge, London and New York, pp 343–356.

Chan, K.M.A., Satterfield, T., and Goldstein, J. (2011). Rethinking ecosystem services to better address and navigate cultural values. *Ecological Economics*, vol 74, pp 8–18.

Chassot, E., Bonhommeau, S., Dulvy, N., et al. (7 authors) (2010). Global marine primary production constrains fisheries catches. *Ecology Letters*, vol 13, no 4, pp 495–505.

Daszak, P., Cunningham, A. A., and Hyatt A. D. (2000). Wildlife ecology – Emerging infectious diseases of wildlife – Threats to biodiversity and human health. *Science*, vol 287, no 5452, pp 443–449.

Diaz, S., Purvis, A., Cornelissen, J.H.C., et al. (8 authors) (2013). Functional traits, the phylogeny of function, and ecosystem service vulnerability. *Ecology and Evolution*, vol 3, no 9, pp 2958–2975.

Finn, J. A., Kirwan, L., Connolly, J., et al. (32 authors) (2013). Ecosystem function enhanced by combining four functional types of plant species in intensively managed grassland mixtures: a 3-year continental-scale field experiment. *Journal of Applied Ecology*, vol 50, no 2, pp 365–375.

Fisher, J., Frank, K., and Leggett, W. (2010). Global variation in marine fish body size and its role in biodiversity–ecosystem functioning. *Marine Ecology Progress Series*, vol 405, no 1, pp 1–13.

Garcia-Llorente, M., Martin-Lopez, B., Iniesta-Arandia, I., Lopez-Santiago, C. A., Aguilera, P. A., and Montes, C. (2012). The role of multi-functionality in social preferences toward semi-arid rural landscapes: an ecosystem service approach. *Environmental Science & Policy*, vol 19–20, pp 136–146.

Isbell, F., Calcagno, V., Hector, A., et al. (15 authors) (2011). High plant diversity is needed to maintain ecosystem services. *Nature*, vol 477, no 7363, pp 199–202.

Isbell, F.I. and Wilsey, B.J. (2011). Increasing native, but not exotic, biodiversity increases above g round productivity in ungrazed and intensely grazed grasslands. *Oecologia*, vol 165, no 3, pp 771–781.

Jackson, L. E., Pulleman, M. M., Brussaard, L., et al. (16 authors) (2012). Social-ecological and regional adaptation of agrobiodiversity management across a global set of research regions. *Global Environmental Change-Human and Policy Dimensions*, vol 22, no 3, pp 623–639.

Jax, K. (2016a). Ecosystem functions: a critical perspective. In Potschin, M., Haines-Young, R., Fish, R. and Turner, R.K. (eds) *Routledge Handbook of Ecosystem Services*. Routledge, London and New York, pp 42–44.

Jax, K. (2016b). Ecosystem services and ethics. In Potschin, M., Haines-Young, R., Fish, R. and Turner, R.K. (eds) *Routledge Handbook of Ecosystem Services*. Routledge, London and New York, pp 301–303.

Kibblewhite, M. G., Ritz, K., and Swift, M. J. (2008). Soil health in agricultural systems. *Philosophical Transactions of the Royal Society B-Biological Sciences*, vol 363, no 1492, pp 685–701.

Landers, D.H., Nahlil, A.M., and C.R. Rhodes (2016). The beneficiary perspective – benefits and beyond In Potschin, M., Haines-Young, R., Fish, R. and Turner, R.K. (eds) *Routledge Handbook of Ecosystem Services*. Routledge, London and New York, pp 74–87.

Loreau, M. (2010). *From populations to ecosystems: Theoretical foundations for a new ecological synthesis*. Princeton University Press, Princeton NJ.

Loreau, M. and Hector, A. (2001). Partitioning selection and complementarity in biodiversity experiments. *Nature*, vol 412, no 6842, pp 72–76.

Luck, G. (2016). Ecosystem service providers. In Potschin, M., Haines-Young, R., Fish, R. and Turner, R.K. (eds) *Routledge Handbook of Ecosystem Services*. Routledge, London and New York, pp 60–61.

Luck, G. W., Harrington, R., Harrison, P. A., et al. (22 authors) (2009). Quantifying the contribution of organisms to the provision of ecosystem services. *BioScience*, vol 59, no 3, pp 223–235.

MA (2005). *Ecosystems and Human Well-being: Synthesis*. Island Press, Washington DC.

Mace, G.M., Norris, K. and Fitter, A.H. (2012). Biodiversity and ecosystem services: a multilayered relationship. *Trends in Ecology & Evolution*, vol 27, no 1, pp 19–26.

Nagendra, H., Lucas, R., Honrado, J. P., et al. (7 authors) (2013). Remote sensing for conservation monitoring: assessing protected areas, habitat extent, habitat condition, species diversity, and threats. *Ecological Indicators*, vol 33, pp 45–59.

Ostfeld, R. S., and Keesing, F. (2012). Effects of host diversity on infectious disease. *Annual Review of Ecology, Evolution, and Systematics*, vol 43, no 43, pp 157–182.

Quijas, S., Jackson, L. E., Maass, M., Schmid, B., Raffaelli, D., and Balvanera, P. (2012). Plant diversity and generation of ecosystem services at the landscape scale: expert knowledge assessment. *Journal of Applied Ecology*, vol 49, no 4, pp 929–940.

Raffaelli, D. (2016). Ecosystem structures and processes: characterising natural capital stocks and flows. In Potschin, M., Haines-Young, R., Fish, R. and Turner, R.K. (eds) *Routledge Handbook of Ecosystem Services*. Routledge, London and New York, pp 62–73.

Schindler, D. E., Hilborn, R., Chasco, B., et al. (7 authors) (2010). Population diversity and the portfolio effect in an exploited species. *Nature*, vol 465, no 7298, pp 609–612.

Tallis, H., Mooney, H., Andelman, S., et al. (13 authors) (2012). A global system for monitoring ecosystem service change. *BioScience*, vol 62, no 11, pp 977–986.

UK NEA (2011). *The UK National Ecosystem Assessment Technical Report*. UNEP-WCMC, Cambridge.

Wilkes, G., Ruecker, N. J., Neumann, N. F., et al. (10 authors) (2013). Spatiotemporal analysis of Cryptosporidium Species/Genotypes and relationships with other zoonotic pathogens in surface water from mixed-use watersheds. *Applied and Environmental Microbiology*, vol 79, no 2, pp 434–448.

Wilsey, B. J., and Polley, H. W. (2002). Reductions in grassland species evenness increase dicot seedling invasion and spittle bug infestation. *Ecology Letters*, vol 5, no 5, pp 676–684.

Worm, B., Barbier, E. B., Beaumont, N., et al. (14 authors) (2006). Impacts of biodiversity loss on ocean ecosystem services. *Science*, vol 314, no 5800, pp 787–790.

Yachi, S. and Loreau, M. (1999). Biodiversity and ecosystem productivity in a fluctuating environment: the insurance hypothesis. *Proceedings of the National Academy of Sciences of the United States of America*, vol 96, no 4, pp 1463–1468.

Briefing Note 4.1 Service providing units

Gary Luck

In 2003, Gretchen Daily, Paul Ehrlich and I (Luck et al., 2003) introduced the concept of service providing units (SPUs) to underscore the importance of maintaining population diversity for the sake of species conservation and the provision of ecosystem services. Changes in populations can have major impacts on ecosystems independent of any changes in species diversity. We also aimed to promote SPUs as a heuristic tool to focus attention on the key components of species populations that would impact the population's capacity to contribute to ecosystem service provision.

At the time, research on the services provided by plants and animals was in its infancy and centred mostly on identifying which species may or may not be contributing to service provision. We argued that a more complete understanding was needed, one which recognised that the number of species populations (population richness), the size and density of each population, the distribution of populations and their genetic differentiation were all vital to the level of contribution a species may make to a given ecosystem service in a particular location.

The SPU concept was introduced as a new approach to link a species population explicitly with the services that it provided. Moreover, the concept can be used as another way to delineate population boundaries, similar to related concepts such as evolutionary or demographic units (see Luck et al., 2003). For example, a population of native pollinating bees confined to an isolated patch of remnant rainforest may provide pollination services to an adjacent coffee crop, but not to other, more distant crops. This population may be nested within a single evolutionary unit or even demographic unit, but for the sake of the population's contribution to ecosystem service provision it should be considered a discrete SPU and managed as such.

Various examples of SPUs now exist in the literature, even if not recognised as such by the authors themselves. Critically, these researchers have attempted to quantify at least one of the key population characteristics that are vital for understanding how a population may contribute to service provision. For example, Mols and Visser (2007) demonstrated the capacity of *Parus major* – an insectivorous bird species – to provide a pest control service in apple orchards by substantially reducing caterpillar damage to the crop. The authors showed that at a density of 1–6 breeding pairs per 2 ha, caterpillar damage is reduced by up to 50% compared to control sites with no breeding pairs. The SPU in this example is the density of breeding pairs within the orchard needed to deliver the service at the required level. At least one breeding pair of *Parus major* every 2 ha within the apple orchard would be a bare minimum. Another example of an SPU comes from Hougner et al. (2006), who demonstrated that the maintenance of oak forest in the National Urban Park of Stockholm required a minimum of 12 resident Eurasian jay (*Garrulus glandarius* – a major dispersal agent of oaks) pairs present each year for 14 years.

At a broader level, Schindler et al. (2010) provide a wonderful example of the importance of managing populations as discrete units in the context of ecosystem-service provision. In Bristol Bay, Alaska, sockeye salmon (*Oncorhynchus nerka*) exist in hundreds of locally adapted populations distributed among tributaries and lakes. These populations are reproductively isolated, have capacity for micro-evolution and are collectively adapted to a wide range of conditions. This population diversity and asynchronous dynamics leads to less variability in returns to the fishing industry that relies on this species. Schindler et al. (2010) estimated that variability in returns is over two times lower than it would be if the species existed in a single homogeneous population, and that closures of fisheries would be ten times more common if the current diversity was reduced to a single homogeneous population.

The notion of quantifying the ecological units that provide ecosystem services applies beyond simply species populations and can include multi-species functional groups, entire ecological communities, habitat types or landscapes. In Luck et al. (2009), we united the concepts of SPUs and ecosystem service providers to demonstrate how different organisational levels can contribute to services in different ways. For example, maintaining the functional diversity of native pollinating bees can be crucial to delivering pollination services to watermelon crops in California (Kremen et al., 2004).

There is much to do to further understanding of the demographic, functional and genetic traits required to provide a particular service at a particular level in any given context. Quantifying these traits raises many challenges, but these must be tackled head on so that society is well informed about what needs to be protected to maintain the provision of vital services into the future.

References

Hougner, C., Colding, J., and Söderqvist, T. (2006). Economic valuation of a seed dispersal service in the Stockholm National Urban Park, Sweden. *Ecological Economics*, vol 59, pp 364–374.

Kremen, C., Williams, N.M., Bugg, R.L., Fay, J.P., and Thorp, R.W. (2004). The area requirements of an ecosystem service: crop pollination by native bee communities in California. *Ecology Letters*, vol 7, pp 1109–1119.

Luck, G.W., Daily, G.C., Ehrlich, P.R. (2003). Population diversity and ecosystem services. *Trends in Ecology and Evolution* vol 18, pp 331–336.

Luck, G.W., Harrington, R., Harrison, P.A., et al. (21 authors) (2009). Quantifying the contribution of organisms to the provision of ecosystem services. *Bioscience*, vol 59, pp 223–235.

Mols, C.M.M., and Visser, M.E. (2007). Great tits (*Parus major*) reduce caterpillar damage in commercial apple orchards. *PLoS One*, vol 2, pp e202.

Schindler, D.E., Hilborn, R., Chasco, B., et al (7 authors) (2010). Population diversity and the portfolio effect in an exploited species. *Nature*, vol 465, pp 609–612.

5

ECOSYSTEM STRUCTURES AND PROCESSES

Characterising natural capital stocks and flows

Dave Raffaelli

Introduction

The term 'ecosystem services' has been around since at least the 1970s, and the concept of Nature providing for human needs even longer (Gómez-Baggethun et al., 2010). However, only relatively recently have ecosystem services started to become included in practical policy appraisal and decision-making. The motivation for this has come largely from an acknowledgement of the value of the natural capital locked up in ecosystems not only for providing market goods but also for underpinning the delivery of a range of benefits for which there is no market and which are difficult to monetise and therefore have little or no value attached to them (e.g. Costanza et al., 1997; Daily, 1997; Stern, 2006; TEEB, 2010). In addition, it is now unarguable that the processes involved in the extraction of goods leads to declines in other services, biodiversity in particular (MA, 2005). Whilst it is clear that levels of natural resource stocks (natural capital) will affect the flows and values of both goods and services, the linkages between stocks and flows are complex (Daly and Farley, 2011), so that the integration of this knowledge and these concepts into practical environmental management and decision-making has proven challenging. Estimating changes in the rate at which services would be delivered (their flow) under different policy options, and whether this is sustainable given the levels of various ecosystem components that generate those services (natural capital stocks), remains a thorny problem for practical ecosystem ecologists and is the focus of considerable research effort world-wide (e.g. the VNN and BESS research programmes in the UK). This chapter clarifies some of these challenges, specifically the concepts of stocks and flows in an ecosystem services context, how both stocks and flows might be evaluated and monitored, and possible ways forward for a more holistic, integrated approach.

Natural capital, stocks and flows: some clarifications

The concept of a capital stock generating a flow of things useful to humankind has great resonance with policymakers, politicians and financiers as well as with practical environmental managers. Natural capital stocks have analogies with financial capital in domestic bank accounts: for a given capital, if one spends more than the interest yielded on that capital, then the capital is eaten into and there will be less future yield from a depleted stock. The extension argument for the conservation of biodiversity is that whilst it is entirely possible to continue to extract high yields of natural resources from managed ecosystems, such as fisheries and agriculture, if more is taken that can be replaced by biological production they will not be sustainable in the long-term because one is likely to be eating into the underlying capital stock at a faster rate than it can be replenished. This common-sense notion is indeed supported by the world-wide downward trends in wild fisheries and few would argue with the basic logic. However, any newcomer to this field will soon discover that these simple concepts of capital and flow are not always clearly or consistently expressed, perhaps reflecting more their power as metaphors for engagement and rhetoric than as useful (at present) operational terms.

The term "natural capital" was first used by Schumacher (1973) as a powerful metaphor in his critique of Western capitalism and the sustainability crisis, and has been subsequently picked up by many authors wishing to bridge the gaps between ecosystem science and economics (Raffaelli and White, 2013). More recently, the natural capital concept has proven useful for constructing frameworks for national resource accounting; Voora and Venema (2008) provide an excellent account of the development of this area. Figure 5.1 provides an illustration of one such scheme in the Canadian context. The Canadian boreal forest occupies 58% of Canada's land area and provides a range of ecosystem services, many of which are not accounted for in traditional natural resource management frameworks that focus on a single good, such as timber. The Boreal Ecosystem Wealth Accounting System (BEWAS) (Anielski and Wilson, 2006) allows a balance sheet to be developed for decision-makers so they are able to take into account the full value of resources and services and are forced to confront the challenge that policy interventions will involve trade-offs between kinds of services. However, within natural capital frameworks such as BEWAS, it is common to find that the relationships between natural stocks, flows and ecosystem services are not at all well-represented, are in different parts of the framework space and may even be confounded and confused. This blurring of what are stocks, what are flows, what are goods and what are services is a feature of much of the policy literature, and the concept of natural capital is often poorly and misleadingly expressed (Raffaelli and White, 2013). For example, the UNEP Natural Capital Declaration (UNEP, 2011) confounds stocks and flows by stating that: "Natural Capital comprises Earth's natural assets (soil, air, water, flora and fauna), and the ecosystem services resulting from them, which make human life possible."

Whilst the "capital-stock-that-generates-interest" analogy might appear to be a common-sense way to approach the sustainable management of ecosystem services, this view is an

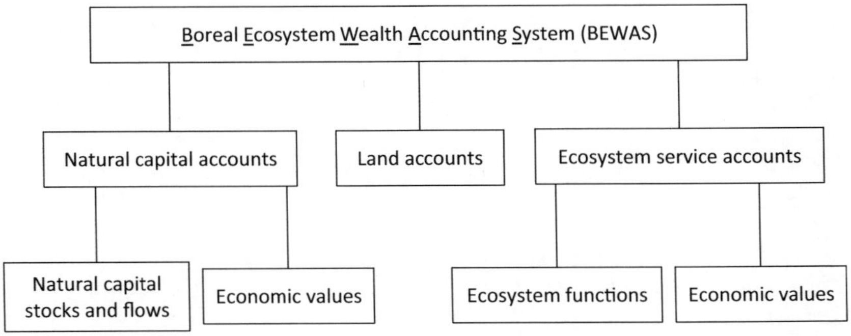

Figure 5.1 The Boreal Ecosystem Wealth Accounting System (BEWAS) framework for natural resource accounting, proposed for tracking the stocks and total value of natural resources and their flows, including land and ecosystem services, for the Canadian boreal ecosystem.

Source: after Anielski and Wilson (2006)

over-simplification. The practical application of this concept is not at all straightforward, even for the simplest situation of a single-species fish stock and fish yields, as testified by world-wide fisheries collapsing despite the existence of advanced management institutions. Ecosystems deliver multiple services and benefits, so that for a proper application of the Ecosystem Approach (CBD, 2009) to environmental management and decision-making using an Ecosystem Services Framework (Turner and Daily, 2008), many different kinds of stocks and flows will need to be traded off against one another to deliver the desired portfolio of benefits, greatly increasing the complexity of the challenge. Stakeholders can usually identify the goods, services and benefits they wish to obtain from their landscape, but specifying and quantifying the stocks that underpin those is far from straightforward. The many biophysical components of stocks that could deliver, for example food, recreation or flood mitigation, can be identified and quantitative models of service flows derived (e.g. Jones et al., 2013), but the feedbacks between extractions of services and levels of stocks (the equivalent of spending more than the interest on the capital account) are harder to establish for landscapes delivering multiple benefits. One of the difficulties is the often overlooked fact that stock-flow relationships come in two distinct forms: traditional stock-flow and fund-service.

Stock-flow and fund-service resources

Stock-flow and fund-service resources are fundamentally different. Stock-flow resources produce more of themselves, can be used up (by humans) at any desired rate, and can be stockpiled for use in the future (Table 5.1).

Good examples would be a stock of trees (woodland or forest) that supplies a flow of trees for timber, or a stock of animals such as cattle or sheep that supplies a flow of meat, leather and wool. Such production services, *sensu* the MA (2005) classification, result from the biomass growth of individuals and new recruits into the population. Restricting the yield to taking the surplus production of biomass and recruits (which are the fastest biomass producers) should, in theory, lead to a sustainable flow. It would not make sense for a farmer to slaughter their breeding stock, but it is noteworthy that it is those very individuals which are the target for most wild fisheries: fishing with a mesh (net) cannot help but select the largest individuals. It should also be noted that for wild fisheries, and probably for many other exploited wild species, stock-recruit relationships have often proven elusive, almost certainly due to the high mortalities of early life stages from density-independent and other unknown factors, so that the simple stock-flow concept becomes complex and challenging for management.

Table 5.1 The salient features of, and differences between, stock-flow and fund-service resources.

Stock-flow resources	*Fund-service resources*
Are materially transformed into what they produce (material cause).	Are not materially transformed into what they produce (efficient cause).
Can be used at virtually any rate desired (subject to the availability of fund-service resources required for their transformation), and their productivity is measured by the number of physical units of the product into which they are transformed.	Can only be used at a given rate, and their productivity is measured as output per unit of time.
Can be stockpiled.	Cannot be stockpiled.
Are used up, not worn out.	Are worn out, not used up.

Source: after Daly and Farley (2011)

For other kinds of ecosystem services – regulatory and cultural, as well as supporting – this simple stock-flow model is not appropriate, and they are better captured by the concept of fund-service resources. Fund-service resources are those where the stock does not actually produce more of itself as a flow, where the flows are delivered at a given rate which cannot be determined by humans, and which cannot be stockpiled for the future (Table 5.1). Fund-service resources cannot be "used up" by extracting a yield or flow, although they can deteriorate or become worn out (Daly and Farley, 2011) (Table 5.1). For instance, a stock of water in a lake can be used by canoeists or swimmers, thereby supplying a recreation service. The swimmers and canoeists do not use up the lake stock in any sense and the lake is not depleted as a function of how many swimmers and canoeists are present. The experience of swimmers and canoeists might be affected by how many are present, but the lake itself (the stock) is not compromised unless it becomes polluted or degraded by activities other than swimming or canoeing. Neither can the recreational service be stockpiled for future use in the same sense that fish and meat can be frozen and timber stored.

In general, biological stock-flow resources are associated with the structural properties of an ecosystem, and fund-service resources with functional properties (Farley et al., 2005). It should be noted that whether a resource is classed as stock-flow or fund-service depends on what it is used for rather than any intrinsic properties: water used for drinking or irrigation is a stock-flow resource, whilst water used for swimming is a fund-service resource. A further example would be the beautiful view that people enjoy of a landscape such as a mountain. The mountain might be perceived by beneficiaries of this cultural service from a considerable distance, but the number of viewers does not deplete the landscape.

Of course, these two kinds of resources are intimately linked: over-exploitation of stock-flow resources will inevitably deplete fund-service resources. Logging a forest for its timber will have significant implications for carbon storage and hence climate regulation as well as for recreation, water quality and quantity regulation, sediment run-off and, depending on whether it is exotic or natural, wild species biodiversity. Managing a landscape or seascape for multiple benefits will clearly require analytical approaches by resource economists that go beyond those traditionally used in order to ensure an appropriate level of resource extraction, one that does not impact ecosystem services (Figure 5.2).

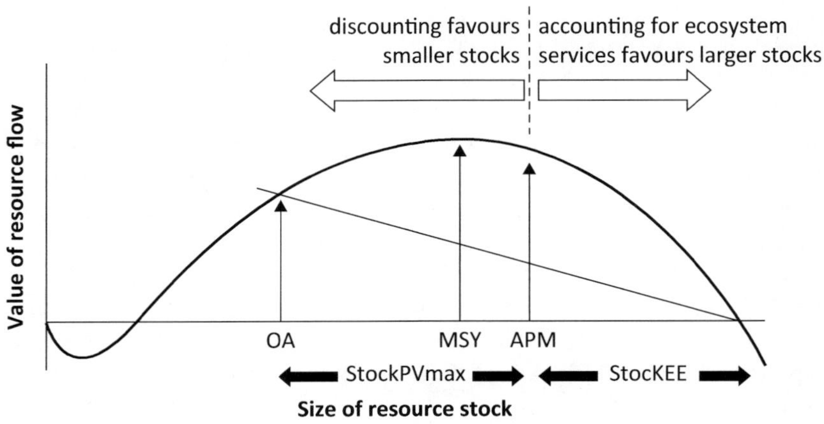

Figure 5.2 Optimal harvest levels for renewable resources, depending on whether managing for a flow from a single stock (left-hand side of curve) or for both stocks and services (right-hand side). OA is the open access equilibrium; APM is the annual profit-maximizing stock; StockPVmax is the stock at which net present value is maximised; StockEE is the objective of seeking a sufficient, rather than the maximum, amount, to gain the joint benefits of both flow and service.

Source: modified from Daly and Farley (2011)

Estimating natural stocks and ecosystem services

The components of natural stocks that underpin ecosystem service flows include soil, water, air and biodiversity. Estimating the quantity and quality of each of these for landscapes, and how they might change under different policy options, is a huge challenge, so it is not surprising that managers and policymakers have sought indicators or proxies for individual components. For example, for the UK, Linstead et al. (2008) identified 46 potential indicators (which they termed environmental compartments) for air, 36 for water, 15 for soils and 194 for biodiversity (some of which appear later, in Table 5.2). Many of these were also indicators of pressures and ecosystem services (e.g. air quality, soil organic matter, water quality and species and habitat diversity), rather than capital stocks as such (how much soil or water there is). The areal extent of major service-providing habitats of conservation importance (Biodiversity Action Plan (BAP) habitats) and measures of their health status could be derived from the UK Countryside Survey (DETR, 2000). For several of the habitat types (agricultural pasture and the marine habitats of cliffs, salt marsh and tidal flats) no indicators were reported by Lindstead et al. (2008), and they do not appear to have been covered by the Countryside Survey. However, stocks could be monitored as aerial extent and suitable biodiversity indicators developed, as for other habitats.

An increasingly common way of linking stocks to flows, based on the amount of habitat present, is through the use of a GIS-type model. Here, biophysical elements are classified and their spatial extent represented, usually as a map, and specific services are derived as a function of the amount of that landscape element, based on locally estimated production functions (good, but involves a lot of work) or some kind of scalar (poorer, but easier). Services may be represented as physical amounts or as monetary values. Often scalars and values are based on previous estimates made elsewhere by applying a benefits transfer approach (e.g. Eade and Moran, 1996; Naidoo and Ricketts, 2006; Troy and Wilson, 2006). The pros and cons of such extrapolations have been widely rehearsed elsewhere (Kareiva et al., 2011), and the "generalization error" of this approach is assumed significant (Plummer, 2009). For some purposes only a ball-park figure may be needed for illustrative purposes. For instance, if the aim of the exercise is to merely establish a broad conversation between different stakeholders about different policy options, then a highly precise estimate of value, with all the research effort that would entail, could be seen as unnecessary. On the other hand, once the conversation turns into real decisions about which policy to adopt, better local estimates are clearly required, or at least the confidence in transferred estimates presented. One difficulty with using generalised values is that services produced will be heterogeneous across different spatial scales. In their analysis of the ecosystem services of biodiversity, recreation and carbon storage in the UK, land cover based proxies agreed poorly for areas where primary data on those services were available, especially for "hotspots" of services (Eigenbrod et al., 2010).

Scale dependency in stock and flow estimates is a non-trivial problem and one not usually recognised or acknowledged by those methods using some kind of scalar for the linkages. Flows from stocks will scale differently with the spatial extent and arrangement of biodiversity elements and with the trade-offs made between services. For instance, a 1000 km^2 block of woodland could reasonably be expected to fix the same carbon per unit area as 100 distinct 10 km^2 blocks. In other words, the relationship between stock and flow is more or less scale independent. In contrast, the service of recreation or soil erosion mitigation per unit area of a 1000 km^2 forest will be much higher than that provided by the smaller blocks. The relationship between spatial extent (block size) and flow is likely to be a step function for these two services, whereas for carbon sequestration it is likely to be more linear. In addition, some services may

Table 5.2 Indicators for major ecosystem services in the UK.

	Indicator	*Ref*
Climate regulation	Greenhouse gas emissions	1, 2
	Ammonia emissions from farmland	3
	Methane emissions	1,2
	Nitrous oxide	1,2
	Central England Temperature Index	4
	Rainfall	5
	Flood occurrence trends	6, 7
	Change in woodland sequestration rate	2
	Change in soil carbon	8
	Net energy delivered from agricultural biomass	3
	New woodland planting	3, 9
Pollination	Plant diversity in fields and margins	3, 8
Water quantity regulation	Groundwater levels	10
	Peak river levels	6,7
Water quality regulation	Biological quality of rivers	11
	Chemical quality of rivers	11
	Nitrate and phosphate levels in rivers	3
	Pesticides in water	3
Erosion prevention	None available	
Food production	Farmland under different production types	3, 23
	Agricultural resilience	12
	Soil quality; soil organic matter	13
	Area of land which is bio-productive	14
Potable water	Water abstracted for agricultural use	3
	Water use for irrigation	3
	Water use in food and drinks industry	15, 16
	Domestic water consumption	11
	Industrial water consumption	16
Genetic resources/biodiversity	Status of farmland BAP priority habitats	3
	Status of farmland BAP priority species	3
	Population trends for common pipistrelles	3, 17
	Population trends for brown hares	3, 18
	Farmland bird populations	3, 18
	Population trends for farmland butterflies	3, 19
	Plant diversity in fields and at field margins	3, 8
	Effective population size for native breeds	20,21
	Improved local biodiversity	22
Raw materials	Commercial forestry	9
	Wool	3, 23
Aesthetics	Landscape features	8
	New woodland planting	9
	Quality of place	28
	% of coastline and rivers with hard defences	
	Extent of sky without light/noise/air pollution	24, 25, 26
	Tranquillity	26
Heritage	Heritage at Risk	27

Sources for ecosystem services indicators

1. Greenhouse gas emissions by local authority http://www.decc.gov.uk/en/content/cms/statistics/climate_change/gg_emissions/uk_emissions/uk_emissions.aspx

2. Greenhouse gas emissions related to land use and land use change http://ecosystemghg.ceh.ac.uk/
3. Defra Observatory Monitoring Network http://www.defra.gov.uk/evidence/statistics/foodfarm/enviro/observatory/indicators/index.htm
4. Central England Temperature data http://hadobs.metoffice.com/hadcet/
5. Rainfall data http://www.metoffice.gov.uk/climate/uk/datasets/
6. Defra/Environment Agency. Flood and coastal defence R&D programme http://evidence.environment-agency.gov.uk/FCERM/Libraries/FCERM_Project_Documents/FD2311_1038_TRP_pdf.sflb.ashx
7. HiFlows-UK flood peak data from river flow monitoring stations http://www.environment-agency.gov.uk/hiflows/91727.aspx
8. Countryside Survey http://www.countrysidesurvey.org.uk/
9. Forestry statistics http://www.forestry.gov.uk/statistics
10. National Water Archive (groundwater) http://www.ceh.ac.uk/data/nrfa/groundwater.html
11. Sustainable development indicators http://www.defra.gov.uk/sustainable/government/progress/data-resources/regional.htm
12. Indicators for a sustainable food system fact sheets http://www.defra.gov.uk/evidence/statistics/foodfarm/general/foodsystemindicators/documents/foodsystemindicators-factsheet.pdf
13. Land Information System http://www.landis.org.uk/index.cfm
14. Natural environment indicators – internal Yorkshire Futures document
15. Federation House Commitment http://www.fhc2020.co.uk
16. Industrial water use http://www.scpnet.org.uk/downloads/Environmental_Data/water-use-166.pdf
17. National Bat Monitoring programme http://www.bats.org.uk/pages/nbmp.html
18. British Trust for Ornithology http://www.bto.org/bbs/index.htm
19. Butterfly indicators http://www.ukbms.org/butterflies_as_indicators.htm
20. Rare breed herd sizes http://www.defra.gov.uk/fangr/documents/ukcountry-rpt2002.pdf
21. Indicators for a sustainable food system http://www.defra.gov.uk/evidence/statistics/foodfarm/general/foodsystemindicators/index.htm
22. National Indicator Set information http://www.audit-commission.gov.uk/localgov/audit/nis/pages/default.aspx
23. Livestock numbers http://www.defra.gov.uk/evidence/statistics/foodfarm/landuselivestock/junesurvey/results.htm
24. Extent of sky not subject to light/noise/air pollution http://www.cpre.org.uk/campaigns/landscape/light-pollution
25. UK National Air Quality Archive http://www.airquality.co.uk/
26. Council for the Protection of Rural England http://www.cpre.org.uk/campaigns/landscape/tranquillity/national-and-regional-tranquillity-maps
27. Heritage at risk register http://www.english-heritage.org.uk/protecting/heritage-at-risk/
28. Progress in the Region http://www.yorkshirefutures.com/articles/progress-region-2008

Source: from Raffaelli et al. (2010)

be produced at very local scales, such as field crop pollinators, whereas others may be more appropriately managed at much larger scales, such as flood alleviation at the catchment scale. Managing landscapes for multiple benefits at a single scale may have unknown consequences for overall service delivery and is especially relevant where biodiversity stocks within a landscape are highly fragmented (Kremen, 2005, Turner and Chapin, 2005). It should also be noted that the values attached to services by stakeholders are also scale-dependent. Institutions at different levels (international, national, state/provincial, municipal, household and individual) differ in the importance they attach to different kinds of services. Thus, for the ecosystem services supplied by the Netherland's De Wieden wetlands, reed for cutting and fisheries were most relevant at the municipal and provincial levels, recreation at the municipal level and nature conservation at the national and international levels (Hein et al., 2006).

Kareiva et al. (2011) are highly critical of the use of benefits transfer and advocate a production function approach, involving models of local ecosystem service supply driven by variables such as land use and cover, ecosystem attributes and the demand for services by people (Pagiola et al., 2004, NRC, 2005, Barbier, 2007). Their book illustrates how this can be achieved using one particular modelling tool, InVEST (Integrated Valuation of Ecosystem Services and Trade-offs), developed by the Natural Capital Project (www.naturalcapitalproject.org) for multiple ecosystem services. Other models such as LUCI (Land Utilisation and Capability Indicator),[1] itself an extension of Polyscape (Jackson et al., 2013), and ARIES (ARtificial Intelligence for Ecosystem Services),[2] take different approaches to obtain similar ends. Such models are likely to prove powerful as decision-making support tools for landscape-scale management. However, our understanding of the fundamental ecological processes that generate the flows of services from stocks and the feedbacks between stocks and flows is still poor (Jones et al., 2013).

Direct measures do not exist for many ecosystem services, such as erosion prevention or climate regulation, and proxies or indicators are needed if we wish to evaluate the effects of policy options and decisions (Layke, 2009). When choosing indicators, a number of features should be taken into account, including: existence of a reference value; the availability of time-series; ability to adapt to different spatial scales of assessment; sensitivity to change (responsiveness, but not too responsive); cost of recording; ease of understanding by all managers and stakeholders; measurability. Most of the countries that have adopted the Ecosystem Approach and an Ecosystem Services Framework have retro-fitted the indicators they were already using and routinely collecting information on. Whilst this is understandable in terms of resources and the desire to utilise historical data in order to detect temporal trends, there is unlikely to be a close match between what is easily collected and what is actually desired (information on levels of and trends in services). The issue is clearly illustrated by Linstead et al.'s (2008) analysis for the UK. They concluded that 80 such indicators have potential for monitoring ecosystem services and 65 of those are currently in use within the UK. However, many of these indicators are measures of yields (of trees, crops, livestock or water) and their proper use and interpretation would require knowledge of the stocks from which these yields are extracted in order to evaluate the sustainability of policy options. This is also an issue for many of the national indicators listed by Layke (2009). For instance, if the service of recreation (measured as number of visitors) is to be managed sustainably, then it might at some point be necessary to limit access to the wider countryside or to nature reserves so that their carrying capacity is not exceeded. Similarly, yields of water from groundwater or yields of produce from agricultural and forestry land can be greatly enhanced through over-exploitation of the underlying capital stocks or, in the case of trees and crops, by the additions of artificial fertilisers, unlikely to be sustainable in the long term. Thus, if such indicators are to be used, they need to be related to the underlying stocks.

Holistic approaches to managing stocks and flows

Understanding the relationships between a single service and its underlying stock is a complex enough process. Extending this approach to many stocks and services, some of which are stock-flow, others fund-service, will be a challenge. Also, evaluating how near the value of an individual indicator might be to an environmental or socially acceptable limit involves huge effort. Extending this to a suite of indicators for an entire system has, to the author's knowledge, not been attempted. At a practical level, managers of landscapes and seascapes need to know if particular policy options that simultaneously change several services and their underlying stocks will push the system being managed beyond its safe operating space (whether any new system state will be sustainable). One possible way to achieve this is to adopt a holistic systems approach to derive system-level measures of resilience, rather than attempting to examine the independent trajectories of all indicators. This is the province of Systems Ecology.

Systems ecology is the science of stocks and flows and has a long history in ecology (reviewed in Raffaelli and White, 2013; Raffaelli and Frid, 2010; Jorgensen et al., 2007), but has yet to be fully applied to the ecosystem service agenda. Norris (2012) has claimed that "a mechanistic understanding of biodiversity change and its consequences for ES can only be addressed using systems approaches", a view echoed by Loreau (2010). Modern readers may be more familiar with the approach as Network Analysis. Networks have holistic properties which are determined by their internal architecture, the arrangement and magnitude of the network elements (components of stocks) and the connections (flows) between them. Network analysis is

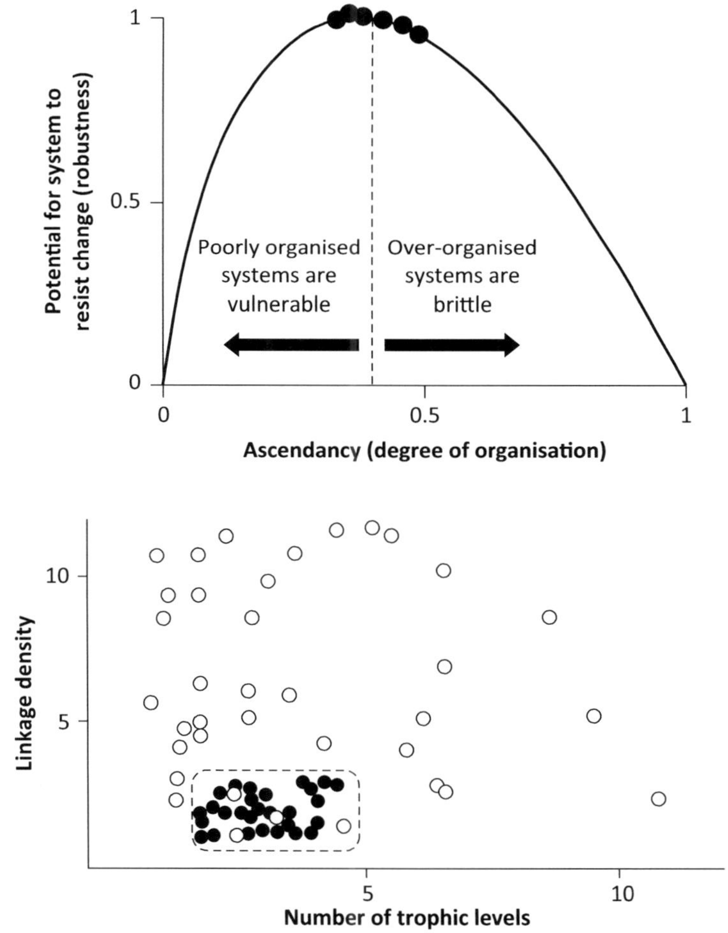

Figure 5.3 (a): Ecological networks are only robust for limited bounds of ascendancy; (b): The normal operating zone (delineated by dotted lines) defined by ascendency considerations for real (solid circles) and random networks (open circles), may be captured by two simple topological properties of food webs: linkage density and number of trophic levels.

Source: modified from Ulanowicz 2011

well-suited for exploring the consequences of simultaneous changes in the different elements that underpin ecosystem services as a result of different policy options. Importantly, such analyses can potentially identify the normal operating space within which changes in different system elements can be allowed to vary as a result of selecting a particular policy without compromising system resilience. Systems-level measures, as opposed to species component measures of stability (Donohue et al., 2013), are most relevant when considering the resilience of the wider system and several such measures are available from network analysis (Heymans et al., 2012; Borrett, 2013; Ulanowicz, 2011).

For instance, Ulanowicz (2011) has argued that the network metric "ascendancy", defined as the size of a network (the sum of all the flows, termed total system throughput) scaled by the network's organisation (how the flows are arranged, termed the average mutual information), has a restricted set of values for real-world ecosystems, around 40% (Figure 5.3a). Too low a value of ascendancy and the system tends to disorder, having insufficient cohesiveness, too high a value and the system becomes vulnerable to perturbations, both externally and internally generated through "self-organising catastrophe" (Holling, 1986; Bak 1996). If such properties can be adequately captured using more easily measured metrics, such as the topological properties of networks (Figure 5.3b), then trade-offs between different services and the network configurations that support them can be explored experimentally by stakeholders in relation to a normal operating space for different policy options. Once the present network is constructed, the stocks which underpin ecosystem services can be identified. Stakeholders and policymakers can then reconfigure the stocks to generate the services demanded under different policy options to establish which, if any, of the new configurations remain in the safe operating space. Raffaelli and White (2013) provide an illustration for ecosystem services in a temperate estuarine system. In addition, the trajectory of the system can be monitored over time with respect to safe operating bounds and policy interventions applied as necessary.

Taking a more holistic systems approach to the management of stocks and flows, one which attempts to maintain the system within sustainability bounds, is potentially attractive and has a strong scientific theory basis. Many of the empirical mapping approaches detailed above are more reductionist and do not offer measures of resilience or sustainability for landscapes delivering, and being managed for, multiple benefits. One critique of network analysis is the large data effort required for network construction, but in reality this may be on a par with that needed for other approaches. As in the case of the more mechanistic mapping approaches, it is possible to include data of varying pedigrees, ranging from local-specific to obtained from a similar system elsewhere to an educated guesstimate. But at least for network analysis models it is possible to openly code for the level of pedigree and thus convey confidence as to whether those data might apply locally.

Notes

1 http://unstats.un.org/unsd/envaccounting/ceea/meetings/tenth_meeting/BK10a.pdf
2 www.ariesonline.org/about/approach.html

References

Anielski, M., and Wilson, S. (2006). *Counting Canada's Natural Capital: Assessing the Real Value of Canada's Boreal Ecosystems*. Canadian Boreal Initiative and Pembina Institute, Ottawa, ON.

Bak, P. (1996). *How Nature Works: The Science of Self-Organised Criticality*. Copernicus Press, New York.

Barbier, E. B. (2007). Valuing ecosystem services as productive inputs. *Economic Policy*, vol 22, no 49, pp 178–229.

Borrett, S. R. (2013). Network Ecology. In *Encyclopedia of Environmetrics*. John Wiley & Son, New York.

Convention on Biological Diversity (2009). Available at: http:/www.cbd.int/ecosystem/ (accessed January 2009).

Costanza, R., d'Arge, R., deGroot, R., et al (13 authors) (1997). The value of the world's ecosystem services and natural capital. *Nature*, vol 387, pp 253–260.

Daily, G. C. (1997). *Nature's Services: Societal Dependence on Natural Ecosystems*. Island Press, Washington DC.

Daly, H. E., and Farley, J. (2011). *Ecological Economics. Principles and Applications*. Island Press, Washington DC.

DETR (2000). *Accounting for Nature: Assessing Habitats in the UK Countryside*. Department of the Environment, Transport and the Regions, London.

Donohue, I., Petchey, O., Montoya, J. M., Jackson, A. L., McNally, L., Viana, M. Healy, K., Lurgi, M., O'Connor, N. E., and Emmerson, M. C. (2013). On the dimensionality of ecological stability. *Ecological Letters*, vol 16, pp 421–429.

Eade, J.D.O., and Moran, D. (1996). Spatial economic valuation: benefits transfer using geographical information systems. *Journal of Environmental Management*, vol 48, pp 97–110.

Eigenbrod, F., Armsworth, P. R., Anderson, B. J., Heinemeyer, A., Gillings, S., Roy, D.B., Thomas, C. D., and Gaston, K. J. (2010). The impact of proxy-based methods on mapping the distribution of ecosystem services. *Journal of Applied Ecology*, vol 47, pp 377–385.

Farley, J., Erikson, J. D., and Daly, H. (2005). *Ecological Economics. A Workbook for Problem-Based Learning*. Island Press, Washington DC.

Gómez-Baggethun, E., de Groot, R., Lomas, P. L., and Montes, C. (2010). The history of ecosystem services in economic theory and practice: from early notions to markets and payment schemes. *Ecological Economics*, vol 6, pp 1209–1218.

Hein, L., van Koppen, K., de Groot, R. S., and van Ierland, E. C. (2006). Spatial scales, stakeholders and the valuation of ecosystem services. *Ecological Economics*, vol 57, pp 209–228

Heymans, S., Coll, M., Libralato, S., and Christensen, V. (2012). Ecopath theory, modelling and application to coastal ecosystems. In McLusky, D. and Wolanski, E. (eds) *Treatise on Estuarine and Coastal Science*. Elsevier, Amsterdam.

Holling, C.S. (1986). The resilience of terrestrial ecosystems: local surprise and global change. In Clark, W. C. and Munn, R. E. (eds) *Sustainable Development of the Biosphere*. Cambridge University Press, Cambridge UK.

Jackson, B., Pagella, T., Sinclair, F., Orellana, B. M., Henshaw, A., Reynolds, B., et al. (9 authors) (2013). Polyscape: a GIS mapping framework providing efficient and spatially explicit landscape-scale valuation of multiple ecosystem services. *Landscape and Urban Planning*, vol 112, pp 74–88.

Jones, L., et al. (15 partners) (2013). Scale dependence of stocks and flows in the valuation of ecosystem services. Available at: http://www.valuing-nature.net/projects/stocks-flows

Jorgensen, S. E., Fath, B. D., Bastianoni, S., et al. (9 authors) (2007). *A New Ecology. Systems Perspective*. Elsevier, Amsterdam.

Kareiva, P., Tallis, H., Ricketts, T. H., Daily, G.C., and Polasky, S. (2011). *Natural Capital. Theory and Practice of Mapping Ecosystem Services*. Oxford University Press, Oxford.

Kremen, C. (2005). Managing ecosystem services: what do we need to know about their ecology? *Ecology Letters*, vol 8, pp 468–479.

Layke, C. (2009). *Measuring Nature's Benefits: A Preliminary Roadmap for Improving Ecosystem Service Indicators*. WRI Working Paper. World Resources Institute, Washington DC. Available at: http://www.wri.org/project/ecosystem-service-indicators.

Linstead, C., Barker, T., Maltby, E., Kumar, P., Mortimer, M., Plater, A., and Wood, M. (2008). *Reviewing Targets and Indicators for the Ecosystem Approach*. Final Report. Defra Project Code NR0119.

Loreau, M. (2010). *From Populations to Ecosystems: Theoretical Foundations for a New Ecological Synthesis*. Princeton University Press, Princeton NJ.

MA (2005). *Ecosystems and Human Well-Being: Synthesis*. Millennium Ecosystem Assessment, Island Press, Washington DC.

Naidoo, R., and Ricketts, T. H. (2006). Mapping the economic costs and benefits of conservation. *Plos Biology*, vol 4, pp 2153–2164.

Norris, K. (2012). Biodiversity in the context of ecosystem services: the applied need for systems approaches. *Philosophical Transactions of the Royal Society B*, vol 367, pp 191–199.

NRC (National Research Council) (2005). *Valuing Ecosystem Services: Towards Better Environmental Decision-Making*. National Academies Press, Washington DC.

Pagiola, S., von Ritter, K., and Bishop, J. T. (2004). *Assessing the Economic Value of Ecosystem Conservation*, TNC-IUCN-WB, Washington DC.

Plummer, M. L. (2009). Assessing benefit transfer for the valuation of ecosystem services. *Frontiers in Ecology and the Environment*, vol 7, pp 38–45.
Raffaelli, D., and Frid, C.L.J. (2010) *Ecosystem Ecology: A New Synthesis*. Cambridge University Press, Cambridge UK.
Raffaelli, D., and White, P.C.L. (2013). Ecosystems and their services in a changing world: an ecological perspective. In Woodward, G. and O'Gorman, E. (eds) *Advances in Ecological Research*, vol 48, pp 1–70.
Raffaelli, D., White, P.C.L.W., and MacGillivray, A. (2010). *Applying an Ecosystem Services Approach to the Yorkshire and Humber Region*. Yorkshire Futures, Leeds.
Schumacher, E.F. (1973). *Small is Beautiful: Economics as if People Mattered*. Harper & Row, New York.
Stern, N. (2006). *The Economics of Climate Change*. HM Treasury, London.
TEEB (2010). *The economics of Ecosystems and Biodiversity: Mainstreaming the Economics of Nature*. UNEP.
Troy, A., and Wilson, M. A. (2006). Mapping ecosystem services: practical challenges and opportunities in linking GIS and value transfer. *Ecological Economics*, vol 60, no 2, pp 435–449.
Turner, K. and Daily, G. (2008). The ecosystem services framework and natural capital conservation. *Environmental and Resource Economics*, vol 39, pp 25–35.
Turner, M. G., and Chapin, F. S. (2005). Causes and consequences of spatial heterogeneity in ecosystem function. In Lovett, G. M., Jones, C. G., Turner, M. G., and Weathers, K. C. (eds) *Ecosystem Function in Heterogeneous Landscapes*. Springer, New York.
Ulanowicz, E. (2005). Ecological network analysis: an escape from the machine. In Belgrano, A., Scharler, U. M., Dunne, J. and Ulanowicz, R. E. (eds.) *Aquatic Food Webs*. Oxford University Press, Oxford.
Ulanowicz, R. E. (1997). *Ecology, the Ascendent Perspective*. Columbia University Press, New York.
Ulanowicz, R. E. (2011). Quantitative methods for ecological network analysis and its application to coastal ecosystems. *Treatise on Estuarine and Coastal Science*, vol 9, pp 35–57.
UNEP (2011). *United Nations Environment Programme Finance Initiative: Natural Capital Declaration*, October 2011.
Voora, V. A. and Venema, H. D. (2008). *The Natural Capital Approach: A Concept Paper*. IISD, Winnipeg.

6

THE BENEFICIARY PERSPECTIVE

Benefits and beyond

Dixon H. Landers, Amanda M. Nahlik and Charles R. Rhodes

Introduction

Humans are dependent on the environment, and the ecosystem services concept helps us see how much. Within the academic and policy communities, subtle differences in definition can greatly affect how we identify and measure ecosystem services and what analysts report to policymakers. Efforts to define, measure, map, and value ecosystem services have met with varying success, but to date, no comprehensive framework or approach has emerged that can consistently inform policymakers in terms useful at different spatial scales. Natural scientists interested in physical measures must directly collaborate with social scientists concerned with valuation and decision-making. This chapter argues that a paradigm shift in how scientists articulate the differences between deeper natural processes and the products of those processes that people use or care about can help to resolve many issues within the ecosystem service community. We pose that great care in separating intermediate ecosystem services from Final Ecosystem Goods and Services will narrow research focus away from inconceivably large values for broad single-metric indicators (e.g., all carbon, all biodiversity in an area), and toward values that are human-scale and clearly affected by specific choices (e.g., Will conversion of wetlands to a new housing development raise the cost of a city's water for decades?). Our approach does not ignore intermediate ecological processes, but instead focusses measurement attention on ecosystem services that people more directly use or appreciate, and on who these people are – because the value of something changes with the user and with the context of their choice.

The objectives of this chapter are threefold. We first explore distinctions between nature's benefits and human beneficiaries of nature, and how seemingly slight differences in the definition of ecosystem services can lead to vastly different outcomes. We adopt a *beneficiary approach* by expressing ecosystem services in terms of use-user combinations, as opposed to a *benefits approach*, which lists myriad potential uses of ecosystem services while implying, but in no way empirically identifying, users. We propose that a beneficiary approach promotes the practicality and feasibility necessary for achieving ecosystem service research and policy objectives, including measuring, mapping, and valuing ecosystem services. A beneficiary perspective may also offer much-needed momentum in moving ecosystem services from concept to implementation, and eventually, to influencing societies (i.e., communities) and economies by refining and highlighting decision-making that affects the environment. Second, we introduce the use of Final Ecosystem Goods and Services (FEGS) as the underlying definition and concept for a beneficiary approach. Distinguishing intermediate ecosystem services from FEGS is a useful foundation for classifications systems that serve both natural and social science objectives. Finally, we present a Final Ecosystem Goods and Services Classification System (FEGS-CS) that appears to meet the needs of a variety of different users interested in quantifying and valuing nature's benefits in a defined and non-duplicative way, by incorporating the perspectives of beneficiaries and, therefore, directly linking to human well-being. Merging the FEGS-CS with an economic approach allows the mapping of flows of FEGS through economic production functions to human well-being. We introduce and briefly describe such an economic approach, the National Ecosystem Service Classification System (NESCS; more easily voiced when slightly rephrased as 'nexus').

The millennium ecosystem assessment and our diversion from it

Since 2005, many natural and social scientists have turned to the ecosystem service definition and general categories that the Millennium Ecosystem Assessment (MA) offers (MA, 2005). Multiple approaches for implementing the MA vision have been proposed – in our view, not necessarily with clarity and consistency (Carpenter *et al.*, 2009; Nahlik *et al.*, 2012). Despite on-going efforts, none of these has successfully met the objective of moving ecosystem services forward into practice by devising a system that supports measuring, mapping, and valuing ecosystem services using the MA categories. This follows from the fact that there is virtually no attempt in the MA approach to explicitly link ecosystem services derived from ecological components to the people that use the services, or beneficiaries. Potential users of the MA's 'benefit approach' (thus so because MA equates services and benefits) are required to separate some ecosystem services from others and to select – without the perspective of how humans perceive them – which of the near-infinite list of environmental things should be measured when characterizing the provision of ecosystem services for policymakers. Even if all these obstacles could be overcome, a further issue associated with a benefit approach remains: How can ecosystem services be prioritized or valued outside of the beneficiary's context? This issue confounds economists and social scientists attempting to estimate values for any ecosystem service.

Benefits versus beneficiaries

For practical reasons, a stronger coupling between ecosystem service uses and users is required. Potential use must connect with an actual user for an ecosystem service to be discretely identified. Only by explicitly matching the uses of nature's services, the *benefits*, to known users, the *beneficiaries*, do we gain the means to measure and account for how much of which of nature's services are used or appreciated, and by whom. Beneficiaries are 'users' of nature's services, but in the broadest sense – 'use' includes consumption, appreciation, and *any* way a person might value the ecosystem service – not excluding valuing that it exists, or benefits others, or in the opinion of the 'user' should be passed unspoiled to future generations, even if the 'user' never touches or even sees it. Without explicit matching of benefits to beneficiaries, there is no uniquely identified value to any *human*'s well-being that we can hope to measure accurately and consistently, and nature's services may continue to be dramatically underappreciated in policy debates and elsewhere.

Humans, whether we recognize it or not, are dependent on the environment for multitudes of benefits that sustain us, our societies, and our economies. How then did our society develop the environmental blind spot that motivated Gretchen Daily and others to compose *Nature's Services*, in which the authors promote the use of ecosystem services to counteract 'the near total lack of public appreciation of societal dependence upon natural ecosystems ... [which presents] a major hindrance to the formulation and implementation of policy designed to safeguard earth's life-support systems' (Daily, 1997, p. xv)?

Development of the environmental blind spot

In any academic field, the first great work attempting to break from, yet build upon, earlier concepts often demonstrates context and sensitivity that later acolytes and practitioners fail to appreciate at some risk to effective progress. This is true of Adam Smith's *Wealth of Nations*, the American Founding Fathers' attempts to define a sustainable and functioning democracy in a new Age of Enlightenment, and Daily's *Nature's Services* (Daily, 1997). The failure of the general public to understand how deeply they rely on nature's processes is in part the result of an Industrial Age worldview that evolved from common readings of Adam Smith's work and ideas tried by the United States' Founding Fathers, among others – a worldview that favors the value of the individual abstractly, as well as favoring the values of productivity, efficiency, and the happiness of individuals relative to society and to the larger ecosphere. In *Nature's Services*, Costanza and Folke note that when societies prioritize goals, the goal of efficiency is naturally coupled with individual preferences, the goal of fairness is naturally coupled with social preferences, and the goal of sustainability is naturally coupled with 'whole system' preferences (see Table 4.1 and surrounding text on p. 57 of Daily, 1997). Favoring individual preferences over 'whole system' preferences would thus place economic production-unit level efficiency before sustainability in decision-making. There is ample evidence that exactly this has happened – MA (2005) presents evidence of cumulative conversion of ecosystem-service-producing areas to commercial use.

Over the centuries, economists have built an idealistic but plausible structure rooted in specific assumptions, which holds that individuals rationally and efficiently pursuing their own interests will also reach a fair, and even sustainable, outcome for society. Much modern economic work: a) explores how reality deviates from the ideals and assumptions, b) seeks to understand by how much, and c) seeks to determine whether and how to correct incentives so that economic dynamics shift toward ideal processes and outcomes. Regardless of one's opinion about how far actual markets are from an ideal process, one must admit that the underlying perspective favoring individual preference action and happiness has come to dominate political-economic thinking.

Many stakeholders and policymakers remain relatively deaf to arguments that do not assume that pursuit of company-level economic goals will also best serve society over the long run. Despite ample evidence that human disruption of natural ecosystems now threatens ecological functions and yields that people care about (Daily, 1997; MA, 2005), overcoming 'the near total lack of public appreciation of societal dependence upon natural ecosystems' (Daily, 1997, p. xv) must involve the dominant individual/economic perspective, if policies needed to safeguard natural systems are to succeed. The hard truth is that the term ecosystem services (in using the word 'services') suggests that natural science, with its objective disciplines, *must* be grafted to social science, with its often subjective disciplines, when determining how and how much humans value the processes and provisions of nature. If the value of nature is not made clearer to them, people will continue to ignore the provisions of nature at direct risk to themselves. We offer an approach that natural and social scientists can agree on, one that defines and identifies measureable ecosystem services, benefits, and beneficiaries.

Why economics matters to everyone who wants ecosystem services to influence policy

Economics includes the study of how to efficiently use limited resources to meet objectives – a seemingly eternal problem for decision-makers, individual, national, or global. In classical economics, limited resources constrain choices, so tracking how resources are produced, processed, and distributed is a primary interest. Under conditions that hold for many limited resources and the goods and services made with them, markets are quite adept at coordinating buyer and seller exchange through price signals. Prices, or some bundle of goods judged to be equivalent in worth, reflect the value people place on non-infinite goods and services.

Classical economics begins from the perspective and with the premise that individual consumer choice as exercised through complete and fair markets will leave everyone as well off as they can be, given the distribution of resources with which everyone began. Because almost all quantitative economic analysis must be formulated in a framework whose foundation is individual choice and actions to raise individual satisfaction, arguments not founded on individual choices must fight an uphill battle for acceptance as an alternative formulation. Economic production functions match inputs, producers, and processes with outputs, and the outputs become inputs for individuals to meet their goals.

Welfare economics – how economics links to measures of well-being and policy choices

Welfare economics is used to compare how different economic states of being affect the well-being of a population, and is thus the basis for policy recommendations by economists. Welfare economics all but demands that a user be able to place (or estimate) a value for goods or services. Values should be comparable to purchase choices in a market (Bockstael *et al.*, 2000), and work best if they can be summed and compared across individuals. 'Values' for things that do not have prices attached to them – as for most ecosystem services – are at a structural disadvantage in analysis that favors a single common scale of comparison (like a price or known bundle of goods). Most ecological processes and resources are un-priced, so they are ignored or subtly embedded in models of economic production. In practice, a price-based perspective assumes away complex ecological production dynamics and any chance that small overlooked effects may accumulate to disrupt the continuing flow of ecosystem services. The default presumption is that if any resource becomes scarce enough, a price will rise and significant effects will be captured in market transactions.

Market failures and how economists can have trouble seeing ecosystem services

Economists are trained to be forthright about the limitations of the assumptions that must be made for basic economic theory to work. Markets are unlikely to work best, or even well, when *market failures* exist – conditions under which markets from their own dynamics will not reach an efficient and mutually beneficial equilibrium. Market failures include the existence of public goods, environmental externalities (good or bad), and asymmetries of information (under which not all parties are sufficiently well-informed to most effectively pursue their own best interests). There should be no surprise that ecosystem services very often correlate with conditions associated with exactly these market failures. Indeed, the ecosystem services concept was defined as a means of identifying some of the un-market-able, but vital, processes and products of the environment otherwise often ignored in everyday economic decision-making (Daily, 1997).

It is precisely because people benefit from many ecosystem services without having to expressly buy them that simply listing potential ecosystem services is not argument enough to include them as certain elements in the set of an individual's *realized* economic choices – meaning the individual actually derives satisfaction (or *utility*) from the ecosystem service, whether they ever use it (e.g., drinking water from snow pack) or just enjoy knowing it exists (e.g., snow leopards). Ecosystem processes and fundamental materials (e.g., soil in the lithosphere) in some objective way exist in an 'ecological production function' that yields ecosystem goods or services that someone values, but without prices there may be little motivation to discover or account for that ecological production function. This is a blind spot between economic theory and practice that the notion of ecosystem services is meant to redress.

Absolute stock versus relative flow: biodiversity as an example

To actually be an ecosystem service, a candidate must have recognizable value to someone. From this understanding, 'biodiversity' might seem to be an ecosystem service/benefit that an individual would value. But without identifying elements of a larger context such as geographical area, or estimated number of species relative to pre-industrial influence, simply naming 'biodiversity' does not evoke the type of trade-off decision-making that an individual might undertake daily (e.g., a *marginal* choice between condition A and condition B, all else being equal, that will make the individual more satisfied).

Without a larger context, the individual can only choose 'biodiversity = yes', or 'biodiversity = no'. Without a biodiverse production environment, the human species cannot fulfill our biological need for a varied diet. Because there can be no rational 'biodiversity = no' choice, 'biodiversity' cannot be 'used/valued' in a *choice* an individual can reasonably make. Economists that model relevant (marginal) choices, rather than economists helping to measure stocks of limited resources, may turn their backs on the discussion. Biodiversity is extremely valuable, but not in ways that individuals consciously choose, as whether one eats a $2 cookie after lunch.

Biodiversity may be depicted in crass financial terms as a *capital stock*, like the capital stock of a firm, where having and using more of it well translates to higher productivity and higher yields for the same level of other inputs. In this sense biodiversity is like the machinery in a factory (physical capital), but in another sense is like the principal in a bank account. Either implies that when biodiversity exists in adequate quantity, it contributes to a *flow* of 'harvestable' income that need not degrade the physical capital/capital stock to the point of exhaustion. Once biodiversity helps generate the *flow* of ('harvestable'/useable/appreciable) goods or services, policy choices may affect these flows, and individuals may derive value from them.

Context, value, and marginal choices

The value a human places on anything depends on context. The context for valuing any particular thing includes: its relative scarcity/abundance; the relative sustainability of the systemic production of the good or service in question; and the degree to which it serves a unique function or the degree to which the functions and the characteristics for which the good is valued can be substituted by a good or goods with similar characteristics. Ultimately, whether a potential substitute is naturally or synthetically produced matters, because the *marginal* cost of using or switching to the second option also factors into the *marginal value calculation*.

For valuing ecosystem services, aspects of context that are likely to matter include:

- the ecological scale needed to maintain the ecosystem service;
- the geographically specific (environmental) context;
- the local economic context (e.g., a beautiful lake may anchor a local economy); and
- elements that affect how someone assigns value, including their culture and place in society, as well as decision-making factors described in behavioral economics and decision theory.

Any attempt to estimate anyone's or everyone's value for a flow of ecosystem services must be made assuming contextual factors defining all these aspects of context. Identifying the flows of ecosystem services is important, but important barriers remain to the transfer of estimations of the value for an ecosystem service from one environmental context to another.

There is no reasonable money-value for 'world carbon' or 'all biodiversity', because the chain of assumptions we make to get monetary values breaks in numerous places (Bockstael *et al.*, 2000). This is why economists focus on *marginal* analysis – how one thing compares to another in a given context – 'on valuing ecological changes, rather than on valuing entire ecosystems' (USEPA, 2009, p.15). Only by comparing changes can there be clear trade-offs, where choices affect outcomes in ways that manifest in different levels of satisfaction for the affected population, whether money values are involved or not.

Ecosystem service flows: distinguishing between intermediate *and* final

Recognizing that biodiversity is not a flow of goods or services does not diminish its vital structural role in an ecosystem's production and resilience. The marginal *level* of biodiversity endemic to a location may be included in any ecological production function that models physical stock, inputs, processes, and outputs in the environment as one might model them for a factory. Biodiversity may support a variety of zooplankton and pelagic amphipods upon which salmon thrive, where the species consumed by salmon are *intermediate*, just as the production of steel is *intermediate* for new cars (Boyd and Banzhaf, 2007, p.619). Cars are the *final* economic product a consumer buys, so in accounting for economic flows, counting the cars also counts all of the inputs and processes that comprise them. In this way, calling an ecosystem *structure or process* (terms as used by de Groot *et al.*, 2002) like biodiversity an ecosystem service would lead to counting it twice when we try to keep track of what people value (the ecosystem service, or in this case, the salmon), because the *intermediate* effects will already have affected production of the *final* ecosystem services that we do count. If something has an important ecological function, but this function cannot be expressly connected to someone's value for it, it is not a final ecosystem service.

The separation of intermediate from final ecosystem services when modeling the transmission of ecosystem products into the human value chain (or 'economy', where any *thing* may be used, transformed, or directly appreciated) is a critical step, distinguishing intermediate ecological processes, structures, and production functions from final uses and users (Boyd and Banzhaf, 2007). This separation more clearly identifies uses with users and minimizes double counting of ecosystem services (see Box 6.1). It does not solve all problems associated with trying to convert a complex holistic cyclical and circular (i.e., ecological) system into economic production functions and uses, but it is an important stride forward.

Box 6.1 Double counting ecosystem services

Many now recognize substantial overlap among the broad MA categories to be a potential pitfall of the MA approach (Boyd and Banzhaf, 2007; Fisher *et al.*, 2008; Ott and Staub, 2009; Haines-Young and Potschin, 2013; Staub *et al.*, 2011; Nahlik *et al.*, 2012). Supporting and Regulating services in the MA context are generally biogeochemical processes, functions, or ecological structures that necessarily create and underpin the Provisioning and Cultural service categories – implicitly inducing double counting, because a single ecosystem service may depend upon elements in multiple MA categories. Double counting should be minimized or avoided in classification or accounting schemes. An ecosystem services classification system that minimizes or avoids double counting of both stocks and flows of ecosystem services must categorize both uses and users. A listing of uses (benefits) *cannot* be tallied because it is exploratory or qualitative, whereas users (beneficiaries) and the incremental amounts of FEGS the users are identified as using or appreciating *can* be tallied, because one or more users or a larger amount is quantifiable. While perhaps frustratingly difficult to enumerate, this philosophical connection between ecosystem services and users is key to practical application of the ecosystem services concept – for identifying relevant indicators and metrics, for highlighting direct and indirect uses, and for valuation.

Final Ecosystem Goods and Services (FEGS) conceptually and theoretically link a product derived from a particular sector of the environment with a specific user or beneficiary. FEGS are 'the components of nature, directly enjoyed, consumed, or used to yield human well-being' (Boyd and Banzhaf, 2007, p. 619). FEGS explicitly allow one to separate the infinite list of ecosystem services into intermediate ecosystem services (i.e., most ecological processes and functions that we recognize are important to understand, but that do not directly influence human well-being) and FEGS (Figure 6.1). Clear and concise separation between intermediate and final ecosystem services has five advantageous characteristics that support the development and use of an ecosystem service classification system:

1 FEGS can be associated with readily definable ecological production functions;
2 FEGS link to biophysical measures;
3 FEGS count only direct interactions (uses) between a user (or beneficiary) and the ecosystem, thereby minimizing double counting;

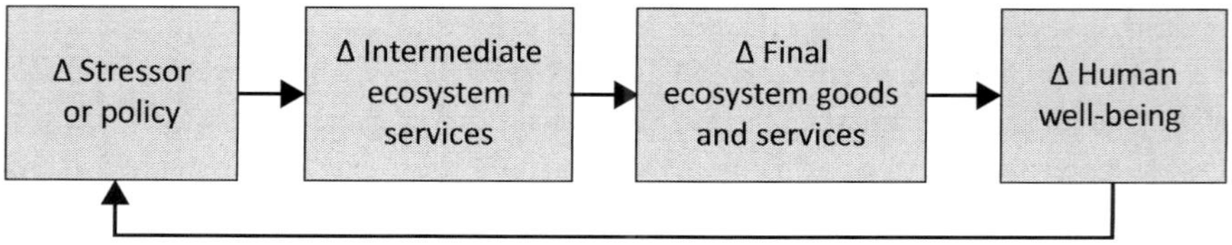

Figure 6.1 Flowchart illustrating how Final Ecosystem Goods and Services (FEGS) relate environmental stressors or policy changes, intermediate ecosystem services, and human well-being.

Source: adapted from P. Ringold (pers. comm)

4 FEGS relate clearly to human beneficiaries and connect to human well-being; and,
5 FEGS facilitate direct communication and collaboration between natural and social scientists (Nahlik *et al.*, 2012).

The use of the FEGS definition provides order within the previously discussed broad notion of ecosystem services. The FEGS approach avoids much of the ambiguity inherent in other ecosystem services definitions and approaches, while providing the insight necessary for ecologists to measure ecosystem services, and for economists to value ecosystem services by connecting a specific beneficiary to what they directly utilize, consume, or enjoy from the environment. In short, FEGS are the intersection of nature and people, or what is produced by the environment and what is valued in the economy.

The use or appreciation by a beneficiary of an environmental element makes that element a FEGS. Remove the user, the environmental element, or the relationship between the user and that element, and there is no FEGS – as when one lists ecosystem-service 'benefits' for which users are implied but not identified. In any classification system for potential FEGS, the potential of the user to use or appreciate a candidate ecosystem service must exist. In any accounting of FEGS, there must be a distinction between potential use (i.e., stocks of FEGS) and actual use (i.e., flows of FEGS to beneficiaries), for which a quantity of FEGS should be specified (or a proxy for use or quantity).

Limits of economics do not limit contribution of the 'marginal decision' perspective

Economists model and track marginal changes in stocks and flows because these are their tools of trade, so the intermediate-versus-final distinction, which better enables marginal analysis, is critical to how policies will be determined. Similarly, the intermediate-versus-final distinction helps national accounting of the environmental or 'green' variety to track total changes and flows of ecosystem services based on exchange values for services in an accounting period (calculated to reflect what would have been paid had there been a market for these ecosystem services). Defining ecosystem services in terms economists and accountants can accommodate may seem impure to many ecologists, but policymakers will continue to employ economists and accountants who rely on information that favors precise definitions, distinct *single-direction* flows, and dollar-specific bottom lines. Certain characteristics of environmental elements we call ecosystem services do challenge conventional uses of standard definitions and metrics used in economics and accounting. This does not mean that economic and accounting systems are incapable of accommodating the more subtle productions of ecosystems – it means that economics and accounting must adapt.

The need to adapt is no surprise to many economists or accountants. Ecosystem services are, by definition, resources that are increasingly recognized as limited and are 'used' in the broadest sense to raise human welfare, so they *should* be used efficiently. Within ecosystem services, FEGS fall squarely within the natural purview of economics, although not as squarely within the set of things that markets allocate or even price well. Ecosystem services are not used efficiently for the same reasons that so few of them are marketable items with a unit price – they often exist as public goods, are diminished by negative externalities of the market system, are loosely defined, are poorly understood, are not well-modeled or quantified, and are notoriously difficult to value in money terms.

Economists will need more of this missing information to formally bring FEGS into decisions about efficient use of resources, including benefit-cost analyses. Bringing FEGS into formal analyses works toward the ideal of estimating the true value, or Total Economic Value (TEV), of alternative outcomes. TEV includes direct use, indirect use, the option to use, and a range of values a person may attach to the existence of a thing that they will not directly or indirectly use now or later. When benefit-cost analyses are requested or required (USEPA, 2010), quantified and dollar-valued figures born of this process generally exclude most elements currently associated with ecosystem services – not because they are not limited resources, but because they have been poorly defined, modeled, and measured. Formal inclusion of ecosystem services in decision-making may help make clear the societal interest in affordable and sustainable flows of FEGS relative to a single industry's profit motive.

The clearer we can define resources, quantities, system dynamics, and use-user combinations, the more accurately we can identify which people value which FEGS, the type of value, and how much they value the FEGS – and the more efficiently economists can model the contextual trade-offs that individuals, firms, or governments may responsibly consider. Careful economic approaches may favour the ease of monetizable values first, quantifiable values second, and qualitative values third, but should respect all three types in valuation research and reporting, precisely because decision-makers would be unwise to focus on only monetized values in their decision-making about ecosystem services (given challenges within the science that are unlikely to be overcome in the foreseeable future). Issues of fairness and the choice of which stakeholders' priorities should be treated as dominant are not directly addressed by the tools of economics. Nonetheless, the formalization required to address issues implicit in economic modeling will offer greater information and context for these and other decision problems.

Why targeted ecological measurements matter to everyone who wants ecosystem services to influence policy

To explore the practical difference between a beneficiary approach and a benefits approach, consider the example of fresh water. Water is often considered a benefit provided by the ecosystem. How would an ecologist measure fresh water? Would they measure water quality, quantity, clarity, temperature, or one (or all) of the myriad measurements that exist for water? Starting from the benefit – water in this case – is not particularly informative because, while fresh water will always be a vital provision for many uses and people, its own ubiquitous usefulness means that vague categorization offers little insight to ecologists about how to measure it, or to economists estimating its value. What to measure and value depends on who is using the water (i.e., the beneficiary), how the water is being used, and in what larger context.

Consider water from a stream in the context of three specific interests: a farmer irrigating crops, a subsistence user drinking water for survival, and a steel mill using water for cooling. Knowing the farmer's needs, an ecologist might measure water quantity during the growing season, salinity, certain chemical concentrations, and pathogens that could harm crops or those who eat them. For the subsistence user, the degree to which the water is safe to drink may be the most important criterion to measure (e.g., laboratory screening for pathogens, inorganic chemicals, and organic compounds). The industrial use of water for cooling will be affected by the volume or quantity of water available and its temperature – measurements ecologists can provide. From a practical perspective, using a beneficiary approach to hypothesize or identify the interests of beneficiaries (i.e., users or individuals) is necessary to select a biophysical measurement that aptly captures a specific benefit for a specific user. Linking beneficiaries to a specific sector of the environment from which they ultimately receive benefits also: a) divides potential benefits into concise, minimally-overlapping components based on uses, and b) minimizes overlap in beneficiaries. The beneficiary approach is also imperative for connecting these measurements to what people care about, and, ultimately, for relating these characteristics to human well-being.

Final ecosystem goods and services classification system (FEGS-CS)

FEGS are the basis of the Final Ecosystem Goods and Services Classification System (FEGS-CS). The fundamental goal in developing the FEGS-CS was to organize ecosystem services in a comprehensive, consistent, and meaningful manner. To meet these goals, FEGS that pertain explicitly both to the environmental landscape and to specific beneficiaries were identified and defined. A series of guiding questions, outlined in the United States Environmental Protection Agency (USEPA) FEGS-CS Report (Landers and Nahlik, 2013), was used to hypothesize how beneficiaries use, interact with, or perceive benefits derived from nature. These questions helped to direct the identification and definition of FEGS by describing their importance to the beneficiary. Combining a classification of environmentally derived goods and services with a requirement that a human beneficiary be part of the definition of FEGS separates the FEGS-CS from other ecosystem service classification approaches. So far, FEGS-CS efforts have identified 338 FEGS, which Landers and Nahlik (2013) place into 21 different categories of FEGS.

National ecosystem services classification system (NESCS)

The FEGS-CS defines elements integral to the production of ecosystem services in a geographic space, and fixes the 'use' of these to actual users/beneficiaries of the ecosystem products. While this fusion of use and user is perfectly acceptable for many applications of FEGS-CS, it can constrain economists attempting to map the flow of FEGS into the economic sector. The National Ecosystem Services Classification System (NESCS) begins from an economic perspective and offers more flexibility on the beneficiary side by separating uses from users. In NESCS different users may employ the same 'use', or any particular user may employ the same FEGS to different uses. NESCS maps flows of *environmental end-products*, which like FEGS are 'final', being only those ecosystem services at the pass-point from the environment to human use or appreciation of them. By mapping flows of environmental end-products through specific users and uses, NESCS can identify and represent relevant FEGS fully, completely, and uniquely in any user's utility function. This approach satisfies microeconomic theory, helps minimize double counting, and links directly to users comparable to categories in standard industrial accounting frameworks like the North American Industry Classification System (United States Census Bureau, 2014). It is these capacities of NESCS that may help ecosystem services find their proper weight in policy decisions about how to manage renewable environmental resources.

Figure 6.2 demonstrates how an economic way of considering production and human well-being may be expanded to include how products or services from nature (that may never be marketed or priced in a discrete way) may nonetheless influence human well-being. Beginning in the upper left with physical capital (e.g., machinery) and labor resources (i.e., employees), we see that services from capital and labor factor into intermediate and final economic production functions. Final *economic* goods or services are those that a household may buy directly (e.g. not the motor housing or factory labor, but the whole refrigerator, which, when added to the household stock, can provide a flow of food-cooling services over time). Households buy things to raise their *utility*, or level of satisfaction, contributing to individual human well-being. Final *ecosystem* goods and services provide explicit environmental inputs to intermediate or final economic production functions (first circles, top row), or to level of satisfaction (third circle, top row). FEGS enter the human economy as 'final' in any of these places, because by definition they are 'final' at the point the FEGS become used or appreciated.

Considering the natural environment as a productive resource for humanity follows directly from Daily and others' choice of the word 'services'. From this point of view, natural capital (the

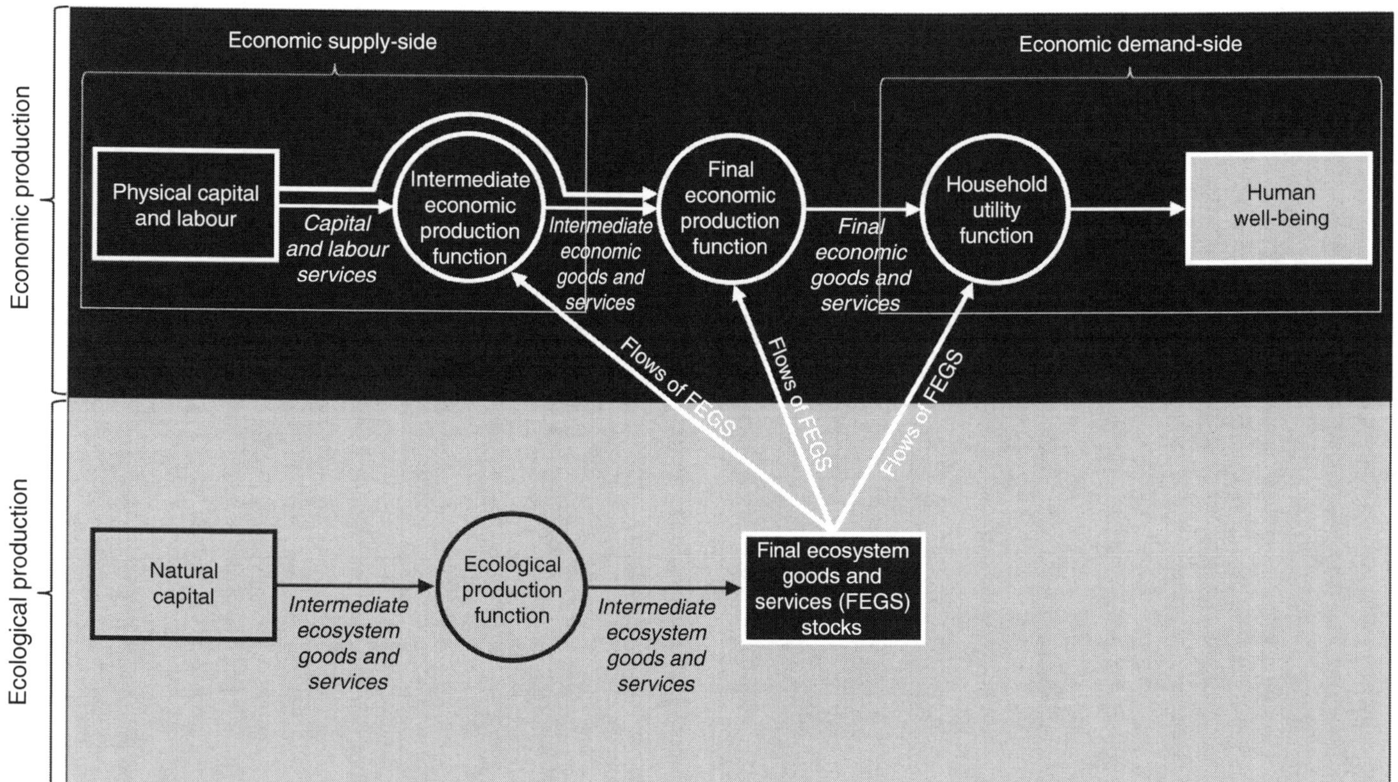

Figure 6.2 National Ecosystem Services Classification System (NESCS) conceptual framework: relationships between ecological and economic systems, with Final Ecosystem Goods and Services (FEGS) as the 'hand-off' between the two systems.

Source: adapted from USEPA (2015)

dynamic elements that in harmony generate the Supporting and Regulating services described in the MA) provides inputs for ecological production functions. Stocks of FEGS do not necessarily raise utility (e.g. the sea cucumber that lives and dies beyond human attention), but flows of FEGS (along the arrows from the ecological box to the economic box) *do* raise utility. This critical distinction between potential and actual use is made possible only by identifying the use and user of the ecological end-product. FEGS can pass into the human value chain as economic inputs (e.g., fish caught for a cannery), as final economic products (e.g., fresh fish at a restaurant), or directly to end consumers (e.g., catch-and-eat recreational fishing). Note that the end-consumer example especially involves no market transaction for the fish – a more explicit problem of economic measurement than the commercial fishermen not 'giving money to Mother Nature' for the wild fish they extracted to sell in the market. Both cases defy standard market measures of value.

NESCS begins with the same final ecosystem services perspective as the FEGS-CS, making possible a unique and non-duplicative mapping of ecosystem service flows, something difficult to achieve using earlier versions of MA-based categorization where ecological processes (i.e., potentially intermediate ecosystem services) exist on the same level as Provisioning and Cultural services. The further ability of NESCS to uniquely identify uses and users of FEGS makes it possible to gauge values for the same FEGS according to the different contexts of use that different users face. Without the ability to specify context, 'valuation' means little because the value of something changes with the user and with the context of the choice.

Next steps and conclusions

Classification is a discretionary process, important to science because each classification system poses a platform for common-standard definitions. Still, classification ultimately remains an iterative and evolving exercise in theory and in educated judgment, to make scientific discussions and strategies more efficient. The FEGS-CS and the NESCS provide improvements on previous ecosystem service classification systems by recognizing that without pairing them to specific users (beneficiaries), uses (benefits) cannot be tabulated in discussions of one policy choice versus another. FEGS-CS and NESCS are each a set of rules intended to define unique combinations of classification layers and candidate ecosystem services.

Specific applications of the FEGS-CS and NESCS will pose challenges, and thereby inform the development and expansion of these systems for later applications. A priority will be to identify which among all possible FEGS use-user combinations are relevant for the specific place and scale of a specific policy question, then to determine appropriate metrics to quantify not only the stocks of FEGS, but also their flows in the NECS context.

Determining appropriate metrics will likely prove easier for the production side than for the valuation side, because it is easier to confidently measure what *is* than what *might be* – yet both parts are necessary. Part of TEV includes someone valuing the existence of a thing, or a desire to preserve it for posterity, even if they themselves will never see or physically interact with it. Such values are hard to assign to a specific person, and harder to quantify or put in dollar terms even when one can. The less information one has about a thing or about the way someone values it, the harder it will be to put an amount or price on it. Ecosystem services remain difficult in this way.

A beneficiary-based classification system is an important step forward in helping ecosystem services hold deserved gravitas in policy discussions. The beneficiary approach makes it easier to focus on environmental measurements relevant to actual users, while also making it easier to distinguish actual beneficiaries of ecosystem services from prospective ones. Better identification does not solve questions of the best way to perform valuation of ecosystem services. But the ability to designate specific uses and users is progress – a milestone on the road to the quantification and valuation that policymakers will continue to demand.

Acknowledgements

The authors would like to thank Carl Obst (consultant to System of Environmental-Economic Accounting at United Nations Statistics Division), Mike Plummer (United States National Oceanic and Atmospheric Administration), Susan Preston (Environment Canada), and Mike Papenfus (United States Environmental Protection Agency, Office of Research and Development) for their comments and insight on earlier drafts of this chapter. The views expressed in this chapter are those of the authors and do not necessarily reflect the views or policies of the United States Environmental Protection Agency.

References

Bockstael, N.E., Freeman III, A. M., Kopp, R.J., Portney, P.R., and Smith, V.K. (2000). On measuring economic values for nature. *Environmental Science and Technology*, vol 34, pp 1384–1389.

Boyd, J.W., and Banzhaf, S. (2007). What are ecosystem services? The need for standardized environmental accounting units. *Ecological Economics*, vol 63, pp 616–626.

Carpenter, S.R., Mooney, H.A., Agard, J., et al. (15 authors) (2009). Science for managing ecosystem services: Beyond the Millennium Ecosystem Assessment. *Proceedings of the National Academy of Sciences*, vol 106, pp 1305–1312.

Daily, G.C. (ed.) (1997). *Nature's Services: Societal Dependence On Natural Ecosystems*, Island Press, Washington DC.

de Groot, R. S., Wilson, M. A., and Boumans, R. M. (2002). A typology for the classification, description and valuation of ecosystem functions, goods and services. *Ecological Economics*, vol 41, no 3, pp 393–408.

Fisher, B., and Turner, R.K. (2008). Ecosystem services: Classification for valuation. *Biological Conservation*, vol 141, pp 1167–1169.

Fisher, B., Turner, K., Zylstra, M., et al. (19 authors) (2008). Ecosystem services and economic theory: Integration for policy-relevant research. *Ecological Applications*, vol 18, pp 2050–2067.

Haines-Young, R. and Potschin, M. (2013). *Common International Classification of Ecosystem Services (CICES)* (V4.3). Report to the European Environment Agency. Available at: www.cices.eu.

Jax, K., Barton, D.N., Chan, K.M.A., et al. (26 authors) (2013). Ecosystem services and ethics. *Ecological Economics*, vol 93, pp 260–268.

Landers, D.H., and Nahlik, A. M. (2013). *Final Ecosystem Goods and Services Classification System (FEGS-CS)*, Report Number EPA/600/R-13/ORD-004914, United States Environmental Protection Agency, Washington DC.

MA (2005). Hassan, R., Scholes, R. J., and Ash, N. (eds) *Ecosystems and Human Well-Being: Current State and Trends, Volume 1: Findings of the Condition and Trends Working Group of the Millennium Ecosystem Assessment*, Island Press, Washington, DC.

Nahlik, A. M., Kentula, M.E., Fennessy, M.S., and Landers, D.H. (2012). Where is the consensus? A proposed foundation for moving ecosystem service concepts into practice. *Ecological Economics*, vol 77, pp 27–35.

Ott, W., and Staub, C. (2009). *Welfare-Significant Environmental Indicators. A Feasability Study on providing a Statistical Basis for the Resources Policy: Summary*, Environmental Studies Number 0913, Federal Office for the Environment, Bern, Switzerland.

Pearce, D., Atkinson, G., and Maourato, S. (2006). *Cost-Benefit Analysis and the Environment: Recent Developments*, Organisation for Co-operation and Development (OECD) Publishing, Paris.

Staub, C., Ott, W., Heusi, F., et al. (7 authors) (2011). *Indicators for Ecosystem Goods and Services: Framework, Methodology and Recommendations for a Welfare-Related Environmental Reporting*, Environmental Studies Number 1102, Federal Office for the Environment, Bern.

Turner, R. K., and Daily, G.C. (2008). The ecosystem services framework and natural capital conservation. *Environmental and Resource Economics*, vol 39, pp 25–35.

United Nations (2014). *System of Environmental-Economic Accounting*. Available at: www.unstats.un.org/unsd/envaccounting/ (accessed 16 April 2014).

United States Census Bureau (2014). *North American Industry Classification System*. Available at: www.census.gov/eos/www/naics/ (accessed 16 April 2014).

United States Environmental Protection Agency (2009). *Valuing the Protection of Ecological Systems and Services*. United States Environmental Protection Agency, Washington DC.

United States Environmental Protection Agency (2010). *National Center for Environmental Economics Guidelines for Preparing Economic Analyses*, Report Number EPA-240-R-10–001, United States Environmental Protection Agency, Washington, DC

United States Environmental Protection Agency (2013). *Green Accounting*. Available at: www.epa.gov/sustainability/analytics/green-accounting.htm (accessed 16 April 2014).

United States Environmental Protection Agency (2015). *National Ecosystem Services Classification System (NESCS): Framework Design and Policy Application*, Report Number EPA- 800-R-15-002, United States Environmental Protection Agency, Washington DC.

7
A SOCIAL-ECOLOGICAL PERSPECTIVE ON ECOSYSTEM SERVICES

Lasse Loft, Alexandra Lux and Thomas Jahn

Introduction

For an illustration of social-ecological relationships, we will take you on a short journey to a Yemenite island. After sailing through the Indian Ocean, we have arrived at the Socotra Archipelago, an island system 150 km off the horn of Africa. Its long geological isolation and unique climate have combined to create coastal and terrestrial ecosystems characterized by a large number of endemic species, turning the area into a biodiversity hotspot often referred to as 'the Galápagos islands of the Indian Ocean'. Even though the archipelago was recognized as a UNESCO world heritage site in 2008, pressure is mounting on its ecosystems and the services they provide for local communities. Overgrazing, habitat destruction through unsustainable fishing practices, and the introduction of alien invasive species are among the drivers of biodiversity and ecosystem services loss. In addition, climate change affects the monsoon patterns, leading to a change in species composition, while ocean acidification threatens the existence of local coral reefs. This combination poses a high risk of losing one of the most unique biodiversity hotspots in the world, as well as the basis for the livelihoods of the local communities.

This example is one of many around the globe that show how humankind as an integrated part of the earth's systems dynamics affects ecosystems, climate and the hydrosphere in an unprecedented manner, both through direct drivers such as land-use change and (anthropogenic) climate change and underlying indirect drivers like socio-political, economic and cultural change. According to Crutzen (2002), the rates, scales, types and combinations of changes caused by people are comparable to geological forces. Crutzen thus argues that the current period should be known as the 'Anthropocene'. Rockström et al. (2009) point out that we are currently leaving humanity's safe operating space, as we are exceeding critical thresholds regarding biodiversity loss, climate change and the nitrogen cycle. These perspectives point to the significance of interactions between society and nature. Furthermore, Anthropocene and planetary boundary theses imply two important tasks for science: first, a better understanding of human-nature interactions, and second, the need for developing practical measures to meet the challenge of sustaining a viable earth system. Drawing on the example above, we need to better understand how anthropogenic changes of ecosystems affect human well-being and how human interventions can mitigate these effects. In this context, ecosystem services (ES) are a well-established, but contested, concept for emphasizing the multiple benefits of ecosystems to humans (see e.g. Schröter et al., 2014). However, if the ES concept is to be used to help science and policy find solutions for the global challenges, it needs to better recognize human-nature interactions and explore feedback loops more closely.

Human-nature interactions as the core of a social-ecological perspective

Recent studies have shifted towards more comprehensive conceptualizations based on the analysis of 'coupled human and natural systems' (Liu et al., 2007) or 'human environment systems' (Turner et al., 2003), thereby acknowledging that people and nature are linked across multiple scales. In fact, it could be argued that the concept of social-ecological systems (SES) has become central to discourses on human-nature interactions (e.g. Ostrom, 2009). There is a variety of definitions of what a SES is. We will follow Glaser et al. (2012, p.4) in their understanding that 'a social-ecological system consists of a bio-geo-physical unit and its associated social actors and institutions' and that 'social-ecological systems are complex and adaptive and delimited by spatial or functional boundaries surrounding particular ecosystems and their problem context'. The island of Socotra can be conceptualized as such an SES in which the island ecosystems form the bio-geo-physical unit and the people with their fishing practices and management rules are the social actors that shape the system directly at the local level. Additionally, the island ecosystem system is affected by climate change, which is largely caused by societies in developed countries.

Capturing ecosystem services in social-ecological systems

Economists and ecologists increasingly apply the ES concept as a way of conceptually bridging human and natural systems. Potschin and Haines-Young (2011; 2016) and others use the metaphor of a 'production chain' to represent the link between biophysical structures and processes, human benefits and values, and, ultimately, human well-being. ES are essential for societal and individual viability, as their availability (or unavailability) affects material welfare and societal or individual capacities. Framed this way, the concept focuses on the supply of ecosystem services to humans. However, issues such as the demand for and access to ecosystem services must also be considered, as should the feedback effects of people's actions on ecosystems (Reyers et al., 2013). For our Socotra example, it is not enough to measure and quantify important ES, such as fish biomass. If we are aiming at developing strategies for a sustainable transformation of the island system, we need to incorporate rules, institutions and local management practices in our analysis, such as banning dynamite fishing or enforcing strict regulations on plant and coral collection. An increasing number of publications now include the feedback loop between social and natural systems – from human well-being to indirect socioeconomic and cultural drivers of change via direct drivers of ecosystem change. In this conceptualization, ES are considered an integral part of SES (e.g. Carpenter et al., 2009).

Although SES are abstractions of real-world situations, processes and structures, they can be used as an analytical entity that allows a formalized description and analysis of societal relations to nature (Hummel et al., 2011). Figure 7.1 shows a model that uses ES as a bridging concept within a hybrid SES. The SES is a framework that can be used to structure key factors, concepts or variables, and the assumed relationships between them, such as the spatial boundaries of systems, units of analysis, time horizons, inputs and drivers.

Characteristics of social-ecological systems

We have described the SES approach as a way to summarize the hybrid elements and relations of human-nature interactions (Figure 7.1). Behind this statement lies a complex system approach that allows for the analysis of interactions and cross-scale linkages and supports the integration of different types of knowledge (Hummel et al., 2011). Numerous relationships exist in these coupled systems, and a number of specific properties of SES have been identified as follows:

- The core components of an SES are ecosystem functions and actors. While the former represent the capacity of natural processes and structures to provide services, actors are seen as the direct or indirect beneficiary of those ES, or the party affected by an 'ecosystem disservice'. The actors directly or indirectly influence the development of a SES through management activities or unintended side effects (e.g. de Groot et al., 2002, Liehr et al., 2015).
- As mentioned above, the social-ecological structures and processes that form the analytical core of the SES are hybrid constructs – thus, we cannot identify their ultimate natural and social elements. Against this background, particular, context-specific social-ecological structures can be identified in terms of (1) *societal practices* of natural resource use; (2) *knowledge* of human-nature interactions; (3) *institutions for natural resource use* represented by economy, politics, law and culture; and (4) *technology* as man-made material structures to exploit natural resources (Liehr et al., 2015).
- Human-nature interactions represented in SES vary across space and time because they are path-dependent and culturally variable. Thus, SES can evolve or transform over time and across space. The interactions between people and nature in such transformations are neither straightforward nor unidirectional – causes may become effects and SES's behavior is hardly predictable through time lags or spatial effects. This is why we speak about complex and non-linear feedback loops. In addition, SES vary in their capacity to cope with disturbances in terms of resilience (Liu et al., 2007, Hummel et al., 2011). In sum, this creates a challenging complexity, which may lead to unintended and surprising outcomes if not understood in its entirety.

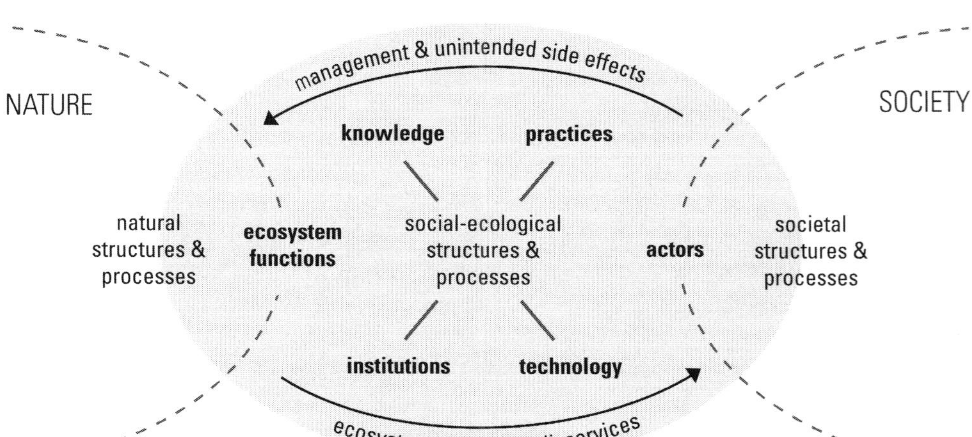

Figure 7.1 The concept of social-ecological systems.

Source: modified according to Hummel et al., 2011

A transdisciplinary research approach

Applying the SES framework for analyzing ES and assessing future actions, needs and options is anything but simple. Research in this hybrid field requires integrated theoretical approaches and methods that go beyond simply employing existing concepts from natural and social sciences in a multidisciplinary approach. One major challenge is the integration of different bodies of knowledge including knowledge from various scientific disciplines in natural sciences, social sciences and the humanities, technical professional knowledge and traditional knowledge. One way of structuring research that incorporates the different bodies of knowledge and actors while enabling problem-oriented, integrative research on ES is through transdisciplinary approaches (Jahn et al., 2012 and Hauck, 2016).

Conclusion

The analysis of ES can be an important element in developing responses to the challenges of global change at the local, regional and global level. However, describing, analyzing and modeling ES in human-nature interactions requires a wider framing to capture the beneficial or adverse relations between natural and societal structures and processes and their specific contexts. We have shown that the analysis of SES can fulfill this conceptual requirement as they can aid in structuring existing knowledge from different scientific disciplines as well as from different stakeholders. Applying a transdisciplinary research process allows for a better understanding of ES in their broader context.

References

Carpenter, S. R., Mooney, H. A., Agard, J., Capistrano, D., DeFries, R. S., Díaz, S., Dietz, T., Duraiappah, A. K., Oteng-Yeboah, A., Pereira, H. M., Perrings, C., Reid, W.V., Sarukhan, J., Scholes, R. J., and Whyte, A. (2009). Science for managing ecosystem services: beyond the Millenium Ecosystem Assessment. *PNAS*, vol 106, no 5, 1305–1312.

Crutzen, P.J. (2002). Geology of mankind. *Nature*, vol 415, p 23.

de Groot, R. S., Wilson, M., and Boumans, R. (2002). A typology for the description, classification and valuation of ecosystem functions, goods and services. *Ecological Economics*, vol 41, no3, pp 393–408.

Glaser, M., Ratter, B.M.W., Krause, G., Welp, M. (2012). New Approaches to the Analysis of Human-Nature Relations. In Glaser, M., Krause, G., Ratter, B., and Welp, M. (eds) *Human Nature Interactions in the Anthropocene* (pp 3–12). London: Routledge.

Hauck, J. (2016). Transdisciplinarity. In Potschin, M., Haines-Young, R., Fish, R. and Turner, R.K. (eds) *Routledge Handbook of Ecosystem Services*. Routledge, London and New York, pp 92–93.

Hummel, D., Jahn, T., and Schramm, E. (2011). Social-ecological analysis of climate induced changes in biodiversity – outline of a research concept. BiK-F Knowledge Flow Paper, Nr. 11. Frankfurt am Main. Available at: http://www.bik-f.de/files/publications/kfp_nr-11_neu__71c3b9.pdf (accessed 18 May 2015)

Jahn, T., Bergmann, M., and Keil, F. (2012). Transdisciplinarity: between mainstreaming and marginalization. *Ecological Economics*, vol 79, pp 1–10.

Liehr, S., Röhrig, J., Mehring, M., Kluge, T. (2015). Addressing Water Challenges in Central Northern Namibia: How the Social-Ecological Systems Concept Can Guide Research and Implementation. Unpublished manuscript.

Liu, J., Dietz, T., Carpenter, S. R., Alberti, M., Folke, C., Moran, E., Pell, A. N., Deadman, P., Kratz, T., Lubchenco, J., Ostrom, E., Ouyang, Z., Provencher, W., Redman, C. L., Schneider, S. H., and Taylor, W. W. (2007). Complexity of coupled human and natural systems. *Science*, vol 314 pp 1513–1516.

Ostrom, E. (2009). Social-ecological systems: a general framework for analyzing sustainability of social-ecological systems. *Science*, vol 325, pp 419–422.

Potschin, M. B., and Haines-Young, R. H. (2011). Ecosystem services: exploring a geographical perspective. *Progress in Physical Geography*, vol 35, no 5, pp 575–594.

Reyers, B., Biggs, R., Cumming, G. S., Elmqvist, T., Hejnowicz, A. P., and Polasky, S. (2013). Getting the measure of ecosystem services: a social-ecological approach. *Frontiers in Ecology*, vol 11, no 5, pp 268–273.
Rockström, J., Steffen, W., Noone, K., Persson, Å., Chapin, F. S., Lambin, E. F., Lenton, T. M., Scheffer, M., Folke, C., Schellnhuber, H. J., Nykvist, B., de Wit, C.A., Hughes, T., van der Leeuw, S., Rodhe, H., Sörlin, S., Snyder, P. K., Costanza, R., Svedin, U., Falkenmark, M., Karlberg, L., Corell, R. W., Fabry, V. J., Hansen, J., Walker, B., Liverman, D., Richardson, K., Crutzen, P., and Foley, J. A. (2009). A safe operating space for humanity. *Nature*, vol 461, pp 472–475.
Rounsevell, M. D. A and Harrison, P. A. (2016). Drivers of change for ecosystem services. In Potschin, M., Haines-Young, R., Fish, R. and Turner, R. K. (eds) *Routledge Handbook of Ecosystem Services*. Routledge, London and New York, pp 94–98.
Schröter, M., van der Zanden, E. H., van Oudenhoven, A. P. E., Remme, R. P., Serna-Chavez, H. M., de Groot, R. S., and Opdam, P. (2014). Ecosystem services as a contested concept: a synthesis of critique and counter-arguments. *Conservation Letters*, vol 7, no 6, pp 514–523.
Turner, B. L., Matson, P. A., McCarthy, J. J., Corell, R. W., Christensen, L., Eckley, N., Hovelsrud-Broda, G. K., Kasperson, J. X., Kasperson, R. E., Luers, A., Martello, M. L., Mathiesen, S., Naylor, R., Polsky, C., Pulsipher, A., Schiller, A., Selin, H., and Tyler, N. (2003). Illustrating the coupled human-environment system for vulnerability analysis: three case studies. Proceedings of the U.S.

Briefing Note 7.1 Transdisciplinarity

Jennifer Hauck

Influenced by Anglo-American scientific debates in the 1960s and 1970s, Jantsch (1972) was one of the first to use the term 'Transdisciplinarity' (TD). During the 1980s the term was introduced in the European scientific community by Mittelstraß (1992) as a type of research which crosses disciplinary borders and which is based on and meant to solve real world problems. According to the synthesis of Lang et al. (2012), TD research approaches share a number of key characteristics, namely that they:

- focus on problems that are relevant to society;
- enable mutual learning among researchers from different disciplines as well as with and among non-academic research participants; and
- seek to generate knowledge that is solution-oriented, socially robust, and transferable to both research and social practice.

Based on their research and experience, Lang et al. (2012, p.27) devised a structure for an ideal-typical research process, conceptualised as a sequence of three phases:

1. collaboratively framing the problem and building a collaborative research team (Phase A);
2. co-producing solution-oriented and transferable knowledge through collaborative research (Phase B); and
3. (re-)integrating and applying the produced knowledge in both scientific and societal practice (Phase C).

This process of knowledge production takes place at the intersection of science and non-science, and the role of science changes from simply providing technical information to assisting in the process of governance (Funtowicz et al., 2000). In order to facilitate TD research processes, boundary objects (Star, 2010) or boundary concepts (Mollinga, 2010) can help. Boundary concepts are words that function as concepts referring to the same object (e.g. ecosystem services), phenomenon, process, or quality of these, but can carry very different meanings (Mollinga, 2010). The ES concept (see also Potschin and Haines-Young, 2016, this volume) is such a concept, enabling ecologists, economists, sociologists, and researchers from other disciplines, along with policymakers and stakeholders, to speak a common language when talking about the usefulness of ecosystems to human society (Hauck et al., 2014).

Interest in TD is currently high and it is also on the political agenda, as the discussions in the context of IPBES show, mirroring current demand for 'knowledge exchange', 'joint knowledge production' or 'knowledge co-production' in environmental governance. The first refers to overarching and interactive processes that 'generate, share and/or use knowledge through various methods appropriate to the context, purpose and participants involved' (Fazey et al., 2014). The latter two focus particularly on participatory knowledge development processes, aiming at creating knowledge that is acceptable for all participants (van Enst et al., 2014).

While there are many benefits associated with TD knowledge production, two challenges remain. For example, approved quality standards that guide transdisciplinary researchers, programme managers and donors are missing, and the question remains: 'Does TD research bring *better* solutions to the problem addressed, given the extra resources needed?'

One possible way forward to assess TD, proposed by Lang et al. (2012), is to jointly define success criteria both in terms of the desired effects as well as in terms of scientific innovations. Criteria for both aspects are used to evaluate the project throughout the research process by an extended peer group (comprising experts from science and practice). A number of dimensions for evaluation are important here. First, many evaluation criteria focus on the TD process, such as competence of the project partners, adequacy of the problem formulation, flexibility of the project management (Bergmann et al., 2005), legitimacy, or fairness. Second, Walter et al. (2007) propose to assess the societal impacts of TD as well. These include impacts related to the TD process (network building, trust in others, understanding of others, community identification) and to created products (system knowledge, goal knowledge, transformation knowledge). The final impact of a TD process sits between these two categories and concerns the interaction between process and products, namely the distribution of knowledge. In addition, Walter et al. (2007) suggest evaluating changes or improvements of the decision-making capacity of the participants. Therefore, a regular evaluation of the TD research processes by participating stakeholders is recommended. Bergmann et al. (2005), Walter et al. (2007), and Fazey et al. (2014) provide examples of such frameworks.

References

Bergmann, M., Brohmann, B., Hoffmann, E., et al. (7 authors) (2005). Quality criteria of transdisciplinary research. A guide for the formative evaluation of research projects. *ISOE-Studientexte*, no 13. Available at: http://www.isoe.de/ftp/evalunet_guide.pdf (12/4/13).

Fazey, I., Bunse, L., Msika, J., et al. (10 authors) (2014). Evaluating knowledge exchange in interdisciplinary and multi-stakeholder research. *Global Environmental Change*, vol 25, pp 204–220.

Funtowicz, S., Shepherd, I., Wilkinson, D., and Ravetz, J. (2000). Science and governance in the European Union: a contribution to the debate. *Science and Public Policy*, vol 27, pp 327–336.

Hauck, J., Görg, C., Werner, A., et al. (8 authors) (2014). Transdisciplinary enrichment of a linear research process: experiences gathered from a research project supporting the European Biodiversity Strategy to 2020. *Interdiscip. Sci. Rev.*, vol 39, no 4, pp 376–391.

Jantsch, E. (1972). Towards interdisciplinarity and transdisciplinarity in education and innovation. In *Problems of Teaching and Research in Universities*. Organization for Economic Cooperation and Development, Paris.
Lang, D., Wiek, A., Bergmann, M., et al. (8 authors) (2012). Transdisciplinary research in sustainability science – practice, principles, and challenges. *Sustainability Science*, vol 7, pp 25–43.
Mittelstraß, J. (1992). Auf dem Wege zur Transdisziplinarität. *GAIA*, vol 1, no 5, p 250.
Mollinga, P. (2010). Boundary work and the complexity of natural resources management. *Crop Science*, vol 50, no 1, pp 1–9.
Potschin, M., and Haines-Young, R. (2016). Defining and measuring ecosystem services. In Potschin, M., Haines-Young, R., Fish, R., and Turner, R. K. (eds) *Routledge Handbook of Ecosystem Services*. Routledge, London and New York, pp 25–44.
Star, L. (2010). This is not a boundary object: reflections on the origin of a concept. *Technology & Human Values*, vol 35, no 5, pp 601–617.
van Enst, W. I., Driessen, P.P.J., and Runhaar, H. (2014). Towards productive science-policy interfaces: a research agenda. *Journal of Environmental Assessment Policy and Management*, vol 16, pp 1450007.
Walter, A. I., Helgenberger, S., Wiek, A., and Scholz, R. (2007). Measuring societal effects of transdisciplinary research projects: design and application of an evaluation method. *Eval Program Plann*, vol 30, pp 325–338.

Briefing Note 7.2 Drivers of change for ecosystem services

Mark D. A. Rounsevell and Paula A. Harrison

Drivers are the underlying causes of environmental change that act on ecosystem services in the past, present and future (MA, 2005; Rounsevell and Metzger, 2010). They may refer to changes in society, technology, the economy, the physical environment or policy governance, sometimes known as the STEEP driver categories (Bradfield et al., 2005). Knowing the nature and magnitude of drivers and how these affect ecosystems and natural capital, both directly and indirectly, is critical in appraising the flow of ecosystem services. Many ecosystem services evaluation methods such as scenario development, modelling and stress testing of policy alternatives rely fundamentally on understanding how multiple drivers interact across sectors. Table 7.1 provides a non-exhaustive overview of the diverse range of types of drivers that can affect ecosystem service provision or demand.

Drivers cause changes in the properties (known as state variables) of socio-ecological systems (SES) and can affect both the supply of ecosystem services (e.g. climate change) and the demand for ecosystem services (e.g. changes in social preferences). In some cases, changes in the state variable of one SES can become the driver for another SES, with land use change being a prime example of this. Land use change is caused by changes in the economy, climate, society and policy, amongst other drivers, but land use change is also a driver of ecosystem service provision, such as for food or biodiversity. Some drivers can have a direct effect on ecosystem services, e.g. atmospheric CO_2 levels affecting crop yields (food provision), whereas other drivers have an indirect affect, e.g. people's dietary preferences affecting how the land is used, which in turn affects ecosystem service provision. Indirect effects are especially important when considering the consequences of drivers across sectors (Harrison et al., 2015), in which changes in one sector could have important consequences for another sector. An example of this is changes in water abstraction policy affecting irrigation and therefore agricultural land use (food provision), which will have a knock-on effect on farmland bird species (biodiversity and cultural services).

Table 7.1 Categorisation according to the STEEP framework of examples of environmental change drivers that affect ecosystem drivers.

STEEP category	*Generic driver*	*Sub-driver*	*Consequences: example impacts on ES and NC*
Social	Behavioural change (in consuming natural resources)	Dietary preferences, water consumption, energy usage, appreciation of green-space and cultural ecosystem services (tourism, recreation, hunting)	Changes in agro-ecosystems, landscape management for recreation
	Social networks	Knowledge transfer for land/ecosystem management, community partnerships for nature	Land use change, Ecosystem protection
	Education and awareness raising	Public appreciation of nature	Increases in nature-based tourism and recreation, public support for nature protection
	Residential location preferences	Urbanisation, suburbanisation, peri-urbanisation, counter-urbanisation	Changing patterns of urban land use
	Inequality and social equity	Resource over-exploitation, undervaluing nature	Differential pressures on ecosystems
	Recreational use of green space	Hiking, biking, fishing, hunting	Pressures on ecosystems, landscape protection and management
Technological	Crop breeding and agronomy	Changing crop yields, precision agriculture, genetically modified organisms	Changing agricultural land use
	Water use efficiency	Irrigation efficiency, efficiency of water-using white goods, reservoirs and dams	Changing hydrological cycles; aquatic ecosystems
	Water demand	Demand from the domestic, industrial, agricultural and power generating sectors	Changes in water availability and use in different sectors, affects water available for environmental flows
	Transport development	Fuels, modes and networks	Ecosystem functioning arising from infrastructural development
	Renewable energy	Bioenergy, wind, hydro, wave	Reducing climate change impacts on ecosystems
Economic	Trade flows	Food, fibre, timber, water and energy	Displacement of ecosystem impacts
	Economic crises	Stockmarket crashes, banking system stress, currency values	Changing natural resource use

STEEP category	*Generic driver*	*Sub-driver*	*Consequences: example impacts on ES and NC*
	Direct economic incentives	Incentive-based agri-environment schemes, grant aid, demand-side management	Ecosystem protection and sustainable use
	Market-based schemes for ecosystem services	Payment for Ecosystem Services (PES), habitat banking/offsetting, carbon trading schemes	Sustainable resource use and ecosystem conservation
	The green economy	Full cost accounting, valuing of natural capital	Promotes ecosystem services in the economy
	Voluntary standards	Certification, labelling, quality assurance schemes	Sustainable resource use and ecosystem conservation
Environmental	Climate change	Impacts of temperature, precipitation and CO_2 on ecosystem functioning, including physical and human processes within SES	Direct impacts on ecosystem functioning Changing species distributions Sustainable resource management
		Human responses to climate change impacts (adaptation)	Carbon sequestration
		Land-based, climate mitigation strategies	
	Air pollution	Nitrogen deposition BVOCs/ozone	Ecosystem degradation
	Water quality	Nitrogen, phosphates, pesticides	Ecosystem degradation
	Water resources	Drought, flooding, coastal inundation and saline intrusion	Ecosystem degradation
	Land use change	Deforestation, reforestation/afforestation, agricultural expansion or abandonment, urbanisation, nature conservation, bioenergy production	Habitat disturbance, ecosystem transitions, focus on provisioning services at the expense of others
	Natural hazards	Fire, earthquakes, tsunamis, volcanoes	Species composition and competition, ecosystem functioning
Policy Governance	Aid payments	Agricultural development, education, health, clean water	Supporting selected ecosystem types
	Geopolitical change	Regime shift, changing ideologies	Land use change
	Security (and conflict)	War, resource competition (water, land), land grabbing	Land use and management change

Climate mitigation policy	REDD/REDD+, Clean Development Mechanism (CDM), Land Use and Land Use Change and Forestry (LULUCF)	Conserving certain ecosystem types, e.g. forests
Agricultural policy	Agri-environmental schemes, rural development, compulsory set-aside	Reducing impacts on agro-ecosystems
Environmental policy	Convention on Biological Diversity (CBD), UNFCCC, Kyoto protocol, European Water Framework Directive, Ramsar convention, UN convention on law of the sea, Convention to combat desertification	Conservation natural resources
Spatial planning (integration)	Green infrastructure, urban development, environmental protection areas	Conserving natural resources
Multi-level governance	Centralisation versus devolution	Local decision-making about ecosystem protection
Policy integration (mainstreaming)	Sectoral (silo) policies versus integrated policies	Unintended indirect effects on ES, avoidance of perverse outcomes

Source: economic drivers adapted from Brown et al., 2014

Drivers also have different effects across multiple spatial scales (Rounsevell et al., 2010). A driver can be external to a socio-ecological system (e.g. geopolitics, globalisation, industrialisation) or internal to the system (e.g. social preferences for green space and recreation, air pollution, water consumption). However, the same driver could be external or internal depending on the size and location of the socio-ecological system boundaries.

Drivers have different effects at different moments in time and across geographic space. The relative importance of ecosystem service drivers has evolved from the past to the present and will continue to change. For example, the dominant drivers in the past were new technological developments (through industrialisation, agrarian development and transport infrastructure) and the development of global trade and human migration. These will remain important drivers, but changes in social preferences and public and private policy are likely to become increasingly influential. In analysing the future, qualitative narratives are used within scenario storylines to describe how drivers could play out in alternative future worlds (e.g. Rounsevell and Metzger, 2010). These approaches often seek to combine multiple drivers in a co-evolutionary approach, and can also employ participatory approaches to capture tacit knowledge and experience from key stakeholders. Such methods help to develop insights into the effect of drivers on ecosystem properties, which is a pre-requisite for the sustainable management of natural capital and ecosystem service provision.

Acknowledgements

This work was funded by two projects of the 7th Framework Program of the European Commission: "Operational Potential of Ecosystem Research Applications (OPERAs)" (EC grant agreement no 308393) and "Operationalising Natural Capital and Ecosystem Services (OpenNESS)" (EC grant agreement no 308428).

References

Bradfield, R., Wright, G., Burt, G., Cairns, G.V., and der Heijden, K. (2005). The origins and evolution of scenario techniques in long range business planning. *Futures*, vol 37, pp 795–812.

Brown, I., Harrison, P., Ashley, J., et al. (14 authors) (2014). *UK National Ecosystem Assessment Follow-on. Work Package Report 8: Robust Response Options: What Response Options might Be Used to Improve Policy and Practice for the Sustainable Delivery of Ecosystem Services?* UNEP-WCMC, LWEC.

Harrison, P. A., Dunford, R., Savin, C., et al. (7 authors) (2015). Cross-sectoral impacts of climate change and socio-economic change for multiple, European land- and water-based sectors. *Climatic Change*, vol 1, no 3–4, pp 279–292.

MA (2005). *Ecosystems and Human Well-Being. Synthesis Report.* Island Press, Washington DC.

Rounsevell, M.D.A., Dawson, T. P., and Harrison, P. A. (2010). A conceptual framework to assess the effects of environmental change on ecosystem services. *Biodiversity and Conservation*, vol 19, no 10, pp 2823–2842.

Rounsevell, M.D.A., and Metzger, M. J. (2010). Developing qualitative scenario storylines for environmental change assessment. *Wiley Interdisciplinary Reviews Climate Change*, vol 1, pp 606–619.

8

CONCEPTS AND METHODS IN ECOSYSTEM SERVICES VALUATION

Erik Gómez-Baggethun, David N. Barton, Pam Berry, Robert Dunford and Paula A. Harrison

Introduction

Human pressure on the environment is severely undermining the capacity of ecosystems to sustain long-term conditions for human life (Odum, 1979), safety (Rockström et al., 2009), well-being (MA, 2005), and economic prosperity (Jackson, 2009). Because value systems play a critical role in the way societies relate to, and manage, ecosystems, valuation has gained growing attention in environmental science and policy over recent years (TEEB, 2010).

Valuation of ecosystem services is an increasingly used tool for decision-making and planning whose scope of application can range from awareness raising to policy instrument design, priority setting, and environmental litigations in courts, among others (Gómez-Baggethun and Barton, 2013; Laurans et al., 2013). However, valuation is also a controversial practice whose suitability depends strongly on context and purpose (Kallis et al., 2013). Take the example of monetary valuation, arguably the most controversial form of nature valuation. Monetary valuation has proved to be a useful tool to illustrate the economic gains that corporations make by shifting the costs of their economic activity to society at large (e.g. in terms of pollution abatement) (Kapp, 1983). Such costs are difficult to measure and their societal and ecological impacts are hardly commensurable in money (Martínez-Alier, 2002). Yet economic estimations of abatement and restoration costs can inform and promote policy instruments to claim legal liability for environmental damage (Zografos et al., 2014). By contrast, monetizing ecosystem services that are not expected to be governed by market values and norms (most cultural services) or for which no adequate technical substitutes exist (many supporting and regulating services) can be meaningless, or even counterproductive by way of paving the way for undesirable commodification (Gómez-Baggethun and Ruiz-Pérez, 2011).

If ecosystem services valuation is to serve goals of environmental sustainability, social justice, and long-term economic viability (Daly, 1992), it is paramount that policymakers, scientists and practitioners make good use of the available concepts and methods to capture nature's many ecological, social, and economic values to people (Boeraeve et al., 2015). This chapter reviews concepts and methods in the valuation of ecosystems and biodiversity with a focus on ecosystem services. It discusses the merits and limits of different valuation approaches and examines how they may complement each other in assisting decisions and formulating policies. It further discusses scope and limits of the use of ecosystem services valuation and identifies knowledge gaps and priority areas for the research agenda.

In line with the conceptual framework of the Intergovernmental Platform on Biodiversity and Ecosystem Services (IPBES), the term value is used here in the broader sense of its understanding both as 'importance, worth or usefulness' and as 'held values, principles, or moral duties' towards nature (Díaz et al., 2014). Valuation is used accordingly as the act of assessing, appraising, or measuring value or importance (Dendoncker et al., 2013). This notion of value is consistent with the previous literature on ecosystem services, endorsing the broader definition of value as importance (e.g. de Groot et al., 2002; MA, 2005), but enriching the picture with consideration of principles, virtue, and moral duties to better navigate ecological and sociocultural values (Chan et al., 2012a,b; Luck et al., 2012; Jax et al., 2013). From this perspective, relevant values of, and towards, nature should not been seen to emerge only from individual preferences expressed in markets, but also, and most fundamentally, from principles and convictions guiding the ways humans relate to people and nature on the basis of moral, ethical, and political grounds.

There are various ways in which people ascribe meaning to nature, which is expressed very differently across cultural conceptions, philosophical views, and schools of thought (Goulder and Kennedy, 1997; Erikstad et al., 2008). For example, some nations are more dominated by value systems that prioritize individual rights and market-oriented values, while others rely more heavily on value systems that prioritize collective values (Díaz et al., 2014) and where putting market values on nature can be culturally offensive (Turner et al., 2003; Brondizio et al., 2010). The same is true for different disciplines. While some see the monetization of ecosystem services as a transparent method to support decisions, many others regard putting money tags on ecosystems as totally misconceived.

With a view to accommodating the many disciplines and communities that have engaged in the debate on ecosystem services valuation over recent years, the approach adopted in this chapter emphasizes the notion of value pluralism in ecosystem service assessment (Gómez-Baggethun et al., 2014; Martín-López et al., 2014). Value pluralism is the idea that there are multiple values which in principle may be equally correct and fundamental, and yet conflict with each other (O'Neill, 1997). It assumes there is no single value or metric that can comprehensively capture the overall importance of ecosystems because ecosystems may be valuable for many different reasons (Gómez-Baggethun and de Groot, 2010). Valuation involves dealing with multiple and often conflicting valuation languages, whereby different values may be combined to inform decisions, but not always compressed into a single unit. Different values attached to ecosystem services may be weakly comparable (Martínez-Alier et al., 1998), or incommensurable along a single measurement rod (O'Neill, 1997; Martínez-Alier et al., 1998; Paterson, 1998). Major international initiatives on ecosystem services also acknowledge the importance of recognizing multiple types of value in ecosystems and biodiversity (Box 8.1).

Box 8.1 The case for value pluralism in international ecosystem service initiatives

The Millennium Ecosystem Assessment states that "many people ascribe ecological, sociocultural, or intrinsic values to the existence of ecosystems and species" and recognizes these different paradigms, based on 'various motivations and concepts of value, along with the many valuation methods connected with them' (MA, 2005, p.128). Despite its monetary focus, the initiative Economics of Ecosystem Services and Biodiversity also acknowledges that valuation involves dealing with 'conflicting valuation languages" that can be "in a relation of incommensurability with each other' (TEEB, 2010, p.193). The IPBES pushes the perspective of value pluralism further and recognizes from the outset multiple value systems, including 'intrinsic [. . .], instrumental and relational values' (Díaz et al., 2014, p.11). The working group on ecosystem services valuation of the Ecosystem Services Partnership (ESP) refers in its webpage to the "the multitude of values of nature held by peoples" (Source of quotation: www.es-partnership.org/esp/81931/5/0/50).

Ecological, sociocultural, and economic values in ecosystem services

The literature on ecosystem services includes references to multiple types of value (e.g. ecological, social, cultural, economic, relational, spiritual, symbolic, insurance, intrinsic, eudemonistic, bequest, heritage, and place values). For simplicity and systematization, here we follow the convention followed in previous ecosystem services literature (e.g. Farber et al., 2002; Howarth and Farber, 2002; Limburg et al., 2002; Wilson and Howarth, 2002; de Groot et al., 2002; Dendoncker et al., 2013; Castro et al., 2014), and group values into three broader categories: ecological, sociocultural, and economic values. These categories partially overlap and their boundaries simply stem from convention. They should be seen as ideal analytical categories *sensu* Weber (1949), who believed that we cannot understand a particular phenomenon just by describing the great diversity of ways in which they are manifested in the real world. Functional classifications – Weber believed – should make abstraction from real life diversity to synthesize many concrete individual phenomena into unified analytical constructs.

Ecological values

In the literature on ecosystem services, the term *ecological value* has often been used in relation to the status and condition of the ecosystem functions, processes, and components on which ecosystem service delivery depends (Kontogianni et al., 2010; de Groot et al., 2002). It includes measures of ecological health and the integrity of ecosystem function by using parameters such as complexity, diversity, productivity, and stability (de Groot et al., 2003; Harrison et al., 2014). Ecological values are often related to the importance of causal relationships between species and the performance of regulating services, such as the importance/value of a particular tree species in preventing soil erosion or the importance/value of an aquatic species in water purification (MA, 2005; Gómez-Baggethun et al., 2011). Ecological values strongly associate with supporting and habitat services. The magnitude of ecological values is typically expressed in biophysical terms or through constructed scales and indicators of diversity, richness, rareness, productivity, integrity (or health), and resilience.

Ecological value has also been broadly related to the self-organizing capacity of ecosystems, the stability of ecosystems and ecosystem functions, and the quality of the ecological system in terms of its ability to support itself in a dynamic equilibrium (Gren et al., 1994). In terms of their value for humans, ecological values have been associated with the so-called insurance value embedded in ecosystem resilience (Gómez-Baggethun and Barton, 2013). The idea of biodiversity as insurance and as a source of resilience stems from ecology (McCann, 2000), and reflects the proposition that species and functional diversity increase an ecosystem's capacity to absorb perturbations (Elmqvist et al., 2003; Mori et al., 2013). From the perspective of ecosystem services, insurance value is of critical importance because it reflects the capacity of an ecosystem to provide sustained flows of benefits for human well-being despite disturbance and change (Baumgärtner, 2007; Pascual et al., 2010).

Sociocultural values

Sociocultural values include emotional, affective, and symbolic views attached to nature (Daniel et al., 2012). For many people, some ecosystems are associated with deeply held historical, national, ethical, religious, and spiritual values (MA, 2005). These values are ill-suited to commodity metaphors and in most cases cannot be captured by monetary metrics in any meaningful way (Norton and Hannon, 1997; Williams et al., 1992). Sociocultural values relate to non-material benefits people obtain in their interaction with ecosystems through spiritual enrichment, cognitive development, reflection, recreation, or aesthetic experience (MA, 2005).

Aspects that are commonly analysed within the rubric of sociocultural values include sense of place (Feldman, 1990; Brown et al., 2002), spiritual values (Stokols, 1990), emotional values (Milton, 2002), heritage values (Throsby and Throsby, 2001), and sense of community (Chavis and Pretty, 1999). Sociocultural values can also cover relational values between people and non-human species or things (Takeuchi, 2010) and values associated with principles and virtue in the search for a good life (Chan et al., 2012a). Sociocultural values have often been ill-represented in efforts to characterize ecosystem services. Recent research, however, has made substantial progress in better integrating social perspectives and cultural valuation techniques into ecosystem service assessments (e.g. Chan et al., 2011; Daniel et al., 2012; Hernández-Morcillo et al., 2013).

Monetary values

Monetary values of the environment refer both to the market value of ecosystem services (either direct, indirect, or hypothetical) and the costs of their replacement as they decline or disappear, provided that their substitutability is technologically feasible (Turner et al., 2003). The economic literature refers to a range of monetary values of the environment that typically are added up to derive the so-called Total Economic Value (TEV), a framework that divides the economic value of ecosystem services into use and non-use values (Pearce and Turner, 1990; Turner, 1999; Heal et al., 2005). Use value refers to the value of ecosystem services that are used for consumption or production purposes and include direct use, indirect use, and option values. Direct use values derive from the conscious use of ecosystem services. These may be extractive (e.g. timber) or non-extractive (e.g. aesthetic enjoyment). Extractive direct use values relate to provisioning services and non-extractive direct use values relate to cultural services. Indirect use values are attributed to regulating services (e.g. climate regulation, pollination). Option values are often used to frame TEV under conditions of uncertainty and relate to potential or still unknown uses of biodiversity (e.g. bioprospecting to find new medicinal plants), and as a renewable source, through evolutionary processes, of biological solutions to novel problems. Finally, non-use values reflect the mere satisfaction that individuals derive from the knowledge that biodiversity and ecosystem services continue to exist either in their own right, so called 'existence values'; provide their services to others, altruistic values'; or be there for future generations, 'bequest values'. These associations reflect broader links rather than one to one relations.

Since collapses of ecosystem service flows typically involve direct and indirect economic costs, the notion of insurance value has also been approached from an economic perspective

(Perrings, 1995; Armsworth and Roughgarden, 2003). Some attempts have been made to measure insurance value in money terms, either through the investment required to ensure that an ecosystem remains within a given regime in the face of pressure or the costs of the ecosystem leaving its present regime (Baumgärtner and Strunz, 2014). Others note that when the location of ecological thresholds is uncertain or unknown, monetary metrics are unlikely to provide the signals required to avert regime shifts (Limburg et al., 2002), and make a case for adopting the precautionary principle and early warning indicators (Gómez-Baggethun, 2010).

Methods and techniques in ecosystem valuation

In the context of ecosystem services, valuation can be defined as the act of assessing, appraising, or measuring value, primarily in terms of worth, meaning, and importance, but also in relation to principles and moral duties towards biodiversity. Below, we summarize methods and techniques for quantifying or qualifying ecological, sociocultural, and economic values (Figure 8.1).

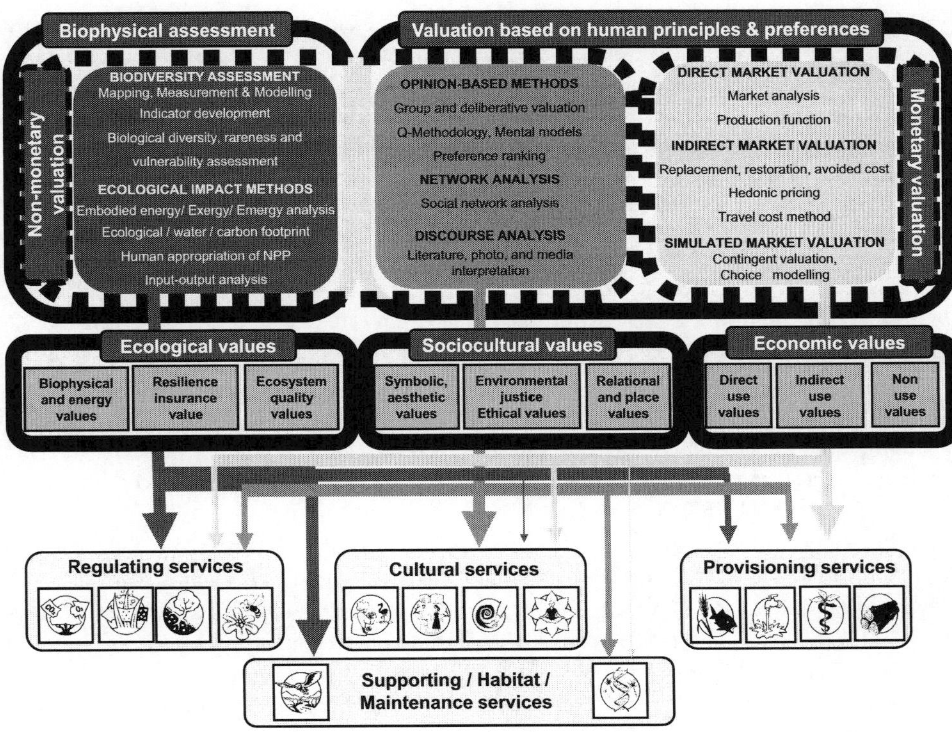

Figure 8.1 Ecosystem services, value types and valuation methods. The upper part of the figure illustrates the divide between i) methods that derive values from biophysical assessments and methods that derive values from human subjectivity (including principles and preferences), and ii) methods based on monetary and non-monetary valuation. The lower part illustrates the connection between methods, value types, and ecosystem services, showing that they do not stand in a one-to-one relation. Arrow width indicates the intensity of the connections.

Source: own elaboration, building on Gómez-Baggethun (2010) with icons by Jan Sasse for TEEB

Ecological valuation

Ecologists and environmental scientists have traditionally used the term *ecological value* to refer to biophysical parameters or impact indicators that are scaled and weighted in order to inform decisions. In relation to ecosystem services, ecologists have focused mostly on assessing the ecosystem components, functions, and biotic attributes that underpin the provision of ecosystem services, including species and functional traits (Luck et al., 2009; Harrison et al., 2014), or in the direct biophysical measurement of ecosystem services (e.g. tons of carbon, cubic meters of timber, or livestock units) for assessing conditions and trends in ecosystem service delivery (MA, 2005). In this sense, ecologists have engaged in assessments of ecosystem function and biophysical accounting more frequently that in valuation *per se*, at least in terms of comparing alternatives to assist policy decisions over ecosystem services.

There is, however, a long tradition of valuation in ecology and environmental assessment (Gosselink et al., 1974; Odum, 1996; Erikstad et al., 2008). For example, ecological valuation has been used to measure the ecological value of a given natural area compared to similar sites (e.g. in terms of its ability to support biodiversity), providing a rational basis for deciding on different management options (Mitsch and Gosselink, 1993). The ecological valuation approach often relies on value indexes and comparison through multicriteria analysis (e.g. Odum, 1979). This type of ecological valuation has been used in decision-making related to contexts such as determining safe minimum standards, environmental impact assessment, and prioritizing the conservation of species and ecosystems. The definition of safe minimum standards for the sustainable use of ecosystem services should be based on information and biophysical parameters on ecological thresholds. Ecological value is of particular importance for 'regulating' and 'supporting' (MA, 2005) or 'habitat' (TEEB, 2010) services, called 'maintenance and regulation' services in the Common International Classification of Ecosystem Services (CICES) (Haines-Young and Potschin, 2013).

Sociocultural valuation

The terms 'non-economic' and 'sociocultural' valuation are increasingly used in ecosystem service assessments (Calvet-Mir et al., 2012; Milcu et al., 2013, Christie et al., 2012) and as part of comprehensive categories of various social scientific and participatory methods (e.g. surveys, interviews, focus groups, citizens' juries, etc.) (Jacobs, 1997). Recently, the term socio-cultural valuation has also been applied as an umbrella term for preference ranking methods analysing human preferences towards ecosystem services in non-monetary terms (Calvet-Mir et al., 2012; Castro et al., 2014; Martín-López et al., 2014).

In the context of ecosystem services, sociocultural valuation has been used in reference to a heterogeneous collection of approaches and methods whose only shared characteristic is not relying on either monetary or biophysical metrics (Chan et al., 2012b; Castro et al., 2014; Lamarque et al., 2011; Calvet-Mir et al., 2012; Martín-López et al., 2012). Following Kelemen et al., (2014) valuation approaches under this umbrella include terms such as 'deliberative valuation' (Howarth and Wilson, 2006; Kenter et al., 2011), 'psycho-cultural valuation' (Kumar and Kumar, 2008), 'social valuation' (James et al., 2013), 'qualitative valuation' (e.g. Zendehdel et al., 2008), and 'subjective assessment' (Aretano et al., 2013).

Economic valuation

Economists have developed methods and tools to value economic impacts on the environment, most often in monetary terms, but sometimes also in terms of contributions to employment

Box 8.2 Valuation of pollution damages in courts: the Chevron-Texaco case in Ecuador

In February 2011, a court in Ecuador ordered oil giant Chevron to pay nearly US$9 billion in damages for polluting the Amazon forest, the second largest ever judgment for environmental contamination in the world after the US$20 billion British Petroleum (BP) has agreed to pay to compensate victims of the Gulf of Mexico oil spill. Allocation of the damages by the judge included a series of environmental and damage valuation studies and a significant part of the money was to be devoted to clean-up and restoration activities. The authors believe this is a good example of where valuation was used to enforce the polluter pays principle in a process that effectively combined ecological, sociocultural, and economic values. The process was conducted in courts and not through direct compensation agreements; the court valued monetarily only those damages directly linked to the cost of reparation of environmental and health impacts (and not compensation, which would have increased the amount of the fine but also tied ecosystem and life loss with money). The plaintiffs maintained throughout the process that they sought compensation only for damages and reparation and that the 'crime committed by Texaco is incalculable'. Chevron was given 15 days to apologize publicly; if not, the fine would double, a clear indication by the judge of the symbolic nature of the punishment, expressing values of recognition, responsibility, and honor.

Source: Kallis et al., 2013

or health. The principal motivation for monetary valuation of ecosystem services is promote awarness of the societal importance of nature, or to enable impacts to be included in cost-benefit analysis (Hanley, 2001; Balmford et al., 2002; Barbier et al., 2009), but, increasingly, it is also used by environmental justice organizations in legal processes conducted in courts (Zografos et al., 2014) (Box 8.2).

Monetary valuation measures 'hidden costs and benefits' in economic activity, often termed externalities. The initial focus was on negative externalities such as pollution (Pigou, 1920/2006), but since the 1960s the scope has extended to cover positive externalities such as environmental amenities (Krutilla and Fisher, 1975) and ecosystem services (TEEB, 2010).

A range of monetary valuation techniques have been developed and increasingly refined for the elicitation of these different value types (reviewed in TEEB, 2010). If available, values are derived from data resulting in real market transactions relating directly to ecosystem services. In the absence of such information, price information is derived from shadow prices obtained in parallel markets that are associated indirectly with the good to be valued (as in hedonic pricing methods), or on observed consumed behaviour (as in travel cost methods). In their absence, valuation relies on expected consumer behaviour in hypothetical markets simulated through surveys (stated preferences), as in contingent valuation and choice modelling methods (Bateman and Turner, 1993; Bateman et al., 2002; Johnson et al., 2015). The three situations described above correspond to a common categorization of the monetary valuation techniques: i) direct market valuation, ii) revealed preference approaches, and iii) stated preferences approaches.

Values from original valuation studies are sometimes applied to other sites through benefit transfer techniques (Barton, 2002; Bateman et al., 2011). Detailed description of these techniques is beyond the scope of this chapter, but comprehensive syntheses of economic valuation are available in the literature (e.g. Heal et al., 2005; Barbier et al., 2009; Pascual et al., 2010; TEEB 2010).

Conclusions

There are many reasons why people attribute importance to nature. To provide comprehensive accounts of these reasons, valuations of ecosystems and their services should rely on a broad range of concepts and methods to capture their ecological, sociocultural, and economic significance. Reducing all values concerning nature into monetary units is considered ethically, politically, and methodologically problematic by many researchers.

Ecosystem service valuation has broadened its scope in recent years, expanding its traditional focus on monetary values to provide more comprehensive pictures of the importance people attribute to biodiversity and ecosystems. The term 'non-monetary' valuation captures this concern and reflects efforts to break the direct association of ecosystem services valuation with money and markets and to promote reflection upon the plurality of values attached to ecosystem services. Once the idea of value pluralism becomes widely accepted in the ecosystem services community, the term 'non-monetary valuation' can be substituted for by affirmative names.

The approach of integrated ecosystem services valuation makes a move in this direction and may be defined as the process of synthesizing, interpreting, and communicating knowledge about the ways in which people conceptualize, understand, and appraise the values of ecosystems services to facilitate deliberation and agreement for decision-making and planning. This chapter adds to this body of literature by synthesizing knowledge from ecology, economics, sociology, and other social sciences for understanding the values of ecosystems and the services they provide to humans. Dealing with multiple values is methodologically challenging and adds layers of complexity vis-à-vis valuations that compress values into a common denominator, such as cost-benefits analysis and various energy-based methods. That different values may not be reduced to a single unit, however, does not mean that there are no ways of using them as the basis of comparison for making sound decisions. Multicriteria evaluation, for example, is one of the methodological approaches that can be used to integrate ecological, sociocultural, and economic values to inform decisions on ecosystem services.

Integrated approaches to valuation can provide more comprehensive pictures of the importance people attribute to biodiversity and ecosystems. Yet it should be noted that ecosystem services is not necessarily the most suitable approach for addressing *every* value in nature. For example, while attempts have been made to stretch ecosystem services to cover intrinsic and other non-anthropocentric values, other complementary concepts and frameworks may capture these aspects of nature's importance in more meaningful ways.

Moreover, values depend on the institutional and distributional settings in which they are expressed. Used in appropriate contexts, valuation can promote awareness of societal dependence on ecosystems and catalyse the use of policy tools for their protection and sustainable use. Used out of context, however, or when using methods and metrics that fail to capture how people ascribe importance to nature in a meaningful way, it may force affected stakeholders to bring their concerns into assessment schemes that do not fit their own values. Ecosystem services valuation is generally more effective for communicating with policymakers and managers concerned with practical aspects of priority-setting than with peasant, indigenous, and other communities for whom nature is deeply interwoven with spiritual values, and where anthropocentric framing of nature's values can thus be misleading.

Further tasks for the research agenda on ecosystem services valuation include finer analyses to identify the most suitable values and metrics that fit particular ecosystem services and decision-making contexts, and further understanding of how different values can be consistently combined to assist decisions. The latter aspect is of critical importance since mounting research shows that values do not add up in a linear fashion, but rather interact in complex and sometimes unexpected ways. For example, empirical research shows that depending on its design, economic valuation, and incentives for ecosystem protection can reinforce or otherwise crowd out intrinsic motivations to protect ecosystems based on ecological and social considerations.

Finally, ecosystem services valuation has so far dealt more satisfactorily with the conception of value as the 'importance, worth, or usefulness' that humans attribute to nature than with the conception of value in terms of 'convictions, principles, or moral duties' towards nature. A promising and yet largely underexplored research area that can make progress in filling this gap is the emerging application of valuation to support litigation processes in courts to claim liability for environmental damage (Phelps et al., 2015). This approach connects ecosystem services valuation with environmental law, right-based perspectives, and concerns of environmental justice.

Acknowledgements

This research was funded by the 7th Framework Program of the European Commission project "Operationalizing Natural Capital and Ecosystem Services (OpenNESS)" (EC grant agreement no 308428).

References

Aretano, R., Petrosillo, I., Zaccarelli, N., Semeraro, T., and Zurlini, G. (2013). People perception of landscape change effects on ecosystem services in small Mediterranean islands: a combination of subjective and objective assessments. *Landscape and Urban Planning*, vol 112, pp 63–73.

Armsworth, P. and Roughgarden, J. (2003) The economic value of ecological stability. *Proceedings of the National Academy of Sciences*, vol 100, pp 7147–7151.

Balmford, A., Bruner, A., Cooper, P., Costanza, R., et al. (19 authors) (2002). Economic reasons for conserving wild nature. *Science*, vol 297, pp 950–953.

Barbier, E.B., et al. (11 authors) (2009). The valuation of ecosystem services. In Naeem, S., Bunker, D., Hector, A., Loreau, M., and Perrings, C. (eds) *Biodiversity, Ecosystem Functioning, and Human Wellbeing: An Ecological and Economic Perspective*. Oxford University Press, Oxford.

Barton, D. (2002). The transferability of benefit transfer: contingent valuation of water quality improvements in Costa Rica. *Ecological Economics*, vol 42, pp 147–164.

Bateman, I. J., Brouwer, R., Ferrini, S., et al. (13 authors) (2011). Making benefit transfers work: deriving and testing principles for value transfers for similar and dissimilar sites using a case study of the non-market benefits of water quality improvements across Europe. *Environmental and Resource Economics*, vol 50, pp 365–387.

Bateman, I. J., Carson, R. T., Day, B., et al. (8 authors) (2002). *Economic Valuation with Stated Preference Techniques: A Manual*. Edward Elgar, Cheltenham/Northampton.

Bateman, I. J., and Turner, R.K. (1993). Valuation of the environment, methods and techniques: the contingent valuation method. In Turner, R. K. (ed.) *Sustainable Economics and Management: Principles and Practice*. Belhaven Press, London.

Baumgärtner, S. (2007). The insurance value of biodiversity in the provision of ecosystem services. *Natural Resource Modeling*, vol 20, pp 87–127.

Baumgärtner, S. and Strunz, S. (2014). The economic insurance value of ecosystem resilience. *Ecological Economics*, vol 101, pp 21–32.

Boeraeve, F., Dendoncker, N., Jacobs, S., Gómez-Baggethun, E., and Dufrêne, M. (2015). How (not) to perform ecosystem service valuations – Pricing gorillas in the mist. *Biodiversity and Conservation* (in press). DOI: 10.1007/s10531-014-0796-1

Brondizio, E. S., Gatzweiler, F., Zagrafos, C., and Kumar, M. (2010). Socio-cultural context of ecosystem and biodiversity valuation. In Kumar, P. (ed.) *The Economics of Ecosystems and Biodiversity (TEEB)*. Earthscan Press, London.

Brown, G.G., Reed, P. and Harris, C. (2002). Testing a place-based theory for environmental evaluation: an Alaska case study. *Applied Geography*, vol 22, pp 49–76.

Calvet-Mir, L., Gómez-Baggethun, E., and Reyes-García, V. (2012). Beyond food production: home gardens' ecosystem services. A case study in Vall Fosca, Catalan Pyrenees. *Ecological Economics* vol 74, pp 153–160.

Castro, A., García-Llorente, M., Martín-López, B., Palomo, I., and Iniesta-Arandia, I. (2014). Multidimensional approaches in ecosystem service assessment. In Alcaraz-Segura, D., Di Bella, C. D., and Straschnoy, J.V. (eds) *Earth Observation of Ecosystem Services*. CRC Press, Boca Raton.

Chan, K., Goldstein, J., Satterfield, T., et al. (8 authors) (2011). Cultural services and non-use values. In Kareiva, P., Tallis, Ricketts, T., Daily, G., and Polasky, S., (eds) *Natural Capital: Theory and Practice of Mapping Ecosystem Services*. Oxford University Press, Oxford.

Chan, K., Guerry, A., Balvanera, P., et al. (17 authors) (2012b). Where are cultural and social in ecosystem services? A framework for constructive engagement. *BioScience*, vol 62, no 8, pp 744–756.

Chan, K., Satterfield, T., and Goldstein, J. (2012a). Rethinking ecosystem services to better address and navigate cultural values. *Ecological Economics*, vol 74, pp 8–18.

Chavis, D.M. and Pretty, P. (1999). Sense of community: advances in measurement and application. *Journal of Community Psychology*, vol 27, no 6, pp 635–642.

Christie, M., Fazey, I., Cooper, R., Hyde, T., and Kenter, J. O. (2012). An evaluation of monetary and non-monetary techniques for assessing the importance of biodiversity and ecosystem services to people in countries with developing economies. *Ecological Economics*, vol 83, pp 67–78.

Costanza, R. (2000). Social goals and the valuation of ecosystem services. *Ecosystems*, vol 3, pp 4–10.

Daily, G. C., and Matson, P.A. (2008). Ecosystem Services: from theory to implementation. *Proceedings of the National Academy of Sciences*, vol 105, no 28, pp 9455–9456.

Daly, H.E. (1992). Allocation, distribution, and scale: towards an economics that is efficient, just, and sustainable. *Ecological Economics*, vol 6, pp 185–193.

Daniel, T. C., Muhar, A., Arnberger, A., et al. (23 authors) (2012). Contributions of cultural services to the ecosystem services agenda. *Proceedings of the National Academy of Sciences*. vol 109, pp 8812–8819.

de Groot, R. S., van der Perk, J., Chiesura, A., and van Vliet, A. (2003). Importance and threat as determining factors for criticality of natural capital. *Ecological Economics*, vol 44, no 2–3, pp 187–204.

de Groot, R. S., Wilson, M. A., and Boumans, R.M.J. (2002). A typology for the classification, description and valuation of ecosystem functions, goods and services. *Ecological Economics*, vol 41, no 3, pp 393–408.

Dendoncker, N., Keene, H., Jacobs, S., and Gómez-Baggethun, E. (2013). Inclusive ecosystem services valuation. In Jacobs, S., Dendoncker, N., and Keene, H. (eds) *Ecosystem Services: Global Issues, Local Practices*, Elsevier, San Diego, CA and Waltham, MA.

Díaz, S., Demissew, S., Carabias, J., et al. (85 authors) (2014). The IPBES Conceptual Framework – connecting nature and people. *Current Opinion in Environmental Sustainability*, vol 14, pp 1–16.

Elmqvist, T., Folke, C., Nyström, C. M., et al. (7 authors) (2003). Response diversity, ecosystem change, and resilience. *Frontiers in Ecology and the Environment*, vol 1, pp 488–494.

Erikstad, L., Lindblom, I., Jerpåsen, G., et al. (7 authors) (2008). Environmental value assessment in a multidisciplinary EIA setting. *Environmental Impact Assessment Review*, vol 28, pp 131–143.

Farber, S. C., Costanza, R., and Wilson, M.A. (2002). Economic and ecological concepts for valuing ecosystem services. *Ecological Economics*, vol 41, no 3, pp 375–392.

Feldmann, R.M. (1990). 'Settlement identity: psychological bonds with home places in a mobile society', *Environment and Behavior*, vol 22, no 2, pp 183–229.

Gómez-Baggethun, E. (2010). To ecologise economics or to economise ecology: theoretical controversies and operational challenges in ecosystem services valuation. PhD Thesis, Universidad Autónoma de Madrid.

Gómez-Baggethun, E., Alcorlo, P., and Montes, C. (2011). Ecosystem services associated with a mosaic of alternative states in a Mediterranean wetland: case study of the Doñana Marsh (southwest Spain). *Hydrological Sciences Journal*, vol 56, pp 1374–1387.

Gómez-Baggethun, E., and Barton, D. (2013). Classifying and valuing ecosystem services for urban planning. *Ecological Economics*, vol 86, pp 235–245.

Gómez-Baggethun, E., and de Groot, R. S. (2010). Natural capital and ecosystem services: the ecological foundation of human society. In Hester, R. E., and Harrison, R. M. (eds) *Ecosystem Services, Issues in Environmental Science and Technology*. Royal Society of Chemistry, Cambridge UK.

Gómez-Baggethun, E., and Martín-López, B. (2015). Ecological Economics perspectives on ecosystem services valuation. In Martinez-Alier, M. and Muradian, R. (eds) *Handbook for Ecological Economics*. Edward Elgar, in press.

Gómez-Baggethun, E., Martín-López, B., Barton, D., et al. (15 authors) (2014). EU FP7 OpenNESS Project Deliverable 4.1., State-of-the-art report on integrated valuation of ecosystem services, European Commission., EC grant agreement no 308428.

Gómez-Baggethun, E., and Ruiz-Pérez, M. (2011). Economic valuation and the commodification of ecosystem services. *Progress in Physical Geography*, vol 35, pp 613–628.

Gosselink, J.G., Odum, E.P., and Pope, R.M. (1974). *The Value of the Tidal Marsh*. Centre for Wetland Resources Publication No. LSU-SG-74–03. Louisiana State University, Baton Rouge.

Goulder, L. H., and Kennedy, D. (1997). Valuing ecosystems: philosophical bases and empirical methods. In Daily, G. C. (ed.) *Nature's Services: Societal Dependence on Natural Ecosystems*. Island Press, Washington DC.

Gren, I.-M., Folke, C., Turner, R. K., and Batman, I-J. (1994). Primary and secondary values of wetland ecosystems. *Environmental and Resource Economics*, vol 4, no 4, pp 55–74.

Haines-Young, R., and Potschin, M. (2013). *Common International Classification of Ecosystem Services (CICES)*: Consultation on Version 4, EEA Framework Contract No EEA/IEA/09/003.

Hanley, N. (2001). Cost-benefit analysis and environmental policy-making. *Environment and Planning C*, vol 19, pp 103–118.

Harrison, P.A., Berry, P. M., Simpson, G., et al. (14 authors) (2014). Linkages between biodiversity attributes and ecosystem services: a systematic review. *Ecosystem Services*, vol 9, pp 191–203.

Heal, G. M., Barbier, E., Boyle, K., et al. (11 authors) (2005). *Valuing Ecosystems Services: Toward Better Environmental Decision-making*. National Research Council, Washington DC.

Hernández-Morcillo, M., Plieninger, T., and Bieling, C. (2013). An empirical review of cultural ecosystem service indicators. *Ecological Indicators*, vol 29, pp 434–444.

Howarth, R., and Farber, S. (2002). Accounting for the value of ecosystem services. *Ecological Economics*, vol 41, no 3, pp 421–429.

Howarth, R. B., and Wilson, M. A. (2006). A theoretical approach to deliberative valuation: aggregation by mutual consent. *Land Economics*, vol 82, no 1, pp 1–16.

Jackson, T. (2009). *Prosperity without Growth. Economics for a Finite Planet*. Earthscan, London.

Jacobs, M. (1997). Environmental valuation, deliberative democracy and public decision-making institutions. In Foster, J. (ed.) *Valuing Nature? Economics, Ethics and Environment*. Routledge, London.

James, G.K., Adegoke, J., Osagie, S., et al. (6 authors) (2013). Social valuation of mangroves in the Niger Delta region of Nigeria. *International Journal of Biodiversity Science, Ecosystem Services & Management*, vol 9, no 4, pp 311–323.

Jax, K., Barton, D., Chan, K., et al. (26 authors) (2013). Ecosystem services and ethics. *Ecological Economics*, vol 93, pp 260–268.

Johnston, R. J., Rolfe, J., Rosenberger, R.S., Brouwer, R., Eds. (2015). *Benefit Transfer of Environmental and Resource Values: A Guide for Researchers and Practitioners*. The Economics of Non-Market Goods and Resources, Springer Science+Business Media, Drodrecht.

Kallis, G., Gómez-Baggethun, E., and Zografos, C. (2013). To value or not to value? That is not the question. *Ecological Economics*, vol 94, pp 97–105.

Kapp, W. (1983). Social costs in economic development. In Ullmann, J. E. (ed.) *Social Costs, Economic Development, and Environmental Disruption*. University Press of America, Lanham.

Kelemen, E., García-Llorente, M., Pataki, G., Martín-López, B., and Gómez-Baggethun, E. (2014). Non-monetary valuation of ecosystem services. OpenNESS synthesis paper No 6. European Commission.

Kenter, J.O., Hyde, T., Christie, M. and Fazey, I. (2011). The importance of deliberation in valuing ecosystem services in developing countries – Evidence from the Solomon Islands. *Global Environmental Change*, vol 21, no 2, pp 505–521.

Kontogianni, A., Luck, G.W., and Skourtos, M. (2010). Valuing ecosystem services on the basis of service-providing units: a potential approach to address the "endpoint problem" and improve stated preference methods. *Ecological Economics*, vol 69, pp 1479–1487.

Krutilla, J.V., and Fisher, A. (1975). *The Economics of the Natural Environment: Studies in the Valuation of Commodity and Amenity Resources*. John Hopkins Press for Resources for the Future, Baltimore, MD.

Kumar, M. and Kumar, P. (2008). Valuation of the ecosystem services: a psycho-cultural perspective. *Ecological Economics*, vol 64, no 4, pp 808–819.

Lamarque, P., Tappeiner, U., Turner, C., et al. (8 authors) (2011). Stakeholder perceptions of grassland ecosystem services in relation to knowledge on soil fertility and biodiversity. *Regional Environmental Change*, vol 11, no 4, pp 791–804.

Laurans, Y., Rankovic, A., Billé, R., Pirard, R., and Mermet, L. (2013). Use of ecosystem services economic valuation for decision-making: questioning a literature blindspot. *Journal of Environmental Management*, vol 119, pp 209–219.

Limburg, K. E., O'Neil, R., Costanza, R., and Farber, S. (2002). Complex systems and valuation. *Ecological Economics*, vol 41, no 3, pp 409–420.

Luck, G.W., Chan, K., Eser, U. et al. (7 authors) (2012). Ethical considerations in on-ground applications of the ecosystem services concept. *BioScience*, vol 62, no 12, pp 1020–1029.

Luck G. W., Harrington, R., Harrison, P. A., et al. (22 authors) (2009). Quantifying the contribution of organisms to the provision of ecosystem services. *Bioscience*, vol 59, no 3, pp 223–235.

MA (2005). *Ecosystem and Human Well-being: Synthesis*, Island Press, Washington DC.

Martín-López, B., Gómez-Baggethun, E., García-Llorente, M., and Montes, C. (2014). Trade-offs across value-domains in ecosystem service assessment. *Ecological Indicators*, vol 37, pp 220–228.

Martín-López, B., Iniesta-Arandia, I., García-Llorente, M., et al. (15 authors) (2012). Uncovering ecosystem services bundles through social preferences: experimental evidence from Spain. *PloS One*, vol 7, no 6, pp 1–11.

Martínez-Alier, J. (2002). *The Environmentalism of the Poor: A Study of Ecological Conflicts And Valuation*. Edward Elgar, Cheltenham.

Martínez-Alier, J., Munda, G., and O'Neill, J. (1998). Weak comparability of values as a foundation for ecological economics. *Ecological Economics*, vol 26, no 3, pp 277–286.

McCann, K. S. (2000). The diversity-stability debate. *Nature*, vol 405, no 6783, pp 228–233.

Milcu, A.I., Hanspach, J., Abson, D., and Fischer, J. (2013). Cultural ecosystem services: a literature review and prospects for future research. *Ecology & Society*, vol 18, no 3, pp 565–598.

Milton, K. (2002). *Loving Nature: Towards an Ecology of Emotion*. Routledge, London.

Mitsch, W.J., and Gosselink, J. G. (1993). *Wetlands*. Van Nostrand Reinhold, New York.

Mori, A. S., Furukawa, T., and Sasaki, T. (2013). Response diversity determines the resilience of ecosystems to environmental change. *Biological Reviews*, vol 88, pp 349–64.

Norton, B.G., and Hannon, B. (1997). Environmental values: a place-based theory. *Environmental Ethics*, vol 19, no 3, pp 227–245.

Norton, B.G., and Noonan, D. (2007). Ecology and valuation: big changes needed. *Ecological Economics*, vol 63, no 4, pp 664–675.

Odum, E. P. (1979). The value of wetlands; a hierarchical approach. In Greeson, P. E., Clarck, J. R., and Clarck, J. E. (eds) *Wetland Functions and Values: The State of Our Understanding*. American Water Resources Association, Bethesda, MD.

Odum, E.P. (1989). *Ecology and Our Endangered Life Support System*. Sinauer Association, Sunderland.

Odum, H.T. (1996). *Environmental Accounting: Energy and Decision Making*. John Wiley, New York.

O'Neill, J. (1997). Value pluralism, incommensurability and institutions. In Foster, J. (ed.) *Valuing Nature?: Economics, Ethics and Environment*. Routledge, London.

Pascual, U., Muradian, R., Brander, L., et al. (15 authors) (2010). The economics of valuing ecosystem services and biodiversity. In Kumar, P. (ed.) *The Economics of Ecosystems and Biodiversity: Ecological and Economic Foundations*. Earthscan, London.

Paterson, M. (1998). Commensuration and theories of value in ecological economics. *Ecological Economics*, vol 25, no 1, pp 105–125.

Pearce, D.W., and Turner, R. (1990). *Economics of Natural Resources and the Environment*. John Hopkins University Press, Baltimore, MD.

Perrings, C. (1995). Biodiversity conservation as insurance. In Swanson, T. M. (ed.) *The Economics and Ecology of Biodiversity Decline: The Forces Driving Global Change*. Cambridge University Press, Cambridge, UK.

Phelps, J., Jones, C. A., Pendergrass, J. A., Gómez-Baggethun, E. (2015). Environmental liability: a missing use for ecosystem services valuation. *Proceedings of the National Academy of Sciences (PNAS)*, in press: doi: 10.1073/pnas.1514893112

Pigou, A. C. (2006) *The Economics of Welfare*. Cossimo Classics, New York. (First published in 1920).

Rockström, J., Steffen, W., Noone, K., et al. (29 authors) (2009). A safe operating space for humanity. *Nature*, vol 461, pp 472–475.

Stokols, D. (1990). Instrumental and spiritual views of people–environment relations. *American Psycologist*, vol 45, no 5, pp 641–646.

Takeuchi, K. (2010). Rebuilding the relationship between people and nature: the Satoyama Initiative. *Ecological Research*, vol 25, no 5, pp 891–897.

TEEB (2010). *The Economics of Ecosystems and Biodiversity: Ecological &Economic Foundations*. Kumar, P. (ed.). Earthscan, London.

Throsby, D., and Throsby, C. D. (2001). *Economics and Culture*. Cambridge University Press, Cambridge, UK.

Turner, K. R., Paavola, J., Cooper, P., et al. (6 authors) (2003). Valuing nature: lessons learned and future directions. *Ecological Economics*, vol 46, pp 493–510.

Turner, R. K. (1999). The place of economic values in environmental valuation. In Bateman, I. J. and Willis, K.G. (eds) *Valuing Environmental Preferences: Theory and Practice of the Contingent Valuation Method in the US, EU, and Developing Countries*. Oxford University Press, Oxford.

Weber, M. (1949). *The methodology of the social sciences*. Free Press, New York.

Williams, D.R., Patterson, M., Roggenbuck, J., and Watson, A. (1992). Beyond the commodity metaphor: examining emotional and symbolic attachment to place. *Leisure Sciences*, vol 14, no 1, pp 29–46.

Wilson, M. A., and Howarth, R. (2002). Discourse-based valuation of ecosystem services: establishing fair outcomes through group deliberation. *Ecological Economics*, vol 41, no 3, pp 431–443.

Zendehdel, K., Rademaker, M., De Baets, B., and Van Huylenbroeck, G. (2008). Qualitative valuation of environmental criteria through a group consensus based on stochastic dominance. *Ecological Economics*, vol 67, no 2, pp 253–264.

Zografos, C., Rodríguez-Labajos, B., Aydin, C.A., Cardoso, A., Matiku, P., et al. (12 authors) (2014). *Economic Tools for Evaluating Liabilities in Environmental Justice Struggles*. The EJOLT experience. EJOLT Report No. 16.

9
A CRITICAL PERSPECTIVE

Mark Sagoff

Weyerhaeuser, one of the world's largest producers of timber products, manages 20.5 million acres (32,500 square miles) of forests in North America, according to its website. This is about the size of Maine and three times the size of Maryland. To measure the value of the ecosystem services its vast plantations provide, including carbon sequestration, the corporation has adopted the categories used by the Millennium Ecosystem Assessment, i.e., provisioning services, regulating services, supporting services, and cultural services. The webpage does not give a total but its accounting suggests that the land it manages contributes each year more than $100 billion to society in ecosystem services.

Should Weyerhaeuser be compensated for providing these ecosystem services and, if so, by whom? It could use the cash. According to the *Seattle Times* (April 17, 2009), "Weyerhaeuser is in a financial crisis so deep the largest U.S. lumber producer turns down the heat in its offices to save money . . . ." With the rebound in the housing market, the company has done better, but nothing like what it could earn with payments for the ecosystem services its plantations possibly provide. Weyerhaeuser has "priced" these services not to raise revenue, however, but to greenwash its website.

In an insightful article, Sant Anna and Nogueira (2012) have written that "the term 'ecosystem services' emerged in the early 1980s to describe a framework for structuring and synthesizing biophysical understanding of ecosystem processes in terms of human well-being." The basic idea is that ecologists, economists, and other experts can measure the contribution that forests and other ecosystems, along with the biodiversity they contain, make to human welfare. According to Sant Anna and Nogueira, "there has been an emergence (or resurrection) of attempts to develop more 'objective' assessments . . . that would not suffer from the 'subjectivism' of the exchange value of an ecosystem service (its 'price') that is constantly changed due to market conditions, as it is for any other good or service."

At least five problems beset attempts to develop an objective economic assessment of the benefits of ecosystem services. First, costs are a lot easier to quantify than benefits. Consider agriculture. Weed, fungal, and insect pests impose costs that are relatively easy to measure in terms of crop losses and pesticide payments. Some of the greatest scourges to crops in the United States, such as the locust, the passenger pigeon, and the screwworm, have been controlled or eliminated. More than two-thirds of the soy and maize planted in the United States relies on Bt and "Round-up Ready" genetic technologies to reduce pesticide costs further. Nevertheless, many pests remain; pigweed, for example, often forces farmers to purchase more expensive herbicides than glyphosate or hand-weed at a cost of about $150 an acre. The costs of nature's disservices are relatively easy to identify and measure.

How would one measure the benefits of biodiversity and ecosystems to agriculture? Markets provide prices for crops; these, however, represent the products but not the processes or services that support them. Ten or more ecosystem services may all be needed to support a crop such as apples. To attach the entire value of the crop to each would be absurd, because then the value of the services would be ten times that of the crop. How would one divide the value of the crop among these services?

The concept of "willingness to pay" (WTP) is no help because it has no known connection to human welfare. No one has shown – or could show – that WTP correlates with well-being because, to do so, one would have to measure welfare and WTP separately. Attempts do this typically demonstrate that after basic needs are met, WTP does not correlate with subjective well-being because, for example, desires are often misinformed or even self-destructive, have unforeseen consequences, and are often motivated by ethical, spiritual, and political commitments rather than by self-interest. Besides, to measure the maximum WTP for one good while holding the price of everything else constant is not only an imponderable but also a meaningless exercise. The resulting amount – represented by a "partial equilibrium" – could not be compared to the market prices of other goods, which represent a general competitive equilibrium in which price is driven down to cost. It would be comparable only with the maximum WTP for each and every other good taken seriatim. What meaning could attach to such measures?

Second, many ecosystem services are in such great supply that it is impossible to assign a scarcity value to them. Most major crops – wheat, maize, soy, and grape, for example – are wind pollinated. It makes no sense to assign a value to the pollination service of the wind, however, because we can take it for granted. Worms do a lot to regulate the soil, but there is no shortage of them. Likewise microbes; these seem to be everywhere. In order to identify economic value one must find scarcity, and this is often hard to do with services that are not under threat and that nature abundantly supplies.

Third, since technology harnesses and magnifies the services nature provides, it is not clear how one would parse the contribution each makes. About 100 plants are cultivated intensively worldwide, and of that number fewer than 20, such as rice, maize, wheat, and rapeseed, provide 90 percent of food crops. Many of these 20 species have been subjected to genetic manipulation for millennia so that even before the advent of genetic engineering, they diverged dramatically in genotype and phenotype from their wild ancestors, which may be unknown or have disappeared. How much value should be attributed to natural biodiversity and how much attributed to human management and intervention?

This question is particularly vexing in the context of fisheries. According to the National Marine Fisheries Service, "aquaculture provides more than 50 percent of all seafood produced for human consumption," a percentage that is expected to rise. Even apparently wild fish are often produced by industrial methods. In California during the spring of 2014, 30 million young salmon reared in the Coleman National Fish Hatchery on a tributary of the upper Sacramento River were trucked to the Sacramento-San Joaquin Delta and released to support the so-called wild fishery. The hatching and trucking were services provided by the California Department of Fish and Wildlife. How would one tease out the services nature provides?

Fourth, even if authorities were able to agree on the economic value of an ecosystem service, such as any of those Weyerhaeuser touts, it would seem to be nearly impossible to force anyone either to provide or to pay for it. One is free to make or not make the exchanges one likes as long as one respects the same freedom of others. It is true that a government may take by eminent domain private property for a public use, but this generally requires that compensation to the landowner be paid. The money paid for ecosystem services must come from somewhere else, from health care or education, for example. There should be more study of the tradeoffs or opportunity costs involved, say, in the decision to spend less to support education, including ecological research, in order to spend more to support ecosystem services.

Fifth, academic scientists or, more generally, PhDs, are not likely to know more about ecosystem services in any particular case than those whose livelihoods depend on them. Only those with skin in the game have a sufficient incentive to know and keep up with the relevant and fluctuating market conditions along with the changing technologies of production. This is the "epistemic" problem von Mises and Hayek advanced during the "socialist calculation debate" a century or so ago. Socialism posited that experts should set prices because only they are trained to calculate the *in natura* or objective bio-eco-geophysical value, for example, a forest makes to human well-being, as opposed to the artificial, ephemeral, and subjective prices that markets generate for the same land. According to Hayek, the relevant information about demand, supply, technology, regulations, and the vagaries of nature, among a thousand other factors, is so ephemeral or transitory, so local, implicit, and contextual, that only those who have a stake in the result can set meaningful prices for any service, ecological or otherwise. What do the professors know that the stakeholders do not?

Elinor Ostrom has shown brilliantly in a series of case studies how those who share a common pool resource are able to organize themselves to take care of it even without the facilitation of academic experts. Perhaps with her work in mind, many ecological economists have called for ecosystems service assessments that "involve more landowners and stakeholders" (Seppelt et al., 2012). The idea that academics can involve stakeholders in the assessment of ecosystem services faces three challenges. First, they have to know who the stakeholders are. How? Second, focus groups are expensive. How would they pay for them? The third challenge is more daunting. How would they get the stakeholders to return their calls?

References

Sant Anna, A.C., and Nogueira, J.M. (2012). Economic valuation of environmental services: increasing the effectiveness of PES schemes in developing countries? *Journal of Agricultural Science & Technology A*, vol 2, issue 9A, pp 1048–1057.

Seppelt, R., Fath, B., Burkhard, B., et al (11 authors) (2012). Form follows function? Proposing a blueprint for ecosystem service assessments based on reviews and case studies. *Ecological Indicators*, vol 21, pp 145–154.

10

ECONOMICS AND ECOSYSTEM SERVICES

A positive contribution to environmental management

R. Kerry Turner

Introduction

In terms of the modern history of economic theory and applications, the notion of human welfare benefits derived from ecosystems first came to prominence in the 1970s. Attempts were made to classify ecosystems from an anthropocentric utilitarian perspective in terms of their structural and functional characteristics and their usefulness (instrumental value) to humans (Westman, 1977). The main motivation at the time was to bring the need for more biodiversity conservation up the public and political agendas. Wetlands were among the first ecosystems focused on, but ranking wetlands in terms of conservation value was, even in the 1970s, controversial if only because of the scientific uncertainties surrounding the extent of current and future functional services produced by any given site. Furthermore, it was recognised that value in nature was a multidimensional concept and the assigned ecosystem services value of a wetland area could be conditioned by a range of ethical, religious, aesthetic or recreational private and public preferences and needs. But there was a general consensus that at the very least efforts to quantify and value ecosystem services were heuristically worthwhile (Westman, 1977).

Economists then focused their attention on the conceptual and methodological problems that attaching monetary values to ecosystem services posed. A range of monetary valuation methods were rapidly developed and the published literature grew exponentially. During the 1980s two 'camps' emerged with the 1988 formation of the International Society for Ecological Economics, which sought to distinguish itself from the more conventional environmental and resource economics tradition.

Environmental economics and ecological economics

Since the 1980s, the evolution of the economic thinking about ecosystems and their valuation and sustainable management has continued within the two traditions, but the precise differences in the approach taken are a matter of interpretation and perception – especially since the Ecological Economics camp is composed of a wide range of analysts from a number of disciplinary backgrounds, including natural science (Turner, 1999; Spash, 1999; Gómez-Baggethun et al., 2010). The distinctive features of Ecological Economics are the acceptance of a pluralistic interdisciplinary methodology which challenges some of the axioms of conventional economics, especially in terms of economic rationality and the dominance of self-interested human preference assumptions over other-regarding preferences. There is also a focus on the macro-systems-level scale, with an emphasis on the dangers posed by system threshold/tipping points if resource and pollution assimilative capacity 'limits' are breached because of continued exponential economic growth. Economic efficiency is seen as only one decision criterion for environmental management, with equity and fairness both within contemporary society and across future generations being equally important to foster trust and accountability in the decision-making process. Justice across generations of time becomes a matter of ethics and morality and not just economics, with many favouring a deontological Kantian approach to sustainable development. This imposes a moral duty on current generations to pass on a capital base and opportunities no less than those currently enjoyed to future generations. Therefore, decision support systems that are enabled by a range of tools and not just formal economic cost-benefit analysis are favoured by ecological economists. A strong sustainability strategy is advocated and contrasted with the weak sustainability thinking prevalent in conventional economics. Within the emerging sustainability science tradition during the 1980s and 1990s, ecosystem services and valuation of society's related benefits became more and more topical in the published literature.

The mainstreaming of ecosystem services and their valuation

This mainstreaming process, some ecological economists have argued, came to be dominated by conventional economic thinking, with a particular focus on monetary valuation of services and economic incentives. A further development was the idea to use conservation incentives through the incorporation of ecosystem service values into markets and so-called payments for ecosystem services (PES) schemes (Wunder et al., 2008; Strassburg et al., 2009).

In the wider context, an Ecosystem Services Framework (ESF) began to evolve from an earlier natural science-based and more holistic analytical approach known as the 'Ecosystem Approach', as detailed by the 1992 Convention on Biological Diversity. The next step was to augment the systems-based science by the inclusion of social science and humanities thinking, to link ecosystem functioning and its outcomes to the provision of services and well-being benefits. For economic and social valuation purposes, the definition proposed by Fisher et al. (2009) clarifies the distinction between ecosystem services and benefits in terms of human well-being. Ecosystem services include ecosystem organisation or structure as well as ecosystem processes and functions (the way in which the ecosystem operates). The processes and functions (abiotic and biotic) become services only if there are humans that (directly or indirectly) benefit from them. The key feature of this definition is the separation of ecosystem processes and functions into intermediate and final services, with the latter yielding welfare benefits. It is changes in the provision of final ecosystem services that analysts are interested in measuring and incorporating into economic and social analysis.

But the ecosystem structure and processes configurations can be complex and not well understood. A wide range of linked 'bundles' of the final ecosystem services (and therefore the valued goods and benefits) can be at risk, and, further, the values themselves are often contested. The assessment and valuation of ecosystem stock and flow situations is therefore not a straightforward task. The monetary valuation of stocks and flows has to rely on a range of accounting and socio-economic approaches, together with an underlying natural science understanding. The incomplete understanding of the complex functioning relationships between ecosystem structure and process and the delivery of ecosystem services poses a dilemma. Is this knowledge gap sufficient to cause the abandonment of ecosystem services-based management? Or does enough knowledge exist for an extension of such management to selective contexts? The position taken here is that the latter is the case and that ecosystem services perspectives and related tools have a positive conservation management role to play.

Ecosystem services monetary valuation and its critics

Understanding the economic value of nature and the services it provides to society is important for all levels of policy and decision-making. But economic methods and tools have to be accompanied by contributions from a range of social and natural science disciplines if a robust decision support system (DSS) is to be constructed. Drawing on thinking from both conventional and ecological economics, the position taken in this review is that the core decision criterion in conventional economics, i.e. economic efficiency, is an important (but not meta-ethical) criterion, especially as environmental resource scarcity increases over time. Cost-benefit analysis can, we would argue, still play an important role in multi-criteria assessment DSSs, but it must be suitably adjusted for equity concerns. The full commodification of all ecosystem services through the assignment of monetary values to all aspects of ecosystem complexity is not meaningful or possible and does not provide a sound scientific or moral basis for sustainable management.

Ecological economics critics of the ecosystem services approach have warned against the use of the approach as an over-arching framework for policy if it is applied without recognising ecosystem complexity, scientific uncertainty and the existence of environmental limits and threshold effects (Norgaard, 2010). It is also argued that the ecosystem services approach tends to rely on conventional project appraisal perspectives and methods which are set within a partial equilibrium framework. For Norgaard (2010), what is required is a general equilibrium approach with a strong sustainability development path objective. Current valuation methods, he claims, only help society 'see' ecosystem services from within our unsustainable economy and therefore underestimate their importance. The economic concept of total economic value which encompasses both human use and non-use environmental values is therefore still not a full measure of the total systems value, which also contains what some have labelled 'glue' or 'primary' value (Turner et al., 2003).

Some ecological economist critics go further and argue that the history of conceptual and methodological problems with monetary valuation of the environment going back to the 1960s indicates that fundamental problems are unlikely to be resolved and that ecosystem services thinking should be abandoned (Baveye et al., 2013). But it can equally well be argued that the body of research that has been accumulated has contributed to new insights into human behaviour. While these new insights have challenged established economic axioms, they have also served to enrich behavioural economics and institutional economics by putting the spotlight on new research challenges. These include equity and compensation in environmental change contexts, social networks and valuation, relative values and other-regarding preferences, and deliberation and multi-criteria decision support tools (see Turner, 2016). Nevertheless, in real-world policy contexts, trade-offs are continually made between conservation and development options, and monetary and opportunity cost calculations can and do play a useful role. While complete 'commodification' of nature should be avoided, many (but not all) ecosystem services can be meaningfully expressed in monetary terms. This type of calculus has 'political' purchase, which can be used to supplement traditional conservation efforts which so far have failed to reverse the global biodiversity loss trend.

References

Baveye, P.C., Baveye, J., and Gowdy, J. (2013). Monetary valuation of ecosystem services: it matters to get the timeline right. *Ecological Economics*, vol 95, pp 231–235.

Fisher, B., Turner, R. K., and Morling, P. (2009). Defining and classifying ecosystem services for decision making. *Ecological Economics*, vol 68, pp 643–653.

Gómez-Baggethun, E., de Groot, R., Lomas, P. L., and Montes, C. (2010). The history of ecosystem services in economic theory and practice: from early notions to markets and payment schemes. *Ecological Economics*, vol 69 pp 1209–1218.

Norgaard, R.B. (2010). Ecosystem services: from eye-opening metaphor to complexity blinder. *Ecological Economics*, vol 69, pp 1219–1227.

Spash, C. L. (1999). The development of environmental thinking in economics. *Environmental Values*, vol 8, pp 413–435.

Strassburg, B., Turner, R. K., Fisher, B., Schaeffer, R., and Lovett, A. (2009). Reducing emissions from deforestation – The "combined incentives" mechanism and empirical simulations, *Global Environmental Change*, vol 19, pp 265–278.

Turner, R.K. (1999). The place of values in environmental valuation. In Bateman, I. J. and Willis, G. K. (eds) *Valuing Environmental Preferences*. Oxford University Press, Oxford.

Turner, R. K. (2016). The 'balance sheet' approach within adaptive management for ecosystem services. In Potschin, M., Haines-Young, R., Fish, R. and Turner, R.K. (eds) *Routledge Handbook of Ecosystem Services*. Routledge, London and New York, pp 289–298.

Turner, R. K., Paavola, J., Cooper, P., Jessamy, V., and Georgiou, S. (2003). Valuing nature: lessons learned and future research directions. *Ecological Economics*, vol 46, pp 493–510.

Westman, W. (1977). How much are nature's services worth. *Science*, vol 197, pp 960–964.

Wunder, S., Engel, S., and Pagiola, S. (2008). Taking stock: a comparative analysis of payments for environmental services in developed and developing countries. *Ecological Economics*, vol 65, pp 834–852.

PART II

Ecosystem services

Methods and techniques for decision support – introduction

As we saw in Part I, the methodological and conceptual progress that has been made over the last decade or so has provided a solid foundation for understanding and assessing ecosystem services. But to install this thinking into the mainstream of environmental planning and management we also require a practical and policy relevant Decision Support System (DSS). The DSS process, which includes a toolbox of methods and techniques (tools), enables the implementation of ecosystem services-based measures. This part of the *Handbook* provides a number of analytical reviews and insights covering the main components of a practical DSS tailored for ecosystem services thinking, and it ends with an assessment of some real applications around the globe.

The DSS is guided by reference to a set of over-arching strategies, and their principles and goals, which in the environmental context include, among others, Sustainable Development, Holistic Ecosystem Approach and Adaptive Management (AM). The set of functional methods and tools must fit into policy and decision-making cycles alongside the overarching strategies. AM highlights the need to make policies and decisions that allow us to change our responses as our knowledge grows and we learn from our successes and failures. AM promotes flexibility in the face of prevailing environmental and socio-economic uncertainties. The initial phase of an AM process takes the form of a scoping exercise which characterises the problems and issues needing to be addressed. So we need to know, for example, the spatial location of the ecosystem in question, the pressures and drivers forcing environmental changes, the type and magnitude of changes in the state of the ecosystem, the time period over which change might occur, the significance of the changes for human welfare and potential policy responses to change.

Pulling together as much information as possible about the ecosystem context and its services is a formidable challenge and requires a systematic interdisciplinary approach. The DSS is therefore comprised of a number of linked components, which include the utilisation of findings from models of key processes that underpin and condition the relevant natural and social capital of an area. To cope with future uncertainty, scenarios can be constructed to reveal possible management options. This accumulated knowledge base then helps to set long-term objectives for managing ecosystems and services, in partnership with stakeholders. The implementation of measures to achieve the objectives can be carried out across the entire ecosystem, or through a number of pilot interventions that can be scaled up later on as required. Outcomes should be carefully monitored and feedback information shared with stakeholders. Finally, the long-term objectives should be reviewed from time to time as the stock of knowledge expands.

In summary, a Decision Support System for ecosystem services management should:

- establish and model baseline conditions and trends for ecosystems and their services;
- identify key policy issues;
- horizon scan, through the use of scenarios, for example;
- construct indicators for the ecosystem stock state and flow of services over time;
- enable a scientific, economic and socio-political-cultural valuation and appraisal of policy options using various tools and models, including participatory and deliberative approaches;
- interrogate and present data and analysis using appropriate methods and formats; and
- facilitate good monitoring and review procedures.

The individual chapters in this part of the *Handbook* cover the different components of the DSS in more detail. Potschin and Haines-Young provide an overview framework for ecosystem assessment and the challenges posed. Three major assessment frameworks are identified, based on habitats, system elements that deliver the service and understanding of places. Each has strengths and weaknesses, as different situations require different perspectives. The authors concentrate on a place-based approach with its links to multi-functionality, the valuation of natural capital and the role of landscape in framing debates about ecosystem services and sustainability. They also highlight, among other things, the need to consider so-called bundles of services that may be provided by ecosystems in a complex way and therefore require landscape scale management. The variety of benefits derived by humans from ecosystem services is very wide, and in some cases supply of one type of service to a group of beneficiaries may preclude the supply of another service to a competing interest group. How trade-off decisions are considered and choices presented to stakeholders is an important process, even more so when the values of different services are 'contested' and mired in political controversy. Balzan and colleagues focus on the particular assessment problems posed by island ecosystems and their beneficiary communities as an example of a place-based assessment. The island societies and ecosystems co-evolve, with linkages being made more close because of insularity. Context and cross linkages become key factors. Difficult trade-off decisions are posed by the juxtaposition of, for example, tourism developments or mangrove shrimp farms generating income and employment, and the conservation of the in situ habitats occupying the same real estate and providing a number of ecosystem services.

Formal models provide a key platform for ecosystem services assessment, and Kienast and Helfenstein present a useful typology covering all the available models suitable for this purpose. They point out that, in ideal circumstances, models should be able to quantify the service supply capacity of a site, assess the demand for services across beneficiary groups and measure the actual supply of services and its sustainability. This is a very tall order and only suites of models can meet this challenge if they can be feasibly coupled together. But judging whether something is sustainable or not also means that a time horizon, stretching out into the future, needs to be specified. Scenario construction and analysis can complement models and enable sustainability science thinking. Cork explains that scenarios are not predictions; rather they are internally consistent narratives (with quantitative and/or qualitative aspects) designed to explore future uncertainties. Co-evolution between the environment and society/economy is throwing up complex interactions with uncertain consequences. Scenario analysis can help us to engage with uncertainty and provide a coping strategy by considering multiple plausible futures, leading to better informed decisions. Policymakers are therefore confronted by a mass of information, and part of the DSS's role is to provide indicative guidance on the available management options.

Indicators, or rather sets of indicators, provide the required signals that can relay complex messages to decision-makers about the consequences of and responses to environmental change, in a simplified way. They are a key component in the DSS, and Müller and colleagues provide an overview of the different approaches to ecosystem services indicator classifications. They also highlight the methodological challenges that still need to be met if indicators are to fulfil their full potential. The information has to be of high quality and provided on a consistent basis. Mapping, as Maes and colleagues show, provides intuitive and simple methods for communicating information and is one necessary response to meet this requirement, supplying information on where ecosystem services are generated and utilised. Land cover data, the spatial distribution of species population data and socio-economic data, for example, can all be turned into map form. The chapter by Liquete and her co-authors, together with the briefing by Foody, shows that when information is needed for large spatial regions remote sensing can be utilised, allowing the acquisition of data about the environment without being in direct contact with it. The typical approach is to use images produced by sensors mounted on air-borne and/or space-borne platforms.

Natural capital, including ecosystems and their services, is a fundamental part of the overall capital stock (along with physical capital, human capital and social capital) which underpins the wealth creation process in all economies. A long established System of National Accounting (SNA) is in place around the world to measure and monitor economic activity over time. More recently, efforts have been made to provide a systematic framework to include ecosystem services stocks and flows in an accounting framework in order to supplement or augment the SNA. Hein and colleagues emphasise that while Gross National Product (GNP) and Gross Domestic product (GDP) measures are standardised and therefore comparable across countries, they are not adequate to capture all facets of natural capital and its contribution to not just economic activity but also wealth and wellbeing. Hein and colleagues focus on the System of Environmental-Economic Accounts (SEEA), which has been developed by the United Nations as a satellite set of accounts to sit alongside national income accounts. Haines-Young and Turner provide more information on some of the other approaches to natural capital and welfare accounting. All accounting methods have their limitations; in particular, they do not adequately encompass early warning signals about the risk of breaching thresholds or passing tipping points when ecosystem changes lead to severe environmental damage or system collapse, with consequent significant costs.

Below the international/national governance levels, there is a need for individual private business firms or sectors to illustrate how their practices impinge on ecosystems and services. The aim would be to change business practices in order to more efficiently manage dependence on ecosystem services and reduce negative impacts on ecosystems. Houdet and colleagues examine this private business practice context and assess the prospects for a harmonised Natural Capital Protocol.

National Income and other related forms of accounting give us a picture of the overall macro-economy, but this picture is built up from the micro-economic scale inhabited by firms and consumers. So how do individuals (consumers, producers and both as citizens) conceive of ecosystem services and benefits and how do they value them? Badura and colleagues examine the advantages and limitations of the conventional micro-economic approach to ecosystem services valuation. These economic values are 'marginal' measures, i.e. they relate to limited changes in the provision of ecosystem services and benefits and not the wide-scale loss of whole habitats/ecosystems. The chapter reviews the range of economic valuation methods that are available and the options open to policymakers if there is a desire to transfer results across locations and decisions.

During the 1980s, a different economic perspective, known as 'Ecological Economics', developed, and Turner uses this interdisciplinary lens to examine the ecosystem services concepts and applications. Ecological Economics thinking puts more emphasis on the macro-scale of the economy and advocates a precautionary approach in which due weight is given to possible threshold effects/tipping points linked to resource depletion and pollution assimilation capacity 'limits'. A 'strong sustainability' position is favoured, which requires economic growth to be constrained. Less faith is put in innovation and technical change and its ability to facilitate capital substitution, as resource and other constraints to economic growth manifest themselves. Ecological economists warn that much natural capital is non-substitutable and its loss may be irreversible, passing on to future generations a significant welfare loss burden.

The value of nature is a multidimensional concept, interpreted differently across interest groups, communities and societies, and therefore decision-making involving trade-offs between ecosystem conservation and development options should be informed by a multi-criteria DSS. The DSS can include economic cost-benefit analysis, with its emphasis on resource efficiency, but equity, fairness and other ethical norms also need to be encompassed. The chapter by Fish and colleagues elaborates the different grounds on which stakeholder participation matters in ecosystem service-based decision-making as well as general theoretical issues and considerations arising from identifying and involving stakeholders in a decision process. The contribution also asks how an ecosystem services perspective changes the way participation is thought about and done with respect to environmental issues. This chapter is complemented by Kenter, who looks at the different types of individual and shared values that exist and how best to capture them in a pluralistic evaluation procedure.

Turner sets out a case for the so-called Balance Sheet Approach (BSA), which provides a new format for the pluralistic presentation of evidence for decision-making. It is a different way of compiling, interrogating and presenting evidence in the decision-making process. It can combine cost-benefit analysis with distributional analysis, i.e., who gains or loses in any trade-off situation. It highlights impacts on different stakeholder groups and localities and reviews feasible compensation measures for losers. Finally, it gives due recognition to the fact that many environmental change contexts are highly 'contested', with competing interest groups involved. Fairness and justice concerns tend to be very significant in these 'contested' contexts, which are sometimes further complicated by opposing ethical norms. Sikor and colleagues point out that justice notions are contextual and experiential, depending as they do on the particular political and historical setting. They argue that decision-making on ecosystem services provision must be not only efficient and effective but be perceived as just. Jax reminds us that ethics as a theory of morality deals with the way humans interact with each other and with non-human nature. He argues that a case can be made for intrinsic value in nature and that this is largely missing from the ecosystem services approach. But recognition of intrinsic value in nature also raises questions about moral rights and interests which may or may not be held by non-human nature. Deciding on a morally acceptable way for humans to deal with the utilisation of ecosystem services remains an open debate.

To close this part of the *Handbook*, Daily describes some of the real-world attempts to implement the ecosystem services framework into decision-making. In terms of pace and scale of policy intervention, China and Latin America stand out. In China, following the widespread droughts and floods in 1997–98, major forestry and conservation initiatives were undertaken. China also established the first National Ecosystem Assessment (2000 to 2010) and set up a zoning network of 'ecosystem conservation areas'. In Latin America the focus has been on water security for rapidly growing urban agglomerations. One response has been the creation of 'water funds', which finance mechanisms through which downstream water consumers and other interests pay for upstream changes in land cover and use.

Key discussion and debating points

- While indicators are a vital component of any DSS, there are a very large number of single ecosystem service indicators. Ecological systems are multifunctional and can provide an array of linked services as 'bundles'. Is the way forward to aggregate single indicators to better represent 'bundles'?
- Expanding National Income Accounts to better reflect changes in the stock of wealth and well-being levels, including the contribution of natural capital, is a requirement of any sustainable development strategy. But should conventional accounts be internally modified or separate complementary green accounts, monetary and /or physical, be constructed?
- It has been argued that in a world in which resource scarcity exists and budgets to enable actions to be taken are limited, economic efficiency must remain an important decision criterion. Is this really the case?
- If we don't have economic valuation or payments for ecosystem services schemes, what other options are there to halt or reverse the on-going global loss of biodiversity?

11

FRAMEWORKS FOR ECOSYSTEM ASSESSMENTS

Marion Potschin and Roy Haines-Young

What is a framework?

A framework is a way of organising our thinking or working practice. In the context of ecosystem assessment, such frameworks are particularly important because the field brings together researchers, resource managers and decision-makers, each with their own skills and backgrounds, and as they embark on a common task it is essential to develop a shared perspective of how things should be approached and what outcomes are required. As Munns et al. (2015) have observed, a major barrier to successful transdisciplinary projects is the misunderstandings and inconsistencies that arise when different disciplines come together. Whether some kind of 'standard lexicon' can be constructed – or is ultimately really necessary – remains to be seen. In this chapter we therefore review these issues, and consider some of the different frameworks for ecosystem assessments that have been discussed in the literature and what has been achieved with them.

The role of conceptual frameworks

In 2012, the Intergovernmental Platform on Biodiversity and Ecosystem Services (IPBES) was established as an independent body open to all member countries of the United Nations, with the aim of 'strengthening the science-policy interface for biodiversity and ecosystem services'. The goal was to better conserve biodiversity, promote its sustainable use and hence secure long-term human well-being. Díaz et al. (2015) provide a useful summary of some of its early work, which was directed at developing a common conceptual framework for the platform.

Given the focus of their concerns IPBES took a conceptual framework to be 'a concise summary in words or pictures of relationships between people and nature . . .' (Díaz et al., 2015, p.3), the purpose of which is to set out the key social and ecological components of the system to be studied and the relationships between them. Díaz et al. (2015) argue that while the resulting framework (Figure 11.1) was 'highly simplified' in the way it depicted the interactions between people and nature, the work that led up to it was innovative both in the open and transparent way it was undertaken and in the way it combined different scientific traditions and cultural perspectives. The simplification was justified, they suggest, because of the interdisciplinary and cross-cultural understanding that needed to be achieved in setting up the Platform.

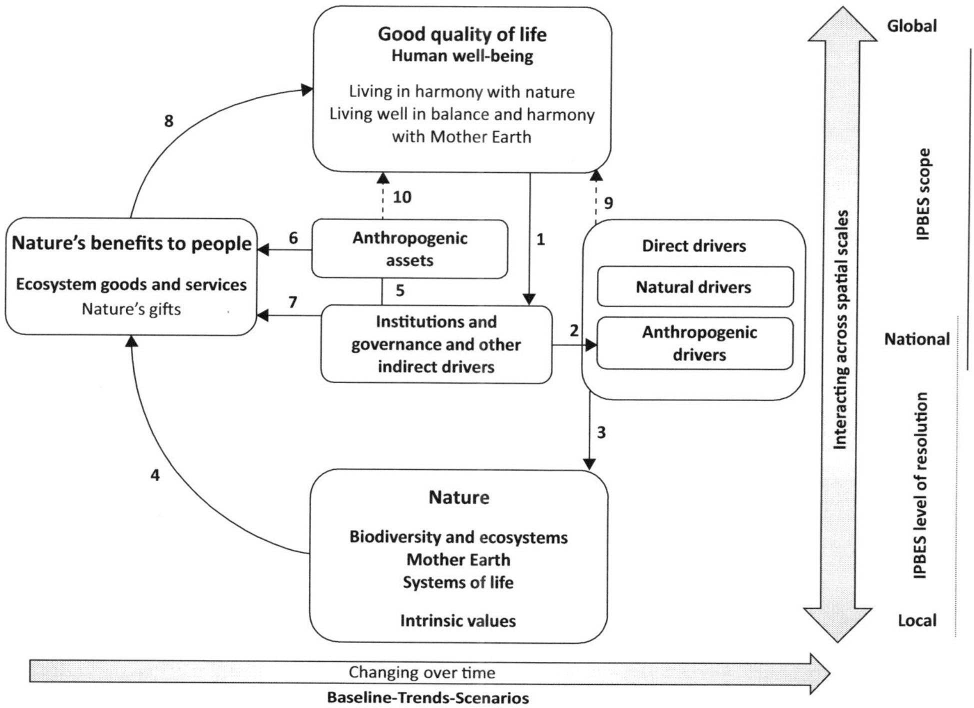

Figure 11.1 The IPBES Conceptual Framework.

Source: after Díaz et al. 2015

In the conceptual framework shown in Figure 11.1, the major headings in the boxes set out broader, more inclusive concepts agreed upon by the participants. However, the diagram also shows the way the language and concepts used by different knowledge systems are related both to them and each other. By way of illustration, consider the way different terminologies are combined under the heading of 'good quality of life'. In the context of the 'western' ecosystem services paradigm we interpret this concept through notions of 'human well-being' (see, for example, the iconic diagram from the Millennium Ecosystem Assessment; MA, 2005). In contrast, 'living-well in balance and harmony with Mother Earth' is a perspective shared by many indigenous peoples worldwide. In considering the differences between them under the general heading of 'good quality of life' it is important to note that we might not just be looking at different labels for the same thing. The idea of 'Mother Earth', for example, is not simply a synonym for 'nature' as used in western science. Rather, it is taken to denote a holistic entity that not only sustains all living things, including people, physically and spiritually, but which is also *entitled to rights as a collective subject of interest* (Pacheco, 2014; Díaz et al., 2015). Indeed, the concept has been incorporated into the constitutional and legal frameworks of Ecuador and Bolivia (Radcliffe, 2012; Satterfield et al., 2013).

The example of the IPBES illustrates that conceptual frameworks are not just a matter of conceptual discussion but also the basis for *action*. Such participative and practical characteristics are in fact essential ingredients of the idea of an *ecosystem assessment* as a 'social process' (Fish et al., 2016). The Millennium Ecosystem Assessment saw this social process as a way of enabling scientific understandings about the causes of ecosystem change and their consequences for human well-being to be made relevant to the needs of decision-makers (MA, 2005). As a result there must be an interaction between participants so that the needs of users of scientific knowledge can be identified and refined, and the content and character of scientific knowledge evaluated. For any ecosystem assessment to be successful, it must be 'credible', 'salient' and 'legitimate' (Cash et al., 2002; Tomich et al., 2010; MacDonald et al., 2014). In the context of the particular set of problems to which the assessment is directed, credibility concerns the *adequacy* of the scientific knowledge, saliency its *applicability*, and legitimacy the fact that outcomes are co-constructed in a fair and un-biased way that respects the different values, beliefs and circumstances of the various stakeholder groups involved.

If one accepts that ecosystem assessments are indeed social processes then it follows that conceptual frameworks cannot simply be taken 'off the shelf'; the collaborative effort in building

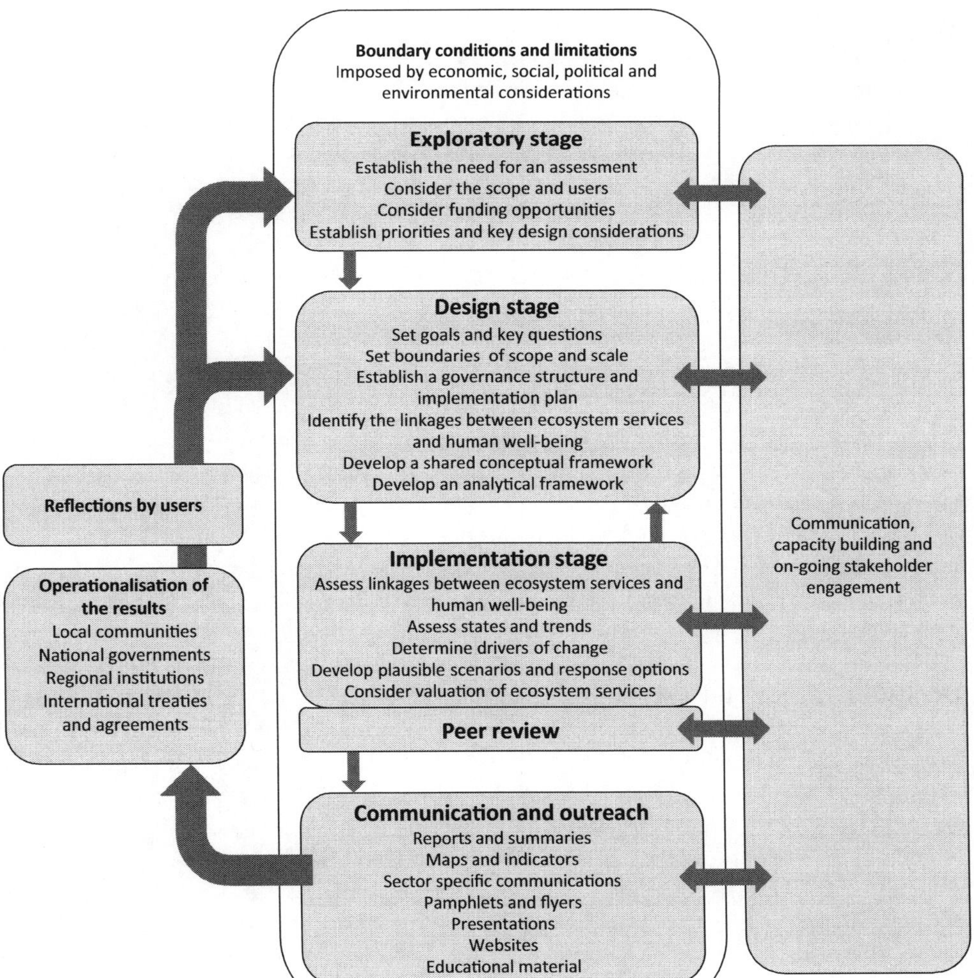

Figure 11.2 An ecosystem assessment process.

Source: adapted from Ash et al., 2010, by the SGA network, http://www.ecosystemassessments.net/about/ecosystem-assessments.html

one is part of the social learning that all participants need to go through if they are to begin to understand each other (Axelsson et al., 2013; see also Figure 11.2). Thus 'blueprints' of the kind described by Seppelt et al. (2012), or the standardised lexicons suggested by Munns et al. (2015) cannot be seen as easily transferrable or uniformly applicable approaches to the problem of assessment. At best, for example, the 'PARSIM' template of Seppelt et al. (2012) merely proposes that in describing an ecosystem study one needs to document its *P*urpose, *S*cope, *A*nalysis, *R*ecommendations and the methods used for *M*onitoring its outcomes. Instead, it is the procedural aspects of building assessments that need to be emphasised.

To use conceptual frameworks effectively in the context of ecosystem assessment, it is worth reflecting in more detail on the different types of purpose to which they are put within the social process that surrounds their construction. In its preliminary documentation, IPBES (United Nations, 2012; see also Díaz et al., 2015) suggested that conceptual frameworks could be viewed as having four purposes:

- tools to make complex systems as simple as they need to be for their intended purpose;
- providing support to structure and prioritise work;
- helping to clarify and focus thinking about complex relationships, supporting communication across disciplines, knowledge systems and between science and policy; and,
- allowing buy-in from a variety of stakeholders by involving them in the development of the framework, and thus increase policy relevance.

While none of these purposes are mutually exclusive, this chapter has mainly focussed on the last two points, which deal more with the tasks of ensuring *engagement* and *advocacy* than with the more theoretical and methodological dimensions implied by the first two items in this list. The need to represent complex systems 'as simple as they need to be' resonates strongly with the spirit and purpose of using *models* in science, which is fundamentally a theoretical undertaking (Kienast and Helfenstein, 2016). Whether these models be qualitative or quantitative in character, if they are to be credible then the important point is that they should be *testable* in some way; conceptual frameworks are therefore not altogether arbitrary but ideally based on knowledge which has been or can be critically evaluated. Nor is the point about structuring and prioritising work atheoretical. Problem recognition, together with an understanding of what might provide a solution to that problem, is essentially conjectural in character. This is especially so when confronted with the kind of complex, 'wicked problems' that surround the interactions between people and nature.

The numbered links between the main elements in the IPBES schema (Figure 11.1) illustrate just how accepted theoretical understandings are used to build a conceptual framework. For example, Díaz et al. (2015, p. 9) argue that 'the evidence so far suggests that causal links between nature and benefits to people [i.e. link 4, Figure 11.1] are strongly scale-dependent, and also straddle over several scales'. Similarly, the proposition in the diagram that intrinsic values and anthropocentric values are separate and distinct is essentially a theoretical one. According to the IPBES framework, intrinsic values 'have no relationship with possible benefits to humans or their quality of life', so they 'fall outside the scope of anthropocentric values and valuation methods' (Díaz et al., 2015, p. 11). This is a particular theoretical position in the sense that it can be argued that there can be an overlap between pure anthropocentric and intrinsic values in relation to some cultural ecosystem services (see for example Schröter et al., 2014). Finally, it is also clear that the conceptual framework is used to identify analytical priorities: according to Díaz et al. (2015), in the diagram solid arrows are used to show the principal influences between the elements of concern to IPBES, while the dotted arrows show those recognised as important, but not the main focus of the Platform.

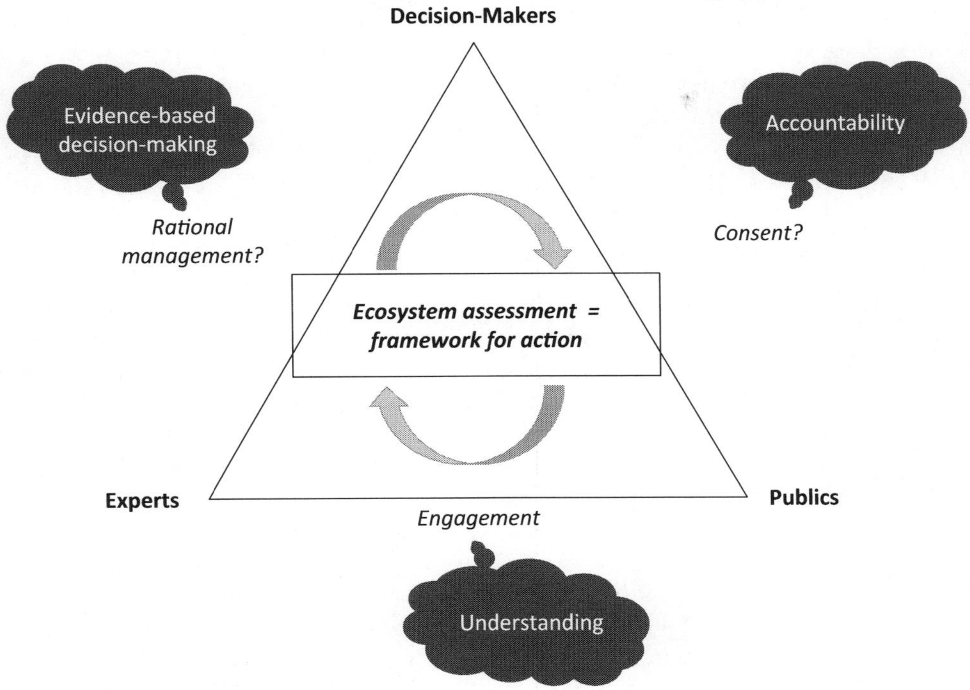

Figure 11.3 The 'post-normal' character of ecosystem assessments.

In the context of an ecosystem assessment the development of a conceptual framework is therefore not a trivial matter, or merely an emblematic task designed to communicate what is distinctive about the particular initiative. Their co-creation is, in a sense, part of the job of defining the scope of a study, and of defining and agreeing the 'boundary conditions' which is a necessary first step in any assessment process (see Figure 11.2). In this chapter we have focussed on the work of IPBES, but there are many other assessments could be used to make the same point (see for example the case of the UK National Ecosystem Assessment, UK NEA; Mace et al. 2011). That they play such a key role in assessment exercises, in fact, illustrates the transdisciplinary, 'post-normal' character (cf. Funtowicz and Ravetz, 1993) of the 'science of ecosystem services', in which different experts, decision-makers and publics come together to deal with a common set of issues. While success in this post-normal world depends on the attempt to base decisions on evidence in rational ways, sustainable solutions are also dependent on the other types of relationships between the key actors, involving such things as consent, engagement and understanding (Figure 11.3).

Operational approaches

Although we have argued that conceptual frameworks are the basis for action, in the sense that they bring experts, decision-makers and publics of different kinds together to address a common issue, it is useful to make a distinction between conceptual frameworks themselves and the approaches actually used to 'operationalise' the analysis. While the large and still growing body of literature on the topic makes generalisation difficult, there are at least three generic 'lines of attack' that people have used in practice. For convenience we will describe them as involving a 'habitat', 'systems' or 'place-based' approach (Potschin and Haines-Young, 2013). While, in considering them, we are clearly focussing on the 'implementation' stages of an assessment (cf. Figure 11.2), the choice of an approach depends on how participants 'see the world' in terms of their conceptual understanding and the emphasis they give to particular issues, such as the analysis of the status and tends of ecosystem services, trade-offs and scale, and especially the ways people and nature are linked.

Habitat or land cover approaches to ecosystem assessment

The habitat approach is probably one of the most widely used in the assessment community. Essentially, it involves viewing the world in terms of ecological habitats and using these units as the basis for looking at the relationship between people and nature. Clearly many different systems for classifying habitats exist, and so the approach covers a broad range of practice. Some assessment exercises choose to think in terms of 'land cover' or 'land use' rather than habitats (see, for example, Helfenstein and Kienast, 2014), but fundamentally all employ the same basic tactic, namely of basing the assessment on an 'ecological unit' of some kind.

The habitat approach was the one employed in the Millennium Ecosystem Assessment; in this instance the units were 'global biomes'. At the sub-global scale other assessments have begun by identifying biotope or habitat units that are more meaningful at regional, national or local scales. In the case of the UK NEA, for example, the so-called 'broad habitats' defined by the UK Biodiversity Action Plan formed the basis of the analysis; the choice was determined partly on the basis of available data but also on the policy relevance of the analytical framework. A similar framework was used in Spain (Spanish National Ecosystem Assessment, 2013); here the selection was based on such things as the need to represent the way 'nature was expressed' at the national scale, the links to well-being and the 'influence of human control', along with the need to integrate the classification used into the broader European framework. As a result, this kind of approach generally provides a cross-tabulation of services against the different types of habitat units as a way of expressing the importance of the links between them, along with such things as the status and trends of services either historically or with respect to some set of scenarios. Figure 11.4 provides an example of output from the UK NEA.

The habitat approach is clearly an efficient way of organising data and presenting the status and trends of services. The assessment of the links between the assessment units and the services can be based on empirical data or, in the case of the UK NEA, expert judgement informed by the available evidence. The approach is also easily extended in the context of the need to map services, because spatially explicit habitat and land cover data are often widely available. Matrix approaches have been widely employed for mapping and modelling ecosystem services based on land use or land cover data, and expert opinion (see for example Maes et al., 2016 and Kienast and Helfenstein, 2016). Jacobs et al. (2015) have made a number of suggestions about how the quality of such methods can be assured by including the confidence of opinions in the matrix and being clear about whose view is included and what assessment methods were used, as well as information on testing and the reliability of assessments.

Table 11.1 provides a summary of some of the main advantages and disadvantages of the habitat approach. On the one hand, it is valuable methodology because it can emphasise the link with existing nature frameworks and so strengthen the arguments for conserving particular habitats; it is also helpful because it can emphasise their multifunctional character and hence help in understanding the kinds of trade-off between ecosystem services that might arise if habitat

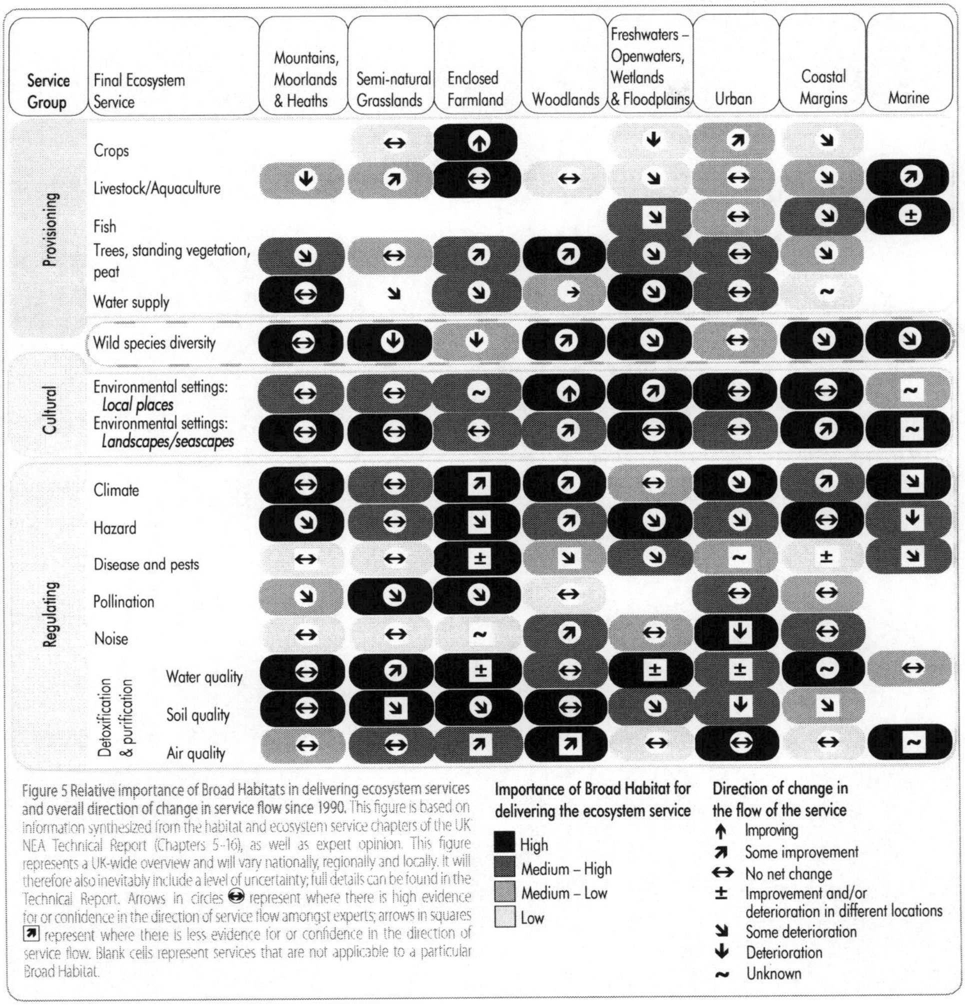

Figure 11.4 Example output based on a habitat approach to assessment from the UK NEA.

Source: UK NEA, 2011

change occurs. Such approaches are also useful because they often also make use of existing data resources and add value to them by linking them to other information about the importance to people. On the other hand, the disadvantages of using habitats as assessment units is that, for services that depend on whole landscapes, it is unclear how the different scales relate to each other or how the contributions of each habitat to overall output can be determined. In the past a disadvantage of the approach has been that this kind of analysis tends to emphasise the capacity of ecosystems to supply services rather than the societal demand for them, although as Burkhard et al. (2012) have shown, this may no longer be the case. These demand-side issues are especially important because one of the major drawbacks of the habitat approach is the difficulty of communicating the ideas to 'the public', who may be unfamiliar with the habitat units used and simply do not 'see the world' in these ways.

Table 11.1 Characteristics, strengths and weaknesses of three assessment approaches.

Approach	Characteristic	Advantages	Disadvantages
Habitat based	Assessment of services made on the basis of stock and condition of components of biodiversity, usually habitat, ecotopes or biomes, land cover etc., and potentially their change over time	• Clear links with exiting conservation frameworks and approaches • Multi-functional character of 'ecosystems' evident • Can often make use of existing biodiversity or habitat monitoring data • Focuses more easily on potential (capacity) of ecosystem units to supply a service, although demand orientated methods are now being developed.	• Unclear how different habitats should be weighted to make some overall assessment of services for a region • Unclear how habitat combinations influence overall service output across whole landscape or land cover mosaics • By focussing on supply-side issues, may be difficult to look at societal demand for the service, but methods are being developed to overcome this • Communication of key messages to publics may be difficult because habitats units are unfamiliar • Unclear how issues of multi-functionality can be addressed
Systems or process-based	Assessment of services is based on structural and functional relationships that determine service output, usually for some defined dynamic process-response unit (e.g., catchment, aquifer etc. or some defined 'service providing unit' that captures key elements of social-ecological system)	• Allows overall assessment of service state and trend to be made • Impacts of alternative assumptions explored easily allowing tests of sensitivity to assumptions and potentially scenario modelling • Generalisation easier, assumptions simpler to test. Depending on how process-response unit is defined, may be possible to look at demand and supply balances	• Systems modelling is complex and present understandings may be limited – especially in the context of predicting spatial pattern • May be difficult to calibrate and test models at local scales due to lack of data • Not quick . . . Not cheap . . . Often assumptions not transparent
Place-based	Services assessed as a bundle across units that have strong social relevance or resonance. Deliberative and temporally sensitive by giving attention to past change.	• Allows better understanding of local contexts, and therefore priorities and values • Can be used to look at patterns of use and demand as well as adequacy of supply of service • Allows issues of trade-offs and any associated conflicts to be identified and potentially resolved, and local scenarios to be examined as part of developing management visions • Allows implications of alternative management of policy options to be tested easily through participatory methods; can support adaptive management approaches • Stimulates social learning • Local buy-in	• Difficult to generalise results because of the uniqueness of place • Difficult to measure or model services at local scales because of uncertainties and lack of base-line data • Needs many different kinds of skills and competences to be combined to accomplish the inter- and transdisciplinary challenges required by the analysis of place • Time-consuming • Expensive

Source: modified from Potschin and Haines-Young, 2013

A systems approach

In contrast to the habitats approach, more system- or process-orientated methods can be used to make assessments. While the habitat framework seeks to identify ways services are generated or used, this is often done in an implicit or indicative way and the processes that underpin the service relationships are often not fully specified. A systems approach generally seeks to provide a more complete picture. Since the topic of modelling ecosystem services is covered elsewhere in this *Handbook* we will not attempt to provide a complete review here, but will highlight the importance of the general approach as an assessment tool.

A systems approach often involves modelling some quantitative 'ecological' and 'economic' production functions that expresses the way service outputs and benefits vary with changes in the various direct and indirect drivers of change. Jonsson et al. (2014) provide an example of how production functions can be constructed for predicting the impact of land-use on the biological control of pests by natural enemies. These kinds of tools can be especially helpful in understanding the changes in value that might result from modifications to the different factors that influence service supply and demand, and eventually of characterising 'Service Providing Units' (SPUs) and of identifying and understanding 'Ecosystem Service Providers' (ESPs) (Luck et al., 2003, 2009, 2016; Kremen, 2005); both broadly refer to the species or other ecological entities that generate services. Model-based approaches are also especially useful in the exploration of future scenarios.

Examples of system-based assessments ranging across different scales and issues include the global modelling work done in the context of the scenarios developed during the Millennium Ecosystem Assessment (see, for instance, Alcamo et al., 2005), the national-scale integrated modelling work done as part of the UK NEA (Bateman et al., 2013, 2014) and the more service-specific work illustrated by Lonsdorf et al. (2009), who applied the 'mobile-agent-based ecosystem service' model for modelling pollinators proposed by Kremen et al. (2007). The work by McVittie et al. (2015) provides an interesting application of modelling approaches based on Bayesian Belief Networks to regulating services provided by riparian buffer strips (Figure 11.5), which also illustrates the growing trend of recent work towards more participatory styles of working with systems concepts. The network, which was based on the cascade model (see Potschin and Haines-Young, 2016), was developed with a range of stakeholders who, through workshops, helped characterise the links between policy objectives, ecosystem services and ecological processes.

The example of the BBN highlights an important issue that is arising through the novel styles of participatory modelling work being done in the context of ecosystem assessments. As McVittie et al. (2015) observe, while the approach is promising we need to better understand the consequences of the trade-offs between realism and precision, on the one hand, and the advantages that collaborative work of this kind brings in terms of shared understandings. Indeed, as we look at the relative advantages and disadvantages of the systems-based approach to ecosystem assessment it is transparency and ease of understanding that probably stands out as one of the major issues (Table 11.1). While model-based approaches are richer theoretically than say habitat approaches, and allow the impacts of different assumptions and management options more easily to be explored, modelling is often demanding in terms of time and resources. Key disadvantages are that systems modelling is complex and usually depends on particular skill sets. Thus it may not be possible to apply such approaches in every situation, not least because the data needed to calibrate models may not be available at local scales. In addition, unless the models used are complex, then the problem of trade-offs between services may be difficult to address. However, it is interesting to note that integrated approaches are emerging; the example of McVittie et al.

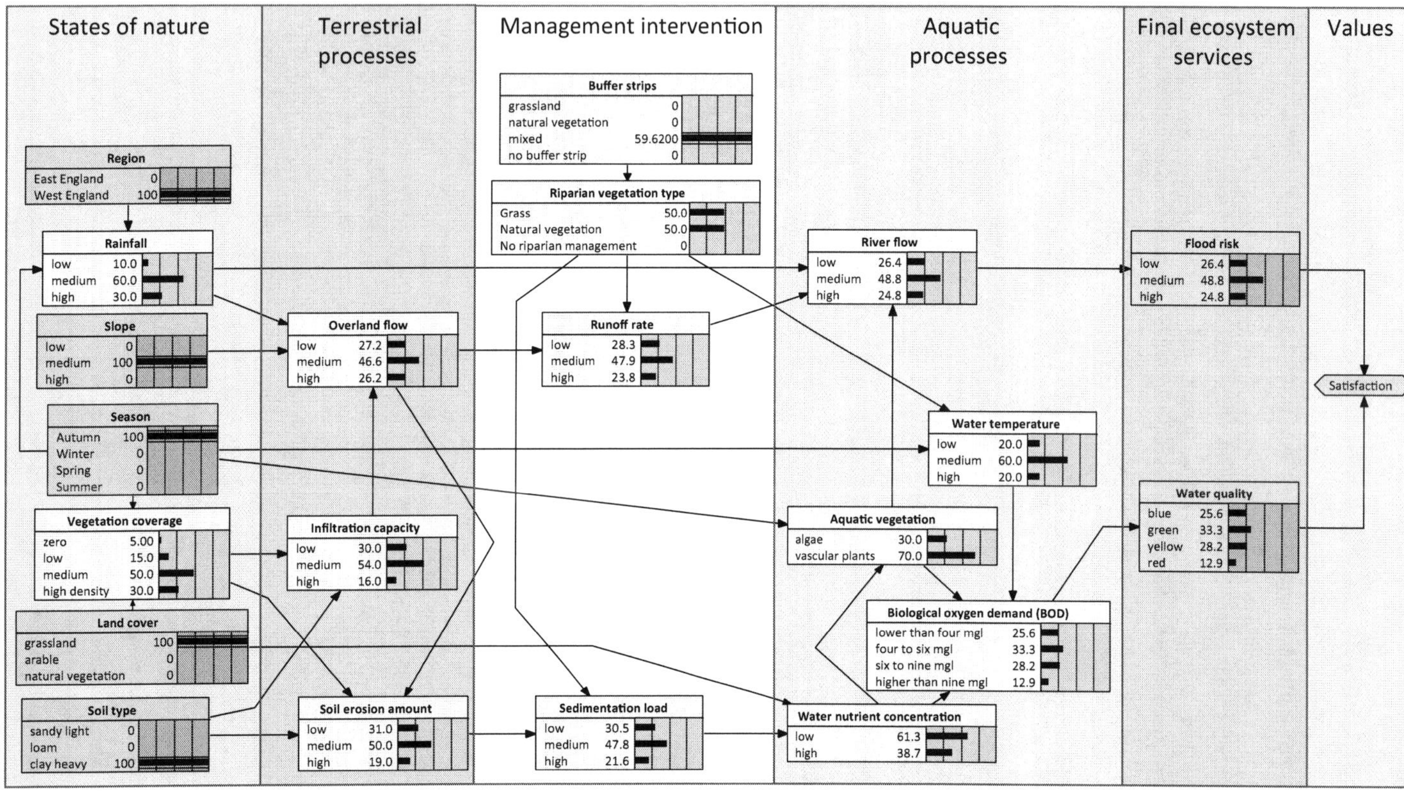

Figure 11.5 A Bayesian Belief Network model for riparian buffer strip management.

Source: McVittie et al., 2015

(2015) includes both flood and water quality regulation, while that of Bateman et al. (2014) looked at the implications for land cover change for food, timber, greenhouse gas emissions, recreation, water quality and biodiversity.

Place-based approaches

The key assumption of our third approach to ecosystem assessment is that *context matters*. Thus a place-based assessment is one which seeks to look at 'bundles' of services in an integrated way, for an area that has some strong social relevance or meaning. It aims to create an understanding of that place, usually through a deliberative process involving the people who know or use the area. The claim is that by articulating the visions and values of the different groups the significance of past and future change in the ecosystem services associated with that place can be more fully gauged.

As we have argued elsewhere (Potschin and Haines-Young, 2013) the emergence of a place-based approach to ecosystem assessment reflects the increasing emphasis given to trans-disciplinary styles of working, and the need to focus on achieving 'solutions' that are sustainable in social and economic terms, as well as being ecologically sound. It also reflects the need to look at ecosystem services at broader landscape scales. The convergence of thinking from landscape ecology about 'multi-functionality', and that of the assessment community about the ways that within heterogeneous habitat mosaics, trade-offs can occur in bundles of ecosystem services, is a particularly helpful outcome of taking a place-based approach.

The assessment literature contains a growing body of case-study material that illustrates the power and contribution of a place-based approach. For example, Plieninger et al. (2013) have demonstrated the importance of a place-based approach for assessing, mapping, and quantifying cultural ecosystem services at community level for an area of Saxony in Germany, while Palomo et al. (2013) used participatory approaches to map ecosystem service flows to the areas surrounding two protected areas in Spain. Kopperoinen et al. (2014) show how a placed-based approach can be used to look at the contribution that green infrastructure makes to the provision of ecosystem services around two cities in southern Finland. A further interesting feature of the emerging literature is an acknowledgement of the importance of the place-based approach in the context of marine studies. Klain et al. (2014), for example, have argued that interviews used in conjunction with maps can help identify the 'meaning of place' and greatly facilitate value articulation by stakeholders on North Island, Vancouver, while Levin and Möllmann (2015) have argued for the importance of a place-based approach within ecosystem based management in understanding regime shifts in marine systems. These studies suggest that the development of participatory mapping is often a key component of a place-based approach (Brown and Fagerholm, 2014) and offers important entry-points for the analysis of stakeholder perspectives (Raymond et al., 2009). Important operational outcomes include a better understanding of where to target management interventions (Bryan et al., 2010), given the inevitable trade-offs that are likely to occur when dealing with a set of ecosystem services associated with a particular area.

Although mapping is an important tool for undertaking a place-based study, as we have emphasised elsewhere, such work has to be undertaken systematically, and is perhaps best done by considering a series of questions that might be explored with stakeholders in any discussions about a particular locality (Table 11.2). Although these questions will need to be rephrased to take account of the different languages and understandings in any particular study, and even broken down in order to simplify, we suggest that taken as a whole they capture many of the key issues that need to be considered when trying to use ecosystem services to identify what connects people to their environment in a particular place. The rationale for the suggested questions is also set out in Table 11.2.

The questions shown in Table 11.2 are designed to elicit both what ecosystems are important and to whom as well as generate an understanding about where they come from. In this respect, question 2 is important because it helps establish the spatial context of the particular place being investigated. Clearly, maps can be a valuable tool in gathering different stakeholders views on these issues. However, the exploration of the questions about the importance

Table 11.2 Framework and rationale for developing a place-based assessment of ecosystem services.

	Question	*Rationale*
1.	What are the ecosystem services associated with this place that matter to peoples' well-being?	Helps in setting the conceptual and spatial boundaries to the assessment; defines the place of concern
2.	How are these services generated? Do they arise locally or are they generated outside the place or area being considered?	Identification of dependencies and cross-scale issues in relation to the supply of services; helps explore the links between the place of interest and other places
3.	How important is each of these services to which individuals or groups, and for what reasons? Do people outside the area also depend on these services?	Helps to identify who has a stake in the deliberations about the place and their needs, and develops understanding of the spatial relationships between one place and other places
4.	How can the importance of these services be prioritised or valued?	Opens up discussions about how values should be assessed and compared (e.g., using individual vs community values; monetary vs non-monetary)
5.	Do we expect to have enough of each of these services either here or elsewhere in the future?	Highlights the issues surrounding the notion of living with environmental limits and questions about sustainability of natural capital
6.	What, if anything, could replace or substitute for each of the benefits obtained from these services, either here or elsewhere?	Links to question 4, and further explores the nature of criticality, compensation and substitutability of benefits
7.	What kinds of management or policy actions are needed to protect or enhance these services and in particular how might actions directed towards one service impact or enhance another?	Helps in understanding the acceptability of management or policy interventions to different stakeholder groups and the identification of potential trade-offs and conflicts and how they might be resolved

Source: after Potschin and Haines-Young, 2013

that different individuals or groups attach to services, and whether people feel provision of services is sufficient at present and likely to be so in the future (questions 3, 4 and 5), probably requires other deliberative techniques, such as monetary and non-monetary valuation and scenarios; the co-construction of a conceptual framework for the assessment may also have a role. Question 6 is designed to examine the values of stakeholders further by looking at the question of substitutability, while question 7 explores the implications of any management interventions, and especially those that might arise in the context of trade-off and synergies between services.

The advantages of a place-based approach (Table 11.1) arise from the way it starts from a social perspective rather than a biophysical one, and so allows a better understanding of local contexts, which often determine people's priorities and values (Kenter, 2016). There is an explicit focus on looking at the bundle of services provided by a particular locality and so discussions about the implications of trade-offs between services might more easily be had. Such a deliberative approach can stimulate social learning amongst those involved in the discussion and 'buy-in' to the recommendations arising from the assessment. As Haines-Young and Potschin (2014) have shown, agreement on what constitutes evidence and indeed the generation of that evidence by joint-working amongst stakeholders can be an important dimension of building trust in a place-based context. The problems associated with the approach are that while the analysis may work for a particular locality, conclusions are often difficult to transfer to other areas or to develop into generalisations that others might use. Moreover, while it recognised that a focus on a particular place can stimulate stakeholder involvement, lack of local data may undermine the credibility of any outcomes. It is also clear that even if such data can be found or generated by the place-based process, the approach can be time-consuming and potentially expensive because it may require the involvement of a range of skills and competences to make it successful (see for example, Fish et al., 2016).

Conclusion

Ecosystem assessments can be undertaken at a range of geographical scales. Whether they are global or local in scope, however, they generally require a conceptual framework of some kind to serve as a platform for discussion and analysis, because assessments are fundamentally 'social processes'. As this chapter has shown, conceptual frameworks provide the basis for developing a common understanding of the problems and issues that need to be explored, relevant types of evidence about services and benefits, and the way changes in service supply and demand might be valued. Such frameworks show how the science of assessments is done in a social context.

Although there is no single way of doing things, because all assessments are unique, there is a growing body of literature that suggests that some broad generalisations about assessment methods can be made. In this chapter we have, for example, outlined the 'habitat', 'systems' and 'place-based' approaches to assessment. While they are not mutually exclusive, and can be combined in the context of any particular study, they do start to map out some distinct routes to understanding how people and nature are linked via the concept of ecosystem services. Most importantly, they identify some important strategies for ensuring that, when working with decision-makers and publics, the science underpinning ecosystem assessments is not only seen as relevant to people's needs, but also accepted as a credible and legitimate source of evidence in public debates.

References

Alcamo, J., van Vuuren, D., Ringler, C., et al. (7 authors) (2005). Changes in nature's balance sheet: model-based estimates of future worldwide ecosystem services. *Ecology and Society*, vol 10, no 2, p 19.

Ash, N., Bennett, K., Reid, W., et al. (11 authors) (2012). Assessing ecosystems, ecosystem services, and human well-being. In Ash, N., Blanco, H., Brown, C., et al. (12 authors) (eds) *Ecosystems and Human Wellbeing: a Manual for Assessment Practitioners*. Island Press, Washington DC.

Axelsson, R., Angelstam, P., Myhrman, L., et al. (13 authors) (2013). Evaluation of multi-level social learning for sustainable landscapes: perspective of a development initiative in Bergslagen, Sweden. *Ambio*, vol 42, no 2, pp 241–253.

Bateman, I., Day, B., Agarwala, M., et al. (27 authors) (2014). *UK National Ecosystem Assessment Follow-on. Work Package Report 3: Economic Value of Ecosystem Services*. UNEP-WCMC, LWEC.

Bateman, I., Harwood, A. R., Mace, G. M., et al. (25 authors) (2013). Bringing ecosystem services into economic decision-making: land use in the United Kingdom. *Science*, vol 341, no 6141, pp 45–50.

Brown, G., and Fagerholm, N. (2014). Empirical PPGIS/PGIS mapping of ecosystem services: a review and evaluation. *Ecosystem Services*, vol 13, pp 119–133.

Bryan, B.A., Raymond, C.M., Crossman, N.D., and Macdonald, D.H. (2010). Targeting the management of ecosystem services based on social values: where, what, and how? *Landscape and Urban Planning*, vol 97, no 2, pp 111–122.

Burkhard, B., Kroll, F., Nedkov, S., and Müller, F. (2012). Mapping ecosystemservice supply, demand and budgets. *Ecological Indicators*, vol 21, pp 17–29.

Cash, D., Clark, W.C., Alcock, F., et al. (6 authors) (2002). Salience, credibility, legitimacy and boundaries: linking research, assessment and decision making. *Proceedings of the National Academy of Sciences*, vol 100, pp 8086–8091.

Díaz, S., Demissew, S., Carabias, J., et al. (29 authors) (2015). The IPBES Conceptual Framework – connecting nature and people. *Current Opinion in Environmental Sustainability*, vol 14, pp 1–16.

Fish, R., Saratsi, E., Reed, M., and Keune, H. (2016). Stakeholder participation in ecosystem service decision making. In Potschin, M., Haines-Young, R., Fish, R., and Turner, R. K. (eds) *Routledge Handbook of Ecosystem Services*. Routledge, London and New York, pp 256–270.

Funtowicz, S. O., and Ravetz, J. R. (1993). Science for the post normal age. *Futures*, September, pp 739–755.

Haines-Young, R., and Potschin, M. (2014). The ecosystem approach as a framework for understanding knowledge utilisation. *Environment and Planning C: Government and Policy*, vol 32, no 2, pp 301–319.

Helfenstein, J., and Kienast, F. (2014). Ecosystem service state and trends at the regional to national level: a rapid assessment. *Ecological Indicators*, vol 36, pp 11–18.

Jacobs, S., Burkhard, B., Van Daele, T., Staes, J., and Schneiders, A. (2015). 'The Matrix Reloaded': a review of expert knowledge use for mapping ecosystem services. *Ecological Modelling*, vol 295, pp 21–30.

Jonsson, M., Bommarco, R., Ekbom, B., et al. (8 authors) (2014). Ecological production functions for biological control services in agricultural landscapes. *Methods in Ecology and Evolution*, vol, 5, no 3, pp 243–252.

Kenter, J. O. (2016). Deliberative and non-monetary valuation. In Potschin, M., Haines-Young, R., Fish, R. and Turner, R. K. (eds) *Routledge Handbook of Ecosystem Services*. Routledge, London and New York. This volume.

Kienast, F. and Helfenstein, J. (2016). Modelling ecosystem services. In Potschin, M., Haines-Young, R., Fish, R. and Turner, R. K. (eds) *Routledge Handbook of Ecosystem Services*. Routledge, London and New York. This volume.

Klain, S.C., Satterfield, T.A., and Chan, K.M. (2014). What matters and why? Ecosystem services and their bundled qualities. *Ecological Economics*, vol 107, pp 310–320.

Kopperoinen, L., Itkonen, P., and Niemelä, J. (2014). Using expert knowledge in combining green infrastructure and ecosystem services in land use planning: an insight into a new place-based methodology. *Landscape Ecology*, vol 29, no 8, pp 1361–1375.

Kremen, C. (2005). Managing ecosystem services: what do we need to know about their ecology? *Ecology Letters*, vol 8, no 5, pp 468–479.

Kremen, C., Williams, N.M., Aizen, M.A., et al. (19 authors) (2007). Pollination and other ecosystem services produced by mobile organisms: a conceptual framework for the effects of land-use change. *Ecology Letters*, vol 10, no 4, pp 299–314.

Levin, P.S., and Möllmann, C. (2015). Marine ecosystem regime shifts: challenges and opportunities for ecosystem-based management. *Philosophical Transactions of the Royal Society B: Biological Sciences*, vol 370, no 1659, pp 20130275.

Lonsdorf, E., Kremen, C., Ricketts, T., et al. (6 authors) (2009). Modelling pollination services across agricultural landscapes. *Annals of Botany*, vol 103, no 9, pp 1589–1600.

Luck, G.W., Daily, G.C., and Ehrlich, P.R. (2003). Population diversity and ecosystem services. *Trends in Ecology & Evolution*, vol 18, no 7, pp 331–336.

Luck, G.W., Harrington, R., Harrison, P.A., et al. (22 authors) (2009). Quantifying the contribution of organisms to the provision of ecosystem services. *Bioscience*, vol 59, no 3, pp 223–235.

Luck, G.W. (2016). Service providing units. In Potschin, M., Haines-Young, R., Fish, R,. and Turner, R. K. (eds) *Routledge Handbook of Ecosystem Services*. Routledge, London and New York. This volume.

MA (2005). *Ecosystems and Human Well-Being: Current State and Trends*. Island Press, Washington DC.

MacDonald, D.H., Bark, R.H., and Coggan, A. (2014). Is ecosystem service research used by decision-makers? A case study of the Murray-Darling Basin, Australia. *Landscape Ecology*, vol 29, no 8, pp 1447–1460.

Mace, G., Bateman, I., Albon, S., et al. (11 authors) (2011). Conceptual framework and methodology. In *The UK National Ecosystem Assessment Technical Report*. UK National Ecosystem Assessment, UNEP-WCMC, Cambridge UK.

Maes, J., Crossman, N. D., and Burkhard, B. (2016). Mapping ecosystem services In Potschin, M., Haines-Young, R., Fish, R., and Turner, R. K. (eds) *Routledge Handbook of Ecosystem Services*. Routledge, London and New York, pp. 188–204.

McVittie, A., Norton, L., Martin-Ortega, J., Siameti, I., Glenk, K., and Aalders, I. (2015). Operationalizing an ecosystem services-based approach using Bayesian Belief Networks: an application to riparian buffer strips. *Ecological Economics*, vol 110, pp 15–27.

Munns, W. R., Rea, A. W., Mazzotta, M. J., Wainger, L. A., and Saterson, K. (2015). Toward a standard lexicon for ecosystem services. *Integrated Environmental Assessment and Management*. DOI: 10.1002/ieam.1631

Pacheco, D. (2014). *Living-Well in Harmony and Balance with Mother Earth. A Proposal for Establishing a New Global Relationship between Human Beings and Mother Earth*. Universidad de la Cordillera, La Paz. Available at: http://ucordillera.edu.bo/descarga/livingwell.pdf

Palomo, I., Martín-López, B., Potschin, M., Haines-Young, R., and Montes, C. (2013). National parks, buffer zones and surrounding lands: mapping ecosystem service flows. *Ecosystem Services*, vol 4, pp 104–116.

Plieninger, T., Dijks, S., Oteros-Rozas, E., and Bieling, C. (2013). Assessing, mapping, and quantifying cultural ecosystem services at community level. *Land Use Policy*, vol 33, pp 118–129.

Potschin, M., and Haines-Young, R. (2013). Landscapes, sustainability and the place-based analysis of ecosystem services. *Landscape Ecology*, vol 28, no 6, pp 1053–1065.

Potschin, M., and Haines-Young, R. (2016). Defining and measuring ecosystem services. In Potschin, M., Haines-Young, R., Fish, R., and Turner, R. K. (eds) *Routledge Handbook of Ecosystem Services*. Routledge, London and New York, pp 25–44.

Radcliffe, S.A. (2012). Development for a postneoliberal era? Sumak kawsay, living well and the limits to decolonisation in Ecuador. *Geoforum*, vol 43, no 2, pp 240–249.

Raymond, C. M., Bryan, B. A., MacDonald, D. H., Cast, A., Strathearn, S., Grandgirard, A., and Kalivas, T. (2009). Mapping community values for natural capital and ecosystem services. *Ecological Economics*, vol 68, no 5, pp 1301–1315.

Satterfield, T., Gregory, R., Klain, S., Roberts, M., and Chan, K.M. (2013). Culture, intangibles and metrics in environmental management. *Journal of Environmental Management*, vol 117, pp 103–114.

Schröter, M., Zanden, E.H., Oudenhoven, A.P., et al. (7 authors) (2014). Ecosystem services as a contested concept: a synthesis of critique and counter-arguments. *Conservation Letters*, vol 7, no 6, pp 514–523.

Seppelt, R., Fath, B., Burkhard, B., et al. (11 authors) (2012). Form follows function? Proposing a blueprint for ecosystem service assessments based on reviews and case studies. *Ecological Indicators*, vol 21, pp 145–154.

Spanish National Ecosystem Assessment (2013). Ecosystems and biodiversity for human wellbeing. Synthesis of the key findings. Biodiversity Foundation of the Spanish Ministry of Agriculture, Food and Environment. Madrid, Spain.

Tomich, T.P., Argumedo, A., Baste, I., et al. (17 authors) (2010). Conceptual frameworks for ecosystem assessment: their development, ownership, and use. In Ash, N., Blanco, H., Brown, C., et al. (12 authors) (eds) *Ecosystems and Human Wellbeing: a Manual for Assessment Practitioners*. Island Press, Washington DC.

UK NEA (2011). *The UK National Ecosystem Assessment: Synthesis of the Key Findings*. UNEP-WCMC, Cambridge UK.

United Nations (2012). Outcome of an informal expert workshop on main issues relating to the development of a conceptual framework for the Intergovernmental Science-Policy Platform on Biodiversity and Ecosystem Services. Plenary of the Intergovernmental Science-Policy Platform on Biodiversity and Ecosystem Services, First session, Bonn, Germany, 21–26 January 2013. IPBES/1/INF/9

Briefing Note 11.1 Place-based assessment of small islands' ecosystem services

Mario V. Balzan, Marion Potschin and Roy Haines-Young

Insularity is a distinguishing feature of islands, but the diversity of their climatic, geomorphological, biotic, sociocultural, political and economic characteristics makes generalisations about them difficult. Most formal definitions tend to place an upper limit in terms of size (Wong *et al.*, 2005), but what counts as an island is difficult to specify because their number on a map increases without limit as the scale gets larger (Rackham, 2012). The UNESCO Man and the Biosphere Programme takes *small* islands to be 10,000 km^2 or less, with 500,000 or fewer residents (Hess, 1990). For the European Union regional policy they are territories having: a minimum surface of 1 km^2; a minimum distance between the island and the mainland of 1 km; a resident population of more than 50 inhabitants; and no fixed link between the island and the mainland (Dijkstra and Poelman, 2011). These definitions effectively introduce a lower limit for island size, and the condition that they are populated, but there are other definitions that do not feature these considerations (e.g. the UN Convention on the law of the seas Article 121 of Part VIII).

Small islands have often been presented as a special case for sustainable development because of their small populations and economies, restricted usable land areas, isolation but dependence on the external markets, high costs of transportation, vulnerability to natural disasters and climate change and limited adaptation capacity (Briguglio, 1995). In terms of biodiversity they also often exhibit a high degree of endemism. Critically, they also support people's well-being in a number of important ways. Marine and coastal ecosystems provide food through fisheries, regulate erosion rates, deliver spiritual and religious values and are often important recreation and tourism sites. Terrestrial ecosystems are similarly important for ensuring freshwater, a scarce resource in many small islands, and wood and non-timber products for local communities, but also for the prevention of erosion, the maintenance of nursery populations and supporting habitats, pollination and cultural ecosystem services in the form of traditional and scientific knowledge systems, ecotourism and recreation (Wong *et al.*, 2005). Socio-economic and environmental insularity often strengthens the linkages between ecosystem services and island communities. For example, in the Mediterranean region islands' rich biodiversity (Medail and Quezel, 1999) has existed for millennia with subsistence production activities that have shaped their landscapes (Plate 11.1).

Placed-based assessments are especially appropriate for islands, where context and cross-linkage matter (Potschin and Haines-Young, 2013; and this chapter). In a study carried in the Solomon Islands, customary land-ownership and strong traditional culture, subsistence gardening and water quality were highly valued by locals who were also willing to pay significantly for improvement in derived ecosystem goods. Locals associated water quality with a healthy ecosystem and the long-term sustainability of water quality regulation was more important than simple use (Kenter *et al.*, 2011). Similarly, in another study from Milne Bay Province, Papua New Guinea, locals identified freshwater and food provisioning as the most important ecosystem services, but also predicted a decline in these and identified overfishing and human population growth as the key drivers. Improved garden and agricultural productivity and population control were consequently the highest ranked management strategies (Butler *et al.*, 2014). In a study within the Azores archipelago aimed at exploring locals' views of marine environment and its conservation, stakeholders were aware of the strategic value of the marine environment for the island's economy, and of key pressures and current trends, and had their own goals and expectations for the establishment of a marine protected area

Plate 11.1 (above) The multifunctional coastal landscape of San Blas, located in the island Gozo (Malta, Central Mediterranean), is an important recreational site to tourists and locals during the summer period, and supports small-scale subsistence agriculture and ecosystems of conservation value; (bottom) The rural landscape of the Fawwara area in the island of Malta is a protected area supporting several habitats and species of European importance whilst being an important agricultural, recreational and eco-tourism site.

(Abecasis *et al.*, 2013). These studies demonstrate that locals are aware of the pressures arising from human activities, and suggest the need for a strong stakeholder involvement in conservation and management strategies aimed at enhancing biodiversity and ecosystem services.

Given the vulnerability of island ecosystems and communities, sustainable management must aim to increase resilience and enhance the ability to recover from disturbance. These goals are often difficult to achieve because of the links between ecosystems: damage to one often gives rise to reverberating effects elsewhere, such as when the removal of forest cover results in declining land and stream fauna, increasing soil erosion and sedimentation and consequently also has adverse impacts on estuarine and marine resources (Hess, 1990). Similarly, the loss of coral reefs and mangroves due to infrastructural and economic developments can threaten human safety, degrade water quality and result in loss of biodiversity. Elsewhere, rapid and unregulated growth in tourism in modern island communities has often been associated with long-term impacts on the sustainability of islands (Aretano *et al.*, 2013). Thus a key challenge for inhabitant communities is that of balancing economic benefits with environmental pressures (van der Velde *et al.*, 2007).

References

Abecasis, R. C., Schmidt, L., Longnecker, N., and Clifton, J. (2013). Implications of community and stakeholder perceptions of the marine environment and its conservation for MPA management in a small Azorean island. *Ocean & Coastal Management*, vol 84, pp 208–219.

Aretano, R., Petrosillo, I., Zaccarelli, N., Semeraro, T., and Zurlini, T. (2013). People perception of landscape change effects on ecosystem services in small Mediterranean islands: a combination of subjective and objective assessments. *Landscape and Urban Planning*, vol 112, pp 63–73.

Briguglio, L. (1995). Small island developing states and their economic vulnerabilities. *World Development*, vol 23, no 9, pp 1615–1632.

Butler, J. R., Skewes, T., Mitchell, D., Pontio, M., and Hills, T. (2014). Stakeholder perceptions of ecosystem service declines in Milne Bay, Papua New Guinea: is human population a more critical driver than climate change? *Marine Policy*, vol 46, pp 1–13.

Dijkstra, L., and Poelman, H. (2011). Regional typologies: a compilation. *European Union Regional Policy*, vol 01, pp 1–16.

Hess, A. (1990). Overview: sustainable development and environmental management of small islands. In D'Ayala, W., and Hein, P. (eds) *Sustainable Development and Environmental Management of Small Islands*. UNESCO, Paris.

Kenter, J. O., Hyde, T., Christie, M., and Fazey, I. (2011). The importance of deliberation in valuing ecosystem services in developing countries – Evidence from the Solomon Islands. *Global Environmental Change*, vol 21, no 2, pp 505–521.

Medail, F., and Quezel, P. (1999). Biodiversity hotspots in the Mediterranean basin: setting global conservation priorities. *Conservation Biology*, vol 13, no 6, pp 1510–1513.

Potschin, M., and Haines-Young, R. (2013). Landscapes, sustainability and the place-based analysis of ecosystem services. *Landscape Ecology*, vol 28, no 6, pp 1053–1065.

Rackham, O. (2012). Island landscapes: some preliminary questions. *Journal of Marine and Island Cultures*, vol 1, no 2, pp 87–90.

van der Velde, M., Green, S. R., Vanclooster, M., and Clothier, B. E. (2007). Sustainable development in small island developing states: agricultural intensification, economic development, and freshwater resources management on the coral atoll of Tongatapu. *Ecological Economics*, vol 61, no 2–3, pp 456–468.

Wong, P. P., Marone, E., Lana, P., and Fortes, M. (2005). Island systems. In Hassan, R., Scholes, R., and Ash, N. (eds) *Ecosystems and Human Well-being: Current State and Trends*. Island Press, pp. Washington DC.

12
MODELLING ECOSYSTEM SERVICES

Felix Kienast and Julian Helfenstein

Introduction

Despite widespread interest in ecosystem services, land managers are frequently left to their own judgment when it comes to selecting ecosystem models for quantifying and mapping the services of a given area. Hence there is a pressing need for a practical framework to guide people's choices. Such a framework should help managers select a credible model suitable for the given knowledge level, geography, and scale. These models should then be able to assess: the capacity of a site to deliver certain services; the demand for ecosystem services; and the actual flow (supply). As these three pillars of ecosystem service provision are rarely found in one and the same model, suites of models that complement each other in terms of drivers, data, scale and knowledge are usually needed. The delicate interplay of capacity, demand, and flow has been repeatedly described by Haines-Young et al. (2012), Burkhard et al. (2012), and others. As summarized in Kienast et al. (2009), their cascade model recognizes natural factors, policy, and economy as significant drivers for the capacity of a site to deliver ecosystem services. "These stocks manifest themselves in the form of landscape structures (e.g. mountains, woodlands, cities) and ecosystem processes and functions (e.g. net primary productivity). 'Goods and services' on the other hand represent the flows of benefits to society from these stocks. These flows (e.g. timber or food production) depend upon both the *capacity* of the landscape to supply these services and the *demand* from society for the benefits they provide" (Kienast et al., 2009, p. 1100).

In our practice-oriented review of modelling approaches, we provide a decision framework to enable managers to select modelling approaches that fit the level of information of their specific management area. The core of the framework is a 6-point checklist that can be used to select appropriate models. We illustrate the use of this checklist with examples. They highlight a broad range of models used to approximate ecosystem services and vary in input data, scale, and model assumptions. Models range from meta-models to process-based numerical simulations, and from top-down approaches to models with high expert and public participation.

Current modelling concepts

Model typology

Before presenting the checklist and illustrating its use, we give a brief summary of modelling concepts currently used for the qualitative and quantitative assessment of ecosystem services

(Crossmann et al., 2013; Seppelt at al., 2012; Chan et al., 2006; Troy and Wilson, 2006; Lesta et al., 2007; Naidoo et al., 2008).

Quantitative models are frequently divided into static and dynamic types (see Turner et al., 2001, p. 48–69). Dynamic models simulate time-dependent characteristics of a system, whereas static models calculate equilibrium states and are time-invariant. In line with Jones et al. (2013), we further distinguish between process-based and empirical approaches. Process-based models aim at modelling a system in a mechanistic way, by, for example, simulating eco-physiological processes with close-to-real physical processes. Empirical models, on the other hand, are based on correlative relationships between drivers and dependent variables. As suggested by Helfenstein and Kienast (2014) and Kienast et al. (2009), these broad categories of models need to be supplemented with "look-up tables", which are a kind of static model linking landscape attributes to ecosystem service provision. They are frequently applied to generate maps if systems knowledge is basic and no process modelling can be applied (see Kienast et al., 2009).

Tiered approaches

The model typology mentioned above has been successfully implemented using so-called tiered approaches (Maes et al., 2013, Grêt-Regamey et al., 2013, TEEB, 2010, Nelson et al., 2009; Tallis et al., 2008). These were originally used in climate modelling to downscale global circulation models and have now been revived in the field of ecosystem service modelling and mapping. They try to match desired model complexity with the available data quality and system knowledge of a region (Tallis and Polasky, 2009). A good example of a tiered approach is promoted by InVEST (Integrated Valuation of Environmental Services and Tradeoffs), a software environment for spatially explicit ecosystem service assessment (Nelson et al. 2009). As a rule, tier 0 and 1 are relatively coarse level assessments with, for example, look-up tables or regression techniques, relating land properties to ecosystem services, whereas tier 2 and 3 involve detailed process models. Schägner et al. (2013) estimate that tier 0 and 1 models make up roughly 70% of the ecosystem service assessments. Consequently, many models in the InVEST software are tier 0 and 1, especially the ones involving cultural services and benefits such as recreation or aesthetics. Tier 2 models involve services where process knowledge is more advanced, as in the case of carbon sequestration, agricultural production and forest yield.

The selection of tier 0 to 3 models depends on the aim of the study, data availability, and the level of knowledge. Furthermore, these model families have – as a rule – very different characteristics in the way they mimic the real world. Some depict it precisely and realistically, but their results are not generalizable. As suggested by Turner et al. (2001), and the classic model classifications of Levins (1966) and Sharpe (1990), there is a clear trade-off between the model attributes "reality", "precision", and "generality"; no currently known model is able to maximize the three properties at the same time (Figure 12.1). Mechanistic models optimize reality and generality, whereas empirical-statistical models optimize reality and precision. Ecosystem service models of tier 0 and 1 likely fall into the category of the statistical, empirical models. They are relatively precise in the area for which they have been calibrated, but it is difficult to generalize from them.

Indicator-based assessments

Increasingly, indicators are used to estimate ecosystem services. Indicators are benchmarks that describe the state of social and ecological systems (Müller and Lenz, 2006). To be useful, a potential metric has to be interpretable, measurable, sensitive to external change, and representative for a process and a specific geographic region (Niemeijer and de Groot, 2008; Bottero, 2011). There

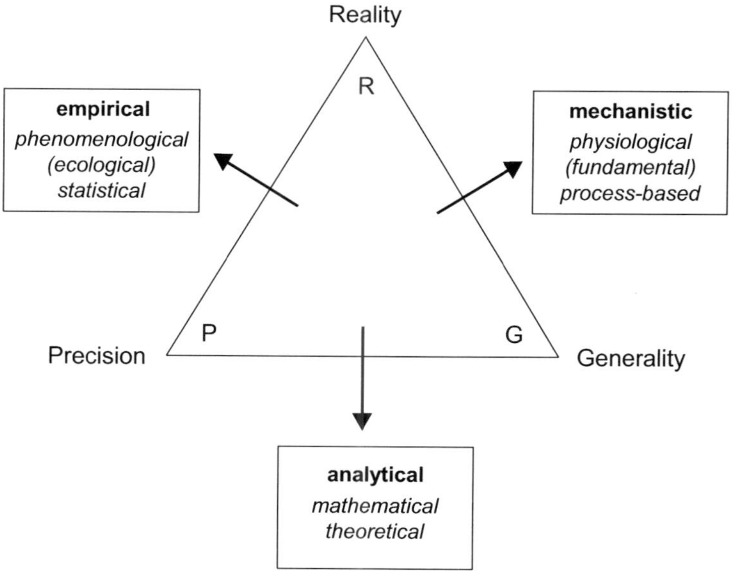

Figure 12.1 Trade-offs in model development.

Source: modified according to Levins, 1966; Sharpe, 1990

is no reason to consider indicators as specifying model outputs. If they are not too aggregated and normative, they can easily be used as input parameters in suites of models.

Egoh et al. (2012) analysed roughly 70 articles dealing with indicators for estimating ecosystem services. They distinguished *primary indicators*, which represent a proxy measure of a service (e.g. "recreation") and *secondary indicators*, which provide the necessary information used to compose the primary indicator (e.g. for recreation, the secondary indicators could be "natural areas", "accessibility" or "population density"). Egoh et al. (2012) found that the number and type of indicators varies strongly and depends on the ecosystem service considered. Whilst pollination, for example, is often measured by a single indicator (the alternative cost of bees), erosion prevention can be approximated by more than six different secondary indicators. Most of the indicators exist for regulating and provisioning services, whilst cultural and supporting services have fewer. In particular, for tier 0 and 1 assessments, indicators are helpful for evaluating the state and trend of service provision, because they can combine detailed and coarser level information (thematic, spatial, and temporal) (Helfenstein and Kienast, 2014). Thus, indicators are a straightforward way of mimicking ecosystem services and of communicating results in an understandable way to end-users. Another advantage is the possibility of using them to generate understandable maps, as in the case of an example from New Zealand (Ausseil et al., 2013). Since mapping is covered elsewhere in the *Handbook* (Maes et al., 2016) and by Maes et al. (2013), the topic is not discussed further here.

Landscape models

The term "landscape" is used here to describe a medium-scale excerpt of the globe's surface, shaped by nature and humans, as perceived by people (Kienast et al., 2007; Müller et al., 2011). There are a number of reasons why this intermediate scale is relevant for assessments, and some have even suggested that the term "landscape services" is more appropriate than "ecosystem services" (Termorshuizen and Opdam, 2009; Kienast et al., 2009). The arguments for considering the landscape scale are threefold. First, service provision does not only depend on the site conditions at the patch scale but also on the *spatial interactions* between ecosystem patches and human influences at intermediate spatial scales. Second, landscapes are an inherent part of the cultural and perceived environment, and play an important role for people's place attachment (Hunziker et al., 2007). Finally, the landscape approach is a widely accepted planning concept for conservation management and zoning (Kienast et al., 2007; Swanwick, 2002, European Landscape Convention ELC (Council of Europe, Web Resource)). However, for simplicity and to limit confusion, we do not use the term landscape services here but rather use the expression "ecosystem services at the landscape scale", to emphasize that this scale is highly relevant for modelling ecosystem services.

The proposed decision framework

Practice-oriented projects frequently encounter the following issues when applying models:

- The level of knowledge about services and the quality and spatio-temporal resolution of the available data are often not properly assessed prior to selecting a model.
- The type of output as well as the stakeholder involvement is not clearly formulated, which leads to frustration for all parties involved in the project.

To address these issues we have designed a simple 6-point decision framework, to help to reduce project failures and manage expectations (Figure 12.2).

Component 1 deals with the level of knowledge that is available. This can be either very basic, narrative or based on experience, or it can be a detailed, process-oriented knowledge ready to be input into an analytical model. If multiple ecosystem services are to be modelled, available knowledge might differ considerably among the services, and might force the modeller

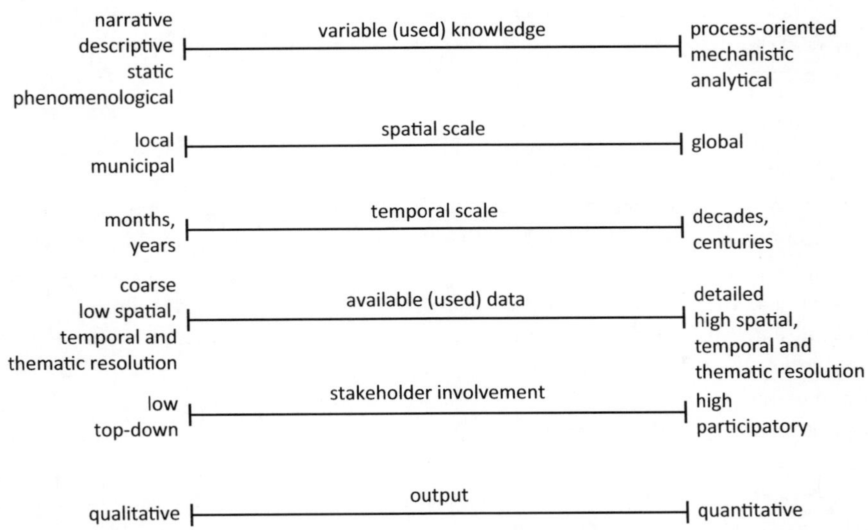

Figure 12.2 Assessment scheme for modelling ecosystem services.

to group services. Components 2 and 3 are concerned with the spatial and the temporal scale of the study. Should results be framed at the global scale over time spans of centuries and decades, or is it a more immediate local assessment at, for example, the level of a municipality? The scale issue seems a trivial one, but if a study is commissioned without a clear spatial scale in mind, at the end of the project the scale may be either too coarse or too fine to allow a sound assessment. Further components of the decision framework deal with the available data and the desired stakeholder involvement. If the data have low spatial and temporal resolution, then different models are required compared to when data of high spatial or thematic resolution are available. The same holds true for stakeholder involvement. If no stakeholder involvement is needed, a top-down approach is usually sufficient. However, if the stakeholder involvement is important, more bottom-up, participatory tools are probably required.

Often expert consultation is used instead of broad participation. As shown by Kianicka et al. (2006) and Buchecker et al. (2010), experts can have biased views of what the public thinks, especially if cultural services such as the visual landscape are at stake. Thus a public hearing or a questionnaire is often helpful, especially in the outreach phase of a project. The last component seems as trivial as the scale issue, but needs to be explicitly addressed to prevent project failure. If a quantitative output is needed, such as exact figures of sequestered CO_2 or the appreciation of a landscape on a Likert scale, the models ought to be mathematical. If single indicators are to be used to mimic ecosystem services, the choice of a model is different compared to a comprehensive study involving multiple indicators or outputs.

Exemplifying the decision framework

The 6-point decision framework proposed in Figure 12.2 was used to describe the main characteristics of selected published ecosystem service assessments and to illustrate how the decision framework could be used. However, this specific decision framework was not actually employed as a selection principle during the design of these studies. Consequently, this "a posteriori" assessment simply shows how the selection of a model (suite) would have been supported by the decision framework. All assessments differ considerably in scale, input data, systems knowledge, and participatory character. They cover multiple ecosystem services and range from the local to the regional/national scale. Continental to global approaches are excluded since there is too much overlap with the mapping section of this handbook, and since a majority of ecosystem service assessments are at the regional and local scale.

Local scale assessments

Consistency of the data for nations or regions is not an issue as long as no benchmarking is performed across spatial-analytical units. At local scales, expectations from the public and from land managers concerning prediction accuracy are high, hence adherence to reality and the precision of models are more important than generality (see Figure 12.1). We illustrate the use of the decision framework with two examples, one from the Netherlands and one from Zanzibar, Tanzania. Figure 12.3 shows the major determinants of the two examples on our 6-dimensional decision framework.

The Dutch example (Willemen et al., 2012) is actually a multi-scale assessment where service demand is calculated at regional level, and supply at both the regional and local level, i.e. for a 750km^2 area. The authors report three ecosystem services: plant habitat, arable production, and cultural heritage. The level of knowledge and the available data are high, and thus mechanistic process-oriented models could actually be applied, at least for arable production. The authors

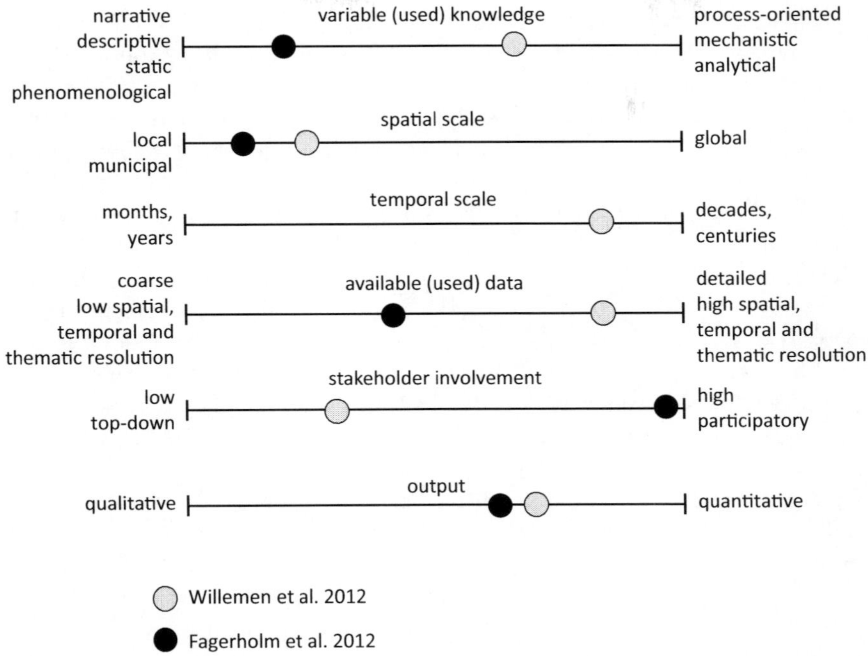

Figure 12.3 Assessment scheme for the two examples at the local scale.

chose, however, a static regression-type solution, where they relate physical landscape properties such as soil characteristics, groundwater level, and farm size to ecosystem service provision. Calculations are finally mapped. Stakeholder involvement is low, and the study mainly fits the needs of regional planners. This example clearly shows that despite very detailed input data, it is not mandatory to mimic the ecosystem services with sophisticated process models. The high-quality data can be utilized to control and improve the regression type models.

The example from Zanzibar (Fagerholm et al., 2012) shows how limited mechanistic and process-oriented knowledge, and limited available environmental data, can be compensated for using indigenous knowledge gained through participation. The paper yields a fully spatially explicit ecosystem service inventory at the community level, including maps of where the services are located. Eleven services are considered, ranging from provisioning to cultural.

Regional scale assessments

While modelling can be kept at a minimum at the local scale, it is a vital component of ecosystem assessment at the regional scale. Regional scale assessments usually profit from broad institutional interest and support from, for example, county planners, regional planning agencies, and environmental protection agencies. Because larger areas are at stake, the level of detail of the input data is generally lower compared to local assessments. The reduced resolution requirements favour remote sensing techniques to gather land cover / land-use information. At this scale, a mix of process-based models and regression type approaches are desirable, preferably with a GIS link. Most of the well-tested software packages for modelling ecosystem services are designed for the regional scale, such as InVEST (Nelson et al., 2009, 2013), ARIES (Artificial

Intelligence for Ecosystem Services, Bagstad et al., 2011; Villa et al., 2014), or Solves (Social Values for Ecosystem Services, Sherrouse et al., 2014). InVEST contains biophysical models at different levels of complexity and currently covers 17 ecosystem services. It was developed for the country-to-state level. A database featuring world-wide application examples is available.[1] ARIES uses probabilistic Bayesian type models, and is tuned to data-scarce conditions. In contrast to deterministic modelling, where equations have to be provided from many empirical observations, ARIES learns relationships between drivers and response variables via artificial intelligence.

Modelling cultural services is a challenge, yet it is of vital importance at this spatial scale. One problem has to do with the costs: participatory approaches or national questionnaires become very costly (large sample size). Recently, however, tools have been developed for the assessment of cultural benefits like scenic beauty (Grêt-Regamey et al., 2008), community values (Raymond et al., 2009), or recreation (Kienast et al., 2012). The package Solves (Social Values for Ecosystem Services) has been successfully applied to mimic scenic beauty and place identity. Sherrouse et al. (2014) give a detailed description of the program.

The two examples that we present for the regional scale are from New Zealand (Ausseil et al., 2013) and Switzerland (Helfenstein and Kienast, 2014). Figure 12.4 shows the major determinants of the two assessments on our 6-dimensional decision framework. Both aim at assessing ecosystem services at the national scale (Switzerland has an area of ca. 40,000 km^2; New Zealand 268,000 km^2), but with two completely different approaches. The New Zealand example restricts the analysis to six provisioning and regulating services (including regulation of climate and provision of food and fibre) with a high degree of knowledge and detailed cartographic data. Sophisticated process-driven models are applied, followed by a reduction to a few key indicators per ecosystem service. The indicators are consistent with the international literature, and are subsequently used to generate national maps and to perform scenario calculations at the

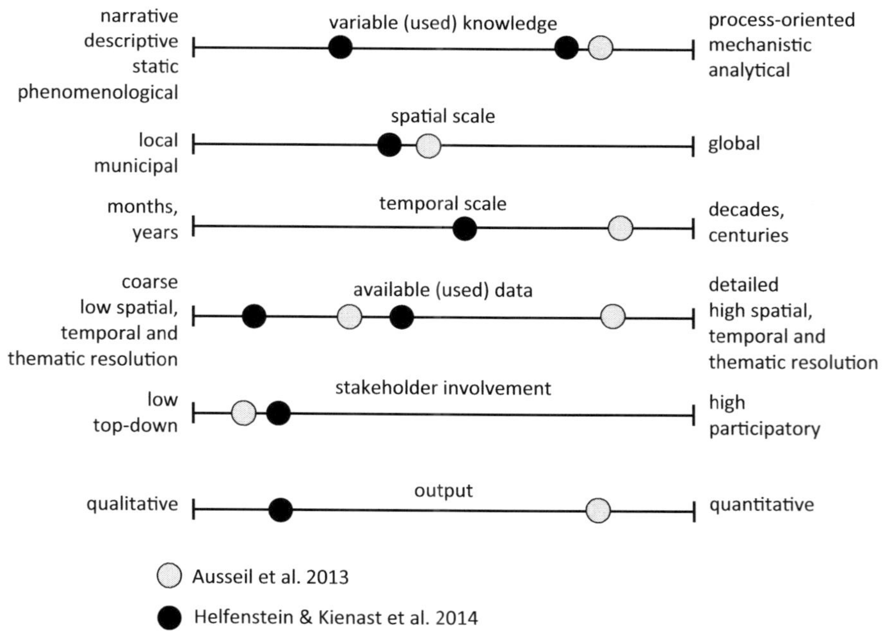

Figure 12.4 Assessment scheme for the two examples at the regional scale.

catchment scale. The approach is an interesting two-stage assessment, with process models firstly being driven with highly detailed temporal and spatial data, and secondly, simple indicators derived from the process models being used to communicate the results to decision-makers, and to perform scenario calculations. The model output is quantitative and can be used in national accounting.

The Swiss example (Helfenstein and Kienast, 2014) is promoted as a rapid assessment, and is a meta-model. Hence, it is considerably different from the New Zealand example. It is characterized by having: highly variable input data and systems knowledge, ranging from simple narratives to detailed spatio-temporal datasets; using results from a variety of modelling approaches assembled from the literature; being based exclusively on indicators; and producing highly qualitative output. Eight provisioning, regulating and cultural services are evaluated with respect to whether service provision is supported or hampered under today's land-use and a trend-scenario future. Theoretically, the analysis can be applied to any given geographical unit, but the regional level appears to be the most appropriate spatial resolution. Provision of ecosystem services by land-use type is assessed from a service/disservice perspective. The core of the approach is a meta-analysis (in the form of a flower diagram, Foley et al., 2005) highlighting the state and trend of ecosystem service provision for forests, agro-ecosystems, and aquatic systems. The straightforward and condensed, but highly informative, way of presenting state and trend of ecosystem service provision at the national scale is the major achievement of the assessment. The authors (Helfenstein and Kienast, 2014, p. 11) found "that the simple but systematic approach is more flexible than traditional mapping approaches, i.e. it allowed combining a variety of spatially non-explicit but highly detailed indicators with spatially explicit indicators." The assessment also proceeded "faster than with a mapping approach, where many known and unknown spatial inaccuracies arise. This flexible incorporation of spatially explicit and non-explicit data provides high quality information on the state and trends of ecosystem services at regional to national extents."

Continental scale assessments

The majority of continental scale assessments are simple look-up tables or very simple process models that are able to use coarse input data from remote sensing, national and continental-wide surveys, and census data. Mapping is of extraordinary importance here and therefore we refer to the chapter on ecosystem service mapping in this book.

Cultural services – a special case for modelling

Out of all ecosystem services, cultural services are probably the ones that are the most difficult to model, as they are less prominently driven by physical properties of ecosystems. Although cultural services are dependent on the physical space, equally, if not more important are people's perceptions and the meanings of places, two variables that are strongly dependent on cultural determinants. Due to this difficulty, some authors even question whether cultural services are ecosystem services at all. Attempts to model cultural services are reported by Willemen et al. (2012), who applied simple indicators such as area of unchanged land since 1900 to mimic cultural heritage areas. In another study undertaken at the national level, Swiss residents (N=2814; return rate 35%) were asked to rate the landscape of their municipality according to several landscape perception concepts. The concepts included in the survey were: complexity; coherence; legibility; and mystery (introduced by Kaplan and Kaplan, 1989), together with the strongly culturally determined concepts of authenticity and fascination (Twigger-Ross and

Uzzell, 1996; Breakwell, 1986; Gehring, 2006). Correlation analysis by Kienast et al. (2015) revealed that there is a relationship between the preference indicators and physical properties of the land. As a rule, they are weak but bear some predictive power. The authors found negative correlations between most preference assessments and, for example, the impervious area, light emissions, or urban permeation. Bearing this in mind, it seems feasible but risky to predict cultural preference features.

In another study, recreation behaviour of people living in five Swiss towns with 10,000–100,000 inhabitants was analysed using questionnaires (Kienast et al., 2012). The focus was on recreation areas within a 10 to 15 minute walking or driving distance. Residents indicated: their outdoor activities; time spent; type of transportation used; preference for given landscape properties; and preferred locations on maps with a cell size of $1km^2$. Kienast et al. (2012) applied statistical models to relate reported presence/absence of people in the $1km^2$ cells to the characteristics of the landscape found there. The surveys revealed that cultural backgrounds – approximated by language – do not significantly change the recreation behaviour of people, and that the landscape characteristics found to significantly influence nearby recreation in the model include distance to residence, open water, forests, summits with an overview, and avoidance of major roads. Based on these relationships, the supply of "nearby recreation" was estimated for the whole of Switzerland. A more detailed description can be found in Buchecker et al. (2013).

Conclusion

The proposed decision framework can serve as a check-list for land managers seeking the 'right' model to assess ecosystem services. Whatever the exact descriptions of the six dimensions are in a specific project, it is important that they are addressed explicitly. Below we highlight the key points of the decision framework that are decisive for proper model selection:

1 *Level and type of knowledge about the ecosystem service to be modelled*: knowledge differs quite strongly among different ecosystem services, which is not a problem if only one service is to be modelled. In full inventories with all ecosystem services included, the issue has to be addressed, especially if trade-offs are calculated. Some assessments solve this problem by building groups of ecosystem services with similar knowledge levels. Others try to compensate for such things as missing spatially explicit data by using alternative data sources. The Tanzania example is such a case; it shows how indigenous knowledge can be gathered in a participatory way and used as a source of information to effectively model ecosystem service provision. It was shown that indigenous knowledge can be given a spatial component and even used for mapping. This ability of laypeople to geographically locate ecosystem services (e.g. recreation, sacred places, wood provision etc.) was observed in many studies both in the industrialized world (Kienast et al., 2012) and in developing countries (Fagerholm et al., 2012). This ability should be used as a source of information in addition to remote sensing and GIS technology.

2 *Spatial and temporal scale*: all examples discussed here profit from clearly defined scales or range of scales for which the assessments are valid. We conclude that it is extremely important to clearly state the expected spatio-temporal framework at the *beginning* of a study, as it is difficult to alter it later.

3 *Available data*: this is one of the most crucial aspects in every project. In many projects there is a constant change of model types as new data becomes available. This can be a reason why financial constraints are sometimes exceeded and project results are inconsistent. It is advisable to "negotiate" the data issue clearly at the beginning of a project, and, after a

"point of no return", additional data should not be included. Instead, such additional data can be used in a next update of an assessment, clearly separated from current results.

4 *Stakeholder involvement*: In addition to what was discussed under (1), including people's view of ecosystem services requires a participatory planning culture. This involves, first, the willingness of the public to voice their views and to participate in the surveys, and second, the willingness of land managers to accept that there is both an expert and a lay view. The challenging part for land managers and conservation specialist is that in modern democracies both expert and lay views are important. Stakeholder and public participation can – to a certain degree – be substituted by expert consultation. However, keep in mind that experts can have biased views of what the public thinks, especially if cultural services such as the visual landscape are at stake. Thus, if finances do not allow for a representative survey, a public hearing or a simple questionnaire can often help to tune the expert view.

5 *Output*: This is a crucial aspect of any assessment. If output types are not clearly stated (tabular result, map, indicator values etc.), model selection is difficult and the risk of using a non-optimal model and generating the wrong output is high. This can be avoided by checking what type of result and output is expected from a model or a software package.

Keeping these crucial points in mind can considerably reduce the risk of selecting inappropriate models. Prior to model selection a project consortium has to clearly answer the following three pedagogic questions:

- ***why*** should an ecosystem service assessment be performed?
- ***what*** is the ***target audience***?
- ***what*** type of ***products*** are envisaged?

Our decision framework can be applied as soon as these general but critical project determinants are clarified.

Note

1 http://www.naturalcapitalproject.org/models/models.html (accessed 23.5.2014).

References

Ausseil, A.-G.E., Dymond, J. R., Kirchbaum, M.U.F., Andrew, R.M., and Parfitt, R. L. (2013). Assessment of multiple ecosystem services in New Zealand at the catchment scale. *Environmental Modelling and Software*, vol 43, pp 37–48.

Bagstad, K. J., Villa, F., Johnson, G.W., and Voigs, B. (2011). *ARIES–Artificial Intelligence for Ecosystem Services: a guide to models and data, version 1.0.* ARIES report series.

Bottero, M. (2011). Indicator assessments systems. In Cassatella, C. and Peano, A. (eds) *Landscape Indicators*. Springer, Dordrecht.

Breakwell, G.M. (1986). *Coping with Threatened Identity*. London, Methuen.

Buchecker, M., Meier, C., and Hunziker, M. (2010). Measuring the effects of consensus-building processes with methods of intervention research. *Eur. Plan. Stud.*, vol 18, pp 259–280.

Buchecker, M., Salvini, G., Baldassarre, G. D., Semenzin, E., Maidl, E., and Marcomini, A. (2013). The role of risk perception in making flood risk management more effective. *Natural Hazards and Earth System Science*, vol 13, no 11, pp 3013 –3030.

Burkhard, B., Kroll, F., Nedkov, S., and Müller, F. (2012). Mapping ecosystem service supply, demand and budgets. *Ecological Indicators*, vol 21(Special Issue), pp 17–29.

Chan, K.M.A., Shaw, R., Cameron, D., Underwood, E. C., and Daily, G. C. (2006). Conservation planning for ecosystem services. *Public Library of Science Biology*, vol 4, no 11, pp 2138–2152.

CICES (2013). *CICES (Common International Classification of Ecosystem Services)* V4.3. Available at: http://cices.eu/

Council of Europe (no date). *The European Landscape Convention*. Available at: http://conventions.coe.int/Treaty/Commun/QueVoulezVous.asp?NT=176&CL=ENG (accessed April 2, 2014).

Crossmann, N. D., Burkhard, B., Nedkov, S., et al (14 authors) (2013). A blueprint for mapping and modelling ecosystem services. *Ecosystem Services*, vol 4, pp 4–14.

Egoh, B., Drakou, E. G., Dunbar, M. B., Maes, J., and Willemen, L. (2012). *Indicators for Mapping Ecosystem Services: A Review*. Report EUR 25456.

Eigenbrod, F., Armsworth, P. R., Anderson, B. J., Heinemeyer, A., Gillings, S., Roy, D. B., Thomas, C. D., and Gaston, K. J. (2010). Error propagation associated with benefits transfer-based mapping of ecosystem services. *Biological Conservation*, vol 143, no 11, pp 2487–2493.

Fagerholm, N., Käyhkö, N., Ndumbaro, F., and Khamis, M. (2012). Community stakeholders' knowledge in landscape assessments – Mapping indicators for landscape services. *Ecological Indicators*, vol 18, pp 421–433.

Foley, J. A., Defries, R., Asner, G. P., Barford, C., et al. (2005). Global consequences of land use. *Science*, vol 309, no 5734, pp 570–574.

Gehring, K. (2006). *Landscape Needs and Notions. Preferences, Expectations, Leisure Motivation, and the Concept of Landscape from a Cross-Cultural Perspective*. Report: Eidg. Forschungsanstalt WSL, Birmensdorf.

Grêt-Regamey, A., Bebi, P., Bishop, I. D., and Schmid, W. A. (2008). Linking GIS-based models to value ecosystem services in an Alpine region. *Journal of Environmental Management*, vol 89, no 3, pp 197–208.

Grêt-Regamey, A., Celio, E., Klein, T. M., and Wissen Hayek, U. (2013). Understanding ecosystem services trade-offs with interactive procedural modeling for sustainable urban planning. *Landscape and Urban Planning*, vol 109, pp 107–116.

Haines-Young, R., and Potschin, M. (2013). *CICES V4.3 – Revised report prepared following consultation on CICES Version 4*. EEA Framework Contract No EEA/IEA/09/003

Haines-Young, R., Potschin, M., and Kienast, F. (2012). Indicators of ecosystem service potential at European scales: mapping marginal changes and trade-offs. *Ecological Indicators*, vol 21 (Special Issue), pp 39–53.

Harrison, P., et al. (2010). Identifying and prioritising services in European terrestrial and freshwater ecosystems. *Biodiversity and Conservation*, vol 19, no 10, pp 2791–2821.

Helfenstein, J., and Kienast, F. (2014). Ecosystem service state and trends at the regional to national level: a rapid assessment. *Ecological Indicators*, vol 36, pp 11–18.

Hunziker, M., Buchecker, M., and Hartig, T. (2007). Space and place – two aspects of the human-landscape relationship. In Kienast, F., Wildi, O., and Ghosh, S. (eds) *A Changing World: Challenges for Landscape Research*. Springer, Dordrecht.

Jones, K. B., Zurlini, G., Kienast, F., et al. (8 authors) (2013). Informing landscape planning and design for sustaining ecosystem services from existing spatial patterns and knowledge. *Landscape Ecology*, vol 28, pp 1175–1192.

Kaplan, R., and Kaplan, S. (1989). *The Experience of Nature. A Psychological Perspective*. Cambridge University Press, Cambridge UK.

Kianicka, S., Buchecker, M., Hunziker, M., and Müller-Böker, U. (2006). Locals' and Tourists' sense of place. A case study of a Swiss alpine village. *Mt. Res. Dev.*, vol 26, no 1, pp 55–63.

Kienast, F., Bolliger, J., Potschin, M., et al. (8 authors) (2009). Assessing landscape functions with broad-scale environmental data: insights gained from a prototype development for Europe. *Environmental Management*, vol 44, pp 1099–1120.

Kienast, F., Degenhardt, B., Weilenmann, B., Wäger, Y., and Buchecker, M. (2012). GIS-assisted mapping of landscape suitability for nearby recreation. *Landscape and Urban Planning*, vol 105, pp 385–399.

Kienast, F., Frick, J., van Strien, M. J., and Hunziker, M. (2015). The Swiss landscape monitoring program – a comprehensive indicator set to measure landscape change. *Ecological Modelling*, vol 295, pp 136–150.

Kienast, F., Ghosh, R., and Wildi, O. (eds) (2007). *A Changing World: Challenges for Landscape Research*. Landscape Series, Springer.

Koschke, L., Fürst, C., Frank, S., and Makeschin, F. (2012). A multi-criteria approach for an integrated land-cover-based assessment of ecosystem services provision to support landscape planning. *Ecological Indicators*, vol 21, pp 54–66.

Lautenbach, S., Kugel, C., Lausch, A., and Seppelt, R. (2011). Analysis of historic changes in regional ecosystem service provisioning using land use data. *Ecological Indicators*, vol 11, no 2, pp 676–687.

Lesta, M., Mauring, T., and Mander, Ü. (2007). Estimation of landscape potential for construction of surface-flow wetlands for wastewater treatment in Estonia. *Environ. Manage.*, vol 40, pp 303–313.

Levins, R. (1966). The strategy of model building in population biology. *American Scientist*, vol 54, pp 421–431.

Maes, J., Crossman, N. D., and Burkhard, B. (2016). Mapping ecosystem services. In Potschin, M., Haines-Young, R., Fish, R. and Turner, R. K. (eds) *Routledge Handbook of Ecosystem Service*. Routledge, London and New York, pp 188–204.

Maes, J., Teller, A., Erhard, M., et al. (43 authors) (2013). *Mapping and Assessment of Ecosystems and their Services. Indicators and Guidelines for Ecosystem Assessments under Action 5 of the EU Biodiversity Strategy to 2020.* Publications office of the European Union, Luxembourg.

Müller, F., de Groot, R., and Willemen, W. (2011). Ecosystem services at the landscape scale: the need for integrative approaches. *Landscape Online*, vol 23, pp 1–11.

Müller, F., and Lenz, R. (2006). Ecological indicators: theoretical fundamentals of consistent applications in environmental management. *Ecological Indicators*, vol 6, pp 1–5.

Naidoo, R., Balmford, A., Costanza, R., Fisher, B., Green, R. E., Lehner, B., . . . & Ricketts, T. H. (2008). Global mapping of ecosystem services and conservation priorities. *Proceedings of the National Academy of Sciences*, vol 105, no 28, pp 9495–9500.

Nelson, E., Bhagabati, N., Ennaanay, D., et al (6 authors) (2013). Modeling terrestrial ecosystem services. *Encyclopaedia of Biodiversity*, 2nd edition. March 2013.

Nelson, E., Mendoza, G., Regetz, J., et al. (14 authors) (2009). Modeling multiple ecosystem services, biodiversity conservation, commodity production, and trade-offs at landscape scales. *Frontiers in Ecology and the Environment*, vol 7, pp 4–11.

Niemeijer, D. and de Groot, R. S. (2008). A conceptual framework for selecting environmental indicator sets. *Ecological Indicators*, vol 8, pp 14–25.

Raymond, C. M., Bryan, B. A., MacDonald, D. H., Cast, A., Strathearn, S., Grandgirard, A., and Kalivas, T. (2009). Mapping community values for natural capital and ecosystem services. *Ecological Economics*, vol 68, no 5, pp 1301–1315.

Schägner, J. P., Brander, L., Maes, J., and Hartje, V. (2013). Mapping ecosystem services' values: current practice and future prospects. *Ecosystem Services*, vol 4, pp 33–46.

Seppelt, R., Fath, B., Burkhard, B., et al. (11 authors) (2012). Form follows function? Proposing a blueprint for ecosystem service assessment studies based on reviews and case studies. *Ecological Indicators*, vol 21, pp 145–154.

Sharpe, P.J.A. (1990). Forest modeling approaches: compromises between generality and precision. In Dixon, R. K., et al. (eds) *Process Modeling of Forest Growth Responses to Environmental Stress.* Timber Press, Portland, Oregon.

Sherrouse, B. C., Semmens, D. J., and Clement, J. M. (2014). An application of social values for ecosystem services (SolVES) to three national forests in Colorado and Wyoming. *Ecological Indicators*, vol 36, pp 68–79.

Swanwick, C. (2002) *Landscape Character Assessment. Guidance for England and Scotland.* Report Scottish Natural Heritage.

Tallis, H., Kareiva, P., Marvier, M., and Chang, A. (2008). An ecosystem services framework to support both practical conservation and economic development. *PNAS*, vol 105, pp 9457–9464.

Tallis, H., and Polasky, S. (2009). Mapping and valuing ecosystem services as an approach for conservation and natural-resource management. *Annals of the New York Academy of Sciences*, vol 1162, pp 265–283.

TEEB (The Economics of Ecosystems and Biodiversity) (2010) *Ecological and Economic Foundations*. Kumar, P. (ed.). Earthscan, Oxford.

Termorshuizen, J.W., and Opdam, P. (2009). Landscape services as a bridge between landscape ecology and sustainable development. *Landscape Ecology*, vol 24, pp 1037–1052.

Troy A., and Wilson, M. A. (2006). Mapping ecosystem services: practical challenges and opportunities in linking GIS and value transfer. *Ecological Economics*, vol 60, pp 435–449.

Turner, M. G., Gardner, R. H., and O'Neill, R.V. (2001). *Landscape Ecology in Theory and Practice: Pattern and Process*. Springer, New York.

Twigger-Ross, C., and Uzzell, D.L. (1996). Place and identity processes. *Journal of Environmental Psychology*, vol 16, pp 205–220.

van Oudenhoven, A.P.E., Petz, K., Alkemade, R., Hein, L., and de Groot, R.S. (2012). Framework for systematic indicator selection to assess effects of land management on ecosystem services. *Ecological Indicators*, vol. 21 (special issue), pp110–122.

Villa, F., Bagstad, K. J., Voigt, B., et al. (7 authors) (2014). A Methodology for Adaptable and Robust Ecosystem Services Assessment. *PLOS ONE* 9: DOI: 10.1371/journal.pone.0091001

Vitousek, P., Aber, J.D., Howarth, R.W., et al. (8 authors) (1997). Human alteration of the global nitrogen cycle: sources and consequences. *Ecological Applications*, vol 7, no 3, pp737–750.

Willemen, L., Veldkamp, A., Verburg, P. H., Hein, L., and Leemans, R. (2012). A multi-scale modelling approach for analysing landscape service dynamics. *Journal of Environmental Management*, vol 100, pp 86–95.

Zhang, W., Ricketts, T. H., Kremen, C., Carney, K., and Swinton, S. M. (2007). Ecosystem services and dis-services to agriculture. *Ecological Economics*, vol 64, no 2, pp253–260.

13

INDICATORS FOR ECOSYSTEM SERVICES

Felix Müller, Benjamin Burkhard, Ying Hou,
Marion Kruse, Liwei Ma and Peter Wangai

Introduction

Indicators are depictions of system qualities, quantities or states, which are not directly accessible by the observer. Ecological indicators provide aggregated information on phenomena within human-environmental systems to characterize environmental management options. They provide signals that relay complex messages in a simplified and useful manner, providing communication tools in environmental management (Dale and Beyeler, 2001, Müller and Wiggering, 2004, Turnhout et al., 2007, Niemeijer and de Groot, 2008, ten Brink et al., 2011, Müller and Burkhard, 2012).

Basic features of ecosystem indication

As the target of ecological indication is the provision of quantitative information for decision-making processes, ecosystem-based indicators have to represent the complex interactions between biotic and abiotic components. The indicandum requires a holistic approach, linking structures and functions of environmental entities. As a consequence, indicator sets are necessary to cope with the enormous complexity and to provide essential arguments for trade-offs throughout planning processes. Indicator derivation, application and valuation for the representation of ecosystem states and viabilities are thus complex, interdisciplinary and multidimensional procedures.

Derivation and definition of indicators

The first step of the process is indicator selection. As we are operating at a complex systems level it is advisable to use the subsequent steps of systems analysis for this purpose. In the beginning the problem and the purpose have to be defined in detail because they provide the indicandum – the object of indication (see Figure 13.1). Thereafter the developer has to decide upon the spatial and temporal boundaries of the indication, the scales and the aspired level of complexity. These decisions are followed by determining the influencing elements, relations and constraints. Figure 13.1 shows a conceptual diagram that illustrates the subsequent steps.

The result is a better comprehension of the indicandum, which now can be used to make indicator proposals and optimize these proxies with reference to scientific correctness, applicability and data availability. These procedures will include changes in the whole derivation

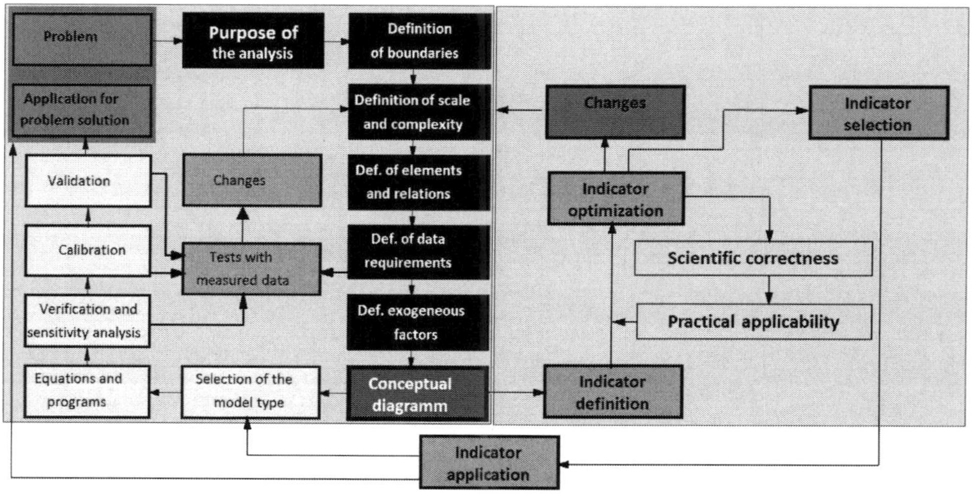

Figure 13.1 A systems-based scheme for general indicator derivation and definition.

layout and will finally lead to a selection and application of the indicator for modelling purposes, monitoring interpretations, scenario outcomes or directly for the solution of the initial problem.

Quality of indicators

The resulting indicator set can be characterized by different suitabilities and qualities. In general, good indicators should communicate information of high "scientific correctness" and high "practical applicability" with an optimal applicability (Kandziora et al., 2013, Dale and Beyeler, 2001, Müller, 2005, van Oudenhoven et al., 2012). Each indicator should be clear, understandable and comprehensive. Due to the fragility of the indicandum, an indicator should be sensitive to the slightest change of the indicated system. Thus, the scalability of an indicator is vital for purposes of measuring and mapping changes with spatio-temporal explicitness. An indicator should also be integrative, have a known response to stress and disturbance and a low variability. Good indicators should therefore follow several criteria, which are summarized in Table 13.1.

Recent approaches to indicate ecosystem services

In general, ecosystem services are most often derived from biophysical ecosystem properties, and discerned utilizing ecological data. Studies referring to the ecosystem service potential are mainly based on landscape data to reflect the capacity of ecosystems to provide specific services. The ecosystem service flows are mainly based on direct measurements, questionnaires and statistics, while the demand for ecosystem services is often clarified through the application of economic methodologies. Besides these empirical methods, expert opinions and systems play an important role in ecosystem service assessments, and in recent years there has been an increasing application of modelling and mapping procedures. Value-transfer and economic techniques are often used for applying monetary values (see also Badura et al., 2016; Gomez-Baggethun et al., 2016; Potschin and Haines-Young, 2016b; Turner, 2016).

Table 13.1 Some attributes of indicator quality.

Scientific correctness	*Practical applicability*
A clear representation of the **indicandum**	A high **political relevance** concerning management options
A clear proof of relevant **cause–effect relations**	**Direct relationship** to respective management actions
An optimal **sensitivity** of the representation	A high **comprehensibility** and public transparency
A high **transparency** of the derivation strategy	Strong **acceptability** by users/stakeholders
A high v**alidity**, accuracy, precision, representativeness	An orientation towards (quantitative) **environmental targets**
An optimal degree of **aggregation**	A capacity to **communicate** information clearly
An appropriate measurability and high **data availability**	A link with information on the **normative loadings** in the applied indicators systems
A good fulfilment of **statistical requirements**	A relationship to **long–term** trends and applicability for **early warning** purposes

Source: after Müller and Wiggering, 2004

Frequently used ecosystem service indicators

Considering the frequencies with which ecosystem service indicators are used in the literature, some basic trends are visible: regulating ecosystem services, especially global climate regulation and nutrient regulation, are the most frequently assessed services (Feld et al., 2009). Although their determination is linked with complicated ecophysiological methodologies, the translation of specific functions into ecosystem services, as well as the recent problems in environmental policy, supports these applications. In terms of provisioning ecosystem services, food supply and water use are the most common. Cultural services are often less clear in their definitions and quantification methods (Hernández-Morcillo et al., 2013).

Ecosystem service indicator sets

One significant feature of approaches to ecosystem service indication is that single indicators are of limited use. It is necessary to work with indicator sets or bundles in order to fulfil the requirements of complexity on the one hand and trade-off suitability on the other. The resulting indicator sets are very diverse as they include descriptive aspects as well as evaluative items (Müller and Burkhard, 2012). This diversity, and the high number of different service categorizations, results in a multitude of different indicator sets. The most commonly applied indicator sets are provided by The Economics of Ecosystems and Biodiversity (TEEB) initiative, (TEEB, 2010) and the Common International Classification of Ecosystem Services (CICES) (Haines-Young and Potschin, 2013).

As a global study to assess the economic impact of global loss of biodiversity, TEEB proposes a classification system based on four categories: provisioning, cultural, regulating and habitat services. Since the specific focus of TEEB is on economic assessment, indicators are needed that are suitable to indicate the economic value in relation to the changes in ecosystems and biodiversity. These indicators must be convertible into economic terms.

The CICES framework is the most comprehensive ecosystem service indicator set currently proposed. CICES has been framed around human needs and primarily describes ecosystem outputs directly contributing to human well-being. Nevertheless, information about supporting or intermediate ecosystem services can be integrated. Abiotic outputs from nature (such as mineral/non-mineral abiotic substances, renewable abiotic/non-renewable energy sources) have been defined as not fundamentally depending on living ecosystem processes and are thus not ecosystem services. But due to their importance for trade-off analyses and mapping, they have been included in CICES as abiotic outputs from natural systems.

Another comprehensive ecosystem service indicator set ('Kiel' column in Table 13.2) has recently been presented by Burkhard et al. (2014), including indicators for 11 regulating, 14 provisioning and six cultural ecosystem services. Their assessment framework distinguishes between indicators for ecosystem service potentials, flows (*de facto* used services) and societal demands for ecosystem services. The information such indicators provide is highly relevant for ecosystem service trade-off assessments. Table 13.2 gives a simplified overview of the differences between the classifications mentioned before.

Table 13.2 Overview of selected ecosystem service indicator systems.

	TEEB	*CICES*	*Kiel*
General distinctions			
Ecosystem service categories	Provisioning, Regulating, Habitat or Supporting, Cultural	Provisioning, Regulation & Maintenance, Cultural, Abiotic outputs from natural systems	Regulating, Provisioning, Cultural
Amount of ecosystem services considered	17	53 classes in 22 groups	28
Indicator quantification units suggested	Yes	No	Yes
Indicator peculiarities	Indicators should be convertible into economic values	Focusing on outputs directly contributing to human well-being	Distinction between ecosystem potentials, flows and demands
Exemplary services			
Regulating ecosystem service: climate regulation	Climate regulation by carbon sequestration: net annual rate of atmospheric carbon stored (t C/ha*yr)	Global climate regulation by reduction of greenhouse gas concentrations by amount, concentration or climatic parameter	Amount of methane, carbon dioxide and water vapor taken up by vegetation, soils and marine systems (t CO_2/ha per year)
Provisioning ecosystem service: timber	Timber production: dry matter productivity (kg DM/ha*yr)	Material by amount, type, use, media (land, soil, freshwater, marine)	Ecosystem service flow: Harvested wood (solid m^3/a; volume/a); Yield (€/ha per year)
Cultural ecosystem service: tourism	Nature-related outdoor tourism: protected area visitor numbers (n/yr)	By visits/use data, plants, animals, ecosystem type	Number of facility visitors (n/facility per year); Turnover from tourism (€/ha per year)

Source: based on TEEB 2010; Haines-Young and Potschin, 2013; Burkhard et al., 2014

Ecosystem service indicators for different ecosystem types

Indicators for terrestrial ecosystem services

Globally, terrestrial ecosystems cover 147x10^6 km^2 of total surface area (Vitousek et al., 1986), including forests, grasslands, rangelands, mountainous landscapes, arid and desert ecosystems, swamps and marshes. The respective indicators for provisioning services consequently show a widespread range of products such as energy biomass, timber, fiber and resins, food crop production per hectare, wildlife and livestock production (honey, meat, skins, bones, trophies), pharmaceutical products, colour pigments, volumes of water in cubic meters or numbers of employment positions in various related sectors (Burkhard et al., 2009; Layke et al., 2012; Müller and Burkhard, 2012; Fagerholm et al., 2012; de Wit et al., 2012; Busch et al., 2012). Indicators for regulating services include fluxes in atmospheric gases (CO_2, NO_2, SO_2), carbon stock in standing forests vegetation and soils, micro-climate regulation (temperature, precipitation, evapotranspiration, wind), top soil protected from erosion, reduced sedimentation in nearby rivers and dust particles filtered by vegetation, groundwater quality and recharge. Major indicators for cultural services in terrestrial ecosystems include the numbers of sacred sites, heritage sites, public parks and reserves, and measures for the aesthetics of natural ecosystems. As different ecosystem types comprise different particularities in their ecological settings, some of these special features are highlighted in the following paragraphs.

Urban ecosystems. In most urbanized areas of the world, goods and services are supplied from the surrounding natural environments. They are harnessed for industrial, social, economic, commerce and trade development. Provisioning ecosystem services in urban areas are mainly indicated by urban crop production, clean water supply, wood/biomass fuel, biotic construction materials (timber, rafts), art and craft materials and reeds (Fagerholm et al., 2012; de Wit et al., 2012; Layke et al., 2012). Most urban activities, such as transportation, mining or manufacturing, can cause significant environmental degradation and pollution. Therefore, regulating ecosystem services in urban areas is crucial. Predominant indicators for urban regulating ecosystem services include the amount of soil loss by erosion, fluxes of atmospheric gases, loads of solid particles in air, ground water purity, temperature and precipitation fluctuations, magnitude and lapse period of characteristic storms. Due to pollution, high population density and congestion in the urban ecosystems (Pickett and Grove, 2009), the demand for cultural ecosystem services is ever increasing. Respective indicators include features of natural areas and urban greenspace, their accessibility and their recreation potential.

Agricultural ecosystems. Agricultural ecosystems are expanding globally due to high human demand on provisioning services (food, fiber, fuels). Consequently, many other terrestrial ecosystems have been converted into agricultural areas (MA, 2005). Their intensive use often affects regulating and cultural services negatively (e.g. nutrient loss, erosion, decrease in visual quality), calling for trade-off assessments. Since agricultural ecosystems are the main sources of provisioning services, specific indicators are related to food production, e.g. crop production, livestock production and fuel production. Many of these provisioning services can be measured in monetary units, as they are tangible and traded on various markets.

Forest ecosystems. Forest ecosystems account for 22.24% of the earth's terrestrial area (Vitousek et al., 1986), providing multiple ecosystem services. The associated indicators are mainly related to timber production, harvested biomass for energy, wild food provision, climate regulation, fire management, air quality regulation, noise reduction, water purification and recreational and aesthetic values. In urban forests, regulating services such as air filtration, micro climate regulation, noise reduction and rainwater drainage, and cultural services such as recreational and aesthetic values, provide special significance (Bolund and Hunhammar, 1999, Gamfeldt et al., 2013, Ninan and Inoue, 2013).

Indicators for aquatic ecosystem services

Marine ecosystems. The marine ecosystem is composed of the major oceans and their connectors and extensions (Costanza et al., 1997) and covers approximately 70% percent of the Earth's surface. This vastest ecosystem of the world contributes more than 60% of the total economic value of the biosphere, of which the coastal zone is particularly significant (Costanza et al., 1997; Martínez et al., 2007). Scientists have paid most attention to food provision, particularly fisheries (Alcamo et al., 2005). Typical indicators of marine food provision include total landings or catch per unit effort and number of viable fisheries (Liquete et al., 2013). Water storage and provision are another two major provisioning services of marine ecosystems. Biotic materials and biofuels are further important provisioning services, which can be indicated for example, by the amount of fuel and amount of timber used from mangroves and the amount of ornamental sources used (Bohnke-Henrichs et al., 2013). The indicators of water purification focus on the presence of excess nutrients (eutrophication) or suspended particulate matter (Hyytiäinen et al., 2008). Coastal protection indicators generally refer to the presence of biotic structures that disrupt water movement. Another critical regulating service, climate regulation, is typically indicated by CO_2 uptake and accumulation or sequestration rates. The main cultural services, recreation and tourism, can be indicated by the number of visits to an area (Tomlinson et al., 2011).

Freshwater ecosystems. Freshwater ecosystems include standing-water or lentic ecosystems (e.g. lakes and ponds), running-water or lotic ecosystems (e.g. streams and rivers) and inland wetlands, whose water levels fluctuate seasonally or annually. Although freshwater ecosystems cover a relatively small proportion of the earth's surface compared with marine and terrestrial ecosystems, the ecosystem services they provide to humans exceed their relative area.

River and lake ecosystems: Freshwater supply is the principal service delivered by river and lake ecosystems and can be indicated by, for example, water withdrawal amount (Burkhard et al., 2014). Humans capture and harvest a large amount of food from rivers and lakes. For this service, the amount of freshwater food harvested serves as a typical indicator. Many rivers make key contributions to energy supply by generating hydropower. The respective indicators include the amount of hydropower generation and the respective market value (Wang et al., 2010). For the most important regulating services of river and lake ecosystems, water purification and waste treatment, the associated indicators include biochemical degradation capacity of COD and the amount of N and P removed (Bohnke-Henrichs et al., 2013). Natural rivers and lakes attract people due to their aesthetic values, coupled with a sense of wildness (Harrison et al., 2010, Everard et al., 2010). Rivers and lakes are often sites of various kinds of recreational activities, such as swimming, bathing, canoeing or angling, and the numbers of visitors are used as indicators.

Inland wetlands. Freshwater wetlands are areas permanently covered by shallow freshwater within part of the annual cycle. Although wetlands cover only about two percent of the surface of the earth, they play a relatively important role in the global cycling of sulfur, nitrogen and phosphorus, as well as carbon, due to the aerobic and anaerobic stratification of wetland. Wetlands are important carbon sinks and may account for as much as 40% of the global reserve of terrestrial carbon (Sheng et al., 2004). This property enables wetlands to regulate global climate, which can be indicated by, for example, the amount of carbon retained per year (Fennessy and Craft, 2011). Another important regulating service is natural hazard regulation, which is mainly dependent on a functioning floodplain (Harrison et al., 2010). The applied indicators include the number of floods causing damages and the reduction of peak discharges (Burkhard et al., 2009). Wetlands also function in water purification by retaining water pollutants. Indicators such as elements removed from water and pollutant concentrations of the river downstream the wetland can be used to represent this service. Besides regulating services, inland wetlands provide some cultural and provisioning services, such as recreation and tourism, aesthetic values and food supply (e.g. rice from paddy fields, fish, shrimps or ducks).

Indicator concepts in development

Ecosystem services are highly complex and embedded in adaptive human-environmental systems. Therefore, one-size-fits-all solutions for ecosystem service assessments are not easy to be found, and several further concepts are under discussion that will impact indicator development. Three salient aspects of conceptual debate are described below.

- **Distinguishing ecosystem service potentials, flows and demands**

Each ecosystem has a certain potential to supply services based on its characteristic properties and resulting functions. Humans can harness these ecosystem service potentials by different actions, such as fostering land use change. These activities trigger flows of ecosystem services from nature to society, satisfying multiple human demands. Trees in a forest (standing biomass), for example, hold high potential for timber supply. Only if the trees are harvested and further processed are flows of timber and related products activated to benefit humans in the form of goods. In order to gain a more realistic picture of real processes and existing flows, it is important to distinguish between ecosystem service potentials, flows (*de facto* use) and demands (Bastian et al., 2013; Villamagna et al., 2013) and to provide separate indicators for their assessment (Burkhard et al., 2014).

- **Distinguishing intermediate and final services**

The supply of ecosystem services results from complex interactions among species and their abiotic environment initiated or supplemented by human activities. Different processes and actors are normally involved in the generation and use of an ecosystem service at different stages. Food provisioning ecosystem services are, for instance, based on numerous regulating (e.g. nutrient cycling, water and climate regulation, pollination), provisioning (freshwater) and cultural (knowledge systems) ecosystem services. Thus, the final food provision is based on several intermediate services (after Fisher et al., 2009; Boyd and Banzhaf, 2007). When summing up all these services' values to a total (economic) value, risks of double counting and erroneous indication may occur. Therefore, the distinction between different stages of ecosystem service delivery and respective indicators can provide relevant aspects for assessments (Burkhard et al., 2014).

- **Distinguishing ecosystem and environmental services**

There are several terms and definitions under discussion, as the classification of ecosystem services is not fully consistent and a flexible approach to deal with case study-specific features is necessary. Environmental services are for example understood as having no direct ecological processes responsible for their generation (Haines-Young and Potschin, 2010). However, abiotic energy resources and mineral extraction are included in CICES (see above) and in Kandziora et al. (2013) are regarded as having impacts on decision-making and planning. In a land cover-based ecosystem services assessment method, Burkhard et al. (2014) apply and discuss abiotic energy resources and mineral extraction, while other authors exclude these (abiotic) provisions from the group of ecosystem services.

Strategic position of ecosystem service indicators

Indicators in the ecosystem service cascade

The 'ecosystem service cascade', introduced by Haines-Young and Potschin (2010; see also Potschin and Haines-Young, 2016a) links biodiversity, ecosystem services and human well-being to underline the coupled human-environmental system in which the concept of ecosystem services is placed. Therefore, indicators are needed to assess the ecological functions (e.g. by ecological integrity; see Müller, 2005, Müller et al., 2012), ecosystem services and human well-being (benefits and values, OECD, 2011). Therefore, an overall indication is demanding for indicators of economic, social and personal well-being. Respective indicators are proposed in Kandziora et al. (2013).

In Figure 13.2, a linkage with the Driver Pressure State Impact Response (DPSIR) indicator framework is demonstrated. Thereby the variables of the subsystem "ecosystems and biodiversity" are assigned to the indication of the system's state. A change in these variables will provoke a modification in the impact section, which initially is represented by the provision of ecosystem

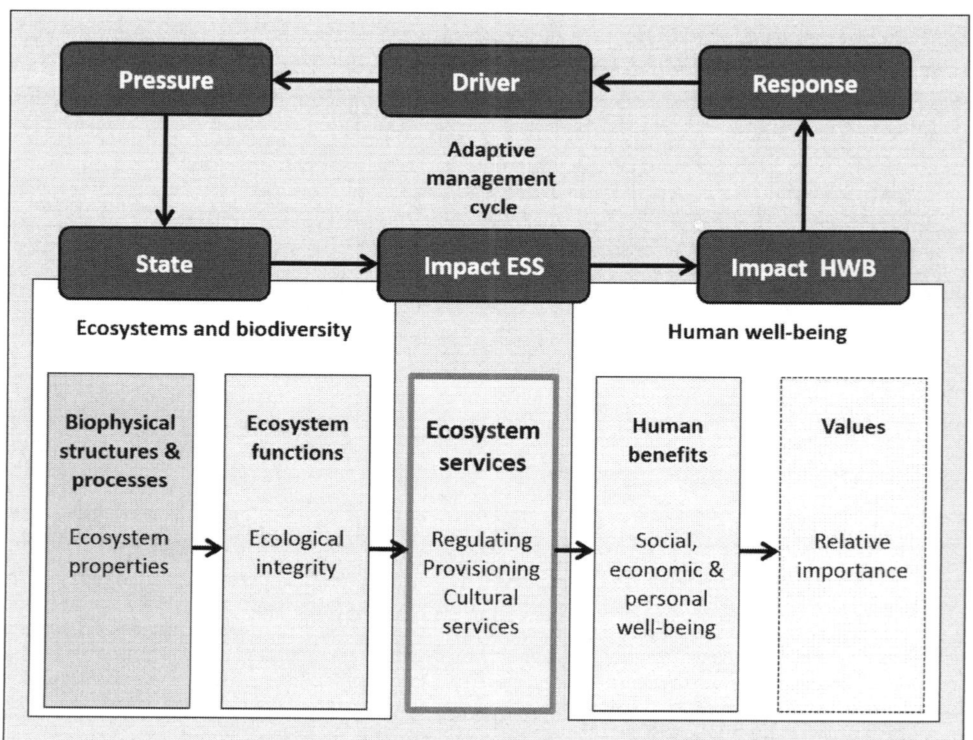

Figure 13.2 The "ecosystem service cascade" embedded in an adaptive management cycle.
Source: after Potschin and Haines-Young (2011), and Müller and Burkhard (2012)

services. If they are changed, by definition, there must be a further impact on the variables of human well-being. Once the impacts on human well-being are perceived, a management measure (response) may take place. The response can change the drivers and pressures, resulting in an adaptive management cycle.

Indicator aggregations and indicator dimensions

Ecosystem service indicators are constructed to provide information for decision-making processes in human-environmental systems. This information is demanded on different levels, by, for example, local managers, stakeholder groups working on planning trade-offs, regional administrations or national policymakers. All of these groups have different demands for specific indicator characteristics and feasibilities. Therefore, the results of single variables in indicator sets often have to be aggregated to provide a smaller and more manageable group of values. The respective steps of aggregation are connected with several problems, because there is a loss of information and precision with each abstraction or indicator combination. Therefore there is a developmental conflict between transparency and theoretical correctness on the one hand and simplicity and empirical reliability on the other.

Furthermore, there is a problem of multiple dimensions. While provisions can be quantified in t/ha*a or in monetary currencies, regulations enfold many different dimensions (tons fixed carbon, number of pollinators, retention of soil in t/ha*a, etc.), and cultural services are characterized by numbers of visitors, travel distances or estimations of landscape beauty. Thus the question is if the developer will translate the results into one currency (money, time, relative change, relative difference from a reference state, relative scores) or if the user will be confronted with several dimensions. Of course, suitable illustrations (e.g. ecosystem service profiles, radar diagrams, maps) can be helpful, but also in these depictions there is a hidden weighting problem. Do all indicators have the same significance? Does the change of the value of indicator A by 10% have the same relevance as the 10% change of the value of indicator B? These questions are discussed fervently; there is no theoretical principle, and thus the solutions are case-study-specific.

The role of scales in ecosystem service indication

In ecology, scale refers to the spatial or temporal dimension of an ecological phenomenon. The dynamics of ecological patterns and processes can be most effectively studied at their individual, characteristic scales. In ecosystem service assessments, a change of the study scales may cause changes in the evaluation results (Wu and Li, 2006). Therefore, the selection of the research scale is considerably important.

Some ecosystem services, such as most provisioning services, can be assessed at multiple spatial scales, from local to global. However, some others, such as local climate regulation and spiritual inspiration, are subject to distinctive local or regional scales. Special indicators are needed to indicate some services at distinct spatial scales. Different regions can, for example, have different characteristic air pollutants. Thus, distinctive air quality indicators are needed to represent the air quality regulation service of the respective region. These roles of scale also apply to temporal scale issues. For example, the delivery of most provisioning services depends on the growing seasons of organisms. A further example refers to the reliance of local climate, air quality and natural hazard regulations on the annual cycle of temporal fluctuations of, for example, vegetation cover, weather and phenology. Assessments with annual average values may conceal the temporal fluctuations (seasonal, monthly, daily) of the service provisions and overlook the temporal mismatches of ecosystem service supply and demand (e.g. demands for air quality regulation, recreation and tourism throughout the year but few supplies of the services in winter). Different temporal scales of indication may also need distinctive indicators. For instance, to indicate flood regulation at small temporal scales (e.g. one or two years), the reduction of peak discharges is an appropriate measurement, but to deal with large temporal scales such as tens of years, some other indicators, e.g. the number of annual floods, are more effective.

Discussion and conclusion

This volume as a whole demonstrates the enormous potential of the ecosystem service approach in environmental science and application. Referring to the role of ecosystem service indicators, there are – of course – several limitations and problems which hopefully can be reduced and solved in the near future.

One class of these limitations arises from the multitude of sources of uncertainty in ecosystem service indication. A study by Hou et al. (2013) has pointed out several uncertainties related to ecosystem service mapping and assessments and which in an important sense are reflected in the way definitions are established and articulated by the developer throughout the systems analytical process (see Figure 13.1). Has the correct scale been chosen? Does the delineation of the research area fit the purpose and the complexity of the problem? Is the definition of elements and relations suitable? Are the correct exogeneous factors chosen? What is the accuracy of the chosen land cover data? Do they provide satisfactory resolutions and extents? Is the quality of the input data satisfactory? Are the translation algorithms for defining ecosystem service potential correct? Where do uncertainties lie in expert opinions? Which information is lost throughout indicator aggregation? The list could be extended.

Yet all of these uncertainties can also be understood as challenges. If we want to indicate ecosystem services, we have to cope with the complexity of the respective human-environmental system, we have to be aware of the high amounts and strengths of interrelations and potential trade-offs, we have to face the different components of the ecosystem service cascade and consider the indication of ecosystem functions and structures which should be conducted in parallel. Finally, we can come back to some of the features of good indicators from Table 13.1 and derive research demands from those requirements:

- We should intensify the check for a clear and sufficient representation of the indicandum by the indicator and we should be aware that very often the attained information is insufficient, e.g. by measuring crop production and using the outcome to indicate the overall potential of an ecosystem for provisioning services. At least the user should know that that indication is incomplete.
- We should use systems analytical techniques for indicator definition because this ensures that the main causal relations between the indicator and the regulating variables are taken into account.
- We should make tests of sensitivity of the chosen indicators, because they can either "over-react" by showing very high variabilities or they could be well-buffered, not reacting at all to actions or interventions in the system.
- The user should be informed about the scientific background and the technique of the derivation strategy in detail. This will enable an adequate valuation of the results to be determined.
- The developer should seek an adapted degree of indicator aggregation, and the users should be informed about the resulting information loss and impreciseness.

- We should be aware that data availability plays an enormous practical role. But this should not mean that we retreat from theoretical analysis. Only if the optimal indicator setting is known can we assess the quality of the resulting indicator set.
- In spite of the system's complexity, the resulting indicator set should be as comprehensible for the user as possible. The results should be transparent, but not over-simplified. To find the optimal degree of aggregation and transparency, participatory indicator development represents a logical part of indicator development.
- Whenever possible, quantitative targets should be defined. Then an aggregated indication of the distances from these targets is a logical outcome and easy to understand.
- Finally, the developer should inform the user about the normative assumptions which are implied by the indicator set. This does not only include the obvious normativeness but also the basic philosophy of the indicator development.

These long lists of methodological challenges and aspirations suggest that the indication of ecosystem services is still a very embryonic enterprise, that many problems have to be faced and that several developmental steps and decisions are still ahead. Nevertheless, the available results show that ecosystem service indication has an enormous potential to translate the basic results of research into practical application.

References

Alcamo, J., van Vuuren, D., Ringler, C., Cramer, W., Masui, T., Alder, J., and Schulze, K. (2005). Changes in nature's balance sheet: model-based estimates of future worldwide ecosystem services. *Ecology and Society*, vol 10, p19.

Bastian, O., Syrbe, R.-U., Rosenberg, M., Rahe, D., and Grunewald, K. (2013). The five pillar EPPS framework for quantifying, mapping and managing ecosystem services. *Ecosystem Services*, vol 4, pp 15–24.

Bohnke-Henrichs, A., Baulcomb, C., Koss, R., Hussain, S. S., and de Groot, R. S. (2013). Typology and indicators of ecosystem services for marine spatial planning and management. *Journal of Environmental Management*, vol 130, pp 135–145.

Badura, T., Bateman, I., Agarwala, M. and Binner, A. (2016). Valuing preferences for ecosystem-related goods and services. In Potschin, M., Haines-Young, R., Fish, R. and Turner, R.K. (eds) *Routledge Handbook of Ecosystem Services*. Routledge, London and New York, pp 228–242.

Bolund, P., and Hunhammar, S. (1999). Ecosystem services in urban areas. *Ecol. Econ.*, vol 29, pp 293–301.

Boyd, J., and Banzhaf, S. (2007). What are ecosystem services? The need for standardized environmental accounting units. *Ecological Economics*, vol 63, pp 616–626.

Burkhard, B., Kandziora, M., Hou, Y. and Müller, E. (2014). Ecosystem service potentials, flows and demands – concepts for spatial localisation, indication and quantification. *Landscape Online*. Available at: http://www.landscapeonline.de/wp-content/uploads/DOI103097-LO201434.pdf .

Burkhard, B., Kroll, F., and Müller, E. (2009). Landscapes' capacities to provide ecosystem services – a concept for land-cover based assessments. The official journal of the international association for landscape ecology, chapter Germany. *Landscape Online*, vol 15, pp 1–22.

Burkhard, B., Kroll, F., Nedkov, S., and Müller, F. (2012). Mapping supply, demand and budgets of ecosystem services. *Ecological Indicators*, vol 21, pp 17–29.

Busch, M., Gee, K., Burkhard, B., Lange, M., and Stelljes, N. (2012). Conceptualizing the link between marine ecosystem services and human well-being: the case of offshore wind farming. *International Journal of Biodiversity Science, Ecosystem Services & Management*, vol 7, no 3, pp 190–203.

Costanza, R., D'Arge, R., de Groot, R., Farberk, S., and Grasso, M. (1997). The value of the world's ecosystem services and natural capital. *Nature*, vol 387, pp 253–260.

Dale, V.H., Beyeler, S.C. (2001). Challenges in the development and use of ecological indicators. *Ecological Indicators*, vol 1, pp 3–10.

de Wit, M., Van Zyl, M., Crookes, D., Blignaut, J., Jayiya, T., Goiset, V., and Mahumani, B., (2012). Including the economic value of well-functioning urban ecosystems in financial decisions: evidence from a process in Cape Town. *Ecosystem Services*, vol 2, pp 38–44.

Everard, M., Jones, L., and Watts, B. (2010). Have we neglected the societal importance of sand dunes? An ecosystem services perspective. *Aquatic Conservation: Marine and Freshwater Ecosystems*, vol 20, pp 476–487.
Fagerholm, N., Käyhkö, N., Ndumbaro, F., Khamis, M. (2012). Community stakeholders' knowledge in landscape assessments – mapping indicators for landscape services. *Ecological Indicators*, vol 18, pp 421–433.
Feld, C. K., da Silva, P. M., Sousa, J. P., de Bello, F., Bugter, R., Grandin, U., Hering, D., Lavorel, S., Mountford, O., Pardo, I., Paärtel, M., Roämbke, J., Sandin, L., Jones, K. B., and Harrison, P. (2009). Indicators of biodiversity and ecosystem services: a synthesis across ecosystems and spatial scales. *Oikos*, vol 118, pp 1862–1871.
Fennessy, S., and Craft, C. (2011). Agricultural conservation practices increase wetland ecosystem services in the Glaciated Interior Plains. *Ecological Applications*, vol 21S, pp S49-S64.
Fisher, B., Turner, R. K., and Morling, P. (2009). Defining and classifying ecosystem services for decision making. *Ecological Economics*, vol 68, no 3, pp 643–653.
Gamfeldt, L., Snäll, T., Bagchi, R., Jonsson, M., Gustafsson, L., Kjellander, P., Ruiz-Jaen, M. C., Fröberg, M., Stendahl, J., and Philipson, C. D. (2013). Higher levels of multiple ecosystem services are found in forests with more tree species. *Nat. Commun.*, vol 4, p 1340.
Gómez-Baggethun, E., Barton, D.N., Berry, P., Dunford, R. and Harrison, P.A. (2016). Concepts and methods in ecosystem services valuation. In Potschin, M., Haines-Young, R., Fish, R. and Turner, R.K. (eds) *Routledge Handbook of Ecosystem Services*. Routledge, London and New York, pp 99–111.
Haines-Young, R. and Potschin, M. (2010). The links between biodiversity ecosystem services and human well-being. In Raffaelli, D. and Frid, C. (eds) *Ecosystem Ecology: A New Synthesis*. Cambridge University Press, Cambridge UK, pp 110–139.
Haines-Young, R. and Potschin, M. (2013). *Common International Classification of Ecosystem Services (CICES)*: Consultation on Version 4, August-December 2012. EEA Framework Contract No EEA/IEA/09/003. Download at www.cices.eu.
Harrison, P. A., Vandewalle, M., Sykes, M. T., Berry, P. M., Bugter, R., de Bello, F., Feld, C. K., Grandin, U., Harrington, R., Haslett, J. R., Jongman, R., Luck, G. W., Da Silva, P. M., Moora, M., Settele, J., Sousa, J. P. and Zobel, M. (2010). Identifying and prioritising services in European terrestrial and freshwater ecosystems. *Biodiversity and Conservation*, vol 19, pp 2791–2821.
Hernández-Morcillo, M., Plieninger, T., Bieling, C. (2013). An empirical review of cultural ecosystem service indicators. *Ecological Indicators*, vol 29, pp 434–444.
Hou, Y., Burkhard, B., and Müller, F. (2013). Uncertainties in landscape analysis and ecosystem service assessment. *Journal of Environmental Management*, vol 127, supplement, pp S117-S131.
Hyytiäinen, K., Ahtiainen, H., and Heikkilä, J. (2008). An integrated simulation model to evaluate national measures for the abatement of agricultural nutrients in the Baltic Sea. *Agricultural and Food Science*, vol 18, pp 440–459.
Kandziora, M., Burkhard, B., and Müller, F. (2013). Interactions of ecosystem properties, ecosystem integrity and ecosystem service indicators e a theoretical matrix exercise. *Ecological Indicators*, vol 28, pp 54–78.
Layke, C. (2012). Indicators from the global and sub-global Millennium Ecosystem Assessments: an analysis and next steps. *Ecological Indicators*, vol 17, pp 77–87.
Liquete, C., Piroddi, C., Drakou, E. G., Gurney, L., Katsanevakis, S., Charef, A., and Egoh, B. (2013). Current status and future prospects for the assessment of marine and coastal ecosystem services: a systematic review. *PLOS ONE*, 8.
MA (2005). *Ecosystems and Human Well-being: Synthesis*. Island Press, Washington DC.
Martínez, M. L., Intralawan, A., Vázquez, G., Pérez-Maqueo, O., Sutton, P., and Landgrave, R. (2007). The coasts of our world: ecological, economic and social importance. *Ecological Economics*, vol 63, pp 254–272.
Müller, F. (2005). Indicating ecosystem and landscape organisation. *Ecological Indicators*, vol 5, no 4, pp 280–294.
Müller, F., and Burkhard, B. (2012). The indicator side of ecosystem services. *Ecosystem Services*, vol 1, pp 26–30.
Müller, F., Burkhard, B., Kandziora, M., Schimming, C., and Windhorst, W. (2012). Ecological indicators: ecosystem health. In Jørgensen, S.E. (ed.) *Encyclopedia of Environmental Management, Volume 1*. CRC Press, Baton Rouge, LA.
Müller, F., and Wiggering, H. (2004). Umweltziele und Indikatoren. In *Umweltziele und Indikatoren*. Springer, Berlin and Heidelberg.
Niemeijer, D., and de Groot, R. (2008). A conceptual framework for selecting environmental indicator sets. *Ecological Indicators*, vol 8, pp 14–25.

Ninan, K. N., and Inoue, M. (2013). Valuing forest ecosystem services: what we know and what we don't. *Ecol. Econ.*, vol 93, pp 137–149.

OECD (2011). *How's Life?: Measuring Well-being*. OECD Publishing. Available at: http://dx.doi.org/10.1787/9789264121164-en

Pickett, T., and Grove, J. (2009). Urban ecosystems: what would Tansley do? *Urban Ecosystem*, vol 12, pp 1–8.

Potschin, M., and Haines-Young, R. (2011). Ecosystem Services: exploring a geographical perspective. *Progress in Physical Geography*, vol 35, no 5, pp 575–594.

Potschin, M. and R. Haines-Young (2016a). Defining and measuring ecosystem services. In Potschin, M., Haines-Young, R., Fish, R. and Turner, R. K. (eds) *Routledge Handbook of Ecosystem Services*. Routledge, London and New York, pp 25–44.

Potschin, M. and R. Haines-Young (2016b). Frameworks for ecosystem assessments. In Potschin, M., Haines-Young, R., Fish, R. and Turner, R. K. (eds) *Routledge Handbook of Ecosystem Services*. Routledge, London and New York, pp 125–143.

Sheng, Y., Smith, L. C., MacDonald, G. M., Kremenetski, K.V., Frey, K. E., Velichko, A. A., Lee, M., Beilman, D.W., and Dubinin, P. (2004). A high-resolution GIS-based inventory of the west Siberian peat carbon pool. *Global Biogeochemical Cycles*, vol 18, GB3004, doi:10.1029/2003GB002190.

TEEB (2010). *The Economics of Ecosystems and Biodiversity: Mainstreaming the Economics of Nature: A Synthesis of the Approach, Conclusions and Recommendations of TEEB*. Available at: http://www.teebweb.org/our-publications/teeb-study-reports/synthesis-report/

ten Brink, P., Gantioler, S., Gundimeda, H., Sukhdev, P., Tucker, G., and Weber, J.-L. (2011). Strengthening indicators and accounting systems for natural capital. In ten Brink, P. (ed.) *The Economics of Ecosystems and Biodiversity in National and International Policy Making*. Earthscan, London.

Tomlinson, B., Sastre, S., Blasco, D., and Guillén, J. (2011). The systems approach framework as a complementary methodology of adaptive management: a case study in the urban beaches of Barcelona. *Ecology & Society*, vol 16, no 4. Available at http://dx.doi.org/10.5751/ES-04484-160428

Turnhout, E., Hisschemöller, M., and Eijsackers, H. (2007). Ecological indicators: between the two fires of science and policy. *Ecological Indicators*, vol 7, issue 2, pp 215–228.

Turner, R. K. (2016). Ecological economics and ecosystem services In Potschin, M., Haines-Young, R., Fish, R. and Turner, R. K. (eds) *Routledge Handbook of Ecosystem Services*. Routledge, London and New York, pp 243–255.

van Oudenhoven, A., Pets, P., Alkemade, R., Hein, L., and de Groot, R. (2012). Framework for systematic indicator selection to assess effects of land management on ecosystem services. *Ecological Indicators*, vol 21, pp 110–122.

Villamagna, A. M., Angermeier, P. L., and Bennet, E. M. (2013). Capacity, pressure, demand, and flow: a conceptual framework for analyzing ecosystem service provision and delivery. *Ecological Complexity*, vol 15, pp 114–121.

Vitousek, P. M., Ehrlich, P. R., Ehrlich, A. H., and Matson, P. A. (1986). Human appropriation of the products of photosynthesis. *BioScience*, vol 36, pp 368–373.

Wang, G., Fang, Q., Zhang, L., Chen, W., Chen, Z., and Hong, H. (2010). Valuing the effects of hydropower development on watershed ecosystem services: case studies in the Jiulong River Watershed, Fujian Province, China. *Estuarine Coastal and Shelf Science*, vol 86, pp 363–368.

Wu, J., and Li, H. (2006). Concepts of scale and scaling. In Wu, J., Jones, K. B., Li, H., and Loucks, O. (eds) *Scaling and Uncertainty Analysis in Ecology*. Springer, Dordrect NL.

14
USING FUTURES-THINKING TO SUPPORT ECOSYSTEM ASSESSMENTS

Steven Cork

Introduction

The literature on futures-thinking includes diverse philosophical approaches and methods. Futures-thinking draws on a range of disciplines to acknowledge and address the psychological and other barriers to thinking openly and creatively about future possibilities and their implications for planning over multiple timeframes. A key component of most approaches to futures-thinking is the development of scenarios (internally consistent narratives designed to explore future uncertainties), which have proved to be particularly useful in ecosystem studies because they allow researchers and managers to explore the interactions among social, technological, ecological, economic, political and other factors influencing possible future human needs and the capacity of ecosystems to meet some of those needs.

The past decade and a half has seen an explosion of attempts to explain and categorise approaches to futures-thinking and the types of scenarios produced. These attempts have revealed a wide range of disciplinary viewpoints, which can be confusing to the uninitiated. This chapter attempts to distil key messages and guidelines from this literature in order to help researchers, policymakers and land managers decide when and how to apply futures-thinking when considering ecosystem services.

Systematic thinking about multiple futures is important for ecosystem assessments

The concept of ecosystem services draws on both economics and ecology to aid the consideration of human dependence on nature in decision-making (Daily, 1997). The main focus of ecosystem services research has been on defining and measuring the services and assessing the value of benefits that flow from them (e.g. Fisher *et al.*, 2009). Increasingly, however, the concept is being used to facilitate dialogue, social learning and better decisions at all levels (Pahl-Wostl, 2009; Australia21, 2014).

Thinking about the future is required in almost all applications of the ecosystem-services paradigm, whether economists are asking about marginal changes in ecosystem services or communities are asking about the consequences of alternative development trajectories for their local area. It is not surprising, therefore, that futures-thinking is found to be valuable in many projects considering links between ecosystems and human well-being (Bennett *et al.*, 2003; Peterson *et al.*, 2003; Alcamo, 2009c). It is surprising, however, how many projects and policies that influence, or are influenced by, human demands on ecosystems do not employ systematic futures-thinking (Cork, 2010; Bengston *et al.*, 2012; Australia21, 2014).

Futures-thinking is more than scenarios

Various terms are used to refer to the processes involved in thinking systematically about multiple possible futures (see 'Definitions' below). For consistency, 'futures-thinking' is used here to denote the full range of approaches. The methods associated with futures-thinking explore how external forces have interacted with human thinking, values and behaviours to shape the past and present and how such interactions might shape the future. These methods draw on a

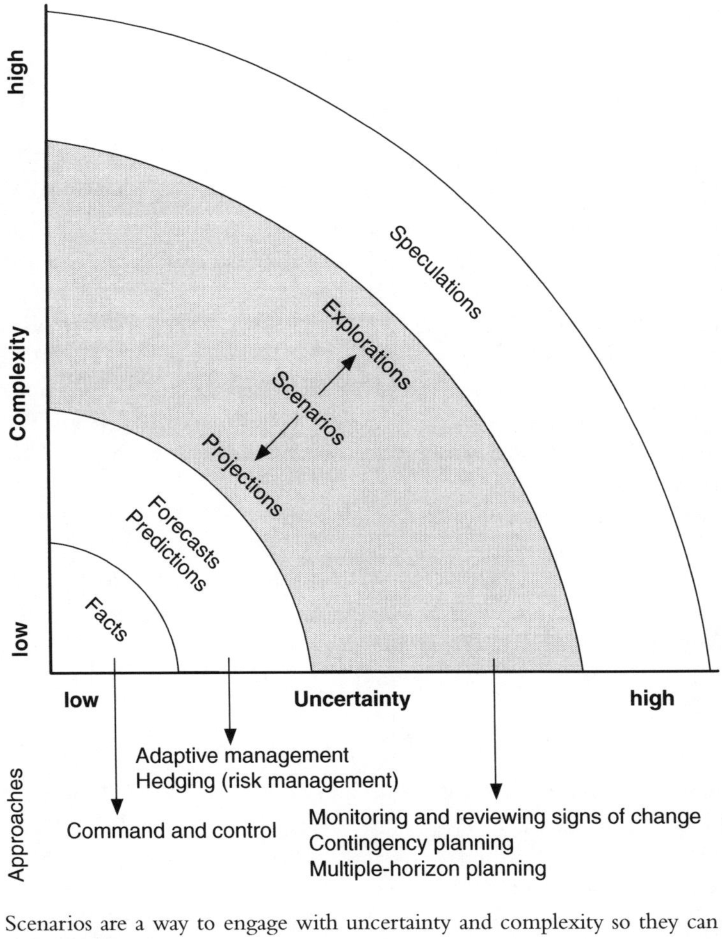

Figure 14.1 Scenarios are a way to engage with uncertainty and complexity so they can be taken into account in decision-making.

Source: adapted from Zurek and Henrichs, 2007 and Peterson et al., 2003

range of disciplines, including systems analysis, psychology and other social sciences, operations research, statistics, economics and others. As discussed later, the depth of thinking about the 'human condition' varies among approaches to futures-thinking. Development of scenarios is a common way to bring this thinking together, while encouraging interaction and social learning among stakeholders and connecting longer-term possibilities with immediate actions (Curry, 2009, Figure 14.1). The quality of scenarios is dependent on the quality of the analysis and thinking that precede them and the scenarios themselves can range in form from simple to elaborate; text-based to picture, video or performance based; qualitative to quantitative or, most often, mixtures of all of these. Rushing to scenarios is a key risk in futures-thinking (Slaughter, 2002).

Quantitative or semi-quantitative scenarios have been the main focus of ecosystem studies to date, because such studies usually involve comparisons between multiple possible future social-ecological situations, leading to conclusions about economic and social implications. Many futures-thinkers would argue that the deeper qualitative aspects of futures-thinking have not been employed as much as they could or should in relation to relationships between humans and the environment.

Definitions and origins

Definitions

Since the 1960s there have been attempts to create theory and a discipline around thinking about the future (Marien, 2002; Slaughter, 2002; Hines and Gold, 2013). This hoped-for discipline is most often called 'futures-studies', 'foresight' or 'strategic foresight' (Hines and Gold, 2013). The process is often referred to as 'scenario planning', although this is sometimes taken to mean the particular style of futures-thinking promoted by groups like the Global Business Network (Scearce et al., 2004). French futurists have coined the terms 'futuribilia' (from 'future' and 'possibilities') (Malaska and Virtanen, 2009) and 'la prospective' (Godet, 2012).

According to Slaughter (2009, p. 8), two factors distinguish futures studies from other research: 'Futures studies examine not only possible, but also probable, preferable, and wildcard futures, and, typically attempt to gain a holistic or systemic view based on insights from a range of different disciplines'.

Scenario development is a key tool used within futures-thinking (Börjeson *et al.*, 2006). Raskin *et al.* (2005, p. 36) define scenarios as: '. . . plausible, challenging, and relevant stories about how the future might unfold, which can be told in both words and numbers. Scenarios are not forecasts, projections, predictions, or recommendations. They are about envisioning future pathways and accounting for critical uncertainties'.

Origins

People have always used stories to explore the future (Molitor, 2009). Learned writing about futures-thinking appeared as early as the 16th century (Malaska and Virtanen, 2009; Bengston et al., 2012). But it appears that, while individuals have some raw capacity, organisations and societies do not display 'foresighting ability' unless it is deliberately cultivated (Slaughter, 2006). The last half of the 20th century has, in fact, been described as a journey 'from forecasting to foresight' (Curry, 2009), as futures-thinkers have recognised and addressed the illogical and risky tendency of humans to rely on prediction of the most likely future (van der Heijden, 1996; Taleb, 2007).

Modern refinement of techniques for developing scenarios from futures-thinking is generally said to have been pioneered by Herman Kahn, working with the RAND corporation and the US Military in the 1950s (Chermack *et al.*, 2001; Mietzner and Reger, 2005). Subsequently, scenario planning has been rooted in systems analysis and operations research (Molitor, 2009). In the 1960s and 1970s these ideas were developed and applied by various groups, including Royal Dutch Shell, the Club of Rome, the Stanford Research Institute, the French futures-thinkers and various universities.

In the years since, many approaches to futures-thinking have been developed (Bishop *et al.*, 2007; Alcamo, 2009c; Varum and Melo, 2010). Two major development paths are important to consider: (1) increasingly sophisticated quantitative modelling to engage with the complexity and uncertainty of the biophysical world (Alcamo, 2009a); (2) a focus on social learning and deep thinking about worldviews and human consciousness and how these affect how societies evolve (Inayatullah, 2008; Slaughter, 2008; Stewart, 2008). Increasingly, these approaches have been merged to consider environment-human interactions (Raskin *et al.*, 2005; Rothman, 2009).

Approaches

When might futures-thinking be successful (and when might it not)?

Scearce *et al.* (2004) developed a decision tree to diagnose when 'scenario thinking' (their term) is the most appropriate tool to use (Figure 14.2). Two key insights are that:

- Scenarios (or futures-thinking generally) are rarely successful without support from leaders and enthusiasm from those managing the project; and,
- The value of developing scenarios is as much, or more, in the process as in the final product.

Senior managers might be unsupportive if they fear that futures-thinking might question and/or side-step their judgment (Wack, 1985) or if they don't want to look far into the future (Molitor, 2009). Key risks include failure to challenge established ideas sufficiently, inadequate time or other resources, an inappropriate mix of participants, confusion about the best approach to take, and facilitators not having adequate talent or experience (Bradfield *et al.*, 2005; Mietzner and Reger, 2005; Molitor, 2009).

To address these risks, rigorous analytical processes have been documented (Bishop *et al.*, 2007; see methods section below) and theory from systems-thinking and social sciences has been incorporated into futures-studies (Stewart, 2008; Alcamo, 2009c). Scenarios embedded in high quality futures-thinking and participatory approaches are seen as powerful tools for integrating knowledge, scanning for future possibilities, and taking account of the different worldviews that people bring to complex issues (Swart *et al.*, 2004; Mietzner and Reger, 2005; UK National Ecosystem Assessment, 2011; Raupach *et al.*, 2013; Judge, 2008). Scenarios can play as much a role in making sense of the present as in anticipating futures (Curry, 2009).

As discussed later in this chapter, most concerns about futures-thinking can be avoided by being clear about how it will be used and having realistic expectations – futures-thinking cannot predict the future but can encourage mental preparedness, flexibility and innovation for possible eventualities (Royal Dutch Shell, 2013).

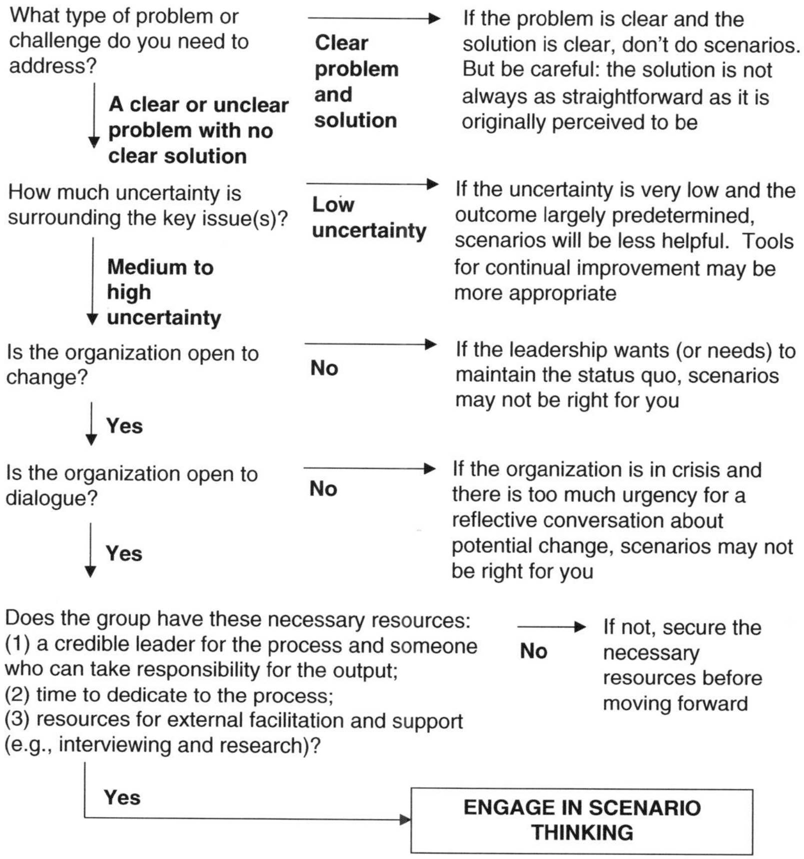

Figure 14.2 Steps in determining whether 'scenario thinking' is an appropriate tool.

Source: Scearce et al., 2004

Typologies

Many classifications of futures-thinking approaches and scenario types have been proposed, based on diverse criteria, including: social interests of futurists; focus on worldviews/consciousness versus drivers of change external to individuals; thinking approach (intuitive, deductive, probabilistic); type of future (probable, possible, preferred etc.); types of knowledge used; design; time-horizon; quantitative versus qualitative scenarios; complexity; geographic scale; approach to risk; forward from the present versus backward from the future; participative versus expert-driven; breadth of driving forces and key uncertainties explored; and modes of communication and action (Marien, 2002; van Notten *et al.*, 2003; Stewart, 2008; Rothman, 2009; Varum and Melo, 2010). Some key typologies are considered below.

Different philosophical approaches to futures-thinking

During the evolution of futures-thinking, several different philosophical approaches have emerged, distinguished by the degree of focus on human consciousness versus external driving forces (Figure 14.3).

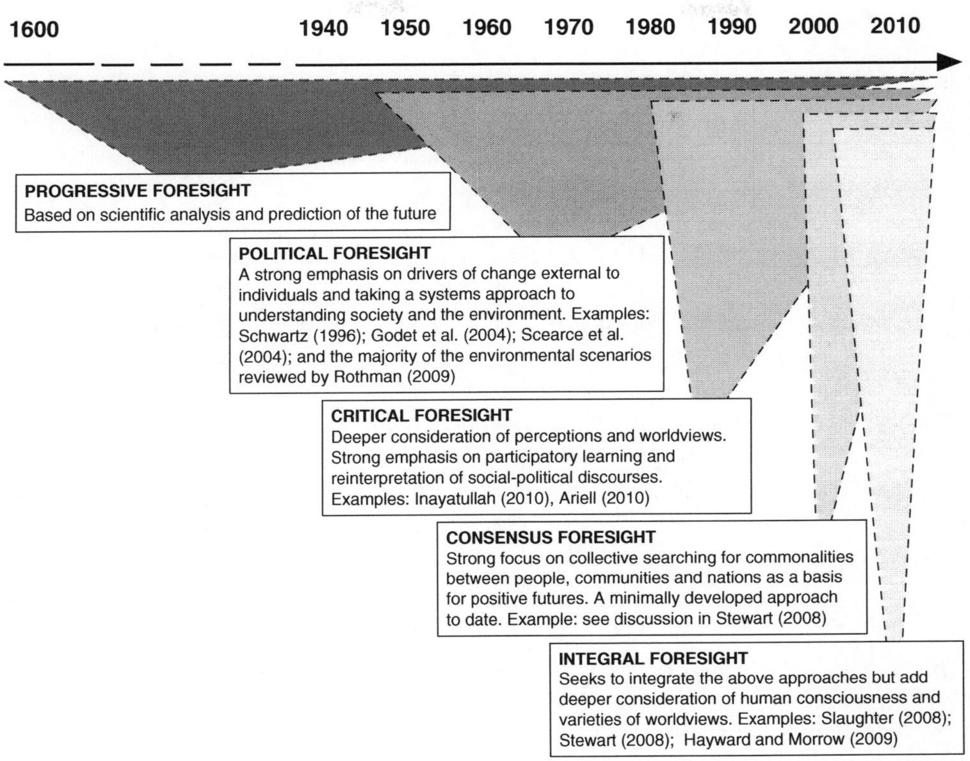

Figure 14.3 Evolution of philosophic approaches to futures thinking.

Source: interpreted from Stewart, 2008

All of these approaches are in current usage, although progressive foresight is becoming rare and, so far, there are very few examples of consensus foresight. The application of critical and integral foresight is increasing, but the majority of published futures-studies dealing with relationships between humans and nature are of the political foresight type (Raskin *et al.*, 2005; Rothman, 2009). Box 14.1 illustrates how critical and political foresight approaches lead to different visions of the future but also how different methods for political foresight yield different visions. All of the approaches illustrated in Figure 14.3 and Box 14.1 could be useful in ecosystem studies, depending on the social-ecological context.

Box 14.1 Examples of some approaches to developing scenarios and how they draw on different philosophies of futures-thinking

Curry and Shultz (2009) compared four approaches to building scenarios about the future of European civil society. Causal Layered Analysis is an example of 'critical foresight'. The other three are examples of 'political foresight' (as explained in Figure 14.3).

The 2x2 matrix approach develops scenarios around the four combinations of two 'critical uncertainties' (in this case: (1) 'top down' versus 'bottom up' relationships between individuals and institutions; and (2) a focus on individual versus collective interests in society).

Causal Layered Analysis (CLA) was used to build scenarios around different dominant worldviews within society: (1) confidence in global institutions as 'guardians of the future'; (2) an individual focus on 'my world' rather than 'our world', leading to a fragmented world of low cooperation; (3) a European 'enlightenment tradition', in which civil society protects us against powerful interests.

The Manoa approach assumes that futures are generated by the turbulent intersection of multiple trends. The one scenario produced here considered interactions among new spiritual processes, transformations in conflict resolution, public safety, and security, and increases in the length and quality of life.

Archetype scenario (or 'incasting') approaches are discussed in more detail later. In this study participants considered five archetype futures: 'Continued Growth'; 'Ideological Exclusionism'; 'Environmental Sustainability'; 'High Technology Transformation'; and 'Spiritual Transcendence'.

Approaches to futures-thinking also have been distinguished by the 'social interests' of practitioners. In Australia, 'pragmatic' interests ('carrying out today's business better') are much more common than 'progressive' ('contains some sort of explicit commitment to systemic improvement') or 'civilisational' ('seeks to understand the possible characteristics of the next level of civilisation') interests (Ramos, 2004; Slaughter, 2009). A similar conclusion might be valid for ecosystem services studies.

A purpose-focused typology

Börjeson *et al.* (2006) built on a typology of scenarios by van Notten *et al.* (2003) with a more explicit focus on purpose. This strongly operational focus is likely to be useful to those wishing to consider how ecosystem processes, services and benefits might be influenced by future trends and/or how societal goals might be achieved in uncertain futures (Figure 14.4).

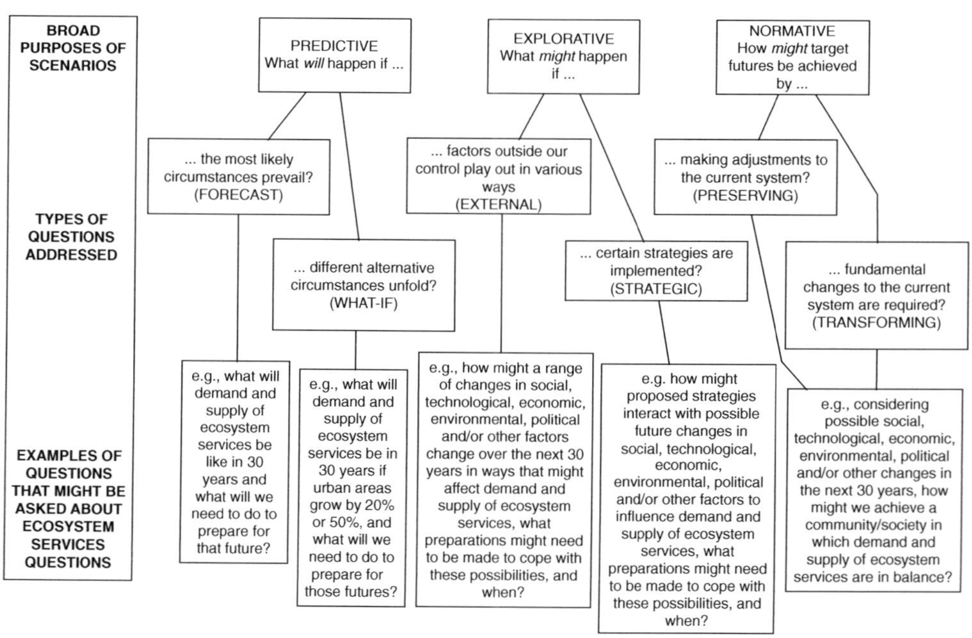

Figure 14.4 Based on Börjeson et al.'s (2006) scenario typology with examples relevant to ecosystem services approaches. Note that many futures-thinkers disagree with including predictive scenarios as they do not fulfil criteria for robust futures-thinking.

The 'explorative-external' and 'explorative-strategic' distinction is similar to that between 'inquiry-driven' and 'strategy-driven' (Alcamo, 2009a). Increasingly, participatory environmental futures-thinking combines these two approaches (Bennett *et al.*, 2003; Swart *et al.*, 2004; Raskin *et al.*, 2005; Alcamo, 2009b; Rothman, 2009).

Other distinctions that could be helpful

The 'archetype' scenarios approach emerged in the mid-1970s (Bezold, 2009) and has become popular again recently (Hunt *et al.*, 2012). It is particularly useful for generating dialogue and imaginative thinking quickly. The idea is that most scenarios fall into a few categories (see Box 14.1) and that dialogue about possible futures can be generated by asking how different archetype futures might emerge.

Two recent projects in Australia have used archetype scenario approaches to facilitate conversations about Australia's future (Raupach *et al.*, 2013; Alford *et al.*, 2014; Cork *et al.*, 2014; Costanza *et al.*, in press).

Methods

Principles

The following summary of principles for effective futures-thinking with a focus on scenarios draws on key reviews over several decades (Godet and Roubelat, 1996; Schwartz, 1996; van der Heijden, 1996; Inayatullah, 2008; Scearce *et al.*, 2004; Mietzner and Reger, 2005; Slaughter, 2008; Stewart, 2008; Alcamo and Henrichs, 2009; Rothman, 2009):

- Futures-thinking is not about prediction but about engaging with uncertainty, considering multiple plausible futures and making better-informed decisions.
- Scenarios are part of a bigger process of futures-thinking that includes a virtuous cycle of learning.
- A key element of the process is removing obstacles to creative and strategic thinking (including thinking flaws and institutional constraints).
- It is vital to be clear about objectives and ensure the approach suits these.
- Scenarios should; be relevant, coherent, plausible, internally consistent, sufficiently different to explore relevant extremes, and challenge conventional wisdom.
- The time-horizon should be sufficient for the key trends and drivers of change to have a chance to have their impacts.
- It is important to consider alternative worldviews and assumptions in depth.
- Support from senior management, key stakeholders and 'decision-owners' is vital.

Key steps

The methodological steps in developing scenarios under most approaches are similar, although the emphasis will vary (see citations in previous sub-section). Key steps include:

- Problem analysis/framing (e.g. interviews with relevant stakeholders and experts, analysis of perceptions and worldviews).
- Analysis of assumptions about the past, present and future, and their validity and/or criticality.

- Gathering and analysis of necessary information (typically involves analysis of past events and trends over varying time scales from a few years to decades or millennia and 'horizon scanning' to consider emerging trends and other factors that might influence the futures being explored).
- Analysis of how these factors might interrelate (e.g. influence, structural (systems), cross-impact, consistency analyses).
- Identification of foci for scenarios (e.g. critical uncertainties).
- Development of 'scenario logics' (uncertain and relatively certain aspects of the future and how they might differ between scenarios – see examples in Cork *et al.*, 2006 and Haines-Young *et al.*, 2011).
- Exploration of logics using qualitative approaches (e.g. informed imagination, experts and challenges from experts) and various quantitative approaches (e.g. data-driven models).
- A set of scenarios (these might be very brief or very detailed narratives and they might include numbers, pictures, videos, sounds, objects and/or live drama, depending on the intended audiences and purpose).
- Consideration of consequences, which might include 'wind tunnelling' (e.g. imagining how we might function in different futures), and wild cards (disruptive events, surprises).
- Identification of signposts, branching points and other ways to get early warning of emerging future trajectories.
- Strategies and actions, usually over several time horizons (e.g. 1, 5, 10, or more years).

Two preparatory steps – horizon scanning and assumption/worldview analysis – deserve particular emphasis because they provide the platform for successful futures-thinking.

Horizon scanning and other collection and analysis of contemporary and historical data to understand context and processes stimulate thinking about what is possible and provide a basis for assessing plausibility (Galtung and Inayatullah, 1996; Voros, 2003; Schultz, 2006; Sutherland *et al.*, 2011).

Analysing assumptions is important because people tend to base their views of the world on untested, and often illogical, thinking. Many 'thinking flaws' have been identified, most involving some sort of selective gathering, recalling or 'mental filtering' of information to avoid complexity and uncertainty, and their effect is to limit imagination about what is possible (Taleb, 2007; Judge, 2008). Ways to address thinking flaws include: systematic analysis and testing of assumptions (Godet *et al.*, 2004; Dewar, 2002); considering deeper aspects of human consciousness in horizon scanning (Voros, 2003); exploring values and beliefs in depth (Inayatullah, 2008); and systems-thinking and analysis (Mietzner and Reger, 2005).

Deductive versus inductive methods

A distinction has been drawn between 'deductive' and 'inductive' methods for developing scenarios (Rothman, 2009; Godet, 2012). In deductive methods, drivers of change and 'critical uncertainties' are identified first, and then scenarios explore how combinations of these uncertainties might evolve to create alternative futures. Examples include the IPCC climate change scenarios (Nakićenovic and Swart, 2000), the global scenarios of the Millennium Ecosystem Assessment (Millennium Ecosystem Assessment, 2005), and the analysis-driven 'la Prospective' approach (Godet, 2012). In 'inductive' methods, drivers of change and critical uncertainties are allowed to emerge as the dialogue evolves. Rothman (2009) gives the United Nations Environment Program's GEO3 and GEO4 scenarios as examples. Both approaches could be relevant in ecosystem assessments, depending on the objectives and focus of the project and the preferences of those involved.

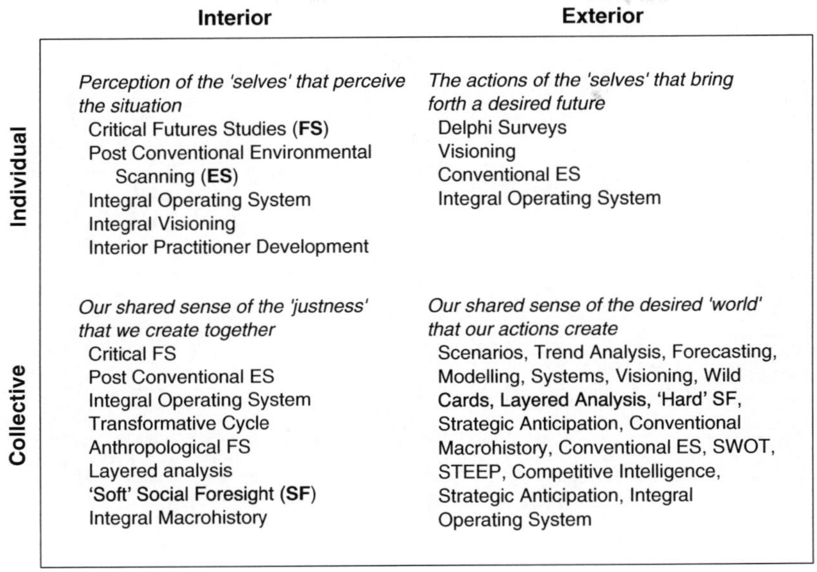

Figure 14.5 'Integral' approaches to futures-thinking combine traditional 'exterior' methods with 'interior' thinking.

Source: drawing on Slaughter, 2006, Hayward and Morrow, 2009

Critical and integral approaches

'Critical' and 'integral' approaches might be particularly useful for understanding how societal processes might influence demand for ecosystem services in alternative futures (Figure 14.5). Like ecosystem assessments, critical and integral futures approaches aim to improve societies' ability to think more deeply about social as well as bio-physical futures (Inayatullah, 2008; Slaughter, 2006; Stewart, 2008; Box 14.1).

Applications

Decisions about managing human-ecosystem interactions are taken at many levels, and are ideally informed by: an understanding of human ethics, morality and needs, attitudes and worldviews; an understanding of how ecological and social systems work together; and insights about how social, technological, economic, political and other factors might influence all of the above in the future (Figure 14.6). The following sub-sections discuss how different types of futures-thinking have been applied to consider these complex interactions in ecosystem assessments (summarised in Table 14.1).

Choosing futures-thinking approaches

Table 14.2 summarises examples of futures-thinking studies that have focused on ecosystem services. These are discussed below with reference to the components of Figure 14.6. Many other relevant studies were reviewed by Rothman (2009).

All studies in Table 14.2 explored worldviews to some degree. Social learning through participatory scenario development was a particular focus of studies 11, 12, 16, 19, 20, 22 and 26.

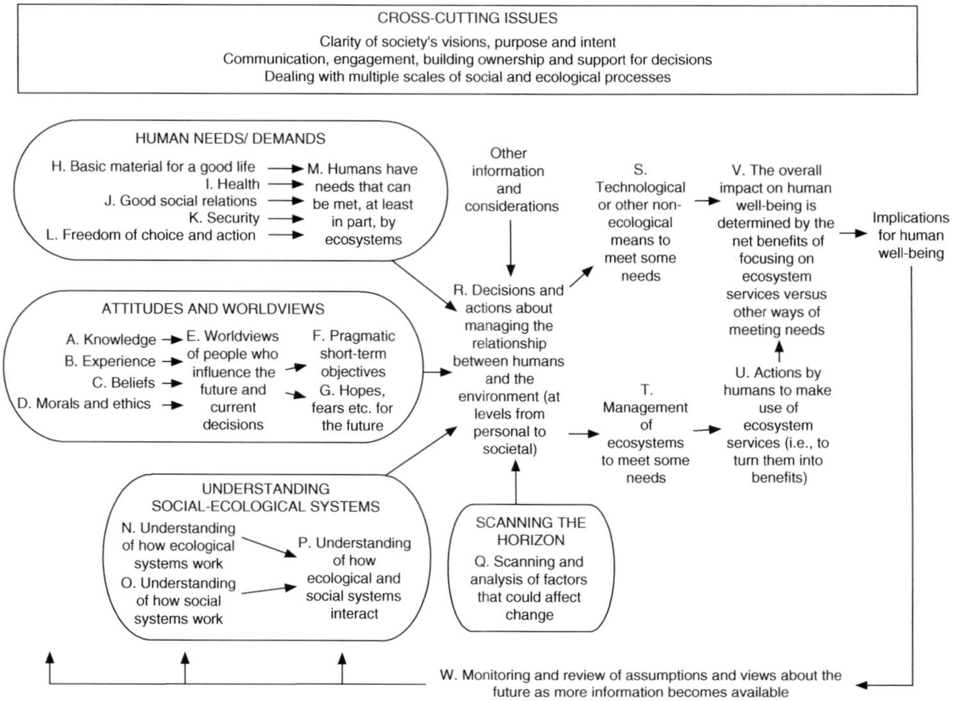

Figure 14.6 Some of the factors influencing societal decision-making about management of ecosystem services and implications for human well-being (letter codes A–W against factors are referred to in Table 14.1).

Table 14.1 Contributions futures-thinking can make to the processes depicted in Figure 14.6 (letter codes A–W in the table refer to the codes against factor names in Figure 14.6).

Process	*Potential contributions from futures-thinking*
Considering worldviews that people bring to assessing the worth of ecosystem services (A–D)	Assumption analysis and revealing of worldviews through dialogue. In-depth exploration of perceptions and underlying worldviews (especially 'critical' and 'integral' approaches)
Understanding and anticipating human needs (H, L and Q)	Exploring where and how people might live in the future In-depth thinking about relationships between humans and the environment (especially 'critical' and 'integral' approaches)
Understanding how ecological and social systems work and interrelate (N–P)	Providing reasons and platforms that bring diverse disciplines and stakeholders together to improve scientific understanding and/ or explore how social and ecological processes might interact in the future
Exploring the possible future implications of decision options (S–U)	Asking 'what if' questions about decision-options in structured and systematic ways, adding breadth and depth to analyses of risks and opportunities
Monitoring and review of assumptions and views (W)	Social learning, review of assumptions, updating of information, early-warning indicators of future trajectories and triggers for action

Table 14.2 Examples of futures-studies directly or indirectly considering ecosystem services.

	Scale	Authors	Focus	Scenario details				
				Existing/ new	*Philosophy*	*Type*	*Participants*	*Quantitative/ qualitative*
1	G	(Raskin *et al.*, 2002)	Pathways of world development; worldviews	New	polFor	nor	Exp	Qmix
2	G	(Nakićenovic and Swart, 2000)	Social, economic and technological impacts on greenhouse gas emissions	New	polFor	expExt	Exp	Qmix
3	G	(Moss *et al.*, 2010)	As above	ExS	polFor	expStrat	Exp	Qmix
4	G	(Millennium Ecosystem Assessment, 2005)	Ecosystem services, human well-being, policy and management consequences	New	polFor	expExt	Exp	Qmix
5	G	(Holmgren, 2009)	Climate change and peak oil	New	polFor	expExt	Exp	Qmix
6	G	(Sala *et al.*, 2000)	Global drivers of change in biodiversity	New	polFor	expExt	Exp	Quan
7	G	(Novacek and Cleland, 2001)	As above	ExS	polFor	nor	Exp	Qmix
8	NS	(Office for Science and Technology, 2002)	Challenges and policy for flood and coastal defence in the UK	ExA	polFor	expStrat	Exp	Quan
9	NS	(Haines-Young *et al.*, 2011)	Ecosystem services and human well-being in UK futures	New	polFor	expExt	Part	Qmix
10	NS	(Metzger *et al.*, 2006)	Land use change and vulnerability of human-environment systems in Europe	ExP	polFor	expStrat	Exp	Quan
11	NS	(Kok *et al.*, 2006)	Understanding drivers of desertification at multiple scales (Europe, Mediterranean, local)	ExP	polFor	nor	Part	Qual
12	NS	(Kok *et al.*, 2011)	'Fast track' thinking about Europe's freshwater futures	ExA	polFor	nor	Part	Qmix
13	NS	(Alford *et al.*, 2014, Cork *et al.*, 2014)	Understanding worldviews and engaging society in conversations about alternative Australian futures	New	critFor	nor	Part	Qual
14	NS	(Ariell, 2010)	Worldviews and myths underpinning policies for managing carbon emissions from forests		critFor		Exp	
15	LR	(Plieninger *et al.*, 2013)	Local perceptions about drivers of change in cultural landscapes and ecosystems (Swabian Alb reserve, Germany)	New	polFor	expExt	Part	Qual
16	LR	(Palomo *et al.*, 2011)	Future management of ecosystems and human well-being (Doñana social-ecological system, Spain)	New	polFor	norPres	Part	Qual
17	LR	(Higgins *et al.*, 1997)	Effects of clearing alien plants on ecosystem services (mountain fynbos ecosystems, South Africa)	ExS	progFor	pred	Exp	Quan

(*Continued*)

Table 14.2 (Continued)

	Scale	*Authors*	*Focus*	*Scenario details*				
				Existing/ new	*Philosophy*	*Type*	*Participants*	*Quantitative/ qualitative*
18	LR	(Bohensky *et al.*, 2006)	Gariep River basin futures (South Africa)	ExP	polFor	expExt	Part	Qmix
19	LR	(Peterson *et al.*, 2014)	Future uses of ecological services (Northern Highlands Lake District, Wisconsin, USA)	New	polFor	expExt	Part	Qmix
20	LR	(Johnson *et al.*, 2012)	Collective imagination of adaptation in social and environmental communities (Minnesota, USA)	New	polFor	nor	Part	Qual
21	LR	(Estoque and Murayama, 2012)	Land-use change and ecosystem services (Baguio City, Philippines)	ExS	progFor	pred	Exp	Quan
22	LR	(Cork and Delaney, 2009)	Future challenges and opportunities for regional catchment communities (Namoi catchment, NSW, Australia)	New	polFor	expExt	Part	Qmix
23	LR	(Abel *et al.*, 2003)	Value of ecosystem services in several development trajectories (Goulburn Broken catchment, Victoria, Australia)	ExS	progFor	pred	Part	Qmix
24	LR	(Wang *et al.*, 2006)	Values, aspirations and plausible irrigation futures for Goulburn Broken catchment	New	polFor	expExt	Part	Qmix
25	LR	(Bryan *et al.*, 2011)	Environmental, economic and social impacts of alternative environmental targets (Lower Murray River, South Australia)	ExS	polFor	expStrat	Part	Qmix
26	LR	(Pearson *et al.*, 2010)	Create and demonstrate a generic sustainable land-use scenario framework (Rocky Point region, south-east Queensland, Australia)	New	polFor	nor	Part	Qmix
27	LR	(VanWynsberghe *et al.*, 2003)	Reconciling global carrying capacity with human well-being (Georgia Basin, British Columbia, Canada)	ExA New	polFor	expExt	Part	Quan

Scale: G=Global; NS=National or sub-global; LR=Local or regional

Existing scenarios: ExA=archetypes; ExP=previously developed scenarios; ExS=strategies/targets

Futures-thinking philosophy: progFor=Progressive foresight; polFor=Political foresight; critFor=critical foresight (Figure 14.3)

Scenario type: pred=Predictive; expExt=Explorative external; expStrat=Explorative strategic; nor=Normative (either preserving or transforming) (Figure 14.4)

Experts or participatory: Process primarily run by experts (Exp) or a mixture of experts and stakeholders (Part)

Qualitative or quantitative: Whether the inputs were primarily one or the other or a roughly equal mixture

Studies 11 and 18 (and several subsequent studies by these authors) used participatory approaches to link futures-thinking across geographical scales.

Studies 13 and 14 are examples of worldview-focused ('critical' foresight) approaches. 'Critical' and 'integral' foresight are potentially important for helping communities reflect on their relationships with the natural environment and/or significant cultural changes (Barber, 2010) or where progress is hindered by different interpretations of issues (Ariell, 2010; Inayatullah, 2010).

Most studies in Table 14.2 consider how drivers of change might influence human demands on ecosystem services. The Millennium Ecosystem Assessment (study 4) sought to make explicit the relationships between ecosystem services and human needs. Studies like 21 help stakeholders identify which ecosystem services are important to them (see also Maynard *et al.*, 2010). Exploration of future demands on ecosystems is not often seen in ecosystem services projects generally (Bengston *et al.*, 2012) or in strategic thinking across society (Cork, 2010; Australia21, 2014). The UK's National Ecosystem Assessment is a notable exception (study 9).

Studies indicated by 'Qmix' are examples of combining quantitative and qualitative modelling to deepen understanding of how ecological and social systems interrelate and to identify and address knowledge gaps. Alcamo (2009b) and Alcamo *et al.* (2009) outline the steps and types of models used in such approaches.

Most studies in Table 14.2 employed horizon scanning within a range of budgetary and time constraints. Studies 1 and 4 illustrate detailed scanning.

The distinction between exploratory and strategic scenarios (there is a mix in Table 14.2) is not always clear, as most studies use exploratory scenarios to consider the implications of current or future strategies anyway. Several have taken 'fast track' approaches by: starting with archetype scenarios and 'backcasting' to consider how those futures might come about (e.g. studies 12 and 13); using previously developed scenarios as their starting point (8, 10, 11, 23); or using strategies or targets as their 'scenarios' (17, 23). A variation on normative backcasting is exploring pathways to sustainable futures (1, 7, 20). In study 27, archetypes were established initially by experts and then developed further by stakeholders.

Studies 17, 21 and 23 have been labelled 'progressive foresight', because they 'predict' the most likely future in some detail. Mainstream futures-thinkers would probably not regard these as foresight studies but they do fit onto some broader conceptions of scenario-use.

Conclusions

This chapter has synthesised a large body of literature on futures-thinking and shown how it can contribute to ecosystem assessments. The main benefits are:

- Exploring and communicating worldviews that influence assessment of worth.
- Exploring needs for, and potential demands on, ecosystems services in alternative futures.
- Bringing scientists and others together to better understand how coupled social-ecological work and what ecosystem services they might supply in alternative futures.
- Exploring the implications of alternative strategies for managing the interrelationships between humans and ecosystems.
- Considering how imagined, desirable futures might be achieved and undesirable ones avoided.
- Monitoring and reviewing assumptions about, and images of, alternative futures as more information becomes available.

The potential of futures-thinking has not been fully realised in ecosystem services projects to date, despite the strong focus on change and future generations inherent in this concept. Although the literature on methods is complex, core principles emerge and guidelines are provided in this chapter for matching types of scenario approaches and methods to different purposes of ecosystem services projects.

References

Abel, N., Cork, S., Gorddard, R., *et al.* (11 authors) (2003). *Natural Values: Exploring Options For Enhancing Ecosystem Services In The Goulburn Broken Catchment.* CSIRO, Canberra.

Alcamo, J. (2009a). The case for scenarios of the environment. In Alcamo, J. (ed.) *Environmental Futures – The Practice of Environmental Scenario Analysis.* Elsevier, Amsterdam. www.sciencedirect.com/science/article/pii/S1574101X08004018 (accessed 30 June 2014).

Alcamo, J. (2009b). The SAS approach: combining qualitative and quantitative knowledge in environmental scenarios. In Alcamo, J. (ed.) *Environmental Futures – The Practice of Environmental Scenario Analysis.* Elsevier, Amsterdam. www.sciencedirect.com/science/article/pii/S1574101X08004067 (accessed 30 June 2014).

Alcamo, J. (ed.) (2009c). *Environmental Futures – The Practice of Environmental Scenario Analysis.* Elsevier, Amsterdam.

Alcamo, J., and Henrichs, T. (2009). Towards guidelines for environmental scenario analysis. In Alcamo, J. (ed.) *Environmental Futures – The Practice of Environmental Scenario Analysis.* Elsevier, Amsterdam.

Alcamo, J., Kok, K., Busch, G., and Priess, J. (2009). Searching for the future of land: Scenarios from the local to global scale. In: Alcamo, J. (ed.) *Environmental Futures – The Practice of Environmental Scenario Analysis.* Elsevier, Amsterdam. www.sciencedirect.com/science/article/pii/S1574101X08004043 (accessed 30 June 2014).

Alford, K., Cork, S., Finnigan, J., *et al.* (6 authors) (2014). The challenges of living scenarios for Australia in 2050. *Journal of Futures Studies*, vol 18, pp 115–1266.

Ariell, A. (2010). Forest futures: a causal layered analysis. *Journal of Futures Studies*, vol 14, pp 49–64.

Australia21 (2014). *Considering the environment in national strategic thinking*, Australia21, Canberra. www.australia21.org.au/wp-content/uploads/2014/06/ALConsidering-the-environment-in-national-strategic-thinking_V11_Approved.pdf (accessed 6 July 2014).

Barber, M. (2010). Solving the futures challenge – all you need is a 3LA. *Futures*, vol 42, pp 170–173.

Bengston, D.N., Kubik, G.H., and Bishop, P.C. (2012). Strengthening Environmental Foresight: Potential Contributions of Futures Research. *Ecology and Society*, vol 17. www.ecologyandsociety.org/vol17/iss2/art10/ (accessed 25 June 2014).

Bennett, E., Carpenter, S., Peterson, G., *et al.* (6 authors) (2003). Why global scenarios need ecology. *Frontiers in Ecology and the Environment*, vol 1, pp 322–329.

Bezold, C. (2009). Jim Dator's alternative futures and the path to IAF's aspirational futures. *Journal of Futures Studies*, vol 14, pp 123–134.

Bishop, P., Hines, A., and Collins, T. (2007). The current state of scenario development: an overview of techniques. *Foresight*, vol 9, pp 5–25.

Bohensky, E.L., Reyers, B., and Van Jaarsveld, A.S. (2006). Future ecosystem services in a Southern African River Basin: a scenario planning approach to uncertainty. *Conservation Biology*, vol 20, pp 1051–1061.

Börjeson, L., Höjer, M., Dreborg, K.-H., Ekvall, T., and Finnveden, G. (2006). Scenario types and techniques: towards a user's guide. *Futures*, vol 38, pp 723–739.

Bradfield, R., Wright, G., Burt, G., Cairns, G., and Van Der Heijden, K. (2005). The origins and evolution of scenario techniques in long range business planning. *Futures*, vol 37, pp 795–812.

Bryan, B.A., Crossman, N.D., King, D., and Meyer, W.S. (2011). Landscape futures analysis: assessing the impacts of environmental targets under alternative spatial policy options and future scenarios. *Environmental Modelling & Software*, vol 26, pp 83–91.

Chermack, T.J., Lynham, S.A., and Ruona, W.E.A. (2001). A review of scenario planning literature. *Futures Research Quarterly*, vol 17, pp 7–32.

Cork, S. (2010). Ways forward in the population and environment debate. www.aph.gov.au/About_Parliament/Parliamentary_Departments/Parliamentary_Library/Publications_Archive/~/

media/05%20About%20Parliament/54%20Parliamentary%20Depts/544%20Parliamentary%20Library/Pub_archive/PopulationEnvironDebate.ashx (accessed 6 May 2011).

Cork, S., Alford, K., Grigg, N., *et al.* (6 authors) (2014). Helping a nation think about its futures. *Solutions*, vol 4, pp 9–14. www.thesolutionsjournal.org/node/237114 (accessed 6 July 2014).

Cork, S., and Delaney, K. (2009). *Scenario Planning for Sustainable Landuse in the Namoi Catchment*, Namoi Catchment Management Authority, Tamworth. http://grapevine.net.au/~anncork/Namoi.htm (accessed 16 September 2015).

Cork, S.J., Peterson, G.D., Bennett, E.M., Petschel-Held, G., and Zurek, M. (2006). Synthesis of the storylines. *Ecology and Society*, vol 11. www.ecologyandsociety.org/vol11/iss2/art11/ (accessed 6 July 2014).

Costanza, R., Kubiszewski, I., Cork, S., *et al.* (11 authors) (2015). Scenarios for Australia in 2050: a synthesis and proposed survey. *Journal of Futures Studies*, vol 19, no 3, pp 49–76.

Curry, A. (2009). From foresight to insight: using scenarios well. *Journal of Futures Studies*, vol 13, pp 119–122.

Curry, A., and Schultz, W. (2009). Roads less travelled: different methods, different futures. *Journal of Futures Studies*, vol 13, pp 35–60.

Daily, G.C. (1997). *Nature's Services – Societal Dependence on Natural Ecosystems*. Island Press, Washington DC.

Dewar, J.A. (2002). *Assumption-Based Planning*. Cambridge University Press, Cambridge UK.

Estoque, R.C., and Murayama, Y. (2012). Examining the potential impact of land use/cover changes on the ecosystem services of Baguio city, the Philippines: a scenario-based analysis. *Applied Geography*, vol 35, pp 316–326.

Fisher, B., Turner, R., and Morling, P. (2009). Defining and classifying ecosystem services for decision making. *Ecological Economics*, vol 68, pp 643–653.

Galtung, J., and Inayatullah, S. (1996). *Macrohistory and Macrohistorians*. Praeger, Westport, CT.

Godet, M. (2012). To predict or to build the future? Reflections on the field and differences between foresight and la Prospective. *The Futurist*, May-June 2012, pp 46–49.

Godet, M., Bassaler N., Monti, R., and Richou, S. (2004). *Guideline for Strategic Prospective workshops*, LIPSOR, Paris. http://en.laprospective.fr/dyn/anglais/ouvrages/trameatelier7english2004.pdf (accessed 16 September 2015).

Godet, M., and Roubelat, F. (1996). Creating the future: the use and misuse of scenarios. *Long Range Planning*, vol 29, pp 164–171.

Haines-Young, R., Paterson, J., Potschin, M., Wilson, A., and Kass, G. (2011). The UK NEA scenarios: development of storylinesand analysis of outcomes. In *The UK National Ecosystem Assessment Technical Report*, UK National Ecosystem Assessment, UNEP-WCMC, Cambridge UK.

Hayward, P., and Morrow, R. (2009). An integral approach to scenarios. *Journal of Futures Studies*, vol 13, pp 115–118.

Higgins, S.I., Turpie, J.K., Costanza, R., *et al.* (7 authors) (1997). An ecological economic simulation model of mountain fynbos ecosystems – Dynamics, valuation and management. *Ecological Economics*, vol 22, pp 155–169.

Hines, A., and Gold, J. (2013). Professionalizing foresight: why do it, where it stands, and what needs to be done. *Journal of Futures Studies*, vol 17, pp 35–54.

Holmgren, D. (2009). *Future Scenarios. How Communities Can Adapt to Peak Oil and Climate Change*. Chelsea Green Publishing, White River Junction, VT.

Hunt, D.V.L., Lombardi, D.R., Atkinson, S., *et al.* (27 authors) (2012). Scenario archetypes: converging rather than diverging themes. *Sustainability*, vol 4, pp 740–772.

Inayatullah, S. (2008). Six pillars: futures thinking for transforming. *Foresight*, vol 10, pp 4–21.

Inayatullah, S. (2010). Multiple narratives of the futures of the global financial crisis. *Journal of Futures Studies*, vol 14, pp 109–128.

Johnson, K.A., Dana, G., Jordan, N.R., *et al.* (7 authors) (2012). Using participatory scenarios to stimulate social learning for collaborative sustainable development. *Ecology and Society*, vol 17, www.ecologyandsociety.org/vol17/iss2/art9/ (accessed 6 July 2014).

Judge, A. (2008). Stepping into, or through, the Mirror: embodying alternative scenario patterns. *Journal of Futures Studies*, vol 13, pp 129–138.

Kok, K., Rothman, D.S., and Patel, M. (2006). Multi-scale narratives from an IA perspective: part I. European and Mediterranean scenario development. *Futures*, vol 38, pp 261–284.

Kok, K., van Vliet Mathijs, M., Bärlund Ilona, I., Dubel, A., and Sendzimir, J. (2011). Combining participative backcasting and exploratory scenario development: experiences from the SCENES project. *Technological Forecasting and Social Change*, vol 78, pp 835–851.

MA (2005). *Ecosystems and Human Well-Being, Volume 2: Scenarios*. Island Press, Washington DC.

Malaska, P., and Virtanen, I. (2009). Theory of futuribles and historibles. *Futura*, vol 1/09, pp 65–84.
Marien, M. (2002). Futures studies in the 21st century: a reality-based view. *Futures*, vol 34, pp 261–281.
Maynard, S., James, D., and Davidson, A. (2010). The development of an ecosystem services framework for south east Queensland. *Environmental Management*, vol 45, pp 881–895.
Metzger, M.J., Rounsevell, M.D.A., Acosta-Michlik, L., Leemans, R., and Schroter, D. (2006). The vulnerability of ecosystem services to land use change. *Agriculture, Ecosystems & Environment*, vol 114, pp 69–85.
Mietzner, D., and Reger, G. (2005). Advantages and disadvantages of scenario approaches for strategic foresight. *International Journal of Technology Intelligence and Planning*, vol 1, pp 220–239.
Molitor, G.T. (2009). Scenarios: worth the effort. *Journal of Future Studies*, vol 13, pp 75–86.
Moss, R.H., Edmonds, J.A., Hibbard, K.A., *et al.* (19 authors) (2010) The next generation of scenarios for climate change research and assessment. *Nature*, vol 463, pp 747–756.
Nakićenovic, N., and Swart, R. (eds) (2000). *Emissions Scenarios 2000: Special Report of the Intergovernmental Panel on Climate Change.* Cambridge University Press, Cambridge UK.
Novacek, M.J., and Cleland, E.E. (2001). The current biodiversity extinction event: scenarios for mitigation and recovery. *Proceedings of the National Academy of Sciences of the United States of America*, vol 98, pp 5466–5470.
Office for Science and Technology (2002). *Foresight Futures 2020.* Department of Trade and Industry, London, futuresavvy.net/wp-content/uploads/scenario%20documents/Foresight-Futures-2020-SPRU.pdf (accessed 6 July 2014).
Pahl-Wostl, C. (2009). Participation in building environmental scenarios. In Alcamo, J. (ed.) *Environmental Futures – The Practice of Environmental Scenario Analysis.* Elsevier, Amsterdam.
Palomo, I., Martín-López, B., López-Santiago, C., and Montes, C. (2011). Participatory scenario planning for protected areas management under the ecosystem services framework: the Doñana social-ecological system in Southwestern Spain. *Ecology and Society*, vol 16, www.ecologyandsociety.org/vol16/iss1/art23/ (accessed 25 June 2014).
Pearson, L.J., Park, S., Harman, B., and Heyenga, S. (2010). Sustainable land use scenario framework: framework and outcomes from peri-urban south-east Queensland, Australia. *Landscape and Urban Planning*, vol 96, pp 88–97.
Peterson, G.D., Beard Jr, T.D., Beisner, B.E., *et al.* (8 authors) (2014). Assessing future ecosystem services: a case study of the Northern Highlands Lake District, Wisconsin. *Conservation Ecology*, vol 7. www.consecol.org/vol7/iss3/art1 (accessed 6 July 2014).
Peterson, G.D., Cumming, G., and Carpenter, S.R. (2003). Scenario planning: a tool for conservation in an uncertain world. *Conservation Biology*, vol 17, pp 358–366.
Plieninger, T., Bieling, C., Ohnesorge, B., *et al.* (6 authors) (2013). Exploring futures of ecosystem services in cultural landscapes through participatory scenario development in the Swabian Alb, Germany. *Ecology and Society*, vol 18. www.ecologyandsociety.org/vol18/iss3/art39/ (accessed 6 July 2014).
Ramos, J.M. (2004). *Foresight Practice in Australia: A Meta-Scan of Practitioners and Organisations.* Australian Foresight Institute, Swinburne University, Melbourne.
Raskin P., Banuri T., Gallopin G., *et al.* (7 authors) (2002). *Great Transition: The Promise and Lure of the Times Ahead.* Stockholm Environment Institute, Boston.
Raskin, P., Monks, F., Ribeiro, T., Van Vuuren, D., and Zurek, M. (2005). Global scenarios in historical perspective. In Carpenter, S. R., Pingali, P. L., Bennett, E. M., and Zurek, M. B. (eds) *Ecosystems and Human Well-being, Volume 2: Scenarios. Findings of the Scenarios Working Group of the Millennium Ecosystem Assessment.* Island Press, Washington DC.
Raupach, M.R., McMichael, A.J., Alford, K.J.S., *et al.* (11 authors) (2013). Living scenarios for Australia as an adaptive system. In Raupach, M. R., McMichael, A. J., Finnigan, J. J., Manderson, L., and Walker, B. H. (eds) *Negotiating Our Future: Living Scenarios for Australia to 2050, Volume 1.* Australian Academy of Science, Canberra. www.science.org.au/sites/default/files/user-content/resources/file/australia_2050_v1_ls_0.pdf (accessed 6 July 2014).
Rothman, D.S. (2009). A survey of environmental scenarios. In Alcamo, J. (ed.) *Environmental Futures – The Practice of Environmental Scenario Analysis.* Elsevier, Amsterdam.
Royal Dutch Shell (2013). *40 Years of Shell Scenarios.* Royal Dutch Shell. http://s05.static-shell.com/content/dam/shell-new/local/corporate/corporate/downloads/pdf/shell-scenarios-40years book080213.pdf (accessed 20 December 2014)
Sala, O.E., Chapin, F.S., Armesto, J.J., *et al.* (19 authors) (2000). Global biodiversity scenarios for the year 2100. *Science*, vol 287, pp 1770–1774.

Scearce, D., Fulton, K., and Global Business Network Community (2004). *What if? The Art of Scenario Thinking for Non-Profits*, Global Business Network, Emeryville, CA. www.monitorinstitute.com/downloads/what-we-think/what-if/What_If.pdf (accessed 6 July 2014).

Schultz, W.L. (2006). The cultural contradictions of managing change: using horizon scanning in an evidence-based policy context. *Foresight*, vol 8, pp 3–12.

Schwartz, P. (1996). *The Art of the Long View*. Currency Doubleday, New York.

Slaughter, R. (2009). The state of play in the futures field: a metascanning overview. *Foresight*, vol 11, pp 6–20.

Slaughter, R.A. (2002). Futures studies as a civilizational catalyst. *Futures*, vol 34, pp 349–363.

Slaughter, R.A. (2006). *Pathways and Impediments to Social Foresight. Monograph Series 2003–2006 No. 10*, Strategic Foresight Program, Swinburne University, Melbourne.

Slaughter, R.A. (2008). What difference does 'integral' make? *Futures*, vol 40, pp 120–137.

Stewart, C.C. (2008). Integral scenarios: reframing theory, building from practice. *Futures*, vol 40, pp 160–172.

Sutherland, W.J., Bardsley, S., Bennun, L., *et al.* (10 authors) (2011). Horizon scan of global conservation issues for 2011. *Trends in Ecology & Evolution*, vol 26, pp 10–16.

Swart, R.J., Raskin, P., and Robinson, J. (2004). The problem of the future: sustainability science and scenario analysis. *Global Environmental Change*, vol 14, pp 137–146.

Taleb, N.N. (2007). *The Black Swan: The Impact of the Highly Improbable*. Allen Lane, Camberwell.

UK National Ecosystem Assessment (2011). *The UK National Ecosystem Assessment: Synthesis of the Key Findings*. UNEP-WCMC, Cambridge, UK.

van der Heijden, K. (1996). *Scenarios. The Art of Strategic Conversation*. John Wiley & Sons, Chichester.

van Notten, P.W.F., Rotmans, J., van Asselt, M.B.A., and Rothman, D.S. (2003). An updated scenario typology. *Futures*, vol 35, pp 423–443.

VanWynsberghe, R., Moore, J., Tansey, J., and Carmichael, J. (2003). Towards community engagement: six steps to expert learning for future scenario development. *Futures*, vol 35, pp 203–219.

Varum, C.A., and Melo, C. (2010). Directions in scenario planning literature: a review of the past decades. *Futures*, vol 42, pp 355–369.

Voros, J. (2003). A generic foresight process framework. *Foresight*, vol 5, pp 10–21.

Wack, P. (1985). Scenarios: shooting the rapids. *Harvard Business Review*, November-December, pp 3–14.

Wang, Q.J., Soste, L., Robertson, D., Handley, S., and Chaffe, R. (2006). Scenario planning for irrigation futures of the Goulburn Broken region. In Petheram, R. J., and Johnson, R. C. (eds) *Practice change for Sustainable Communities: Exploring Footprints, Pathways and Possibilities: APEN 2006 International Conference*. La Trobe University, Victoria. www.regional.org.au/au/apen/2006/refereed/1/3035_wangqj.htm#TopOfPage (accessed 6 July 2014).

Zurek, M.B., and Henrichs, T. (2007). Linking scenarios across geographical scales in international environmental assessments. *Technological Forecasting and Social Change*, vol 74, pp 1282–1295.

15

MAPPING ECOSYSTEM SERVICES

Joachim Maes, Neville D. Crossman and Benjamin Burkhard

Introduction

The continued and sustainable provision of ecosystem services, and the protection of natural capital, is increasingly recognized by decision-makers as both a strategy to cope with a changing environment and as a cost-effective alternative to building infrastructure. Exemplary cases involve the deployment of green infrastructure in and around cities, which is linked to public health through the provision of increased air quality, regulation of the micro-climate through the cooling effects of vegetation, and recreation; the sustainable management of farmland, ensuring that pollination and biological control of pests maintain or enhance the production crops; the purification of wastewaters in constructed wetlands which provide additional services such as urban storm water management and maintenance of bird habitats; the restoration of floodplains to protect downstream communities; or the financial compensation of land owners for conserving or managing their land to provide sustained levels of ecosystem services (i.e. payment for ecosystem services). Many of these cases, if not all, share a common requirement for high-quality and consistent information on the condition of ecosystems and the services provided at different scales. Mapping ecosystems and their services is thus increasingly seen as a response to this requirement. This trend is evidenced by a rapidly expanding scientific literature and the expanding availability of different indicators, mapping tools, models, and decision support systems (Martínez-Harms and Balvanera, 2012; Egoh et al., 2012; Crossman et al., 2013b; Pagella and Sinclair, 2014).

Maps are a powerful way to convey information to users (Wood, 2010). Maps provide intuitive and simple methods for communicating information amongst stakeholders (scientists, policy makers, resource managers, and citizens) about the complex interactions between ecosystem services at a range of spatial and temporal scales (Burkhard et al., 2013; Cowling et al., 2008). Maps can be used to visualize trade-offs and synergies among ecosystem services; they may help identify spatial congruence or mismatches between supply, flow, and demand of ecosystem services or between ecosystems providing services and beneficiaries receiving services (Burkhard et al., 2012, 2014). For example, global trade is increasingly leading to beneficiaries of provisioning services being located in one hemisphere and supply of services located in the other (Meyfroidt et al., 2010). At smaller scales, urban dwellers benefit from many services supplied by agro-ecosystems some distance away (Kroll et al., 2012; Seto et al., 2012). Maps can help better understand these spatial relationships. Maps can support the selection, planning, and management of areas for conservation and green infrastructure (Palomo et al., 2013).

This chapter provides an overview of the latest developments in mapping ecosystem services. It aims to provide an answer to the following two questions: What aspects of ecosystem services can be mapped and what methods and tools are available for mapping? The chapter concludes with some future challenges related to mapping.

The cascade model as a framework for mapping ecosystem services

A number of classification systems for ecosystem services are available (Millennium Ecosystem Assessment, 2005; TEEB, 2010); the Common International Classification of Ecosystem Services (CICES)[1] is one of the most recent and comprehensive (Potschin and Haine-Young, 2016). While these classification systems are useful for broad organization, reporting, and auditing of ecosystem services, they need to be extended for mapping so as to include different aspects of ecosystem services. Using the example of freshwater supply, a provisioning ecosystem service, river basin managers may need to know how climate change or deforestation will impact on the supply of water in their jurisdiction. The manager would need a map of water supply and its change given a land use or climate change scenario. However, there is the question of what exactly a water supply map is. Would it be a map of lakes and reservoirs with indication of their volume; or a gridded map of water yields based on the difference between rainfall and evapo-transpiration by vegetation; or a map of the realized annual water abstractions in the region of interest; or a map of the water users? All these maps may provide the necessary information because together they help the river basin manager to assess where water is supplied, where it is consumed, and, importantly, how forests, wetlands, lakes and groundwater reserves contribute to this ecosystem service by maintaining a continuous flow of water from the ecosystems to the consumers.

This example shows why many ecosystem service researchers use the ecosystem services "cascade model" (Haines-Young and Potschin, 2010) as a framework for reporting and communicating results. This model conceptualizes how ecosystem services are generated and how they flow to people to meet their demands and provide benefits and value to society. The cascade model, or variants thereof, was adopted by The Economics of Ecosystems and Biodiversity study (TEEB, 2010) and is a popular framework to organize and summarize ecosystem services indicators (de Groot et al., 2010; Crossman et al., 2013b), maps (Pagella and Sinclair, 2014; Burkhard et al., 2012; Drakou et al., 2015), models and decision support systems (Bagstad et al., 2013b), and assessments (Santos-Martín et al., 2013).

The cascade model disentangles the ecosystem services supply chain. It contains a biophysical supply side and human demand side. Landscapes hold a certain ecosystem service supply capacity or potential (Bastian et al., 2012; Burkhard et al., 2014) which flow to people when they use or exploit them to satisfy their needs. Distinguishing between these components offers greater precision in understanding how different pressures and management decisions affect ecosystems and human well-being (Schröter et al., 2014; Villamagna et al., 2013; Tallis et al., 2012). A mapping project should thus carefully consider what components of the ecosystem services supply chain to map. Table 15.1 provides three examples of ecosystem services and associated metrics for mapping the supply chain.

Mapping ecosystem service supply

The type (e.g. woodland, grassland or wetland), spatial arrangement (well-connected vs. patches in an agricultural landscape), productivity (nutrient-rich vs. nutrient-poor systems), and

Table 15.1 Examples of ecosystem services and associated mapping indicators along the ecosystem service supply chain.

Ecosystem services supply chain		*Ecosystem service mapping indicators*		
Cascade	*Synonyms*	*Timber production*	*Water purification*	*Recreation*
Structure and process ↓	Supply Capacity Stock of natural capital Potential	Forest land cover Net primary production	Wetland area Denitrification	Protected areas Proximity to water Naturalness
Function ↓		Forest standing stock Annual timber growth	Removal capacity	Outdoor recreation potential
Service ↓	Flow Use Actual service Intermediate service	Harvested timber	Changes to nutrient and pesticide loads	Recreation opportunity spectrum
Benefit ↓	Demand Good Final service	Construction wood Biofuel Paper pulp	Clean water Improved water quality	Improved health Physical activities
Value (economic, health, social)		Net economic value of forest products	Avoided costs for water treatment Willingness to pay for improved water quality	Travel cost value

condition (healthy vs. degraded) of ecosystems have a profound influence on their capacity to deliver ecosystem services. Hence, these properties of ecosystems are frequently used to map the supply of ecosystem services (Martínez-Harms and Balvanera, 2012; Maes et al., 2012; Pagella and Sinclair, 2014).

Primary data, measurements, and observations stemming from remote sensors or collected in the field are the most accurate for maps of ecosystem service supply, and process models offer the most reliable method of estimating change to supply from a management intervention (Crossman et al., 2013a). However, this data is not always available, often limited to provisioning services, or based on samples which require up-scaling, down-scaling, or interpolation to cover for areas with missing data. Models may also not be available where system understanding is relatively poor (Crossman et al., 2013a). Researchers are thus in many cases reliant on proxies to map ecosystem service supply (Eigenbrod et al., 2010).

A common proxy method to map ecosystem services in absence of primary data and models is the use of ecosystem types often derived from land cover data. In their simplest form, ecosystem service maps use the proportion of a certain land cover type in a study area as an indicator for the supply of an ecosystem service. For example, the proportion of wetlands is assumed to approximate the potential supply of typical wetland services such as storm water regulation and water purification.

Land cover data is also used in combination with other datasets or expert-based opinions to add more detail (Jacobs et al., 2015). This method compares the potential supply among different land cover types to deliver ecosystem services and scores each class accordingly on a relative scale (Burkhard et al., 2009; 2012). If available, other spatial information, such as soil, hydrological, or vegetation data, help fine-tune the spatial assessment units or update the scores. The result is a land cover map which identifies higher or lower areas of ecosystem service supply. An example is mapped high or low above-ground vegetation carbon stocks as proxy for climate regulation.

Luck et al. (2003) suggested that species populations represent the fundamental unit in the provision of ecosystem services. Therefore quantifying the spatial distribution of species, vegetation, and habitat is important for mapping ecosystem service supply, especially of dominant taxa, which contribute significantly to total supply; or for those ecosystem services which depend strongly on species diversity, such as wild food provision or pollination. In addition, certain species contribute more than others to supplying ecosystem services because of specific characteristics or functional traits. For example, tree species with a big crown and a dense pack of leaves are likely to sequester more carbon, capture more air pollutants and provide more shade than species with an open crown structure. Similarly, pollinators with a larger body size tend to fly further than small-sized pollinator species and are thus able to pollinate crops over larger distances. Therefore mapping supply of ecosystem services will be improved by adding species trait information to distribution maps.

Ecosystem service mapping using process-based and/or empirical models that simulate ecosystem functioning or ecological production functions are likely to provide more accurate estimates of ecosystem services supply and changes to supply from an intervention (Crossman et al., 2013a). However, this approach requires significant investments in terms of data acquisition and expertise because the aim is to understand and quantify underlying mechanisms which drive the supply of ecosystem services.

Mapping ecosystem service demand and value

Quantifying and mapping ecosystem service demand is less common when compared to studies on ecosystem service supply (Honey-Rosés and Pendleton, 2013; Burkhard et al., 2013). Mapping demand for ecosystem services requires spatial information on where benefits are received or enjoyed and what values are attributed to these benefits by beneficiaries. Accordingly, studies have taken different approaches which vary from mapping the dependencies and needs of people on ecosystem services (García-Nieto et al., 2013; Nedkov and Burkhard, 2012; Mubareka et al., 2013), to mapping monetary and non-monetary values associated to the benefits provided by ecosystems (Crossman et al., 2011; Bryan and Crossman, 2013; Palomo et al., 2013).

While mapping ecosystem service supply is dependent on biophysical data and models (land cover, other environmental information, process models), mapping demand is dependent on socio-economic data (such as land use and demographic and population statistics). Cities, rural communities, and farms are dependent on ecosystem services such as storm water regulation, recreation or pollination. Hence, land use data have been used to map and quantify the demand for ecosystem services (Burkhard et al., 2012). Land use data are useful to down-scale aggregate statistics that quantify demand, including production statistics of agriculture, forestry and fisheries or data on tourism (Kroll et al., 2012). For example, coarse-scale (e.g. national) data on timber harvests can be downscaled to finer-scale units (e.g. watersheds, municipality, land parcels) by using the ratio of total harvestable forest area between the smaller and larger spatial units.

Land cover data can also be used to map demand. An indicator for ecosystem service demand is the economic value of ecosystem services, and benefit or value transfer functions (Richardson et al., 2014) are commonly used to measure this demand. The hypothesis is that primary, case study-based values which are assigned to individual services, to bundles of ecosystem services, or to ecosystems, can be transferred to assess the value provided by other, similar sites (Troy and Wilson, 2006; Richardson et al., 2014). For example, under a benefit transfer approach, the total economic value (TEV) delivered by a land cover type, such as a mountain grassland, is an approximate estimate for the TEV of any other grassland situated in mountain areas within a similar bio-geographic region. Using land cover data, an elevation map and a map of climatic zones, TEV can thus be mapped. A more robust benefit transfer, such as a meta-analysis (Brander et al., 2013), incorporates variables that describe the socio-economic context. Schägner et al. (2013) provide a review of methods and studies that map economic values of ecosystem services.

Not all ecosystem service values can be described in economic or monetary terms (Chan et al., 2012), especially values associated with cultural and spiritual ecosystem services. Several methods are available to assign other types of values to ecosystem services. These include mapping social and community (or stakeholder) values (Sherrouse et al., 2011; Palomo et al., 2014; Raymond et al., 2009; Plieninger et al., 2013) and landscape, environmental, or conservation values (Bryan et al., 2011; Cimon-Morin et al., 2013). Participatory GIS (Fagerholm et al., 2012) is a technique increasingly used to map social and community ecosystem service values, while mapping conservation values of biodiversity and ecosystem services has a deep history in the conservation planning literature (Chan et al., 2006; Margules and Pressey, 2000).

Mapping ecosystem service flows

Ecosystem services are realized and have value when humans benefit directly or indirectly from them. At this point, supply meets demand and ecosystem goods and services "flow" from where they are generated (Service Providing Units, SPUs) to where they are received (Service Benefitting Areas, SBAs). This includes a dynamic temporal dimension which is often difficult to capture in a map. In contrast, maps visualize the spatial mismatches between service providers and service receivers well. Several spatial configurations that characterize the flow of ecosystem services are recognized (Costanza, 2008; Villamagna et al., 2013; Serna-Chavez et al., 2014). The flow of ecosystem services to beneficiaries can be *in situ* (e.g. maintenance of soil quality), local (e.g. pollination), regional or global (e.g. climate regulation). The flow can be directional (e.g. upstream ecosystems providing downstream flood control), or the beneficiary can move to the service (e.g. recreation). Decoupled ecosystem service supply and demand patterns are typical for goods and services traded over long distances, such as food or timber, but mapping these flows is uncommon (Crossman et al., 2013a) despite significant consequences (Meyfroidt et al., 2010).

Two mapping approaches have been used to visualize flows of ecosystem services. One approach, described in the two previous sections, maps the ecosystem service supply (i.e. SPUs) and demand (i.e. SBAs) separately. A few studies have mapped supply and demand on a single chart to demonstrate how ecosystem services flow in the landscape. This has been done for flood regulation services (Nedkov and Burkhard, 2012; Stürck et al., 2014). Hotspot maps have been used to visualize ecosystem service flows in areas where supply and demand overlap (García-Nieto et al., 2013). The other approach aims to visualize the networks that connect ecosystems to their beneficiaries, a central theme of the Artificial Intelligence for Ecosystem Services (ARIES) tool (Bagstad et al., 2013a; Johnson et al., 2012).

Approaches and tools for mapping ecosystems services

A tiered approach to mapping ecosystem services

Common to other scientific disciplines that use maps and models, including climate research, spatial economics, and ecosystem modelling, the mapping of ecosystem services involves a trade-off between simple and more complex approaches (Kareiva et al., 2011). Simple mapping approaches require fewer data, can be easier to explain and understand, and may generate greater transparency and trust among users. Simpler approaches also allow for incorporation of local data, models, and stakeholder knowledge, but their accuracy and precision may be lower. More complex approaches require more data and expertise, but may be more credible and accurate (Kareiva et al., 2011). Tier 1 maps are typically derived from land cover or land use datasets or from other data that allow direct mapping of ecosystem services. Tier 2 maps add more complexity to tier 1 maps, making use of additional data to add further detail within a land cover or use class. Tier 3 maps integrate process models and detailed understanding of the systems under investigation.

Tier 1 maps

The simplest form of mapping is based on land cover and/or land use data or other environmental data that approximate ecosystem services. The relative area of certain ecosystem types or land use classes derived from land cover data is frequently used to map ecosystem service supply and demand. In particular, maps of vegetation, potential vegetation, habitats, or ecosystems are useful to map the ecosystems that deliver services and hence, to make inferences about the relative quantity of the services provided. Biodiversity monitoring data is also a useful source to map ecosystem services or ecosystem service providers. Consider for instance the harvest of wild foods for consumption (Schulp et al., 2014a). The simplest approach would be to map the ecosystems or habitats where the species of interest occur.

Tier 2 maps

The Tier 1 approach can be further improved if data at different levels of aggregation are used as a base to derive more complex indicators, which are combined to estimate ecosystem services. Land use data is linked to different datasets according to known relationships between land use, location-based information, and ecosystem service provision from the literature. Based on these relationships, the capacities of different land uses to provide ecosystem services can be quantified at different locations and aggregated at different scales. For example, in order to estimate wild food production, literature data or expert based scores on production can be linked to different forest types and mapped at the country scale (Schulp et al., 2014a). Likewise, national consumption or catch statistics of wild food (e.g. Schulp et al., 2014a), if available, can be downscaled using the area of different forest types as a spatial surrogate to obtain a map of wild food production. This procedure requires basic GIS skills.

Tier 3 maps

The Tier 2 approach can be further refined by modelling biophysical processes. For example, wild food production may be assessed by modelling the spatial distribution of appropriate flora and fauna species using climate data as well as other environmental data which limit the distribution of species (e.g. Schulp et al., 2014a). In a second step, process-based data can be used to assess annual production and, in combination with forest types, the result is a spatially resolved model of wild food production. Constructing a model is on one hand time-consuming and requires expert knowledge on modelling. Adjusting an existing model to local conditions, on the other hand, is much easier. Models can be extended by integrating expert knowledge (for example using Bayesian Belief networks), and can be used to assess uncertainty in quantification and valuation.

Example

Figures 15.1–6 show examples of tier 1–3 maps of the supply and demand of pollination services in south-western France. All maps are rescaled between 0 and 1 to allow comparison. Both tier

1 maps for supply and demand present the share of land which supports or requires pollination, respectively. The tier 2 maps are based on the tier 1 maps but an additional score gives more weight to those land types which support or require pollination more than other land types do. Finally, the tier 3 maps are based on a pollination model.

It is clear that different approaches to mapping produce quite different results. For example, the very simple tier 1 pollination supply map (Figure 15.1) is substantially different to the tier 2 (Figure 15.2) and tier 3 (Figure 15.3) maps. The tier 1 map treats all land types which supply pollination equally. Adding knowledge about differences in capacity to supply pollination, through expert judgment (tier 2) or through a model (tier 3), considerably impacts the outcome. The tier 2 and tier 3 maps are relatively similar, but notable differences can be observed, particularly in the most southern area of the region of interest. This is a densely forested area in the Pyrenees, a mountain chain on the border between France and Spain. The tier 3 approach gives more weight to those land types which are in a pollinator's flight range from crops. These are typically forest edges, riparian areas, or linear elements in an agricultural landscape. The tier 2 map therefore overestimates the potential of forests to provide pollination services to cropland. When observations for validating maps are not available, it is important that mapmakers are transparent about the indicators used, the methods applied, and the related uncertainties. Showing where different maps agree and disagree can assist stakeholders in map interpretation (Schulp et al., 2014b).

On the demand side, the tier 1 (Figure 15.4) and tier 2 (Figure 15.5) maps are relatively similar, but they differ notably from the tier 3 map (Figure 15.6), which is based on actual land use by

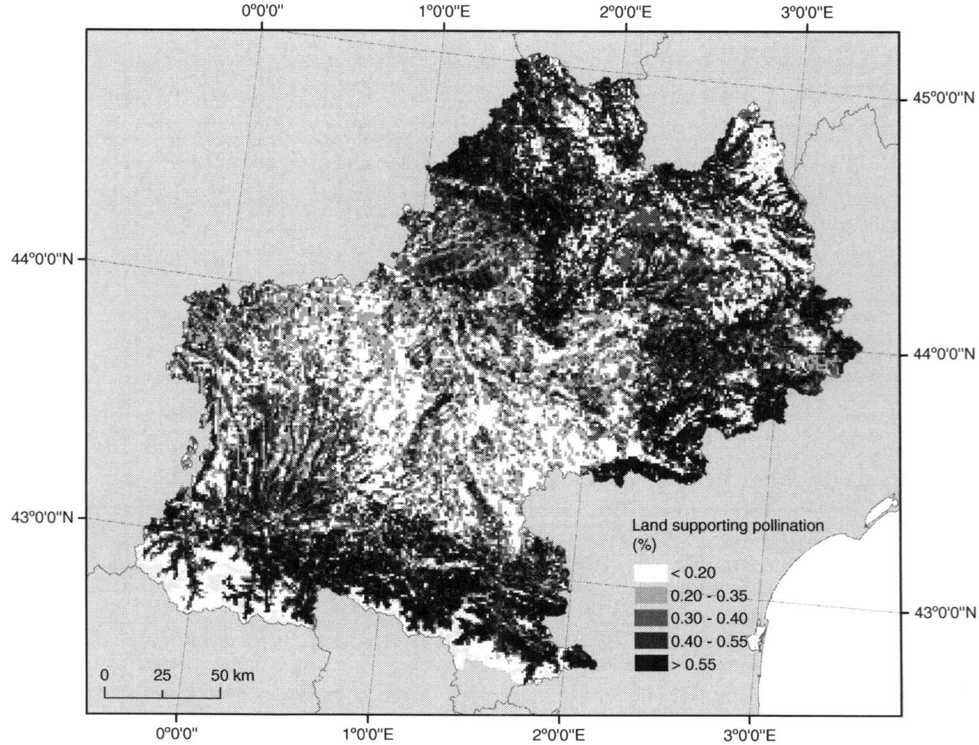

Figure 15.1 Tier 1 ecosystem service supply map of pollination. This map was produced using data on the share of land pixels (1ha) in 1 km² cells which support pollinators.

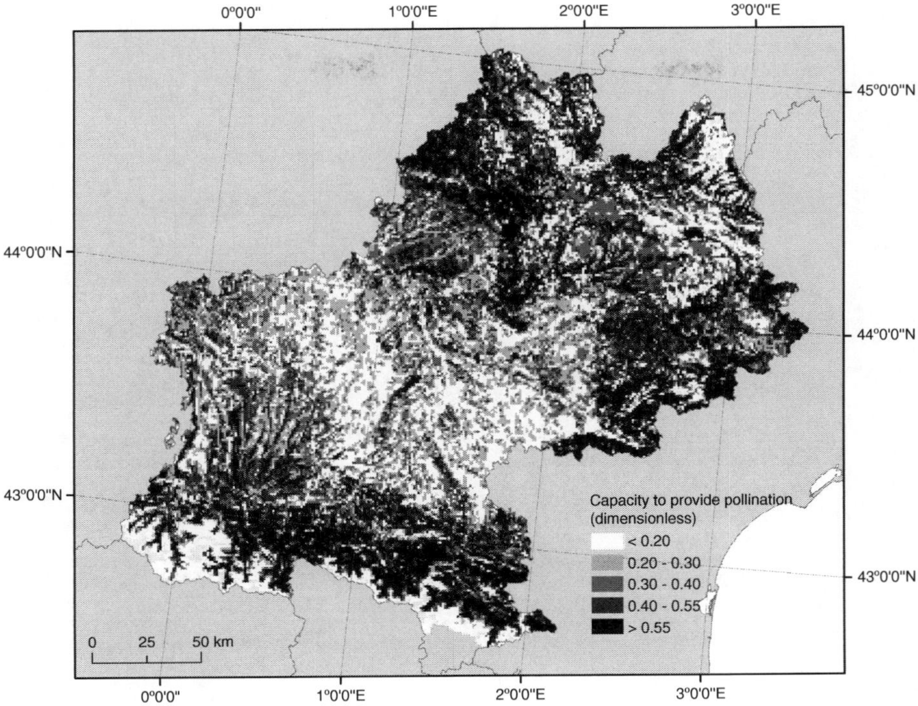

Figure 15.2 Tier 2 ecosystem service supply map of pollination. This map was produced using a mixture of land cover and human judgement inputs and shows the average capacity of land cover types to supply pollination, using the Burkhard et al. (2014) methods.

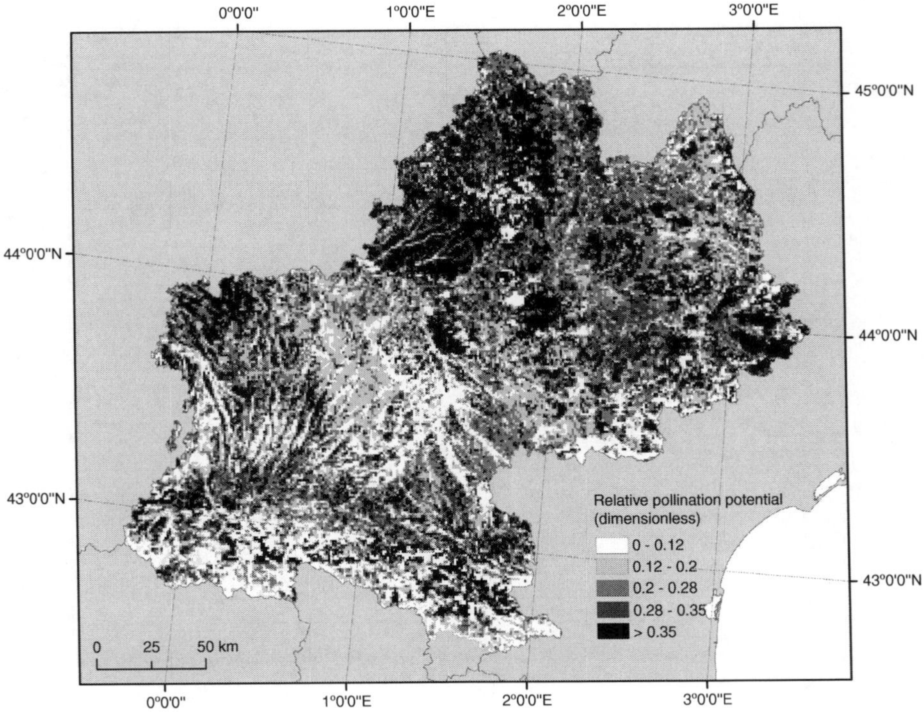

Figure 15.3 Tier 3 ecosystem service supply map of pollination. This map was produced using a relative pollination potential model which scores land pixels on their nesting and flower availability making use of multiple datasets and linked to the flight distance of wild bees (Zulian et al., 2014).

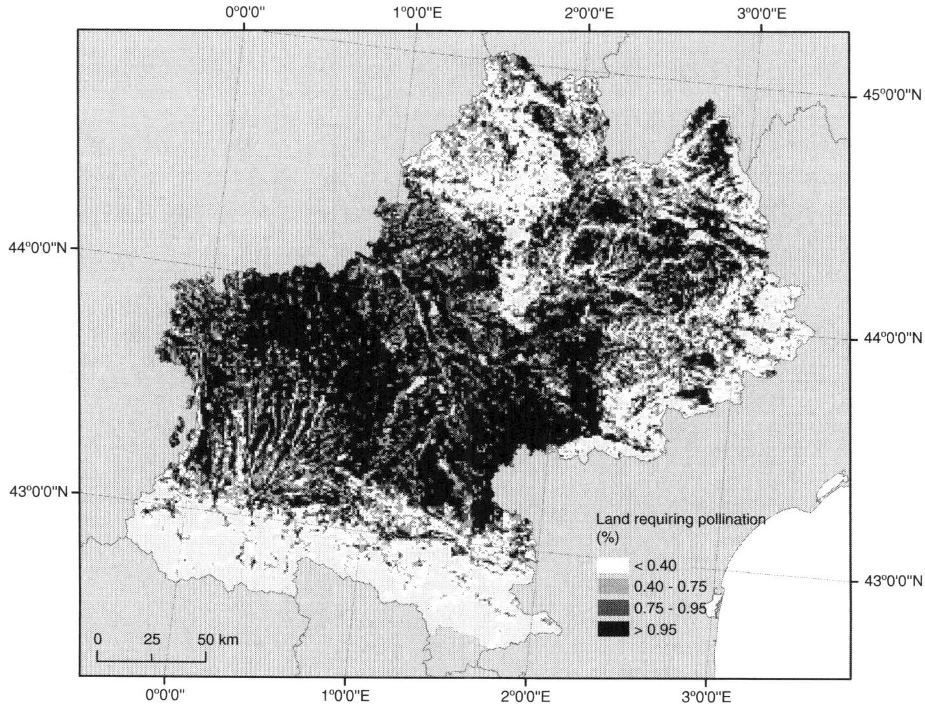

Figure 15.4 Tier 1 ecosystem service demand map of pollination. This map was produced using data on the share of land pixels (1ha) in 1 km² cells which require pollination.

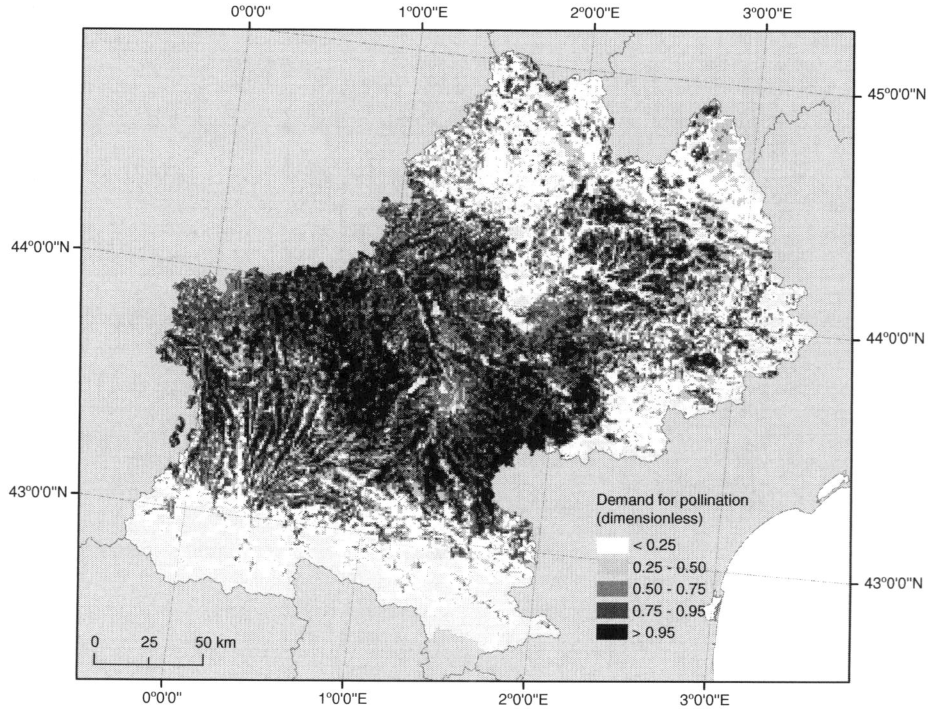

Figure 15.5 Tier 2 ecosystem service demand map of pollination. This map was produced using a mixture of land cover and human judgement inputs and shows the average demand pollination by land cover types, using the Burkhard et al. (2014) methods.

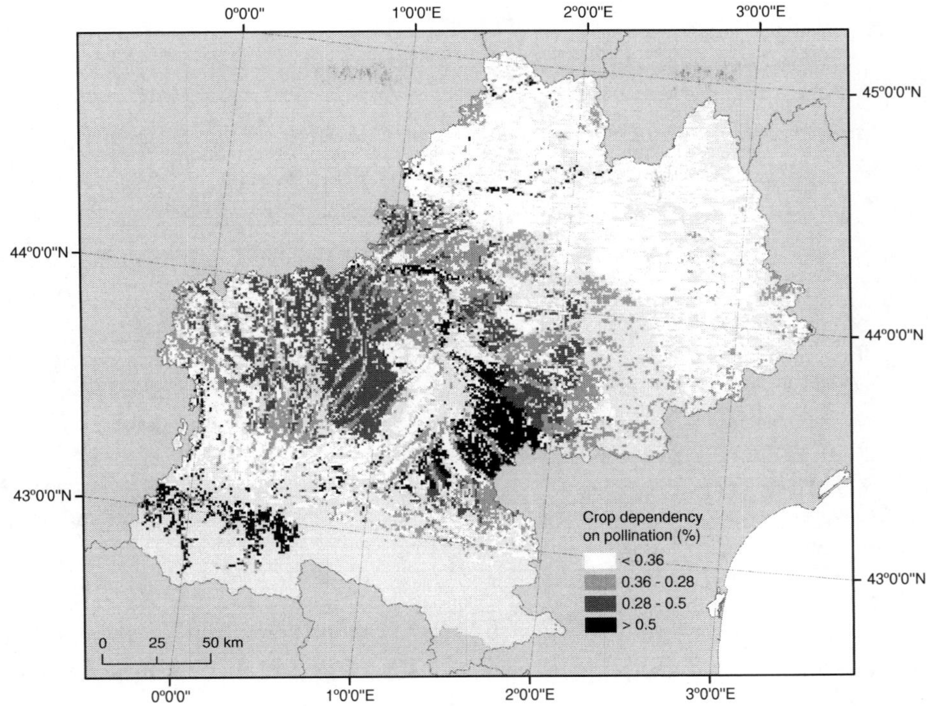

Figure 15.6 Tier 3 ecosystem service demand map of pollination. This map was produced by estimating demand of crops for pollination services by taking crop land use data and the dependency of different crops on insect pollination (Zulian et al., 2013).

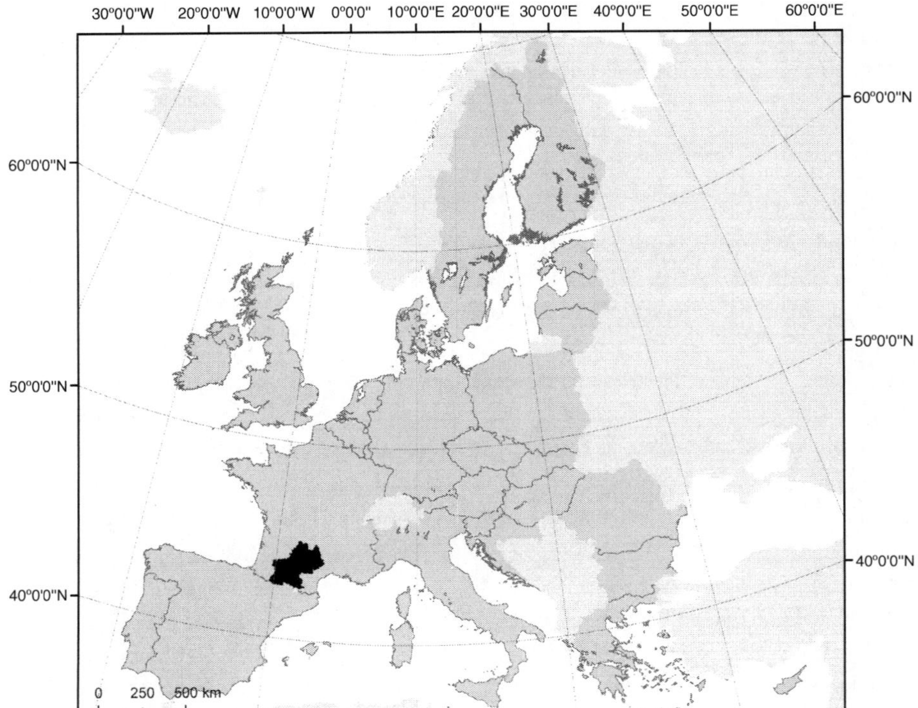

Figure 15.7 Study area.

different crop types with different dependencies on pollination. Areas where grain and maize are the dominant crop do not require pollination, whereas areas where sunflowers and fruits are cultivated are highly dependent. This difference is not made in the tier 1 and tier 2 demand maps.

Indicators for mapping ecosystem services

Ecosystem service indicators are pieces of information that efficiently communicate the characteristics and trends of ecosystem services, making it possible for policymakers to understand the condition, trends, and rate of change in ecosystem services. Ecosystem service indicators can be mapped if they contain geographic information such as coordinates, or if they are based on information that is collected for a well-defined geographic location, confined by geological, geographic, or administrative boundaries. Several recent reviews (Martínez-Harms and Balvanera, 2012; Egoh et al., 2012; Crossman et al., 2013b) summarize the recent literature on indicators that have been used to map ecosystem services. More information on ecosystem service indicators can be found in Müller et al. (2016).

State-of-the-art ecosystem service mapping tools

The mainstreaming of ecosystem services into public and private decision-making (see Box 15.1) has led to the development of numerous tools and decision support systems for quantifying and mapping ecosystem services. Current tools range from simple spreadsheet models to complex software (Bagstad et al., 2013b). Spreadsheet models use lookup-tables to link the quantity of an ecosystem service, such as one derived from field experiments or expert judgment (Jacobs et al., 2015), to other data which are readily available, such as land cover or species occurrences. These simple models document ecosystem services and their inter-relationships within a region, possibly including the underlying ecosystem functions and biodiversity. The advantage of these models is they can be applied simply and relatively quickly by stakeholders where limited empirical data are available (Burkhard et al., 2009, 2012; Kienast et al., 2009).

Box 15.1 Mapping and assessment of ecosystems and their services in the European Union (MAES): a dedicated action of the EU Biodiversity Strategy

The mapping and assessment of ecosystems and their services is an essential part of the EU Biodiversity Strategy to 2020 and a necessary condition to make ecosystems and their services key parameters informing planning and development processes and decisions. In particular, Action 5 of the Strategy requires Member States, with the assistance of the European Commission, to map and assess the state of ecosystems and their services in their national territory by 2014, assess the economic value of such services, and promote the integration of these values into accounting and reporting systems at EU and national level by 2020.

The European working group on Mapping and Assessment of Ecosystems and their Services (MAES), which includes experts of the European Commission, the Member States, and the research community, has been instrumental in providing an analytical framework, a typology of ecosystems and ES, and a first set of indicators for mapping and assessment (Maes et al., 2013; 2014). Importantly,

ESMERALDA, a research project funded under the European Commission's framework program for research (Horizon 2020) will provide much more detailed guidance to various stakeholders for mapping and assessing ecosystem services.

The work being carried out on the mapping and assessment of ecosystems and ecosystem services is not only important for the advancement of biodiversity objectives, including the development of Europe's green infrastructure, but also for informing the development and implementation of related policies on water, climate, agriculture, forest, and regional planning.

More information is available at: http://biodiversity.europa.eu/maes

More advanced tools for modelling and mapping are user-driven and are typically designed for a specific application or policy decision. Examples of such models are ESTIMAP (Zulian et al., 2014) and CLIMSAVE (Harrison et al., 2016).

A new generation of generic ecosystem service modelling and mapping tools have been designed for replicable and quantifiable ecosystem service analyses, independent of scale, site or policy context. Bagstad et al. (2013b) reviewed 17 of these generic-type tools, many of which are provided as GIS toolboxes or as stand-alone applications. Arguably the most commonly used tools are InVEST, ARIES, SolVES, and GUMBO. InVEST, the open-access GIS-based Integrated Tool to Value Ecosystem Services and their trade-offs[2] (Kareiva et al., 2011), is currently the most prominent ecosystem service mapping tool. InVEST contains separate ecosystem service models and uses land cover data as the major input in a tiered approach. Model outputs include ecosystem service estimates in biophysical or monetary units. ARIES, the web-based Artificial Intelligence for Ecosystem Services[3] (Villa et al., 2009) is also a more common ecosystem service mapping and valuation tool. ARIES applies Bayesian Belief networks to analyse ecosystem service flows from supply areas to beneficiaries. SolVES, the Social Values for Ecosystem Services,[4] is a GIS tool focussing on assessing, quantifying and mapping perceived social values for ecosystems (Sherrouse et al., 2011). GUMBO, the Global Unified Metamodel of the Biosphere,[5] simulates global interactions of natural, built, social, and human capital and their dynamics (Boumans et al., 2002).

Mapping standards and catalogues

Achieving common standards and adopting agreed procedures is of paramount importance to operationalize the growing body of knowledge on ecosystem services. Yet, as shown by several studies, consistency in mapping approaches is still a major challenge. Crossman et al. (2013b) addressed this challenge by proposing a template to assist the quantification and mapping of ES. Although each mapping study will necessarily be driven by the available data, the blueprint proposed by Crossman et al. (2013b) provides a checklist of information needed to model and map ecosystem services. Furthermore, it is a first step towards the compilation of templates summarizing ecosystem service mapping studies in a standardized format, so that maps can be stored in on-line databases for further use by the community of practitioners.

A major initiative has emerged that provides a centralized platform to store and share mapped ecosystem service data that follows consistent standards. The Ecosystem Services Partnership Visualization tool[6] (ESP-VT) is a visualization and data-sharing tool for ecosystem service maps. ESP-VT is an online tool that stores, organizes, visualizes, and shares ecosystem service maps and associated documentation and metadata (Drakou et al., 2015). It is a joint development of the thematic working group on Mapping Ecosystem Services of the Ecosystem Services Partnership, the Joint Research Centre of the European Commission, and the Commonwealth Scientific and Industrial Research Organisation (CSIRO) in Australia. The ESP-VT supports communication within and beyond the community of ecosystem service mappers by providing a common framework and set of standards for cataloguing and displaying ecosystem service spatial data. The ESP-VT is unique among the current ecosystem service tools because it visualizes, compares, and makes ecosystem service maps available for sharing within the ES community. The ESP-VT can be used to store and share potentially any ecosystem service map, whether produced for a small case study area using simple spread-sheet and/or tier 1 methods, or for a larger geographic area using more complex modelling and tools (e.g. InVEST, ARIES).

Challenges

Despite the recent advancements with respect to mapping ecosystem services, including mapping typologies (Pagella and Sinclair, 2014), common standards (Crossman et al., 2013b; Drakou et al., 2015), and modelling tools and decision support systems (Bagstad et al., 2013b), several major challenges remain to be addressed. Two challenges frequently identified in current literature are the need for enhanced collaboration with stakeholders to incorporate their knowledge and perspectives, and a better assessment of uncertainty (Maes et al., 2012; Pagella and Sinclair, 2014; Alkemade et al., 2014; Burkhard et al., 2013; Grêt-Regamey et al., 2013; Liu et al., 2013; Schulp et al., 2014b). These interrelated challenges are concerned with the scientific quality and accuracy of maps to support decisions in natural resources management, river basin management, land use planning activities, or conservation. Stakeholders, decision-makers and general users of mapping products have often-legitimate questions and concerns about maps. Maps are powerful visual communication tools with an "air of authority" (Hauck et al., 2013) that may lead to improper use. For example, the use of tier 1 ecosystem service supply maps to support decisions on investment in either further exploitation, or restoration, of specific ecosystem services should be done with much caution because of the simplicity and therefore high uncertainty of the models and data used to derive the maps.

Ecosystem services, which link ecosystems to people and economies, are inherently complex and uncertain, but novel participatory decision-making techniques such as Bayesian Networks (Haines-Young, 2011) and Multi-Criteria Analysis (Liu et al., 2013), have recently been used to bring in wider knowledge domains and reduce uncertainty. By tapping into local knowledge, participatory decision making techniques offer ways to better bridge the science-policy interface and bring together end-users in science and decision- and policy-making in order to match what can be offered and what is really needed from mapped ecosystem services.

Data sharing capabilities need to be enhanced and data gaps such as the incomplete coverage of, in particular, marine areas, are further challenges that need to be overcome. The ESP-VT (Drakou et al., 2015), should go some way toward identifying data gaps and improving the sharing of ecosystem services spatial data. Furthermore, the Marine Ecosystem Service Partnership[7] is working on reducing gaps in marine ecosystem services data.

There is also a need to increase our capacity to measure ecosystem services in order to provide more accurate ecosystem services maps. For example, Crossman et al. (2013a) argue for improved integration of biophysical (biotic, abiotic), land use (intensity) and socio-economic data.

Conclusions

The ecosystem services framework is increasingly popular to support decision-making, planning, and policy in both the public and private sectors, from local to global scales (World Business Council for Sustainable Development, 2012; World Bank, 2012; UK National Ecosystem Assessment, 2011). Critical to this growth in application of ecosystem services are robust quantification, valuation and mapping of supply, and demand and flows of goods and services (Crossman et al., 2013b; Burkhard et al., 2014). The scientific literature abounds with maps of ecosystem services, and many, in combination with an untold number of unpublished maps, are being used to assist decision-makers to prioritize investments in land, water, and ocean management. Ecosystem services maps visualize and quantify where and to what extent ecosystems contribute to human well-being and therefore maps operationalize the ecosystem services framework and concepts.

Examples where maps of ecosystem services support decisions include conservation planning, green infrastructure development, ecological restoration, and carbon sequestration. Ecosystem service maps also underpin the achievement of many targets across a number of policy areas (such as the European Union Biodiversity Strategy to 2020 see Maes et al., 2012), which need spatial information of ecosystem services including agriculture, urban planning, and river basin management.

To understand ecosystem service provision in a spatial context, there is a need to identify both where services are generated and where they are used. Accordingly, several methods have been developed to map supply of and demand for ecosystem services, with recent attention to the mapping of ecosystem service flows. This has resulted in a growing number of tools and decision support systems which can be used to map ecosystem services.

Key to high-quality ecosystem services maps are scientific accuracy, reproducibility, and credibility. This can be achieved by incorporating stakeholder knowledge, by the use of common mapping standards and by improving our systems to monitor ecosystem services.

Notes

1 http://cices.eu/
2 http://www.naturalcapitalproject.org/InVEST.html
3 http://www.ariesonline.org/
4 http://solves.cr.usgs.gov/
5 http://ideas.repec.org/a/eee/ecolec/v41y2002i3p529-560.html
6 esp-mapping.net
7 http://www.marineecosystemservices.org/about

References

Alkemade, R., Burkhard, B., Crossman, N.D., Nedkov, S., and Petz, K. (2014). Quantifying ecosystem services and indicators for science, policy and practice. *Ecological Indicators*, vol 37, part A, pp 161–162

Bagstad, K.J., Johnson, G.W., Voigt, B., and Villa, F. (2013a). Spatial dynamics of ecosystem service flows: a comprehensive approach to quantifying actual services. *Ecosystem Services*, vol 4, pp 117–125.

Bagstad, K.J., Semmens, D.J., Waage, S., and Winthrop, R. (2013b). A comparative assessment of decision-support tools for ecosystem services quantification and valuation. *Ecosystem Services*, vol 5, pp 27–39.

Bastian, O., Haase, D., and Grunewald, K. (2012). Ecosystem properties, potentials and services – The EPPS conceptual framework and an urban application example. *Ecological Indicators*, vol 21, pp 7–16.

Boumans, R., Costanza, R., Farley, J., et al. (8 authors) (2002). Modeling the dynamics of the integrated earth system and the value of global ecosystem services using the GUMBO model. *Ecological Economics*, vol 41, pp 529–560.

Brander, L., Brouwer, R., and Wagtendonk, A. (2013). Economic valuation of regulating services provided by wetlands in agricultural landscapes: a meta-analysis. *Ecological Engineering*, vol 56, pp 89–96.

Bryan, B.A., and Crossman, N.D. (2013). Impact of multiple interacting financial incentives on land use change and the supply of ecosystem services. *Ecosystem Services*, vol 4, pp 60–72.

Bryan, B.A., Raymond, C.M., Crossman, N.D., and King, D. (2011). Comparing spatially explicit ecological and social values for natural areas to identify effective conservation strategies. *Conservation Biology*, vol 25, pp 172–181.

Burkhard, B., Crossman, N., Nedkov, S., Petz, K., and Alkemade, R. (2013). Mapping and modelling ecosystem services for science, policy and practice. *Ecosystem Services*, vol 4, pp 1–3.

Burkhard, B., Kandziora, M., Hou, Y., and Müller, F. (2014). Ecosystem service potentials, flows and demands – concepts for spatial localisation, indication and quantification. *Landscape Online*, vol 34, pp 1–32.

Burkhard, B., Kroll, F., Müller, F., and Windhorst, W. (2009). Landscapes capacities to provide ecosystem services – a concept for land-cover based assessments. *Landscape Online*, vol 15, pp 1–22.

Burkhard, B., Kroll, F., Nedkov, S., and Muller, F. (2012). Mapping ecosystem service supply, demand and budgets. *Ecological Indicators*, vol 21, pp 17–29.

Chan, K.M.A., Satterfield, T., and Goldstein, J. (2012). Rethinking ecosystem services to better address and navigate cultural values. *Ecological Economics*, vol 74, pp 8–18.

Chan, K.M.A., Shaw, M.R., Cameron, D.R., Underwood, E.C., and Daily, G.C. (2006). Conservation planning for ecosystem services. *Plos Biology*, vol 4, pp 2138–2152.

Cimon-Morin, J., Darveau, M., and Poulin, M. (2013). Fostering synergies between ecosystem services and biodiversity in conservation planning: a review. *Biological Conservation*, vol 166, pp 144–154.

Costanza, R. (2008). Ecosystem services: multiple classification systems are needed. *Biological Conservation*, vol 141, pp 350–352.

Cowling, R.M., Egoh, B., Knight, A.T., O'Farrell, P.J., et al. (9 authors) (2008). An operational model for mainstreaming ecosystem services for implementation. *Proceedings of the National Academy of Sciences of the United States of America*, vol 105, pp 9483–9488.

Crossman, N.D., Bryan, B.A., de Groot, R.S., Lin, Y.-P., and Minang, P.A. (2013a). Land science contributions to ecosystem services. *Current Opinion in Environmental Sustainability*, vol 5, pp 509–514.

Crossman, N.D., Bryan, B.A., and Summers, D.M. (2011). Carbon payments and low-cost conservation. *Conservation Biology*, vol 25, pp 835–845.

Crossman, N.D., Burkhard, B., Nedkov, S., et al. (14 authors) (2013b). A blueprint for mapping and modelling ecosystem services. *Ecosystem Services*, vol 4, pp 4–14.

de Groot, R.S., Alkemade, R., Braat, L., Hein, L., and Willemen, L. (2010). Challenges in integrating the concept of ecosystem services and values in landscape planning, management and decision making. *Ecological Complexity*, vol 7, pp 260–272.

Drakou, E.G., Willemen, L., Crossman, N.D., et al. (8 authors) (2015). A visualization and data-sharing tool for ecosystem service maps: lessons learned, challenges and the way forward. *Ecosystem Services*, vol 13, pp 134–190.

Egoh, B., Drakou, E.G., Dunbar, M.B., Maes, J. and Willemen, L. (2012). *Indicators for Mapping Ecosystem Services: A Review*. Publications Office of the European Union, Luxembourg.

Eigenbrod, F., Armsworth, P.R., Anderson, B.J., et al. (8 authors) (2010). The impact of proxy-based methods on mapping the distribution of ecosystem services. *Journal of Applied Ecology*, vol 47, pp 377–385.

Fagerholm, N., Käyhkö, N., Ndumbaro, F., and Khamis, M. (2012). Community stakeholders' knowledge in landscape assessments – Mapping indicators for landscape services. *Ecological Indicators*, vol 18, pp 421–433.

García-Nieto, A.P., García-Llorente, M., Iniesta-Arandia, I., and Martín-López, B. (2013). Mapping forest ecosystem services: from providing units to beneficiaries. *Ecosystem Services*, vol 4, pp 126–138.

Grêt-Regamey, A., Brunner, S.H., Altwegg, J., Christen, M., and Bebi, P. (2013). Integrating expert knowledge into mapping ecosystem services trade-offs for sustainable forest management. *Ecology and Society*, vol 18, no 3. Available at: http://dx.doi.org/10.5751/ES-05800-180334

Haines-Young, R. (2011). Exploring ecosystem service issues across diverse knowledge domains using Bayesian Belief Networks. *Progress in Physical Geography*, vol 35, pp 681–699.

Haines-Young, R. and Potschin, M. (2010). The links between biodiversity, ecosystem services and human well-being. In Raffaelli, D. G. and Frid, C.L.J. (eds) *Ecosystem Ecology. A New Synthesis*. Cambridge University Press, Cambridge UK.

Harrison, P.A., Holman, I.P., and Berry, P. M. (2015). Assessing cross-sectoral climate change impacts, vulnerability and adaptation: an introduction to the CLIMSAVE project. *Climatic Change*, vol 128, pp 153–167.

Hauck, J., Görg, C., Varjopuro, R., et al. (7 authors) (2013). "Maps have an air of authority": potential benefits and challenges of ecosystem service maps at different levels of decision making. *Ecosystem Services*, vol 4, pp 25–32.

Honey-Rosés, J., and Pendleton, L.H. (2013). A demand driven research agenda for ecosystem services. *Ecosystem Services*, vol 5, pp e160-e162.

Jacobs, S., Burkhard, B., Van Daele, T., Staes, J., and Schneiders, A. (2015). "The Matrix Reloaded": a review of expert knowledge use for mapping ecosystem services. *Ecological Modelling*, vol 295, pp 21–30.

Johnson, G.W., Bagstad, K.J., Snapp, R.R., and Villa, F. (2012). Service path attribution networks (SPANs): a network flow approach to ecosystem service assessment. *International Journal of Agricultural and Environmental Information Systems*, vol 3, pp 54–71.

Kareiva, P., Tallis, H., Ricketts, T.H., Daily, G.C., and Polasky, S. (2011). *Natural Capital: Theory and Practice of Mapping Ecosystem Services*. Oxford University Press, Oxford.

Kienast, F., Bolliger, J., Potschin, M., et al. (8 authors) (2009). Assessing landscape functions with broad-scale environmental data: insights gained from a prototype development for Europe. *Environmental Management*, vol 44, pp 1099–1120.

Kroll, F., Muller, F., Haase, D., and Fohrer, N. (2012). Rural-urban gradient analysis of ecosystem services supply and demand dynamics. *Land Use Policy*, vol 29, pp 521–535.

Liu, S., Crossman, N. D., Nolan, M., and Ghirmay, H. (2013). Bringing ecosystem services into integrated water resources management. *Journal of Environmental Management*, vol 129, pp 92–102.

Luck, G.W., Daily, G.C., and Ehrlich, P.R. (2003). Population diversity and ecosystem services, *Trends in Ecology and Evolution*, vol 18, pp 331–336.

MA (2005). *Ecosystems and Human Well-Being: Synthesis*. Island Press, Washington DC.

Maes, J., Egoh, B., Willemen, L., et al. (14 authors) (2012). Mapping ecosystem services for policy support and decision making in the European Union. *Ecosystem Services*, vol 1, pp 31–39.

Maes, J., Teller, A., Erhard, M., et al. (42 authors) (2013). *Mapping and Assessment of Ecosystems and their Services. An Analytical Framework for Ecosystem Assessments under Action 5 of the EU Biodiversity Strategy to 2020*. Publications Office of the European Union, Luxembourg.

Maes, J., Teller, A., Erhard, M., et al. (42 authors) (2014). *Mapping and Assessment of Ecosystems and their Services. Indicators for Ecosystem Assessments under Action 5 of the EU Biodiversity Strategy to 2020*. Publications Office of the European Union, Luxembourg.

Margules, C.R., and Pressey, R.L. (2000). Systematic conservation planning. *Nature*, vol 405, pp 243–253.

Martínez-Harms, M.J., and Balvanera, P. (2012). Methods for mapping ecosystem service supply: a review. *International Journal of Biodiversity Science, Ecosystem Services and Management*, vol 8, pp 17–25.

Meyfroidt, P., Rudel, T.K., and Lambin, E.F. (2010). Forest transitions, trade, and the global displacement of land use. *Proceedings of the National Academy of Sciences of the United States of America*, vol 107, pp 20917–20922.

Mubareka, S., Maes, J., Lavalle, C., and De Roo, A. (2013). Estimation of water requirements by livestock in Europe. *Ecosystem Services*, vol 4, pp 139–145.

Müller, F., Burkhard, B., Hou, Y., Kruse, M., Ma, L. and Wangai, P. (2016). Indicators for ecosystem services. In Potschin, M., Haines-Young, R., Fish, R. and Turner, R.K. (eds) *Routledge Handbook of Ecosystem Services*. Routledge, London and New York, pp 157–169.

Nedkov, S., and Burkhard, B. (2012). Flood regulating ecosystem services – Mapping supply and demand, in the Etropole municipality, Bulgaria. *Ecological Indicators*, vol 21, pp 67–79.

Pagella, T.F., and Sinclair, F.L. (2014). Development and use of a typology of mapping tools to assess their fitness for supporting management of ecosystem service provision. *Landscape Ecology*, vol 29, pp 383–399.

Palomo, I., Martín-López, B., Potschin, M., Haines-Young, R., and Montes, C. (2013). National Parks, buffer zones and surrounding lands: mapping ecosystem service flows. *Ecosystem Services*, vol 4, pp 104–116.

Palomo, I., Montes, C., Martín-López, B., et al. (7 authors) (2014). Incorporating the social–ecological approach in protected areas in the anthropocene. *Bioscience*. Available at: doi: 10.1093/biosci/bit033

Plieninger, T., Dijks, S., Oteros-Rozas, E., and Bieling, C. (2013). Assessing, mapping, and quantifying cultural ecosystem services at community level. *Land Use Policy*, vol 33, pp 118–129.

Potschin, M., and Haines-Young, R. (2016). Defining and measuring ecosystem services. In Potschin, M., Haines-Young, R., Fish, R. and Turner, R.K. (eds) *Routledge Handbook of Ecosystem Services*. Routledge, London and New York, pp 25–44.

Raymond, C.M., Bryan, B.A., Macdonald, D. H., et al. (7 authors) (2009). Mapping community values for natural capital and ecosystem services. *Ecological Economics*, vol 68, pp 1301–1315.

Richardson, L., Loomis, J., Kroeger, T., and Casey, F. (2014). The role of benefit transfer in ecosystem service valuation. *Ecological Economics*, vol 115, pp 51–58.

Santos-Martín, F., Martín-López, B., García-Llorente, M., et al. (6 authors) (2013). Unraveling the relationships between ecosystems and human wellbeing in Spain. *PLoS ONE*, vol 8, p e73249.

Schägner, J.P., Brander, L., Maes, J., and Hartje, V. (2013). Mapping ecosystem services' values: current practice and future prospects. *Ecosystem Services*, vol 4, pp 33–46.

Schröter, M., Barton, D.N., Remme, R.P., and Hein, L. (2014). Accounting for capacity and flow of ecosystem services: a conceptual model and a case study for Telemark, Norway. *Ecological Indicators*, vol 36, pp 539–551.

Schulp, C.J.E., Burkhard, B., Maes, J., Van Vliet J., and Verburg, P. H. (2014b). Uncertainties in ecosystem service maps: a comparison on the European scale. *PloS One*, vol 9, no 10, p e109643.

Schulp, C.J.E., Thuiller, W., and Verburg, P.H. (2014a). Wild food in Europe: a synthesis and mapping of the supply, demand and benefits of terrestrial wild food as an ecosystem service. *Ecological Economics*, vol 105, pp 292–305.

Serna-Chavez, H.M., Schulp, C.J.E., Van Bodegom, P.M., et al. (6 authors) (2014) 'A quantitative framework for assessing spatial flows of ecosystem services', *Ecological Indicators*, vol 39, pp 24–33.

Seto, K.C., Reenberg, A., Boone, C.G., et al. (10 authors) (2012). Urban land teleconnections and sustainability. *Proceedings of the National Academy of Sciences*, vol 109, pp 7687–7692.

Sherrouse, B.C., Clement, J.M., and Semmens, D.J. (2011). A GIS application for assessing, mapping, and quantifying the social values of ecosystem services. *Applied Geography*, vol 31, pp 748–760.

Stürck, J., Poortinga, A., and Verburg, P.H. (2014). Mapping ecosystem services: the supply and demand of flood regulation services in Europe. *Ecological Indicators*, vol 38, pp 198–211.

Tallis, H., Mooney, H., Andelman, S., et al. (13 authors) (2012). A global system for monitoring ecosystem service change, *Bioscience*, vol 62, pp 977–986.

TEEB (2010). *The Economics of Ecosystems and Biodiversity: Ecological and Economic Foundations*. Earthscan, London and Washington DC.

Troy, A., and Wilson, M.A. (2006). Mapping ecosystem services: practical challenges and opportunities in linking GIS and value transfer. *Ecological Economics*, vol 60, pp 435–449.

UK National Ecosystem Assessment (2011). *The UK National Ecosystem Assessment: Synthesis of the Key Findings*. UNEP-WCMC, Cambridge UK.

Villa, F., Ceroni, M., Bagstad, K., Johnson, G., and Kriviv, S. (2009). ARIES (ARtificial Intelligence for Ecosystem Services): a new tool for ecosystem services assessment, planning, and valuation. *11th International BioECON Conference on Economic Instruments to Enhance the Conservation and Sustainable Use of Biodiversity*, Venice, Italy.

Villamagna, A. M., Angermeier, P.L., and Bennett, E.M. (2013). Capacity, pressure, demand, and flow: a conceptual framework for analyzing ecosystem service provision and delivery. *Ecological Complexity*, vol 15, pp 114–121.

Wood, D. (2010). *Rethinking the Power of Maps*. The Guilford Press, New York.

World Bank (2012). *Wealth Accounting and the Valuation of Ecosystem Services*. Available at: http://www.wavespartnership.org/waves/

World Business Council For Sustainable Development (2012). *Biodiversity and Ecosystem Services: Scaling Up Business Solutions. Company Case Studies that Help Achieve Global Biodiversity Targets*. World Business Council for Sustainable Development, Geneva.

Zulian, G., Polce, C., and Maes, J. (2014). ESTIMAP: a GIS-based model to map ecosystem services in the European Union, *Annali di Botanica*, vol 4, pp 1–7.

16

A PRACTICAL APPROACH TO MAPPING OF ECOSYSTEMS AND ECOSYSTEM SERVICES USING REMOTE SENSING

Camino Liquete, Eva Haas, Torsten Bondo, Christina Hirzinger, Melanie Schnelle, Dominik Reisinger, David Lyon, John Finisdore and Michael Ledwith

Introduction

Earth observation (EO) via remote sensing technologies is increasingly being used for ecosystem assessment and valuation (Andrew et al., 2014; Potschin and Haines-Young, 2016a; Gomez-Baggethun et al., 2016). Processed satellite data enables development of spatially explicit and geographically broad assessments of ecosystems around the globe. These often cost less and take less time than field surveys, even when including resources required for ground-truthing of remote data. As satellites revisit the same areas at regular intervals, ecosystem change over time can be captured. Processing this data with information about users of ecosystem services (ES) allows decision-makers to understand not only ecological functions but also how they deliver benefits to people (Boyd and Banzhaf, 2007; Landers and Nahlik, 2013). As a result, remotely sensed or EO products can support defining, measuring and assessing ecosystem services effectively.

EO products measure ecosystems services using proxies or indicators based on key habitat characteristics. Satellite data can monitor one or more elements of the "cascade model" (see Potschin and Haines-Young, 2016b; Maes et al., 2016), such as the natural capacity of ecosystems to deliver services (ecosystem function in the cascade model) or the actual human use of them (ecosystem flow). Ideally, EO products should also address other elements of the cascade model, in particular the human benefit and valuation of ecosystem flows. This can be done by integrating in situ socio-economic data into EO products.

In this chapter we describe some practical examples of how EO products can be used for the assessment of specific ecosystem services around the globe. A more general briefing on remote sensing approaches and methods can be found in Foody (2016).

Mapping ecosystems and their services from space

Two European Space Agency funded projects demonstrate the value of EO products for ecosystem service evaluation.[1] The *Geographic Ecosystem Monitoring and Assessment Service* (G-ECOMON) project is focused on ecosystem services assessments in multiple sectors, and the *Earth Observation Services for Ecosystem Valuation* (ECOSERVE) project is demonstrating the value of EO products for economic or social valuation.

These projects led to the development of an EO4ES Matrix (Tables 16.1, 16.2 and 16.3) where a list of ecosystem services' indicators (classified following *The Economics of Ecosystems and Biodiversity*[2] and the cascade model, and based on sources such as Egoh et al., 2012 and Maes et al., 2014) is linked with remote sensing and derived products. The Matrix speeds practitioners' identification of EO products that meet their needs. It also shows that any one EO product can be used for different ecosystem services and one ecosystem service indicator can be estimated with various EO products. The list of ecosystem service indicators is not exhaustive, but illustrative.

Table 16.1 Matrix linking EO based products and ecosystem service indicators, here for the example of disaster risk reduction in Afghanistan.

ECOSYSTEM SERVICES				*EO BASED PRODUCTS*			
Section	*Ecosystem service*	*Type of indicator*	*Example of indicator*	*DSM/ DTM/ nDSM*	*Land cover characterisation & change*	*Land use characterisation & change*	*Erosion potential mapping*
Provisioning services	Food	Capacity	Food and feed crop area				
			Managed grassland area				
		Flow	Food and feed crop yield				
			Managed grassland yield				
	Raw materials	Capacity	Wood fuel stock and change				
Regulating services	Erosion prevention and maintenance of soil fertility	Capacity	Vegetal cover				
			Soil erodibility area				
			Forested area				
			Slope				
		Flow	Universal Soil Loss Equation (USLE)				

Note: the squared cells point to direct relationships, where EO information covers the indicator's needs; while the hatched cells imply indirect relationships, where additional information is usually required. DSM/DTM/nDSM= Digital Surface Model/Digital Terrain Model/Normalized Digital Surface Model

Table 16.2 Matrix linking EO based products and ecosystem service indicators, here for the example of biomass potential for energy production in Liberia.

ECOSYSTEM SERVICES				*EO BASED PRODUCTS*			
Section	*Ecosystem service*	*Type of indicator*	*Example of indicator*	*Land cover characterisation & change*	*Land use characterisation & change*	*Forest extent, type and density*	*Biomass estimate*
Provisioning services	Raw Materials	Capacity	Wood fuel stock and change				
			Presence/ distribution of commercial tree species				
		Flow	Timber harvesting rates				
			Wood fuel production				
Regulating services	Carbon sequestration & storage	Capacity	Above-ground biomass				

Note: the dark grey squared cells point to direct relationships, where EO information covers the indicator's needs; while the hatched cells imply indirect relationships, where additional information is usually required

Table 16.3 Matrix linking EO based products and ecosystem service indicators, here for the example of monitoring habitat in a national park in Indonesia.

ECOSYSTEM SERVICES				*EO BASED PRODUCTS*			
Section	*Ecosystem service*	*Type of indicator*	*Example of indicator*	*Land cover characterisation & change*	*Land use characterisation & change*	*Forest extent, type and density*	*Biomass estimate*
Provisioning services	Raw Materials	Capacity	Wood fuel stock and change				
			Fire occurrence data				
		Flow	Timber harvesting rates				
Habitat services	Habitat for species	Capacity	Habitat mapping				
			Fragmentation of the natural habitats				
Cultural services	Tourism	Capacity	Naturalness				
			Distribution of habitats for emblematic species				
			Accessibility				

Note: the squared cells point to direct relationships, where EO information covers the indicator's needs; while the hatched cells imply indirect relationships, where additional information is usually required

Examples

A core component of the G-ECO-MON project was the use of eleven trials to demonstrate the utility of several EO products. The applications included impact assessment, production assessment, payment for ecosystem services and ecosystem accounting among others. This section illustrates the results from this work.

Ecosystem services for disaster risk reduction, Afghanistan

The valleys in the Koh-e Baba Mountains, located in the western Hindukush Mountains of Afghanistan, are prone to a multitude of natural hazards, including avalanches, flash floods and droughts. EO products allowed the mapping and monitoring of the mountain ecosystems and their services, providing information for risk reduction (Table 16.1).

The main ecosystem services that occur in Koh-e Baba are regulating services from rangelands that moderate extreme events like landslides and avalanches through erosion prevention. A digital surface model and a land use classification consisting of agriculture areas, grassland, forest, rangelands, shrublands, bare soil, snow and ice areas as well as settlements (Figure 16.1) provided UNEP Eco-DRR (United Nations Environment Programme Ecosystem-based Disaster Risk Reduction) a baseline for use in a more detailed ecosystem survey and valuation. The final products will guide decisions for the sustainable management, conservation and restoration of ecosystems for risk reduction.

Figure 16.1 This sub meter digital surface model was derived from three Pléiades scenes (triplet) acquired in July 2014. This time of the year is ideally suited to map vegetation cover, and the black areas show rangelands that have important provisioning and regulating functions.

Source: map produced by GeoVille, Austria, including material of Pléiades © CNES 2014, Distribution Airbus DS

Identification of sustainable biomass, Liberia

Liberia's agricultural and forestry production systems can deliver sustainable rural electrification by utilising residues such as old rubber trees for small-scale electricity generation. EO products allow for the detection of the areal distribution and relative and age of rubber trees. The amount of biomass available can be estimated as well as the travel time necessary for collection.

The Kwendin, Liberia study focused on the provisioning ecosystem services of tree type, combined with an age differentiation, providing the amount of fuelwood available. Using a multi-temporal analysis, the harvesting rate and the amount of fuel production was detected. Furthermore, the calculation of above-ground biomass allowed Winrock International, the user in this particular case, to calculate carbon sequestration and storage in the fuelwood areas (Table 16.2).

Wildlife conservation in Indonesia

The forested wetlands in the Berbak National Park (Figure 16.2) – extending over 2400 km^2 in western Indonesia – teem with the chatter of migratory birds and IUCN Red-listed species, such

Figure 16.2 Land cover classification map of Berbak, Indonesia derived from a 2013 Landsat-8 image. Most of the area is covered by primary forest but there are large recent burn scars in the western side of the park. Notice the extension of logging activity along human-made canals into the protected area of the park. Map coordinate system: WGS 1984 UTM Zone 48S. Graticule: GCS WGS 1984.

Original source: land cover/use map (1:125 000) of the Berbak National Park, Sumatra, Indonesia. Created by Metria AB as part of the G-ECO-MON project funded by ESA. © 2013

as the Sumatran tiger. At the time of writing forest fires were threatening these ecosystems. Local population expansion and associated need for croplands and fuel have resulted in encroachment into these protected areas. As a result, a payment for ecosystem services (PES) scheme was introduced to limit the intrusion and to abate the impact on Sumatra's biodiversity.

The only realistic manner to verify the PES, considering the jungle terrain and lack of infrastructure, is through EO-based monitoring. Using satellite data, land cover classification maps for the years 2003, 2007, 2010 and 2013, as well as change detection maps for all image pairs, were produced for the Zoological Society of London (Table 16.3).

Conclusion and outlook

EO products are powerful and affordable tools to map and assess ecosystems services, as well as to analyse their potential degradation through time. Revisit and processing times allow for relatively quick assessments over time. This study shows that EO products can be directly linked to specific ecosystem services used in established conceptual frameworks. Their application in developing countries at different scales has been demonstrated in the presented case studies.

Both G-ECO-MON and ECOSERVE projects have produced applied reports. An article showing the full EO4ES Matrix and reviewing EO products that can map and assess ecosystem services is under preparation.

Notes

1 http://www.space4ecosystems.com/
2 TEEB classification: http://www.teebweb.org/resources/ecosystem-services/

References

Andrew, M. E., Wulder, M. A., and Nelson, T.A. (2014). Potential contributions of remote sensing to ecosystem service assessments. *Progress in Physical Geography*, vol 38, no 3, pp 328–353.

Boyd, J. W., and Banzhaf, S. (2007). What are ecosystem services? The need for standardized environmental accounting units. *Ecological Economics*, vol 63, pp 616–626.

Egoh, B., Drakou, E. G., Dunbar, M. B., Maes, J., and Willemen, L. (2012). *Indicators for Mapping Ecosystem Services: A Review*. Publications Office of the European Union, Luxembourg.

Foody, G. (2016). Remote Sensing. In Potschin, M., Haines-Young, R., Fish, R. and Turner, R.K. (eds) *Routledge Handbook of Ecosystem Services*. Routledge, London and New York, pp 211–212.

Gomez-Baggethun, E., Barton, D. N., Berry, P., Dunford, R., and Harrison, P. A. (2016). Concepts and methods in ecosystem services valuation. In: Potschin, M., Haines-Young, R., Fish, R., and Turner, R. K. (eds) *Routledge Handbook of Ecosystem Services*. Routledge, London and New York, pp 99–111.

Landers, D. H., and Nahlik, A. M. (2013). *Final Ecosystem Goods and Services Classification System (FEGS-CS)*. EPA/600/R-13/ORD-004914. U.S. Environmental Protection Agency, Office of Research and Development, Washington, DC.

Maes, J., Crossman, N. D., and Burkhard, B. (2016). Mapping ecosystem services. In: Potschin, M., Haines-Young, R., Fish, R., and Turner, R. K. (eds) *Routledge Handbook of Ecosystem Services*. Routledge, London and New York, pp 188–204.

Maes, J., Teller, A., Erhard, M., Murphy. P., Parching, M. L., et al. (46 authors) (2014). *Mapping and Assessment of Ecosystems and their Services. Indicators for Ecosystem Assessments under Action 5 of the EU Biodiversity Strategy to 2020*, 2nd report. Publications Office of the European Union, Luxembourg.

Potschin, M. and Haines-Young. R. (2016a). Frameworks for ecosystem assessments. In: Potschin, M., Haines-Young, R., Fish, R., and Turner, R. K. (eds) Routledge Handbook of Ecosystem Services. Routledge, London and New York, pp 125–143.

Potschin, M., and Haines-Young. R. (2016b). Defining and measuring ecosystem services. In: Potschin, M., Haines-Young, R., Fish, R., and Turner, R. K. (eds) *Routledge Handbook of Ecosystem Services*. Routledge, London and New York, pp 25–44.

Briefing Note 16.1 Remote sensing

Giles M. Foody

Remote sensing is a means of acquiring data about the environment without being in direct contact with it (Lillesand et al., 2009). In the context of ecosystem services, the key focus is typically on the use of images acquired by sensors mounted on airborne and/or spaceborne platforms. In most cases the sensor is pointed vertically downwards and acquires data for the entire field of view. As a result, the images acquired have a map-like format and provide spatially continuous data for the imaged area. Since the sensors that are widely used also typically acquire digital images, the data are in a form that is relatively easy to integrate with other spatial data sets within a geographical information system. This allows the remotely sensed imagery to be linked with other available information on the region and perhaps its human population to provide a suite of spatial information to inform studies of ecosystem services.

The fundamental basis of remote sensing is that the way electromagnetic radiation (e.g. sunlight) interacts with a target on the Earth's surface is a function of the target's properties. A remotely sensed image simply conveys information on the interaction that can be analysed to yield key ecosystem variables. The radiation used may be characterised by its wavelength, and most attention has been paid to those wavelengths that can pass through the atmosphere relatively unimpeded, ensuring that the signal recorded by the sensor is a function of the target's properties and not those of the atmosphere; cloud, however, can obscure the ground for all except microwave remote sensing.

Most remote sensing has made use of visible and near-infrared wavelength radiation. For example, aerial photography has been used to acquire data on the environment for decades. Standard aerial photographs can be interpreted to yield a vast array of information such as the type and condition of vegetation present and, especially if stereoscopic photography, the relief of the landscape. Although photography is still widely used, remote sensing has developed greatly since the launch of the first major Earth resources satellite programme, Landsat, in 1972. Building on the successful use of false colour infrared photography, which is especially informative on vegetation type and amount, sensors have been developed to capture reflected solar radiation in different parts of the spectrum. Much initial research was undertaken using multi-spectral sensors that measured the reflected radiation from the Earth's surface in a small number of broad spectral wavebands. For example, the pioneering Landsat multispectral scanning system (MSS), launched in 1972, measured the spectral response of the Earth's surface in four wavebands: green, red and two near-infrared. The data in each waveband could be analysed separately, but could also be combined to, for example, form a false colour composite image that was similar to colour infrared photography. Since then, sensing systems have advanced greatly. Hyperspectral sensors, for example, measure the reflectance in numerous narrow spectral wavebands. The detailed spectral response acquired by such sensors is useful in providing information on specific environmental features (e.g. chlorophyll content). Sensors have also been developed to explore other parts of the spectrum, notably at thermal infrared wavelengths, as well as at microwave wavelengths. The former provide information on the thermal properties of the environment (e.g. land surface temperature) while the latter is especially informative on its structure and amount (e.g. vegetation biomass). Microwave remote sensing also differs greatly from conventional methods, involving an active sensor, long wavelength radiation and a typically sideways looking sensor. A key attraction of microwave remote sensing is the ability of the radiation used to pass through the atmosphere with little if any attenuation, even in the presence of cloud. In addition, given that it is an active system, it is not dependent on sunlight, making it an all-day, weather-independent data source. There has also been a growth in laser (LiDAR) remote sensing which provides useful information, especially in relation to the vertical dimension helping, for example, to provide 3D characterisation of the environment which can be especially valuable for forested environments. Given that remote sensing systems, especially those in satellite orbits around the Earth, can be acquired repeatedly, remote sensing is now a major source of 4D data on the environment.

Remote sensing technology has evolved considerably in the last few decades. There are a vast array of remote sensing systems in operation, with many more planned. While some long-term programs, such as Landsat, have developed to facilitate environmental monitoring and so provide a relatively continuous set of observations in time, other systems have been developed to exploit new opportunities with, for example, wavebands positioned to help capture very specific information on the environment. There has also been considerable development of airborne systems, including the use of sensors mounted on unmanned aerial vehicles (UAVs), which allow data to be acquired for very specific locations. The various sensors deployed can be characterised by their spatial, spectral, temporal and radiometric properties. A user simply needs to focus on the attributes most suitable for the task at hand when selecting a remotely sensed data set. However, it is important to note that there are often trade-offs in key data set properties. For example, there is typically a trade-off between the spatial and temporal resolution of a remote sensing system. However, it is common to find that landscape to regional scale studies make use of data from sensors such as those carried on the Landsat series of satellites (typically with a spatial resolution in the 15–80 m range and acquired potentially every 16–18 days) while global scale studies would typically make use of data from a sensor such as MODIS (with a spatial resolution of 250m – 1 km but acquired daily); the exact details of the data available will vary with the specific system used and site characteristics.

Reference

Lillesand, T., Kiefer, R. J., and Chipman, J. (2009). *Remote Sensing and Image Interpretation*, sixth edition. Wiley, New York.

17

AN INTRODUCTION TO ECOSYSTEM ACCOUNTING

Lars Hein, Bram Edens,[1] *Carl Obst, Roy Remme, Matthias Schröter and Elham Sumarga*

Introduction

In this chapter, we provide an introduction to ecosystem accounting, its key properties and niche in the overall system for environmental-economic accounting, and some examples of its application. Ecosystem accounting provides a systematic framework to include ecosystem services and ecosystem assets in an accounting structure, aligned with the System of National Accounts (SNA, United Nations et al., 2009). The SNA is a comprehensive system for measuring and organising information on economic activities in a country. National accounts comprise several specific accounts, such as the production account recording the value of domestic production; capital accounts that describe investment, depreciation, and other changes in assets; and balance sheets which record economic and financial wealth. Because the SNA provides an international standard, indicators produced according to the national accounts (e.g. GDP) are comparable across countries.

It is widely acknowledged (e.g. Costanza et al., 2014) that national accounts do not adequately capture all facets of natural capital. In response, the System of Environmental-Economic Accounts (SEEA) has been developed over the course of 20 years as a satellite system to the SNA (United Nations et al., 2009). The SEEA describes various types of accounts, such as water accounts, energy accounts, land accounts, and economic accounts for environmental taxes and subsidies. Ecosystem accounts are a recent development, differing from traditional environmental accounts in the sense that they aim to analyse natural capital from the perspective of spatially explicit ecosystem assets. A first guideline was developed in 2013 (European Commission et al., 2013), and further development and testing of ecosystem accounting methodologies is on-going (Edens and Hein, 2013; Obst et al., submitted).

Characteristics of ecosystem accounting

A first important consideration in ecosystem accounting is that ecosystem services are distinguished from ecosystem benefits (TEEB, 2010; European Commission et al., 2013), with the service capturing the contribution from the ecosystem, and the benefit representing the way people use the ecosystem service. For instance, standing stock of timber represents a service, and harvested wood a benefit. Connecting an ecosystem account to the national accounts requires linking each service and benefit to production and consumption, and to specific economic sectors.

Two key elements of ecosystem accounting are ecosystem assets and ecosystem services flow. An asset represents the 'capacity of the ecosystem to generate services, now and in the future' (European Commission et al., 2013). In order to understand the future flow of services, it is important to understand the 'condition' or 'state' of the ecosystem. For instance, soil degradation may affect agricultural production in future years, or overfishing may affect either fish yields or the effort required to catch fish. For each service, there may therefore be a need to identify specific condition indicators as part of an ecosystem account. Given the spatial variability of ecosystem use, ecosystem service flows, and ecosystem assets and condition, a spatial approach is taken in ecosystem accounting (European Commission et al., 2013). To the extent possible, ecosystem accounting should consider ecological thresholds in ecosystem services supply (Hutchings and Myers, 1994; Scheffer and Carpenter, 2003). Figure 17.1 below presents the main elements of an ecosystem account. Even though the eventual aim of an ecosystem account may well be to capture all ecosystem assets and ecosystem services relevant to an economy, in a pilot phase it is likely that a more focussed approach is appropriate. The focus of the pilot phase may be determined by policy needs and priorities, as well as data availability and the feasibility of modelling specific services with currently available models.

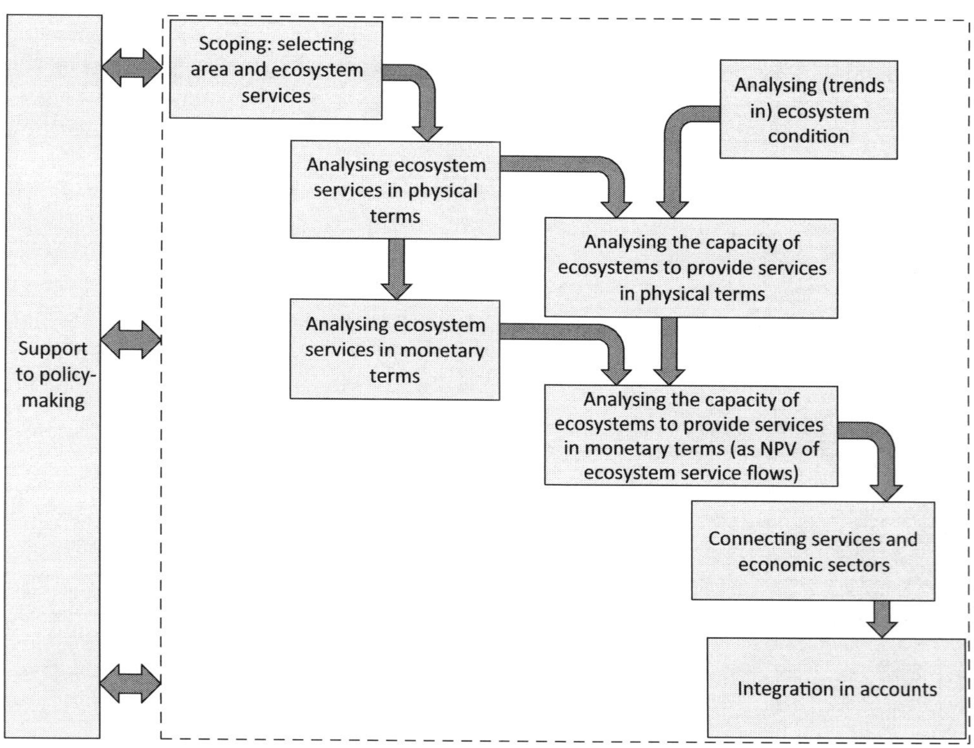

Figure 17.1 Compilation process of ecosystem accounting. The flow diagram presents the main steps required for developing an ecosystem account. The policy setting determines the ecosystem services to be included, and on the basis of this selection capacity and condition indicators need to be identified and assessed.

Monetary valuation in an accounting context focusses on measuring production. A key difference between welfare-based valuation (e.g. Bateman et al., 2011) and production-based valuation approaches is that the latter includes the value of production only and excludes consumer surplus. Production is valued through exchange values, which can be related to market values for those ecosystem services that have representative market values (cf. United Nations et al., 2009). The monetary value of the asset represents the net present value of the expected flow of ecosystem services, under current management. Further details on valuation in an accounting context are provided in European Commission et al. (2013).

Applications

Ecosystem accounts have a number of applications. Ecosystem accounts (i) present a comprehensive overview of ecosystem assets including the services provided; (ii) indicate interdependencies between ecosystems and economic activities; (iii) allow measuring changes in ecosystem assets over time; and (iv) have a number of other potential applications that can support environmental management. An example of how ecosystem services flows can be included in an ecosystem account is provided in Table 17.1. A full ecosystem account would, in line with Figure 17.1, also include information on ecosystem assets (opening and closing stocks, ecosystem condition), and services flows and assets would be expressed in both physical and monetary units (European Commission et al., 2013).

An example of how ecosystem accounts provide insight in the sustainability of landscape management is provided in Figure 17.2. Figure 17.2 shows that free-ranging sheep consume

Table 17.1 Basic ecosystem account of Limburg province, the Netherlands, for 2010. Mean annual flows per ha or km^2 and total annual flows (per land cover unit) of three ecosystem services are shown for five land cover types. The standard deviation (SD) shown in brackets reflects the spatial variability of mean flows (based on Remme et al., 2014).

Land cover type	*Ecosystem service*					
	Hunting		*Air quality regulation*		*Recreational cycling*	
	Mean (SD) kg meat km^{-2} yr^{-1}	*Total kg meat yr^{-1}*	*Mean (SD) kg PM$_{10}$ km^{-2} yr^{-1}*	*Total kg PM$_{10}$ yr^{-1} (x 1000)*	*Mean (SD) trips ha^{-1} yr^{-1}*	*Total trips yr^{-1} (x 1000)*
Pasture	21 *(17)*	9,100	909 *(528)*	404	103 *(77)*	1,863
Cropland	20 *(17)*	14,732	956 *(535)*	717	98 *(72)*	2,611
Forest	23 *(20)*	8,100	2,001 *(1,228)*	700	128 *(95)*	1,565
Heath	32 *(25)*	678	2,056 *(1,116)*	45	84 *(61)*	30
Peat	13 *(3)*	70	968 *(347)*	7	84 *(43)*	3
Total		32,680		1,873		6,072

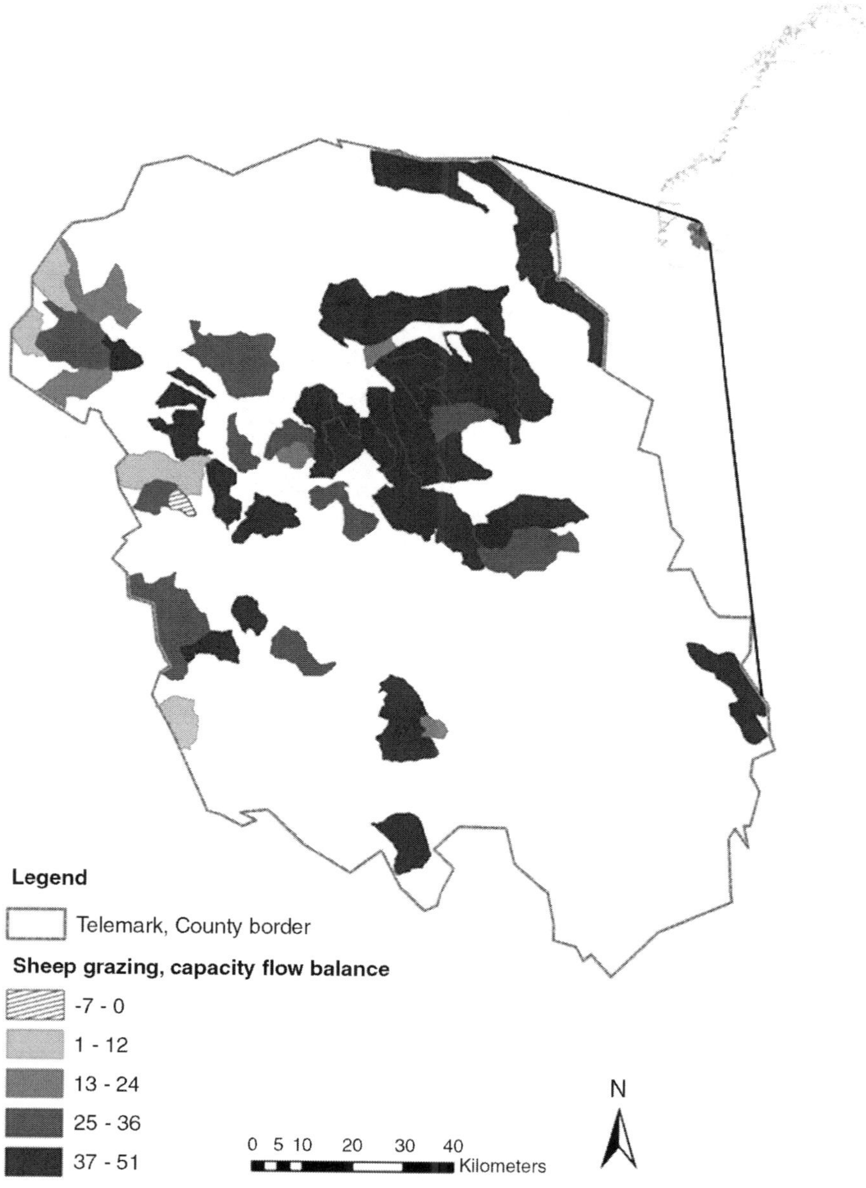

Figure 17.2 Differences between actual sheep grazing pressure and the capacity for generating the ecosystem service, expressed in number of animals per km^2, in Telemark, Norway (Schröter et al., 2014, modified). Positive values indicate a pressure lower than capacity.

only a small part of the available pasture resource in Telemark County, Norway (Schröter et al., 2014), and that this activity is sustainable in most of the county, in the sense that extraction rates are considerably lower than the regenerative capacity of the ecosystem.

Ecosystem accounts can also indicate which parts of a landscape should be maintained and protected in a (semi-)natural state in order to sustain the generation of regulating services that are critical for generating other ecosystem services. Figure 17.3 provides an example involving

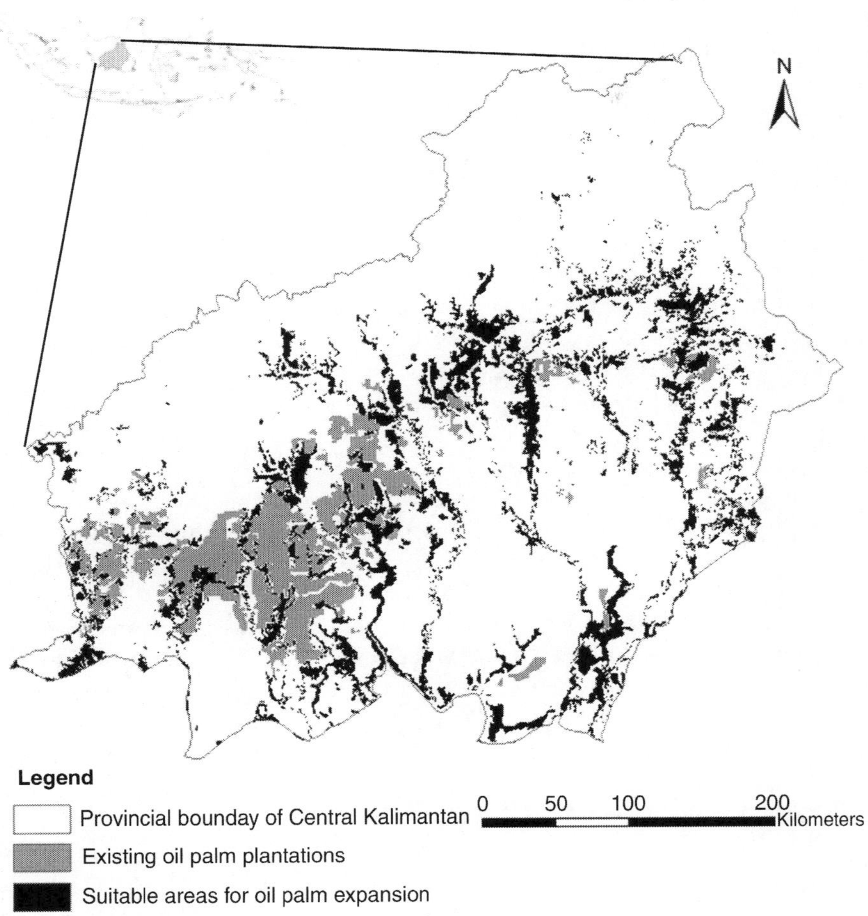

Figure 17.3 Central Kalimantan: areas where oil palm expansion would lead to low impacts on ecosystem services, based on criteria of local stakeholders (Sumarga and Hein, 2014, modified).

the identification of areas potentially suitable for oil palm expansion in Central Kalimantan, Indonesia. Areas that should not be converted in order to maintain the generation of ecosystem services (e.g. carbon sequestration, rice production, prime wildlife habitat) were identified using selection criteria identified in a stakeholder workshop (Sumarga and Hein, 2014). Ecosystem accounting can also indicate opportunity costs of preserving ecosystems and potential compensation through 'Payment for Ecosystem Services' mechanisms.

Discussion and conclusions

Ecosystem accounting has important limitations. First, it is complex to set up a full account, including condition, capacity, and service flows, in both physical and monetary units. There are still important data shortages in most parts of the world. Furthermore, not all aspects of ecosystem use (e.g. cultural and spiritual aspects) can be meaningfully integrated into an accounting framework. A particular challenge is how issues such as resilience and the non-linear dynamics of ecosystems can be best captured in ecosystem accounts. Compared to standard national accounts and other satellite accounts, the relation between assets and services is more complex in ecosystem accounting because of the diversity of ecosystem services and the complexity of ecosystem processes governing the impacts of ecosystem use on the condition of the ecosystem.

Second, national accounts and therefore ecosystem accounts measure production and not welfare. Hence, an ecosystem account will provide a partial indication of the economic value of ecosystems, which will often be an underestimate of the total economic value since consumer surpluses are excluded. Accounting methods which are welfare-based, such as Comprehensive Wealth Accounting (e.g. Ferreira et al., 2008) and Inclusive Wealth Accounting (Duraiappah and Muñoz, 2012), have a number of limitations as well, in particular the gap between data required for such methods and datasets that can in practice be measured or modelled.

Third, ecosystem accounting does not provide a tool to understand and design measures to deal with the long-term effects and risks of ecosystem and climate change. Risks are not made explicit in accounts, and long-term effects only have a small effect on current values of ecosystem capital because of limited discounting periods and due to discounting itself. Finally, having more information on the status of and trends in ecosystem assets does not necessarily imply that this information will lead to improved governance of ecosystems. Better ecosystem management is hampered not only by a lack of information on ecosystem change and ecosystem assets, but also by other characteristics, related to the economic and institutional structures governing resource use (e.g. Ostrom, 2008).

Although a consensus is gradually emerging on the conceptual framework for ecosystem accounting (e.g. Boyd and Banzhaf, 2007; Mäler et al., 2009; European Commission et al., 2013; Edens and Hein, 2013), not all aspects of ecosystem accounting have been tested, and further experimentation is required. A number of steps remain. First, there is a need for technical capacity-building in the field of ecosystem accounting, a process that has already started with recent international initiatives such as the World Bank WAVES program and the 'Advancing Natural Capital Accounting' project of the UN Environment Programme in collaboration with the UN Statistics Division. Second, further research on how earth observation systems can support the development of ecosystem accounts has to be carried out. Third, there is a need to further test how ecosystem accounting can support environmental and resource management through, for instance, supporting land use planning. Finally, based on these experiences, ecosystem accounting needs to be further developed and standardised, under the mandate of the UN Statistics Commission, and involve a collaboration between statistical agencies, UN agencies working on environmental and accounting themes, and the research community.

Note

1 The views expressed in this paper are those of the authors and do not necessarily reflect the policies of Statistics Netherlands.

References

Bateman, I., Mace, G., Fezzi, C., Atkinson, G., and Turner, K. (2011). Economic analysis for ecosystem service assessments. *Environmental and Resource Economics*, vol 48, pp 177–218.

Boyd, J., and Banzhaf, S. (2007). What are ecosystem services? The need for standardized environmental accounting units. *Ecological Economics*, vol 63, pp 616–626.

Costanza, R., Kubiszewski, I., Giovannini, E., et al. (10 authors) (2014). Time to leave GDP behind. Comment in *Nature*, vol 505, pp 283–285.

Duraiappah, A. K., and Muñoz, P. (2012). Inclusive wealth: a tool for the United Nations. *Environment and Development Economics*, vol 17, pp 362–367.

Edens, B., and Hein, L. (2013). Towards a consistent approach for ecosystem accounting. *Ecological Economics*, vol 90, pp 41–52.

European Commission, Organisation for Economic Co-operation and Development United Nations, World Bank (2013). *System of Environmental-Economic Accounting 2012, Experimental Ecosystem Accounting*. UN DESA, New York.

Ferreira, S., Hamilton, K., and Vincent, J. R. (2008). Comprehensive wealth and future consumption: accounting for population growth. *World Bank Economic Review*, vol 22, no 2, pp 233–248.

Hutchings, J. A., and Myers, R. A. (1994). What can be learned from the collapse of a renewable resource? Atlantic Cod, *Gadus Morhua*, of Newfoundland and Labrador. *Canadian Journal of Fisheries and Aquatic Sciences*, vol 51, pp 2126–2146.

Mäler, K.-G., Aniyar, S., and Jansson, Å. (2009). Accounting for ecosystems. *Environmental and Resource Economics*, vol 42, pp 39–51.

Obst, C., Hein, O., and Edens, B. (submitted). National accounting and the valuation of ecosystem assets and their services.

Ostrom, E. (2008). Institutions and the environment. *Economic Affairs*, vol 28, pp 24–31.

Remme, R. P., Schröter, M., and Hein, L. (2014). Developing spatial biophysical accounting for multiple ecosystem services. *Ecosystem Services*, vol 10, pp 6–18.

Scheffer, M., and Carpenter, S. R. (2003). Catastrophic regime shifts in ecosystems: linking theory to observation. *Trends in Ecology and Evolution*, vol 18, pp 648–656.

Schröter, M., Barton, D. N., Remme, R. P., and Hein, L. (2014). Accounting for capacity and flow of ecosystem services: a conceptual model and a case study for Telemark, Norway. *Ecological Indicators*, vol 36, pp 539–551.

Sumarga, E., and Hein, L. (2014). Mapping ecosystem services for land use planning, the case of central Kalimantan. *Environmental Management*, vol 54, pp 84–97.

TEEB (2010). *The Economics of Ecosystems and Biodiversity. Ecological and Economic Foundations*. Kumar, P. (ed.) Routledge, Oxford.

United Nations, European Commission, International Monetary Fund, Organisation for Economic Co-operation and Development and World Bank (2009). *System of National Accounts 2008*. United Nations, New York.

18

ACCOUNTING FOR ECOSYSTEM SERVICES IN BUSINESS

Joël R. A. Houdet, John Finisdore, Julia Martin-Ortega, Helen Ding, John K. Maleganos, James Spurgeon, Tobias Hartmann and David Steuerman

From impact mitigation to ecosystem services-based approaches

Firms have traditionally viewed biodiversity and ecosystem services (ES) as external constraints on their activities because of regulations and laws imposed on them (e.g. through Environmental Impact Assessments). This perception still holds true for many firms. ES are routinely quantified, monitored and accounted for by companies when directly used, especially in the agricultural, mining, fishing and pharmaceutical sectors, or where the organisational goal is to avoid waste (e.g. loss of provisioning ES) and hence costs (Schaltegger *et al.*, 2000). However, where ES are used indirectly, especially those interacting with production processes in the supply chain, they are often ignored. This is often due to lack of awareness and lack of assessment or measurement tools tailored to the needs of businesses (Houdet *et al.*, 2009). Yet managing ecosystems for the sustained or *ad hoc* delivery of specific ES, especially those which generate higher returns on investment, may lead to unforeseen ecosystem change, degradation or even collapse. For instance, managing the biomass and productivity of tree plantations to maximize CO_2 sequestration can lead to diminished stream flows, increased soil salinization and acidification (Jackson *et al.*, 2002).

Pressures from stakeholders such as neighbouring communities, NGOs, other firms and government agencies to better manage business dependencies on ES and reduce impacts on ecosystems is, however, progressively changing business practice (Houdet, 2008; TEEB, 2010; WBCSD, 2012; WRI, 2008). As a result, new business strategies and practices are emerging around the notion of ES, as for instance in water-intensive beverages production (Kissinger, 2013). These strategies and practices are seen as key assets that can generate or sustain positive outcomes.

As resources become scarcer, and ES availability becomes more unreliable, changes in ES quality, supply and price are increasingly impacting business sustainability. As a consequence, companies are now more likely to explore strategies which reduce risks and costs, secure raw material supply, foster positive stakeholder relationships or even develop new business opportunities. These typically involve going beyond impact avoidance, mitigation and offset measures, and towards sustainable ES interdependency management or stewardship (Houdet, 2010; Houdet *et al.*, 2012; Mulder *et al.*, 2012; TEEB, 2010; WBCSD, 2012; WRI, 2008). Such strategies, and the associated tools, skills and processes, can be described as 'ecosystem services-based approaches' as defined by Martin-Ortega *et al.* (2015).[1]

Yet to successfully manage both impacts and dependencies on ES, companies need to have the right set of information. *You can only manage what you measure.* This means firms have to invest in accounting tools and systems to measure, monitor and hence help secure their ecological infrastructure: all key interactions with ES need to be managed to maintain or improve competitive advantage (Houdet *et al.*, 2012).

What tools can business use to account for ecosystem services?

Companies have started accounting for ES to improve management practices and even restore natural capital (Hartmann, 2014; Houdet *et al.*, 2012; TEEB, 2010; WBCSD, 2012; WRI, 2008). Accounting for ES may be undertaken for different purposes, depending on whose needs and aspirations it aims to satisfy (Houdet *et al.*, 2009). Box 18.1 summarises the potential benefits to companies of adopting a Natural Capital Accounting approach. The objectives of internal stakeholders (management, staff members, majority shareholders) can be very different from external ones (environmental NGOs, trade unions, regulation bodies, neighbouring communities and businesses), hence the potential for contradictory outcomes when undertaking ES biophysical assessment and economic valuation. Two complementary approaches to account for ES from a business perspective are therefore currently being explored.

Box 18.1 Potential benefits to companies from adopting ES accounting in business

- General decision-making
- – Improved sustainability decision-making
- – Integrated thinking and reporting
- – Better relations with stakeholders and regulators
- – Gain and develop experience and capacity around ES
- Operational
- – Cost savings
- – Improving / securing raw material supply and quality of raw materials
- – Inform selection of optimum impact mitigation measures
- – Provide programmed certainty
- – More engaged workforce
- – Meet client/government requirements
- – Long-term raw material stock security
- Financial
- – Maintain and enhance revenues
- – Understand and make most of new environmental markets
- – Maintain social license to operate
- Risks
- – Manage and reduce risks
- – Help ensure security of supply of natural resources
- – Become more aware of potential price increases
- – De-risk future uncertainty
- – Flattening investment risks (reducing risks of sunk costs/ false investments)

• Reputation
– Maintain and enhance reputation
– Demonstration of creating shared value
• Strategy
– Increase external uptake by standard-setters and national governments
– Help inform internal and external communications
– Help prioritise risks and opportunities
– Map stakeholders and engage them in managing ES too
– Inform strategic biodiversity action plans

Source: adapted from Spurgeon, 2014

First, existing environmental or sustainability tools are being improved at the production, organizational (information systems for decision-making) and institutional (tools for engaging external stakeholder) levels (Houdet *et al.*, 2012; Natural Capital Coalition, 2014; Waage and Kester, 2014). These include impact assessments, mitigation and offset measures, environmental management, accounting and valuation systems, life-cycle assessment (LCA) methodologies, product labels or certification schemes, and sustainability reporting guidelines. Among these tools, some use and produce biophysical information (e.g. LCA, offset measures), while others focus on estimating monetary values (e.g. management accounting and economic valuation) and sometimes couple biophysical data with monetary information (e.g. impact assessment, eco-efficiency analysis).

Second, various organizations have been developing new tools that focus on addressing ES issues specifically (Houdet *et al.*, 2012; Waage and Kester, 2014; WBCSD, 2013), including:

- Tools for raising awareness of ES, and their values, risks and opportunities for the business community (e.g. ESB – Ecosystem Services Benchmark, ESR – Ecosystem Services Review, BBII – Business & Biodiversity Interdependency Indicator),
- Tools for mapping ES at the broad landscape level for scenario purposes to aid decision-making for policy and planning (e.g. ARIES – Artificial Intelligence for Ecosystem Services, InVEST – Integrated Valuation of Ecosystem Services and Tradeoffs) (see also Maes *et al.*, 2016, and Kienast and Helfenstein, 2016);
- Tools used for finer-scale, local assessments at the land asset level, which can involve any type of landowner (e.g. EROVA – Environmental Risk, Opportunity and Valuation Assessment, EcoAIM—Ecological Asset Information Management; EcoMetrix; Wildlife Habitat Benefits Estimation Toolkit); and
- Tools used for finer-scale assessments for specific business perimeters for internal management (e.g. WBCSD, 2012, Gonzalez and Houdet, 2009) and external reporting purposes (e.g. Houdet *et al.*, 2010; PUMA, 2010).

While these tools may help firms to incorporate biodiversity and ES concerns into their activities, most are at the development stage and there has been limited assessment of their effectiveness and business uptake (Waage and Kester, 2014; WBCSD 2013). Currently there is no standardized and widely-recognized set of tools for measuring ES impacts and dependencies in business. Box 18.2 explains briefly the key questions companies need to answer when attempting to undertake an ES assessment.

Box 18.2 The 7 main questions for businesses to consider when undertaking a successful ES assessment

1. **What aspect of the business do you want to target for analysis?** For instance, this may be a product or a set of products; projects or operational site(s); the whole company or a business unit; the value chain, or part of it, upstream and / or downstream; a region or landscape in which the business operates.
2. **What is your business aspect's main relationship to functioning ecosystems and environmental issues?** Directly through landholdings/ownership of land? Directly but not on your own land? Indirectly through the value chain?
3. **What is your company's main intended use of undertaking an ES assessment?** For internal (e.g. supply chain management, risk and opportunity analysis) or external (e.g. reporting and disclosure, compliance, mergers and acquisitions, labeling) applications?
4. **What components of natural capital and the environment are you interested in assessing?**
5. **What form of measurement are you most interested in considering?** Qualitative, biophysical and / or monetary (financial, or economic values)?
6. **Whose perspective are you most interested in?** Management? Shareholders? External stakeholders and society?
7. **To what extent do you want to limit the resources (e.g. time, budget and skills) to undertake an ES assessment?**

Source: adapted from Spurgeon, 2014

Towards a standardised ES accounting protocol: what criteria need to be taken into account?

A lack of standardisation means that ES cannot be consistently accounted for from the perspective of a whole business, a business unit, its supply chains or its products and project(s). This prevents ES accounting from being 'mainstreamed'. The need for the development of standardized sets of indicators, measurement and assessment protocols to address business impacts and dependencies on ES has therefore been widely recognised (Houdet, 2010; Houdet and Germaneau, 2014; Houdet *et al.*, 2015; Mulder *et al.*, 2012; Natural Capital Coalition, 2014; Waage and Kester, 2014). Such a standardised protocol requires many key issues to be addressed, including:

- **Purpose – inward or outward looking**: Would the protocol focus on the impacts of ES on business activities, on the impacts of business on ES and their users (third parties) or on both? Who are information users? Would the information be used for internal (decision-making, planning, performance monitoring) or external (to satisfy voluntary or regulatory third-party disclosure) purposes?
- **Scope – of which there are a number of dimensions:**
  - **Thematic**: Which ES shall be covered by the protocol? Shall all be accounted for or only the most material ones? Who defines materiality (i.e. from which perspective, that of the company or that of its stakeholders)? Can one develop a methodology for material analysis that is flexible enough to account for intra- and inter-sectoral differences?

  - **Organisational**: Where does the protocol start and stop accounting for ES from an organisational perspective? Only for activities fully controlled by the company? What about ES dependencies and impacts of the supply chains and that of the life-cycles of its products?
  - **Temporal**: ES changes over time as a result of coupled natural and anthropogenic impacts. To precisely measure the interactions between the changes of ES and past and present business activities, better understand the associated benefits and costs for business and society as well as identify potential opportunities for risk reduction, it is essential to choose the baseline year and target year relevant for the overall scope of the accounting exercise. So key questions remain to be addressed: e.g. how should the protocol account for ES over time? For only specific periods (e.g. as is done under the GHG Protocol) or past and future events, as typically done in environmental management accounting? For instance, Houdet et al. (2014) argue that choosing a periodic approach can be misleading, because it does not account for past impacts and dependencies (e.g. land clearance which took place before construction of a factory).
  - **Spatial**: Tracing ES from their sources to their points of use is likely to be a challenge. Should it make reference to both the supply and demand side, and so require accounting of ecological infrastructure and ecosystem assets (stocks), functions, processes and the flows responsible for the availability and delivery of the ES? How can one avoid double counting? For instance, should a water bottling company account exclusively for its final water use at the point of its consumption (i.e. scope 1 as per GHG Protocol principles) and ignore the sources, delivery channels and associated stakeholders in its catchment areas? Would the company have to account for the dependencies and impacts on ecosystem services linked to suppliers of the materials used to produce the water bottles? What's more, the sources of ES information can be expected to be diverse, especially at the landscape and supply chain levels. How can the protocol prescribe data quality requirements for disparate data sources? When ES information is spatially located, to what degree should spatial information be part of indicators for ES?
- **Indicator and metric selection:**
  - Fundamental attributes of sound data for the protocol should at least include completeness, validity, relevance, timeliness, consistency and transparency. Comparability should also be a priority, especially for processed or aggregated data, to allow for intra- and inter-sectoral comparisons.
  - Which ES classification should be used? Although standardised ES classifications exist, such as CICES (Haines-Young and Potschin, 2013, see also Potschin and Haines-Young, 2016) or FEGS-CS (Landers and Nahlik, 2013, see also Landers *et al.*, 2016), there is a question of whether they are relevant for business. Would these or a combination of them be sufficient to account for all relevant spatial dimensions of ES? The challenge is to design one that: (a) is unambiguous for people from different disciplines; (b) is suitable to all business applications (e.g. life-cycle analysis, site management, risk analysis, budgeting, reporting to external stakeholders); (c) allows for biophysical measurement at multiple scales (geographic scalability); (d) allows values to be aggregated and disaggregated (economic scalability); (e) has exhaustive and mutually exclusive categories; and, (f) avoids overlap or double counting of ecosystem services when valued qualitatively, quantitatively or monetarily.
  - Moreover, at what point does something 'leave' the natural environment and become a good or service within the human economy? At what point does human intervention in an ecosystem render its outputs as manmade rather than a product of the natural

environment? And, subsequently, at what point are the users of an ES identified taken into account? In that context, the approach followed in FEGS-ES by Landers and Nahlik, (2013) is likely to exclude many business and ecosystem interactions, with the risk of causing disinterest of companies operating upstream of value chains (e.g. retailers, banks, insurance providers).

- ES indicators or metrics could potentially be qualitative, quantitative, aggregated, biophysical or monetary, and even coupled (e.g. biophysical amount per unit of good sold). Should the protocol describe their use in absolute (e.g. total volumes of ES used) or relative terms (ES impact per worker, per unit of product, per hectare), or from a time-relevant perspective? Should they also cover human and socio-economic phenomena, and include methodologies for assessing them in economic terms? If so, would this be based on strong or weak sustainability principles (i.e. opportunity costs versus restoration cost approaches)? To what extent should intangible business values of ES (e.g. the social dimension of ES attributed to business) be taken into account in decision-making and what would be the right scale for value aggregation?

Most importantly, the ES accounting protocol should be flexible enough to be usable by different stakeholder groups, with varying set of skills and interests, to achieve mainstreaming into any relevant business policy, strategy, activity or practice. Guidelines and mechanisms for quality assurance are also required to generate the trust needed for all stakeholders to make use of it.

Conclusion

Accounting for ES in a standardised, comparable and transparent manner has become a critical challenge for mainstreaming ES in business decision-making and applications. We have identified some key criteria for the development of such a business oriented framework. All these have in common the need to define the specific purposes and boundaries of the accounting exercise, which needs to be addressed by the firms themselves as well as by their internal and external stakeholders. While tools for identifying, quantifying, mapping and valuing ecosystem services are being developed, these have not always been designed for the business world, making their direct use difficult. Moreover, they do not yet have the widespread multi-stakeholder support required for effective uptake. In partnership with a growing network of organisations, the Natural Capital Coalition (previously The Economics of Ecosystems and Biodiversity (TEEB) for Business Coalition), is supporting the development of a harmonised Natural Capital Protocol. Let us hope that the exercise is successful and gains the support of all stakeholders. Indeed, the extent of its uptake by all business sectors will largely depend on supportive government policies, widespread understanding of its need and uses in specific sectors and timely access to the right set of skills and resources.

Note

1 Ecosystem services-based approaches are a way of understanding the complex relationships between nature and humans in order to support decision-making, with the aim of reversing the declining status of ecosystems and ensuring the sustainable use/ management/conservation of resources. They are based on four core elements: (1) the focus on the status of ecosystems, and the recognition of its effects on human wellbeing; (2) the understanding of the biophysical underpinning of ecosystems in terms of service delivery; (3) the integration of natural and social sciences and other strands of knowledge for a comprehensive understanding of the service delivery process; and (4) the assessment of the services provided by ecosystems for incorporation into decision-making.

References

Gonzalez, G., and Houdet, H. (2009). *Accounting for Biodiversity and Ecosystem Services from a Management Accounting Perspective. Integrating Biodiversity into Business Strategies at a Wastewater Treatment Plant in Berlin.* Veolia Environnement, Orée, Paris.

Haines-Young, R., and Potschin, M. (2013). *CICES V4.3 – Revised Report Prepared Following Consultation on CICES Version 4.* August-December 2012. EEA Framework Contract No EEA/IEA/09/003.

Hartmann, T. (2014). *How Companies Value Natural Capital – Taking Stock and Looking Forward*, Global Nature Fund, Bonn. Available at: http://www.naturkapitalbilanzierung.de/ (accessed 29 May 2014).

Houdet, J. (2008). *Integrating Biodiversity into Business Strategies. The Biodiversity Accountability Framework.* FRB, Orée, Paris.

Houdet, J. (2010). *Entreprises, biodiversité et services écosystémiques. Quelles interactions et strategies? Quelles comptabilités?* PhD Thesis, AgroParisTech.

Houdet., J., Burritt, R., Farrell, K., Martin-Ortega, J., Ramin, K., Spurgeon, J., Atkins, J., Steuerman, D., Jones, M., Maleganos, J., Ding, H., Ochieng, C., Naicker, C., et al. (16 authors) (2014). What natural capital disclosure for integrated reporting? Designing & modelling an Integrated Financial – Natural Capital Accounting and Reporting Framework. Synergiz – ACTS, Working Paper 2014–01, France – Kenya.

Houdet, J., and Germaneau, C. (2014). Accounting for biodiversity and ecosystem services from an EMA perspective: towards a standardised biodiversity footprint methodology. In Jones, M. (ed.) *Accounting for Biodiversity*. Routledge, Abingdon and New York..

Houdet, J., Johnstone, A., and Germaneau, C. (2015). Ecosystem services-based approaches to water management: what opportunities and challenges for business? In Martin-Ortega, Ferrier, R. C., Gordon. I. J., and Khan, S. (eds) *Water Ecosystem Services: A Global Perspective*. Cambridge University Press, Cambridge UK.

Houdet, J., Pavageau, C., Trommetter, M., and Weber, J. (2009). Accounting for changes in biodiversity and ecosystem services from a business perspective. Preliminary guidelines towards a Biodiversity Accountability Framework. Ecole Polytechnique, Department of Economics, France.

Houdet, J., Trommetter, M., and Weber, J. (2010). *Promoting Business Reporting Standards for Biodiversity and Ecosystem Services. The Biodiversity Accountability Framework.* Orée – FRB, Paris.

Houdet, J., Trommetter, M., and Weber, J. (2012). Understanding changes in business strategies regarding biodiversity and ecosystem services. *Ecological Economics*, vol 73, pp 37–46.

Jackson, R. B., Banner, J. L., Jobbágy, E. G., Pockman, W. T,. and Wall, D. H. (2002). Ecosystem carbon loss with woody plant invasion of grasslands. *Nature*, vol 418, pp 623–626.

Kienast, F., and Helfenstein, J. (2016). Modelling ecosystem services. In Potschin, M., Haines-Young, R., Fish, R. and Turner, R.K. (eds) *Routledge Handbook of Ecosystem Services*. Routledge, London and New York, pp 144–156.

Kissinger, G. (2013). *Reducing Risk. Landscape Approaches to Sustainable Sourcing. SAB Miller Case Study.* Eco-Agriculture Partners, Washington DC.

Landers, D. H., and Nahlik, A. M. (2013). *Final Ecosystem Goods and Services Classification System (FEGS-CS).* EPA/600/R-13/ORD-004914. U.S. Environmental Protection Agency, Office of Research and Development, Washington DC.

Landers, D. H., Nahlil, A. M., and Rhodes, C. R. (2016). The beneficiary perspective – benefits and beyond. In Potschin, M., Haines-Young, R., Fish, R. and Turner, R.K. (eds) *Routledge Handbook of Ecosystem Services*. Routledge, London and New York, pp 74–87.

MA (2005). *Ecosystems and Human Well-Being: Synthesis.* Island Press, Washington DC.

Martin-Ortega, J., Ferrier, R. C., Gordon, I. J., and Khan, S. (eds) (2015). *Water Ecosystem Services: A Global Perspective.* Cambridge University Press, Cambridge UK.

Mulder, I., Mitchell, A.W., Peirao, P., Habtegaber, K., Meneses, L., and Simonetti, R. (2012). *The NCD Roadmap: 4 Steps to Implement the Natural Capital Declaration for the Financial Industry.* UNEP Finance Initiative, Geneva and Global Canopy Programme, Oxford.

Natural Capital Coalition (2014). Valuing natural capital in business: towards a harmonized framework. Natural Capital Coalition – ICAEW. Available at: http://www.naturalcapitalcoalition.org/js/plugins/filemanager/files/Valuing_Nature_in_Business_Part_1_Framework_WEB.pdf (accessed 29 May 2014).

Potschin, M. and Haines-Young, R. (2016) Defining and measuring ecosystem services. In Potschin, M., Haines-Young, R., Fish, R., and Turner, R. K. (eds) *Routledge Handbook of Ecosystem Services*. Routledge, London and New York, pp 25–44.

PUMA (2010). PUMA's environmental profit and loss account for the year ended 31 December 2010. Available at: http://about.puma.com/wp-content/themes/aboutPUMA_theme/financial-report/pdf/EPL080212final.pdf (accessed 29 May 2014).

Ruhl, J. B., Kraft, S. E., and Lant, C. L. (2007). *The Law and Policy of Ecosystem Services*. Island Press, Washington DC.

Schaltegger, S., Hahn, T., and Burritt, R. L. (2000). *Environmental management accounting: Overview and main approaches*. Centre for Sustainability Management at the University of Lueneberg, Lueneberg and Canberra.

Spurgeon, J.P.G. (2014). *Natural Capital Accounting for Business: Guide to Selecting an Approach*. Final report to the EU Business and Biodiversity Platform, performed under ICF contract.

TEEB (2010). *The Economics of Ecosystems and Biodiversity: Report for Business*. Bishop, J. (ed.). Earthscan, London and Washington DC.

Waage, S., and Kester, C. (2014). Making the invisible visible: analytical tools for assessing business impacts & dependencies upon ecosystem services. BSR's Environmental Services, Tools & Markets Working Group. Available at: http://www.bsr.org/fr/our-insights/report-view/making-the-invisible-visible-analytical-tools-for-assessing-business-impact (accessed 29 May 2014).

WBCSD (2012). *Guide to Corporate Ecosystem Valuation. A Framework for Improving Corporate Decision-Making*. WBCSD, Geneva.

WBCSD (2013). *Eco4Biz – Ecosystem Services and Biodiversity Tools to Support Business Decision-Making*. WBCSD, Geneva.

WRI (2008). *The Corporate Ecosystem Services Review. Guidelines for Identifying Business Risk and Opportunities Arising from Ecosystem Change*. World Resources Institute, World Business Council for Sustainable Development and Meridian Institute, Washington DC.

19

VALUING PREFERENCES FOR ECOSYSTEM-RELATED GOODS AND SERVICES

Tomas Badura, Ian Bateman, Matthew Agarwala and Amy Binner

Introduction: rationale for valuation of ecosystem related goods and services

If all resources were infinite there would be no need for us to value their different contributions to human welfare; indeed there would be no need for economics. Regrettably, we do not live in such a world. Indeed human ingenuity has devised multiple ways for us to exceed the capacity of the planet to provide all the resources which our rapidly expanding and increasingly affluent population demands (cf. Costanza, 2016). Given this situation, a fundamental responsibility of good governance is to encourage the allocation of scarce resources such that they best satisfy society's requirements and aspirations. Economic analyses can help evaluate the myriad options for resource allocation in terms of their inherent trade-offs (that is, in terms of what must be given up to pursue them) and identify that course of action which delivers the highest net benefit (i.e. which maximizes the gap between the benefits of an option and its costs, including the opportunity costs of forgone alternatives).

When the costs and benefits of each course of action are readily observable, the task of identifying the option with the greatest net benefit is relatively straightforward. However, as is the case with the natural environment, when the effects of an option are imperfectly understood, or relevant costs and benefits are difficult to assess, then identifying which course of action is the best poses a major challenge. From an economic perspective, the natural environment is a value-generating resource which should be fully integrated into decision-making systems (Atkinson et al., 2012). However, this integration is often far from straightforward. The natural environment can be viewed as a repository of a variety of ecosystem processes, arrayed in a complex and interlinked web where multiple processes link together to determine the inputs to further processes. From an economic perspective, these processes become of relevance to human welfare when they deliver 'final ecosystem goods and services' (Landers and Nahlik, 2013), otherwise known as 'final ecosystem services' (Fisher et al., 2009). While these have been the subject of much academic debate and reconceptualization (Fisher and Turner, 2008; Fisher et al., 2009; Haines-Young and Potschin, 2013; Bateman et al., 2011a; Johnston and Russell, 2011; Potschin and Haines-Young, 2011 and 2016; Staub et al., 2011), an apt early definition is that ecosystem services are those "components of nature, directly enjoyed, consumed or used to yield human well-being" (Boyd and Banzhaf, 2007, p. 619). The ecosystem service concept is therefore inherently anthropocentric, focussing upon nature's contributions to human well-being. While it is obviously true that underpinning natural processes are vital to the provision of ecosystem services, the two should not be confused. Indeed, attempts to value the former processes are liable to result in double counting errors if such values are then added to those of ecosystem services. This concentration on well-being also means that common classifications of ecosystem services into "supporting", "regulating", "provisioning" and "cultural" categories (MA, 2005), while being "heuristically relevant", are of less pertinence to economic analysis where the focus is on those welfare-bearing (tangible) goods and (intangible) services which are directly related to final ecosystem services.

An important distinction therefore is that economic analysis does not attempt to directly value ecosystem services, but rather assesses the value of the contribution of ecosystem services to related welfare-bearing goods and services. Occasionally these will be effectively identical. Natural landscapes are both final ecosystem services and valued economic services when viewed by those who consider them aesthetically pleasing. However, many ecosystem services only generate welfare-bearing goods when they are used as inputs in a production process and combined with other human-derived inputs (such as labour, machinery, expertise, etc.). For example, while natural processes provide wild fish, it is only when combined with fishing expertise, boats, nets, and so on, that fish are converted to food. Nature still provides essential and highly valuable inputs to such production processes, but to confuse ecosystem services with related goods over-values the former and undermines the validity of decision-support analyses.

As overviewed throughout this chapter, valuation of ecosystem service related goods can involve a number of complexities; not least of which is that many of these goods are not traded in markets and so lack market prices (cf. Turner, 2016). Of course some of these goods are priced (e.g. food and timber), but many others are not (e.g. an equable climate, clean water, natural hazard regulation). Furthermore, the goods and services to which the natural environment contributes are frequently measured and reported in a broad range of units. For instance, greenhouse gas sequestration is reported in tons of carbon equivalent sequestered, water purification in cubic metres of water purified, and recreation in the number of visits to a site. Attempting to draw meaningful comparisons and evaluate trade-offs between these diverse units, especially when we consider the diversity of inter-related effects which arise from, say, land use change, is a task of Herculean proportions. Nevertheless, decision-makers are routinely faced with precisely this challenge, which is only made more complicated by the fact that ecosystem services are not always complementary (e.g. an increase in food production sometimes comes at the expense of a decrease in water quality). Consequently, there is an obvious advantage in making the various trade-offs inherent in decisions comparable through a common unit. Arguably, any common unit could be used, but there are particular advantages associated with the use of money; it is a pure unit of exchange with no inherent value. Furthermore, it is of course the unit which is most familiar to decision-makers. Importantly, money units place the value of spending on welfare-bearing ecosystem services on a level playing field with other investment options. Failure to include the economic value of ecosystem services within decision-making has led to their long term abuse and worldwide decline (MA, 2005; SCBD, 2010; TEEB, 2011; UK NEA, 2011). Reversal of this trend requires radical change, rejecting the status quo in decision-making and ensuring genuine comparability of the value of maintaining ecosystem services with that of other investments. Economic valuation provides a key element in delivering this change. Van Beukering et al. (2015) highlight a number of further advantages of the use of economic valuation of ecosystem services, including:

- The role of valuation as an advocacy tool which helps place the value of ecosystem services on the planning agenda by highlighting the importance of ecosystems for private sector profitability as well as for the provision of public goods and human health and security (e.g. air purification and natural hazards protection).

- Ecosystem valuation is also an important tool for assisting transparent and better-informed decision-making.
- Valuation of ecosystems can also be used for damage assessment in terms of either setting up a compensation fee for potential environmental damage or resolving legal disputes between conflicting parties.
- Valuation can also be used for determining taxes, fees, or charges for use of ecosystem services, which can discourage their usage and support conservation.

Methodological developments and related empirical studies on valuation have proliferated over the past three decades, with a notable increase in academic output over the past 10 years (Fisher et al., 2009). As a result, economists can draw upon a substantial body of research and a range of tools (see section 3) for assessing the contribution of the natural environment to human wellbeing across a range of contexts (e.g. Costanza et al., 1997, 2014; Bateman et al., 2013; UK NEA, 2011, 2014; Goldstein et al., 2012). Valuation research is becoming ready to use for policy application.

The following sections examine three areas of ecosystem valuation research. We first discuss the main concepts underlying this area of research, outlining the problems in our understanding of the natural world and how this influences our ability to assess environmental values. Second, we overview the basic methods of ecosystem valuation, examine options for generalising and transferring results across locations and decisions, and review pioneering integrated decision-support tools. Finally, we discuss the future prospects for environmental valuation, considering the main challenges faced by researchers and practitioners in this area, and how valuation can act as an institution of change for society-environment relationships.

Basic concepts of environmental valuation

All ecosystem management decisions (and any environmental degradation) have intertemporal implications (e.g. Mäler, 2008). Given this, we can conceptualize the natural environment as a collation of stocks of ecosystem assets, generating flows of ecosystem services over time (e.g. Atkinson et al., 2012; Barbier, 2007; Bateman et al., 2011a). In this context, economic appraisal of ecosystem service-related goods involves the valuation of service flows over time. The value of an ecosystem asset is therefore the net present value of future flows of ecosystem services. The important question, then, is how the ecosystem asset value changes in response to human interventions; how its future prospects change in response to what is happening to ecosystems and biodiversity now.

The economic definition of value is based on the choices and trade-offs people make in relation to the good or service in question. The most commonly applied indicator of value is a good's market price. But when no such price exists, values can be derived either from related markets or from stated behaviour in hypothetical situations (see methods, section 3). Value is then estimated in terms of four measures (Hicks, 1943): (1) an individual's maximum willingness to pay (WTP) to obtain an increase in the provision of a welfare-bearing good or service; (2) their maximum WTP to prevent a reduction in such provision; (3) the minimum amount that an individual is willing to accept in compensation for the loss of a welfare-bearing good or service (their willingness to accept; WTA) and; (4) the amount they are WTA to forego a welfare gain from increased provision of such a good. WTP and WTA are reflected in the preferences and choices people express in either existing or hypothetical markets. Summing individual preferences over the relevant population provides an estimate of the corresponding aggregate value of a given change in the provision of an ecosystem-related economic good or service. Note that correctly identifying the relevant population is of crucial importance to avoiding over- or under-estimation of related values (Bateman et al., 2006).

The economic concept of value can be divided into use and non-use values (Pearce and Turner, 1990). These are not mutually exclusive, as people can hold both types of values for the same good. Further, use values can be divided into three categories: direct, indirect, and option values. Direct use values arise from a direct interaction with an environment and include extractive (e.g. fisheries, timber) and non-extractive (e.g. recreation or aesthetic value of a natural view) values. Indirect use values stem from ecosystem service-related goods which are not used directly (e.g. water and air purification or natural hazards protection). Option values arise from the potential future use of ecosystems (e.g. medical research).

Non-use values are not related in any way to current or future use of ecosystem goods or services by the individual expressing the value in question; they arise simply from knowing the continued and maintained existence of an ecosystem (or elements thereof) is secure. Such values are often related to charismatic species and rare habitats (e.g. the Sumatran tiger, Bateman et al., 2010; or the Brazilian Amazon, Horton et al., 2003; Morse-Jones et al., 2012). Non-use values can be divided into three categories: existence, bequest, and altruistic. Existence value relates to the satisfaction people obtain from the existence of ecosystems and biodiversity, quite separate from its use. Bequest value relates to the welfare people gain from knowing that ecosystems and biodiversity will be passed on to future generations. Finally, altruistic or other-regarding value (e.g. Ferraro et al., 2003) relates to the satisfaction individuals gain from ensuring that ecosystems and biodiversity are available for other people in their generation. Note that this nomenclature deliberately eschews the use of the term "intrinsic value" (Bateman et al., 2011a).

In practical terms, it is also important to highlight the difference between exercises attempting to reflect the total accounting value of the services provided by an ecosystem and economic analyses of the unit (marginal) value of relatively small changes in those services and their related goods. While there have been attempts to calculate the total value of world/region/country's ecosystems (e.g. the influential work of Costanza et al., 1997, 2014), this approach is considered problematic in economic theory terms (e.g. Bockstael et al., 2000; Heal et al., 2005). One line of critique argues that any such (total) value of ecosystems is an underestimate of infinity, as humanity relies on the natural world for its own existence. Most importantly, it is argued that very few policies concern total loss of ecosystems and such exercises are of no (or little) use for practical decisions. We take a somewhat intermediate position. Accounting exercises, whether they are for market or non-market goods, necessarily rely upon certain strong assumptions. Most particularly they typically ignore the increase in marginal values which generally arise as stocks are depleted and related services fall. This is why an assessment of the total accounting value of the world's ecosystem services can be some finite sum while economic intuition points to the real value being infinite. Nevertheless, accounting exercises do play an important role in raising the issue of environmental degradation and biodiversity loss by bringing it to wider policy and public attention. Green national accounts become of greater relevance when they are considered over time, as they can flag trends in stocks and highlight potential areas for policy action (e.g. ONS, 2014). However, accounting studies cannot identify the optimal efficient response to those trends. This is where economic marginal values come into their own, as they can single out the most efficient courses of action in response to some ecosystem service concern and finite resources.

Economic analyses of ecosystem services therefore examine marginal values, reflecting the fact that most policies consider changes in ecosystem services provision on a limited rather than absolute scale. Multiplying the marginal value by the relevant unit of change implied by the policy change then provides the change in ecosystem service values for the policy in question. However, it is also necessary to take into account the fact that in many cases marginal values are not constant (e.g. Brander et al., 2006); this could result, for example, from changes in the supply of ecosystem services, availability of substitutes, or changes in societal norms.

While an economic analysis of ecosystem services can be undertaken at a particular site for local decision-making, economic analyses can also be part of larger ecosystem service assessments across differing scales. Bateman et al. (2011a) provides a general framework and nomenclature for integrating ecosystems and economic analyses in such assessments. Two types of assessment can be made. Sustainability analyses typically focus on past changes in natural capital stocks and seek to determine whether past development was on a sustainable path in terms of increasing or decreasing capital assets. Conversely, programme evaluation analyses are forward-looking and aim to evaluate development options. This often comprises use of policy scenarios, forecasts of environmental change, and trends in domestic and world markets (e.g. UK NEA, 2011, 2014).

Contextualising values

Values can be highly context-dependent, and therefore valuation frameworks need to recognize how space, time, biological factors, and institutions influence the values assessed. The spatial element of valuation studies is of key importance for many ecosystem services. For example, the benefits of water related services (e.g. water purification and provision) are often experienced downstream from where they were generated, while natural hazard regulation services (e.g. flood and storm protection) can be positioned a considerable distance from the human populations and infrastructure they protect. Moreover, some ecosystem values, particularly use values, decrease as the distance between the asset and the valuing individual increases (a phenomena known as the distance decay effect see e.g. Bateman et al., 2006). Bateman et al. (2011b) show how the location of outdoor recreation sites matters, reporting values between £1,000 and £65,000 per annum for recreation, depending on the proximity to significant conurbations. The spatial dimension of ecosystem service values, including the distance decay effect, is increasingly being incorporated into valuation studies with the use of Geographical Information Systems (GIS) tools and significantly contributes to valuation research development (e.g. see Goldstein et al., 2012; Bateman et al., 2013, 2014).

Ecosystem-related values are also influenced by the time profile over which they are measured and accounted for. Many of the values of ecosystem services occur over long time scales (e.g. carbon sequestration in peat soils and its impact on climate). To account for the temporal dimension of benefits, economists use (social) discount rates, which reflect the theoretical and empirical observation that, for a wide variety of reasons, people prefer receiving benefits sooner than later. There is, however, no clear consensus on the choice of discount rates, and it is a source of considerable debate (see Gowdy et al., 2010, for an extensive discussion).

The crucial role of natural science input

Natural science input lies at the heart of any ecosystem service-related valuation. Economics is well-suited to assess the link between quantified ecosystem services and people's well-being. However, it is crucial to understand the "production function" of ecosystem services from a natural sciences perspective, e.g. how different biological factors influence ecosystem functions. A particular challenge is the still-limited understanding of the role biodiversity plays in ecosystem functioning and the provision of ecosystem services (Naeem et al., 2012; Mace, 2014). For many ecosystems we lack sufficient understanding of natural thresholds and tipping points to reflect in subsequent economic analyses (Mäler et al., 2003; Rockström et al., 2009).

Ideally, analyses of potential tipping points should appraise the value of maintaining resilience within natural systems. Ensuring that the natural environment maintains the capacity to self-equilibrate following shocks is clearly vital, especially if the frequency and intensity of such shocks is liable to increase in line with a general degradation of natural systems caused by pressures such as climate change. However, the natural science information required to adequately assess (and thereby value) the maintenance of resilience levels is, to date, rarely available. In the absence of such information, concerns regarding the maintenance of resilience may be better handled through the adoption of precautionary approaches such as Safe Minimum Standards (SMS). Here economic valuation and decision-making operate as usual until a threshold in ecosystem functioning is identified. SMS are then employed to ensure the resilience of resources with economics being confined to the identification of cost-effective solutions for delivery of those standards (Bateman et al., 2011a).

Methods for valuing ecosystem related goods and services

There are a number of methods available for estimating the value of ecosystem service related economic goods and services. In some cases market prices can be relevant, although adjustments may have to be made for any distortions in those prices (e.g. to allow for the impact of price subsidies or constraints). However, many important goods and services, including a large number associated with ecosystem services, arise outside market contexts and lack private property rights. This problem has been long recognized, and from just a few path-breaking studies in the middle of the last century a burgeoning literature has developed, detailing methods for the economic valuation of non-market goods (e.g. Bateman et al., 2002a; Freeman et al., 2014; Bouma and van Beukering, 2015). This section outlines the major methods employed for valuing ecosystem related goods and services and provides illustrative examples of each.

Broadly speaking, valuation methods can be categorized as:

- Market valuation methods where market price information is used to indicate the value of related non-market goods. Approaches include the direct use of market prices (adjusted as necessary) for ecosystem service related goods, production function approaches, and, where appropriate, the use of cost information (for example in assessing the marginal costs of abating greenhouse gases).
- Revealed preference methods where analysis of the purchase of market priced goods is used to indicate the implicit value of strongly related non-market goods. Approaches include hedonic pricing (e.g. assessing the uplift in property prices attached to houses in quieter locations) and variants of the travel cost method (where an individual's willingness to incur costs reveals their value for the recreational sites they visit).
- Stated preference methods, where hypothetical markets are developed and survey or experimental participants make choices regarding the provision and associated costs of non-market goods.

Market valuation methods

The market price of certain goods provides useful information regarding the value of related ecosystem services. However, two practical problems need to be addressed before an unbiased estimate of underlying non-market values can be obtained. First, prices need to be adjusted for any market distortion (such as government subsidies or taxes) as well as for non-competitive practices. Second, ecosystem services may constitute only a portion of the value of inputs underpinning the production of a given marketed good, as other inputs, such as labour, expertise, and manufactured or social capital are also required to produce such goods. These various contributions to value can be disentangled through the estimation of production functions, revealing the value added by each input. For example, Fezzi et al. (2014) examine the contribution of climatic conditions (specifically temperature and precipitation) to agricultural output in the UK. By controlling for the contribution of other inputs (such as fertilizer) and policy interventions (such as subsidies), they isolate the effects of climate and use this to examine the likely impact of future climate change. These effects are expressed both in terms of changes in land use and, crucially, the economic value implications of climate change. Applying this method requires collection of data on, and understanding of how, changes in the quality and quantity of ecosystem services affect the costs of production of the final good, and the supply and demand for that good and for the other factors of production (Koetse et al., 2015). This method can in principle be applied to the valuation of inputs from a variety of ecosystem services, ranging from the maintenance of beneficial species such as pollinators to coastal protection from tropical storms (Barbier, 2007).

Some studies have used costs as approximation of the value of ecosystem service inputs. One approach is to look at the damage costs avoided by not allowing an ecosystem service to degrade (e.g. storm and flood protection; Badola and Hussain, 2005). Similarly, it is possible to analyse the expenditure and behaviour people incur to avoid such damage (Rosado et al., 2000). Some studies also look at the cost of replacement or restoration of an ecosystem service. However, the latter two are considered problematic as these costs might have little relationship to the values they aim to approximate (e.g. Barbier, 2007).

One obvious limitation of the above methods is that they can only apply to ecosystem services which are directly related to the production of market-price goods. Alternative approaches are required in cases where that relation is more indirect.

Revealed preference methods

Many ecosystem services are associated with non-market, unpriced goods. Here revealed preference approaches such as the travel cost or hedonic pricing method can prove useful. The travel cost method is a commonly used approach for valuing recreational benefits. It relies on the premise that values of recreation benefits are implicitly shown in people's behaviour in travel markets. Through analysis of travel expenses in terms of actual travel costs, time costs, and admittance fees, it is possible to assess the implicit price of access to the recreation site and incur the value of recreational benefits. Employing the travel cost method Egan et al. (2009) combined information from a household survey on recreational usage of 129 lakes in Iowa, US, with the detailed information on lakes' water quality to estimate the demand function for recreational trips to Iowa's lakes conditioned on their water quality.

Another route towards revealed preference valuation is provided by the hedonic pricing method. This measures the implicit price of an ecosystem service-related good as revealed in the observed price of an associated, market-priced good. The most common application of the hedonic method is via the property market, as house prices, once they are stripped of the influence of structural factors (number of bedrooms, garden size, etc.), neighbourhood variables (local unemployment rates, etc.), and accessibility characteristics (access to places of work, high quality schools, etc.), reveal clear associations with local environmental quality. Common applications include the valuation of aesthetic views, air quality, flood risk, and many other local amenities (e.g. Day et al., 2007).

Stated preference valuation methods

All the methods listed so far rely on observed behaviour either directly or indirectly occurring in extant markets. However, an alternative approach is to generate hypothetical markets through which survey or experimental respondents can be asked to express their WTP or WTA for changes in the provision of ecosystem service related goods. This is particularly useful for assessing preferences regarding situations which have not yet occurred (e.g. Metcalfe et al., 2012). Stated preference methods are also the only approach available for estimating pure non-use values such as those associated with the survival of endangered species (Morse-Jones et al., 2012). Debate continues, however, about the robustness of such methods if respondents do not have well-formed preferences for such changes.

Two variants of the stated preference approach are in common usage, the contingent valuation method (CVM) and choice modelling (CM) technique. As can be inferred from its name, the CVM approach elicits individuals' values contingent on there being a (hypothetical) market within which those values can be expressed. All four WTP/WTA measures can be elicited using CVM, provided that the market can be conveyed in a manner which respondents find credible. As the first of the stated preference methods to be widely applied, CVM has been subject to considerable critical appraisal, and it is certainly true that the questionnaire framing effects long recognized by psychologists in other contexts and the preference anomalies identified by experimental economists in the laboratory frequently translate to stated preference applications. However, this has generated a wider awareness across the economics profession of the complexity of human preferences within both environmental and other contexts. In response to this, CVM applications have undergone stringent examination, with increasing emphasis being placed upon the crucial role of design and implementation in applications. Carson et al. (1994) conducted a CVM study to estimate the monetary measure of the compensation for the negative impact of chemicals on wildlife species in California, USA. Employing a referendum format of the CVM, where respondents vote for or against a particular policy, the study estimated the WTP for decreasing the recovery period of the four affected species from 50 to five years to be around $575 million (with a standard error of $27 million). Notably, the development of the survey used in this study was conducted over the course of 32 months and provides a fine example of a comprehensive CVM study.

CM approaches are, in many respects, similar to the CVM. Again, they can, in theory, be applied to almost any ecosystem-related good and rely upon hypothetical markets to elicit respondents' choices. However, while CVM typically asks respondents about a single change in provision of a good, CM approaches elicit choices regarding multiple such changes. This is achieved by noting that many goods are composed of multiple attributes. For example, a given land use might involve different areas (or "levels") of woodland, farmland, and conservation land. By varying these levels we define multiple goods. CM respondents are then asked to choose between some set of these goods. Multiple definitions of these goods can be created, so respondents can be asked to answer many such choice questions, generating large amounts of preference data. By adding a variable cost to each of these goods, the analyst can observe how respondents trade off money against changes in the levels of each attribute.

Value transfer and its variants

In many cases, the costs of undertaking high-quality integrated natural science and economic valuation assessments is vastly outweighed by the net benefits generated from the improved decision-making facilitated by such studies. However, in some instances the resources necessary for such studies are unavailable. Provided that some estimate of the physical impacts of an intervention is available then values may be approximated through value transfer methods.

In essence, value transfer takes information from previously assessed "study" sites and utilizes this to estimate values for some alternative "policy" sites, or different changes at the same site. It is not a valuation method per se, as it is based on results from previous valuation studies. It reflects a pragmatic approach, recognising that it is not possible (or necessary) to value all ecosystems and their services when we have enough base studies from which values can be (robustly!) extrapolated.

A key requirement for the correct use of value transfer methods is that, on the assumption that precise matches between policy and study sites are unlikely, any differences are understood, quantified, and incorporated within the transfer process. Common adjustments between sites are to reflect differences in standards of living, varying levels of population, different spatial configurations and substitute availability, or different levels of provision change. The general aim of adjusting for the differences between the sites when using value transfer approaches is to minimize the "transfer errors" (the difference between the transferred values and the "actual" values which a particular site or change generates).

Value transfer methods embrace a variety of techniques, varying from the simple transfer of adjusted mean values (univariate transfer) to sophisticated applications of *value functions* (multivariate transfer) specifically developed for transfer purposes. Simple value transfer (e.g. Muthke and Holm-Mueller, 2004) takes values from primary valuation study sites (or a pool of such studies) and transfers these to the focal policy site(s). The values transferred are either (adjusted) means or unit values. It is important to note here that poor or incomplete adjustments can exhibit bigger errors than simple mean transfers (Brouwer and Bateman, 2005).

The value function transfer approach (e.g. Bateman et al., 2011c), in contrast, employs statistical methods to estimate a relationship between the study site characteristics and values estimated. The derived function is then used to predict values for the policy site(s), using the data from the policy sites as predictors in the value function. This method explicitly incorporates the difference between the sites, as the actual characteristics of the policy site determine the final value obtained from the function. Clearly, identifying the most appropriate variables and specifications for such value functions becomes a central issue. Potential approaches include meta-analyses of the extant valuation literature (e.g. Brander et al., 2006) or simply using statistical methods to identify relevant variables. However, Bateman et al. (2011c) argue that for value transfer purposes it is preferable to use function specifications which conform to economic theory (and possibly omit some context-specific variables) rather than simply rely upon the best statistical fit. This is because, while best fit models may incorporate site-specific variables unique to study sites, theoretically derived functions are more likely to incorporate generic variables, applicable to both study and policy sites.

Bateman et al. (2011c) also provide a performance comparison between the univariate and multivariate approaches to value transfer in a controlled multi-site experiment. Results show that for heterogeneous sets of sites, value function transfer exhibits lower transfer errors than using mean value transfer. In contrast, when transferring between similar sets of sites, mean value transfer performed better.

Recent methodological developments in value transfer include the adoption of the GIS techniques (see e.g. Bateman et al. 2002b, 2006; Troy and Wilson, 2006; Brander et al., 2012). This trend holds the promise of further methodological refinement for value transfer, reflecting the – in many cases crucial – spatial dimension of ecosystem service and related goods provision (see section 2). Incorporating GIS into valuation studies facilitates the construction of spatially explicit value functions. Sen et al. (2014), for example, provide a novel methodology and application for spatially sensitive prediction of outdoor recreation visits and values for different ecosystems. Using data on recreation trips in the UK from over 40,000 households, geographical and environmental data, and meta-analysis of recreation values the authors derive a spatially explicit estimation of visit numbers and recreation values under present and potential future land use in the UK.

Despite its promise, value transfer techniques are necessarily constrained by the availability of primary valuation studies. A range of initiatives aim to collate already existing valuation evidence and organize it in searchable database form – e.g. Environmental Valuation Reference Inventory (EVRI, www.evri.ca). Such databases can be a useful tool for initial screening for value transfer purposes and help to systematize the available evidence. However, coherent coverage of valuation studies is far from complete in terms of both the quality and quantity of applications.

Decision-support tools and integrated ecosystem analyses

A range of decision-support tools aiming to support systematic ecosystem assessments have emerged recently. These tools integrate ecology, economics, and geography, with some employing ecosystem valuation tools (Bagstad et al., 2013). A number of these rely on an integrated set of models, which can support policy and decision-making with a spatially explicit advice (see e.g. InVEST, Tallis et al., 2013; ARIES, Bagstad et al., 2011; or TIM, Bateman et al., 2014). These tools vary in their ability to address differing scales (spatial and temporal) as well as data and computational constraints, but most seek to inform decision-making by mapping the impacts of, say, climate and land use change on the provision of ecosystem-related goods and services. Employing the Integrated Valuation of Environmental Services and Tradeoffs (InVEST) tool, Goldstein et al. (2012) analyze seven land use scenarios for a private land development in Hawaii with the aim of balancing both private and public interests on a local level, while taking into account carbon storage, water quality, and financial return over a 50-year horizon. Bateman et al. (2014) in UK NEA (2014) provide analysis of optimal land use policy in Great Britain for the period of the next 50 years in terms of spatially explicit advice for forestry policy under different optimizing rules. The Integrated Model (TIM), developed for this study, takes into account monetary estimates of agriculture and timber production, recreation, agricultural and forestry Green House Gasses emissions, and non-monetary measures of water quality and bird diversity.

The integrated modelling tools make use of different combinations of valuation methods, represent an integration of knowledge and models across disciplines, and often require several years of development and data harmonization, but nevertheless show promise of further development in ecosystem valuation research.

Conclusions: the future of valuation research

Failure to include the value of nature in decisions can lead to inefficient resources allocation and further environmental degradation. Carefully applied, economics and monetary valuation can fill this gap. They provide information that is understood by stakeholders at all levels and facilitate comparative assessment across different policy areas. Increasingly, the value of ecosystem-related goods and services has become a central issue in environmental policy debates. Such attention is appropriate given the importance of economic analysis for understanding and addressing many of the trade-offs underlying environmental decision-making. However, the policy demands on valuation research are immense. Current European Union (EU) and Convention on Biological Diversity (CBD) biodiversity targets aim to account for and incorporate the values of ecosystem services and biodiversity in decision-making frameworks by 2020. While initial progress towards these targets has been made (SCBD, 2014), actually delivering on these objectives still requires urgent and extensive action in terms of knowledge generation, but also through further mainstreaming across society.

Further work is required to strengthen the scientific knowledge base surrounding ecosystem services assessments and accompanying economic analyses. This concerns both work within and across the involved scientific disciplines. Both economics and natural science have a number of areas urgently needing further research in order to strengthen the ecosystem assessments required for policy use. As economists continue developing valuation methods, they must improve their understanding of preference formation and dynamics, integrating advances from behavioural and experimental economics as well as subjective well-being research and contributions from disciplines such as psychology or cognitive science. In turn, natural science's further understanding of ecosystem functioning, its relation to biodiversity, and the role and occurrence of thresholds and non-linearities are key bases for further advancement of ecosystem assessment research. The challenge for ecosystem science is indeed of sizeable proportions, requiring "a new kind of interdisciplinary science . . . to build understanding of social-ecological systems" to support decisions with robust advice (Carpenter et al., 2009, p.1309).

Further, valuation of ecosystem-related goods and services requires building up a strong and robust evidence base. A co-ordinated effort is crucial to develop a sufficient basis to support broad-scale value transfer, as well as to build a set of best practice examples for replication in different contexts. Importantly, rigorous monitoring systems for mapping the outcomes of policy interventions are crucial for understanding valuation-based (and broadly speaking ecosystem-related) policies, their improvement and further refinement. Environmental economics, as any other research with policy implications and on-the-ground results, needs evaluation programmes similar to ones present in development studies, which evaluate interventions, their results and effectiveness. In the case of valuation research and applications, this can take the form of follow-up studies, monitoring how the results of analyses influenced on-the-ground reality as well as how the values estimated transpire in reality. This can have benefits for both knowledge base-building and further understanding of what works on the ground and what does not. In addition, such evaluation can help us understand the dynamics behind the formation of values and their change through time.

Co-ordination of related policy processes and research initiatives is crucial in order to exploit synergies and strengthen the evidence base. Environmental valuation exercises can inform numerous policy and research processes, including those related to climate change, mitigation, and adaptation (e.g. including reduced emissions from deforestation and forest degradation); moving beyond GDP and natural capital accounting; water and agricultural policies; and natural hazard policies. Moreover, co-operation across sectors can highlight pressing problems and identify priority research needs. Importantly, valuation research is well-positioned for a productive co-operation with the private sector, which can play a key role in both changing the impact businesses have on the environment and driving consumer behaviour change. While numerous areas exist where businesses can build on valuation research (e.g. TEEB, 2012), there are still only isolated cases of cooperation between valuation researchers and private sector.

Valuation research and ecosystem research more broadly has gathered momentum and holds the potential to change the way nature has been disregarded in our decisions throughout the past century, causing major environmental destruction. Valuation research can contribute to ecosystem related research and, further, to broader "sustainability science". However, bridging across disciplines and sectors is vital in order for this information to turn into changes in decisions. This remains a major challenge, indeed, "[s]uch a massive effort in social-ecological science is unprecedented in human history, yet it is commensurate with the problems we face and with the potential of sustainability science" (Carpenter et al., 2009, p.1311).

References and further readings

Atkinson, G., Bateman, I., and Mourato, S. (2012). Recent advances in the valuation of ecosystem services and biodiversity. *Oxford Review of Economic Policy*, vol 28, no 1, pp 22–47.

Badola, R., and Hussain, S. A. (2005). Valuing ecosystem functions: an empirical study on the storm protection function of Bhitarkanika mangrove ecosystem, India. *Environmental Conservation*, vol 32, no 1, pp 85–92.

Bagstad, K. J., Semmens, D. J., Waage, S., and Winthrop, R. (2013). A comparative assessment of decision-support tools for ecosystem services quantification and valuation. *Ecosystem Services*, vol 5, pp 27–39.

Bagstad, K. J., Villa, F., Johnson, G., and Voigt, B. (2011). *ARIES—Artificial Intelligence for Ecosystem Services: A Guide to Models and Data*, Version1.0 Beta. The ARIES Consortium, Bilbao.

Barbier, E.B. (2007). Valuing ecosystem services as productive inputs. *Economic Policy*, vol 22, no 49, pp 177–229.

Bateman, I. J., Abson, D., Beaumont, N., Darnell, A., Fezzi, C., Hanley, N., Kontoleon, A., Maddison, D., Morling, P., Morris, J., Mourato, S., Pascual, U., Perino, G., Sen, A., Tinch, D., Turner, R.K., Valatin, G., *et al.* (50 authors) (2011b). Economic values from ecosystems. In: *The UK National Ecosystem Assessment Technical Report, UK National Ecosystem Assessment*, UNEP-WCMC, Cambridge UK. Available at: http://uknea.unep-wcmc.org/

Bateman, I. J., Brouwer, R., Ferrini, S., Schaafsma, M., Barton, D. N., Dubgaard, A., Hasler, B., Hime, S., Liekens, I., Navrud, S., De Nocker, L., Ščeponavičiūtė, R., and Semėnienė, D. (2011c). Making benefit transfers work: deriving and testing principles for value transfers for similar and dissimilar sites using a case study of the non-market benefits of water quality improvements across Europe. *Environmental and Resource Economics*, vol 50, no 3, pp 365–387.

Bateman, I. J., Carson, R. T., Day, B., Hanemann, W. M., Hanley, N., Hett, T., Jones-Lee, M., Loomes, G., Mourato, S., Özdemiroglu, E., Pearce, D.W., Sugden, R., and Swanson, J. (2002a). *Economic Valuation with Stated Preference Techniques: A Manual*. Edward Elgar, Cheltenham.

Bateman, I., Day, B., Agarwala, M., Bacon, P., Badura, T., Binner, A., De-Gol, A., Ditchburn, B., Dugdale, S., Emmett, B., Ferrini, S., Carlo Fezzi, C., Harwood, A., Hillier, J., Hiscock, K., Hulme, M., Jackson, B., Lovett, A., Mackie, E., Matthews, R., Sen, A., Siriwardena, G., Smith, P., Snowdon, P., Sünnenberg, G., Vetter, S., and Vinjili, S. (2014). *UK National Ecosystem Assessment Follow-on. Work Package, Report 3: Economic Value Of Ecosystem Services*. UNEP-WCMC, LWEC.

Bateman, I. J., Day, B. H., Georgiou, S., and Lake, I. (2006). The aggregation of environmental benefit values: welfare measures, distance decay and total WTP. *Ecological Economics*, vol 60, no 2, pp 450–460.

Bateman, I. J., Fisher, B., Fitzherbert, E., Glew, D., and Naidoo, R. (2010). Tigers, markets and palm oil: market potential for conservation. *Oryx*, vol 44, no 2, pp 230–234.

Bateman, I. J., Harwood, A. R., Mace, G. M., Watson, R. T., Abson, D. J., Andrews, B., Binner, A., Crowe, A., Day, B. H., Dugdale, S., Fezzi, C., Foden, J., Hadley, D., Haines-Young, R., Hulme, M., Kontoleon, A., Lovett, A. A., Munday, P., Pascual, U., Paterson, J., Perino, G., Sen, A., Siriwardena, G., van Soest, D., and Termansen, M. (2013). Bringing ecosystem services into economic decision-making: land use in the United Kingdom. *Science*, vol 341, no 6141, pp 45–50.

Bateman, I. J., Jones, A. P., Lovett, A. A., Lake, I. R., and Day, B. H. (2002b). Applying geographical information systems (GIS) to environmental and resource economics. *Environmental and Resource Economics*, vol 22, no 1–2, pp 219–269.

Bateman, I. J., Mace, G. M., Fezzi, C., Atkinson, G., and Turner, K. (2011a). Economic analysis for ecosystem service assessments. *Environmental and Resource Economics*, vol 48, no 2, pp 177–218.

Bockstael, E. N., Freeman, A. M., Kopp, R. J., Portney, P. R., and Smith, V. K. (2000). On measuring economic values for nature. *Environmental Science & Technology*, vol 34, no 8, pp 1384–1389.

Bouma, J., and van Beukering, P.J.H. (eds) (2015). *Ecosystem Services: From Concept to Practice*. Cambridge University Press, Cambridge UK.

Boyd, J., and Banzhaf, S. (2007). What are ecosystem services? The need for standardized environmental accounting units. *Ecological Economics*, vol 63, pp 616–626.

Brander, L. M., Brauer, I., Gerdes, H., Ghermandi, A., Kuik, O., Markandya, A., Navrud, S., Nunes, P.A.L.D., Schaafsma, M., Vos, H., and Wagtendonk, A. (2012). Using meta-analysis and GIS for value transfer and scaling up: valuing climate change induced losses of European wetlands. *Environmental and Resource Economics*, vol 52, no 3, pp 395–413.

Brander, L. M., Florax, R.J.G.M., and Vermaat J. E. (2006). The empirics of wetland valuation: a comprehensive summary and a meta-analysis of the literature. *Environmental and Resource Economics*, vol 33, pp 223–250.

Brouwer, R., and Bateman, I. J. (2005). Benefits transfer of willingness to pay estimates and functions for health-risk reductions: a cross-country study. *Journal of Health Economics*, vol 24, no 3, pp 591–611.

Carpenter, S. R., Mooney, H. A., Agard, J., Capistrano, D., Defries, R. S., Díaz, S., Dietz, T., Duraiappah, A. K., Oteng-Yeboah, A., Pereira, H. M., Perrings, C., Reid, W. V, Sarukhan, J., Scholes, R. J., and Whyte, A. (2009). Science for managing ecosystem services: beyond the Millennium Ecosystem Assessment. *Proceedings of the National Academy of Sciences of the United States of America*, vol 106, no 5, pp 1305–1312.

Carson, R.T., Hanemann, W. M., Kopp, R. J., Krosnick, J. A., Mitchell, R. C., Presser, S., Ruud, P. A., and Smith, V. K. (1994). *Prospective Interim Lost Use Value Due to PCB and DDT Contamination in the Southern California Bight*. Natural Resource Damage Assessment, Report to National Oceanic and Atmospheric Administration.

CBD – Convention on Biological Diversity (2010). *Strategic Plan for Biodiversity 2011–2020*, COP 10 Decision X/2. Aichi, Japan. Available at: http://www.cbd.int/sp/targets/

Costanza, R. (2016). Ecosystem services in theory and practice. In Potschin, M., Haines-Young, R., Fish, R. and Turner, R. K. (eds) *Routledge Handbook of Ecosystem Services*. Routledge, London and New York, pp 15–24.

Costanza, R., dArge, R., de Groot, R., Farber, S., Grasso, M., Hannon, B., Limburg, K., Naeem, S., Oneill, R.V., Paruelo, J., Raskin, R. G., Sutton, P., and van den Belt, M. (1997). The value of the world's ecosystem services and natural capital. *Nature*, vol 387, pp 253–260.

Costanza, R., de Groot, R., Sutton, P., van der Ploeg, S., Anderson, S.J., Kubiszewski, I., Farber, S., and Turner, R.K. (2014). Changes in the global value of ecosystem services. *Global Environmental Change*, vol 26, pp 152–158.

Day, B., Bateman, I. J., and Lake, I. (2007). Beyond implicit prices: recovering theoretically consistent and transferable values for noise avoidance from a hedonic property price model. *Environmental and Resource Economics*, vol 37, no 1, pp 211–232.

Egan, K. J., Herriges, J. A., Kling, C. L., and Downing, J. A. (2009). Valuing water quality as a function of water quality measures. *American Journal of Agricultural Economics*, vol 91, no 1, pp 106–123.

Ferraro, P. J., Rondeau, D., and Poe, G. L. (2003). Detecting other-regarding behavior with virtual players. *Journal of Economic Behavior & Organization*, vol 51, no1, pp 99–109.

Fezzi, C., Bateman, I. J., Askew, T., Munday, P., Sen, A., and Harwood, A. (2014). Valuing provisioning ecosystem services in agriculture: the impact of climate change on food production in the United Kingdom. *Environmental and Resource Economics*, vol 57, pp 197–214.

Fisher, B., and Turner, R. K. (2008). Ecosystem services: classification for valuation', *Biological Conservation*, vol 1, no 2007, pp 8–10.

Fisher, B., Turner, R. K., Morling, P. (2009). Defining and classifying ecosystem services for decision making. *Ecological Economics*, vol 68, no 3, pp 643–653.

Freeman III, M. A., Herriges, J. A., and Kling, C. L. (2014). *The Measurement of Environmental and Resource Values: Theory and Methods*, 3rd edition. RFF Press, Washington DC.

Goldstein, J. H., Caldarone, G., Duarte, T. K., Ennaanay, D., Hannahs, N., Mendoza, G., Polasky, S., Wolny, S., and Daily, G. C. (2012). Integrating ecosystem-service tradeoffs into land-use decisions. *Proceedings of the National Academy of Sciences*, vol 109, no 19, pp 7565–7570.

Gowdy, J., Howarth, R., and Tisdell, C. (2010). Discounting, ethics, and options for maintaining biodiversity and ecosystem services. In: Kumar, P. (ed.) *The Economics of Ecosystems and Biodiversity: Ecological and Economic Foundations*. Earthscan, London.

Haines-Young, R., and Potschin, M. (2013). Common International Classification of Ecosystem Services (CICES). Report to the European Environment Agency EEA/BSS/07/007 (download: www.cices.eu).

Heal, G. M., Barbier, E. B., Boyle, K. J., Covich, A. P., Gloss, S. P., Hershner, C. H., Hoehn, J. P., Pringle, C. M., Polasky, S., Segerson, K., and Shrader-Frechette, K. (2005). *Valuing Ecosystem Services: Toward Better Environmental Decision Making*. The National Academies Press, Washington DC.

Hicks, J. R. (1943) The four consumer surpluses. *Review of Economic Studies*, vol 8, pp 108–116.

Horton, B., Collarullo, G., Bateman, I. J., and Peres, C. (2003). Evaluating non-user willingness to pay for a large scale conservation program in Amazonia: a UK/Italian contingent valuation study. *Environmental Conservation*, vol 30, no 2, pp 139–146.

Johnston, R. J., and Russell, M. (2011). An operational structure for clarity in ecosystem service values. *Ecological Economics*, vol 70, pp 2243–2249.

Koetse, M. J., Brouwer, R., and van Beukering, P.J.H. (2015). Economic valuation. In Bouma, J. and van Beukering, P.J.H. (eds) *Ecosystem Services: From Concept to Practice*. Cambridge University Press, Cambridge UK.

Landers, D. H., and Nahlik, A. M. (2013). *Final Ecosystem Goods and Services Classification System (FEGS-CS)*. EPA/600/R-13/ORD-004914. U.S. Environmental Protection Agency, Office of Research and Development, Washington DC.

MA (2005). *Ecosystems and Human Well-Being: Synthesis*, Island Press, Washington DC.

Mace, G.M. (2014). Biodiversity: its meanings, roles, and status. In: Helm, D. and Hepburn, C.J. (eds) *Nature in the Balance: The Economics of Biodiversity*. Oxford University Press, Oxford.

Mäler, K.-G. (2008). Sustainable development and resilience in ecosystems. *Environmental and Resource Economics*, vol 39, no 1, pp 17–24.

Mäler, K.-G., Xepapadeas, A., and De Zeeuw, A. (2003). The economics of shallow lakes. *Environmental and Resource Economics*, vol 26, no 4, pp 603–624.

Metcalfe, P., Baker, B., Andrews, K., Atkinson, G., Bateman, I. J., Butler, S., Carson, R. T., East, J., Gueron, Y., Sheldon, R., and Train, K. (2012). An assessment of the non-market benefits of the water framework directive for households in England and Wales. *Water Resources Research*, vol 48, Article Number: W03526.

Morse-Jones, S., Bateman, I. J., Kontoleon, A., Ferrini, S., Burgess, N. D., and Turner, R. K. (2012). Stated preferences for tropical wildlife conservation amongst distant beneficiaries: charisma, endemism, scope and substitution effects. *Ecological Economics*, vol 78, pp 9–18.

Muthke, T., and Holm-Mueller, K. (2004). National and international benefit transfer testing with a rigorous test procedure. *Environmental and Resource Economics*, vol 29, pp 323–336.

Naeem, S., Duffy, J. E., and Zavaleta, E. (2012). The functions of biological diversity in an age of extinction. *Science*, vol 336, pp 1401–1406.

ONS (2014). *UK Natural Capital – Initial and Partial Monetary Estimates*. Office for National Statistics. Available at: http://www.ons.gov.uk/ons/rel/environmental/uk-natural-capital/initial-estimates/art-article.html

Pearce D., and Turner, R. K. (1990). *Economics of Natural Resources and the Environment*. Harvester Wheatsheaf, London.

Potschin, M., and Haines-Young, R. (2011). Ecosystem services: exploring a geographical perspective. *Progress in Physical Geography*, vol 35, no 5, pp 575–594.

Potschin, M. and R. Haines-Young (2016). Defining and measuring ecosystem services. In Potschin, M., Haines-Young, R., Fish, R. and Turner, R. K. (eds) *Routledge Handbook of Ecosystem Services*. Routledge, London and New York, pp 25–44.

Rockström, J., Steffen, W., Noone, K., Persson, A., Chapin, F. S., Lambin, E. F., Lenton, T. M., Scheffer, M., Folke, C., Schellnhuber, H. J., Nykvist, B., de Wit, C. A., Hughes, T., van der Leeuw, S., Rodhe, H., Sörlin, S., Snyder, P. K., Costanza, R., Svedin, U., Falkenmark, M., Karlberg, L., Corell, R.W., Fabry, V. J., Hansen, J., Walker, B., Liverman, D., Richardson, K., Crutzen, P., and Foley, J. A. (2009). A safe operating space for humanity. *Nature*, vol 461, no 7263, pp 472–475.

Rosado, M., Cunha-e-Sa, M. A., Ducla-Soares, M. M., and Rainho Catela Nunes, L. M. (2000). Combining averting behavior and contingent valuation data: an application to drinking water treatment. FEUNL working paper no. 392. Available at: http://ssrn.com/abstract=880458

SCBD – Secretariat of the Convention on Biological Diversity (2010). *Global Biodiversity Outlook 3*. Montréal.

SCBD – Secretariat of the Convention on Biological Diversity (2014). *Global Biodiversity Outlook 4*. Montréal

Sen, A., Harwood, A. R., Bateman, I. J., Munday, P., Crowe, A., Brander, L., Raychaudhuri, J., Lovett, A. A., Foden, J., and Provins, A. (2014). Economic assessment of the recreational value of ecosystems: methodological development and national and local application. *Environmental and Resource Economics*, vol 57, no 2, pp 233–249.

Staub C., Ott W., Heusi F., Klingler G., Jenny A., Hacki M., and Hauser, A. (2011). Indicators for ecosystem goods and services: framework, methodology and recommendations for a welfare-related environmental reporting, *Environmental Studies No. 1102*. Federal Office for the Environment, Bern.

Tallis, H. T., Ricketts, T., Guerry, A., Wood, S. A., Sharp, R., Nelson, E., Ennaanay, D., Wolny, S., Olwero, N., Vigerstol, K., Pennington, D., Mendoza, G., Aukema, J., Foster, J., Forrest, J., Cameron, D., Arkema, K., Lonsdorf, E., Kennedy, C., Verutes, G., Kim, C., Guannel, G., Papenfus, M., Toft, J., Marsik, M., Bernhartdt, J., and Griffin, R. (2013). *InVEST 2.5.3 User's Guide*. The Natural Capital Project, Stanford, CA.

TEEB (2011). *The Economics of Ecosystems and Biodiversity in National and International Policy Making*. ten Brink, P. (ed.). Earthscan, London.

TEEB (2012). *The Economics of Ecosystems and Biodiversity in Business and Enterprise*. Bishop, J. (ed.). Earthscan, London.

Troy, A., and Wilson, M. A. (2006). Mapping ecosystem services: practical challenges and opportunities in linking GIS and value transfer, *Ecological Economics*, vol 60, no 2, pp 435–449.

Turner, R. K. (2016). Ecological economics and ecosystem services. In Potschin, M., Haines-Young, R., Fish, R. and Turner, R.K. (eds) *Routledge Handbook of Ecosystem Services*. Routledge, London and New York, pp 243–255.

UK National Ecosystem Assessment (2011). *The UK National Ecosystem Assessment: Synthesis of the Key Findings*. UNEP-WCMC, Cambridge UK.

UK National Ecosystem Assessment (2014). *The UK National Ecosystem Assessment: Synthesis of the Key Findings*. UNEP-WCMC, LWEC.

van Beukering, P.J.H., Brouwer, R., and Koetse M. J. (2015). Economic values of ecosystem services. In Bouma, J. and van Beukering, P.J.H. (eds) *Ecosystem Services: From Concept to Practice*. Cambridge University Press, Cambridge UK.

20

ECOLOGICAL ECONOMICS AND ECOSYSTEM SERVICES

R. Kerry Turner

Introduction: nature's value

During the 1970s a debate involving natural science, economic and social science perspectives emerged around the issue of the 'value' of ecosystems and biodiversity. Given the inherent complexity of nature, it is not surprising that the concept of 'value' is open to multiple interpretations and meanings. A number of different dimensions of nature-based value (see Potschin and Haines-Young, 2016) can be discerned and evaluated in different ways, e.g. in monetary terms via economic analysis and the concept of total economic value (TEV, where TEV = use value + non-use value (existence and bequest values)); in biophysical and geochemical terms via natural science; in more qualitative terms via sociology, cultural geography and the arts and humanities; etc. Each of these value dimensions has validity in its own domain. Environmental philosophers have constructed a generic value typology with four categories: *anthropocentric instrumental value*, which maps closely on to the economic concepts of use and most of non-use values; *anthropocentric intrinsic value*, a culturally dependent concept which is linked to human stewardship of nature motivation and which needs a human to ascribe intrinsic value to non-human nature – the economist's idea of existence value may overlap into this value category. The other two value categories – *non-anthropocentric instrumental value* and *non- anthropocentric intrinsic value* – are less directly relevant to the policy initiatives, unless in the latter value category's case a radical ethical position is accepted as the societal norm, which is currently not the case (Hargrove, 1992). But, while this academic thinking and debate was necessary and heuristically worthwhile, it was not in itself sufficient to mitigate the on-going global loss of ecosystems and habitats and species (Westman, 1977). Practical and implemented management responses were key requirements if the rate and extent of environmental change was to be moderated.

In terms of the modern history of economic theory and applications, while the notion of human benefit in welfare terms derived from ecosystems first came to prominence in the 1970s, consensus proved elusive. A significant amount of research attention was then focused on the conceptual and methodological problems that assigning monetary values to ecosystem services posed. A range of monetary valuation methods were rapidly developed and the published literature grew exponentially (e.g. Pearce and Turner, 1990; Freeman, 1993). During the 1980s, two 'camps' emerged with the1988 formation of the International Society for Ecological Economics which sought to distinguish itself from the more conventional environmental and resource economics tradition. This chapter focuses on the ecological economics tradition.

Ecological economics

The Ecological Economics camp is composed of a wide range of analysts from a number of disciplinary backgrounds, including natural science (Spash, 1999; Turner, 1999; Gómez-Baggethun et al., 2010). The distinctive features of Ecological Economics are the acceptance of a pluralistic interdisciplinary methodology. Much of the analysis produced within Ecological Economics poses challenges to some of the axioms of conventional economics, especially in terms of economic rationality and the dominance of self-interested human preference assumptions over other-regarding preferences. There is also a focus on the macro-systems level scale with an emphasis on the dangers posed by system threshold/tipping points linked to resource and pollution assimilative capacity 'limits' caused by continued exponential economic growth. At the practical level, real world policy and project appraisal has also been criticised in terms of often being overly technocratic. According to Ecological Economics the policy process should be undertaken in as open and transparent a way as is feasible and should not be dominated by a narrow interpretation of cost-benefit analysis. Project and policy appraisal should be undertaken on the basis of a multi-criteria approach in all but the most straightforward of policy contexts. Economic efficiency is therefore seen as only one decision criterion for environmental management. Equity and fairness both within contemporary society and across future generations are equally important to foster trust and accountability in the decision-making process. Justice is not to be seen as exclusively a matter of income distribution, it is also a question of procedures which determine it. Both quantitative and more qualitative and deliberative decision support tools and approaches may be encompassed within the policy process. In short, a strong sustainability strategy is advocated and contrasted with the weak sustainability thinking prevalent in conventional economics (Turner, 1988; Daly and Cobb, 1989; Neumayer, 1999).

Weak and strong sustainability

Sustainable economic development has been characterized as a process of change in an economy that ensures that welfare is non-declining over the long run (Pearce et al., 1990). Both economics camps have used capital theory to define sustainable development in terms of the value of a capital stock over time. The capital stock includes a natural capital component which encompasses ecosystems and their services together with the abiotic environment. So, a society's wealth creation potential is governed by its stock of manufactured, human, natural and institutional/social capital. Some ecological economists have highlighted problems in the aggregation of natural and produced capital, but the most salient issue of contention is related to the feasibility of substitution across the capital stock components. Weak sustainability thinking, favoured by many conventional economists, assumes that there are and will be in the future almost limitless possibilities for capital substitution due to innovation and technological change. Economic growth, it is believed, can continue indefinitely as long as the 'Hartwick rule' is observed, that is, that the economic rents derived from the exploitation of exhaustible natural resources (fossil fuels and the like) are invested in other forms of capital capable of yielding an equivalent stream of income in the future (Hartwick, 1978). An extension to the 'Hartwick rule' has been proposed, which suggests that in some contexts changes in natural capital assets might be offset by investing in produced capital or social capital (Aldred, 2002; Turner, 2007; Lazaro-Touza and Atkinson, 2013). However, the size of the change in natural assets (e.g. loss of coastal wetlands or tropical forests and their ecosystem services) is important. The more extensive the loss and therefore the more significant the consequences, the less people are prepared to accept compensation in terms of other capital assets and either vote for conservation *per se*, or compensation through equivalent increases in natural capital nearby.

The concept of 'green growth' has recently gained some traction in international policy circles and has to some extent overshadowed the sustainable development dialogue. Green growth has been interpreted as a process in which gross domestic product (GDP, a measure of the rate of economic activity) grows together with a substantive (beyond business as usual) expansion in environmental protection measures. This development path can be guided by either weak or strong sustainability standards, depending on the position taken on natural capital substitution possibilities and therefore on technological innovation. A spectrum of arguments in favour of green growth and investment in natural capital (maintaining/enhancing the stock and flow of ecosystem services) as a prime stimulus for growth have been put forward. They start with advocating short-run Keynesian multiplier impacts that can assist in jerking an economy out of recession. Some then go further, through a policy of extensive investment in environmental protection and the correction of market failures. The claim here is that, while correcting for market failures may be costly, the future damage costs of inaction now will be an order of magnitude greater. Finally, some analysts claim that environmental investments themselves can be a significant driver of the growth process through innovation and the so-called first mover advantage (Porter and van der Linde, 1995). Sustainable delivery and use of ecosystem services can play a part in a green growth strategy.

Nevertheless, strong sustainability thinking advocated by ecological economists is much less optimistic about the possibilities for capital substitution and argues that vital elements of natural capital are 'critical assets' which are subject to irreversible loss constraints. There are no plausible technological substitutes for climate stability, stratospheric ozone, topsoil or species diversity and related ecosystem services (Ayres, 1993). Critical natural capital should therefore be protected by environmental standards and regulations to curb exploitative market forces, with due regard for the opportunity cost such a precautionary approach might entail (Bateman et al., 2011). Many Ecological Economists see development as a process of co-evolution of environment and society with the economy as an open subsystem (maintained by material and energy inflow) of the overall environment. It is then important to remain vigilant over the existence of system thresholds/tipping points leading to possible collapse and limits to growth, and to conserve as far as possible natural capital and future opportunities and options for the generations to come (Howarth, 1995).

The Ecological Economics camp also contains analysts who take a different ethical stance from the consequentialist philosophy (i.e. an act is right or wrong depending on the consequences of the action for human welfare/utility) supported by conventional economics. Deontological philosophy is given much wider support in ecological economics (i.e. an act is right or wrong depending on how well or not it conforms to a moral norm such as Kantian 'categorical imperatives'). Sustainable development goals carry with them an underlying intergenerational equity morality which imposes sustainability rules as prior moral constraints on the maximization of social preferences when welfare gains/losses are being assessed across generations. So, actions that deliver welfare gains to the present generation by imposing plausible risks of significant and irreversible losses on future generations are morally questionable.

Within the emerging sustainability science tradition during the 1980s and 1990s, ecosystem services and valuation of society's related benefits became more and more mainstream in the published literature. A further development was the use of conservation economic incentives through the incorporation of ecosystem service values into markets and so-called payments for ecosystem services (PES) (see chapter 44 in this volume; Wunder et al., 2008; Strassburg et al., 2009; Brouwer, 2016). Even more ambitiously, some claimed that these schemes could also help with other policy objectives, such as poverty alleviation. However, more research into, and practical application of, PES schemes needs to be done before the real equity impact of these schemes in a range of institutional and governance contexts can be determined (Corbera et al., 2007).

Ecosystem services framework

In the wider context, an Ecosystem Services Framework (ESF) began to evolve from an earlier natural science-based and more holistic analytical approach known as the 'Ecosystem Approach' as detailed by the 1992 Convention on Biological Diversity (CBD). The next step was to augment the systems-based science by the inclusion of social science and humanities, to link ecosystem functioning and the delivery of services which contribute to human well-being. Hence, the underlying aim is not so much to solely maximize environmental or biodiversity conservation, but rather to manage the rate of change in ecosystems structure (including species composition) and functioning as socio-economic and ecological systems co-evolve through time.

Many definitions and classification schemes for ecosystem services now exist (Costanza et al., 1997; Daily, 1997; Boyd and Banzhaf, 2007). One of the most widely cited is the Millennium Ecosystem Assessment definition (MA, 2005). Its classification and framework provides a platform for moving towards a more operational classification system which explicitly links changes in ecosystem services to changes in human welfare. By adapting and re-orienting this definition, it can be further tailored to suit the purpose at hand with little loss of functionality. Wallace (2007), for example, has focused on land management, while Boyd and Banzhaf (2007) and Mäler et al. (2009) take national income accounting as their policy context.

For economic and social valuation purposes, the definition proposed by Fisher et al. (2009) clarifies the distinction between ecosystem services and benefits: *ecosystem services are the aspects of ecosystems utilized (actively or passively) to produce human well-being.* Fisher et al. (2009) see ecosystem services as the link between ecosystems and things that humans benefit from, not the benefits themselves. The key feature of this definition is the separation of ecosystem processes and functions into intermediate and final services, with the latter yielding welfare benefits. The term 'intermediate services' should not be interpreted as signifying lesser significance but rather as a necessary signal in order to clearly demarcate (in valuation terms) final services and provide technically correct guidance to avoid double counting when services are valued in economic terms (Fisher et al., 2009). It is changes in the provision of final ecosystem services that economic analysts are interested in measuring and incorporating into policy analysis (see Badura et al., 2016).

Ecosystem services benefits are the 'exports' from the ecosystem sector to the human economic sector (Banzhaf and Boyd, 2012). Complementary assets (e.g. time, energy, finance or skills) also usually have to be combined with the natural capital to yield benefits. Following the UK NEA (2011) and UK NEAFO (2014) conceptual framework for ecosystem services assessment, the outcomes from the functioning of ecosystems have been generically labelled 'goods', which refer to a range of human welfare benefits derived from the flow of final services provided (Balmford et al., 2011; Bateman et al., 2011). But the scope of the delivered final ecosystem services (and therefore the valued goods and benefits) is very wide and the values are contested. The assessment and valuation of ecosystem stock and flow situations is therefore not a straightforward task. The monetary valuation of stocks and flows in particular is complex and has to rely on a range of accounting and socio-economic approaches, together with a sound underlying natural science understanding. This latter requirement is the cause of an on-going debate about the adequacy of ecological knowledge on the relationship between ecological processes and the delivery of ecosystem services.

Valuation and management challenges

Some critics worry that, because the links between biodiversity, biophysical processes and the supply of ecosystem services are complex, our understanding of the 'system' is not always sufficient to be confident about the consequences of ecosystem services-based management (Adams,

2014). Overall biodiversity may not be conserved if management focuses too narrowly on specific processes and services and misses negative impacts on other components of an ecosystem. It is important to distinguish between contexts in which single ecosystem services are being considered and the co- provision of multiple services ('bundles') through so-called service providing units (SPUs). The latter are a combination of a population (with its diversity and redundancy characteristics), biological communities and any network of interacting organisms that provide a service, and are mapped in terms of spatial area (Luck, 2009). Conserving sufficient 'patches' of SPUs and avoiding excessive habitat fragmentation will help to maintain ecosystem services provision and biodiversity (see also Luck, 2016).

Ecological economists warn that some services will not be amenable to monetary valuation because they are linked to socio-cultural entities, with specific historical conditions and symbolic significance. The values expressed for such cultural entities may well manifest themselves through collective social networks such as groups, communities and even nations. They may not be best identified through an individual's monetary valuation, but through group deliberation and shared values in quantitative or qualitative terms, or through other evidence sources, e.g. archives (see UK NEAFO, 2014; Kenter, 2016). Cultural or societal values, as well as communal and group values, include principles and values as well as a shared sense of what is worthwhile held by members of a society, community or group and culturally transmitted and assimilated over time as social norms. These shared non-monetary values signal that human well-being and quality of life is a function of both individual satisfaction and the meeting of a variety of social, health-related and cultural collective needs. Cultural values include shared values often acquired over long periods of time and connected to specific local places. Similarly, valuing the contribution that ecosystem services make to human well-being (or 'happiness', see below) cannot be reduced to individual preferences (WTP) and motivations alone.

Society's acceptance of the reliability and legitimacy of decision-making processes that have been informed by technical evidence and have highlighted trade-off dilemmas can in certain contexts be heavily influenced by whether shared values have or have not been explicitly recognized and accounted for in the political process (Fish et al., 2011). It is also important to note that, while techniques are evolving to better understand shared values, the social learning mechanisms themselves are 'processes to be engaged in', facilitating policy deliberation among equal partners.

Happiness data, human well-being and ecosystem services

With its origins in the 1970s (Easterlin, 1974), scientific interest in the so-called happiness approach and empirical measures of human happiness has increased significantly in the last decade (Layard, 2005; Layard, 2010; Oswald and Wu, 2010). This emerging field of research offers a novel way of valuing environmental goods which models individuals' self-reported happiness as a fraction of their income and the prevailing environmental conditions (Welsch and Kuhling, 2008). The approach seeks to measure individual stated subjective well-being (happiness or life satisfaction) through large-scale survey data collected on a regular basis. The data on subjective well-being has been used in economic research as an empirical approximation for 'experienced utility', which is an *ex post* hedonic quality (satisfaction) associated with an act of choice (Kahneman et al., 1997). Some surveys use verbal categories, while others use a numerical scale (e.g. 1 = dissatisfied to 10 = satisfied). Measures of happiness and life satisfaction correlate well with each other and according to factor analysis represent a single unitary construct. A mix of personal and demographic characteristics, as well as socio-economic factors, is significantly correlated with happiness. Both the degree of urbanization and the provision of environmental amenities, for example, are significant variables (negative and positive, respectively). So, using happiness data for environmental valuation, we may be able to estimate the value of ecosystem service benefits differently than standard cost-benefit analysis and willingness to pay methods. By correlating people's reported subjective well-being with ecosystem service benefits and personal income (including lagged individual income and other people's income) it is possible to identify the environmental quality and income utility link and to estimate the implied utility-constant trade-off between them (e.g. the increase in income necessary to compensate individuals for any decline in environmental benefit provision). So far, air pollution, water pollution, noise, climate conditions and natural hazards have been valued using the happiness approach. To the extent that ecosystem functioning contributes to the maintenance/improvement of, for example, air and water quality, valuation of ecosystem service benefits is possible via this 'happiness approach'.

Stock versus flow values

The distinction between ecosystem services stocks and flows has also to be reflected in the economic valuation approach adopted. A paper in the journal *Nature* by Costanza et al. (1997) estimated the value of global ecosystem services at $33 trillion, which led to a protracted debate and controversy over the 'true' value of the natural environment. This analysis has been attacked on a number of grounds, including that the aggregate value was not necessarily the sum of the parts, and that US$33 trillion was more than global income and therefore people's ability to pay (Heal, 2000). Further work (Howarth and Farber, 2002) sought to defend the Costanza et al. approach by arguing that the estimates of ecosystem services value were analogous to National Income Accounting entities such as GDP with a constant set of value weights. The underlying rationale here is that the aggregate measure is a quantity parameter (the stock concept), and, while it is related to value, it does not directly value the planet's ecosystem services in total. In this sense it is an accounting price measure of the quantity of ecosystem services holding prices constant, where the measures are not based on economic theory but on accounting rules (Costanza et al., 2014). In this stock accounting context, the criticism related to people's budget constraint and ability to pay is not relevant, because the measure is based on virtual (not real) prices and virtual incomes (i.e. incomes adjusted to enable individuals to hypothetically pay for the services). For the income and expenditure accounts to balance, the total expenditure must be less than actual and virtual income.

The current extent of European coastal blue carbon (the carbon storage service provided by salt marshes and sea grasses), for example, has been estimated to have an accounting stock price (value) of about US$180 million (Luisetti et al., 2013). Such total (stock) values can be estimated and compared for two different points in time as a heuristic to help to appreciate the change in natural capital. This viewpoint is, however, controversial and is not supported by many mainstream economists. For them the only relevant measure is the marginal economic value. For economic valuation (as opposed to accounting) it is important to be able to quantify and evaluate gains or losses in stock assets and consequent service flows. Now, instead of holding prices constant, we seek to determine marginal economic value as it relates to an incremental increase in a set of ecosystem services over time and space. When the ecosystem final services value relates only to non-market services, it can be combined with GDP (in the same way as relevant pollution and other externalities are internalized) to yield a more green GDP measure. The present value of a discounted flow of ecosystem services values can contribute to stock of wealth accounts such as the Inclusive Wealth account (UNU-HDP and UNEP, 2012). An important consideration is that the flow and stock values (i.e. the accounting and economic values), as explained above, serve different purposes and should not be added up.

It has been argued by Norgaard (2010) that the whole natural capital/stock and flow approach to environmental management has serious limitations and can serve to obscure the need for more radical and extensive reforms of institutions and governance. Much depends on how pressing the global sustainability constraints really are, but institutional and governance issues are clearly key parameters that need to be addressed in any meaningful sustainability dialogue, and so far progress at the national and international level in this dimension has been very limited (see also Rockstrom et al., 2009; Primmer, 2016). The idea of safe minimum standards (SMS) has long been proposed to constrain development pressures on ecosystems. These rules/regulations/zoning and other measures would, however, fall short of Kantian moral imperatives status, as they would allow for a pragmatic trade-off between current costs of SMS and the future benefits of risk abatement.

A number of issues surround the notion of irreversibility and related concepts of threshold effects and tipping points. While all these terms are now in use in the environmental conservation and economics literature, we refer to thresholds in the context of individual ecosystems or landscape ecology limited to the regional spatial scale. Tipping points, then, describe global scale system and subsystem non-linear and abrupt reactions to environmental change pressures (Rockstrom et al., 2009).

Ecosystems function via feedbacks between different components of structure and process. When the feedback effects are positive, any given initial perturbation (stress or shock) of the system will be amplified, and the prevailing state of the system may be such that a complete switch into a different state is triggered (a classic case is the enrichment of shallow lakes via excessive N & P inputs from the surrounding catchment, causing abrupt change in water quality and aquatic plant and fish communities). The initial ecosystem state prior to the 'flip' is then a threshold or a bifurcation. The capacity of an ecosystem to 'absorb' stress or shock and remain in its prevailing state is known as resilience, which has a natural science and social dimension. It is still a matter of scientific debate whether greater diversity in ecosystems provides a buffering capacity (greater stability or resilience) and in which specific contexts. A further degree of uncertainty surrounds the question of whether the ecosystem state change is reversible or irreversible in the future. This is a far from straightforward question, and we currently lack sufficient scientific and other knowledge to be able to offer robust prescriptions. In the shallow lake example mentioned earlier, it is the case that remedial management actions (such as sediment pumping, and N & P abatement) can restore water quality and other related services. But even in this case, the timing and extent of the necessary abatement programme is not clear-cut, with adverse consequences for the overall costs of action. In more complex co-evolving environmental and socio-economic contexts, resilience and irreversibility are conditioned by multiple factors, including current technological/scientific gaps, or impracticability constraints in the form of significant cost burdens as well as cultural and governance limitations (Folke et al., 2010; The Royal Society, 2014).

Given the uncertainty around contemporary environmental change conditions (local through to global), there is a positive probability that a given ecosystem in a given change context will be pressurized into a thresholds zone and across a point, causing it to flip to a less desirable new state. The probability of flipping is lower as resilience is maintained/increased, and management interventions to 'conserve' resilience are therefore important. Resilience capacity can be regarded as capital stock (natural capital) which yields an insurance service and benefit. In practice, data (time series and other) constraints have so far precluded the monetary valuation of this insurance service, with the one exception of an agro-ecosystem study in Australia.

Two different approaches have been put forward in the environmental economics literature as coping strategies for the irreversibility problem. The first was a modified CBA method, known as the Krutilla-Fisher (K-F) approach. It laid down that, in relevant preservation versus development situations, the benefit of the preservation option should be factored into the CBA equation. Preservation benefits forgone should be treated as part of the costs of development and should be assumed to increase through time because of the relative price effect. The development benefits should have an offsetting discount factor, in addition to the 'basic' discount rate, because of 'technological obsolescence'. It is also the case that the present value of development can be very sensitive to the preservation-relative price effect and the obsolescence factor (Pearce and Turner, 1990). Given the prevailing information gaps, it is recommended that the benefit of the doubt be given to preservation over development in all cases where benefits and costs are reasonable closely balanced. The K-F strategy was designed to cope with ecosystem and landscape asset losses characterized by uniqueness and national/international significance. The dilemma was how to cope with possible irreversible losses of unique assets such as designated national park lands. But it may be the case that some potential ecosystem losses are of 'local' significance and 'uniqueness'. In these cases, a conservation versus development trade-off needs to be addressed.

In these 'local irreversibility' situations, a 'shadow project' approach in which sustainability considerations are integrated into the CBA calculus may be relevant. The decision-maker is asked to consider a range of decisions about development options and impose a sustainability constraint into the decision support system and process (i.e. to keep the stock of natural capital constant over time by suitable compensatory expenditures). The sum of the ecosystem damage done by a whole sequence of development projects would have to be offset by separate projects within the 'portfolio' of decisions being made. These compensatory projects would not have to pass the positive B-C ratio test (Barbier et al., 1990). The precise form of 'acceptable' compensation will vary from context to context and is an under-researched area. Roach and Wade (2006), for example, have examined the use of so-called habitat equivalency analysis, which estimates ecological service loss and then scales restorative ecological compensation to offset the damage impact.

A second line of argument, if irreversibility concerns are relevant, incorporates the notions of the precautionary principle and the safe minimum standard. There is a line of reasoning that can link ecosystem diversity and resilience maintenance (with 'primary'/'glue'/'infrastructure' values in nature, alongside the 'insurance' value noted earlier) together with support for the precautionary principle and strong sustainability (Gren et al., 1994; Turner et al., 2003). The precautionary principle is itself shrouded in ambiguity, but CBA can provide a useful filter for it if a 'safe minimum standards' (SMS) interpretation covering species, habitats and ecosystems is accepted. This goes back to the work of Ciriacy-Wantrup (1952) and Bishop (1978), in which it was advocated that a project should be rejected if irreversible losses of nature could consequently occur, unless the social costs of doing so were prohibitive. Thus, decision-makers facing the prospect of very high preservation or conservation costs might choose to sanction a development option even though it carries a small risk of significant ecosystem damages. Nevertheless, judging what is an 'unacceptably large' or 'tolerably low' social cost can be informed by ecology, economics and risk analysis, but ultimately is a 'political' call. Ethical and political choices will have to be made and deliberatively agreed upon.

Ideally, ecosystems would be managed with sustainable development in mind. In practice, there are a number of acknowledged reasons why ecosystem degradation continues unabated. These reasons include both market failure and poor governance. One of the key causes of market failure is lack of information, and so the provision of information on the economic and social value of ecosystems can only contribute to, but not guarantee, better decision-making. Governance systems are equally important, and management measures such as terrestrial and marine protected areas and marine conservation zoning are often highly contested. But the main driver of environmental change both nationally and internationally is a macro-economic strategy dominated by a 'maximize growth' (GNP) philosophy.

Ecosystem services provision can also extend over considerable periods of time, given appropriate environmental management regimes. The time horizon itself presents the policy process with a number of socio-economic and ethical considerations.

Discounting and equity considerations

It is often necessary to choose between options that differ in temporal patterns of costs and benefits, or that differ in their duration. Discounting provides a common matrix that enables comparison of costs and benefits that occur at different points in time. Use of discounting yields an outcome in which future costs and benefits are valued less highly than those that occur in the present, and the procedure is integral to cost-benefit analysis (CBA) and cost effectiveness analysis (CEA). The choice of the discount rate can have a significant effect on the economic viability of management options and their relative economic ranking. It signals the rate at which future consumption is to be traded against consumption in the present. Use of a 'high' positive rate of discount discriminates against the future and, in project terms, against options that involve high initial costs and a stream of benefits that extends far out into the future. Instead, it tends to favour projects that have immediate benefits and delayed cost burdens (Turner, 2007). But, while a low discount rate favours the future, this may be politically and morally questionable if immediate well-being increases are slowed or compromised altogether and the burden falls disproportionately on the poor.

The discounting question raises a number of much deeper ethical and strategic considerations related to equity and fairness principles and practice. Fairness in contemporary society (intra-generational equity) is sidestepped in conventional applications of CBA via the acceptance of the economic efficiency criterion which weighs all benefits and costs equally, regardless of whether they affect rich or poor in society (Turner, 2007). But a case for actual compensation for losers (financial and/or environmental), especially in contexts with 'contested' environmental change consequences, has much to recommend it.

The standard CBA practice of positive, fixed and short term (<25 years) discounting does not sit easily within policy contexts with pressures and drivers such as climate change and related global economic forces. A growing number of analysts agree that discounting at a constant and relatively high (i.e. determined by reference to market interest rates data) rate of discount over time horizons of 100 years or more is problematic. The effect is to make even large costs/benefits incurred in the distant future seem inconsequential. Asheim (2012) summarized the dilemma in the following way: contemporary society needs to distinguish between what the current generations as a collective should do ethically to serve the interests of all generations from an impartial perspective, and what contemporary countries or individuals should do strategically to serve their own interests when such actions influence the strategic action of other countries and individuals.

The ethical and strategic dilemma is not as straightforward as it may first appear. Zero or negative discounting poses a threat to the least well-off in today's society and can result in large sacrifices from the present for the benefit of later generations, who may actually be better off. A single invariant low rate of discount could, in some circumstances, also allow a greater volume of projects to pass the CBA test and therefore strain resource/environmental capacities. Nevertheless, some modifications to the standard CBA discounting procedure have been adopted in UK public sector project appraisal. A time declining discount rate (DDR) procedure over at least 100 years is now recommended for projects with significant environmental impacts.

A range of reasons have been put forward in support of DDR, including uncertainty about future interest rates and the macro-economic state of the economy. Some empirical evidence exists for 'hyperbolic discounting', indicating that individuals value medium and distant futures on an equivalent basis, i.e. the discount rate falls the longer the time horizon. It may be that individuals revise and re-evaluate plans continually as they age over time. In the discounted utilitarian approach in conventional CBA, the counter-argument would be that social discounting as practiced by governments should not mimic the 'time inconsistent'/'irrational' behaviour of individuals exhibiting hyperbolic discounting behaviour. But, while policy inconsistency at a given period of time is an institutional failure and should be corrected, policy switching over longer periods of time is inevitable and necessary if future uncertainties and 'surprises' are unavoidable. Finally, Knetsch (2005) has claimed that individuals discount future losses at a lower rate than the value of future gains and, therefore, rates reflecting observed individual preferences would give more weight to future environmental losses, justify greater current sacrifices to deal with them and support policies that reduce the risk of future loss.

The ethical versus strategic behaviour dichotomy is the focus for a key set of arguments. If individual people also have social preferences (shared values) and if they trade off their own material interests against the wider interests of society, then some notion of fairness that captures shared values and/or constrains self-interested preferences is required. Roemer (2011) has argued that the utilitarian social welfare function used in conventional CBA assumes that the decision problem for a society with many generations is equivalent to the decision problem of an infinitely-lived consumer. This claim is refuted by Dasgupta (2011), but both agree that the discount rate based on market data and applied to the climate change problem is too large. They disagree on the better ethically defensible sustainability criterion, with Roemer favouring the Rawlsian intergenerational 'maximin' approach, i.e. each generation passes on a non-declining, in per capita welfare terms, capital (human, physical and natural capital components) bequest. Roemer argues that consumption as conventionally defined in economics is not the only component of welfare/well-being. Educated leisure, quality of the local to global environment and knowledge are also direct inputs into 'happiness'/welfare. But intergenerational maximin is not problem-free (e.g. how much sacrifice is a fair burden for the current generation, rich and poor?). Nevertheless, if a spirit of moral pluralism (i.e. there is no meta-ethical criterion and the context/consequences of ethical choices should not be ignored) is accepted, intergenerational maximin may still prove to be a usable ethical guide.

Multi-criteria decision support systems: a possible way forward

Assessing better the economic value of nature and the services it provides to society is important for local, national and global policy and decision-making. But a multidisciplinary toolkit is required if a robust decision support system (DSS) is to be constructed. While the core decision criterion in conventional economics, i.e. economic efficiency, is important, especially if, as expected, the scarcity of environmental goods and services increases over time, it is not a meta-criterion. Cost-benefit analysis can still play an important role in multi-criteria assessment DSSs if suitably adjusted for equity concerns. But the full commodification of all ecosystem services through the assignment of monetary values to all aspects of ecosystem complexity is not meaningful, or possible, and does not provide a sound scientific or moral basis for sustainable management.

DSSs can be better informed, for example, by a more sophisticated understanding of relevant social and policy networks (Borgatti et al., 2009) and also via methods and techniques encompassing multiple values and decision criteria. Economic assessment methodologies can be modified to incorporate equity issues (e.g. via the application of weights to costs and benefits, or the payment of actual compensation), and the economic analysis itself can be augmented by a wider trade-off analysis, such as MCA or deliberative (non-)monetary valuation techniques.

DSSs and their component methods and tools, such as CBA, CEA and MCA, require the acceptance of different assumptions about the capacities and motivations of the individuals involved, and the role the methods play in framing/scoping the assessment process. From an institutional perspective, CBA and other methods can be characterized as value articulating institutions, in the sense of rule structures facilitating value (Vatn, 2009). If the existence of plural rationalities is accepted, the role of such institutions is to signal which rationality is expected. The choice of assessment approach, for example, is related to whether the benefits and costs involved are linked to an exchangeable commodity or some kind of public good; if it is the allocation of a public type good that is being contested, then deliberative methods may play a useful role, although they need to be tailored to suit the situation, e.g. citizens or stakeholder forums or a hybrid. The balance sheet approach (see Turner, 2016) aims to provide an appraisal format based on a pluralistic perspective. The particular sequencing of policy tools (methods/tools) through the balance sheet approach starts with CBA/CEA but then encompasses other complementary 'tools' to apply adaptive management principles. Nevertheless, overly complex DSSs should be avoided. Psychologists have found that in some situations too many choices and comparative evaluation can result in worse choices. People guided by their own intuition and aided by simple heuristics come to decisions with less regret (Gigerenzer, 2007).

References

Adams, W.D. (2014). The value of valuing nature. *Science*, vol 346, pp 549–551.

Aldred, J. (2002). Cost-benefit analysis, incommensurability and rough equality. *Environmental Values*, vol 11, pp 27–46.

Asheim, G.B. (2012). Discounting while treating generations equally. In Hahn, R. W. and Ulph, A. (eds) *Climate Change and Common Sense: Essays in Honour of Tom Schelling*. Oxford University Press, Oxford.

Ayres, R.U. (1993). Cowboys, cornucopians and long run sustainability. *Ecological Economics*, vol 8, pp 189–207

Badura, T., Bateman, I., Agarwala, M., and Binner, A. (2016). Valuing preferences for ecosystem-related goods and services. In Potschin, M., Haines-Young, R., Fish, R. and Turner, R.K. (eds) *Routledge Handbook of Ecosystem Services*. Routledge, London and New York, pp 228–242.

Balmford, A., Fisher, B., Green, R.E., et al. (7 authors) (2011). Bringing ecosystem services into the real world: an operational framework for assessing the economic consequences of losing wild nature. *Environmental and Resource Economics*, vol 48, pp 161–175.

Banzhaf, H.S., and Boyd, J. (2012). The architecture and measurement of an ecosystem services index. *Sustainability*, vol 4, pp 430–461.

Barbier, E.B., Markandya, A., Pearce, D.W., et al. (1990). Environmental sustainability and cost-benefit analysis. *Environment and Planning A*, vol 22, pp 1259–1266.

Bateman, I.J., Mace, G.M., Fezzi, C., Atkinson, G., and Turner, R.K. (2011). Economic analysis for ecosystem service assessments. *Environmental and Resource Eco*nomics, vol 48, pp 177–218.

Bishop, R. (1978). Endangered species and uncertainty: the economics of a safe minimum standard. *American Journal of Agricultural Economics*, vol 60, pp 10–18.

Borgatti, S.P., Mehra, A., Brass, D.J., and Labianca, G. (2009). Network analysis in the social sciences. *Science*, vol 323, pp 892–895.

Brouwer, R. (2016). Payments for ecosystem services. In Potschin, M., Haines-Young, R., Fish, R. and Turner, R.K. (eds) Routledge Handbook of Ecosystem Services. Routledge, London and New York, pp 548–553.

Brouwer, R. (2016). Payments for ecosystem services. In Potschin, M., Haines-Young, R., Fish, R. and Turner, R.K. (eds) *Routledge Handbook of Ecosystem Services*. Routledge, London and New York, pp 548–553.

Boyd, J., and Banzhaf, S. (2007). What are ecosystem services? *Ecological Economics*, vol 63, pp 616–626.

Ciriacy-Wantrup, S. (1952). *Resource Conservation: Economics and Policies*. University of California Press, Berkeley CA.

Corbera, E., Brown, K., and Adger, W.N. (2007). The equity and legitimacy of markets for ecosystem services. *Development and Change*, vol 38, pp 587–613.
Costanza, R., D'Arge, R., de Groot, R.D., et al. (13 authors) (1997). The value of the world's ecosystem services and natural capital. *Nature*, vol 387, pp 253–260.
Costanza, R., de Groot, R., Sutton, P., et al. (7 authors) (2014). Changes in the global value of ecosystem services. *Global Environmental Change*, vol 26, pp 152–158.
Daily, G. (1997). *Nature's Services*. Island Press, Covelo CA.
Daly, H.E., and Cobb, J.B. (1989). *For the Common Good*. Beacon Press, Boston MA.
Dasgupta, P. (2011). The ethics of intertemporal distribution in a warming planet: reply and response to John E. Roemer. *Environmental and Resource Economics*, vol 50, pp 475–493.
Easterlin, R. (1974). Does economic growth improve the human lot? Some empirical evidence. In David, P. A., Reder, M. W., et al. (eds) *Nations and Households in Economic Growth: Essays in Honour of Moses Abramovits*. Academic Press, New York.
Fish, R., Burgess, J., Chilvers, J., et al. (8 authors) (2011). *Participatory and Deliberative Techniques for Embedding an Ecosystems Approach into Decision Making*. Full Technical Report (Project Code: NR0124). Defra, London.
Fisher, B., Turner, R.K., and Morling, P. (2009). Defining and classifying ecosystem services for decision making. *Ecological Economics*, vol 68, pp 643–653.
Folke, C., Carpenter, S.R., Walker, B., et al. (6 authors) (2010). Resilience thinking: integrating resilience, adaptability and transformation. *Ecology and Society*, vol 15, no 4, p 20.
Freeman, M.A. (1993). *The Measurement of Environmental and Resource Values*. Wiley, New York.
Gigerenzer, G. (2007). *Gut Feelings: The Intelligence of the Unconcious*. Penguin Books, New York.
Gómez-Baggethun, E., de Groot, R., Lomas, P.L., and Montes, C. (2010). The history of ecosystem services in economic theory and practice: from early notions to markets and payment schemes. *Ecological Economics*, vol 69, pp 1209–1218.
Gren, I.-M., Folke C., Turner R. K., and Bateman I. (1994). Primary and secondary values of wetland ecosystems. *Environmental and Resource Economics*, vol 4, pp 55–74.
Hargrove, C. (1992). Weak anthropocentric intrinsic value. *The Monist*, vol 75, pp 183–207.
Hartwick, J.M. (1978). Substitution among exhaustible resources and intergenerational equity. *Review of Economic Studies*, vol 45, pp 347–354.
Heal, G. (2000). *Nature and the Marketplace*. Island Press, Washington DC.
Howarth, R. B (1995). Sustainability under uncertainty: a deontological approach. *Land Economics*, vol 71, pp 417–427.
Howarth, R.B., and Farber, S. (2002). Accounting for the value of ecosystem services. *Ecological Economics*, vol 41, pp 421–429.
Kahneman, D., Walker, P., and Sarin, R. (1997). Back to Bentham: exploration of experienced utility. *Quarterly Journal of Economics*, vol 112, pp 373–405.
Kenter, J. O. (2016). Deliberative and non-monetary valuation. In Potschin, M., Haines-Young, R., Fish, R. and Turner, R.K. (eds) *Routledge Handbook of Ecosystem Services*. Routledge, London and New York, pp 271–288.
Knetsch, K. (2005). Gains, losses and the US-EPA economic analyses guidelines: a hazardous product? *Environmental and Resource Economics*, vol 32, pp 91–112.
Layard, R. (2005). *Happiness: Lessons from a New Science*. Penguin Books, London
Layard, R. (2010). Measuring subjective well-being. *Science*, vol 327, pp 534–535.
Lazaro-Touza, L. and Atkinson, G. (2013). Nature, roads or hospitals? An empirical evaluation of 'sustainable development preferences'. *Ecological Economics*, vol 95, pp 63–72.
Luck, G. (2009). Quantifying the contribution of organisms to the provision of ecosystem services', *BioScience*, vol 59, pp 223–235.
Luck, G. (2016). Ecosystem Service Providers. In Potschin, M., Haines-Young, R., Fish, R. and Turner, R. K. (eds) *Routledge Handbook of Ecosystem Services*. Routledge, London and New York, pp 60–61.
Luisetti, T., Jackson, E., and Turner, R.K. (2013). Valuing the European "coastal blue carbon" storage benefit. *Marine Pollution Bulletin*, vol 71, pp 101–106.
MA (2005). *Millennium Ecosystem Assessment: Ecosystems and Human Wellbeing: Synthesis*. Island Press Washington DC.
Mäler, K.-G., Aniyer, S., and Jansson, A. (2009). Accounting for ecosystems. *Environmental and Resource Economics*, vol 42, pp 39–51.
Neumayer, E. (1999). *Weak versus Strong Sustainability: Exploring the Limits of Two Opposing Paradigms*. Edward Elgar, Cheltenham.

Norgaard, R. (2010). Ecosystem services: from eye-opening metaphor to complexity blinder. *Ecological Economics*, vol 69, pp 1219–1227.

Oswald, A., and Wu, S. (2010). Objective confirmation of subjective measures of human well-being: evidence from the USA. *Science*, vol 327, pp 576–579.

Pearce, D.W., Markandya, A., and Barbier, E.B. (1990). *Blueprint for a Green Economy*. Earthscan, London.

Pearce, D.W., and Turner, R.K. (1990). *Economics of Natural Resources and the Environment*. Harvester Wheatsheaf, Hemel Hempstead.

Porter, M.E., and van der Linde, C. (1995). Toward a new conception of the environment-competitiveness relationship. *Journal of Economic Perspectives*, vol 9, pp 97–118.

Potschin, M., and R. Haines-Young (2016). Frameworks for ecosystem assessments. In Potschin, M., Haines-Young, R., Fish, R. and Turner, R. K. (eds) *Routledge Handbook of Ecosystem Services*. Routledge, London and New York, pp 125–143.

Primmer, E. (2016). An institutional perspective. In Potschin, M., Haines-Young, R., Fish, R. and Turner, R. K. (eds) *Routledge Handbook of Ecosystem Services*. Routledge, London and New York, pp 582–585.

Roach, B., and Wade, W. (2006). Policy evaluation of natural resource injuries, using habitat equivalency analysis. *Ecological Economics*, vol 58, pp 421–433.

Rockstrom, J., Steffen W., Noone, K., Persson, Å., Chapin, F. S., Lambin, E. F., . . ., Foley, J. A. (2009). A safe operating space for humanity. *Nature*, vol 461, pp 472–475.

Roemer, J.E. (2011). The ethics of intergenerational distribution in a warming planet. *Environmental and Resource Economics*, vol 48, pp 363–390.

Spash, C.L. (1999). The development of environmental thinking in economics. *Environmental Values*, vol 8, pp 413–435.

Strassburg, B., Turner, R.K., Fisher, B., Schaeffer, R., and Lovett, A. (2009). Reducing emissions from deforestation – The "combined incentives" mechanism and empirical simulations. *Global Environmental Change*, vol 19, pp 265–278.

The Royal Society (2014). *Resilience to Extreme Weather*. The Royal Society Science Policy Centre Report 02/14, London.

Turner, R.K. (1988). Wetland conservation: economics and ethics. In Collard, D. and Pearce, D. W. (eds) *Economics, Growth and Sustainable Environments: Essays in Memory of Richard Lecomber*. The Macmillan Press, London.

Turner, R.K. (1999). The place of values in environmental valuation. In Bateman, I. J. and Willis, G. K. (eds) *Valuing Environmental Preferences*. Oxford University Press, Oxford.

Turner, R.K. (2007). Limits to CBA in UK and European environmental policy: retrospects and future prospects. *Environmental and Resource Economics*, vol 37, pp 253–269.

Turner, R.K. (2016). The 'balance sheet' approach within adaptive management for ecosystem services. In Potschin, M., Haines-Young, R., Fish, R. and Turner, R.K. (eds) *Routledge Handbook of Ecosystem Services*. Routledge, London and New York, pp 289–303.

Turner, R.K., Paavola, J., Cooper, P., Jessamy, V., and Georgiou, S. (2003). Valuing nature: lessons learned and future research directions. *Ecological Economics*, vol 46, pp 493–510.

UK National Ecosystem Assessment (2011). *The UK National Ecosystem Assessment Technical Report*. UNEP-WCMC, Cambridge UK.

UK National Ecosystem Assessment Follow-on (2014) *The UK National Ecosystem Assessment Follow-on: Synthesis of the Key Findings*. UNEP-WCMC, LWEC.

UNU-IHDP and UNEP (2012). *Inclusive Wealth Report 2012. Measuring Progress Toward Sustainability*. Cambridge University Press, Cambridge UK.

Vatn, A. (2009). An institutional analysis of methods for environmental appraisal. *Ecological Economics*, vol 68, pp 2207–2215.

Wallace, K.J. (2007). Classification of ecosystem services: problems and solutions. *Biological Conservation*, vol 139, pp 235–246.

Welsch, H. and Kuhling, J. (2008). Using happiness data for environmental valuation: issues and applications, *Journal of Economic Surveys*, vol 23, pp 385–406.

Westman, W. (1977) 'How much are nature's services worth', *Science*, vol 197, pp 960–964.

Wunder, S., Engel, S. and Pagiola, S. (2008). Taking stock: a comparative analysis of payments for environmental services in developed and developing countries, *Ecological Economics*, vol 65, pp 834–852.

21

STAKEHOLDER PARTICIPATION IN ECOSYSTEM SERVICE DECISION-MAKING

Robert Fish, Eirini Saratsi, Mark Reed and Hans Keune

Introduction

It is very difficult to imagine the application of an ecosystem services perspective to policy and decision-making processes without recourse to the issue of stakeholder engagement and participation. The history of the concept is itself a history of stakeholder agendas being gradually negotiated and aligned with respect to a common purpose, and around which professional norms and standards – conceptual models, data infrastructures, valuation instruments and so forth – have accrued significance. Stakeholders enter the fray of this perspective in a variety of guises: as data inputs into models; as enlightened recipients of information; as well as critics and companions in its work. These roles are based on a much larger set of assumptions about why it matters to speak and engage with people.

The purpose of this chapter is to elaborate the different grounds on which stakeholder participation occurs and general issues and considerations arising from relating stakeholder processes to ecosystem service-based decision-making. Rationales for stakeholder participation are shown to have ethical and practical dimensions and a range of models for stakeholder engagement can be deployed, from information dissemination to active co-production of research and policy. Most importantly, the chapter asks how an ecosystem services perspective changes how we think and do participation. When we reflect on the more general history of environmental management we see precedents and literatures on stakeholder engagement that have developed without the need for, and thus are somewhat indifferent to, an ecosystem services perspective. These precedents and literatures provide us with some important touchstones in making the case for a participatory approach in natural resource management, and understanding its key contours and assumptions (for an overview see Reed, 2008; Reed *et al.*, 2009), but the larger point is how ecosystem services thinking creates new needs, concerns and opportunities with respect to participation in decision-making. In his critical perspective Sagoff (2016) asks rhetorically how would ecosystem services researchers 'get the stakeholders to return their calls?', a question he poses in the context of wider epistemological concerns over advancing this perspective in decision making. This chapter culminates with exploring the risks and opportunities arising from an ecosystem service-driven approach to participation.

The chapter is complementary to the concerns of Kenter (2016), where precise methods and techniques for engagement are set out under the rubric of understanding values and valuation.

It should also be seen as the corollary of all other chapters: the ambitions of this developing agenda rarely, if ever, proceed without some concern or expectation for engagement in decision-making, however tightly and narrowly drawn, or obscured, this may be.

Definitions and rationales for stakeholder participation

Perhaps unsurprisingly for a term that has long passed what Bryson (2004: 22) describes as its 'tipping point' into common discourse, the idea of 'stakeholder' is open to many variants and these definitions 'differ in how inclusive they are'. At its most expansive, stakeholder can be taken to refer to any organisation, group or individual affected by, with an interest in, or influence over, a decision-making issue (Grimble and Wellard, 1997; Warner, 2007; Brown *et al.*, 2015). Under this rubric, stakeholders might be taken to include scientific experts, policy experts and representatives from public, private or third sector organisations in a decision process, along with members of the wider public who do not represent any formal or informal organisation but themselves. The term may be extended further still, to include: non-humans; temporal considerations, such as the stake of future and past generations; or broader and imagined social collectivities such as the membership of a local community, a national public, a global citizenry and so forth.

More prescriptively, usage of the term is sometimes distinguished by a concern for direct interests and forms of influence in decision making, such as an NGO, as opposed to more loose and indirect affiliations with an issue; as in the 'general public' or 'experts' or 'business'. Hence, Renn *et al.* (1995) view participatory processes as '[f]orums for exchange that are organised for the purpose of facilitating communication between *government, citizens, stakeholders and businesses* regarding a specific decision or problem' (Renn *et al.*, 1995; authors' emphasis). Others have conceptualised participation in terms of flows of information and/or knowledge, distinguishing participation as a two-way knowledge exchange process from communication and consultation processes in which there are one-way flows of information or knowledge to or from stakeholders (Rowe and Frewer, 2000). Whichever way we choose to distinguish the term, the parameters that define who participates and on what basis participation occurs remain central. In this chapter we take the former, open and expansive, definition, in order to encapsulate and address a broad and varied concern with involving organisations, groups or individuals in ecosystem service decision-making.

A concern to work with, and learn about, the views of stakeholders within environmental decision-making is a well-established feature of agendas for sustainable development. For example, over two decades have passed since the 1992 Rio Declaration on Environment and Development, adopted by the United Nations Conference on Environment and Development, stated that:

> [e]nvironmental issues are best handled with the participation of all concerned citizens, at the relevant level. At the national level, each individual shall have appropriate access to information concerning the environment that is held by public authorities [. . .] and the opportunity to participate in decision-making processes. States shall facilitate and encourage public awareness and participation by making information widely available.
>
> *[Principle 10]*

A number of international conventions have since served to reinforce and extend these concerns, including the Aarhus Convention, and most notably in the present context, the Convention on Biological Diversity (CBD). The CBD is important, for it sets out the wider principles that have since informed operational applications of the ecosystem services concept. The Convention stresses that the management of ecosystems and their services is a matter of 'societal choice' and that stakeholder involvement is key to the long term sustainability and equitable management of ecosystem services, one reflecting legitimate rights to influence and shape decisions that impact human livelihoods and well-being. The concern reflects practical necessity as well, since pathways to the sustainable management of ecosystem services are rarely self-evident and uncontested. Not only will the underlying purposes of decisions themselves be open to competing claims and perspectives, but decisions will also often be made in circumstances where evidence is highly uncertain or incomplete, and where concern for the management of ecosystems services must be compared alongside a range of other benefits driving and informing decisions. A consequent theme running through advocacy of a participatory approach is that it is a way of building resilience into decision-making; a means of unlocking society's capacity to adapt to, and innovate within, its changing circumstances.

In summary, and drawing on the work of Fiorino (1990) and Chilvers (2009), we might say there are three arguments that have tended to be made for stakeholders' involvement in environmental decision-making:

- *Normative-based arguments*: that participation is important in its own right. Individuals and groups have a legitimate right to influence decision-making processes that have a potential bearing on them or the groups which they represent. Participation is therefore closely related to ideas of empowerment, social justice, equity and indeed democracy;
- *Process-based arguments*: that participation is a better way of achieving particular ends. Where individuals and groups have been involved in the process and reasoning behind decisions, trust and confidence in decisions and institutions is enhanced, social cohesion built and corresponding conflicts surrounding decisions are reduced; and
- *Outcome-based arguments*: that participation leads to better ends. Through participation outcomes and outputs for decision-making processes are more informed. Involving stakeholders not only exposes decision-making to different values and priorities, but also provides a wider base of evidence and practical 'know-how' to guide decisions.

Approaches to the design of participatory processes can be influenced strongly by which rationales are driving the need for participation, but these rationales should best be regarded as *mutually reinforcing*. It can be argued, as was done by the EU in the Aarhus Convention, that there are ethical reasons why citizens should have the right to participate in decisions about their natural environment; given the normative international policy goals of protecting and enhancing ecosystem services, more pragmatic arguments further legitimise participation, making it an ever greater priority in both policy development and implementation.

Of course, this is not to also claim that participation in decisions about ecosystem services is entirely unproblematic. Indeed, there are many examples of participatory processes that did not meet their goals, or led to unintended consequences, where special interest groups were able to bias outcomes or where conflicts were exacerbated or marginalised groups further disempowered (e.g. Cooke and Kothari, 2001). However, there is increasing evidence that effective facilitation and design of participatory processes, including pluralistic representation of stakeholder interests and the management of power dynamics, can avoid many of these failings in many contexts (e.g. Reed, 2008; Reed and Sidoli del Ceno, in press; de Vente *et al.*, under review). We therefore continue by exploring some of the models of participation that have been proposed to ensure effective engagement.

General models and levels of engagement with stakeholders

From a process-orientated starting point, the case for a participatory approach extends across the whole cycle of decision-making, from debating and assigning the priorities that drive decision-making to critically appraising the options and trade-offs that surround pathways to management, and through to designing mechanisms for implementing, monitoring and learning that arise making particular types of decisions. In general, the way that participatory processes are used is dictated by the context and purpose for which they are applied. For example, the level and type of participation may be influenced by contextual factors, such as the level of conflict surrounding an issue, the spatial scale over which the decision has to be made, as well as the extent to which decisions are open to negotiation with those invested with the formal authority to make the relevant decision (such as statutory bodies charged with particular environmental responsibilities and functions) (de Vente *et al.*, under review). Linked to this, the level and type of participation used may depend on the purpose of engaging stakeholders in the decision-making process, for example:

- *Educating/informing stakeholders*: where the objective of engagement is governed by a need for transparency in situations where there exists very limited, or no scope, for negotiating a particular action. In these circumstances, the engagement is non-interactive. It is about disseminating information to constituencies of interested/affected people.
- *Consulting with/learning from stakeholders*: where the objective of engagement is to primarily take account of the views, interests or concerns of stakeholders. In these circumstances, engagement is interactive: a given organisation or group retains full control over decision-making but other individuals and groups can seek to influence aspects of the proposed action in structured and tightly defined ways.
- *Collaborating/working with stakeholders*: where the objective of engagement is to foster shared approaches to and ownership of decision-making among stakeholders, based, for instance, on collective problem framing, priority setting, option evaluation and plan formulation. In these circumstances, the decision-making process rests on a more iterative and active programme of information and knowledge exchange, and debate and argumentation.

For an engagement process to be considered *participatory* it should involve some degree of exchange, interaction and reciprocity of information and ideas between stakeholders and those responsible for the decision process (Rowe and Frewer, 2000). Broadly put, the distinction between 'learning from' and 'working with' stakeholders is one of weaker and stronger forms of participation, though there are significant subtleties within and across this general scheme. For instance, an emphasis on deliberation, where the focus is on extended discussion, debate and scrutiny of decision issues often straddles a consultative and collaborative worldview on participation, but is often a defining characteristic of deeper forms of engagement.

From the practical perspective of the policymaker and analyst, a participatory approach provides information that can complement and extend insight derived from analytic desk-based approaches, that is, approaches to decision-making based on in-house evidence gathering and research. From this starting point, the case for learning and working with stakeholders arises where there are perceived gaps in available evidence or insight, or where relying on 'desk analysis' appears insufficient, such as in decision contexts that suggest high risks and large impacts.

Like the rationales for participation described above, the distinctions between educating, consulting and collaboration in decision-making should be regarded as complementary, rather than simply competing, contexts for engagement. Information dissemination, for instance, can inadvertently open up opposition to a proposed action, thereby creating the conditions for more consultative forms of engagement. Consultative approaches to decision-making have the potential, in turn, to involve more collaborative elements to them, while the results of collaboration will often need to be disseminated through 'at a distance' models of communication such as information/education campaigns. At the same time, information provision is a platform upon which participants can be recruited into, and guided through, consultative and collaborative processes. In other words, these different strategies for stakeholder engagement are directly and indirectly related. Nonetheless, the important overriding point to note is that the case for employing participatory approaches to decision-making arises in situations where courses of action are at least partially open to negotiation and reframing.

Distinguishing stakeholders

The general assumption behind a participatory approach is that there is a need to 'cast the net' widely in terms of who is being involved in decision-making. Inclusive, legitimate and informed responses to issues of natural resource management are likely to emerge where mechanisms are put in place to accommodate a diversity of 'world views' and experiences. This seems particularly significant in the context of ecosystem services decision-making, if the concern is to make decisions with respect to the *full range* of material and social benefits that people derive from the natural environment.

In principle, identifying relevant stakeholders is a formative concern in decision-making though it is often approached in an *ad hoc* way. Systematic analysis of the reasoning behind the inclusion of different groups and individuals in a decision-making process is generally rare. Recent work describing the types of methodologies that can be used to analyse the terms on which stakeholders participate has emerged within the environmental management literature. In particular, *stakeholder analysis* is increasingly being advocated as a systematic way of identifying, differentiating and categorizing stakeholders, and investigating the relationships between them. In this vein Reed *et al.* (2009, p. 1935) remark that, 'in the absence of stakeholder analysis, there is a danger that particularly powerful and well-connected stakeholders can have a greater influence on decision-making outcomes than more marginalised groups'. Thinking through 'who's in and why' helps mitigate stakeholder processes simply reinforcing an accepted way of approaching an issue or problem and prevailing wisdoms about who are the custodians or 'keepers' of a problem/issue (see Jax, 2016).

There nonetheless exists no single accepted or standardised way for distinguishing between and grouping participants. Research by Grimble and Wellard (1997) proposes a seminal formula making analytical distinctions between: stakeholders who affect or are affected by a decision; those who have primary or secondary interests or roles; and stakeholders with contrasting levels of influence and importance (see Box 21.1). Here we offer three further related ways of making distinctions, namely competency, power and how stakeholders are affected by decisions.

In the first case – distinctions based on ideas of competency – the concern is with the different types of knowledge stakeholders are presumed to bring to a decision-making process. This often means the inclusion of specialists, that is, stakeholders who participate because of their formal credentials/ background in an issue, such as scientists occupying roles in public and private sector organisations. Importantly, and often in contrast, it includes stakeholders who represent 'lay', 'indigenous', 'traditional' forms of knowledge: that is, those who participate because of accumulating practical and local experience and understanding about an issue, even if this is not a formally codified specialism (See Box 21.2). And finally, we can point to the idea of procedural competency, which is to say those who participate because of their knowledge of the mechanisms that ultimately foster or impede delivery. This category covers competencies in formal

Box 21.1 Stakeholder analysis

A study by Morris et al. (2009) seeking to understand and predict the relationship between farmer decision-making and bird populations employed a stakeholder decision analysis methodology to identify individuals and groups with an 'interest' in and 'influence' over farmland bird populations. Stakeholders were initially identified using a focus group with researchers, individuals and groups classified in an 'interest–influence' matrix, displaying their attributes and inter-relationships. Distinctions were drawn between:

Key players: those with a high interest and influence over the issue (such as NGOs and farmers)

Context setters: those with a high influence, but little direct interest over the issue (such as the local authority)

Subjects: those with high interest, but little influence (such as bird watchers)

Crowd: those with little interest or influence (such as wider sections of the general public)

A framework based on ecosystem services was used to identify and classify these stakeholders according to their *interest* in the services provided by the regulating, production, habitat, carrier and information functions of agricultural land. At the same time, *influence* was assessed by drawing distinctions between different instruments and sources of power. This formed the basis of a graphical representation/mapping of stakeholders which provided guidance on how stakeholders and institutions could be engaged to halt and reverse the decline of farmland birds (Figure 21.1).

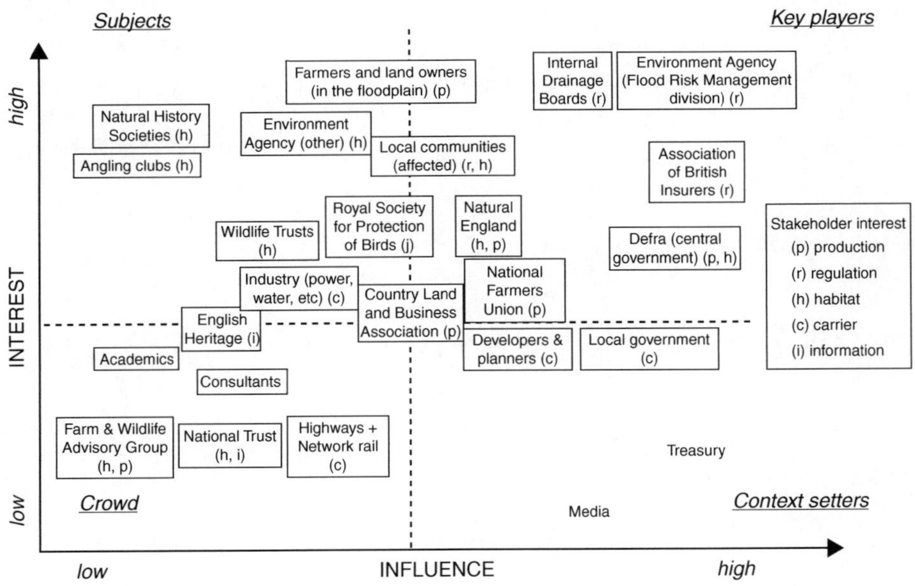

Figure 21.1 Influence and interest matrix.

Source: Morris et al., 2009

procedures (for instance, procedures that guide the actions of state, commercial and third sector actors) but it also includes an understanding of more informal and tacit protocols – sometimes culturally and locally specific – that may lock/unlock decision-making (such as 'how to work' with particular individuals, social groups or delivery partners).

Box 21.2 Knowledges for ecosystem services management

In its classical form, knowledge may be understood as justified true belief; an expression of understanding about something, socially agreed upon, and given in the form of propositions, such as statements of facts and information. There are many ways in which we can construct and interpret different types of knowledge (see Figure 21.2 below) as the basis for thinking about participation in decision-making.

Science provides us with one salient, influential and codified model of knowledge production and the conceptual and analytical bases of ecosystem services research are founded on this. The influence of scientific knowledge can partially be accounted for because it is considered to provide universal knowledge, justified according to replicable and socially agreed upon standards and protocols.

Despite their different histories and origins as terms, the assertion of 'lay', 'traditional' and 'indigenous' knowledge challenge the idea that science holds a monopoly on truth; and thus ecosystems and their management should not stand and fall on scientific bodies of knowledge alone. A common unifying strand of these knowledges is their socially, culturally, historically and spatially embedded character. The valorisation of place and locality based approaches to engagement provides the most salient context in which alternative forms of knowledge within decision-making are advanced and the technocratic rationalities of scientific knowledge tempered. (see also Potschin and Haines-Young, 2016a).

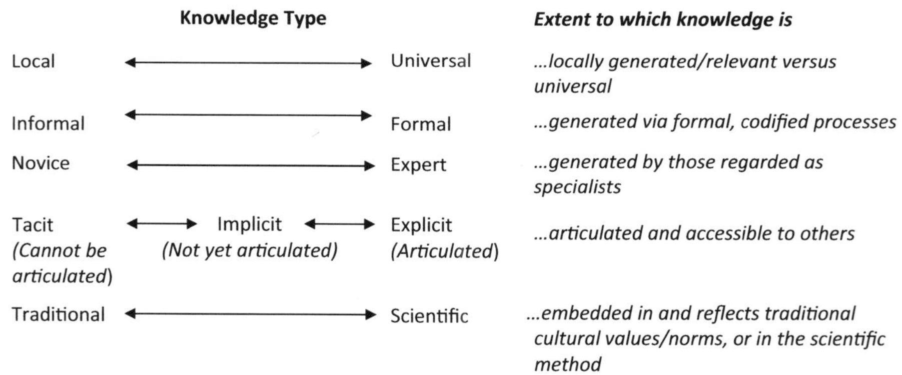

Figure 21.2 Overview of knowledge types.

Source: adapted from Raymond et al., 2010

Second, distinctions may be based on the relative power of the stakeholders. This concerns stakeholder capacities to shape/influence the direction of the process and its outcome. It includes consideration of designated forms of power – that is, those with formal authority to act on behalf of others, from individual and groups working through legal and statutory powers of the state (such as a provincial authority) to non-professional groups acting on behalf of others (such as a community or interest group). It also includes involving those who control the resources that underpin the success of decisions, such as private and public land owners, water utilities, food producers, service providers and so forth. And finally, it includes the more subtle consideration of network power; stakeholders who shape the capacities of things to happen by dint of their strategically important place in a decision-making network, such as individuals who broker decisions at a local level (for example, a land agent) or work at the interface of different stakeholder groups (for instance, a community organiser). Network power can also rely on charismatic authority: those whom others 'follow'. Yet questions of power with respect to participation are not only issues of who holds and wields power, but also, crucially, about consideration of the role of disempowered and hard-to-reach groups: those who sit outside of, or are marginal to, conventional circuits of decision making power. Thus it is common to think of participation in terms of empowerment, which may serve to fundamentally change relationships and social dynamics in a decision situation, and actively transform the terms of debate.

Third, distinctions may be made on assumptions about the impacts and benefits of decisions, that is, thinking about stakeholders in terms of the bearing ecosystem service management has upon them. A common approach here would be to bind impacts/benefits to a particular geographical location of a decision situation – that is impacts/benefits surrounding a discrete locality. However, at least one of the implications of an ecosystem services worldview, with its emphasis on understanding flows of benefit, is that the relationships between service production areas and service benefit are often decoupled spatially (Fisher *et al.*, 2009), suggesting that using locality as a surrogate for impact/benefit may be problematic in many cases (e.g. beneficiaries of carbon sequestration). There is thus a need to explore indirect and cumulative impacts on other contexts and scales, a point to which we return below. What we can say more generally, though, is that issues of impact and benefit often go hand in hand with distributional considerations within participation, that is, the representation of bodies of knowledge and insight according to social cleavages (such as class, age, gender, ethnicity) and thus how they may stand to win or gain from change.

Ecosystem services and the changing vernacular of stakeholder participation

As we remarked earlier, the concept of ecosystem service has developed alongside a rise in normative appeals to stakeholder participation in policy- and decision-making. These appeals are writ large in the principles of the Ecosystem Approach and the CBD. Yet the theory and practice of ecosystem services research does not simply inherit and apply the idea of stakeholder participation. Rather, it actively cultivates new tendencies and preoccupations in participation, and brings with it new corresponding risks and opportunities.

The gradual harmonisation of resource management agendas around the idea of ecosystem services has, firstly, created a new vocabulary and conceptual framing in which the process of stakeholder participation occurs, and we might say that this serves to govern how stakeholders are constructed into participatory processes, and how they talk about human relationships and interactions with the natural world. At its most rudimentary, the issue of language is both a challenge and opportunity. It is a challenge because, as environmental researchers and policy-makers have steadily remade the world in the image of ecosystem services (Robertson, 2012), stakeholders are increasingly obliged to learn a new and distinctive grammar of sustainability, and this itself comes with no guarantees (Box 21.3). The discourse on the one hand presents itself as a holistic and all-encompassing world view, and thus inclusive, but on the other, describes the natural world in a highly specific way. This might be framed as a translational issue arising from the challenge of mainstreaming – a matter of communicating and explaining – lest the ecosystem services agenda gradually retreat within itself: that is, as a private expert discourse with a limited public face.

Box 21.3 Making sense of ecosystem services

Stakeholder understanding of the concepts and framework ecosystem services raises challenges for embedding this thinking in to participatory processes. A public dialogue research project carried out by Fish and Saratsi (2015) in the United Kingdom exposed its guiding logic to a cross section of the general public with no *a priori* awareness of this discourse.[1]

The process found that participants were particularly supportive of the framework's holistic and inter-connected view of the environment and felt it served to challenge preconceived wisdoms about what was considered the remit of the environmental agenda.

In general, the categories of ecosystem service that resonated most strongly with participants were cultural and provisioning services. Cultural services were frequently singled out by participants as an indication of the framework's holistic outlook whilst the logic of provisional services was well understood and articulated by participants who saw the strong connection between environmental processes and economic prosperity.

Participants found the categories of regulating and supporting services less self-explanatory. Supporting services were appreciated by participants as important because their definition captured the idea of human dependencies on a deeper, more fundamental, nature. Participants understood this category as embracing all background processes of nature and were perplexed why specific examples of regulating services belonged in a separate category to that of supporting services (cf. Potschin and Haines-Young 2016b, on ecosystem functions), In addition, the regulating services category tended to be equated with the practice of regulation, as in governance and law-making, along with the practical act of management. Participants were also sceptical of how well supporting services could be addressed in decision-making.

The language was considered too technical and specialist overall, requiring explanation and context to be understood. Participants believed that ecosystem services terminologies should be used with caution if the purpose was to: promote general awareness of environmental issues and challenges among society; convey what government is doing about the environment; or, more specifically, involve wider publics in ecosystem service-based decision-making.

Furthermore, a significant minority of participants were sceptical about advancing the use of the term 'services' to describe human relationships with nature. These participants worried about the long term implications of this way of thinking, expressing concern that people would end up paying for things they currently have the right to access and use freely. They also expressed concern that human responsibilities and duties of care towards nature are obscured by the concept of services and there is a need to bring this aspect out more explicitly if policy and decisions around this framework are to be developed. Some also felt the framework may lead to a rather bureaucratic and 'tick boxy' approach to the management of natural resources.

The language issue is, however, more than a challenge of translation. The concept and framework of ecosystem services is advanced by a language that actively constructs and transforms the way we talk about, ask questions of, and act towards, the natural world. It hardly needs emphasising that the underpinning logic of ecosystem services is to interpret the natural systems from a utilitarian, unapologetically human-centred, starting point, and although this is hardly unusual (just reflect on the term 'natural *resources*') the very language that this agenda propagates has its own effects with respect to willingness and inclinations to engage. What follows from the language of ecosystem services is, for some, the presumption that the intentions of the agenda are related to the essentially economic and neoliberal framing of environmental governance and sustainability debates. In this, the language of ecosystem services absorbs and legitimises wider tendencies that are, at best, naively framed as a pragmatic calculus to make better the case for nature, and at worst, a sinister indication that the moral fabric of environmentalism is being corroded.

The more optimistic assessment of this recalibration of environmental discourse is that a new conversation about human-nature relationships has been started and that is precisely because the language makes these relationships sound 'strange', and demands explanation. In other words a new wave of critical debate has been energised and animated. This might well be a key legacy of the ecosystem services agenda as it matures in relation to wider processes of engagement. We are, in fact, already witnessing a significant evolution and expansion in the terminologies that govern this discourse and corresponding developments in tools and techniques. These require new models and patterns of stakeholder participation, not least in the consideration of shared and cultural values for ecosystems and the challenge these pose to prevailing orthodoxies of neoclassical valuation (Kenter *et al.*, 2015). Can we usefully think of stakeholder participation as a way of maintaining a critical distance from this agenda, through which the idea of ecosystem services may be actively challenged and dismantled? Or should we engage with the methods and mindsets of participation, thus actively transforming and extending the terms on which the ecosystem services agenda is carried forward? In either case ecosystem services is not a static agenda with respect to stakeholder participation.

It is useful at this juncture to further emphasise the distinctive thrust of technical instruments implied by the application of the ecosystem services thinking to participatory models of decision-making. Specifically, we need to consider the role of this agenda in further strengthening the move towards so called 'analytic-deliberative' models of participation. The term analytic-deliberative is borrowed here from the risk management literatures (Stern and Fineberg, 1996; Renn 1999; Chilvers, 2007), and is specifically taken to refer to the emerging hybridity and integration of participatory and technical data infrastructures as the basis for decision making (Fish, 2011). This manifests itself as a whole swath of socio-technical tools (such as deliberative multi-criteria tools, deliberative valuation, participatory modelling, and deliberative mapping) (see Kenter, 2016) that increasingly systematises stakeholder values and preferences with respect to environmental goals, whilst broadening and deepening what counts as evidence.

Importantly, too, it changes how data is imagined and used in the context of its users and beneficiaries. The still-widely asserted, though much discredited, linear model of knowledge production and acquisition pervading the role of scientific knowledge in decision-making processes is fundamentally challenged here, as deliberative and scientific processes start to co-construct, and collapse into, one another (see also Haines-Young and Potschin, 2014). Such an analytic-deliberative tendency is driven only partially by enhanced capabilities and technological platforms to capture and integrate social complexity as valid data. In addition to this, analytic-deliberative approaches have the potential to extend the interpretation and use of ecosystem service data beyond purely processing and representing natural phenomena. The ecosystem services framework accrues significance because it seeks to populate the links between ecosystem services and well-being. However this cascade might be structured and operationalised, we cannot escape the point that understanding the benefits people receive from nature, and how they view and relate to them, is a decisive feature of the paradigm and why researchers came to speak of human-environment relations in terms of 'services'. So stakeholder participation is being actively enmeshed in the process of data construction as the scientific bases of the discourse expands outwards into a transdisciplinary world (Lang *et al.*, 2012; see also Hauck, 2016).

We can push further and say that the analytic-deliberative tendency of the ecosystem services perspective requires new scientific capacities to imagine and inform its complex construction of the world (Cilliers, 2005; Keune and Dendoncker, 2013). By this we mean that the holistic worldview of the ecosystem services framework has an insatiable need for data that can make sense of system interdependencies and chart its way towards an agreed purpose. But how much data is enough? Many usages of the framework are currently heuristic, but what is occurring too is the emergence of new, increasingly open-sourced, capacities for data creation and acquisition that presume new models of stakeholder engagement. The notion of citizen science, where ecosystem service knowledges might be produced through the labour and efforts of a committed, but diffuse, environmental citizenry is part of the process by which the ambitions of this framework are being practically enabled (Hand, 2010).

And what of the effects of this tendency towards analytic-deliberative approaches on the idea and expectations of participation itself? As the distinction between data and participation become obscured, analytic-deliberation implies a more calculating approach to knowledge production within participation. This might be taken to imply that knowledge emerging from within participatory process carries with it greater expectations that knowledge will 'do something', and that the 'talking shop' charge of the critics of participation is thrown into relief. Yet, the very virtues of a dialogic and interpretive perspective, with its focus on debate, argumentation and expanding possibilities, can easily be stifled as much as enhanced if the model of engagement is one driven by a concern primarily to inform the analytics; the stakeholder as a mere data input or data point.

Finally, we return to the placement of stakeholders within a framework where the idea of nature is constructed less as an object to be conserved *per se*, than something yielding flows of service-derived benefits to well-being. This is not only a more relational construction of nature but also a model of human-environment interactions requiring a more flexible and unbounded notion of stakeholder. The idea of 'flows of benefit' raises an important distinction between, on the one hand, the site and process of service production and, on the other, the more dissipated sense in which benefits are created, experienced, consumed and felt. If one of the starting points for participation has been to think of stakeholder engagement as an *in situ* activity that is typically a highly localised process, an ecosystem services perspective challenges the local scales at which participation is usually enacted. Decisions about ecosystem services are necessarily multi-level and spatially complex. The ecosystem processes underpinning different services operate at different spatial and temporal scales, ranging from the microscopic and slope scale to landscape and global scales. These processes may be manipulated to influence flows of services over periods ranging from months (e.g. altering hydrological processes to prevent erosion and improve the provision of clean water) to decades (e.g. in the case of woodland or soil processes that sequester carbon and help regulate the climate). The interlocking and often interacting spatial and temporal scales at which ecosystem services are provided requires an approach to decision-making that can operate at these multiple scales, representing the needs of future generations and considering the priorities of social groups from local to national and international scales, depending on the nature of the services being decided upon. Bache and Flinders (2004) call this 'multi-level governance'; decision-making processes that facilitate knowledge exchange and collaboration between individual and institutional actors across multiple scales (not necessarily respecting administrative boundaries). Multi-level governance creates interfaces between informal and formal institutions, which can empower non-state (often local) actors to participate in and influence governance and policy far beyond their traditional sphere of influence

(Bache and Flinders, 2004; Newig and Fritsch, 2009). To achieve the integration of different knowledge about ecosystem services at different spatial and institutional scales, it is necessary to understand how knowledge is socially situated within the networks of individual actors who are affected by or who can affect environmental decisions/governance arrangements. How these individual actors then self-organise or are organised in formal and informal institutions may explain why some groups are able to mobilise ecosystem service evidence for decision-making more effectively than others, and how decision-makers find and use this evidence.

The self-mobilisation of ecosystem service knowledge via social interaction across social networks and communities of practice has been described as 'social learning' (Reed *et al.*, 2010). Despite the capacity for social learning processes to facilitate knowledge exchange across the sorts of scales required for many decisions about ecosystem services, the extent to which ecosystem service knowledge translates into decisions and behaviours will depend on many factors. Kenter *et al.* (2015) argue that short-term participatory processes may be able to change preferences and what they call 'contextual values' such as the amount of money someone might be willing to pay to support the provision of an ecosystem service. However, to effect change in more deeply-held beliefs (or what they call 'transcendental values'), engagement may be required over much longer periods of time, potentially extending to generational timescales. Although this presents a challenge for the funding and design of participatory processes, this recognises that participation often requires investment in long-term dialogue, and should not be seen as a 'quick fix' for attitudes and behaviours threatening the provision of services, like climate regulation, over the long-term. This might be conceived as purely a technical question of co-ordination, and a need to create new socially, spatially and temporally extended approaches to participation and value elicitation with respect to how ecosystem services are used and prioritised, but in practice we know very little about what it means to navigate priorities as we 'follow the flows'.

Conclusion

The governance of natural resources from an ecosystem services perspective both inherits and continues a tradition of environmental decision-making where the idea of stakeholder participation is viewed as axiomatic to the integrity of its processes and the credibility of its outputs and outcomes. There is an ethical, practical and substantive basis to the idea of stakeholder participation that holds true of the ambitions and concerns of this volume. General rationales for engagement endure under this new model of human-environment relations and practitioners have an enormous companion tradition of work that helps it understand and apply principles of good practice (see Box 21.4).

Box 21.4 Best practice stakeholder participation

In his review of stakeholder participation for environmental management, Reed (2008) distils eight general principles for guiding best practice:

1. Stakeholder participation needs to be underpinned by a philosophy that emphasises empowerment, equity, trust and learning;
2. Where relevant, stakeholder participation should be considered as early as possible and throughout the process;

3. Relevant stakeholders need to be analysed and represented systematically, rather than presuming stakeholders are self-evident and self-construed;
4. Clear objectives for the participatory process need to be agreed among stakeholders at the outset;
5. Methods should be selected and tailored to the decision-making context, considering the objectives, type of participants and appropriate level of engagement;
6. Highly skilled facilitation is essential, and includes not only technical skills, but also empathy and openness to address multiple, often conflicting and entrenched, positions;
7. Local and scientific knowledges should be integrated to produce a more comprehensive and balanced understanding of complex and dynamic natural systems and processes; and
8. Participation needs to be institutionalised; there is a need to create organisational cultures in which participation is actively planned for and supported.

For Reed, these principles should be read less like a toolkit than a contract built out of an underlying philosophy of trust and respect.

Be that as it may, the future of stakeholder participation in ecosystem service decision-making is not immune to a reality that can often depart significantly from normative arrangements. History provides us with plenty of lessons where wider sources of ideas, expertise and creative inspiration is either overlooked or neglected. Broad and deep engagement with stakeholders is an aspiration that few in conventional decision making roles would deny, but we need to recognise it is also a practically difficult process to get right. Many have pointed to the difficulties of building diversity in to decision-making. Too often, critiques claim, participation involves engagement with a 'ready-made' audience of stakeholders with direct responsibilities, interests and expertise in a given decision-making context. Diversity depends on working with 'hard-to-reach' audiences, both in terms of the participation of those who may not readily identify themselves with a given issue, and those without the inclination or confidence to see themselves as sources of authority and insight, and capable of problem solving. Participation is an enabling process and that depends on the upfront commitment of time, material resources and skills. State and non-state actors are often not trained in engagement even if requirements and mandates to do so exist. What is practically achievable in the context of a given approach to participation can often fall wildly short of expectations and serve to alienate rather enable the very community of stakeholders deemed critical to building socially acceptable and robust outcomes. At its very worst, some have argued participation can serve to dis-empower, exclude, oppress and act as a smokescreen behind which decision-making institutions conduct 'business as usual'.

Yet as this emerging epistemic community enters the fray of participation with its own particular preoccupations and penchants, so too its does gradually impart new meanings and terms on which participation occurs and changes its vernacular. Superficially and rhetorically put, this might be said to bring its own opportunities and challenges. These are matters reflecting wider questions of the language and the conceptual framing of the discourse and the conditions and assumptions that surround its needs and understandings of data, as well as its placement in a much wider, more vexed, debate about whose side the framework is on.

Note

1 Findings were based on views of 118 participants from a cross-section of backgrounds (age, gender, occupation, ethnicity, and rural and urban residents), as well as levels of self-reported awareness and interest in environmental issues. In total, 341 participant days were committed to a dialogue process involving nine one-day regional events (Birmingham, Exeter, Glasgow) and one one-and-a-half day national event (London).

References

Bache, I., and Flinders, M. (2004). *Multi-level Governance*. University Press, Oxford.

Brown, G., de Bie, K., and Weber, D. (2015). Identifying public and stakeholder perspectives for implementing place-based land management. *Landscape and Urban Planning*, vol 139, pp 1–15.

Bryson, J. M. (2004). What to do when stakeholders matter: stakeholder identification and analysis techniques. *Public Management Review*, vol 6, no 1, pp 21–53.

Chilvers, J. (2007). Towards analytic-deliberative forms of risk governance in the UK? Reflecting on learning in radioactive waste. *Journal of Risk Research*, vol 10, no 2, pp 197–222.

Chilvers, J. (2009). Deliberative and participatory approaches in environmental geography. In: Castree, N., Demeritt, D., Liverman, D., and Rhoads, B. (eds) *A Companion to Environmental Geography*. Blackwell, Oxford.

Cilliers, P. (2005). Knowledge, limits and boundaries. *Futures*, vol 37, pp 605–613.

Cooke, B. and Kothari, U. (eds) (2001). *Participation: The New Tyranny?* Zed Books, London.

de Vente, J., Reed, M. S., Stringer, L. C., Valente, S., and Newig, J. (in press). How does the context and design of participatory decision-making processes affect their outcomes? Evidence from sustainable land management in global drylands. *Ecology & Society*.

Fiorino, D. (1990). Citizen participation and environmental risk: a survey of institutional mechanisms. *Science, Technology and Human Values*, vol 15, no 2, pp 226–243.

Fish, R. (2011). Environmental decision making and an Ecosystems approach: some challenges from the perspective of social science. *Progress in Physical Geography*, vol 35, no 5, pp 671–680.

Fish, R. and Saratsi, E. (2015). *Naturally Speaking. A Public Dialogue on the UK National Ecosystem Assessment*. Final Report to Sciencewise, University of Exeter, Exeter.

Fisher, B., Kerry Turner, R., and Morling, P. (2009). Defining and classifying ecosystem services for decision making. *Ecological Economics*, vol 68, no 3, pp 643–653.

Grimble, R., and Wellard, K. (1997). Stakeholder methodologies in natural resource management: a review of principles, contexts, experiences and opportunities. *Agricultural Systems*, vol 55, no 2, pp 173–193.

Haines-Young, R., and Potschin, M. (2014). The ecosystem approach as a framework for understanding knowledge utilisation. *Environment Planning C*, vol 32, pp 301–319.

Hand, E. (2010). Citizen science: people power. *Nature*, vol 466, pp 685–687.

Hauck, J. (2016). Transdisciplinarity. In: Potschin, M., Haines-Young, R., Fish, R. and Turner, R. K. (eds) *Routledge Handbook of Ecosystem Services*. Routledge, London and New York, pp 92–94.

Jax, K. (2016). Ecosystem services and ethics. In: Potschin, M., Haines-Young, R., Fish, R., and Turner, R. K. (eds) *Routledge Handbook of Ecosystem Services*. Routledge, London and New York, pp 301–303.

Kenter, J. O. (2016). Deliberative and non-monetary valuation methods. In: Potschin, M., Haines-Young, R., Fish, R. and Turner, R.K. (eds) *Routledge Handbook of Ecosystem Services*. Routledge, London and New York, pp 271–288.

Keune, H. and Dendoncker, N. (2013). Negotiated complexity in ecosystem services science and policy making – ecosystem services. In: Jacobs, S., Dendoncker, N., and Keune, H. (eds) *Ecosystem Services – Global Issues Local Practices*. Elsevier, New York.

Kenter, J. O., O'Brien, L., Hockley, N., et al. (24 authors) (2015). What are shared and social values of ecosystems? *Ecological Economics*, vol 111, pp 86–99.

Lang, D., Wiek, A., Bergmann, M., et al. (7 authors) (2012). Transdisciplinary research in sustainability science – practice, principles, and challenges. *Sustainability Science*, vol 7, supplement 1, pp 25–43.

Morris, J., Posthumus, H., Hess, T., Gowing, D., and Rouquette, J. (2009). Watery land: the management of lowland floodplains in England. In: Winter, D. M. and Lobley, M. (eds) *What is Land for?* Earthscan, London.

Newig, J., and Fritsch, O. (2009). Environmental governance: participatory, multi-level–and effective? *Environmental Policy and Governance*, vol 19, no 3, pp 197–214.

Potschin, M., and R. Haines-Young (2016a). Frameworks for ecosystem assessments. In: Potschin, M., Haines-Young, R., Fish, R. and Turner, R. K. (eds) *Routledge Handbook of Ecosystem Services*. Routledge, London and New York, pp 125–137.

Potschin, M., and R. Haines-Young (2016b). Defining and measuring ecosystem services. In Potschin, M., Haines-Young, R., Fish, R. and Turner, R. K. (eds) *Routledge Handbook of Ecosystem Services*. Routledge, London and New York, pp 25–44.

Raymond, C. M., Fazey, I., Reed, M. S., Stringer, L. C., Robinson, G. M., and Evely, A. C. (2010). Integrating local and scientific knowledge for environmental management: from products to processes. *Journal of Environmental Management*, vol 91, no 8, pp 1766–1777.

Reed, M. S. (2008). Stakeholder participation for environmental management: a literature review. *Biological Conservation*, vol 141, no 10, pp 2417–2431.

Reed, M. S., Evely, A. C., Cundill, G., et al. (11 authors) (2010). What is social learning? *Ecology & Society*, vol 15, no 4, r1. Available at: http://www.ecologyandsociety.org/vol15/iss4/resp1/

Reed, M.S., Graves, A., Dandy, N., et al. (9 authors) (2009). Who's in and why? A typology of stakeholder analysis methods for natural resource management. *Journal of Environmental Management*, vol 90, no 5, pp 1933–1949.

Reed, M. S., and Sidoli del Ceno, J. (in press). Alternative approaches to managing conservation conflicts: from top-down to bottom-up. In: Young, J. and Redpath, S. (eds) *Conflicts in Conservation: Strategies for Coping with a Changing World*. Cambridge University Press, Cambridge UK.

Renn, O. (1999). A model for an analytic-deliberative process in risk management. *Environmental Science and Technology*, vol 33, no 18, pp 3049–3055.

Renn, O., Webler, T., and Wiedemann, P. (1995). *Fairness and Competences in Citizen Participation. Evaluating Models for Environmental Discourse*. Kluwer, Dodrecht.

Robertson, M. (2012). Measurement and alienation: making a world of ecosystem services. *Transactions of the Institute of British Geographers*, vol 37, no 3, pp 386–401.

Rowe, G., and Frewer, L. (2000). Public participation methods: a framework for evaluation in science. *Technology and Human Values*, vol 25, no 1, pp 3–29.

Sagoff, M. (2016). A critical perspective. In: Potschin, M., Haines-Young, R., Fish, R. and Turner, R. K. (eds) *Routledge Handbook of Ecosystem Services*. Routledge, London and New York, pp 112–114.

Stern, P., and Fineberg, H. (eds) (1996). *Understanding Risk: Informing Decisions in a Democratic Society*. National Academy Press, Washington DC.

Warner, J. (2007). The beauty and the beast: multi-stakeholder platform for integrated water management. In: Warner, J. (ed.) *Multi-stakeholder Platform for Integrated Water Management*. Ashgate, Aldershot.

22

DELIBERATIVE AND NON-MONETARY VALUATION

Jasper O. Kenter

Introduction

A sophisticated array of methods has been developed by economists to value ecosystem services. However, there is also increasing recognition of the limitations of these methods. These limitations are in part practical, and in part have to do with the way that neoclassical economists conceptualise value. As chapter 8 pointed out, concerns have been raised that putting a monetary value on the environment and applying market-based thinking to it turns the natural world into a commodity. If nature is commodified, can this be rhymed with notions that other species, or even whole ecosystems, have intrinsic value independent of their benefits to human well-being? Does it make sense to try to monetise our cultural, emotional, and spiritual connections to nature? Economic methods focus on the values of individuals. What about the collective values we share as communities or society as a whole?

Deliberative and non-monetary methods are often put forward as a way to consider values in a broader and more pluralistic way than mainstream economics. This chapter will introduce a wide range of deliberative and non-monetary methods to assess values ascribed to and associated with the natural world, which can be used as either an alternative to or in conjunction with economic methods. First, a summary will be given of the key limitations of economic methods. This is followed by a brief discussion of what values are, and a short conceptualisation of different types of individual and shared values. This will then aid a discussion of how these different types of values can be considered through different, but not mutually exclusive, types of methods: *deliberative, analytical-deliberative, psychometric,* and *interpretive*. I will conclude with a brief discussion of when deliberative and non-monetary methods are particularly advantageous, how different methods can be used together, and where future research might be headed.

Beyond monetary valuation

There exists a wide range of economic methods available to elicit ES values, including methods based on market prices, revealed preferences (where environmental values are revealed by choices expressed through observable behaviour, such as the willingness to travel to a certain place), and stated preferences (where preferences are elicited through surveys). Fundamentally, all of these are based on the notions of willingness-to-pay (WTP) and willingness-to-accept.

Here, the assumption is made that the value that someone has for something is reflected in how much someone is willing to spend on it, or in the amount of money someone would be willing to accept to lose it.

Neoclassical economics, currently the dominant economic paradigm, makes some further assumptions about values (e.g. Lawson, 2013). Values and behaviour are assumed to rely on rational, pre-formed, utilitarian preferences. When someone prefers A over B, this is assumed to be because A would provide more utility to that person than B. Individuals make choices by maximising their utility, rationally trading off the positives and negatives of different options. Originally, utilitarianism considered that utility was about increasing happiness. However, neo-classical economics considers utility in a more general way, as the degree that something satisfies wants or desires. Here, utility (and thus the value of something) can be indirectly, not directly, measured through WTP.

Another important neoclassical assumption is that all explanations can be couched solely in terms of individuals. Thus, the value of something to society is always an aggregate of individual utility. Moreover, individual utility is assumed to be self-regarding. This means individuals are assumed to base their preferences solely on their own utility. The welfare of others (including future generations and non-humans) is assumed to influence individuals only if it increases their own utility.

Finally, mainstream economics makes an epistemological assumption that preferences can be positively and objectively measured (albeit indirectly through WTP). The main way that patterns in these preferences can be understood is through quantitative modelling.

These assumptions limit the scope of the methods based on them. It can be questioned whether all behaviour can really be couched in terms of pre-formed, self-regarding individual preferences that are rationally traded off. In relation to the natural environment, peoples' motivations are very diverse, including rights, duties, virtues, and cultural beliefs, identities and narratives that are hard to translate into measures of utility. For example, people might be willing to pay for something because they feel it is the right thing to do, rather than because it satisfies their individual preferences. Many shared values operate at the level of communities and cultures, rather than individuals. However, economic approaches are unlikely to recognise pluralistic values and conceptions of well-being because the notions of what constitutes valid economic knowledge tend to exclude subjective and qualitative material. Economic studies focus on *what* is valued and *how much*, with little attention given to the *why* people value particular ecosystem services in particular places.

In managing the environment, there are many different dimensions of value to any given decision, which are not easy to trade off against each other. When aggregating preferences, some kind of agreement is needed on how to aggregate within dimensions (i.e. how much does each individual count), and across dimensions of valuation (i.e. how are different value criteria to be made commensurate). Take, for example, appraisal of a hypothetical proposed mining project. Dimensions of value could be the usual costs and benefits (expected revenue, construction and operational costs, etc.), the livelihoods of people, the cultural impact of the project, and impacts on local biodiversity. In conventional economic analysis, if the benefits outweigh the costs after compensation, the project would be 'efficient' and deliver a net value to society (regardless of whether these compensations actually take place). This assumes that, in principle, the ecological, social, and cultural dimensions of value can be compensated fully and justly. In practice, many people feel very uncomfortable trading off values that relate to culture or ethics in monetary terms. However, unless all parties can deliberate together and agree about how different dimensions should be traded off against each other, it is not possible to come to any single conclusion, and some economists themselves have argued that for this reason methods such as cost-benefit analysis have only limited use in these cases (Hockley, 2014). Often people do not have clear pre-formed values and prefer to deliberate on values with others. People also often resist attempts for their values to be converted into monetary amounts, and, in managing ecosystem services, the use of economic approaches can increase conflict rather than resolve it when people feel that their other values are not taken into consideration (O'Brien, 2003). It has been argued that the description of the environment in terms of preferences, utility, WTP, and also ecosystem 'services' at the least fails to reflect the deeper and often shared meanings that places might hold (Daniel et al., 2012; Owen et al., 2009) and at the worst is in itself a political act of commodification and enclosure (O'Neill, 2008).

Economists make value-laden decisions in cost-benefit analysis about the distribution of property rights, and who counts how much when evaluating the economic efficiency of different options. By basing decisions on the status quo, economic analysis often implicitly supports the rich and powerful, because in monetary terms they will have the largest benefits and costs. It is possible to make adjustments on the basis of different assumptions about property rights, but this requires a deliberation on what alternative assumptions and adjustments should be made on aggregate of the preferences of diverse individuals and interests (i.e. who should count how much).

Thus, it is clear that mainstream economic approaches have significant limitations. On their own, they cannot fully reflect the value of ecosystem services. To address these issues, a number of different ways forward have been suggested. Increasingly, deliberative approaches are seen as a way in which preferences can be more clearly formed and expressed (e.g. Alvarez Farizo et al., 2007), but they can also provide an opportunity to engage with a broader array of motivations and moral stances than utilitarian preferences alone (e.g. Kenter et al., 2011). Deliberative approaches can be monetary and non-monetary. Psychological approaches often use statistical techniques similar to economics, but they provide a different avenue to conceptualising well-being, in a multi-dimensional way rather than through a single monetary indicator. Others advocate interpretive or qualitative approaches (e.g. Daniel et al., 2012). These are less focused on general conclusions, concentrating on idiosyncratic knowledge that reveals the meaning of and narratives associated with places. They can provide a rich source of information on why particular places or natural features are important to us. Understanding the 'why' is often very important in the management of these places and in dealing with conflicts between different social groups and interests. All these different approaches can be combined with various types of mapping techniques, where participants indicate (and potentially deliberate) where ecosystem services and their values lie within in the landscape.

An overview of key methods and their relation to different types of values is presented in Table 22.1. As this table shows, particular methods relate to different types of individual and shared values in terms of outcomes of the methods, and in the case of deliberative methods also the process. These different value types will be discussed in the next section.

However, it is important to note that not all non-monetary methods are by definition in juxtaposition to the instrumental paradigm of mainstream economics (Raymond et al., 2014). For example, 'public participation' mapping studies use surveys where individuals evaluate and score the landscape on the basis of a pre-determined categorisation of services or features. These individual values are then aggregated across populations in a similar way to economic studies, using arithmetic and without significant deliberation. While these studies share the advantages of economic approaches that a large sample can be assessed and that differences between groups can be assessed statistically, they also share many of the limitations outlined above. In contrast, 'participatory' mapping or GIS originates from within the participatory action research (PAR) community, and generally involves groups of participants who discuss and deliberate the value of

Table 22.1 Selection of key deliberative and non-monetary methods that can be used to assess different types of values of ecosystem services.

Technique		*Description*	*Types of values that may be elicited*
Deliberative	In-depth discussion groups	Group (usually 4–8 people) discussions (often repeated), during which participants shape the terms of discussion, develop themes in ways relevant to their own needs and priorities.	*Process*: Cultural/societal, communal, transcendental, group, deliberated, other-regarding, values in relation to society. *Outcome*: Deliberated group or individual, transcendental, and/or contextual values.
	Citizen's juries	A small cross-section of the general public who come to a considered judgement about a stated policy issue/problem through detailed exposure to and scrutiny of, the relevant evidence base. Group responds by providing a recommendation or 'verdict'.	*Process*: Cultural/societal, communal, transcendental, other-regarding, values in relation to society. *Outcome*: Deliberated group contextual values and indicators (verdict).
	Deliberative opinion polls	Technique designed to observe the evolution of the views of a large citizen test group as they learn about a topic. Typically the group votes on the issues before and after an extended debate.	*Process*: Cultural/societal, communal, transcendental, group, deliberated, other-regarding, values in relation to society. *Outcome*: Deliberated individual value indicators (vote counts).
Analytical-deliberative	Participatory modelling	The involvement of stakeholders in the design and content of analytical models that represent ES and their benefits under different spatial and temporal conditions.	*Process*: Cultural/societal and communal contextual values. Other-regarding and transcendental values only likely to be made explicit if prompted through reflection/deliberation process. *Outcome*: Deliberated group contextual values and indicators (relative importance of different parameters and their relationships).
	Deliberative monetary valuation	Techniques that use formal methods of group deliberation to come to a decision on monetary values for environmental change. May be allied to survey-based techniques (contingent valuation or choice experiments) or use a non-econometric approach to establish values (e.g. incorporating citizen's juries).	*Process*: Cultural/societal and communal contextual values. Other-regarding and transcendental values only likely to be made explicit if prompted through reflection/deliberation process. *Outcome*: Deliberated and/or group value indicators (deliberated individual or group WTP, deliberated individual or group fair price, deliberated individual or group social WTP).
	Deliberative multi-criteria analysis	Techniques that involve groups of stakeholders designing formal criteria against which to judge the non-monetary and (sometimes) monetary costs and benefits of different management options as the basis for making a decision.	*Process*: Cultural/societal and communal contextual values. Other-regarding and transcendental values only likely to be made explicit if prompted through reflection/deliberation process. *Outcome*: Deliberated contextual individual or group values and indicators (ratings/rankings/scores).

Interpretive, potentially deliberative	Participatory mapping/GIS	A group of stakeholders consider or create a physical or digital map to indicate landscape features that are valuable (and/or problematic). Participants may also rate or rank these features for importance. Map layers can also incorporate photo, video, artwork, poetry, etc.	*Process*: Communal contextual values, if features are important/assessed on a larger scale: contextual cultural/societal values. *Outcome*: As above. If features are deliberated and decided upon or rated/ranked by groups, these take the form of deliberated group contextual values and indicators.
	Storytelling	Participants are asked to tell stories about their experiences of or in relation to places. These may be reflected upon in a group setting to discuss values related to these experiences.	*Process*: Communal contextual values, if features are important/assessed on a larger scale: contextual cultural/societal values. Other-regarding and transcendental values only likely to be made explicit if prompted through reflection/deliberation process. *Outcome*: As process. If stories are deliberated in a group setting, these may take the form of deliberated group values. Number of times particular themes or values are expressed can provide indicators.
	Interviews	Participants are interviewed about their values, beliefs and preferences. Group interviews allow for deliberation and are similar to in-depth discussion groups. However, in group interviews, terms are set by the interviewer rather than the group.	*Process and outcome*: As storytelling.
Interpretive	Media analysis	Use of a range of textual analysis tools (particularly content, frame and discourse analysis) on (mass) media outputs or social media content over a selected period of time.	*Process*: n/a *Outcome*: transcendental, communal, societal and cultural values, other-regarding-values.
	Desk-based cultural history study	This approach can be used effectively as a first option to quickly scan existing literature over a specified period of time to identify values connected with the decision being considered. The study can cover academic and grey literature, as well as creative writing (prose and poetry). Historical analysis can deliver understanding of past value and belief conflicts that can help to better manage present issues and mitigate risks.	*Process*: n/a *Outcome*: transcendental, communal, societal and cultural values, other-regarding values.

(Continued)

Table 22.1 (Continued)

Technique		*Description*	*Types of values that may be elicited*
	Other interpretive methods	A wide range of qualitative techniques including ethnography and participant observation, genealogy, life history methods, dramaturgical analysis, reviewing landscape character descriptions, other textual analysis of various sorts including discourse, content and frame analysis.	*Process*: n/a *Outcome*: Variable, can be particularly suited to transcendental, communal, societal, and cultural values.
Psychometric deliberative	Values compass	This method asks participants to consider which of their individual transcendental values are most important by ranking or rating them, and then asks to discuss the degree to which these values are important for one's community, culture or society. Values can also be ranked or rated on a group basis. It is based on the values typology developed by Schwartz (1990).	*Process*: transcendental individual, communal, cultural, and/or societal values. *Outcome*: As process, plus group and deliberated values.
Psychometric	Subjective well-being indicators	These can be used to assess how and the degree to which places contribute to one's well-being, and are thus highly suitable for assessing the value of cultural ecosystem services using a quantitative non-monetary metric.	*Process*: n/a *Outcome*: communal, societal, and cultural contextual values.
	Other psychometric	Psychometric testing refers to the measurement of psychological phenomena and processes, e.g. knowledge, experience, attitudes, values, worldviews. Psychometric models (e.g. Values-Beliefs-Norms, Theory of Planned Behaviour) can be used to better understand the impact of deliberative processes on values.	*Process*: n/a *Outcome*: standard scales exist for transcendental values, and can be developed on a case-by-case basis for contextual communal, cultural, and social values. Statistical models can be used to relate psychometric variables (e.g. transcendental values) to contextual values and indicators such as WTP.

The content of this table was derived from Kenter et al. (2014; 2015).

different places. PAR focuses on facilitating a process of change within organisations or communities, directly helping to address key issues that are pertinent to participants. There is a particular focus on bottom-up learning, and the role of the researcher is more as a facilitator than an expert (Kumar, 2002). Here the types of values that emerge are usually more idiosyncratic and interpreted post-hoc. A richer picture is likely to emerge, and there is more opportunity for implicit values to be made more explicit through deliberation, though probably with a smaller group of participants than in a public participation mapping exercise. However, there is no definite line drawn between instrumental approaches like public participation GIS and more deliberative and participatory approaches (Brown and Kyttä, 2014), and some authors argue for pragmatically combining elements of the two to integrate some of the benefits of both instrumental and deliberative paradigms (Raymond et al., 2014).

The diverse range of deliberative and non-monetary approaches and their particular advantages and disadvantages will be discussed in more detail below. However, it is useful to first consider what the different types of values might be elicited through them.

A plural perspective on values

As discussed above, there are many different values and dimensions of value associated with management of ecosystem services. An important benefit of using deliberative and non-monetary methods is that they can take a broader perspective on values than conventional monetary approaches. However, different methods are more suitable for eliciting different types of values, and it is important to distinguish between these types to make a more informed choice about what methods are suitable in particular situations and how their results can be interpreted. While there are various frameworks that type values for ecosystem services in a range of different ways, the UK National Ecosystem Assessment developed a framework specifically designed for characterising and better understanding valuation processes (Kenter et al., 2015). The framework distinguishes five *dimensions* of values: (i) the value concept; (ii) the value provider; (iii) the process used to elicit values; (iv) the scale of value; (v) and its intention (Figure 22.1).

In terms of the *concept of value*, a distinction can be made between values in the sense of: '*criteria that people use to select and justify actions and to evaluate people (including the self) and events*' (Schwartz, 1992); opinions about worth or importance; and the worth of something itself, often expressed in monetary terms. Another way of looking at this is that values can be differentiated between guiding principles and goals that transcend specific situations (e.g. fairness, honesty, enjoyment), which are called *transcendental values*; values that are dependent on an object of value and hence contextual and attitudinal, which are called *contextual values* (e.g. clean water); and measures of the worth of something (e.g. WTP of £100 to improve water quality), which are called *value indicators*.

In terms of *providers of value*, we can distinguish individuals, ad-hoc groups, communities, and societies as a whole. Societies, as a whole, share *cultural* and *societal values*, which may be considered shared principles and virtues as well as a shared sense of what is worthwhile and meaningful. Within societies and cultures there is a wide range of social groups that may express distinct *communal values*, including local communities, faith groups, groups of people that share an activity such as recreational users of the environment, communities of practice, etc. In addition, there are the ad-hoc groups associated with research, such as a discussion group of stakeholders or a focus group with members of the public, which can come to collective value outcomes that we term *group values* with techniques such as citizens' juries or multi-criteria analysis, which will be detailed below.

The dimension of *elicitation process* distinguishes between *non-deliberated* and *deliberated values*.

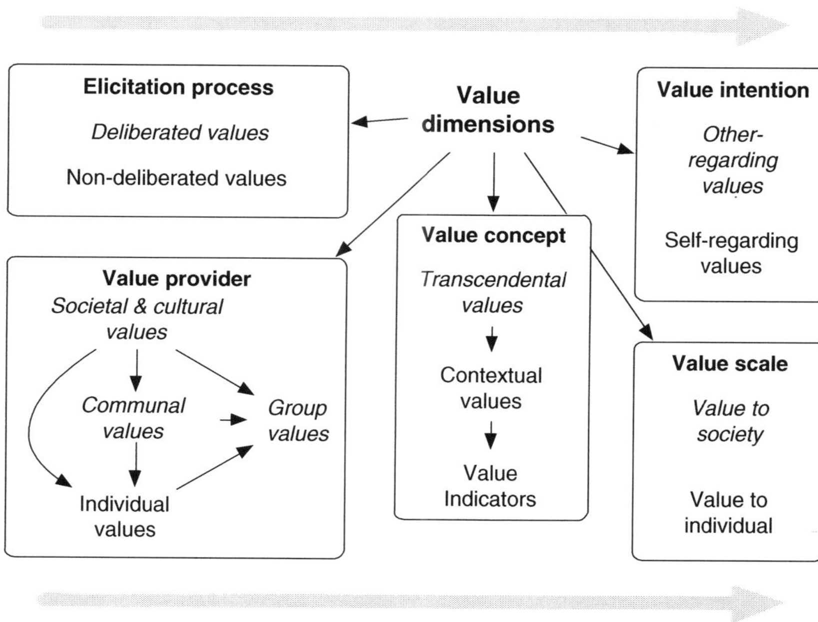

Figure 22.1 Dimensions and types of values.

Source: adapted from Kenter et al. 2015

Bold titles indicate dimensions of value. Italicised titles indicate types of shared and social values; non-italicised titles indicate other types of values. Arrows within boxes indicate directions of influence. Grey arrows signify that the type of elicitation process and value provider strongly influence what value types are articulated along the concept, intention and scale dimensions.

In terms of *scale*, we can distinguish the individual scale, and the social or societal scale, which has bearing on *values to society*, or in relation to society. An example is that one might highly value enjoyment and a varied life for oneself (reflected for example in consumer behaviour), but in relation to society other values such as fairness or responsibility might be more important (reflected for example in voting behaviour). An example at the level of indicators is that one might be willing to pay £10 to improve water quality (individual scale), or think that local government should invest £1 million in a water treatment plant (social/societal scale).

The dimension of *intention* relates to whether values are self-regarding or are other-regarding, altruistic values. For example, I may value my own life enjoyment (self-regarding), but also that of my neighbour or that of future generations. *Intention* differs from the *scale* dimension, as values for others are not necessarily values in relation to society.

Armed with an understanding of the distinctions between these different value types, we can now consider how deliberative, analytical-deliberative, psychometric, and interpretive methods might elicit these values.

Deliberative and analytical-deliberative methods

As discussed above, economists have mostly used survey-based techniques (e.g. questionnaires or structured interviews) without significant deliberative components in valuation. There has been a presumption in standard economic approaches that preferences are pre-existing and stable.

However, it is increasingly argued that preferences and contextual values are not pre-formed but need to be generated through some kind of transformative process of deliberation and learning (Kenter et al. 2011, 2014; Parks and Gowdy, 2012; Spash, 2007, 2008). Participatory and deliberative processes are appealing in that they provide participants of valuation studies with time to learn about the good under investigation, as well as time to reflect upon (and construct or potentially modify) their values (Alvarez Farizo and Hanley, 2006; Christie et al., 2006). If the deliberation is undertaken as a group process, participants have the opportunity not only to express and debate their own knowledge, views, and perspectives, but also to learn about and consider the values of those of others in the group. In particular, discussions might address rights, responsibilities, equity, fairness, and other transcendental values and political considerations. Moreover, deliberation provides a crucial opportunity to better consider issues around uncertainties and risks (Zografos and Howarth, 2010). A group learning process is also particularly important with respect to bringing out cultural and communal transcendental values (Kenter et al., 2011) and coming to decisions on group contextual values (Niemeyer, 2004). Such group values might be expressed as a consensus or majority view on what the group believe to be in the best interest of society. Here, participants can come to agreement on how different dimensions of value and different interests should be aggregated. However, consensus views are not always achievable or desirable (Sagoff, 1998) and a deliberative process could also result in the recognition of a diversity of values, where outcomes are achieved that account for reasonable differences (Lo, 2013).

In terms of classifying deliberative methods, two broad groups can be identified: 'deliberative' and 'analytical-deliberative' methods. Deliberative methods, such as in-depth discussion groups and citizen's juries, include a range of techniques that allow stakeholders to 'confer, ponder, exchange evidence, reflect on matters of mutual interest, negotiate and attempt to persuade each other' (Stern and Fineberg, 1996, p. 73). Through this deliberative process, individuals are encouraged to express and develop their views as different evidence and perspectives are considered. The outcomes of deliberative methods are mostly qualitative and might include priority lists, recommendations, and verdicts. Analytical-deliberative methods, such as deliberative monetary valuation (DMV) and multi-criteria analysis (MCA), tend to involve more elaborative approaches which integrate deliberative-based techniques with more formal decision-support tools. Outcomes from such methods are often expressed in monetary terms or other types of quantitative ranking or rating.

Lo and Spash (2012) provide a useful framework in which they set out three approaches to incorporating deliberation into valuation: preference 'economisation', 'moralisation' and 'democratisation'. Preference economisation primarily seeks to utilise deliberation to ease the respondent's cognitive burden associated with expressing stated preference monetary values. Thus, information and group discussions are primarily focused on nurturing value elicitation at the individual level. Preference moralisation seeks to use deliberation to bring out transcendental values, and deliberation is extended to address non-economic considerations, including social norms, rights, and procedural fairness. This is particularly important for valuation of ecosystem services, because environmental values are often latent and require a moralisation process to be brought out.

Within the values framework describe above, moralisation can be seen as a value construction or translation process where transcendental values are brought in and related to a context, so that contextual values can be formed. Building on our conception of transcendental values as much broader than just ethics, including a wide range of life goals and aspirations, moralisation becomes a broad process. For example, values associated with cultural identity are often intertwined with aspects of environmental settings (Church et al., 2011) and deliberation on the importance of nature for cultural identity can significantly change contextual values and preferences (Kenter et al., 2011).

Lo and Spash (2012) consider that effective deliberation processes contain both information-oriented and moralisation aspects, but should also seek 'choice democratisation' as an approach that is consistent with deliberative democratic principles and value plurality. Rather than following standardised procedures, such an approach centres on key principles and requirements in relation to process.

This is particularly important given some of the key limitations of deliberative methods, which have to do with the legitimacy of the deliberative process and with who is represented around the table. In terms of the former, it is hard to ignore the often unequal social relations and institutions outside of the valuation setting, which will influence participants to voice their opinions and concerns (O'Neill, 2007). There are likely to be differences in terms of social status, political influence, class, education, and experience with deliberation and discussion. This can lead to participants not expressing or adjusting their views under the pressure of power dynamics or as a result of perceived social desirability. While this can be managed to some degree through professional facilitation, this also needs explicit consideration in the process design.

As to representation, deliberative approaches usually (though not always) use smaller samples than instrumental approaches and often select participants on the basis of their political representativeness (e.g. as representative of a stakeholder group) rather than randomly. Questions might also be raised in relation to competence (are participants able to assess the issues at stake?) and whether representation needs to counterbalance the type of political and institutional biases described above (Fish et al., 2013). These issues need to be considered through rigorous stakeholder analysis. While there are well-developed processes for stakeholder selection – for an overview see Reed et al. (2009) and Varvasovszky and Brugha (2000) – selection of participants inevitably to some degree remains a subjective process that can substantially influence outcomes.

An elaborate example of applying different deliberative and analytical-deliberative techniques is given in Box 22.1. This case study illustrates how processes of information sharing, moralisation, deliberative-democratic debate, and eliciting group-based values can generate substantially different outcomes from conventional individual survey approaches, addressing some of the key limitations of economic approaches outlined above.

Box 22.1 Valuing UK marine protected areas

The Convention on Biological Diversity has called for 10% of the seas to be designated as a marine protected area (MPA), but there is only limited knowledge of the ecosystem service benefits that this might generate. This large-scale case study by Kenter et al. (2014, 2013) valued cultural services of potential marine protected areas (MPAs). The study considered whether shared values elicited through deliberative workshops were different from individual values, and provides an example of how monetary, deliberative, and non-monetary methods can be integrated to provide a more comprehensive valuation.

Methods

Data gathering consisted of two phases: an online survey with 1,683 divers and sea anglers, and a series of 11 DMV workshops (130 participants) and 5 MCA workshops (55 participants) across Britain. Both the survey and DMV workshops included contingent valuation (CV) questions that asked about WTP towards protecting sites into the future. The CV tasks considered vulnerable species, marine landscape/habitats, large fish and charismatic species, wrecks and rock formations, access options, management restrictions, size, and travel distance[1].

The DMV workshops consisted of a deliberation stage focused on information exchange, followed by individual and group-based CV stages. A second intervention focused on exchange of experiences and values through storytelling by participants and a 'values compass' (Table 22.1). This was followed by another individual and group valuation stage.

The MCA workshops presented participants with a set of goals important to recreational users and a number of scenarios reflecting different MPA management regimes across different marine settings. Participants assessed the importance of different goals as individuals and as groups and then scored how well different management options realised those goals at different settings.

The survey and workshops also included a psychometric component that asked participants to respond to 15 subjective well-being statements about the sites they visited, and questions on their transcendental values, beliefs, and norms.

Results

WTP substantially decreased as a result of the second deliberative intervention and expressing values as group. Participants clearly scrutinised the sites presented to them and they more explicitly evaluated them against other societal priorities. There were extensive discussions around responsibility and fairness in relation to management restrictions and access, which were reflected in WTP changing between individual and group-based valuation. In the final set of group-deliberated values, there was also a convergence between WTP and the subjective well-being indicators, which previously were uncorrelated. Another change was that in the workshops participants formed values for many types of habitats that they didn't have in the online survey.

Storytelling during the DMV workshops brought up a range of themes that expressed communal values and shared experiences and identity, including connectedness, magical and spiritual experiences, adventure, and both social connections and solitary reflection and escape, with some difference of emphasis between divers and anglers.

In the MCA workshops, focusing the deliberation and scoring on site-based values helped tie values to specific landscapes. As in the DMV workshops, deliberated individual values fell between non-deliberated individual values and deliberated group values. Ranking results reflected trade-offs between other-regarding, transcendental values, particularly environmental protection, and self-regarding, utilitarian values (focused on recreational opportunities). Fairness and proportionality around measures and between different sea users was again an important theme.

Asking participants for how they thought their values around marine sites should be assessed, the majority indicated they preferred the workshop format and most of those preferred group to individual choices. Participants also felt more confident about their answers in workshops than in the online survey.

The mixed method approach used in this case study provided a richer picture of values than any single method approach could have. The monetary and analytical elements of the methodology helped to inform decision-makers about the relative priorities of key user groups, while interpretive techniques conveyed their deeper meaning. Deliberation impacted on values by making them more considered but also made contextual values for marine sites a better reflection of underlying transcendental values and the subjective well-being gained from the places visited.

Interpretive methods

Interpretive methods seek to find meaning and understanding through the subjective identification and analysis of discourses. Subjective (and sometimes 'intersubjective') experience is seen to be crucial to understanding why it is what people do. Thus there is an emphasis on the qualitative. Values are understood as constructed between individuals and institutions, through a socio-cultural process (O'Brien, 2003). In relation to ecosystem services, the focus is on understanding how participants relate to the environment and what the meaning is of different places in the land- or seascape and the services they provide. Interpretive methods have largely been used or proposed in relation to cultural ecosystem services. While interest in ecosystem services and cultural services in particular is a recent phenomenon, there is a long history in using interpretive approaches to understanding landscape history (Robertson and Richards, 2003), and these have provided accounts of the communal, cultural, and transcendental values that are tied to specific places or features therein, such as trees, 'exposing the splendour and secrets' (Rackham, 2003). Interpretive methods have a particular advantage over quantitative and instrumental approaches in that they can be used to understand narratives, which can have a profound influence on how we value things and which may underpin peoples' ethics at least as much as much more abstract notions such as rights, virtues, and utility (O'Neill et al., 2008). For example, as is illustrated in Box 22.2, ancient woodlands and veteran trees are not just highly valued because of their ecological characteristics, but also because of the stories they tell. Their role as a witness to human events or as a cornerstone to communal identity throughout the centuries might give them a very profound and incommensurable value.

Box 22.2 Valuing old-growth forests in Nova Scotia, Canada

Across the world, old-growth forests have substantially declined. Management of the remainder is often mired in conflict, with different stakeholders presenting conflicting values around a range of ecosystem services, including food, shelter, fuel, timber, biodiversity, and cultural heritage. This case study from Nova Scotia, Canada, based on research by Owen et al. (2009), provides an example of combining interpretive and deliberative methods to assess stakeholder values, aiding sustainable forest management.

Methods

Nine field trips involved 76 participants from indigenous groups, forestry professionals, environmental organizations, and the public. In the morning participants were introduced to forest stands that differed in maturity and degree of harvesting and asked to reflect on their feelings in a personal diary. In the afternoon a focus group or talking circle was held and participants completed a rating sheet. The journals and focus group discussion were analyzed through thematic coding based on a combination of a pre-defined coding structure (based on a set of transcendental values) and grounded theory (a way of coding where a coding structure is not pre-defined but developed from the data). The rating sheet asked participants to select which transcendental values (e.g. beauty, naturalness, intrinsic values) and ecosystem services (e.g. life support, aesthetics, education) were most highlighted by the old-growth forests, and how these were affected by different silvicultural treatments.

Results

Participants highlighted a diverse range of services, benefits, and values, including habitat, peace, sacredness, beauty, water quality/quantity, education, wildlife appreciation, and recreation. Old-growth forests were also important for identity and culture. One indigenous participant noted: 'Something more powerful in old growth, the larger trees. What was life like when these large trees were saplings? It gives people a sense of history. What were our ancestors doing at this point in time?' (Owen et al., 2009, p. 244).

Participants put forward over 20 suggestions for improving old-growth management, including more collaborative processes, stronger government policy, incentives, more public access and education, changes in management by industry such as set-asides, and employing more selective treatments.

The integrated use of deliberation, rating sheets, and interpretive analysis combined the advantages of different approaches. Rating sheets provided specific results on individual and overall group preference. Focus groups provided more in-depth information on why particular ecosystem services were important and a forum for bringing different points of view together. The qualitative diary data yielded an in-depth understanding of values, particularly around non-material values associated with old growth such as spirituality, beauty, heritage, and equity. The different threads of data delivered forest managers a rich picture of the value of forest ecosystem services to help inform management decisions.

Music, visual and performance arts, and creative literature often express how we relate to and value the environment. But interpretive methods can also be based on fieldwork, such as through semi- or unstructured individual and group interviews, storytelling sessions, or participatory mapping sessions. Here, the boundary between interpretive and deliberative methods can become blurred, as qualitative methods can have more or less aspects of deliberation, and deliberative methods can have more or less elements of interpretation. Results of deliberation tend to revolve around outcomes (e.g. a verdict in a citizens' jury or monetary values in DMV) whereas interpretation is used to analyse the discourses of the process (e.g. the narratives that are used to justify a verdict). For example, in the case study on the value of potential marine protected areas described in Box 22.1, storytelling was used both as a means to understand the meaning of these places to marine users and as a deliberative tool in DMV workshops, and the process of storytelling and discussing deeper-held values impacted participants' WTP for marine conservation.

Finally, desk-based interpretive methods can provide a rapid and relatively low-cost analysis of values for different ecosystem services. For example, cultural history studies or analysis of creative literature can reveal particular cultural representations and values associated with places. Media analysis can consider public discourses associated with specific ecosystem services, including potentially conflicting communal values and beliefs of different social interest groups. It can be used to gather a 'snapshot' of current public views, but it can also be employed to assess public feeling over longer time periods, to assess shifts in values.

Social-psychological methods

Social psychology and sociology have engaged with environmental values over several decades, and much of this research can inform valuation of ecosystem services. Two areas stand out as of particular relevance: theories on transcendental cultural values and how they relate to our environmental behaviour, and research on the subjective well-being value of green spaces.

In terms of the former, research by Schwartz and associates (1999, 1992) developed a list of 56 key transcendental values that could be measured across a wide range of cultures. These include both ethical principles such as honesty and things that can be characterised as desirable end states, such as 'a varied life', 'family security', or 'mature love'. While cultures adhere to these values to different degrees, the set of values appears to follow a universal structure. For example, cultures that have stronger values relating to tradition also tend to have stronger values around security, and those who are willing to transcend their own interests for others are also more likely to have more strongly pro-environmental values. Further evidence suggests that these values do not directly influence contextual values and behaviour, but are mediated by worldviews, beliefs about what is at stake and who is responsible, and norms (Dietz et al., 2005; Stern, 2000). These theories and the psychometric tests and scales that are associated with them can help ecosystem services researchers better understand how cultural and transcendental values influence more specific contextual values for ecosystem services. They can also be used to bring rigour to the design of deliberative valuation processes and help evaluate them.

Psychological approaches to assessing the relation between the environment and subjective well-being also provide a useful means for evaluating the value of ecosystem services.

There is substantial evidence that proximity of green space and interaction with the natural environment provides benefits to well-being in a multi-faceted way, such as restoring mental well-being, improving cognitive function, providing opportunity for reflection, strengthening one's sense of identity, and providing aesthetic appreciation (e.g. Dallimer et al., 2012; Lindemann-Matthies et al., 2010; Wells, 2000).

Thus, in contrast to the mono-dimensional mainstream economic conceptualisation of welfare as preference-satisfaction, well-being is conceived of as multi-faceted. Typically, potential well-being indicators are assembled on the basis of qualitative work (e.g. interviews) and/or existing literature. Data is gathered using quantitative surveys, where factors are analysed statistically. Through multiple studies with different samples in different contexts, more or less generalizable instruments can then be created. Currently, first attempts are being made at developing such an instrument for assessing cultural services, as is illustrated by Box 22.1.

Choosing a method

Deliberative and non-monetary methods for valuing ecosystem services are wide-reaching and originate across a broad range of disciplines, including the arts and humanities, social psychology, sociology, anthropology, geography, development studies, and ecological economics. This means that epistemological paradigms (philosophies about how we can know things) and perceptions on what are legitimate and valid approaches to research and analysis are also diverse. Valuation methods are 'value-articulating institutions' (Vatn, 2009). In other words, the way we approach valuation impacts the type of values that are highlighted. Embracing values as a pluralistic concept means that, to comprehensively value ecosystem services, we need to embrace a diversity of methods to assess them. Both the marine and old growth forest case studies illustrated a mixed method approach, where different components of the methodology emphasised different types of values, and where participants emphasised the importance of different ecosystem services. For example, in the forest case study, journal interpretation more strongly brought out non-material cultural ecosystem service benefits than the focus group or rating sheets.

The choice of which methods to use depends on five key, interrelated things: (i) whether the proposed policy or management that is to be evaluated is likely to lead to significant conflict or contestation; (ii) the complexity of the system under consideration; (iii) the services under consideration and values one might expect to find; (iv) practical limitations; and (v) the stage of the policy cycle.

In terms of the first, key questions to ask are whether different stakeholders share the same framing of the issue in question. Are they likely to more or less agree on who, and which ecosystem services might be affected and how? Is there a common understanding on what the problem or goal is that the policy or management measure is aiming to address or achieve? Is the evidence around the effects of different measures agreed upon? The more strongly negative the answers to these questions, the more likely it is that the use of an instrumental, non-participatory or overly analytical approach will exacerbate resistance and conflict, and the more important it becomes to understand more precisely in what way different ecosystem services are valued and why. Here, interpretive approaches are particularly useful. In strongly contested contexts deliberation can also be of use, but high-quality process design and facilitation is crucial and will need to focus first on building trust and a degree of shared understanding before moving to valuation.

In terms of complexity, in relatively simple contexts quantitative assessments or desk-based assessments will probably suffice. For complex situations more elaborate analytical-deliberative approaches (e.g. participatory modelling) that focus on group-based or social learning, or approaches that incorporate expert knowledge (e.g. citizens' juries), can be particularly useful.

In terms of the services under consideration, it is clear that some services are more amenable to quantitative and analytical approaches than others. Cultural services are generally more idiosyncratic (i.e. they are expressed differently in different places), which advocates for a more interpretive approach. However, it is important to recognise that many provisional and regulating services also have implicit cultural aspects that may be more or less explicit. Many environmental values are subtle and require a deliberative learning process to surface (Kenter et al., 2011). In terms of value types, in non-contested cases it can be sufficient to focus on contextual individual values, while in more complex and contested cases it is important to understand the relations between different transcendental and contextual values, as these are central to understanding conflicts between frames. Here valuation processes need to be designed with value formation as well as elicitation in mind, which might involve combining different deliberative, analytical-deliberative, and interpretive techniques (for examples see Kenter et al., 2014).

Ultimately, however, the choice of method will often be determined by practical restrictions, such as the timescale, expertise, and resources available. Here monetary and quantitative survey-based approaches have a better reputation than qualitative and deliberative methods, though this is not always justified. There are a variety of relatively rapid and low-cost desk-based interpretive methods, such as media analysis and cultural history studies, and particularly at local scales workshop-based deliberative methods can be implemented more rapidly than quantitative valuation surveys, which are demanding in terms of statistical design and analysis. In complex cases, DMV workshops with 10–20 participants each can be more efficient than the use of individual contingent valuation interviews while delivering a potentially higher quality result.

Finally, certain methods are more suitable for different stages of the policy cycle: gathering ideas, surveying values, assessing policies, planning, delivering and managing, and evaluating. Table 22.2 provides an overview of which methods are most suitable for these different stages.

Table 22.2 Examples of methods suitable for different stages of the policy cycle.

Stage of policy cycle	*Potential tools/methods*
Ideas	Visioning
	Storytelling
Survey	Deliberative monetary valuation
	Participatory mapping/GIS
	Psychometric subjective well-being indicators
	Psychological values and beliefs surveys
	Values compass
	(Social) media analysis
	Desk-based cultural history study
	Storytelling
	Interviews
	In-depth discussion groups
	Deliberative opinion polls
	Review landscape character descriptions
	Existing datasets
Assess	Deliberative monetary valuation
	Deliberative multi-criteria analysis
	Citizens' juries
	Participatory (systems) modelling
Plan	Participatory budgeting
	Scenarios
	Visioning
	Visualisations
Deliver / manage	In-depth discussion groups
	Participatory mapping/GIS
	Participatory budgeting
	Review landscape character descriptions
Evaluate	As under 'assess'

Conclusions

This chapter has reviewed a range of deliberative, deliberative-analytical, psychological, and interpretive methods, illustrated by two mixed-method case studies. While many non-monetary and deliberative methods have only been used explicitly for valuing ecosystem services in a limited number of cases, most have been used successfully to evaluate policies and management in other contexts. Which methods to choose depends on whether the proposed policy or management that is to be evaluated is likely to lead to significant conflict or contestation, the complexity of the system under consideration, the services under consideration and values one might expect to find, and practical limitations. In complex and contested cases deliberative and non-monetary methods may be more appropriate to use than mainstream economic methods, because they can consider a broader spectrum of values and conceptions of well-being and may meet with less resistance. In many other cases they can be used alongside economic methods to provide a more comprehensive valuation, considering not just how much ecosystem services are worth but also what they mean to people.

Note

1 The innovative use of attributes in CV tasks made it possible to associate WTP with specific aspects of sites (in a similar way as in choice experiments); see Kenter et al. (2013).

References

Alvarez Farizo, B., and Hanley, N. (2006). Improving the process of valuing non-market benefits: combining citizens' juries with choice modelling. *Land Economics*, vol 82, p 465.

Alvarez Farizo, B., Hanley, N., Barberán, R., and Lázaro, A. (2007). Choice modeling at the market stall: Individual versus collective interest in environmental valuation. *Ecological Economics*, vol 60, pp 743–751.

Brown, G., and Kyttä, M, (2014). Key issues and research priorities for public participation GIS (PPGIS): A synthesis based on empirical research. *Applied Geography*, vol 46, pp 122–136.

Christie, M., Hanley, N., Warren, J., et al. (6 authors) (2006). Valuing the diversity of biodiversity. *Ecological Economics*, vol 58, pp 304–317.

Church, A., Burgess, J., Ravenscroft, N., et al. (14 authors) (2011). Cultural services. In: UK *National Ecosystem Assessment: Technical Report*. UNEP-WCMC, Cambridge UK.

Dallimer, M., Irvine, K. N., Skinner, A.M.J., et al. (9 authors) (2012). Biodiversity and the feel-good factor: understanding associations between self-reported human well-being and species richness. *Bioscience*, vol 62, pp 47–55.

Daniel, T. C., Muhar, A., Arnberger, A., et al. (21 authors) (2012). Contributions of cultural services to the ecosystem services agenda. *Proceedings of the National Academy of Sciences*, vol 109, pp 8812–8819.

Dietz, T., Fitzgerald, A., and Shwom, R. (2005). Environmental values. *Annual Review of Environment and Resources*, vol 30, pp 335–372.

Fish, R., Winter, M., Oliver, D., et al. (6 authors) (2013). Employing the citizens' jury technique to elicit reasoned public judgments about environmental risk: insights from an inquiry into the governance of microbial water pollution. *Journal of Environmental Planning and Management*, vol 57, pp 1–21.

Hockley, N. (2014). Cost-benefit analysis: a decision-support tool or a venue for contesting ecosystem knowledge? *Environmental Planning C*, vol 32, pp 283–300.

Kenter, J. O., Bryce, R., Davies, A., et al. (13 authors) (2013). The value of potential marine protected areas in the UK to divers and sea anglers. *UK National Ecosystem Assessment Interim Report*. UNEP-WCMC, Cambridge UK.

Kenter, J. O., Hyde, T., Christie, M., and Fazey, I. (2011). The importance of deliberation in valuing ecosystem services in developing countries – Evidence from the Solomon Islands. *Global Environmental Change*, vol 21, pp 505–521.

Kenter, J. O., O'Brien, L., Hockley, N., et al. (24 authors) (2015). What are shared and social values of ecosystems? *Ecological Economics*, vol 111, pp 86–99.

Kenter, J. O., Reed, M. S., Irvine, K. N., et al. (18 authors) (2014). *UK National Ecosystem Assessment follow-on phase. Work Package Report 6: Shared, Plural and Cultural Values of Ecosystems*. UNEP-WCMC, Cambridge UK.

Kumar, S. (2002). *Methods for Community Participation*. Practical Action Publishing, Rugby, Warwickshire.

Lawson, T. (2013). What is this "school" called neoclassical economics? *Cambridge Journal of Economics*, vol 37, pp 947–983.

Lindemann-Matthies, P., Junge, X., and Matthies, D. (2010). The influence of plant diversity on people's perception and aesthetic appreciation of grassland vegetation. *Biological Conservation*, vol 143, pp 195–202.

Lo, A.Y. (2013). Agreeing to pay under value disagreement: reconceptualizing preference transformation in terms of pluralism with evidence from small-group deliberations on climate change. *Ecological Economics*, vol 87, pp 84–94.

Lo, A.Y., and Spash, C. L. (2012). Deliberative monetary valuation: in search of a democratic and value plural approach to environmental policy. *Journal of Economic Surveys*, vol 27, pp 768–789.

Niemeyer, S. (2004). Deliberation in the wilderness: displacing symbolic politics. *Environmental Politics*, vol 13, pp 347–372.

O'Brien, E. A. (2003). Human values and their importance to the development of forestry policy in Britain: a literature review. *Forestry*, vol 76, pp 3–17.

O'Neill, J. (2007). *Markets, Deliberation And Environment*. Routledge, London and New York.

O'Neill, J., Holland, A., and Light, A. (2008). *Environmental Values*. Routledge, London.

Owen, R., Duinker, P., and Beckley, T. (2009). Capturing old-growth values for use in forest decision-making. *Environmental Management*, vol 43, pp 1–12.

Parks, S., and Gowdy, J. (2013). What have economists learned about valuing nature? A review essay. *Ecosystem Services*, vol 3, pp e1-e10.

Rackham, O. (2003). *The Illustrated History of the Countryside*. Weidenfeld & Nicholson, Cambridge UK.

Raymond, C. M., Kenter, J. O., Plieninger, T., and Turner, N. J. (2014). Comparing instrumental and deliberative paradigms underpinning the assessment of social values for cultural ecosystem services. *Ecological Economics*, vol 107, pp 145–156.

Reed, M. S., Graves, A., Dandy, N., et al. (9 authors) (2009). Who's in and why? A typology of stakeholder analysis methods for natural resource management. *Journal of Environmental Management*, vol 90, pp 1933–1949.

Robertson, I., and Richards, P. (2003). *Studying Cultural Landscapes*. Arnold, London.

Sagoff, M., (1998). Aggregation and deliberation in valuing environmental public goods: a look beyond contingent pricing. *Ecological Economics*, vol 24, pp 213–230.

Schwartz, S. H. (1990). Individualism-collectivism critique and proposed refinements. *Journal of Cross-Cultural Psychology*, vol 21, no 2, pp 139–157.

Schwartz, S. H. (1992). Universals in the content and structure of values: theoretical advances and empirical tests in 20 countries. *Advances in Experimental Social Psychology*, vol 25, pp 1–65.

Schwartz, S. H. (1999). A theory of cultural values and some implications for work. *Applied Psychology*, vol 48, pp 23–47.

Schwartz, S. H., and Bilsky, W. (1990). Toward a theory of the universal content and structure of values – Extensions and cross-cultural replications. *Journal of Personality and Social Psychology*, vol 58, pp 878–891.

Spash, C. L. (2007). Deliberative monetary valuation (DMV): issues in combining economic and political processes to value environmental change. *Ecological Economics*, vol 63, no 4, pp 690–699.

Spash, C. L. (2008). How much is that ecosystem in the window? The one with the bio-diverse trail. *Environmental Values*, vol 17, pp 259–284.

Stern, P. C. (2000). New environmental theories: toward a coherent theory of environmentally significant behavior. *Journal of Social Issues*, vol 56, pp 407–424.

Stern, P. C., and Fineberg, H. V. (eds) (1996). *Understanding Risk: Informing Decisions in a Democratic Society*. National Academy Press, Washington DC.

Varvasovszky, Z., and Brugha, R. (2000). How to do (or not do) a stakeholder analysis. *Health Policy and Planning*, vol 15, pp 338–345.

Vatn, A, (2009). An institutional analysis of methods for environmental appraisal. *Ecological Economics*, vol 68, pp 2207–2215.

Wells, N. M. (2000). At home with nature: effects of "greenness" on children's cognitive functioning. *Environment and Behavior*, vol 32, pp 775–795.

Zografos, C., and Howarth, R.B. (2010). Deliberative ecological economics for sustainability governance. *Sustainability*, 2010, pp 3399–3417.

23

THE 'BALANCE SHEET' APPROACH WITHIN ADAPTIVE MANAGEMENT FOR ECOSYSTEM SERVICES

R. Kerry Turner

Introduction: what is adaptive management?

Contemporary global concern regarding climate change, energy supply, food and water security, and the loss of ecosystems and their services, has made clear that a scientifically robust, pluralistic and policy-relevant understanding of how these issues are interrelated is essential if losses of human welfare are to be minimised. The concept of ecosystem services is one construct for understanding how changes to our natural environment impact our welfare (Boyd and Banzhaf, 2007; Fisher et al., 2009; Fisher et al., 2011) and, given that ecosystem services research may play a significant role in linking conservation activities and human welfare, systematic approaches to measuring, modelling and mapping ecosystem services (and their value to society) are urgently needed (Carpenter et al., 2006).

Three principles lie at the core of any systematic attempt to construct and deliver effective ecosystem services management and policy. They are: pluralism, pragmatism and precaution. Pluralism is necessary because the construction of a support process and tools for decision-making incorporating ecosystem services requires collaboration across scientific disciplines. In addition, there is a need to recognise the prevailing and growing diversity in social and cultural values, ethics and norms in contemporary society. These trends make the value of nature a multidimensional concept that is context-dependent. A degree of pragmatism is also needed if awareness of the ecosystem services concept is to be raised and emphasised in governance circles, especially finance ministries, and other sectors of society, including private business. To underpin this strategy the aim should be to manage ecosystems in a way that maintains or enhances their resilience and the valuable flow of services they provide, and not to argue solely for the maximisation of biodiversity conservation *per se* as a moral imperative. Therefore, the methodology adopted should encompass quantification as far as is feasible, with both monetary and non-monetary valuation of ecosystem services. In a world where resources are scarce and budgets limited, trade-off decisions and relative value judgements are inevitable and commonplace. Third, since there is scientific uncertainty about how some ecosystems and their services may be adversely affected by human development, a precautionary approach to decision-making is needed. There is an ever-present danger that thresholds may be breached as economic activity stresses some environmental systems to such an extent that irreversible changes are triggered, with consequent loss of ecosystem services. But many would argue that we cannot wait for more complete information, as this may result in services being further degraded or lost. Decisions should therefore take place within a risk-based framework using an adaptive management (AM) approach and practical decision tools, which emphasises flexibility and 'learning by doing'.

Adaptation is about deliberate action to adjust to risk, and involves decision-making. The AM approach to management recognises the uncertainties that currently shroud issues such as climate change and its future impacts and is based on the following set of guidelines:

- A recognition that people are an integral component of the ecology and functioning of ecosystems, and that a process of co-evolution between human society (and economy) and the environment has now become self-evident due to the scale and intensity of global development and trade;
- Some global change impacts (such as those predicted to follow climate change) require a pragmatic and precautionary adaptive response in advance of long-term mitigation tied to institutional change at the international governance level;
- Environmental management interventions need to be multifunctional rather than focused on single ecosystems/services, with the longer-term aim of understanding and managing 'landscape' level ecological processes and relevant socio-economic driving forces and pressures which reduce resilience;
- To quantify gains and losses from any given policy option choice, it is necessary to assign meaningful monetary values to some ecosystem services once translated into social welfare benefits, and to provide non-monetary evaluation of other (particularly cultural) service benefits;
- Because adaptation often results in winners and losers, there is an increasing need for novel forms of loss compensation (e.g. through habitat compensation/creation/restoration, renewable resource re-stocking or financial payments) and insurance; and
- That the role of the citizen, individuals and groups (social networks) are as important as central decision-making in highly contested environmental change contexts. The concept of acceptable, tolerable and intolerable risk, for example, is conditioned by institutional and social values and is therefore highly contested.

The AM process defines the location of the ecosystem being managed, the time period over which it may be impacted by environmental change and potential responses to change. The maximum feasible amount of knowledge then needs to be produced in a systematic way via models, futures scenarios, mapping, valuation and other methods, to characterise the ecosystem and services. Partnership working with key stakeholders will be an important feature of this management process and the setting of objectives. To meet the objectives, a mix of landscape-level and/or local-scale pilot interventions need to be actioned and monitored. Information sharing with stakeholders should be an on-going feature of the process. As the stock of management information increases, objectives should be reviewed and new measures deployed that are relevant to the new context. To judge the success or otherwise of adaptation options a decision support system (DSS) is needed, which is a combination of process, methods and techniques (tools). The DSS should be based on multiple criteria – efficiency, effectiveness, equity and legitimacy – to try to meet the challenge posed by diverse, contestable and sometimes subjective ecosystem knowledge (Adger et al., 2005).

According to Weiss (1999), there are at least four types of knowledge usage:

- Direct instrumental use of knowledge to inform decision-making;
- Indirect conceptual influence of knowledge on policy;
- Legitimisation of pre-existing policy lines; and
- Knowledge use sanctioned and imposed by hierarchy.

Dunlop (2014) builds on this typology to investigate ecosystem services knowledge and the barriers and opportunities for its use in policy. She distinguishes between epistemic learning, which is based on instrumental knowledge use, and reflexive learning, in which knowledge use is open-ended. Reflexive learning takes place in contexts characterised by uncertainty in which social networks use knowledge to enhance discussion and reconfigure received wisdom. It can also be the case that learning and knowledge are part of a bargaining process between interest groups and the formation of coalitions. Finally, a hierarchy may impose prescribed knowledge and use. The DSS described in the next section seeks to combine both epistemic and reflexive learning potential, and works from where we are now in policymaking circles in an iterative way forward.

The need for a new format for decision support systems

Current, more or less centralised, decision-making struggles to adequately cope with the discovery, transmission and aggregation of ecosystem knowledge (Turnpenny et al., 2014). In the past, too much emphasis has been placed on policy support processes and enabling tools, which are anchored to a rational mechanical worldview that only utilises reductive, quantitative metrics. Today there is increasing evidence of the policy influence that networked publics (social networks) can have, aided by the pervasive spread of networked technology. In some contexts, social networks can respond more quickly to, and with fresh insights into, contested policy issues and their often unintended consequences. Nevertheless, a robust DSS should still contain a strategic decision-making capability with its effectiveness dependent on the presence of both hierarchies and networks. The interplay between the hierarchy and the networks is important and needs to be continuously bolstered by transparent information feedback loops.

Cost benefit analysis (CBA) and related methods have been part of the public policy DSS toolbox for many decades, although their actual impact on policy decisions has been much debated (Ackerman et al., 1974; Hanley and Spash, 1994; Pearce, 1998; Hanley, 2001; Turner, 2007; Hockley, 2014). In the future, the distinction between process and individual technical tools will become less important. For some analysts, CBA should be seen not as a tool but as a venue where ecosystem knowledge is aggregated and contested (Hockley, 2014). In this chapter a different perspective is put forward, which retains CBA as one tool combined with others, which are then embedded in a more flexible and comprehensive evidence gathering, collating and interrogating process, labelled the Balance Sheet Approach (BSA) (Turner, 2007). The BSA is as much about investigating the policy issues and questions as it is about finding answers.

If CBA or related methods are to continue to play a role in the policy process, then a more explicit focus on distributional and fairness issues (i.e. who gains and who loses from environmental change and consequent policy responses) is required (see Jax, 2016; Sikor et al., 2016). The spread of costs and benefits across different affected individuals and groups in society needs to be accounted for and a weighting procedure applied. Project appraisals funded by economic development agencies have routinely included distributional weights, but this practice has not been commonplace in other public sector applications. As a minimum, the way in which the CBA 'accounts' are set out and formatted needs to change to incorporate and highlight financial transfers and the distributional impact of costs and benefits across stakeholders. A format, for example, which disaggregates the benefits and costs of a project or policy among stakeholders and records all inter-stakeholder financial transfers, can also serve to illuminate key issues such as the level of aggregation adopted and the project/policy accounting boundary. Changing the accounts format is a necessary first step, but Kristrom (2005) has gone further and proposed a 'hierarchy of options approach' in which explicit distributional weighting is applied, based on a rule that requires higher weights on all costs and benefits accruing to socially disadvantaged or below-average income groups. Alternatively, explicit distributional weights can be introduced to reflect the degree of inequality aversion present in society by examining past public policy decisions, or the prevailing marginal rates of income tax (Atkinson et al., 2000). We look at a particular way of formatting appraisal data and findings within the BSA below.

The balance sheet approach

Any decision support system that is put in place to assist in evaluating the gains and losses involved in adaptation to climate change and ecosystem services impacts will need to encompass a wide diversity of consequences and different stakeholder perspectives. The BSA (see Figure 23.1) comprises three complementary components (balance sheets) which can be seen as 'roughly comparable' sets of findings with overlaps and linkages. In using them, the aim would therefore be to determine the 'best' combination of data, methods and analysis, depending on the actual activity and context under appraisal (Turner, 2007). Conventional national/strategic policy appraisal relies heavily on standard economic and environmental impact analysis represented by sheet 1, but our environment, economy and society are all changing at an increasingly

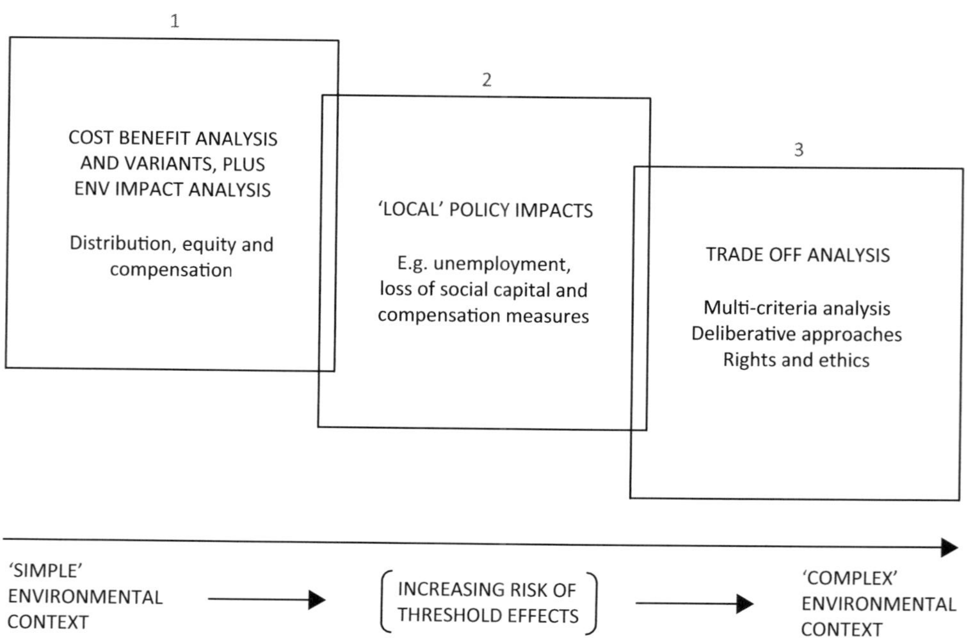

Figure 23.1 Balance sheet approach.

rapid rate, and in more complex ways. This may mean that a more comprehensive and spatially explicit appraisal process involving sheets 2 and 3 will be required.

The move across the figure from left to right, from one stylised polar position to another, illustrates the risk transition from relatively slow and relatively well-understood environmental change towards very dynamic and/or longer-term change that is more shrouded in uncertainty. The information in the BSA progressively encompasses more data and findings depending on the complexity of, and uncertainty around, the policy context under consideration. So, sheet 1 will need to contain evidence drawn from conventional economic and environmental impact analysis, but with added emphasis on equity and fairness considerations. The information on who gains and who loses in project/policy decision-making, and what, if any, compensation should be paid to recompense 'losers', also needs to be highlighted and included. Compensation can take a number of forms and is not restricted to financial payments. Some analysts have argued that individuals may not view money as adequate compensation for loss of certain environmental assets (natural capital), but may accept investments in social capital as an offset (Aldred, 2002; Turner, 2007). Empirical testing of this argument has recently given some support to the proposition that in certain circumstances social capital compensation may be acceptable (Lazaro-Touza and Atkinson, 2013). This distribution and equity focus then forms a key link to evidence presented in sheet 2, and in conceptual terms the DSS process begins to add reflexive learning and knowledge to the instrumental knowledge gathered in sheet 1.

Sheet 2 should contain the results of collecting and drilling into the information on the spatial and socio-economic characteristics of 'winners' and 'losers' down to regional and local scales and the implications for different policy contexts. A novel feature of this section of the evidence should be an up-front review of feasible compensation measures for the 'losers', rather than *ad hoc* responses to stakeholder reactions and political pressure after a decision has been announced. Using this sheet to interrogate regional and local project/policy impacts may reveal not just competing users for an ecosystem service(s) but also 'contesting' groups with profoundly different moral/ethical positions, attitudes to risk and cultural heritages – see Jax (2016). This will make the formulation of any overarching policy or delivery plan more difficult.

The evidence in sheet 3 should therefore specifically address these more 'contested' contexts. It will be drawn from the findings of multi-criteria analysis methods and group-based deliberative methods (ranging from deliberative monetary valuation exercises to cultural ecosystem mapping) which encourage discussion and debate (negotiation and arbitration) among relevant participants. This may or may not lead to a consensus about appropriate actions.

In the BSA, therefore, the three types of complementary assessments (balance sheets) try to give some guidelines for steering a reasonably objective course through these 'contested' policy contexts (see Figure 23.2):

1. Economic (monetary) CBA using a conventional economic efficiency criterion (macro-economy efficiency), but augmented with an assessment of the extent and condition of ecosystems linked to levels of ecosystem services (natural capital asset check (UK NEAFO, 2014)), and a distributional analysis of impacts and possible equity weighting;
2. Regional and local financial impacts and policy analysis, covering impacts like local unemployment, loss of community identity and related financial multiplier effects which often raise issues of compensation and insurance; and
3. Trade-off analysis to support dialogue and negotiations between contending groups, which is better suited to dealing with collective or shared values across wider society such as, for example, intrinsic value in biodiversity, cultural assets value etc. It is also necessary to remind ourselves that ecosystems underpin economies and provide life support systems and not *vice*

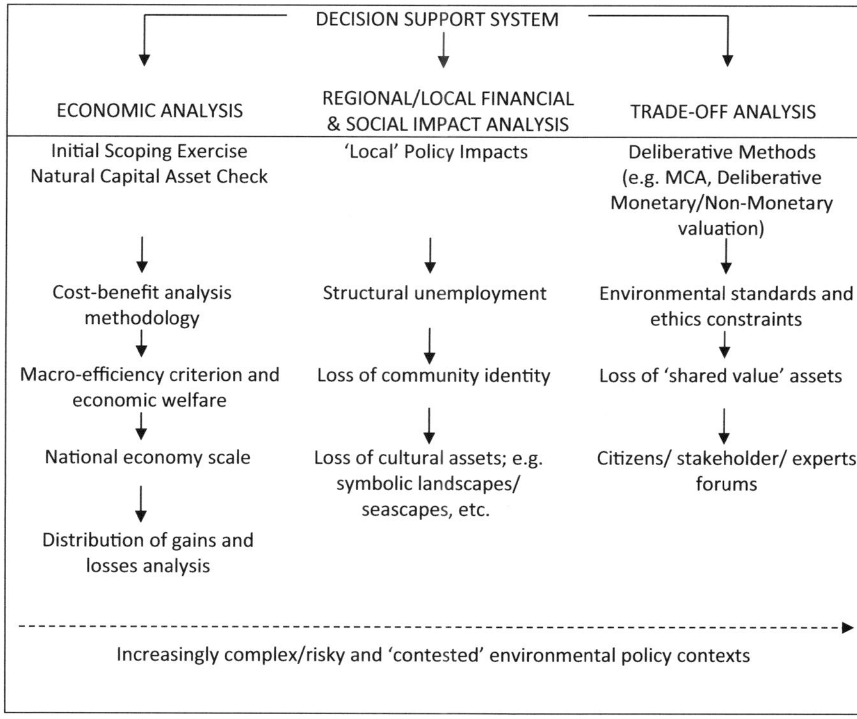

Figure 23.2 Policy Appraisal Balance Sheets.

> *versa*. If we accept a 'strong sustainability' goal, then a 'limits-based' approach with environmental standards is required to protect irreplaceable natural capital from threshold effects which can result in tipping points and system collapse, and to achieve intergenerational equity (Turner, 1999; Norgaard, 2010).

The analytical sequence of the balance sheet approach would begin with an economic cost-benefit and environmental scoping analysis, and then proceed to include the other balance sheets, depending on the context and the type of environmental issue under scrutiny. The aim would not be to aggregate the results of each balance sheet but to present the policy process with the set of linked findings in as transparent a way as possible.

Given the range of data that relates to terrestrial and marine ecosystems and related socio-economic activities, there is a pressing need to agree broad categories of data which can illuminate the economic, social and environmental dimensions of environmental change in specific contexts. The BSA aims to achieve this by separating out, in the first instance, economic data and analysis. So, in the first column of Figure 23.2, economic data is covered and is guided by the criterion of macro-economic efficiency/effectiveness (enabled via CBA or Cost Effectiveness Analysis) and informed by market-based data, willingness to pay (WTP) data and cost data (including second best data such as gross value added (GVA) etc.). A key link to the analysis in the second column in Figure 23.2 is provided at the bottom of the first column when the issue of the distribution of costs and benefits is raised, i.e. who gains and who loses from any change. The method thus includes certain elements of social appraisal in economic appraisal as well as questions of trust and legitimacy. The assessment in the second column of Figure 23.2 expands on the sort of data and issues that are best classified as social effects, with a spatial boundary (local/regional) condition imposed on the analysis. The assessment in the final column of Figure 23.2 continues the social analysis, but now encompasses values and impacts that are often expressed at local to national scales and conditioned by a variety of underlying ethical criteria (and sometimes moral imperatives). Clearly, the information in the columns overlap, but the aim is to define a logical decision support process which is trying to, a priori, scope and analyse real world (often 'wicked') economic and socio-political issues. To provide a simple illustration of the application of the BSA, we focus on coastal ecosystem services and management in the next section.

Coastal ecosystem services and coastal zone management in the UK

The global economy, driven by international finance and trade flows, now impacts almost all the world's ecosystems, not least coastal and marine, through a range of pressures and drivers. Coastal areas in particular are exposed to environmental change pressures because of their spatial location and attributes. They have therefore seen disproportionate increases in human population and economic and trade activities, with consequent natural habitat destruction and coastal hard engineering in response to actual and perceived increased flood and erosion risk. Coastal areas and shelf seas have been the repositories for both flow pollution, leading to eutrophication, and stock pollution through heavy metal contamination. The sources of this pollution have been in adjacent catchments, in estuarine locations or from direct discharges into the seas. Global trade has served to intensify shipping activities, with pollution impacts and the introduction of exotic invasive species into new areas through a combination of technological changes in hull design, ballast water discharges and probably climate change effects (Turner and Schaafsma, 2015). Offshore aggregates mining, increased navigation dredging, wind farm developments and aquaculture have also led to habitat and species loss.

This heavy, extensive and on-going utilisation of coastal and marine resources and the risks posed by climate change to marine and coastal resources and their dependent human communities ensure that management decisions will be contested by competing interests. The goal of a return to good (pristine) conditions is also unlikely to prove practicable, and so the DSS and social dialogue has to focus on the future and feasible future environmental system states.

Coasts provide a number of ecosystem services which generate goods and benefits, such as:

- Formation of dunes and marshland which can help with flood defence and reduce coastal erosion;
- Diverse habitats for wildlife;
- Wild and farmed food such as fish and shellfish;
- Recreation opportunities;
- Seascapes, which have cultural significance, promote well-being and support tourism; and
- Climate regulation service through the uptake of carbon.

The complexities and the non-commensurate values that characterise the real world political economy of 'contested' natural resource allocation and trade-offs are clearly illustrated in coastal zone policy. Adaptation to sea level rise and related climate change impacts highlight coastal protection and sea defence policy options with the 'wicked' characteristics common in many environmental management situations. The dynamic character of the coastal zone, in which areas of land and sea are constantly changing, is also a challenge. There are also potentially conflicting demands on the use of coastal resources to meet different human wants/needs, e.g.

nature conservation versus fishing. In the past, decision-makers considering a change in coastal policy at the strategic level have not always fully connected policymaking and policy delivery down to the local scale. In fact, coastal management needs to achieve a balance between strategic requirements and local schemes; because of the way coastal processes often work, schemes restricted to a given stretch of coast can nevertheless have much wider and sometimes negative consequences. One community's erosion control scheme, for example, can mean loss of beach frontage down the coast.

In the past decade, governments around the world, including the UK government, have sought to re-orientate policy in terms of future investments in coastal management away from a 'hold the line' philosophy and towards a more flexible approach based on cost benefit/cost effectiveness assessments (BSA sheet 1). The new adaptive management approach has included coastal realignment schemes in selected locations, and also a greater recognition of coastal processes such as erosion and subsequent beach replenishment. The realignment schemes can involve the removal of hard engineered sea defences such as sea walls and their replacement with more natural ('soft') defences such as saltmarshes or sand dune complexes. Individual schemes are traditionally appraised on a cost benefit basis and rank ordered in terms of their cost benefit ratio. The scheme rank order is then placed against the available resource budget and projects to be actioned are chosen. But the traditional DSSs and policy planning had not been sufficiently adjusted before the headline strategic policy shift (to more soft rather than hard defences) became widely publicised and stakeholder concerns were raised. Poor policy support sequencing (i.e. little or no recognition of what the BSA sheet 2 could provide) has meant that difficult 'local' policy impacts and controversies have been raised, and policymakers have been slow to respond. Thus, while the switch towards a more flexible coastal management regime can be justified on overall cost grounds and national strategic requirements, together with a precautionary approach to possible climate-related sea level rise and storm intensity and frequency predictions, the impacts and consequences on local communities and the distribution of the impacts across geographical scales and socio-economic groups have not been fully assessed. A hierarchical approach informed by instrumental knowledge does provide an adequate starting point for the policy appraisal process, given that budgets are always limited. But the distributional consequences should have been recognised in advance of the policy switch, and mitigation measures should have been in place, as well as a more targeted information and awareness campaign. Instead, the agencies involved have had to play 'catch up', following numerous stakeholder protests and campaigning over so-called shoreline management plans put forward by the Environment Agency in England, and wide press coverage. So, acceptable 'compensation' measures for the 'losers' in any given coastal scheme (and for that matter flooding risk situations more generally across catchments) have only slowly emerged as controversy has escalated. The pathfinder scheme trialled in East Anglia (UK), for example, has examined a number of compensation measures for householders affected by coastal erosion (Turner and Welters, 2014). It has also been pointed out that the shifting responsibility between state and citizen in adaptation contexts can potentially create vulnerabilities and there is thus a renewed interest in the role that the insurance industry may play as responsibility for risk management changes (Adger et al., 2013). Under the BSA, the distributional impacts and 'local' impacts would have been diagnosed prior to the strategic policy switch, and arguably more effective ameliorative measures would have been put in place.

In contexts where especially prized assets are at risk (e.g. designated nature reserves, symbolic landscapes or buildings or even small communities with a particular cultural heritage) compensation may not be appropriate or feasible. Particularly difficult trade-off choices are posed in such circumstances and decision-making can be assisted through BSA sheet 3, to at least open a social dialogue and highlight moral imperatives held by some stakeholder interests, and therefore to contribute to reflexive learning.

Conclusions

The balance sheet process and tools offer a different way to appraise policy in terms of collating interrogating and presenting evidence. It seeks to generate as robust an evidence base as is feasible in different policy contexts, including ecosystem services and their future maintenance. For some critics it still does not go far enough if a strong sustainability development path is an overriding imperative and the risk of threshold effects should be minimised (Norgaard, 2010). Its reliance on partial equilibrium analysis, some critics would argue, prevents it from playing as full a role as it should, and governance, institutional and ethical reforms are required if sustainable development goals are to be approached (Norgaard, 2010; Baveye et al., 2013). The partial equilibrium position assumes that macro-level limits in terms of pollution or resource depletion are not breached but carry no guarantee that this is the case. The position taken in the balance sheet approach is that system-wide constraints to help avoid thresholds and tipping points are necessary, and that precaution should be the watchword given the uncertainties that currently prevail in our understanding of the consequences of climate change and other environmental change pressures and drivers. However, this does not have to mean complete abandonment of monetary valuation of some parts of nature and ecosystem services thinking (Baveye et al., 2013). Indeed, in practice, the constraints themselves have to be subject to cost effectiveness tests as well as political and ethical scrutiny.

References

Ackerman, B. A., Ackerman, S. R., Sawyer, J. W., and Henderson, D. W. (1974). *The Uncertain Search for Environmental Quality*. The Free Press, New York.

Adger, W.N., Arnell, N.W., and Tompkins, E. L. (2005). Successful adaptation to climate change across scales. *Global Environmental Change*, vol 15, pp 17–86.

Adger, W.N. et al. (5 authors) (2013). Changing social contexts in climate change adaptation', *Nature Climate Change*, vol. 3, pp 330–333.

Aldred, J. (2002). Cost-benefit analysis, incommensurability and rough equality. *Environmental Values*, vol 11, pp 27–46.

Atkinson, G., Machado F., and Mourato, S. (2000). Balancing competing principles of environmental equity. *Environment and Planning A*, vol 32, pp 1791–1806.

Baveye, P.C., Boveye, J., and Gowdy, J. (2013). Monetary valuation of ecosystem services: it matters to get the timeline right. *Ecological Economics*, vol 95, pp 231–235.

Boyd, J., and Banzhaf, S. (2007). What are ecosystem services? *Ecological Economics*, vol 63, pp 616–626.

Carpenter, S. R, Defries, R., Dietz, T., et al. (7 authors) (2006). Millennium Ecosystem Assessment: research needs. *Science*, vol 314, pp 257–258.

Dunlop, C.A. (2014). The possible experts: how epistemic communities negotiate barriers to knowledge use in ecosystem services policy. *Environment and Planning C: Government and Policy*, vol 32, pp 208–228.

Fisher, B., Turner, R.K., Burgess, N.D., et al. (22 authors) (2011). Measuring, modelling and mapping ecosystem services in the Eastern Arc Mountains of Tanzania. *Progress in Physical Geography*, vol 35, pp 595–611.

Fisher, B., Turner, R.K., and Morling, P. (2009). Defining and classifying ecosystem services for decision making. *Ecological Economics*, vol 68, pp 643–653.

Hanley, N. (2001). Cost-benefit analysis and environmental policy making. *Environment and Planning C*, vol 19, pp 103–118.

Hanley, N., and Spash, C. (1994). *Cost-Benefit Analysis and the Environment*. Edward Elgar, Aldershot.

Hockley, N. (2014). CBA: a decision-support tool or a venue for contesting ecosystem knowledge? *Environment and Planning C Government and Policy*, vol 32, pp 283–300.

Jax, K. (2016). Ecosystem services and ethics. In: Potschin, M., Haines-Young, R., Fish, R. and Turner, R.K. (eds) *Routledge Handbook of Ecosystem Services*. Routledge, London and New York, pp 42–44.

Kristrom, B. (2005). Framework for assessing the distribution of financial effects of environmental policies. In: Serret, Y. and Johnston, N. (eds) *The Distributional Effects of Environmental Policy*. Edward Elgar, Cheltenham.

Lazaro-Touza, L., and Atkinson, G. (2013). Nature, roads or hospitals? An empirical evaluation of "sustainable development preferences". *Ecological Economics*, vol 95, pp 63–72.

Norgaard, R.B. (2010). Ecosystem services: from eye-opening metaphor to complexity blinder. *Ecological Economics*, vol 69, pp 1219–1227.

Pearce, D.W. (1998). Environmental appraisal and environmental policy in the European Union. *Environmental and Resource Economics*, vol 11, pp 489–501.

Sikor, T., Martin, A., Fisher, J., and He, J. (2016). Ecosystem services and justice. In: Potschin, M., Haines-Young, R., Fish, R. and Turner, R.K. (eds) *Routledge Handbook of Ecosystem Services*. Routledge, London and New York, pp 299–301.

Turner, R.K. (1999). The place of values in environmental valuation. In: Bateman, I.J. and Willis, G.K. (eds) *Valuing Environmental Preferences*. Oxford University Press, Oxford.

Turner, R.K. (2007). Limits to CBA in UK and European environmental policy: retrospects and future prospects. *Environmental and Resource Economics*, vol 37, pp 253–269.

Turner, R.K., and Schaafsma, M. (eds) (2015). *Coastal Zones Ecosystem Services: From Science to Values and Decision Making*. Springer Ecological Economics Series, Berlin, forthcoming.

Turner, R.K., and Welters, R. (2014). *New Tools to Improve the Management of the Coastal Environment*. LEWC Policy and Practice Policy and Practice Notes No.10, Swindon.

Turnpenny, J., Russel, D., and Jordan, A. (2014). The challenge of embedding an ecosystem services approach: patterns of knowledge utilisation in public policy appraisal. *Environment and Planning C: Government and Policy*, vol 32, pp 247–262.

UK National Ecosystem Assessment Follow-On (2014). *The UKNEAFO: Synthesis of the Key Findings*. UNEP-WCMC, LWC.

Weiss, C. (1999). The interface between evaluation and policy. *Evaluation*, vol 5, pp 468–486.

Briefing Note 23.1 Ecosystem services and justice

Thomas Sikor, Adrian Martin, Janet Fisher and Jun He

There are strong reasons for ecosystem services researchers to take a look at justice. Justice is about what is morally right, being 'the first virtue of social institutions' according to John Rawls. Ecosystem services governance is amenable to justice analysis because it sets up relationships between stakeholders, such as between the providers and users of ecosystem services. Governance affects the distribution of benefits and responsibilities among stakeholders and requires collective decisions about management objectives and the means to achieve these.

More specifically, justice serves in the analysis of the resolution of the trade-offs highlighted in ecosystem services research. This is important from a social perspective, since governance should be not only effective and efficient but also just. It is equally imperative from an ecological perspective, because people's environmental behaviour depends on how they perceive the legitimacy and fairness of ecosystem services governance.

Analysis of ecosystem services governance benefits from taking an empirical approach that can capture the plurality of justice, because more often than not, stakeholders do not agree on a single definition of what is morally right. Another key premise is that notions of justice are contextual and experiential, as they depend on the particular political and historical setting. Empirical approaches complement the normative approaches known from environmental ethics and political philosophy, which tend to start from a particular moral position and then derive certain ethical principles in a deductive manner.

Empirical justice analysis involves two lines of enquiry. The first requires identification of dominant notions of justice that accompany ecosystem services governance, examination of how appropriate these are in the contexts where ecosystem management takes place, and analysis of how these notions are operated in practice. Establishing the interplay between notions, contexts and actual practice is crucial because the moral implications of ecosystem services governance become apparent only if actual practices of governance are studied within particular setting. The second involves characterization of different stakeholders' notions of justice in particular contexts, exploration of their justification in public discourse, and the determination of how some of these notions find or lose support in public discourse. The study of justifications employed in public discourse helps to distinguish between mere statements of self-interest and notions of justice that are argued to possess wider relevance. The underlying premise is that, in reality, there are often tensions between dominant notions of justice and those of local stakeholders, and that interactions between these ideas of justice are a major social dynamic influencing the nature and outcomes of ecosystem services governance.

A conceptual framework helps to characterize notions of justice, organized around three elements (see Diagramme 23.1):

- *Dimensions* refer to how notions emphasize concerns about distribution, participation or recognition. Distribution is about the assignment of rights to resources and responsibilities among stakeholders. Participation, or procedural justice, directs attention to the roles of different stakeholders in decision-making. Recognition is about acknowledging people's distinct identities and histories.
- *Subjects* are the kinds of stakeholders considered to possess rights or bear responsibilities, have a role in decision-making, deserve recognition or are deserving of care from other stakeholders. Common examples in conservation are the local poor, entire local populations, indigenous peoples, global society, future generations, non-human organisms and nature.

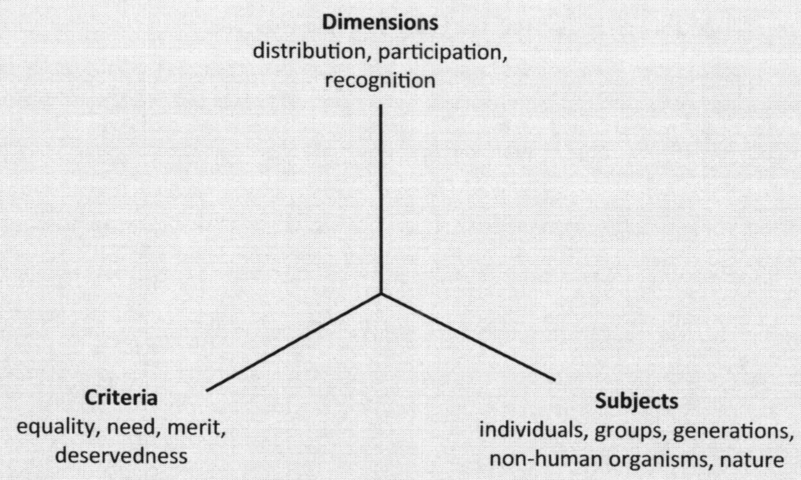

Diagramme 23.1 A simple conceptual framework for characterizing notions of justice

Source: the authors, based on Sikor *et al.* (2014)

- *Criteria* are decision-making guidelines that organize the relationship between subjects for particular dimensions of justice. Typical criteria are equality, need, merit and deservedness.

Ecosystem services governance, in its theoretically ideal form, tends to set up the relationships between stakeholders according to a distinct model, which assigns management duties and rewards according to relationships between providers and users of ecosystem services. Stakeholders are various kinds of people and social groups within a particular locality or country, or even worldwide. Providers and users are assumed to interact with each other on the basis of marginal utility, in which ecosystem services can be secured where at least one user values the service higher than at least one provider expects to receive. This is the ideal model in the sense of a guiding image, even though actual governance practices often depart from it significantly.

There are various Payments for Ecosystem Services projects and programs around the world that illustrate the basic tenets of this model despite variations in their specific designs. For example, an experimental scheme in Rwanda provided financial incentives to reduce human activity in the Nyungwe National Park (Martin *et al.*, 2014). Local people received individual payments distributed equally among households and collective payments for the provision of public goods. In return, they had to refrain from hunting, tree-cutting and mining in the park in order to help provide ecosystem services important to global society.

The ideal form of ecosystem services governance is associated with particular, dominant notions of justice. It tends to prioritize the distributive dimension, in the form of the assignment of duties, rewards, benefits and contributions; it emphasizes contemporary individual users and providers of ecosystem services as the primary subjects of justice (as opposed to particular social groups or future generations); and it employs the criterion of merit, defined as marginal utility, for organizing the relationship between these users and providers. In the ideal form, the overall management objective, the duties and rewards of providers, and the benefits and contributions of users are all defined on the basis of marginal utility, e.g. the compensation that the market determines as sufficient to motivate the provision of certain services.

These dominant notions are compatible with utilitarian theories of distributive justice as long as they improve overall human well-being. A central branch of utilitarianism considers social actions to be just if the generated welfare gains outweigh any connected losses. Such actions are morally right even if they make some stakeholders worse off. Ecosystem services governance in its ideal form meets this moral requirement because voluntary interactions between providers and users lead to the maximization of aggregate utility.

Nevertheless, the appropriateness of utilitarian theories of distributive justice depends on certain preconditions, in particular the absence of significant economic inequalities or power asymmetries leading to coercive market conditions, and general consent to the primacy of utilitarian theories of distributive justice. Such preconditions are often not present in the Global South. The limited appropriateness of utilitarian notions is reflected in the rising popularity of rights-based approaches to conservation. Rights-based approaches in their ideal forms derive from other notions of justice that emphasize meeting a minimum threshold for affected people.

Further readings

Chan, K.M.A., and Satterfield, T. (2013). Justice, Equity, and Biodiversity. In: Levin, S. A. (ed.) *The Encyclopedia of Biodiversity*. Elsevier, Oxford.

He, J., and Sikor, T. (2014). Notions of justice in payments for ecosystem services: insights from China's Sloping Land Conversion Program in Yunnan Province. *Land Use Policy*, vol 53, pp 207–216.

Martin, A., Gross-Camp, N., Kebede, B., McGuire, S. and Munyarukaza, J. (2014). Whose environmental justice? Exploring local and global perspectives in a payments for ecosystem services scheme in Rwanda. *Geoforum*, vol 54, pp 167–177.

Sikor, T. (ed.) (2013). *The Justices and Injustices of Ecosystem Services*. Earthscan, London.

Sikor, T., Martin, A., Fisher, J., and Jun, H. (2014). Towards an empirical analysis of justice in ecosystem governance. *Conservation Letters*, vol 7, pp 524–532.

Briefing Note 23.2 Ecosystem services and ethics

Kurt Jax

"Ecosystem services" is not a neutral term. In all of its definitions it relates to beneficial relations of nature (or ecosystems) to humans. It thus already addresses a value dimension, and carries with it the notion of being desirable. So it is no surprise that the application of the ecosystem services concept entails morally relevant issues. Ethics, as a theory of morality, deals with the ways humans interact with each other and – more recently, with the rise of environmental ethics – how humans interact with non-human nature (O'Neill et al., 2008). The main questions an ethical perspective asks of ecosystem services are:

- which values of nature (and in consequence, which objects) are considered by the ecosystem services concept and which are excluded?

- who has a voice in deciding of what constitutes an ecosystem service?
- who benefits from the use of ecosystem services and who carries the costs?

These questions thus relate not only to the kind of values of nature embraced by people, but also to issues of justice and equity.

Many discussions and concerns about the ecosystem services concept relate to the first question. The ecosystem services concept clearly deals with nature within a utilitarian framework, i.e. one that explicitly aims at the use(fulness) of nature for humans in terms of fostering human well-being. In principle, the types of values of nature for humans to which the ecosystem services concept can relate range from purely instrumental ones (nature as a pure means), via fundamental values (basic conditions for existence and life) through eudaimonistic values, the latter describing a non-instrumental approach to nature, parts of it being valued in themselves as a condition for a good, flourishing human life (Muraca, 2011; Jax et al., 2013). What is not covered by ecosystem services, however, are intrinsic values, understood as values for nature, independent of any human interests (Davidson, 2013). While thus most values of nature embraced by people may be within the scope of an ecosystem services perspective, there is a danger that values here may be perceived much more restricted, namely in purely economic and even monetary terms, leading to a "commodification" of nature, i.e. the transformation of ecosystem components or processes into products or services that can be privately appropriated, assigned exchange values and traded in markets (Gómez-Baggethun and Ruiz-Pérez, 2011), which counteracts the intuitions and values of many conservation-minded people. Bringing all values down to a common evaluation measure like money also overly simplifies the different qualities of value types that may not be easily traded off or substituted against each other. Beyond this, some items (e.g. some species) that are valued by those promoting biodiversity conservation may become invisible or excluded from consideration because they are beyond any utilitarian framework.

The values embraced in ecosystem services protection and in biodiversity conservation certainly have a high overlap but do not completely coincide. The question of what constitutes an ecosystem service is not given by nature as such but is always determined by societal choices. A water plant with mass production in a eutrophic lake may be considered a disservice because it impedes many uses of the lake, but when the same plant is used for bioenergy purposes, it can be characterised as a service (Escobar et al., 2011). The question who decides about which parts of nature are conceptualised and managed as ecosystem services thus becomes also one of whose voice is heard and whose values are being applied. To leave it simply up to scientists and politicians, without including other stakeholders, can violate the interests of many social groups.

Finally, questions of justice can arise with the application of the ecosystem services concept. The distributions of benefits and costs derived from the provision and use of ecosystem services can be distributed very unevenly among different regions and/or social groups, e.g. when the production of energy crops imported to highly industrialised countries threatens the livelihood of people elsewhere. The distribution of benefits and costs associated with the provision of ecosystem services is important across all scales, both spatial (e.g. global gain, local loss or vice versa) and temporal (use by present generation or options for future generations).

The degree to which the ethical issues described above become relevant depends crucially on the ways in which the ecosystem services concept is defined and used. If the concept is used more in a didactic manner, i.e. to generally raise awareness for the value of nature beyond direct market values, few of these problems will be pertinent. If ecosystem services, however, becomes the basis of planning and management, the various values and valuation languages should be respected and possible conflicts and trade-offs be made explicit and mediated (see Luck et al., 2012, Jax et al., 2013).

References

Davidson, M.D. (2013). On the relation between ecosystem services, intrinsic value, existence value and economic valuation. *Ecological Economics*, vol 95, pp 171–177.

Escobar, M.M., Voyevoda, M., Fühner, C., and Zehnsdorf, A. (2011). Potential uses of Elodea nuttallii-harvested biomass. *Energy, Sustainability and Society*, vol 1, p 4. Available at: http://www.energsustainsoc.com/content/1/1/4

Gómez-Baggethun, E., and Ruiz-Pérez, M. (2011). Economic valuation and the commodification of ecosystem services. *Progress in Physical Geography*, vol 35, pp 613–628.

Jax, K., Barton, D.N., Chan, K.M.A., et al. (26 authors) (2013). Ecosystem services and ethics. *Ecological Economics*, vol 93, pp 260–268.

Luck, G.W., Chan, K.M.A., Eser, U., et al. (7 authors) (2012). Ethical considerations in on-ground applications of the ecosystem services concept. *BioScience*, vol 62, pp 1020–1029.

Muraca, B. (2011). The map of moral significance: a new axiological matrix for environmental ethics. *Environmental Values*, vol 20, pp 375–396.

O'Neill, J., Holland, A., and Light, A. (2008). *Environmental Values*. Routledge, London.

24

SECURING NATURE AND PEOPLE

Can we replicate and scale success?

Gretchen C. Daily

There are three pathways to wisdom.
The first is through contemplation, and that is the noblest.
The second is through imitation, and that is the easiest.
The third is through experience, and that is the bitterest.

Confucius

Introduction

Globally, 'natural capital' now appears in society's thinking about agriculture, water, energy, health, fisheries, forestry, mining, cities, and the infrastructure supporting these and other vast sectors – and it is increasingly evident in the ways communities, corporations, governments, and other institutions frame decisions (e.g., Balmford et al., 2002; Li et al., 2005; Ouyang, 2007; MEP and CAS, 2008; Kiesecker et al., 2010; Rapidel et al., 2011; Goldstein et al., 2012; Levy et al., 2012). There is growing demand for rigorous approaches that integrate the values of natural capital into major development decisions in order to reverse the degradation of Earth's life-support systems and enhance human well-being.

How can we help channel and magnify the energy of this movement into large-scale, durable, improved outcomes for nature and people? Clearly a wide range of strategies is needed, as featured in this *Handbook*. I focus my brief remarks on meeting and cultivating further demand from decision-makers now. Including the value of ecosystem services in the decisions of governments, corporations, traditional cultures, and individuals is designed to broaden our understanding of the roles Nature plays in our lives and the reasons – including the moral imperative – for conserving it (e.g., Leopold, 1949; Norton, 1987; Ehrenfeld, 1988; Rolston, 2000; Daily and Ellison, 2002; Dasgupta, 2010).

A strategy

Our urgent challenge today is to move from ideas to action on a broad scale (Carpenter et al., 2006, 2009; Daily et al., 2009; Perrings et al., 2011). Doing so requires understanding the production of services from landscapes and seascapes, together with their biodiversity and human activities; quantifying service flows, from producers to beneficiaries, across space, social class, economic sector, and time; understanding the decision-making processes of individuals, communities, governments, and corporations; integrating research with institutional design and policy implementation; and crafting policy interventions with flexibility for learning and improvement.

Each of these alone is a complex and difficult task. Together they form a daunting but critically important agenda for collaboration, pointed towards four major advances that will help transform how businesses, governments, and individuals interact with nature:

1 *fundamental understanding* of ecosystem services, human well-being, and their inter-linkages, in biophysical, economic, social, institutional, and governance domains;
2 *practical, science-based tools and approaches*, co-developed with users and well-tailored for application in highest-leverage decision contexts;
3 *real-world demonstrations* that implement new approaches in diverse sites and sectors globally, with potential for replication and scaling; and
4 *engaging leaders and building capacity* to achieve transformative and lasting change.

All of this work is accelerating in institutions globally, and involves greatly increasing the interaction between researchers and real-world generators and users of knowledge. Key are tools that help decision-makers visualize the impacts of potential policies – identifying trade-offs and compatibilities between environmental, economic, and social benefits – by modelling and mapping the delivery, distribution, and economic value of ecosystem services under alternative scenarios (Tallis et al., 2011). Many innovative partnerships are now forming around the world to advance this agenda. Demonstrations range from informing the infrastructure development strategies of major development banks and private investors in transportation and other key sectors to working with indigenous communities in strategic planning of land and ocean resource uses to balance conflicting values, to engaging corporations to quantify the risks and opportunities of alternative sourcing options along key supply chains (Ruckelshaus et al., 2013; Kenter et al., 2014).

We need to go much further to engage leaders, build capacity, and magnify the impact of these successes in order to forge and accelerate lasting, transformative change. The focus is on developing tools for mainstreaming natural capital systemically, across high-leverage decision contexts. There remains a long way to go, but some early demonstrations are inspiring.

Real-world demonstrations

Beginning in the late 1990s, larger-scale investment in natural capital for water flow regulation in China – and for a broad suite of ecosystem services in Costa Rica – set pioneering examples that are now being adapted elsewhere and scaled up. Today, China and Latin America, broadly, offer highly innovative models. I briefly present two (of many) cases, operating at different scales and in different kinds of social-ecological systems. In each, there is an acute or looming crisis, innovative leadership, and an experiment underway in pursuit of dual goals: securing natural capital and human well-being.

China's land-use and human development planning

The ecosystem service investments being made in China today are breath-taking in their goals, scale, duration, and innovation. Following massive droughts and flooding in 1997–98, China implemented several national forestry and conservation initiatives, into which investments exceeded 700 billion yuan (about USD100 billion) over 2000–2010 (Zhang et al., 2000; Liu et al., 2008). The larger and older of these initiatives, the Sloping Land Conversion Program (SLCP), involves 120 million farmers directly and is being rigorously evaluated to improve its design and efficacy. These initiatives have goals of securing critical natural capital through targeted investments across landscapes, and alleviating poverty through targeted wealth transfers from coastal provinces to inland regions where many ecosystem services originate (Daily et al., 2013).

China also stands out in strengthening the scientific foundation supporting these public policies. This is illustrated in the development of a first National Ecosystem Assessment, spanning a wide range of ecosystems, services, and spatial scales, over the past decade (2000–2010). Perhaps most ambitiously, China is establishing a new network of 'ecosystem function conservation areas' (EFCAs) (MEP and CAS, 2008). In 2008, the network comprised 50 zones spanning 27% of the country; in 2015 it grew to 58 zones spanning 49% of the country, as shown in Figure 24.1.

EFCAs are a way of zoning land so as to focus conservation and restoration in places with highest return-on-investment for public benefit in order to halt and reverse degradation of vital ecosystems and their services. The zoning is also meant to focus high-impact human activities in places where they will do least damage. EFCAs are also a way of focusing poverty alleviation efforts in places where the stakes are highest, both for local residents and for beneficiaries of ecosystem services living farther away. Implementing EFCAs involves compensation mechanisms, whereby regional beneficiaries – of, for example, water purification and sand storm control for

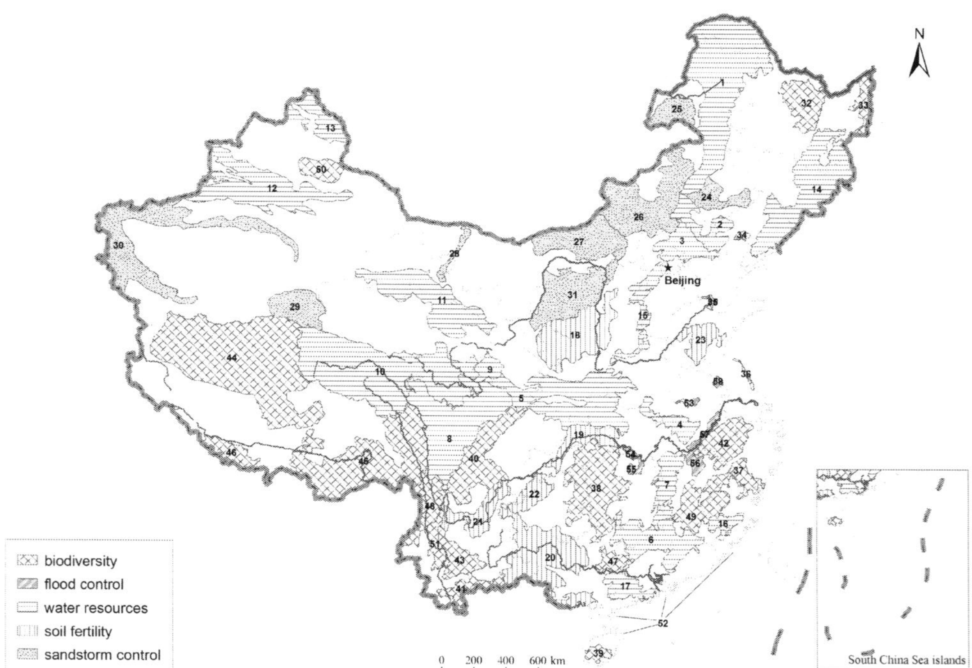

Figure 24.1 China's new system of Ecosystem Function Conservation Areas (EFCAs).

As delineated by the Ministry of Environmental Protection and the Chinese Academy of Sciences, EFCAs span 49% of China's land area in 58 zones. EFCAs have dual goals of securing biodiversity and ecosystem services and alleviating poverty (Ouyang, 2007; Ehrlich et al. 2012). They have been delimited using InVEST and related modelling. Figure courtesy of H. Zheng and Z. Ouyang, Research Center for Eco-Environmental Sciences, Chinese Academy of Sciences

Beijing – invest in the transformation to more sustainable livelihoods and durable improvements in well-being among the landholders producing the services (e.g., Zheng et al., 2013).

While these initiatives represent a massive scientific and policy undertaking, they are very young and there is still little understanding of the local costs of implementation, or their effects on poor and vulnerable populations. The EFCA model represents a new paradigm for integrating conservation and human development, but for this policy innovation to have wide applicability and success, it will be important to assess and improve local livelihoods (e.g., Li et al., 2011; Li et al., 2012; Liang et al., 2012).

Pushing the frontier further, in March 2014 the central government approved development of a new metric – Gross Ecosystem Product – to be implemented highly visibly, alongside Gross Domestic Product, at all levels of government. The potential impacts of ecosystem service investments in China are enormous, both within the country and globally. As a matter of national security, national leaders now speak of 'China's dream' as becoming 'the ecological civilization of the 21st Century.'

Water for cities in Latin America and beyond

In the mid-1990s, New York City made one of the first and most famous investments in ecosystem service provision in recent history, to secure drinking water for the 10 million people living and working in the city (Daily and Ellison, 2002). Today, with rapidly growing urban populations, expanding natural resource extraction in upstream watersheds, and climate change, water security for cities is a growing concern globally.

Water funds are a finance mechanism through which downstream water consumers and other parties (e.g., conservation and human development organizations, public entities) pay for upstream changes in land cover and use. The objectives typically include maintenance or enhancement of: water quality; regular water flows (for dry-season supply and flood control); groundwater recharge; terrestrial and aquatic biodiversity; and livelihoods and well-being in upstream human communities. Other services are also anticipated, such as carbon storage and sequestration, crop pollination, and pest control (Goldman-Benner et al., 2012).

There is now a major effort underway, supported by the Nature Conservancy, the World Bank, the Inter-American Development Bank, and FEMSA Foundation, to replicate and standardize these funds in terms of design, implementation, and monitoring, across more than 40 major cities throughout Latin America. Analysis is focused on predicting the relative feasibility and payoff of potential land-use / cover changes, such as conservation and restoration in headwaters, on steep slopes, and in riparian areas; and shifting to more sustainable pastoral and cropping practices. The modelling software InVEST has been tailored for use in this decision context, to predict where and which potential activities would yield the greatest societal return-on-investment. Investments are then targeted across landscapes to yield the highest return, subject to stakeholder preferences for security and equity.

Since the official establishment of the first water fund in Quito, Ecuador, in 2006, the model has spread rapidly (Figure 24.2). The Nature Conservancy is now exploring establishing the first funds in Africa. The effort is focused on developing standards – in biophysical modelling, financing, governance, and monitoring – that can be sensibly tailored to contrasting conditions, but that guide the process and incorporate lessons from elsewhere.

Water funds are an inspiring example of rapid advances in all areas of the strategy laid out above. Today they are still at an early and vulnerable stage. With strong commitment on the part of key stakeholders, one could envision the emergence of flexible yet durable institutions that help guide growth of cities and management of the natural capital they depend on, more broadly.

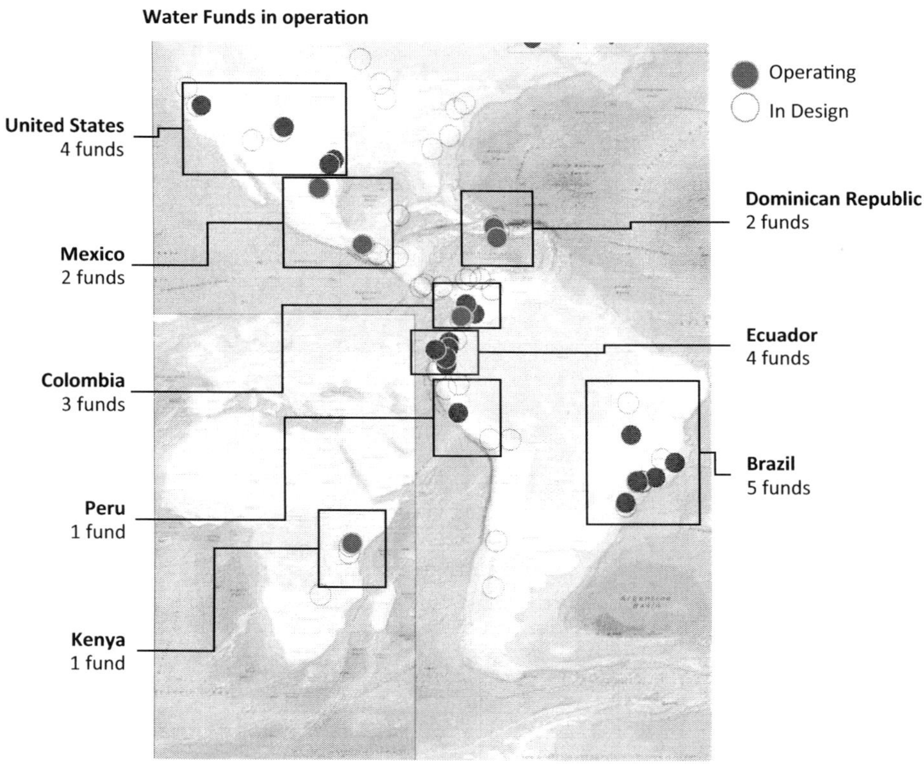

Figure 24.2 Water funds featured by stage of development.

Filled circles indicate that the water fund has been created, with a legal agreement among parties. Open circles designate water funds that are in the process of creation; four more (not depicted) are underway in Latin America since 2014. Figure courtesy of the Latin American Water Funds Partnership Dashboard (Nov 2014) and The Nature Conservancy, Internal Survey of Water Funds (Dec 2014).

Conclusion

Mainstreaming natural capital into decisions is a long-term game plan, requiring co-evolving advances in knowledge, social institutions, and culture. The contributions of many people and places are needed to achieve this. First, governments, businesses, and individuals must find it easy to inculcate ecosystem services and natural capital into their decisions, and the methods for doing so must be transparent, credible, and predictable. In many cases, sectors of society are open to the concepts of ecosystem services and natural capital, but simply do not know how to use them in a tangible way. Second, there need to be examples of projects or enterprises that – as a result of properly valuing ecosystem services and natural capital – end up with improved decisions, institutions, and human well-being. These examples both test our knowledge against real-world problems and produce compelling stories of how an ecosystem services approach made a difference. Lastly, these examples of success must have visibility and charisma, to draw political and thought leaders to them and thereby trigger broader awareness and action.

As daunting as this strategy is, much is within striking distance. The environmental and human development movement has a much more diverse and powerful community behind it now than ever before. Co-development of knowledge with knowledge users is beginning to provide tools and approaches that are rigorous yet practical. There are enough policy experiments underway that compelling demonstrations are already forthcoming – the first step in an iterative process between basic science and application to real-world problems. Science by itself cannot change the world, but science plus the vision and action of leaders can.

Acknowledgements

I am deeply grateful to the many collaborators and partners whose insight, trust, and commitment make this work possible. I thank the Pontifical Academy of Sciences and the Pontifical Academy of Social Sciences Joint Workshop on *Sustainable Humanity, Sustainable Nature: Our Responsibility*, for which I presented a longer version of this essay.

References

Balmford, A., Bruner, A., Cooper, P. et al. (19 authors) (2002). Economic reasons for conserving wild nature. *Science*, vol 297, pp 950–953.

Carpenter, S., DeFries, R., Dietz T. et al. (7 authors) (2006). Millennium Ecosystem Assessment: research needs. *Science*, vol 314, pp 257–258.

Carpenter, S., Mooney, H., Agard, J. et al. (15 authors) (2009). Science for managing ecosystem services: beyond the Millennium Ecosystem Assessment. *Proceedings of the National Academy of Sciences, USA*, vol 106, pp 1305–1312.

Daily, G. C., and Ellison, K. (2002). *The New Economy of Nature: The Quest to Make Conservation Profitable.* Island Press, Washington DC.

Daily, G. C., Ouyang, Z., Zheng, H., et al. (9 authors) (2013). Securing natural capital and human well-being: innovation and impact in China. *Acta Ecologica Sinica*, vol 33, pp 669–676.

Daily, G. C., Polasky, S., Goldstein, J., et al. (9 authors) (2009). Ecosystem services in decision-making: time to deliver. *Frontiers in Ecology and the Environment*, vol 7, pp 21–28.

Dasgupta, P. (2010). Nature's role in sustaining economic development. *Philosophical Transactions of the Royal Society B.*, vol 365, pp 5–11.

Ehrenfeld, D. (1988). Why put a value on biodiversity? In: Wilson, E. O. (ed.) *Biodiversity*. National Academy Press, Washington DC.

Ehrlich, P. R., Kareiva, P., and Daily, G. C. (2012). Securing natural capital and expanding equity to rescale civilization. *Nature*, vol 486, pp 68–73.

Goldman-Benner, R. L., Benitez, S., Boucher, T., et al. (8 authors) (2012). Water funds and PES: practice learns from theory and theory can learn from practice. *Oryx*, vol 46, no 1, pp 55–63.

Goldstein, J. H., Caldarone, G., Duarte, T. K., et al. (9 authors) (2012). Integrating ecosystem service tradeoffs into land-use decisions. *Proc. National Academy of Sciences USA*, vol 109, pp 7565–7570.

Kenter, J. O., Reed, M.S., Irvine, K. N., et al. (18 authors) (2014). *Shared, Plural and Cultural Values: A Guide for Decision-Makers*. UK 41 National Ecosystem Assessment follow-on phase. UNEP-WCMC, Cambridge UK.

Kiesecker, J. M., Copeland, H., Pocewicz, A., and McKenney, B. (2010). Development by design: blending landscape-level planning with the mitigation hierarchy. *Frontiers in Ecology and the Environment*, vol 8, pp 261–266.

Leopold, A. (1949). *A Sand County Almanac*. Oxford University Press, Oxford.

Levy, K., Daily, G. C., and Myers, S. (2012). Human health as an ecosystem service: a conceptual framework. In: Ingram, J. C., Rumbaitis Del Rio, C. and DeClerck, A. (eds) *Integrating Ecology and Poverty Reduction: Ecological Dimensions*. Springer, New York.

Li, C., Li, S., Feldman, M. W., Daily, G. C., and Li, J. (2012). Does out-migration reshape rural households livelihood capitals in the source communities? Recent evidence from Western China. *Asian and Pacific Migration Journal*, vol 21, pp 1–30.

Li, F., Wang, R., Paulussen, J., and Liu, X. (2005). Comprehensive concept planning of urban greening based on ecological principles: a case study in Beijing, China. *Landscape and Urban Planning*, vol 72, pp 325–336.

Li, J., Feldman, M., Li, S., and Daily, G. C. (2011). Rural household income and inequality under payment for ecosystem services: the Sloping Land Conversion Program in Western China. *Proceedings of the National Academy of Sciences USA*, vol 108, pp 7721–7726.

Liang, Y., Li, S., Feldman, M. W., and Daily, G. C. (2012). Does household composition matter? The impact of the Grain for Green Program on rural livelihoods in China. *Ecological Economics*, vol 75, pp 152–160.

Liu, J., Li, S., Ouyang, Z., Tam, C., and Chen, X. (2008). Ecological and socioeconomic effects of China's policies for ecosystem services. *Proceedings of the National Academy of Sciences USA*, vol 105, pp 9489–9494.

MEP (Ministry of Environmental Protection of China) and CAS (Chinese Academy of Sciences) (2008). *National Ecosystem Service Zoning in China*. CAS, Beijing.

Norton, B. (1987). *Why Preserve Natural Variety?* Princeton University Press, Princeton, NJ.

Ouyang, Z. Y. (2007). *Ecological Construction and Sustainable Development*. Science Press, Beijing.

Perrings, C., Duraiappah, A., Larigauderie, A., and Mooney, H. A. (2011). The biodiversity and ecosystem services science-policy interface. *Science*, vol 331, pp 1139–1140.

Rapidel, B., DeClerck, F., Le Coq, J.-F., and Beer, J. (eds) (2011). *Ecosystem Services from Agriculture and Agroforestry: Measurement and Payment*. Earthscan, London.

Rolston III, H. (2000). The land ethic at the turn of the millennium. *Biodiversity and Conservation*, vol 9, pp 1045–1058.

Ruckelshaus, M., McKenzie, E., Tallis, H., et al. (11 authors) (2013). Notes from the field: lessons learned from using ecosystem services to inform real-world decisions. *Ecological Economics*, vol 115, pp 11–12.

Tallis, H. T., Ricketts, T., Guerry, A. D., et al. (26 authors) (2011). *InVEST 2.4.4 User's Guide*. The Natural Capital Project, Stanford.

Zhang, P., Shao, G., Zhao, G., et al. (7 authors) (2000). China's forest policy for the 21st Century. *Science*, vol 288, pp 2135–2136.

Zheng, H., Robinson, B. E., Liang, Y., et al. (9 authors) (2013). The benefits, costs, and livelihood implications of a regional PES (Payment for Ecosystem Service) program. *Proc. National Academy of Sciences USA*, vol 110, pp 16681–16686.

PART III

Ecosystem services in an applied context – introduction

The chapters in this part of the *Handbook* take an 'applied perspective' on ecosystem services. They consider issues across some key global biomes and offer an insight on the status of the major service types found there. In particular, these reviews provide an understanding into the issues and opportunities that experts working in these different ecosystems feel are most relevant to current debates. The chapters also describe some of the management and policy responses that are developing around the particular problems found in these different contexts.

An important message to emerge from the chapters assembled here is that a focus on management questions is a way of coping with the complexity of all these ecosystems and the multiple issues that arise within them. In the case of marine ecosystems, for example, Charles and colleagues describe a demand-driven approach, or 'triage,' designed to help practitioners develop a clear and purposeful process for engaging with all relevant stakeholders. The aim is to help those involved in assessment to understand its scope, the desired outcomes and the metrics and indicators used to inform the decisions that have to be made.

The need for new and context-relevant methodological developments is also a feature of the chapter on freshwater ecosystems. Brauman, for example, argues that the particular nature of services associated with freshwaters presents a significant inter-disciplinary challenge, and so we need to look to new ways of working 'to help people define problems and be transparent about disciplinary assumptions.' She suggests that these novel approaches need to differ from those of traditional hydrology, ecology and economics by focussing more explicitly on landscape change and on new variables or metrics. Brauman claims that land management has too long focussed on optimizing landscapes for non-water-related services; landscape effects on water have only become apparent through unexpected negative consequences. She concludes that we need to be more adept at connecting ecohydrologic processes in the landscape to hydrologic services, so that we can better understand the effects on water quantity, quality, location and timing.

The case for taking a landscape-orientated approach, and the opportunities for stimulating integrated management, is highlighted in the review of regulating services by Fürst and her co-authors; it is also a view supported by Luque and Iverson in their review of forest ecosystem services. Indeed, these authors argue that such an approach is essential if we are to cope with the many pressures that are now evident in these systems as a result of land use and climate change, and changing demands from society. Improving methods to measure and value biodiversity and ecosystem resilience are required to overcome current knowledge gaps. The chapters in this section also demonstrate that establishing the link between the ecosystem and people's livelihoods is especially important step. The case is well-made in the context of forests but also carries over strongly into the chapters on both drylands and grasslands.

Stringer and Dougill observe that drylands are especially vulnerable because they exhibit multiple thresholds and multiple ecological and social states, some of which support the supply of certain ecosystem services, some of which do not. They argue that these dynamics need to be understood because, globally, the livelihoods of a significant number of people depend on the integrity of these ecosystems. The case study material that these authors provide shows that current policies and economic incentives are both drivers of, and potential solutions to, dryland ecosystem service degradation. In Botswana's Kalahari rangelands, for example, policies that favour cattle production and private land tenure lead to critical trade-offs with land use strategies that prioritise wildlife management, biodiversity, the provision of veld products and cultural services. The issue of the conversion of ecosystems is also a key theme in the chapter on grasslands, where Egoh and colleagues describe some of the consequences of the widespread transformations of land use that have been experienced in this biome. Further case studies are provided in the briefings on grasslands in North America by Franzluebbers and Steiner and Australia by MacLeod and McIvor. Collectively, these authors show that worldwide, many types of grassland are threatened by cropping, intensification, afforestation and urbanisation, and that there is an emerging threat from biofuel production. Examples of the new strategies are emerging to cope with these pressures to be found in South Africa; in the past, efforts to conserve grasslands were largely based on their high biodiversity. Today, effective arguments are being framed around the range of ecosystem services they provide and the contribution that these strategies can make to poverty alleviation.

The conversion of ecosystems through cultivation and development and its impacts are key themes of the chapters led by Plieninger and Elmqvist. While Plieninger and colleagues recognise the importance of the provisioning role of cultivated ecosystems for humanity, there are nevertheless opportunities to create land use systems that can more effectively integrate agricultural production and the delivery of other types of ecosystem services. These authors support the need for a landscape approach, and illustrate its potential in the context of agroforestry, organic farming and the creation of high nature value farming methods. As in the case of urban ecosystems, recognition of the opportunities associated with ecosystem services is helping people to rethink the relationships between people and nature. However, the study of ecosystem services in cities is, according to Elmqvist and his co-authors, a particular challenge, because urban landscapes are 'largely unexplored and surprisingly poorly understood.' A key knowledge gap that they identify relates to their multi-functional character, and the way this knowledge informs our understanding and management of trade-offs between services.

The need to consider trade-offs within the bundles of services that different ecosystems provide is a recurring theme across many of the chapters in this part of the *Handbook*. Plieninger and colleagues suggest that in the context of cultivated lands, the question of whether and how intensive or extensive agriculture influences the various ecosystem services is a vital issue, requiring 'targeted research.' But as the general review of provisioning services by Galford and Ricketts demonstrates, these issues are not confined to any particular types of ecosystem. Van der Putten and Wall, for example, provide a valuable insight into the multi-functionality of soil ecosystems and their contribution to sustainable development. Fürst and colleagues make a similar argument about multi-functionality in the relation to regulating services, and argue that this issue now presents a major challenge for planning in general.

The chapter by Elmqvist and his colleagues suggests that despite growing evidence on the links between urban ecosystems and quality of life, the relationships between specific ecosystem services and well-being should not be 'taken for granted' or applied in simplistic ways through urban planning. As this chapter and the associated briefing by Gill on 'green infrastructure' shows, there is an increasingly valuable dialogue developing between the scientific and planning communities, based on a recognition that ecosystem services can be a major source of resilience for cities; green infrastructure is not simply about biodiversity conservation – it can help people cope with both environmental and socio-economic shocks.

All of the chapters in this part of the *Handbook* show that a major consequence of thinking about ecosystem services is that we have to better understand the way ecological and social systems are 'coupled.' This builds on some of the key themes covered in the first part of this *Handbook*. In this context, the material presented here shows that there are, for example, many lessons to be gained from the study of Mediterranean landscapes. Martín-López and colleagues observe that in these areas the co-evolution between human societies and nature has promoted traditional, multi-functional landscapes that can provide a richer 'bundle' of ecosystem services than modern-day ones. As a result, they argue, we need to find ways of 'retrieving the qualities of the former mosaic landscape' in order to avoid the kinds of landscape homogenisation that appear to be the consequence of contemporary development patterns. In short, we need to build new cultural landscapes that retain their multi-functional character.

A greater convergence of thinking between those working with cultural landscapes and those working on coupled socio-ecological systems is a clear message to emerge from the chapters in this part. The richness of the debate that lies ahead is highlighted by the chapter by Chan and Satterfield, who contend that the management of cultural ecosystem services is indeed an essential consideration for sustainability, because cultural ecosystem services are crucial contributors to human well-being and because they may be key to sustainable human-ecological relationships. They observe that, on the face of it, the relationship between ecosystem services and questions of sustainability appears straightforward, because the flows of benefits are like the interest we gain from our stocks of natural capital. As a result, much of the discussion about sustainability is focussed on resource management rather than the cultural aspect of all ecosystem services. Chan and Satterfield conclude, however, that instead we need to see cultural ecosystem services both as a new kind of human-natural capital and something capable of generating new capital: 'not only do positive experiences and attachment with nature yield both well-being and key capacities, they also yield the stewardship attitudes and identities that may be fundamental to local and global sustainability.' The briefing note by Cooper also reinforces this point; he argues that alongside the question of what nature can do for us, consideration of the spiritual also challenges us to ask what we are duty-bound to do for nature.

All of the chapters in this part of the *Handbook* demonstrate that the concept of ecosystem services can change attitudes and approaches, but more still needs to be done to mainstream these ideas in decision-making. Whether we can achieve such a transformation in thinking, given the speed with which many of our ecosystems are being modified, is an issue that must concern us all. Progress will certainly demand a better and richer understanding of how we value nature. Success will also necessitate more sophisticated ways of handling those values if we are to avoid what Elmqvist and others see as 'over-simplistic narratives' on the economic benefits of, for example, restoring ecological infrastructure in cities. As Chan and Satterfield conclude, it will require more deliberative decision-making regarding the values and futures we wish to foster. The contribution of this *Handbook* is to encourage these kinds of debate.

Key discussion and debating points

- Societies often expand provisioning services at the expense of other services. What kinds of management strategy might mitigate any trade-offs?
- The management of regulating services often requires us to take a multi-scalar perspective on the interactions and dependencies in socio-ecological systems. What implications does this have for governance?
- Many ecosystem services depend on a mosaic of different habitats. What might a landscape perspective add to our understanding of how ecosystem services are generated and used?
- The notion of 'green infrastructure' is now often used in the context of urban planning. How does the idea link to that of ecosystem services? Are they conflicting or mutually reinforcing concepts?
- It has been argued understanding cultural ecosystem services is an essential step towards sustainability because they both contribute to human well-being and because they shape people's attitudes and identities around the notion of stewardship. What evidence do we have that this is indeed the case and how can we take better account of these services when managing for sustainability?

25
ECOSYSTEMS SERVICES
Provisioning

Gillian Galford and Taylor Ricketts

Introduction

The food on our plate. The water we drink. The timber that frames our homes. The wood that heats our home. These are some goods and benefits that people gain from the provisioning services that ecosystems contribute. Without these goods and benefits, our lives would be very different, perhaps unimaginably so.

Provisioning services have several distinguishing characteristics. They result in tangible products of an ecosystem process, or goods. Note that "service" is a verb, as in, the action the service does, not the product itself. For example, growing tomatoes is an ecosystem service, not the product (tomato) itself. Provisioning services typically yield ecosystem goods with a readily measurable market value. These ecosystem goods are the essential inputs to all economic production.

The products of provisioning services are typically private goods, meaning that economists consider them to be both rival and excludable. Rival means the use of a good by one person leaves less for use by another person. Excludable means one person's use keeps another from using a good. Fisher *et al.* (2009, p.647) explain these concepts with a chicken: "If I buy a chicken in the market, my use of it precludes your use (rival) and I can keep you from using it by say locking it up, or eating it (excludable)." These characteristics make ecosystems goods amenable to private supply and market trading. Understanding how ecosystems goods are rival and excludable can help to identify typical ways they are – and can be – managed.

Provisioning services are often managed at the expense of others. The Millennium Ecosystem Assessment (MA, 2005) examined changes in global ecosystems services due to human activities. The MA estimated 60% of ecosystem services were being degraded or used unsustainably, including fresh water, fisheries, air purification, and regulation of pests. Provisioning services fared better than other ecosystems services, although often at the expense of other services. For example, degradation of regulating services is often due to efforts to enhance provisioning services, such as carbon storage losses from forest converted to cropland. Throughout the late 20th century, provisioning services grew rapidly as humans expanded farm lands (160% increase in food production), harvested more water (doubled water use), and cleared more timber (tripling of harvests for pulp and paper, timber by 160%). Provisioning services are not necessarily being used in responsible ways, despite being favoured. Nearly half of provisioning services are exploited at unsustainable rates (MA, 2005).

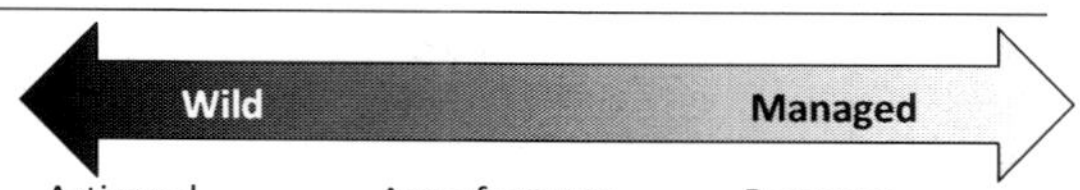

Category	Examples		
Food	Artisanal fisheries	Agro-forestry	Rowcrop agriculture
Water	Wells		Dams
Materials	Grass thatching		Paper products
Fiber	Hand-made rope		Cotton clothing
Fuel	Firewood	Charcoal	Biofuels
Medicine	Teas	Commercially grown plants	Synthetic compounds

Figure 25.1 Categories of provisioning services and examples of these services along a spectrum of wild to manage

The MA findings highlight how mismanagement can enhance provisioning services to the detriment of other ecosystem services. Provisioning services are inexorably linked to other services which are inputs or intermediates to provisioning. For example, tomatoes are a final ecosystem product and water for irrigation, in this case, is an intermediate service. Increasing irrigation could maximize tomato production, but might deplete water resources as a result. Optimizing one service must consider its impacts and interactions with other essential services. The theme of sustainability weaves into this discussion.

In this chapter, we review six main categories of provisioning services: the supply of food, water, timber, fibre, fuel, and medicine. We describe each service, provide examples along a spectrum (Figure 25.1) from wild to highly managed supply, highlight interactions with other types of ecosystems services, and flag sustainability concerns. The spectrum of wild to highly managed services is a function of inputs required to produce or enhance a good, such as capital inputs like labour, fertilizer, and processing. We consider increasing inputs to the natural system to mean the system is more managed. We take wild to mean ecosystems that are not actively managed by humans, although one could argue that almost every ecosystem has been influenced by humans. We conclude with points about the interactions among services.

Food

Food comes from a wide diversity of provisioning ecosystems. Sources range from non-managed ecosystems (deep-sea fisheries) to lightly managed or wild ecosystems (venison, mushrooms, aquaculture, and non-timber forest products) to heavily human-managed row-crop agricultural ecosystems (wheat, corn, etc.).

Wild foods

Wild plant foods can be harvested with little to no human management. Brazil nuts (*Betholletia excela)*, native to northern South America, are among the largest Amazonian trees, often reaching

50m and up to 500 years in age. Brazil nuts are typically collected from forests, not plantations, and generate income without destructive extraction. Intensive or over-harvesting leads to declines in production, as the forest stand declines without nuts to grow new trees (Peres *et al.* 2003).

Blueberries (genus *Vaccinium)* existed in wild or semi-wild communities until cultivation in the 20th century, with lowbush varieties extending along the northern Atlantic coast of the US. This abundant, sweet, and colourful berry was important forage for many indigenous groups. Native Americans preserved this nutrient-dense food as powders to add to porridges, smoked to add to meats, or dried in the sun. Today, commercial cultivation of highbush blueberries produces over 200,000 metric tons in the United States alone (FAO STAT, 2014).

Wildlife serves as a primary source of meat and income for hundreds of millions of people (Milner-Gulland *et al.*, 2003). Bushmeat can play a very important role in nutrition; Golden *et al.* (2011) found that without bushmeat, children in Madagascar would have an almost 30% higher rate of anaemia, up to 300% in poor households. Anaemia leaves children susceptible to other diseases and creates cognitive, motor, and physical deficits. The benefits of wild meats must be considered in balance with the sustainability of their harvest in order to promote production for future generations.

Deep-sea fishing taps a pool of resources largely unmanaged or controlled by humans. While people have indirect impacts on salt-water fisheries, such as through climate change, invasive species introduction, and pollution, typically they do not manipulate these systems as intensively as land-based agroecosystems. The management of fisheries involves controlling human use. Commercial fisheries are controlled by local measures to sustain the fisheries indefinitely and avoiding overfishing, such as with Atlantic Cod. The most common are the restriction of fishing to particular seasons to avoid critical breeding and rearing periods, specific equipment that minimizes bi-catch or other damages to the ecosystem, and specific quotas to restrict the amount of a population removed from an ecosystem (Cochrane *et al.*, 2011).

In general, wild foods can provide a low-cost food source with large impacts on food security. In rural South Africa, households utilize wild-grown spinaches (96% of households), wild fruits (88%), and wild honey (50%) (Shackleton and Shackleton, 2004). These food sources become more important when a head of household is lost or food is scarce. In South African communities heavily affected by HIV/AIDS, for example, 62% of children were found to supplement their diets with wild foods, and for 30% of the surveyed children, wild foods composed over half of their diet (McGarry and Shackleton, 2009). McGarry and Shackleton (2009) found that the most vulnerable children have the highest proportion of wild foods in their diets.

Low-intensity food production systems

Fisheries show an evolution from wild systems to intermediate forms of management. Human appropriation of rivers has caused steady decline of fisheries worldwide. Damming of rivers, popular in the early 19th century, provided needed hydro-electrical power. These dams cut off the native runs of fish, reducing their habitat and populations. Other factors contributed to the decline of wild fish populations, including pollution, nutrient loading, river channelization, and introduction of invasive species and exotic diseases. In no case is this more apparent than with the native salmon of North America. Native peoples of North America used rivers to secure wild salmon, an important food source and cultural icon, particularly to the tribes of the Pacific Northwest (Campbell and Butler, 2010). Today, most salmon in the lower half of North America are raised in fish hatcheries and released into rivers. Some salmon runs have been observed to be made up of as little as 10% wild individuals (Barnett-Johnson et al., 2011). These hatchery-raised fish have much lower success rates, high propensity for disease, and less genetic diversity than their wild counterparts.

Some fish farming remains low-intensity, such as tilapia (cichlids from the genera *Oreochromis, Sarotherodon*, and *Tilapia*), which is cultivated in small ponds around the world. Famous for its fast growing period, tasty meat, and ease of cultivation, tilapia production is being used as a solution to food insecurity. For example, farmers in Haiti have developed small rearing farms for tilapia to sell in local markets. These farms contain a few small tanks for raising fry and a small pond for growing the fish. Tilapia can thrive in murky waters and feed on local plant sources such as dried weeds. For the cost of labour inputs, tilapia can be a valuable nutrition and income source for a community or farmer (SAI, 2011).

In most of the world, pastures are the primary land-use for cultivating domesticated animal products such as beef, pork, and goat. Pastures range from natural grasslands, such as the Cerrado (savannah) of Brazil or the Serengeti Plains of Tanzania, to pastures with managed forages, added nutrients, or irrigation. Low-intensity grazing (less than 1 Animal Unit per ha) works where land is cheap or can be freely grazed, as in open range systems. To maximize production, confined or semi-Confined Animal Feed Operations (CAFOs) are being used. In these systems, animal production (e.g. milk, dairy) occurs almost entirely indoors and can lead to large point-sources of nutrient pollution; yet another example of the delicate trade-off between provisioning of food and sustainability.

Croplands also have a place in the middle of the spectrum with less-intensively managed systems. In subsistence agriculture, food production meets the needs of the farm family with little to modest product for resale (Figure 25.2). These systems may have lower crop yields than more intensive practices, but typically require less management inputs and are easier for smaller-scale farming. Low-intensity systems may use lower rates of fertilizer additions but still have an impact

Figure 25.2 Boys in Mali preparing fields for planting.

Source: photo by G. Galford

on other ecosystems services. For example, lower yields correspond to smaller plants. This leaves the soil vulnerable to erosion, without roots and leaf canopy to lessen the impact of rainfall. Loss of topsoil through erosion negatively affects the provisioning of food and causes negative impacts to other ecosystem services, like water quality (Vitousek *et al.*, 2009, Sanchez, 2002). Many subsistence farmers rely on wild foods to supplement their crops.

Human-dominated food production systems

At the far end of the spectrum are human-dominated production ecosystems. Most calories are produced from human-dominated ecosystems, such as the corn and wheat fields of the United States or the rice paddies of south Asia (FAOSTAT, 2014). Here, food production is largely controlled by farmers' decisions of what crops and varieties to plant and which management practices to use (e.g. using fertilizers or irrigation). People also determine land use of the fields, including management of conservation areas that foster regulating services, such as pollination, water purification, and erosion control.

Production from row-crop agriculture can be intensive, using a large amount of inputs (e.g. fertilizer and water) or growing multiple crops within one year (Figure 25.3), and crop yields, or outputs, tend to be high. In this regard, food production as a provisioning service to humans can be maximized. Maize is possibly the most ubiquitous crop that is intensively farmed. It requires large additions of fertilizer to reach yields needed for sustenance and/or profit (Vitousek *et al.*, 2009).

Figure 25.3 Fertilizer bags being prepared for the growing season, Mato Grosso Brazil (1 bag = 1,000 tons).

Source: photo by G. Galford

Water

Water supply is a provisioning ecosystem service with many different manifestations, including drinking, cooking, bathing, and irrigation, among others. All of these sources of water benefit from purification, as healthy ecosystems remove toxins, nutrients, sediment, and more.

Throughout human history, societies have been built around availability and access to fresh water. Rainwater harvest is a straightforward way to access water and is common where ground water and surface water are limited but rain comes infrequently in large volumes. Ground water pumps allow access to water hundreds of feet below the surface to service households or communities.

Surface water harvest ranges from small to large scales. This generally involves damming water flow to extract water. It relies upon a healthy watershed ecosystem to deliver clean water. One of the largest cities in the world, New York City, depends on a surface water system. A watershed north of New York City spans roughly 5000 km^2 and delivers 5.3 billion litres of water per day to 9 million people. New York City's investment in this resource in 1927 has resulted in the provisioning of clean, copious water for the city over many decades. In 2007, the US government found that the watershed was sufficiently providing high-quality water and could avoid filtering the water – saving the city huge costs (McPhearson *et al.*, 2014).

Building materials

Ecosystems provide us with the timber and building materials, common provisioning goods, needed for homes and buildings. Many cultures rely on natural materials for roofing, siding, stoves, and more. Here we discuss a spectrum of resources, from more wild to more processed goods.

In Malawi, tall grasses grow in abundance at the edge of fields and roads. These grasses are cut during the dry season and used as a thatched roof in traditional building (Figure 25.4). Newer

Figure 25.4 Straw-thatch roof on a house in Malawi contrasted with an aluminum roof in the background. Notice also support beams made from crudely cut tree trunks.

Source: photo by G. Galford

or upgraded homes have more durable aluminium roofs, but the traditional thatched grass roofs remain common. Apparent in the Malawi photo is the use of other building materials sourced from local ecosystems. Hand-cut tree trunks support the thatched roof. Little processing has been done to harvest this tree, clean it of limbs and leaves and put it into its current place – this is a direct provisioning of building materials for construction.

Many cultures use earthen materials for walls and siding on houses, often called adobe, which require more labour and specialized knowledge. Traditional adobe is a mix of clay (less than 30%), sand (roughly 50%) and plant-derived materials such as straw or dung of ruminants (roughly 20%). In some places this mixture is used as a wall board or sheet rock in homes that can be plastered over for a more finished look. In parts of Central America, this same mixture is used to construct a 'Lorena' stove, or enclosed fire box with chimney, for cooking. The Lorena stove improves fuel efficiency of charcoal or wood by up to 50%. It improves health as the cooks, usually women, no longer stoop over a fire or inhale smoke. The most labour- and time-intensive use of adobe materials is creating bricks.

Moving to the other end of the spectrum, the most industrialized products include machine-milled timber and wood-chip products. In North America, vast reaches of forest have been managed with clear-cutting operations that cause water quality problems, habitat loss for endangered species, and a homogenization of forest age and structure. Many public lands are leased to logging corporations for production of timber, wood, and other products. These clear-cut practices involve several decades of forest growth, typically with a single species planted for rapid growth, before clearing. While these practices still exist, other producers are seeking third-party certification programs for environmental and social sustainability in forestry, such as the Forest Stewardship Council (FSC). FSC oversees practices on over 183 million hectares worldwide, almost half of which are public lands used for forestry. FSC-certified products span beyond just building materials and include furniture.

Fibre

Many fibres used as material resources undergo anywhere from minimal to industrialized processing before people use them for clothing, cordage, sails, insulation, paper, and more.

Cordage, such as ropes, uses one of the least processed fibres – hemp (*Cannabis sativa*). Hemp is one of the oldest domesticated plants known, in cultivation for over 12,000 years. Today, small farms in Central America use its raw forms to make ropes, or it may be highly processed to make paper, fibre board, and posts (NAIHC, 1997).

Fibres from animals, such as sheep and alpaca wool, are some of the warmest materials available and have been used for centuries for clothing. In Central Asia, wool fibres are also used as insulation in Mongolian yurts. These large, round structures are covered with cloth and sheep's felted wool, which provide insulation and weatherproofing. Nomadic pastoralists harvest the wool.

Moving to more industrialized production of fibres, we see higher inputs (irrigation, labour, fertilizers, etc.), post-processing, and distance travelled from source to use. Forest resource use can also be industrialized. Tree and plant fibres are used in products such as paper, rope, and mulch. Paper involves large amounts of processing and is an integrated essential in modern culture. Currently, the USA leads the world in wood pulp production (56 million metric tons per year), followed by Canada (23 million), Sweden (11 million), Finland (11 million), and Japan (10 million) (FAOSTAT, 2014).

Cotton is one of the most ubiquitous fibre crops around the world. It requires irrigation, fertilization, pesticides, advanced machinery, and labour. Consider the global journey involved in a cotton shirt. In 2011, the USA was the top cotton exporter in the world, with 2.8 million metric tons of cotton lint exported, followed by India at 1.9 million tons (FAOSTAT, 2014). China is by far the world's largest importer of cotton lint at 3.6 million tons (FAOSTAT, 2014), 2.6 million of which comes from the USA (US Census Bureau, 2014). Cotton is processed into yarn, fabric, and apparel in China. Many of these products are re-imported to countries that produced the cotton. In fact, 34% of all USA imports of cotton products (apparel and non-apparel) come from China (OTEXA, 2014).

Fuel

Humans have gathered fuel from local ecosystems for millennia. Firewood, for example, is one of the most commonly used fuels across space and history. A study in rural South Africa found that over 80% of the population uses firewood as their primary fuel source (Williams and Shackleton, 2002). Firewood can also be processed – sectioned, cut and dried, or processed into charcoal or wood pellets. The production of charcoal allows access to energy for cooking and heating in many urban and semi-urban areas without access to gas or stable heat sources, and comes with environmental trade-offs. The demand for wood for charcoal, particularly for developing urban areas, can lead to increased rates of deforestation (DeFries *et al.*, 2010).

Dung from domestic animals has been widely used for fires, smoking meat, and anything else firewood is used for. Recently, animal manures, plant residues, and solid wastes are being broken down in anaerobic digesters to produce gas that can be burned like propane for cooking, heating, and electricity generation. Biodigesters are being experimented with on farms and in areas where solid waste disposal is lacking or causing environmental degradation (Asam et al., 2011). Cobenefits of these digesters include mitigating greenhouse gas emissions, reducing water pollution, and recovering byproducts, such as cost savings in manure management, nutrient retention, manure odour, and pathogen reduction (Yirode et al., 2009).

Today, we find new ways to convert ecosystems products into combustible energy sources. Crops traditionally used for cooking oils, such as sunflower and canola, are being used as biodiesel. Corn and sugarcane ethanol (alcohol) are produced for use in internal combustion engines in mixture with or as replacement of fossil fuels. As of yet, these crop-based biodiesels and ethanol take almost as much energy to produce as they yield (Bourne, 2007). Second generation biofuels in development use cellulose from crop residues or tree fibres, or algae, to synthesize oils. All represent provisioning services from ecosystems.

Medicine

Ecosystems provide us with a range of medicines, which again span a range from those harvested directly from nature to those that have been domesticated and cultivated, or synthetically replicated.

Herbal medicines from wild and cultivated stocks are an essential part of global healthcare. In Africa, 80% of people use some form of herbal medicines for diseases such as malaria (WHO, 2002). Wilcox and Bodeker (2004) reviewed herbal treatments of malaria and found over 1200 plant species used globally, with a fifth of patients in endemic regions using traditional herbal remedies. In South Africa, studies have found that 65–70% of medicinal plant materials sold in markets come from forests or savannahs (Cocks *et al.*, 2004, Williams, 2004).

Some plant-based medicinals are now intensively cultivated. Opium poppy, morphine's source, was one of the first plants humans cultivated. Echinacea, used in tablets, teas, and tinctures, boosts immune systems and decreases severity of colds and infections. Marijuana is cultivated for treating severe pain, muscle spasms, nausea, cancer, seizures, glaucoma, cachexia, and

HIV/AIDS. Colorado, US, recently legalized the sale of medical marijuana and saw a demand for 130 metric tons per year, exceeding expectations by almost 90% (Light *et al.*, 2014).

Synthetic chemicals now can replicate the plant products that provide medicinal services. Jaborandi, native to Brazil, treats fever, detoxifies the body, and reduces fluid retention. It is used as the active ingredient in glaucoma medicine (Sidha, 2014). Aspirin, used for pain and fever reduction, is now commercially produced, but was originally administered by chewing willow bark.

Important cultural nuances affect the socially acceptable uses of plants for medicine. Coca is an instructive example. Native to South America, the coca plant has been used in teas or chewed without psychological effects associated with cocaine. Uses include mitigating altitude sickness, functioning as an anaesthetic and analgesic, stanching bleeding, treating malaria, and improving digestion (Weil, 1981). In contrast, the commercial cultivation of coca for cocaine production is fought stanchly by the international community, with measures including aerial monitoring and eradication of coca plantations. In tropical regions, spraying herbicides to kill coca plants may have negative health impacts on local people, adversely affect other crops, and contaminate water supplies.

Interactions among services

We have considered a set of specific provisioning services in this chapter, but it is important to note that provisioning services interact with each other, as well as with other types of services. These interactions can be positive (i.e. synergistic) or negative (i.e. represent trade-offs). Positive interactions are common, for example, between provisioning and cultural ecosystems services. Cuisine, or the style of cooking, is largely reflective of the foods available in a region, ranging from the crops grown and animals farmed to wild foods foraged. Architectural styles often reflect the natural materials available for construction, from post and beam homes to straw and clay. Negative interactions are also easy to identify. Intensive food production, for example, often occurs at the expense of other supporting services (e.g. crop pollination), regulating services (e.g. carbon storage), and cultural services (e.g. recreation).

For example, the yield increases and financial gains from intensive farming are not without trade-offs. Water quality degrades as nutrients from fertilizers moves into waterways. These nutrients can cause harmful blooms of algae and create anoxic waters, such as the Dead Zone in the Gulf of Mexico, which is fed by nutrients carried from croplands in the Midwestern United States down the Mississippi River. Algae flourish on the added nutrients and deplete oxygen, which created the infamous Dead Zone (Donner *et al.*, 2004). There can be serious damages to fisheries when waters reach high temperatures in the summer and the waters are anoxic.

Management to sustain the natural ecosystem can increase food production. For example, mangrove habitats are important nurseries and feeding grounds that can increase fisheries landings substantially. One study found that the value of the fisheries associated with mangroves in the Gulf of Mexico is US$37,500 per ha of mangrove fringe. This is 300 times greater than government estimates that do not include accounting for mangroves (Aburto-Oropeza et al., 2008). Elsewhere, Brashares *et al.* (2004) document that inter-annual declines in fishery catch led to increases in capture of bush meat in Ghana. A switch from fish to bush meat is an efficient adaptation and use of wild resources, but was shown to have large negative impacts on populations of native mammals, raising concerns for biodiversity.

Human demands for water provisioning can conflict with demands from other species. One recent example comes from North America, where environmental regulations for the protection of species make smaller hydropower projects less financially viable. The Condit Dam on the White Salmon River in Washington State (Becker, 2006) was completed in 1913. Its first fish ladder was destroyed within a year by a storm. A second fish ladder was destroyed four years later and another was never built. In 2005, PacifiCorp found that bringing the dam into federal compliance with a new fish ladder would incur costs to their ratepayers well beyond increased rates that would result from taking the small dam offline. The Condit Dam was therefore breached on October 26, 2011 and removed by December 2012 (PacifiCorp, 2015). Breeding salmon returned upstream of the former dam site in the summer of 2013 for the first time in over a century (Florip, 2012). While energy generation from the White Salmon has decreased, opportunities for wild salmon and recreation have improved.

Materials from nature provide additional ecosystem services, often cultural. As a result of the local ecosystem and its resources, societies have emerged with variations in architecture – for example, wood floors and walls in forested regions where timber is freely available. Take the

Figure 25.5 Top left: Plants are boiled to great fabric dyes over an open wood fire. Top right: After dying the fabric, stencils are used by the artist (Lower left) who paints on designs using fabric dyes made of plants and mud (Left center). The final product is earth-colored fabrics whose patterns tell traditional stories.

Source: photos by G. Galford

cultural significance of log cabin homes built from timber cleared with the settlement of the American West. This style of building still evokes a cultural style reminiscent of a particular time and place. Modern log cabin homes mimic the original nature-based architecture, with the added benefits of additional insulation and amenities. Foods such as cheese and wine are also emblematic of certain regions of the world, such as Gruyere cheeses of Switzerland or Bordeaux wines of France.

Fibres are useful in everyday living and often become part of our cultural identity. Denim blue jeans are iconic of the West. Originally created by Levi Strauss to provide tough, durable pants for gold miners in California during the mid to late 1800s, today's jeans are popular around the world and are emblematic of American culture. In Mali, bogolan (Figure 25.5) is a traditional fabric of great artistry. This cotton fabric is hand-woven from locally grown cotton and dyed with fermented dyes made of clays and local plants. The N'gallama (*Anogeissus leiocarpa*) leaves impart a yellow colour that is the base of all bogolan patterns. The elaborate bogolan patterns tell stories about historical events, myths, or proverbs. Today, bogolan remains a strong piece of Mali's cultural identity (Blanchart *et al.*, 2010).

Fuels and their extraction, such as firewood collection, shape environmental conditions with further impacts for ecosystems services. Deforested areas have a reduced capacity to provide services such as food and fibre and, ultimately, fuels (Makungwa *et al.*, 2013). Water quality may be degraded without trees holding soils in place (Baker, 2006). Burning wood and charcoal contributes to greenhouse gas emissions and particulates in the air (Andreae and Merlet, 2001) that can trigger asthma.

These examples show an explicit trade-off between one provisioning service – e.g. food production from croplands – and another – food from fisheries. Balancing these positive and negative interactions among services will require a synthetic approach to measuring, monitoring, and understanding multiple forms of ecosystem service. Capitalizing on synergies, and managing trade-offs will be essential to meeting the challenge of sustainability in the 21st century.

References

Aburto-Oropeza, O., Ezcurra, E., Danemann, G., Valdez, V., Murry, J. et al. (6 authors) (2008). Mangroves in the Gulf of California increase fishery yields. *Proceedings of the National Academy of Sciences*, vol 105, no 30, pp 10456–10459.

Andreae, M. O., and Merlet, P. (2001). Emission of trace gases and aerosols from biomass burning. *Global Biogeochemical Cycles*, vol 15, no 4, pp 955–966.

Asam, Z-Z., Poulsen, T. G., Nizami, A. S., Rafique, R., Kiely, G., and Murphy, J. D. (2011). How can we improve biomethane production per unit of feedstock in biogas plants? *Applied Energy*, vol 88, no 6, pp 2013–2018.

Baker, A. (2006). Land Use and Water Quality. *Encyclopaedia of Hydrological Sciences*, pp. 16:188.

Barnett-Johnson, R., Grimes, C. B., Royer, C. F., and Donohoe, C. J. (2011). Identifying the contribution of wild and hatchery Chinook salmon (Oncorhynchus tshawytscha) to the ocean fishery using otolith microstructure as natural tags. *Canadian Journal of Fisheries and Aquatic Sciences*, vol 64, no 12, pp 1683–1692.

Becker, D. H. (2006). The challenges of dam removal: the history and lessons of the Condit dam and potential threats from the 2005 federal power act amendments. *Environmental Letters*, vol 36, no 811, pp 811–866.

Blanchart, P., Dembeléb, A., Dembeléb, C., Pléab, M., Bergströmc, L., *et al.* (10 authors) (2010). Mechanism of traditional Bogolan dyeing technique with clay on cotton fabric, *Applied Clay Science*, vol 50, no 4, pp 455–460.

Bourne Jr., J. (2007). Biofuels, *National Geographic Magazine*, October.

Brashares, J. S., Arcese, P., Sam, M. K., Coppolillo, P. B., Sinclair, A.R.E., *et al.* (6 authors) (2004). Bushmeat hunting, wildlife declines and fish supply in West Africa. *Science*, vol 306, no 5899, pp 1180–1183.

Campbell, S.K., and Butler, V.L. (2010). Archaeological evidence for resilience of pacific northwest salmon populations and the socioecological system over the last ~7,500 years. *Ecology & Society*, vol 15, no 1, pp 1–20.

Cochrane, K. L., Andrew, N. L., and Parma, A. M. (2011). Primary fisheries management: a minimum requirement for provision of sustainable human benefits in small-scale fisheries. *Fish and Fisheries*, vol 12, no 3, pp 275–288.

Cocks, M. L., Dold, A. P., and Grundy, I. M. (2004). The trade in medicinal plants from forests in the Eastern Cape province. In: Lawes, M. J., Eeley, H. C., Shackleton, C. M., and Geach, B. S. (eds) *Indigenous Forests and Woodland in South Africa: Policy, People and Practice*. University of KwaZulu-Natal Press, Pietermaritzburg.

Colorado River Water Users Association (2015). *Colorado River: Agriculture*. CRWUA. Available at: www.crwua.org (accessed 15 March 2015).

DeFries, R., Rudel, T. K., Uriarte, M., and Hansen, M. (2010). Deforestation driven by urban population growth and agricultural trade in the twenty-first century. *Nature Geoscience*, vol 3, pp 178–181.

Department of Energy (2005). Updated Condit Hydroelectric Project Settlement Agreement. *United States Federal Register*, vol 70, no 50, pp 12865(E5–1151).

Donner, S. D., Kucharik, C. J., and Foley, J. A. (2004). Impact of changing land use practices on nitrate export by the Mississippi River. *Global Biogeochemical Cycles*, vol 18, no 1, pp 1028–1049.

Endocrine Society (2011). Evaluation, treatment and prevention of vitamin D deficiency: an Endocrine Society clinical practice guide. *The Journal of Clinical Endoctinology & Metabolism*, vol 96, no 7. Available at: http://press.endocrine.org/doi/abs/10.1210/jc.2011-0385

FAOSTAT (2014). *Food and Agriculture Organization of the United Nations, Statistics Division*. Available at: http://www.faostat3.fao.org (online 24 August 2014).

Fisher, B., Turner, R. K., and Morling, P. (2009). Defining and classifying ecosystem services for decision making. *Ecological Economics*, vol 68, no 3, pp 643–653.

Florip, E. (2012). Fish confirmed above Condit Dam: biologists say critters seen are likely hatchery steelhead. *The Columbian*, 17 July. Available at: http://www.columbian.com/news/2012/jul/17/condit-dam-fish-white-salmon-river/ (accessed15 March 2015).

Golden, C.D., Fernald, L.C.H., Brashares, J. S., Rasolofoniaina, B.J.R., and Kremen, C. (2011). Benefits of wildlife consumption to child nutrition in a biodiversity hotspot. *Proceedings of the National Academy of Sciences*, vol 108, no 49, pp 19653–19656.

Light, M. K., Orens, A., Lewandowski, B., and Pickton, T. (2014). *Market Size and Demand for Marijuana in Colorado*. Report prepared for the Colorado Department of Revenue. Available at: www.colorado.gov (accessed 09 October 2014).

MA (2005). *Ecosystems and Human Well-being: Synthesis*. Island Press, Washington, DC.

Makungwa, S. D., Epulani, F., and Woodhouse, I. H. (2013). Fuelwood supply: a missed essential component in a food security equation. *Journal of Food Security*, vol 1, no 2, pp 49–51.

McGarry, D. K., and Shackleton, C. M. (2009). Children navigating rural poverty: rural children's use of wild resources to counteract food insecurity in the Eastern Cape, South Africa. *Journal of Children and Poverty*, vol 15, no 1, pp 19–37.

McPhearson, T., Hamstead, Z. E., Kremer, P. (2014). Urban ecosystem services for resilience planning and management in New York City. *Ambio*, vol 403, pp 502–515.

Milner-Gulland, E. J., Bennett, E. L., SCB 2002 Annual Meeting Wild Meat Group (2003). Wild meat: the bigger picture. *Trends in Ecology and Evolution*, vol 18, no 7, pp 351–357.

NAIHC (1997). Hemp facts. North American Industrial Hemp Council. www.naich.org (accessed 09 September 2014).

OTEXA (2014). *Trade Data: U.S. Textiles and Apparel Imports by Country*. Office of Textiles and Apparel, International Trade Administration, Department of Commerce, USA. Available at: http://www.otexa.ita.doc.gov/ (accessed 09 September 2014).

PacifiCorp (2015). *Condit: Overview*. Available at: http://www.pacificorp.com/condit (accessed 15 March 2015).

Peres, C. A., Baider, C., Zuidema, P. A., Wadt, L.H.O., Kainer, K. *et al.* (17 authors) (2003). Demographic threats to the sustainability of Brazil nut exploitation. *Science*, vol 302, no 5653, pp 2112–2114.

Sanchez, P. A. (2002). Soil fertility and hunger in Africa. *Science*, vol 295, pp 2019–2020.

Shackleton, C., and Shackleton, S. (2004). The importance of non-timber forest products in rural livelihood security and as safety nets: a review of evidence from South Africa. *South African Journal of Science*, vol 100, pp 658–664.

Sidha, P. (2014). Endangered Jaborandi. *British Dental Journal*, vol 217, pp 2–3.

Sustainable Aquaculture Initiative (2011). *The Sustainable Aquaculture Initiative*. Marine Biological Laboratory, Woods Hole. Reports and information available at: www.mbl.edu/sai (accessed 17 March 2015).

US Census Bureau (2014). TradeStats Express. Office of Trade and Economic Analysis, Industry and Analysis, International Trade Association, US Department of Commerce. Available at:http://tse.export.gov/TSE (accessed 09 September 2014).

Vitousek, P. M., Naylor, R., Crews, T., David, M. B., Drinkwater, L. E. *et al.* (17 authors) (2009). Nutrient imbalances in agricultural development. *Science*, vol 324, pp 1519–1520.

Weil, A.T. (1981). The therapeutic value of coca in contemporary medicine. *Journal of Ethnopharmocology*, vol 3, no 2–3, pp 367–376.

WHO (2002). *WHO Traditional Medicine Strategy 2002–2005*. World Health Organization, Geneva.

Wilcox, M. L. and Bodeker, G. (2004). Traditional herbal medicines for malaria. *The British Medical Journal*, vol 329, p 1156.

Williams, A., and Shackleton, C. M. (2002). Fuelwood use in South Africa: where to in the 21st Century? *Southern African Forestry Journal*, vol 196, pp 1–7.

Williams, V. L. (2004) Trace and socio-economic value of forest and woodland resources within the medicinal plant market in Johannesburg. In: Lawes, M. J., Eeley, H. C., Shackleton, C. M., and Geach, B. S. (eds) *Indigenous Forests and Woodland in South Africa: Policy, People and Practice*. University of KwaZulu-Natal Press, Pietermaritzburg.

Yirode, E. L., Gordon, R., and Brown, B. B. (2009). Nonmarket cobenefits and economic feasibility of on-farm biogas energy production. *Energy Policy*, vol 37, no 3, pp 1170–1179.

26

MANAGING REGULATING SERVICES FOR SUSTAINABILITY

Christine Fürst, Susanne Frank and Justice Nana Inkoom

Introduction

Regulating ecosystem services (ES) are those related to the control of biogeochemical cycles and biophysical structures at different scales (Watanabe and Ortega, 2011; Potschin and Haines-Young, 2011). Their contribution to human well-being is often indirect in that they support the supply with other services, such as provisioning services, or mitigate natural risks that endanger human health and life or cause economic damage. The key regulating services to consider are those which can be changed through management, planning, and policymaking, such as hazard mitigation (e.g. flooding), climate regulation (local and global), water purification, water erosion control, biological control of pests and diseases, and pollination (Sandhu et al., 2012).

Most regulating services depend on biogeochemical processes, such as carbon sequestration or water purification, or on biophysical factors or characteristics such as canopy structure or soil quality. Both biogeochemical processes and biophysical properties can be changed by the type of land use, as well as the type and intensity of land management (Maes et al., 2012). The extent to which management strategies can modify the regulating capacities of different land uses is one of the most critical issues that need to be addressed to improve the assessment of regulating services. At present, land management practices are not fully incorporated in land use planning due to a lack of data; information on management practices cannot easily be derived from remote sensing and even where this is possible its incorporation in land information systems is complex and expensive (Lorenz et al., 2013; Witt et al., 2013).

The impact of biogeochemical processes and of biophysical properties on regulating services needs to be considered at multiple scales: for example, the balance of greenhouse gases storage and emission in the permafrost areas and the global oceans drives the global climate. Organic matter accumulation in agricultural and forest soils supports the regulation of local water and nutrient cycles, but contributes also to the global climate regulation through the sequestration of carbon. Such multi-scale interactions and dependencies necessitate involving multiple actors in strategies for sustaining or enhancing regulating ES. Successful management of biogeochemical processes and biophysical properties at the local and regional scale must be accompanied by governance models that enable the cooperation of providers and consumers of regulating services at the regional and global scales (Ernstson et al., 2010).

In this chapter, we discuss strategies for enhancing regulating services through land management and land use planning. We then analyse challenges related to the adaptation of governance models in land use planning and policies at local, regional, and global scales. We will conclude how to support a better integration of land management and spatial prioritization strategies by scale-specific governance instruments.

Strategies for enhancing regulating services in land management and land use planning

To enhance the provision of regulating services by influencing biogeochemical cycles or biophysical properties of land and vegetation, land management usually implies change at the planning unit or parcel scale. Such change and adaptation often involve techniques that seek to optimize the management intensity expressed as the ratio of outputs to material or energetic inputs (Lorenz et al., 2013; Koschke et al., 2013), where the inputs can be such factors as the energy used in soil management, or materials in the form of fertilizers or pest control agents. Such optimization might also make use of natural self-regulation mechanisms in ecosystems, such as allelopathy, to manage the inter- and intraspecific plant competition and hence the regulation of forest stand densities and regeneration processes, techniques to modify the vertical structuring of vegetation layers to use shadowing effects for weed control and for handling interspecific competition in natural regeneration processes, or the sequencing of crop types to reduce the risk of pest outbreaks in agro-ecosystems. The use of such ecological intensification mechanisms can reduce energy and chemical inputs and reduce on- and off-site impacts of land management on the capacity to regulate water quality or to control pests by biological means (Altieri, 1999; Dale and Polasky, 2007; Bommarco et al., 2013).

Some regulating services cannot be exclusively controlled by adapting the land management, but necessitate additionally to account for spatial interactions, and landscape composition and configuration (Fürst et al., 2013a). For instance, soil erosion regulation (Frank et al., 2014, Box 26.1) or hydrological services (Koschke et al., 2014, Box 26.2) depend likewise on the land use and its spatial pattern.

Box 26.1 Case study on soil erosion protection

On the basis of two projects on regional land management adaptation to global change, land use and management scenarios for reducing water erosion were developed for a study area in middle Saxony, Germany (Figure 26.1; Table 26.1; Frank et al. 2014). The decision support software, GISCAME (Fürst et al., 2010a, b), was adapted to provide a qualitative assessment of the performance of the different scenarios against the initial planning goal and to evaluate potential trade-offs with other ES (Fürst et al., 2011, 2012, 2013b).

The regional planners aimed to focus the tests of alternative land use and management options on preferential discharge paths for surface runoff in agricultural areas. The scenario simulation was therefore focused on an 81 km² part of the foothills of the Eastern Ore Mountains (Figure 26.1) which is highly vulnerable to water erosion due to large and homogeneous agricultural parcels and medium to steep slopes. Increasing intensity of heavy rainfall events in the context of climate change is expected to aggravate the water erosion risk.

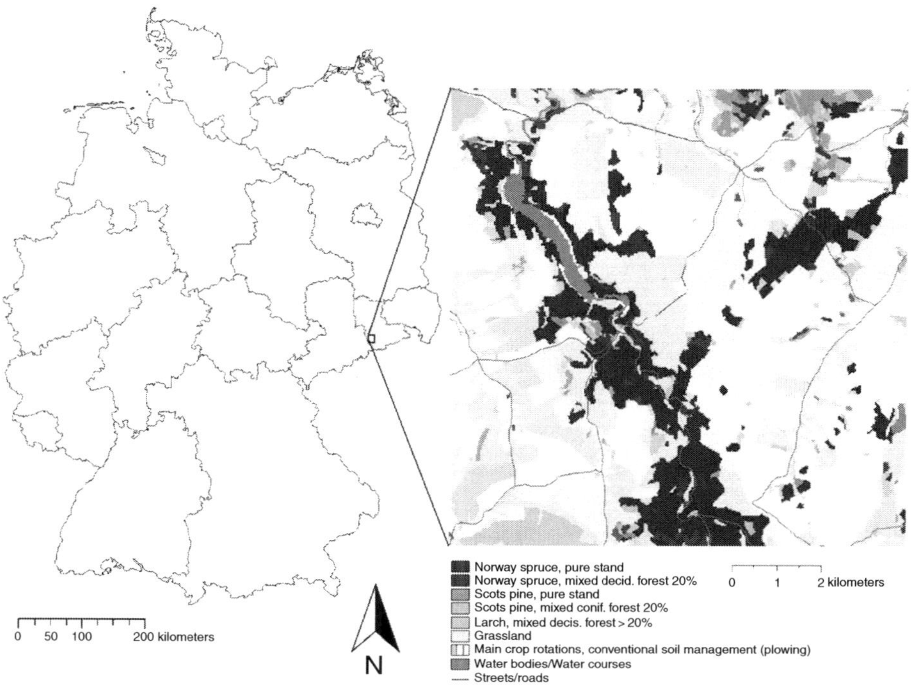

Figure 26.1 Location of the study area in middle Saxony, Germany and of the focus area Eastern Ore Mountains.

Source: Frank et al., 2014

Six most preferable land use and management change scenarios resulted from a consultation process with representatives of the regional planning authority (Table 26.1) and landowners.

The absolute soil loss [t/ha*a] was calculated based on the Revised Universal Soil Loss Equation (RUSLE, Renard et al., 1991) in GISCAME (Figure 26.2). All erosion control measures reduced the potential soil loss in comparison to the BAU scenario. Slight soil erosion reduction by about

Table 26.1 Overview of the land use change and management change scenarios.

No.	*Description*	*Abbreviation*
0	Business as usual. **No changes** in land use	BAU
1	Land use change from field to **grassland** at preferential discharge pathways for surface runoff larger (>4 ha)	Greening I
2	Land use change from field to **short rotation coppices** at preferential discharge pathways for surface runoff larger (>4 ha)	Greening II
3	Land use change from field to **grassland** at preferential discharge pathways for surface runoff larger (>1 ha)	Greening III
4	Land use change from field to **hedgerows** across long, steep slopes	Hedges
5	Land management change from ploughing to a **no-till management**	No-Till
6	**Combination** of Greening II, Hedges, and No-Till.	Combination

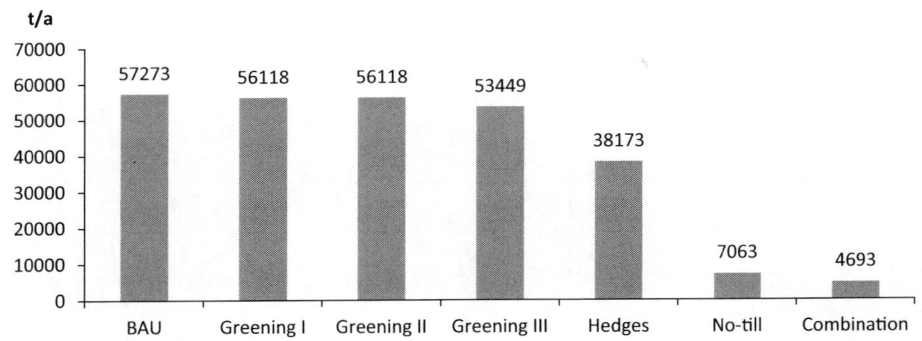

Figure 26.2 Modelled soil losses for the different scenarios.

Source: Frank et al., 2014

2% was found for Greening I and Greening II. In scenario Greening III, the soil loss was reduced by 7%. Hedgerows across the slope were more effective; the potential soil loss was estimated to be decreased by a third. Even more effective was the no-till scenario, which decreased the potential soil losses by 88%. Greening III, Hedges, and No-Till in combination produced a reduction of 92%, which represented the most efficient management option.

At the regional level, six other ES were selected by the regional planning authority and the land owners for which the impact of the water reduction strategies was assessed in GISCAME on a relative scale from 0 (no provision or not relevant) to 100 (highest possible provision in this region), including the additional impact of changes in the landscape composition and configuration (Frank et al., 2012). All measures against water erosion had predominantly positive effects on the other services (Table 26.2): regarding the regulating ES (drought risk regulation, flood regulation, ecological integrity) and cultural services (landscape aesthetics), only positive or no effects were found. In contrast, provisioning services suffered from the measures in most of the scenarios.

Table 26.2 Qualitative assessment of water erosion reduction measures and impact on the potential provision of six other ES compared to the business as usual scenario (BAU). The darker the colour of the cell, the higher the positive (light grey scale) or negative (dark grey) impact on the service provision (Frank et al., 2014). The +/- correctives in the brackets of the services ecological integrity and landscape aesthetics result from the additionally assessed impact of landscape composition and configuration on the overall provisioning potential of these services (see Frank et al., 2012).

Scenario / *ES*	*ES*	*BAU*	*Greening I*	*Greening II*	*Greening III*	*Hedges*	*No-Till*	*Combination*
Regulating services	**Soil erosion protection** (prior service)	81	↗ (82)	↗ (82)	↗ (82)	↗ (82)	↗ (92)	↗ (98)
	Drought risk regulation	45	→ (45)	↗ (46)	→ (45)	↗ (46)	↗ (49)	↗ (49)

(Continued)

Table 26.2 (Continued)

Scenario / *ES*	*ES*	*BAU*	*Greening I*	*Greening II*	*Greening III*	*Hedges*	*No-Till*	*Combination*
Regulating services	**Flood Regulation**	69	→ (69)	↗ (70)	↗ (71)	↗ (70)	↗ (76)	↗ (80)
	Ecological Integrity	34 (+/-0)	→ (34+/-0)	↗ (34+5)	↗ (34+5)	↗ (35+15)	↗ (40+/-0)	↗ (42+15)
Provisioning services	**Provision of food and fodder**	40	→ (40)	↘ (39)	↘ (39)	↘ (37)	↘ (37)	↘ (35)
	Provision of biomass for heating and energy production	54	→ (54)	→ (54)	↘ (53)	↘ (53)	↘ (51)	↘ (48)
Cultural services	**Landscape aesthetics**	43 (-10)	↗ (44–10)	↗ (43–5)	↗ (44–10)	↗ (45+5)	→ (43–10)	↗ (46+5)

Scenario 6, involving a combination of erosion control measures, was the most recommendable and was included in the updating of the regional plan for the model region. The trade-offs for the provision of food, fodder, and biomass were tolerated due to the eminent economic importance of reducing the water erosion.

The array of opportunities to change and adapt land management in favour of regulating services can range from total conservation or non-use of areas to intensification strategies (Tscharndtke et al., 2012). Conservation or non-use can be applicable for areas that are identified as hot spots for globally relevant regulation services. An example is the strategy to conserve tropical forests due to their acknowledged role for global climate regulation (Strassburg et al., 2010). The basic motivation for intensification is to ensure a desired level of supply with food, fodder, or other bio-resources, without opening up new areas for agriculture (Tilman et al., 2011). Intensification is applicable in situations where either not enough land is available for producing basic resources, e.g. in densely settled world regions, or where land should be reserved for other purposes, e.g. for the primary provision of other service groups, such as cultural services (Mortimore, 1993, DeFries et al., 2004). Land sparing by intensification strategies can also be relevant to make improved use of specific capacities of land, depending on its location or on the land use type, to provide regulating services (Hale et al., 2014). An example for place-specific capacities to provide regulating services is land along floodplains and close to gullies to mitigate flood events and water erosion (Poesen and Hooke, 1997). Forests, and particularly old-growth forests, or extensively managed meadows are examples of land use types with high capacities to contribute to regulating services. They can support species diversity and therefore provide biological pest control and can contribute to conserve gene pools for the adaptation of land systems to future situations (Helfenstein and Kienast, 2014).

Both non-use and intensification might imply segregation strategies in land use planning that need to be carefully examined, because they can provoke spatial inequalities by the overuse of some areas in favour of the conservation of others (Castella et al., 2013). Such spatial inequalities can run the risk of ignoring interactions between different scales that drive the overall provision of regulating services in a land system. For instance, protecting areas due to their high biological diversity may not be successful if these areas are not embedded in an overall biotope network that supports species migration and refreshment of gene pools. To avoid negative trade-offs from spatial inequalities, integrative concepts that produce synergy effects from multifunctional (or multi-services) land use are to be preferred (Maes et al., 2012). Examples of how to achieve and make use of synergy effects by multifunctional land use include agro-forestry systems, optimized crop-sequences, and short rotation coppices in agricultural systems. These multifunctional land uses consider in equal measure water and nutrient cycle regulation, increase the carbon sequestration in soils, contribute to biological pest control, and reduce or mitigate water erosion and sediment transfer (Fürst et al., 2013b; Frank et al., 2014).

In a local and regional context, the sustainable provision of regulating services cannot be achieved solely by measures such as low input management at the planning unit scale, or by increasing soil organic matter and management of vegetation cover. On the contrary, a focus on the spatial configuration of different land uses and landscape elements is needed to ensure that the potential of each land use to provide regulating services is fully realised (Frank et al., 2012).

Pollination is an example of a regulating service where landscape configuration is important (Steffan-Dewenter et al., 2002). For animal pollinators, distance to suitable habitats is important, and for wind pollination the configuration of functional elements is a significant controlling factor. Similarly, distance plays a major role in pest and diseases control (Wright, 2002). Examples are the availability and critical mass of predators, natural enemies, and parasites of pests or carriers for human, animal, and plant diseases, or the carrier type-dependent distances between host plants in the development cycle of fungal plant parasites. Also, regulating services such as hazard (flood) mitigation and water erosion and sediment transfer control are determined by landscape structural aspects such as average size, form, and constellation of land use classes with differentiated potential to contribute to the intended regulation (Goldman et al., 2007).

Though the relevance of the landscape configuration for regulating services is broadly acknowledged (Mitchell et al., 2013), the implementation of this knowledge in landscape planning is confronted with the problem of interfering in privately owned land: landscape plans that intend to restructure the land use pattern for enhancing regulating services run the risk of disfavouring single land owners and need to be accompanied by financial instruments to be successfully implemented (Goldman et al., 2007).

We recognize that realisable measures in a given legal frame can be implemented most easily at the local scale. A pre-request for the successful enhancement of regulating ES that are relevant in a regional and global context are strategies that help to nest and bridge scale effects. Such strategies ensure the coordination of actions beyond the scale of the management planning unit, and across administrative or national borders (e.g. Hermann et al., 2014). Also, trade-offs between regulating services, whose relevance for human well-being cannot be directly expressed in monetary terms, and provisioning services should be incorporated in multi-scale planning. Furthermore, there is a need to consider trade-offs and interferences between regulating and cultural ecosystem services: cultural services might often experience the same trend and direction as regulating services, depending on management intensities and spatial constellations, but can also behave contradictorily: infrastructural equipment and development that accompanies "consumable" recreation entails also fragmenting effects that disfavour, for instance, habitat connectivity.

Therefore, emphasis should be given to highly integrative land system development concepts that combine approaches for choosing the most beneficial spatial allocation for each possible land use with the best place-specific management practices (Fürst et al., 2013a).

Challenges related to governance models in land use planning and policies

One of the most relevant challenges in enhancing the provision of regulating services is the question of scale at which services are produced or consumed: most recent publications that have analysed the spatial connectivity and the transfer of services such as global climate regulation and pollination showed that the identification of ES provisioning and consumption hotspots is not easily achieved. Above all, the integration of spatial information on demands in regulating ES and areas which connect provisioning and consumption areas into planning and decision-making is lacking (Syrbe and Walz, 2012).

Another problem of incorporating regulating ES adequately in management, planning, and policy decisions is that they address not only different scales but also highly different actors or stakeholders (Fürst et al., 2012).

The relevance and importance of some services, such as the regulation of local water and nutrient cycles, can be directly observed, experienced, and assessed by a land owner. In this case, shortcomings in the provision of such services can motivate the land owner to adapt the management to reduce or avoid economic trade-offs. Examples are the regulation of vegetation cover over time to avoid water erosion, decrease evaporation, or – in case of forests – optimize the timber mass and quality productivity.

Regulating services that are determined by regional or cross-boundary biogeochemical and biophysical aspects or by the spatial configuration of land uses, however, tend to address largely different groups in society, namely those on whose areas the service is produced and those who gain benefit from its consumption. In this case, the decision if, where, and how to enhance a regulating service by adapted management can create social inequalities: if a regulating service such as flood control cannot be achieved, downstream actors are disfavoured for the sake of not intervening into the land use decisions of upstream land owners. On the other hand, private land owners would be penalized if they were forced to adapt their management exclusively in favour of societal goals (Hart et al., 2013). Without economic compensation, this might provoke adverse effects through rural depopulation or land sales. As a result, cultural landscapes and family land tenure structures can be destroyed and lead to land abandonment or an increase in the average size of land units in the context of industrial agriculture (McGrath, 2012). Both land abandonment and more industrial land use practices can lower a region´s capacity to provide regulating services: an example of negative effects of land abandonment is a higher probability of land slides or avalanches in the mountain regions, where only a careful management of ecosystem structures ensures long-term hazard mitigation (Montanarella, 2006). Industrial land use practices severely change the landscape structure, with impacts for water erosion and functional diversity. Also, the functionality, buffer, and filter capacity of industrially managed agricultural soils can decrease, which endangers the regulation of water and nutrient cycles and lowers the carbon sequestration (Björklund et al., 1999).

Place-specific capacities, in contributing to a regulating service, can also create social inequalities at global scale. The Kumar (2010) study highlighted, for instance, large discrepancies in monetary values of tropical forests for locally relevant provisioning services and regionally or globally relevant regulating services. Residents in world regions that depend economically on the exploitation of land in global hot spots for climate regulation, or land users whose parcels are particularly prone to contribute to flood regulation, should not be punished by greater restrictions than other land users or land owners without adequate compensation.

The question of how to overcome social inequalities expressed as potential trade-offs addresses the problems of scale and relevance in decision-making: though the monetary value of a hectare of tropical forest in the Amazon area for global climate regulation and regional flood control exceeds by far its value for local food, fodder, and timber provision, there is no legal or institutionalized instrument available to make payments to affected land owners for maintaining or increasing highly valuable regulating services. Similarly, public funding mechanisms, as in the context of the Common Agricultural Policy of the EU, are unable to compensate for spatial differences in the capacity to contribute to regulating services (Matthews, 2013). Payments for ES through private funding, or in the context of greening strategies by industrial organisations, could help to solve the problem. Examples of successes already exist in the Netherlands, where a brewery pays for enhancing the ES provision, or in Germany, where a beverage company preferentially purchases fruits for their drinks from farmers who prove that they make an above average contribution to enhancing biodiversity (von Haaren and Bathke, 2008; Steingröver et al., 2010). The potential to transfer such examples to other regions is somehow restricted: industrial actors might not be willing to invest in regions which are not interesting as markets for their products or where the perception of their commitment to conserve nature and biodiversity has no quantifiable impacts for their economic success.

Changes in the governance of land use planning and decision-making could constitute a more sustainable pathway for enhancing the provision of regulating services. A promising development could be, for instance, community-based planning that results in collaborative actions and in community specific mechanisms to balance social inequalities. A benefit of such a model is that the identification of areas where regulating services can best be produced is based upon common knowledge, experience, and consensus-building. Public perception of such valuable areas is raised (Steingröver et al., 2010), and might result in stronger motivation for land owners to contribute by an increased provision of regulating services to societal goals. Likewise, increased public perception can cause higher social pressure on land owners to improve their management and to behave in socially acceptable ways. Another benefit can be that the costs for changing the management can be formulated in a spatially specific way so that compensation measures can be implemented more easily. Social inequalities could be addressed by direct transfer payments from land owners who can continue their business as usual, to those who must adapt or to a community fund.

Although such strategies have been partially implemented, as in the example of the schemes for rural development in Europe (e.g. LEADER / Integrated Rural Development Regions), they are mainly successful in smaller areas or social-ecological systems with a limited number of actors. Nationally different legal regulations can question the applicability of such governance schemes (Fürst et al., 2014): in countries with a strict hierarchical approach for land use planning, such bottom-up processes can contradict standardised decision-making procedures and provoke conflicts between institutionally responsible actors. In such cases, "hybrid" solutions that support consensus-building between institutional and communal responsibilities should be developed. When basing the enhancement of regulating ES exclusively on bottom-up processes, a risk is that bottom-up processes can place too much emphasis on local or regionally relevant services. As a result, globally relevant services or those relevant for future generations might be overseen.

Regulating services can transcend national borders, as in the case of flood regulation in river basins or water purification. In such cases, governance models that support trans-regional partnerships that do not violate different national legal regulations are needed. These could be based on forming new communities that develop action plans and establish cross-border institutionalised coordination. Examples can be found already in the EURO Regions, in river basins (see Box 26.2) or historically grown geopolitical regions with transnational collaboration.

Even in such a broader-scale context, global regulating services cannot always be successfully addressed. To cooperate on enhancing global regulating services would demand voluntary global partnerships and the development of respective funds between stakeholders, or necessitate models where consumers or those impacting regulating services are requested to pay into a fund for compensating social inequalities. Examples therefore can be taken from emissions trading, which could be widened to pay for adapted land management or even for renouncing land use at the globally most efficient places. Though having potential to succeed in enhancing the provision of regulating ES by such payment based schemes, the risk remains that they destroy the cultural identity of and social relationship to the region where affected actors live. This could endanger cultural ES. Consequently, approaches to pay for regulating ecosystem services or to build a global governance of regulating services should be accompanied by an in-depth analysis of potential trade-offs for societal structures, behaviour, and traditions.

Box 26.2 Case study on flood regulation governance at a river basin scale

The Volta River Basin is one of the major West African river basins, shared by six countries: Benin, Burkina Faso, Côte d'Ivoire, Ghana, Mali and Togo (Figure 26.3). Direct and indirect regulating services obtained from the basin include water and nutrients cycling, carbon storage, soil stabilization, and natural hazard regulation (flooding). Throughout the Basin, dams and reservoirs have been constructed to regulate floods, and to store and mobilize water for agriculture, electricity generation, and household purposes.

Flooding in the Volta Basin has a trans-boundary dimension: an increasing use of the water resources due to population growth and decreasing precipitation through climate change, poor agricultural management practices and reservoir construction, and uncontrolled and uncoordinated dam releases from the upper part of the basin threaten the sustainable management of the basin, particularly during and after flooding events (McCartney, Cai, and Smakhtin, 2013). Key trans-boundary concerns of the basin are related to governance, institutional constraints, and disharmonized legal and policy frameworks that undermine effective water resource management at the national and regional levels. Basin states have limited ability and capacity to implement and enforce policies and reforms at all levels of implementation, particularly at the local level. Policy implementation, if attempted, takes place in the context of multiple foci of power and institutions, resulting in duplicated effort and perplexity.

Duality between the legal state and the traditional hierarchies in the basin (land tenure at ethnical group level) impacts land and water resource utilisation. The lack of multi-scale, basin-wide monitoring and modelling results in a limited capacity to predict and mitigate flood propagation through the (sub)catchments.

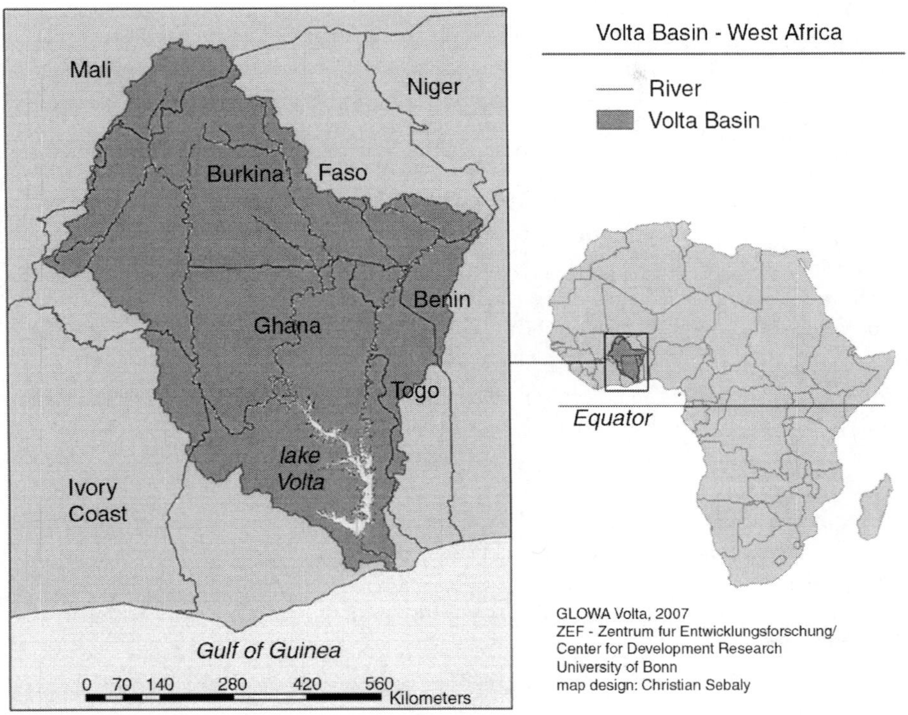

Figure 26.3 The Volta River Basin (LIT).

In 2010, negotiations between Ghana and Burkina Faso, which own the largest part of the Upper Volta Catchment, for a joint Upper Volta Basin Management Agency started to improve trans-boundary flood regulation. The Agency has the mandate to set up action plans and policies to mitigate perennial floods and manage excess water spilled from dams in the upper part of the basin. Other efforts include the strengthening and connection of institutions (e.g. Volta Basin/River Authority) through the establishment of decision support information bases, design of comprehensive housing schemes for flood-prone areas, and monitoring and flood early warning systems to provide prompt and reliable information to the local authorities and citizens, respectively.

However, the regulation of the river by dams and other technical measures led to severe trade-offs for other regulating or provisioning services: the interrupted natural sediment transport endangered traditional floodplain agriculture. Barriers to fish migration severely impacted the natural fish population and diversity and provoked losses for fisheries.

For efficient implementation of flood regulation governance, different needs and perceptions of stakeholders at multiple scales must be respected. Participatory and scenario-based decision support tools supplied with appropriate scientific data and information are required to adequately facilitate evidence-based policy formulation, planning, and decision-making operating laterally and across sectoral scales of riparian countries.

Conclusions

To ensure a sustainable provision of regulating services requires conducting the most appropriate measures or combination of measures at the right scale. This should take into consideration the requirement that measures must be nested if they deal with globally relevant regulating services, where an adapted land management must go hand in hand with accompanying policy interventions and global governance models.

The decision how to identify the right scale and the right measures, or measure combinations, must be made dependent from four key questions, namely:

- Who is concerned or dependent on the provision of a regulating service? By answering this question, the scale at which an impact will result if the regulating service is provided or not can be identified.
- What are the decisive criteria that impact the provision of the regulating service – is it more dependent on biogeochemical processes and biophysical properties of the land cover or are landscape structural aspects dominant (acknowledging that in all cases both criteria will have an impact)?
- What is the scale at which measures for sustaining, enhancing, or restoring a regulating service can be undertaken most efficiently?
- What are the consequences for taking responsibility of these measures and how could these be governed; where will the financial responsibility be located?

Figure 26.4 shows how a decision tree could be structured according to the four questions so that conditional links between its single elements support the selection of appropriate measures, measure alternatives, or measure combinations. The examples marked by the numbers "1" and "2" were selected to demonstrate how to use the decision tree.

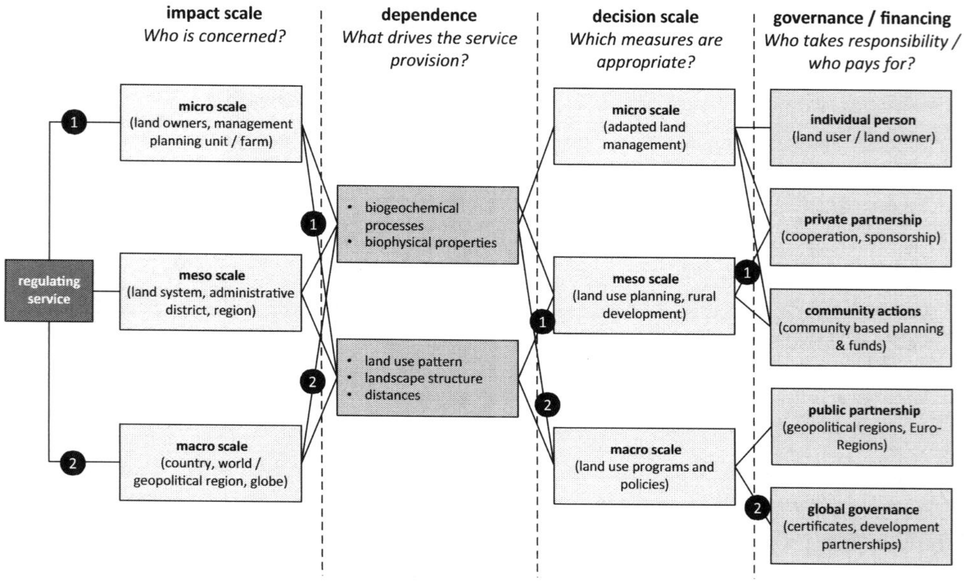

Figure 26.4 Decision tree to identify the best strategies for sustaining and enhancing regulating services.

Example 1: A regulating service, e.g. pollination, is important at local scale for a farmer. This service is dominantly dependent on distances to habitats for pollinators or places where these are actively brought in (beekeeping). In this example, two scales need to be brought together for enhancing a regulating service: sustaining or restoring a habitat quality entails management measures at the local scale, but they are not exclusively in the decision space of one individual person. As the management of this distance dependence can only be undertaken at the meso-scale, but the importance is highest for the farmer, the appropriate measure would be a private partnership either with neighbours to sustain habitat quality for pollinators or with beekeepers.

Example 2: A regulation service such as carbon sequestration is of global importance. This service is dependent on biogeochemical processes. However, as its importance considers humanity in general, not one individual actor whose parcel management should be changed in particular, the appropriate measures can primarily be taken at the policy scale and could be realized by remunerations in the context of the emissions trading or development partnerships, adapted regional development programs and finally adapted land management at the micro scale.

Such a structured decision process must be described in detail for all regulating services, and could facilitate the identification of best strategies or eligible alternatives. For a successful application of the measures and acceptance in practice, the following take-home messages can be summarized:

1 Regulating ES cannot easily be expressed in monetary terms because their contribution to human well-being is often indirect, and most of these services require the consideration of multiple scales in order to understand where they are produced and where they are consumed.
2 To identify the most successful measures to enhance regulating services, their dependence on biogeochemical processes and biophysical properties of the land and on landscape structural aspects and distances needs to be considered.
3 An essential decision criterion to identify the most appropriate scale for actions for enhancing regulating services is the question of where their impact is most relevant. This determines who can take responsibility and how successful measures or measure combinations should be governed.

Acknowledgements

We wish to thank the organizers of the *Handbook* for the opportunity to contribute in an interdisciplinary context experiences from case studies on ES assessment, mapping, and challenges for the implementation in planning and management. We wish to thank the anonymous reviewers for their support, advice, and help. The case studies were taken from the projects RegioPower (22019911(11NF199) funded by the German Federal Ministry of Food and Agriculture in the context of the ERA WoodWisdom / Bioenergy, REGKLAM (01LR0802B) and Glowa Volta (01LW0302A) funded by the German Federal Ministry of Education and Research.

References

Altieri, M. A. (1999). The ecological role of biodiversity in agroecosystems. *Agriculture, Ecosystems and Environment*, vol 74, pp 19–31.

Björklund, J., Limburg, K.E., and Rydberg, T. (1999). Impact of production intensity on the ability of the agricultural landscape to generate ecosystem services: an example from Sweden. *Ecological Economics*, vol 29, no 2, pp 269–291.

Bommarco, R., Kleijn, D., and Potts, S.G. (2013). Ecological intensification: harnessing ecosystem services for food security. *Trends in Ecology and Evolution*, vol 28, no 4, pp 230–238.

Castella, J. C., Lestrelin, G., Hett, C., Bougoin, J., Fitriana Rahma, Y., Heinimann, A., and Pfund, J. L. (2013). Effects of landscape segregation on livelihood vulnerability: moving from extensive shifting cultivation to rotational agriculture and natural forests in northern Laos. *Human Ecology*, vol 41, no 1, pp 63–76.

Dale, V. H., and Polasky, S. (2007). Measures of the effects of agricultural practices on ecosystem services. *Ecological Economics*, vol 64, no 2, pp 286–296.

DeFries, R. S., Foley, J. A., and Asner, G. P. (2004). Land-use choices: balancing human needs and ecosystem function. *Frontiers in Ecology and the Environment*, vol 2, pp 249–257.

Ernstson, H., Barthel, S., Andersson, E., and Borgström, S. T. (2010). Scale-crossing brokers and network governance of urban ecosystem services: the case of Stockholm. *Ecology and Society*, vol 15, no 4, p 28.

Frank, S., Fürst, C., Koschke, L., and Makeschin, F. (2012). Towards the transfer of the ecosystem service concept to landscape planning using landscape metrics. *Ecological Indicators*, vol 21, pp. 30–38.

Frank, S., Fürst, C., Witt, A., Koschke, L., and Makeschin, F. (2014). Making use of the ecosystem services concept in regional planning – trade-offs from reducing water erosion. *Landscape Ecology*, vol 29, no 8, pp 1377–1391.

Fürst, C., Frank, S., Witt, A., Koschke, L., and Makeschin, F. (2013b). Assessment of the effects of forest land use strategies on the provision of Ecosystem Services at regional scale. *Journal of Environmental Management*, vol 127, pp 96-S116.

Fürst, C., Helming, K., Lorz, C., Müller, F., and Verburg, P. (2013a). Integrated land use and regional resource management – A cross-disciplinary dialogue on future perspectives for a sustainable development of regional resources. *Journal of Environmental Management*, vol 127, pp 1–5.

Fürst, C., König, H., Pietzsch, K., Ende, H. P., and Makeschin, F. (2010a). Pimp your landscape – a generic approach for integrating regional stakeholder needs into land use scenario design and sustainable management support. *Ecology and Society*, vol 15, no 3, p 34.

Fürst, C., Lorz, C., and Makeschin, F. (2011). Integrating land management aspects into an assessment of the impact of land cover changes on Ecosystem Services. *International Journal of Biodiversity Science, Ecosystem Services Management*, vol 7, pp 1–14.

Fürst, C., Opdam, P., Inostroza, L., and Luque, S. (2014). A balance score card tool for assessing how successful the ecosystem services concept is applied in participatory land use planning. *Landscape Ecology*, vol 29, no 8, pp 1435–1446.

Fürst, C., Pietzsch, K., Frank, S., Witt, A., Koschke, L., and Makeschin, F. (2012). How to better consider sectoral planning information in regional planning – Example afforestation and conversion. *Journal of Environmental Planning and Management*, vol 55, pp 1–29.

Fürst, C., Volk, M., Pietzsch, K., and Makeschin, F. (2010b). Pimp your landscape! A tool for qualitative evaluation of the effects of regional planning measures on ecosystem services. *Environmental Management*, vol 46, no 6, pp 953–968.

Goldman, R. L., Thompson, B. H., and Daily, G. C. (2007). Institutional incentives for managing the landscape: inducing cooperation for the production of ecosystem services. *Ecological Economics*, vol 64, no 2, pp 333–343.

Hale, I. L., Wollheim, W. M., Smith, R. G., Asbjornsen, A., Brito, A. F., Broders, K., Grandy, A. S., and Rowe, R. (2014). A scale-explicit framework for conceptualizing the environmental impacts of agricultural land use changes. *Sustainability*, vol 6, no 12, pp 8432–8451.

Hart, K., Allen, B., Lindner, M., Keenleyside, C., Burgess, P., Eggers, J., and Buckwell, A. (2013). *Land as an Environmental Resource.* Report Prepared for DG Environment, Contract No ENV.B.1/ETU/2011/0029, Institute for European Environmental Policy, London.

Helfenstein, J., and Kienast, F. (2014). Ecosystem service state and trends at the regional to national level: a rapid assessment. *Ecological Indicators*, vol 36, pp 11–18.

Hermann, A., Kuttner, M., Hainz-Renetzeder, C., Konkoly-Gyuró, E., Tirászi, A., Brandenburg, C., Allex, B., Ziener, K., and Wrbka, T. (2014). Assessment framework for landscape services in European cultural landscapes: an Austrian Hungarian case study. *Ecological Indicators*, vol 37, pp 229–240.

Koschke, L., Fürst, C., Lorenz, M., Witt, A., Frank, S., and Makeschin, F. (2013). The integration of crop rotation and tillage practices in the assessment of ecosystem services provision at the regional scale. *Ecological Indicators*, vol 32, pp 157–171.

Koschke, L., Lorz, C., Fürst, C., Lehman, T., and Makeschin, F. (2014). Assessing hydrological and provisioning ecosystem services in a case study in Western Central Brazil. *Ecological Processes*, vol 3, no 2. Available at: http://www.ecologicalprocesses.com/content/3/1/2 (accessed 13 February 2015).

Kumar, P. (2010). *The Economics of Ecosystems and Biodiversity (TEEB)*. Earthscan, London.

Lorenz, M., Fürst, C., and Thiel, E. (2013). A methodological approach for deriving regional crop rotations as basis for the assessment of the impact of agricultural strategies using soil erosion as example. *Journal of Environmental Management*, vol 127, pp 37–47.

Maes, J., Egoh, D., Willemen, L., Liquete, C., Vihervaara, P., Schägner, J. P., Grizzetti, B., Drakou, E. G., La Notte, A., Zulian, G. Bouraoui, F., Paracchini, M. L., Braat, L., and Bidoglio, G. (2012). Mapping ecosystem services for policy support and decision making in the European Union. *Ecosystem Services*, vol 1, no 1, pp 31–39.

Maes, J., Paracchini, M. L., Zulian, G., Dunbar, M. B., and Alkemade, R. (2014). Synergies and trade-offs between ecosystem service supply, biodiversity, and habitat conservation status in Europe. *Biological Conservation*, vol 155, pp 1–12.

Matthews, A. (2013). Post-2013 EU Common Agricultural Policy, Trade and Development. *ICTSD Issue Paper 39*.

McCartney, M., Cai, X., and Smakhtin, V. (2013). *Evaluating the Flow Regulating Functions of Natural Ecosystems in the Zambezi River Basin*. International Water Management Institute Report 148. doi:10.5337/2013.206 (accessed 13 February 2015).

McGrath, D. M. (2012). *Reintegrating Canadian Agriculture and Ecological Land Management*. Doctoral Thesis, University of Waterloo.

Mitchell, M.G.E., Bennett, E. M., and Gonzalez, A. (2013). Linking landscape connectivity and ecosystem service provision: current knowledge and research gaps. *Ecosystems*, vol 16, no 5, pp 894–908.

Montanarella, L. (2006). Policies for a sustainable use of soil resources. *Geological Society*, vol 266, pp 149–158.

Mortimore, M. (1993). Population growth and land degradation. *GeoJournal*, vol 31, no 1, pp 15–21.

Poesen, J.W.A., and Hooke, J. M. (1997). Erosion, flooding and channel management in Mediterranean environments of southern Europe. *Progress in Physical Geography*, vol 21, no 2, pp 157–199.

Potschin, M. and Haines-Young, R. (2011). Ecosystem services: exploring a geographical perspective. *Progress in Physical Geography*, vol 35, no 5, pp 575–594.

Renard, K. G., Foster, G. R., Weesies, G. A., and Porter, J. P. (1991). RUSLE: Revised Universal Soil Loss Equation. *Journal of Soil and Water Conservation*, vol 46, no 1, pp 30–33.

Sandhu, H. S., Crossman, N. D., and Smith, F.P. (2012). Ecosystem services and Australian agricultural enterprises. *Ecological Economics*, vol 74, pp 19–26.

Steffan-Dewenter, I., Münzenberg, U., Bürger, C., Thies, C., and Tscharntke, T. (2002). Scale dependent effects of landscape context on three pollinator guilds. *Ecology*, vol 83, pp 1421–1432.

Steingröver, E., Geertsema, W., and van Wingerden, W.K.R.E. (2010). Designing agricultural landscapes for natural pest control: a transdisciplinary approach in the Hoeksche Waard (The Netherlands). *Landscape Ecology*, vol 25, no 6, pp 825–838.

Strassburg, B.B.N., Kelly, A., Balmford, A., Davies, R. G., Gibbs, H. K., Lovett, A., Miles, L., Orme, C.D.L., Price, J., Turner, R. K., and Rodrigues, A.S.L. (2010). Global congruence of carbon storage and biodiversity in terrestrial ecosystems. *Conservation Letters*, vol 3, no 2, pp 98–105.

Syrbe, R. U., and Walz, U. (2012). Spatial indicators for the assessment of ecosystem services: providing, benefiting and connecting areas and landscape metrics. *Ecological Indicators*, vol 21, pp 80–88.

Tilman, D., Balzer, C., Hill, J., and Befort, B. L. (2011). Global food demand and the sustainable intensification of agriculture. *PNAS*, vol 108, no 50, pp 20260–20264.

Tscharntke, T., Clough, Y., Wanger, T. C., Jackson, L., Motzke, I., Perfecto, I., Vandermeer, J., and Whitbread, A. (2012). Global food security, biodiversity conservation and the future of agricultural intensification. *Biological Conservation*, vol 151, no 1, pp 53–59.

Von Haaren, C., and Bathke, M. (2008). Integrated landscape planning and remuneration of agri-environmental services. Results of a case study in the Fuhrberg region of Germany. *Journal of Environmental Management*, vol 89, pp 209–221.

Watanabe, M.D.B., and Ortega, E. (2011). Ecosystem services and biogeochemical cycles on a global scale: valuation of water, carbon and nitrogen processes. *Environmental Science & Policy*, vol 14, no 6, pp 594–604.

Witt, A., Fürst, C., and Makeschin, F. (2013). Regionalization of Climate Change: sensitive forest ecosystem types for potential afforestation areas. *Journal of Environmental Management*, vol 127, pp 48–55.

Wright, J. S. (2002). Plant diversity in tropical forests: a review of mechanisms of species coexistence. *Oecologia*, vol 130, no 1, pp 1–14.

27

MANAGING CULTURAL ECOSYSTEM SERVICES FOR SUSTAINABILITY

Kai M. A. Chan and Terre Satterfield

Introduction

It is widely recognized that ecosystems provide people with great non-material benefits "through spiritual enrichment, cognitive development, reflection, recreation and aesthetic experiences" (MA, 2005). These ecological contributions to non-material or 'extra-material' benefits, including both experiences and capabilities, reflect people's interactions with nature and have come to be known amongst ecologists and resource managers as 'cultural ecosystem services' (CES) (Chan et al., 2011). These ecosystem services (ES) are some of the most salient and compelling reasons for people to conserve or restore ecosystems (Chan et al., 2012a). Accordingly, the management of CES is an essential consideration for sustainability, both because CES are crucial contributors to human well-being and because they may be key to sustainable human-ecological relationships.

Whereas CES were initially considered to be one class of ES, alongside other categories (the Millennium Ecosystem Assessment also recognized provisioning, regulating, and supporting ES) (Daily, 1997, MA, 2005), it has become increasingly clear that cultural practices and phenomena and CES are intimately linked and the lens through which many other ES derive meaning (de Groot et al., 2005; Chan et al., 2011; Church et al., 2011; Chan et al., 2012b; Poe et al., 2013). Indeed, in some social-ecological systems, provisioning services are so important locally precisely *because* they are the conduits for CES. In coastal British Columbia, Canada, for example, research has indicated that many extra-material benefits are associated with provisioning services, especially fishing of all kinds (e.g., subsistence or food fish, commercial, or recreational) (Klain et al., 2014; Klain and Chan, 2012). City dwellers appear to be comparatively more focused on non-material CES derived from non-extractive recreational experiences and associated imagining (Martín-López et al., 2012), but CES continue to provide the setting for human relationships with and understandings of nature, shaping the meaning of these relationships (Church et al., 2011; Chan et al., 2012b; Tengberg et al., 2012; Gould et al., 2014a; Matsuoka and Kaplan, 2008). More recently, the centrality of the meaning-making process associated with knowing, perceiving, interacting with, and living within landscapes (Russell et al., 2013) has been coined the 'culturality of ecosystem services' (Pröpper and Haupts, 2014). In short: as a class of ES, CES are foundational: managing CES means managing other ES, and vice versa.

Although the notion of sustainability is present explicitly or implicitly in prominent writing on ES (MA, 2005; TEEB, 2009), the culturality of ES shines a light on an unresolved tension that has yet to be addressed. Whereas it might be simple to imagine the sustainable management of a provisioning or regulating ES, what does it mean to sustainably manage CES? Human needs for protein, fiber, and physical security might seem stable requisites of any system. Likely less stable are cultural and social systems and linked CES, as changes to these systems may predictably or unpredictably change CES in complex ways. This is particularly so when trying to 'manage' for different futures, developments, or generations. What landscape features and configurations will future generations find attractive? What sites and landscapes will they find spiritually fulfilling, saturated with meaning and histories of place, or appropriate setting for social well-being? Clearly, for CES – and so all ES – sustainable management implies more than the maintenance of particular ecosystem stocks or flows.

What sustainability might imply for CES – and so all ES – is the subject of this chapter. We first briefly review the history of research and thinking on CES. We then consider more fully the potential operationalization of connections between biophysical attributes and culture or non-material values. Turning our attention to sustainability, we discuss the implications of competing notions of sustainability for management. On that basis, we integrate considerations of CES and sustainability to advance several propositions regarding appropriate management for CES and so all ES.

Cultural ecosystem services, a brief history: everywhere and nowhere

The number of disciplines and sub-fields that have documented the ways in which people derive non-material benefit from ecosystems is truly dizzying: psychology, education, anthropology, sociology, planning, environmental health, medicine, diverse humanities, etc. (Russell et al., 2013). And yet, despite this tremendous wealth of literature, very few have sought to characterize these ecosystem contributions as CES, that is, to characterize and assess the magnitude of all manner of positive ecosystem contributions to individual social and cultural health (Russell et al., 2013; with a few exceptions, Daniel et al., 2012). And within the ES literature itself (a literature newer than the fields listed above), there is a notable dearth of attention to most CES (Hernández-Morcillo et al., 2013). One possibility is that this is due to their sometimes incommensurate value, intangibility and ontological complexity (Chan et al., 2012b). Ultimately, this amounts to no small irony: CES pervade all ES and similar literatures, but their characterization as such is nascent or in its infancy at best (Chan et al., 2012a; Daniel et al., 2012).

This 'everywhere and nowhere' paradox of CES stems from the central advance in the ES literature, *which is also its greatest limitation*. The ES literature focuses exclusively on quantification of the ecological contribution, with special attention to changes in ES via change in ecosystems (Daily et al., 2009; Kareiva et al., 2011), while admitting but not focusing on a wider array of social and social-ecological interactions. This focus of conventional ES has been enormously productive as concerns economic approaches to cost-benefit analysis and valuation of environmental goods – a productivity also reflected in the ES concept's burgeoning popularity (Costanza and Kubiszewski, 2012). It also privileges a single, incomplete way of thinking about social-ecological interactions (Raymond et al., 2013).

Because some CES are much more conducive to rapid assessment using quantitative approaches, treatment of CES tends to focus on visual aesthetics, and the relative desirability of different landscapes (again, aesthetically). Thus, physical and structural changes have been linked to assessments of desirability (Daniel, 2001; Chamberlain and Meitner, 2012), and/or their worth in monetary terms (Benson et al., 1998; Grêt-Regamey et al., 2008). Similarly, the study of recreation and tourism includes various survey and experimental approaches to isolate the contribution of rock paintings, bird species, and other wilderness attributes to demand or willingness to pay (Boxall et al., 2003; Naidoo and Adamowicz, 2005), such that general models of ecosystems' contribution to recreation and tourism are available for ES modeling (Adamowicz et al., 2011). In general, these treat CES as 'culture' in the high- or popular-art and aesthetics point of view, and not culture in the broader sense of the diverse practices, meanings, identities, and human expressions and organizations that make for unique human communities (Satterfield et al., 2013).

For many researchers of culture, the suggestion implicit in 'cultural ecosystem services' that nature produces culture is both laughable and offensive (Fish, 2011). Disentangling the contribution of particular ecosystem structures and functions to such 'big' ideas as "belonging, cultural heritage, and other symbolic meanings" strikes some as impossible (Kirchhoff, 2012). There are also concerns about the economic metaphor at the core of ES (Raymond et al., 2013; Luck et al., 2012), and about the exclusive utilitarian and anthropocentric focus, which excludes consideration of other ethical perspectives (including principles, virtues, and other relational values) (Jax et al., 2013; Chan et al., accepted) and of the needs of non-human nature (Rees, 1998; Luck et al., 2012). Moreover, there is the problem of quantification in a single metric, which presupposes that diverse values are commensurate (permit trading off), when in fact many people resist such trade-offs in many settings (Ludwig, 2000; Chan et al., 2012b; Satz et al., 2013). The valuation in monetary terms triggers concerns about commodification, in part because dominant socio-political trends and the particular institutional settings for ES (ES markets and payments for ES) push simultaneously toward valuation and commodification (Gómez-Baggethun and Ruiz-Pérez, 2011).

The quandary of 'cultural' nested in ecosystem services

If non-material or extra-material values pervade most if not all ES, and if many such 'cultural' values are also complex and resist quantification, does it follow that the whole project of ES is a boondoggle? Must all ES research dwell on the value complexity of non-material or extra-material considerations? Must studies of agricultural ES also quantify non-material values affected by the altered production or availability of commercial produce? Must studies of the ecological provision of hydropower, for instance, also consider the cultural significance of more abundant energy?

Our position here is that while most ecological or ES changes will affect some non-material experiences or capabilities, some contexts and problems seem to require explicit consideration of value complexity and others do not. First, many ES studies simply quantify ecosystem supply (Tallis et al., 2012), for which biophysical metrics are appropriate. Other studies can interpret the value significance of such biophysical quantities. Second, even when the scope includes benefits and values, and there are important non-material values at play, one-dimensional monetary values may suffice. Tourism and recreation are a case in point: market valuation and travel-cost methods are widely accepted and relatively uncontroversial methods for valuing these benefits associated with ecosystems and particular ecological attributes. On the other hand, few imagine quantifying spiritual benefits associated with ecosystems in monetary terms (Daniel et al., 2012), although context-specific constructed scales or quantitative participatory scoring offers considerable promise (Satterfield et al., 2013).

The applicability of one-dimensional valuation techniques depends on the nature of the value at stake. Chan and colleagues identify eight dimensions of variation in values (preferences vs. principles vs. virtues; market-mediated vs. not; self- vs. other-oriented; individual vs. holistic/group; experiential vs. metaphysical; supporting vs. final / instrumental vs. inherent;

transformative vs. not; and anthropocentric vs. biocentric) (Chan et al., 2011). Each dimension has implications for the appropriate valuation processes (Chan et al., 2012b), but a key consideration for the appropriateness of one-dimensional valuation is whether the benefits in question are mediated mainly by markets. In short, if we have the practice of market exchanges, there are opportunities for respondents' willingness-to-pay or accept estimations to be corrected by experience (Sagoff, 1998). Tourism, although largely non-material, is mostly market-mediated. Some forms of recreation are also mostly market-mediated in the sense that they require considerable purchases or expenses in order to access sites.

When the benefits at stake are mostly market-mediated, a variety of approaches to valuation may be appropriate, including market valuation, travel-cost method, hedonic valuation, and avoided costs (EPA Science Advisory Board, 2009). To the extent that the experiences and capabilities in question are realized partly outside of markets (e.g., entry to a regional park is free because of a community-wide commitment to public access), the results of such monetary valuations are likely under-representations of the worth of benefits to people. Economists recognize that such 'passive-use values' are difficult to valuate accurately, for several reasons (Adamowicz et al., 2011). Consider here the example of aesthetics. To a degree, we purchase access to visual aesthetics through home purchases or room rentals, and these can be valuated using hedonic pricing (National Research Council (U.S.), 2005). But for those who have particular attachments to a place, the assumption that all else is equal among alternative homes/rooms is violated (an assumption implicit in hedonic pricing), and the view is a function not of market transactions but rather of collective action, community-wide commitments, and larger forces (including demographic change). For example, in cultural landscapes aesthetics are inextricably intertwined with multiple values (Klain et al., 2014; Klain and Chan, 2012), including social and educational values (Plieninger et al., 2013b), sense of place, spirituality (Gee, 2010), and triggers for heritage and identity (Tengberg et al., 2012). Furthermore, one is generally not able to purchase substantial changes to one's home landscape (Plieninger et al., 2013a). Thus, while monetary valuation of landscape differences may appropriately represent the aesthetic values to tourists (Grêt-Regamey et al., 2007), one-dimensional valuation may substantially under-represent the worth of landscape differences to locals, and it may be inappropriate (Klain and Chan, 2012; Chan et al., 2012b; Klain et al., 2014).

When the benefits are not primarily mediated through markets, other approaches exist to characterize and even quantify the value dimensions of potential changes. Notable among these are two approaches, each of which includes qualitative dimensions. First, narrative-based approaches are intended to elicit values through field-tested, and even provocative questions (Satterfield, 2001; Gould et al., 2014a; Gould et al., 2014b; Klain et al., 2014). For example, regarding identity and educational values, the interviewer might ask, "Are there places that are important to your sense of identity . . ." or "Have you ever had the experience of a place(s) – or time in the forest or in or on the water – teaching you things?" (Satterfield et al., 2013; Gould et al., 2014b). Maps are helpful in narrative elicitation (Gould et al., 2014b), and the non-monetary values of sites may be spatialized and/or quantified (Klain and Chan, 2012; Satterfield et al., 2013), although this step may be resisted by a substantial number of interviewees (Klain and Chan, 2012). Narrative approaches can also be paired with photo elicitation, which enlists interviewees as active participants in the process by having them take photos of significant features, such that the photos may then be used as prompts in subsequent interviews (Sherren et al., 2010).

Second, although stakeholders may be unwilling to assign quantitative values to benefits on a pre-existing scale, they may be willing to attach relative importance rankings to characterizations of a (cultural) state or quality through 'constructed scales' (Satterfield et al., 2013). Such an approach follows the logic of value-focused decision-making, an approach that has been well honed over decades to facilitate difficult decisions with multi-dimensional value implications (Gregory et al., 2012). In this approach, stakeholders share in the construction of carefully designed quantitative-qualitative scales that can represent anything from the multi-dimensional 'spiritual quality' of a river to the meaning of community stewardship. This might be done, for example, using a five-point scale where each level is coupled with a brief place-specific verbal articulation of the qualities in question: the sound, sight, smell and feel of a river that constitute its spiritual qualities, in one example (Satterfield et al., 2013; Failing et al., 2013). Such measures cannot be transferred to other places, but they can be powerful in the adaptive management of the place in question, and they can be 'aggregated' up into more comprehensive measures such as 'quality of place' where the dimensions of quality might be spiritual quality in one setting, and 'access to beaches' in another, but overall capture locally meaningful attributes of 'place'.

Sustainability in a cultural context

To paraphrase Kitayama (2002), culture is to humans as water is to fish: cultural phenomena are the air we breathe, the norms and practices that constitute or are the currency of identity, the extra-material life of most things that give meaning to the physical world. We are immersed in culture, and it permeates every perception. The quote is also apt in that cultural worlds, like water, are constantly changing; we, like fish, are constantly adapting. If cultural change is the process of substituting new means of satisfying existing needs and preferences (Max-Neef et al., 1992), such change will occur naturally as a process of adaptation, including adaptation to ecosystem change. This adaptation may buffer, or appear to buffer, the negative consequences of ecological and ES change. What then would sustainability imply for CES? Maintaining human well-being despite cultural and ecological change (including via cultural change in response to ecological change)? Maintaining CES as they were? Or, as we will argue, maintaining CES for the sake of cultural continuity and sustainable social-ecological relationships, as decided by affected parties and by those best poised to represent the many without a voice?

These questions intentionally suggest a parallel between CES and the notions of weak and strong sustainability (e.g., Solow, 1974; Hartwick, 1977). Our goal here is not an elaborate treatment of these two perspectives, but to enrich our discussion of CES in reference to one important discourse on sustainability. These two competing visions of sustainability have been articulated in reference to the maintenance of various forms of capital, including human capital (knowledge and skills), financial capital (money), built or manufactured capital (infrastructure), and natural capital (biodiversity and ecosystem structures and functions needed to produce ES). The distinction between strong and weak lies in what must be maintained: whereas strong sustainability implies a non-diminishing stock of natural capital, weak sustainability allows substitution of human-made capital for natural capital (Cabeza Gutés, 1996). Suspending for the moment worthy points about the limitations of either perspective (Bromley, 1998; Robinson, 2004), we first consider how CES might fit into each.

At first glance, the relationship between ES in general and sustainability appears simple. Most ES, as flows of benefits, are effectively the interest or revenue streams from stocks of natural capital. By this logic, there exists a strong argument that sustainability requires describing and protecting the stocks of natural capital necessary for the sustained provision of ES (Potschin and Haines-Young, 2011). At this level, the difference between strong and weak sustainability is crucial. Strong sustainability requires that stocks of natural capital be protected without consideration for substitution of natural with human-made capital. In contrast, weak sustainability allows such substitution. Weak sustainability, however, does not explicitly address the very difficult project of determining when, and to what extent, the substitution of depreciating human-made capital for (often) self-sustaining natural capital can yield sustained stocks of total capital (however that might be measured).

For CES, we are spared that stark choice by another complexity: CES, being unlike other ES, may be considered *capital-producing*. That is, some CES yield capabilities that allow people to access other benefits, including through interactions with nature. In this way, CES produce a portion of human capital (*capabilities*) that is especially important for realizing benefits from nature (both material and non-material). We might call this 'human-natural capital'. Other CES, which produce *experiences* with nature, are essential for the production of the aforementioned human-natural capital. These experience-producing CES yield the analogue of *seed capital*: they produce the capital that enables the human-natural capital (the capabilities). (We acknowledge the irony that CES are not normally associated with capital in the market sense of the word.)

Take, for example, the case of coastal food fishing amongst First Nations off the coast of British Columbia, Canada. Chan et al. (2011) describe the surprising key value placed on black bass (*Sebastes melanops*) by First Nations men in Kyuquot Sound. The men in question quite explicitly taught one of us that only certain experiences were sufficient to captivate young boys as fishermen. The fish had to be predictably abundant enough to yield frequent bites, animated enough to provide a challenging but feasible fight, and tasty enough to produce a tangible reward. In this way, the 'gateway' experience of angling for black bass was used to sustain boys' attention for long enough that they developed the crucial capabilities, and the needed love, to fish for a lifetime and so continue to perpetuate a cultural world dependent both on wild fish and the practice of fishing. While on the water these boys likely also developed key social bonds and social capital, derived activity benefits, spiritual understanding, place attachment, the boat-handling skills and knowledge of the ocean needed for various forms of marine employment, etc. – that is, numerous other non-material benefits via CES.

Unlike natural capital, which – when not subject to large-scale environmental change – is often self-sustaining, human-natural capital depreciates. CES may degrade due to distractions, changes in normative practices, forgetting, the turnover between generations, or forced disruptions and displacement of human communities world-wide. Human-natural capital, that is, must be continually re-created. In many parts of the world, the process of renewal itself is the reconsideration and self-determination of traditional practice. Thus, traditional ecological knowledge (TEK) is renewed and transmitted through practice. Interruptions to that process of transmission can be devastating for the natural capital that is actually partly a result of human-natural coproduction.

The onslaught of colonization for First Nations peoples in Canada illustrates this loss of CES and its associated impacts on other ES and biodiversity (Turner and Turner, 2008). Residential schools programs, forced relocations, suppression of native languages, successive disease transmission and outbreaks, and the reduction of 'native space' to tiny reserves (Harris, 2011) devastated all manner of human-natural capital, be that expressed as estuarine gardens (Deur, 2002) or elaborate social organizations enabled by the sheer wealth of salmon and enormous cultural benefits derived from healthy ecosystems (Masco, 1995). Whereas residential schools and reserves restricted movement greatly, historically, many would have migrated seasonally from one harvesting ground to another (e.g., harvesting edible seaweed on rocky beaches and clams on sandy ones, camas bulbs in the grasslands, salmon in the estuaries), transmitting TEK across time/generations and space, tending the seaweed patches, 'clam gardens', camas fields, and salmon streams to enhance their future productivity (Turner and Turner, 2008; Lepofsky and Caldwell, 2013). Colonization, residential schools, and their many consequences are thus widely understood to have degraded or changed subsistence harvest in a way that eroded CES (including both experiences and capabilities), and fundamentally altered the trajectory for coastal populations and associated landscapes (Turner and Turner, 2008; Turner et al., 2008, Chan et al., 2011). Lest one imagine that this erosive scenario applies only to Indigenous or land-based peoples, one should consider hypotheses of nature-deficit disorders, which holds that loss of time and free play outdoors (Veitch et al., 2007) is impeding creativity, learning, and spirituality, and enhancing the accumulation of stress, and incurring numerous health and behavioural problems (Louv, 2008).

The discussion of capital (however understandably onerous and loaded the term is for many, ourselves included) is helpful to recognize the potential role of CES in capital formation and so sustainability. But what in practice is the sustainability of CES, given the above point about the continual unfolding of cultural practices and norms? Is it letting individuals and collectives dictate the changing nature of CES, assuming that this will maintain CES at a given level, but with desired substitutions? Should we be agnostic if CES change from non-consumptive or low-technology activities involving interactions with organisms and ecosystems (e.g., kayaking, subsistence hunting, or commercial fishing), to technology-intensive activities that treat nature more as a site for thrills (e.g., mud-running, an activity involving driving all-terrain vehicles up creeks at high speeds)? Is it paternalistic to favour some preferences over others?

Several considerations about human behaviour bear mentioning in answering the above question, essentially about CES 'consumer sovereignty' (that preferences are unquestionable) (Norton et al., 1998). First, people, and so their surrounding cultural worlds, adapt to changing circumstances (Norton et al., 1998). This adaptation has normative implications, as in the theory of shifting baselines (Pauly, 1995), which holds that each generation newly imagines the pristine or optimal state as something like what they encountered in nature as youths or young adults.

Second, although such adaptation may obscure the recognition that something is degraded progressively through time, it may nonetheless be the case that much is lost, but not immediately evident (as illustrated with First Nations and the nature-deficit hypothesis above). Agency, or the ability to determine one's fate or collectively and democratically deliberate futures desirable for a community, is fundamentally important to human well-being (Holland et al., 1998; Sparks et al., 2001; Thoits, 2006). However, the apparently free 'choices' that people make can also undermine their long-term well-being (Layard, 2006) for a wide variety of reasons, including imperfect 'rationality' (Kahneman, 2013), individual social traps (by which a behavioral choice with short-term gains acquires significant long-term harms, e.g., overeating, smoking) (Platt, 1973), cultural cognition (the tendency for people to perceive risks in ways that cohere with their political world views and associated social groups) (Kahan et al., 2012), and the compelling power of marketers who exploit these various foibles (McLuhan, 1951; Cialdini, 2007). Unfettered 'free choice' will not necessarily yield positive outcomes for human well-being in the long-term, either for individuals or across generations.

Given the complexities of thinking about capital, both natural and cultural, and the contested nature of the strong vs. weak sustainability conversation, we suggest a definition of sustainability that says nothing explicitly about capital. We favour a definition in line with the Brundtland Commission's (Brundtland, 1987) definition, but generalized so that it can be applied at any spatial scale, and also within a generation (not just between): *sustainability is living within our means, such that our actions and lifestyles do not diminish the capacity of others to live well.* This definition is perhaps aspirational, given the apparent unsustainability of resource use (Wackernagel and Rees, 1996) and our apparent collective transgression of planetary boundaries (Rockström et al., 2009). It is also intentionally broad, to enable the 'constructive ambiguity' that is so powerful at capturing imaginations (Robinson, 2004).

And yet: if sustainability does not imply simply allowing culture and CES to adapt to new conditions and active efforts already in play (e.g., advertising) (Norton et al., 1998), how then do we manage CES for sustainability? Thus far, the management of CES has been a taboo topic for several reasons, one of which is surely an appropriate queasiness about social engineering, paternalism, and like postures regarding knowing what is best for 'people' and 'nature' (Norton et al., 1998). To this, two responses are critical. First, preferences, behaviour, consumption, and so (directly and indirectly) CES are being managed everyday by powerful forces of marketers and in some cases dictatorial governments (McLuhan, 1951; Norton et al., 1998). Second, the principle concern appears to be about manipulation by *outside* agents (Norton et al., 1998). Arguably, concerns of paternalism would be largely mitigated through a deliberative democratic process that questions the legitimacy and appropriateness of different sets of preferences and cultural trajectories. Such a solution is consistent with Sagoff's (1998, 2004) view on citizen preferences and Epstein's (2009) work on reconciling individual choice and the common good.

Managing cultural ecosystem services

The lion's share of discussion about sustainability appears to be about the management of resources or environmental sustainability and *not* about the relationship between ES and CES. Considering CES forces one to grapple with sustainability as a process of managing three different aspects: ecosystems, human preferences and values, and resulting social-ecological interactions. From an economic perspective, these three aspects might be seen as *supply*, *demand*, and *access*. Below, we treat each of these three aspects in turn.

Although people adapt to changing circumstances and baselines, ecosystem quality (*supply*) does seem to matter for the maintenance of CES. The relationship is almost certainly nonlinear, of course, and variable across contexts. For example, coral reefs across the globe have been degraded over the past several centuries and face new threats (Pandolfi et al., 2003), and yet even many of the hardest hit regions continue to enjoy the benefits of coastal tourism, including snorkeling and diving. At some point, however, the resilience of these CES will likely be surpassed, perhaps even before the reefs are replaced with slime (Pandolfi et al., 2005).

In some nations, there has been a widespread shift in *demand* away from local- to national-level nature-based recreation experiences that can be largely explained (statistically) by a rise in video games, home movies, theatre attendance, and internet use (Pergams and Zaradic, 2006). These marked declines appear to be limited largely to the US and Japan, however, as many other nations have enjoyed an increase in nature-based tourism (Balmford et al., 2009) (and the decline in US national park visitation is correlated statistically with a rise in income and in foreign travel, including to protected areas elsewhere (Pergams and Zaradic, 2006)). Thus it is unclear whether there are global declines in demand for CES; it is however quite clear that demand is highly variable and likely subject to change, including intentional.

The concept of environmental identity may be useful here, as it appears to manifest in the quantity of interactions with nature, perceived importance of those interactions, positive experiences in nature, and approval of pro-environmental behavior (Clayton, 2003). Environmental identity is a function of social context, through the values and activities of one's friends and family (Clayton and Opotow, 2003). Accordingly, as alluded to earlier in the chapter, demand for CES cannot be disentangled from the experiences and capabilities that are products of CES (access begets demand and vice versa). Thus, one of the strongest predictors of environmental identity appears to be whether an individual had positive experiences of nature as a child or youth in the company of adult role models (Chawla, 2007). This suggests both that parents have tremendous capacity to encourage (or stunt) an appreciation of nature, and that the quality of nature that enables high CES (positive experiences) is also an important factor in the production of future demand and therefore access. A critical component of environmental identity (as fostered by childhood experiences with role models) is the strong impact on active care for the environment (Chawla, 2007). Thus, experiential CES beget CES, and also the stewardship that sustains nature, which is important for others' enjoyment. By this logic, CES may be the keys to sustainable social-ecological systems.

Because of the centrality of these *interactions* with nature, environmental management for sustainability can no longer be thought of as a process of protecting nature. Rather, management for sustainability must include consideration of how management change will affect demand for, and access to, nature, especially the kind with strong impact on practices and values linked to nature.

A general principle of sustainability might require the mitigation of harm from both social and environmental change. But since 'harm' is a function of preference and values, and these are subject to change via the same social and environmental change, we must attend carefully to identity- and preference-shaping influences. But who should get to decide about these influences? How about those affected, following a principle of truly informed consent, and those who can best stand in their stead? Perhaps sustainability should be understood as a requirement for appropriate procedures rather than for a particular level of resources or material well-being (Robinson and Cole, 2014).

The dynamic nature of values, and CES, demands a thorough consideration of appropriate methods for making all manner of decisions, including about energy futures, urban plans, infrastructure projects, etc. Such thoroughness is beyond the scope of the current chapter, but four points bear mentioning. First, value-focused aka structured decision-making is specifically designed to address the 'constructed' nature of preferences and the complexity of values (Gregory et al., 2012). Second, a critical component of structured decision-making is making explicit the likely consequences of decisions. Above we have argued that changing preferences, values, and relationships may be crucial outcomes, but such outcomes are rarely identified explicitly (the focus is generally instead on material values, and to a lesser extent on non-material impacts on well-being). In fact, in western societies public policy is dominated by an economic body of thinking that conventionally assumes away any impacts on preferences or values (assuming that preferences are given and stable) (Norton et al., 1998). Third, politics and political ecology must always be at the forefront of our thinking about decision-making. Too often policies and infrastructure changes have the effect of benefiting a well-off minority at the expense of the majority and especially the poor (Freudenburg et al., 2008). Fourth, and finally, because many social-ecological losses are of limited reversibility, special consideration should be given to the long-term well-being consequences of losing slow-changing ecosystem features and associated CES. Given the necessarily limited understanding of the well-being implications of recent replacements of technology for interactions with nature, we might be hesitant to sanction yet further erosions of CES. The biophilia hypothesis (Kellert and Wilson, 1993) would suggest that substantial changes in nature and in human-nature relationships are likely to diminish human well-being by a variety of processes, because we co-evolved with nature and have thus become dependent upon it in often-unappreciated ways.

Collectively, we must decide what values we allow to be influenced by which forces. The current laissez-faire approach permits advertisers to manipulate values at will for the sake of economic growth. Yet economic growth, beyond a relatively low threshold of approximately US$20,000/year in per capita gross national product, appears to yield no gain in well-being (Max-Neef, 1995; Layard, 2006). Thus, during the time that the US has seen substantial declines in nature-based recreation, it has seen either a loss (Max-Neef, 1995) or no net gain (Layard, 2006) in well-being, despite substantial increases in income. And via CES, nature contributes greatly to human well-being, albeit in many intangible and invisible ways (Russell et al., 2013). A strong argument can therefore be made that an investment in nature and benign interactions with nature (as in CES) would be in our collective interest.

Conclusion

The study of CES, as nature's benefits to people through non-material processes, is a small subset of a much larger set of interdisciplinary literatures on social-ecological interactions. In the sustainability literature, CES have received short shrift. And yet, such CES can be seen as a newly recognized kind of human-natural capital, and as capital-producing: not only do positive experiences and attachment with nature yield both well-being and key capacities, they also yield the stewardship attitudes and identities that may be fundamental to local and global sustainability.

References

Adamowicz, W.L.V., Naidoo, R., Nelson, E., Polasky, S., and Zhang, J. (2011). Nature-based tourism and recreation values. In: Kareiva, P., Tallis, H., Ricketts, T.H., Daily, G.C. and Polasky, S. (eds.) *Natural Capital: Theory & Practice of Mapping Ecosystem Services*. Oxford University Press, Oxford.

Balmford, A., Beresford, J., Green, J., Naidoo, R., Walpole, M., and Manica, A. (2009). A global perspective on trends in nature-based tourism. *PLoS Biol*, vol 7, e1000144.

Benson, E.D., Hansen, J.L., Schwartz Jr., A.L., and Smersh, G.T. (1998). Pricing residential amenities: the value of a view. *The Journal of Real Estate Finance and Economics*, vol 16, pp 55–73.

Boxall, P.C., Englin, J., and Adamowicz, W.L. (2003). Valuing aboriginal artifacts: a combined revealed-stated preference approach. *Journal of Environmental Economics and Management*, vol 45, pp 213–230.

Brockington, D., Duffy, R., and Igoe, J. (eds) (2008). *Nature Unbound: The Past, Present and Future of Protected Areas*. Earthscan, London.

Bromley, D.W. (1998). Searching for sustainability: the poverty of spontaneous order. *Ecological Economics*, vol 24, pp 231–240.

Brundtland, G.H. (1987). *Report of the World Commission on Environment and Development: "Our Common Future"*. United Nations, New York.

Cabeza Gutés, M. (1996). The concept of weak sustainability. *Ecological Economics*, vol 17, pp 147–156.

Chamberlain, B. C., and Meitner, M.J. (2012). Quantifying the effects of harvest block design on aesthetic preferences. *Canadian Journal of Forest Research*, vol 42, pp 2106–2117.

Chan, K. M.A., Balvanera, P., Benessaiah, K. et al. (20 authors) (accepted). Why protect nature? Rethinking values and the environment. PNAS.

Chan, K.M.A., Goldstein, J., Satterfield, T., Hannahs, N., Kikiloi, K., Naidoo, R., Vadeboncoeur, N., and Woodside, U. (2011). Cultural services and non-use values. In: Kareiva, P., Tallis, H., Ricketts, T.H., Daily, G.C. and Polasky, S. (eds) *Natural Capital: Theory & Practice of Mapping Ecosystem Services*. Oxford University Press, Oxford.

Chan, K.M.A., Guerry, A., Balvanera, P., Klain, S., Satterfield, T., Basurto, X., Bostrom, A., Chuenpagdee, R., Gould, R., Halpern, B.S., Hannahs, N., Levine, J., Norton, B., Ruckelshaus, M., Russell, R., Tam, J., and Woodside, U. (2012a). Where are "cultural" and "social" in ecosystem services: a framework for constructive engagement. *BioScience*, vol 6, pp 744–756.

Chan, K.M.A., Satterfield, T., and Goldstein, J. (2012b). Rethinking ecosystem services to better address and navigate cultural values. *Ecological Economics*, vol 74, pp 8–18.

Chawla, L. (2007). Childhood experiences associated with care for the natural world: a theoretical framework for empirical results. *Children, Youth and Environments*, vol 17, pp 144–170.

Church, A., Burgess, J., Ravenscroft, N., Bird, W., Blackstock, K., Brady, E., Crang, M., Fish, R., Gruffudd, P., and Mourato, S. (2011). Cultural services. In: UK National Ecosystem Assessment (ed.) *The UK National Ecosystem Assessment Technical Report*. UNEP-WCMC, Cambridge UK.

Cialdini, R.B. (2007). *Influence: The Psychology of Persuasion*. Harper Paperbacks, New York.

Clayton, S.D. (2003). Environmental identity: a conceptual and an operational definition. In: Clayton, S.D. and Opotow, S. (eds) *Identity and the Natural Environment: The Psychological Significance of Nature*. MIT Press, Cambridge MA.

Clayton, S.D., and Opotow, S. (2003). *Identity and the Natural Environment: The Psychological Significance of Nature*. MIT Press, Cambridge MA.

Costanza, R., and Kubiszewski, I. (2012). The authorship structure of "ecosystem services" as a transdisciplinary field of scholarship. *Ecosystem Services*, vol 1, pp 16–25.

Daily, G.C. (ed.) (1997). *Nature's Services: Societal Dependence on Natural Ecosystems*. Island Press, Washington DC.

Daily, G.C., Polasky, S., Goldstein, J., Kareiva, P. M., Mooney, H.A., Pejchar, L., Ricketts, T.H., Salzman, J., and Shallenberger, R. (2009). Ecosystem services in decision making: time to deliver. *Frontiers in Ecology and the Environment*, vol 7, pp 21–28.

Daniel, T.C. (2001). Whither scenic beauty? Visual landscape quality assessment in the 21st century. *Landscape and Urban Planning*, vol 54, pp 267–281.

Daniel, T.C., Muhar, A., Arnberger, A., Aznar, O., Boyd, J.W., Chan, K.M.A., Costanza, R., Elmqvist, T., Flint, C.G., Gobster, P.H., Grêt-Regamey, A., Lave, R., Muhar, S., Penker, M., Ribe, R.G., Schauppenlehner, T., Sikor, T., Soloviy, I., Spierenburg, M., Taczanowska, K., Tam, J., and von der Dunk, A. (2012). Contributions of cultural services to the ecosystem services agenda. *Proceedings of the National Academy of Sciences*, vol 109, pp 8812–8819.

de Groot, R., Ramakrishnan, P.S., Berg, A. v. d., Kulenthran, T., Muller, S., Pitt, D., Wascher, D., Wijesuriya, G., Amelung, B., Eliezer, N., Gopal, A. R,. and Rössler, M. (2005). Cultural and amenity services. In: Millennium Ecosystem Assessment (ed.) *Ecosystems and Human Well-being: Current Status and Trends*. Island Press, Washington DC.

Deur, D. (2002). Plant cultivation on the Northwest Coast: a reconsideration. *Journal of Cultural Geography*, vol 19, pp 9–35.

Dowie, M. (2011). *Conservation Refugees: The Hundred-Year Conflict between Global Conservation and Native Peoples*. MIT Press, Cambridge MA.

EPA (U.S. Environmental Protection Agency Science Advisory Board) (2009). *Valuing the Protection of Ecological Systems and Services*. USEPA, Washington DC.

Epstein, R.A. (2009). *Principles for a Free Society: Reconciling Individual Liberty with the Common Good*. Basic Books, New York.

Failing, L., Gregory, R., and Higgins, P. (2013). Science, uncertainty, and values in ecological restoration: a case study in structured decision-making and adaptive management. *Restoration Ecology*, vol 21, pp 422–430.

Fish, R.D. (2011). Environmental decision making and an ecosystems approach. *Progress in Physical Geography*, vol 35, pp 671–680.

Freudenburg, W.R., Gramling, R., Laska, S., and Erikson, K.T. (2008). Organizing hazards, engineering disasters? Improving the recognition of political-economic factors in the creation of disasters. *Social Forces*, vol 87, pp 1015–1038.

Gee, K. (2010). Offshore wind power development as affected by seascape values on the German North Sea coast. *Land Use Policy*, vol 27, pp 185–194.

Gómez-Baggethun, E., and Ruiz-Pérez, M. (2011). Economic valuation and the commodification of ecosystem services. *Progress in Physical Geography*, vol 35, pp 613–628.

Gould, R., Ardoin, N., Woodside, U., Satterfield, T., Hannahs, N., and Daily, G.C. (2014a). The forest has a story: cultural ecosystem services in Kona, Hawai'i. *Ecology & Society*, vol 19, p 55.

Gould, R.K., Klain, S.C., Ardoin, N.M., Satterfield, T., Woodside, U., Hannahs, N., Daily, G.C., and Chan, K.M. (2014b). A protocol for eliciting nonmaterial values using a cultural ecosystem services frame. *Conservation Biology*, vol 2, pp 575–586.

Gregory, R., Failing, L., Harstone, M., Long, G., and McDaniels, T. (2012). *Structured Decision Making: A Practical Guide to Environmental Management Choices*. John Wiley & Sons, Hoboken NJ.

Grêt-Regamey, A., Bebi, P., Bishop, I.D., and Schmid, W.A. (2008). Linking GIS-based models to value ecosystem services in an Alpine region. *Journal of Environmental Management*, vol 89, pp 197–208.

Grêt-Regamey, A., Bishop, I.D., and Bebi, P. (2007). Predicting the scenic beauty value of mapped landscape changes in a mountainous region through the use of GIS. *Environment and Planning B: Planning and Design*, vol 34, pp 50–67.

Happynook, T. (2001). Securing Nuu Chah Nulth. *Canku Ota*, April.

Harris, R.C. (2011). *Making Native Space: Colonialism, Resistance, and Reserves in British Columbia*. UBC Press, Vancouver.

Hartwick, J.M. (1977). Intergenerational equity and the investing of rents from exhaustible resources. *The American Economic Review*, vol 67, pp 972–974.

Hernández-Morcillo, M., Plieninger, T., and Bieling, C. (2013). An empirical review of cultural ecosystem service indicators. *Ecological Indicators*, vol 29, pp 434–444.

Holland, D., William Lachicotte, J., Skinner, D., and Cain, C. (1998). *Identity and Agency in Cultural Worlds*. Harvard University Press, Cambridge MA.

Jax, K., Barton, D.N., Chan, K.M.A., de Groot, R., Doyle, U., Eser, U., Görg, C., Gómez-Baggethun, E., Griewald, Y., Haber, W., Haines-Young, R., Heink, U., Jahn, T., Joosten, H., Kerschbaumer, L., Korn, H., Luck, G.W., Matzdorf, B., Muraca, B., Neßhöver, C., Norton, B., Ott, K., Potschin, M., Rauschmayer, F., von Haaren, C., and Wichmann, S. (2013). Ecosystem services and ethics. *Ecological Economics*, vol 93, pp 260–268.

Kahan, D.M., Peters, E., Wittlin, M., Slovic, P., Ouellette, L.L., Braman, D., and Mandel, G. (2012). The polarizing impact of science literacy and numeracy on perceived climate change risks. *Nature Climate Change*, vol 2, pp 732–735.

Kahneman, D. (2013). *Thinking, Fast and Slow*. Farrar, Straus and Giroux, New York.

Kancewick, M., and Smith, E. (1990). Subsistence in Alaska: towards a Native priority. *UMKC Law Review*, vol 59, pp 645–677.

Kareiva, P., Tallis, H., Ricketts, T.H., Daily, G.C., and Polasky, S. (2011). *Natural Capital: Theory & Practice of Mapping Ecosystem Services*. Oxford University Press, Oxford.

Kellert, S.R., and Wilson, E.O. (1993). *The Biophilia Hypothesis*. Island Press, Washington DC.

Kirchhoff, T. (2012). Pivotal cultural values of nature cannot be integrated into the ecosystem services framework. *Proceedings of the National Academy of Sciences*, vol 109, p e3147.

Kitayama, S. (2002). Culture and basic psychological processes—Toward a system view of culture: comment on Oyserman et al. (2002). *Psychological Bulletin*, vol 128, pp 89–96.

Klain, S.C., and Chan, K.M.A. (2012). Navigating coastal values: participatory mapping of ecosystem services for spatial planning. *Ecological Economics*, vol 82, pp 104–113.

Klain, S., Satterfield, T., and Chan, K.M.A. (2014). What matters and why? Ecosystem services and their bundled qualities. *Ecological Economics*, vol 107, pp 310–320.

Layard, R. (2006). Happiness and public policy: a challenge to the profession. *The Economic Journal*, vol 116, pp C24-C33.

Lee, M. (2002). The cooler ring: urban Alaska Native women and the subsistence debate. *Arctic Anthropology*, vol 39, pp 3–9.

Lepofsky, D., and Caldwell, M. (2013). Indigenous marine resource management on the northwest coast of North America. *Ecological Processes*, vol 2, pp 1–12.

Louv, R. (2008). *Last Child in the Woods: Saving Our Children from Nature-Deficit Disorder*. Algonquin Books, Chapel Hill NC.

Luck, G., Chan, K.M.A., Eser, U., Gómez-Baggethun, E., Matzdorf, B., Norton, B., and Potschin, M. (2012). Ethical considerations in on-ground applications of the ecosystem services concept. *BioScience*, vol 62, pp 1020–1029.

Ludwig, D. (2000). Limitations of economic valuation of ecosystems. *Ecosystems*, vol 3, pp 31–35.

MA (2005). *Ecosystems and Human Well-being: Synthesis*. Island Press, Washington DC.

Martín-López, B., Iniesta-Arandia, I., García-Llorente, M., Palomo, I., Casado-Arzuaga, I., Amo, D.G.D., Gómez-Baggethun, E., Oteros-Rozas, E., Palacios-Agundez, I., Willaarts, B., González, J.A., Santos-Martín, F., Onaindia, M., López-Santiago, C., and Montes, C. (2012). Uncovering ecosystem service bundles through social preferences. *PLoS ONE*, 7, e38970.

Masco, J. (1995). "It is a strict law that bids us dance": cosmologies, colonialism, death, and ritual authority in the Kwakwaka'wakw potlatch, 1849 to 1922. *Comparative Studies in Society and History*, vol 37, pp 41–75.

Matsuoka, R.H., and Kaplan, R. (2008). People needs in the urban landscape: analysis of landscape and urban planning contributions. *Landscape and Urban Planning*, vol 84, pp 7–19.

Max-Neef, M. (1995). Economic growth and quality of life: a threshold hypothesis. *Ecological Economics*, vol 15, pp 115–118.

Max-Neef, M., Elizalde, A., and Hopenhayn, M. (1992). Development and human needs. In: Ekins, P. and Max-Neef, M. (ed.) *Real-Life Economics: Understanding Wealth Creation*. Routledge London.

McLuhan, M. (1951). *The Mechanical Bride: Folklore of Industrial Man*. Vanguard Press, New York.

Naidoo, R., and Adamowicz, W.L. (2005). Biodiversity and nature-based tourism at forest reserves in Uganda. *Environment and Development Economics*, vol 10, pp 159–178.

National Research Council (U.S.) Committee on Assessing and Valuing the Services of Aquatic and Related Terrestrial Ecosystems (2005). *Valuing Ecosystem Services: Toward Better Environmental Decision-Making*. National Research Council, Washington DC.

Norton, B., Costanza, R., and Bishop, R.C. (1998). The evolution of preferences: why "sovereign" preferences may not lead to sustainable policies and what to do about it. *Ecological Economics*, vol 24, pp 193–211.

Pandolfi, J.M., Bradbury, R.H., Sala, E., Hughes, T.P., Bjorndal, K.A., Cooke, R.G., McArdle, D., McClenachan, L., Newman, M.J.H., Paredes, G., Warner, R.R., and Jackson, J.B.C. (2003). Global trajectories of the long-term decline of coral reef ecosystems. *Science*, vol 301, pp 955–958.

Pandolfi, J.M., Jackson, J.B.C., Baron, N., Bradbury, R.H., Guzman, H.M., Hughes, T.P., Kappel, C.V., Micheli, F., Ogden, J.C., Possingham, H.P., and Sala, E. (2005). Are U.S. coral reefs on the slippery slope to slime? *Science*, vol 307, pp 1725–1726.

Pauly, D. (1995). Anecdotes and the shifting baseline syndrome of fisheries. *Trends in Ecology & Evolution*, vol 10, pp 430–430.

Pergams, O.R.W., and Zaradic, P.A. (2006). Is love of nature in the US becoming love of electronic media? 16-year downtrend in national park visits explained by watching movies, playing video games, internet use, and oil prices. *Journal of Environmental Management*, vol 80, pp 387–393.

Platt, J. (1973). Social traps. *American Psychologist*, vol 28, pp 641–651.

Plieninger, T., Bieling, C., Ohnesorge, B., Schaich, H., Schleyer, C., and Wolff, F. (2013a). Exploring futures of ecosystem services in cultural landscapes through participatory scenario development in the Swabian Alb, Germany. *Ecology and Society*, vol 18.

Plieninger, T., Dijks, S., Oteros-Rozas, E., and Bieling, C. (2013b). Assessing, mapping, and quantifying cultural ecosystem services at community level. *Land Use Policy*, vol 33, pp 118–129.

Poe, M.R., Norman, K.C., and Levin, P.S. (2013). Cultural dimensions of socioecological systems: key connections and guiding principles for conservation in coastal environments. *Conservation Letters*, vol 7, pp 166–175.

Potschin, M.B., and Haines-Young, R.H. (2011). Ecosystem services: exploring a geographical perspective. *Progress in Physical Geography*, vol 35, pp 575–594.

Pröpper, M., and Haupts, F. (2014). The culturality of ecosystem services. Emphasizing process and transformation. *Ecological Economics*, vol 108, pp 28–35.

Raymond, C.M., Singh, G., Benessaiah, K., Bernhardt, J.R., Levine, J., Nelson, H., Turner, N.J., Norton, B., Tam, J., and Chan, K.M.A. (2013). Ecosystem services and beyond: using multiple metaphors to understand human–environment relationships. *BioScience*, vol 63, pp 536–546.

Rees, W.E. (1998). How should a parasite value its host? *Ecological Economics*, vol 25, pp 49–52.

Robinson, J. (2004). Squaring the circle? Some thoughts on the idea of sustainable development. *Ecological Economics*, vol 48, pp 369–384.

Robinson, J., and Cole, R.J. (2014). Theoretical underpinnings of regenerative sustainability. *Building Research & Information*, vol 43, pp 133–143.

Rockström, J., Steffen, W., Noone, K., Persson, A., Chapin, F.S., Lambin, E.F., Lenton, T.M., Scheffer, M., Folke, C., Schellnhuber, H.J., Nykvist, B., de Wit, C.A., Hughes, T., van der Leeuw, S., Rodhe, H., Sorlin, S., Snyder, P.K., Costanza, R., Svedin, U., Falkenmark, M., Karlberg, L., Corell, R.W., Fabry, V.J., Hansen, J., Walker, B., Liverman, D., Richardson, K., Crutzen, P., and Foley, J.A. (2009). A safe operating space for humanity. *Nature*, vol 461, pp 472–475.

Russell, R., Guerry, A. D., Balvanera, P., Gould, R.K., Basurto, X., Chan, K.M.A., Klain, S., Levine, J., and Tam, J. (2013). Humans and nature: how knowing and experiencing nature affect well-being. *Annual Review of Environment and Resources*, vol 38, pp 473–502.

Sagoff, M. (1998). Aggregation and deliberation in valuing environmental public goods: a look beyond contingent pricing. *Ecological Economics*, vol 24, pp 213–230.

Sagoff, M. (2004). *Price, Principle, and the Environment*. Cambridge University Press, Cambridge UK.

Satterfield, T. (2001). In search of value literacy: suggestions for the elicitation of environmental values. *Environmental Values*, vol 10, pp 331–359.

Satterfield, T., Gregory, R., Klain, S., Roberts, M., and Chan, K.M. (2013). Culture, intangibles and metrics in environmental management. *Journal of Environmental Management*, vol 117, pp 103–114.

Satz, D., Gould, R.K., Chan, K.M.A., Guerry, A., Norton, B., Satterfield, T., Halpern, B.S., Levine, J., Woodside, U., Hannahs, N., Basurto, X., and Klain, S. (2013). The challenges of incorporating cultural ecosystem services into environmental assessment. *Ambio*, vol 42, pp 675–684.

Sherren, K., Fischer, J., and Price, R. (2010). Using photography to elicit grazier values and management practices relating to tree survival and recruitment. *Land Use Policy*, vol 27, pp 1056–1067.

Solow, R.M. (1974). Intergenerational equity and exhaustible resources. *The Review of Economic Studies*, vol 41, pp 29–45.

Sparks, K., Faragher, B., and Cooper, C.L. (2001). Well-being and occupational health in the 21st century workplace. *Journal of Occupational and Organizational Psychology*, vol 74, pp 489–509.

Tallis, H., Lester, S.E., Ruckelshaus, M., Plummer, M., McLeod, K., Guerry, A., Andelman, S., Caldwell, M.R., Conte, M., Copps, S., Fox, D., Fujita, R., Gaines, S.D., Gelfenbaum, G., Gold, B., Kareiva, P., Kim, C.-K., Lee, K., Papenfus, M., Redman, S., Silliman, B., Wainger, L., and White, C. (2012). New metrics for managing and sustaining the ocean's bounty. *Marine Policy*, vol 36, pp 303–306.

TEEB (2009). *The Economics of Ecosystems and Biodiversity for National and International Policy Makers*. United Nations Environment Programme, Wesseling.

Tengberg, A., Fredholm, S., Eliasson, I., Knez, I., Saltzman, K., and Wetterberg, O. (2012). Cultural ecosystem services provided by landscapes: assessment of heritage values and identity. *Ecosystem Services*, vol 2, pp 14–26.

Thoits, P.A. (2006). Personal agency in the stress process. *Journal of Health and Social Behavior*, vol 47, pp 309–323.

Turner, N.J., Gregory, R., Brooks, C., Failing, L., and Satterfield, T. (2008). From invisibility to transparency: identifying the implications. *Ecology and Society*, vol 13, pp 7.

Turner, N.J., and Turner, K.L. (2008). "Where our women used to get the food": cumulative effects and loss of ethnobotanical knowledge and practice; case study from coastal British Columbia. *Botany-Botanique*, vol 86, pp 103–115.

Veitch, J., Salmon, J., and Ball, K. (2007). Children's perceptions of the use of public open spaces for active free play. *Children's Geographies*, vol 5, pp 409–422.

Wackernagel, M., and Rees, W.E. (1996). *Our Ecological Footprint: Reducing Human Impact on the Earth*. New Society Publishers, Gabriola Island.

Briefing Note 27.1 Ecosystem services and spirituality

Nigel Cooper

Gathered at Assisi, Italy, in 1986 were leaders of several world religions. The World Wide Fund for Nature (WWF) had asked for their support for nature conservation; they provided the Assisi Declarations (World Bank, 2003). The leaders may not have been fully representative or authoritative, but this signaled a recognition among those working for conservation that instead of seeing religion, Christianity in particular, as an obstacle to conservation (as had been the trend following White (1967)), spirituality and religion could be a potential ally. Theologians have continued to develop arguments for caring for nature, often building on the argument that humans ought, are commanded even, to care for what God has created (Gottlieb, 2010).

Meanwhile, in countries of the Global South, where the greatest biodiversity is found, conservationists have been collaborating with local religious groups to build support for protecting nature. Such groups were very aware of the environmental degradation around them. Indigenous peoples in particular are deeply concerned for the protection of their sacred sites; others see their traditional ways of life being undermined. Bodies such as the International Union for the Conservation of Nature (var. dates) and the World Bank (2006) produce advice on how to work with local religious groups on promoting sustainable practices (Pungetti *et al.*, 2012).

The Millennium Ecosystem Assessment, in 2005, adopted a classification of ecosystem services that included 'spiritual services' (mentioned 348 times across all its publications). The Assessment identified, first, the value indigenous peoples put on ecosystems, through providing the sacred sites they revere or as a source of materials used in rituals (e.g. feathers and natural dyes) and, second, ecosystems as a context in which people could receive spiritual experiences apart from any religion, as might be more common in the Global North.

In both these ways, ecosystems provide benefits and so are consistent with the ecosystem services concept. In theory it might be possible to put a monetary value on them, such as using travel-cost methods to price spiritual retreats in the countryside. The inadequacy of this, even from a purely economic point of view, is reflected in the literature, where the difficulties of valuing spiritual ecosystem services are acknowledged but rarely addressed. The ecosystem service approach has drawn attention to spiritual value and so given opportunity for debate over the philosophical inadequacy of restricting spiritual value to just human benefit. The search is therefore on to find methods to bring all types of value into consideration in decision-making.

The first step has been in identifying and characterising the spiritual value of ecosystems, with research in developing regions predominantly investigating the cultic and spiritual approaches of local people. Some of this is revealed through the work of ethnography. Questions about spiritual or religious uses of the local landscape may be included in wider surveys. Narratives are sought from local people to sift out their spiritual dimensions. One technique is to ask people to mark on maps places of spiritual significance alongside other significant places. In this work the spiritual is largely left to participants to self-define, rather than researchers imposing their own definitions of this slippery word.

In regions where it might be assumed religious practice is low, it is even more important to accept a wide definition of the spiritual that is not restricted to the explicitly religious. For example, a British review of the shared values of ecosystems (UK NEA, 2014) revealed several ways in which British people speak of the spiritual dimensions of nature. The popular expression 'recharging the batteries' is closest to the Millennium Ecosystem Assessment's language of spiritual enrichment. This is often experienced in landscapes valued by the Romantic Movement, which saw nature as deeply spiritual. There are traditional sacred sites (from Stonehenge to Salisbury Cathedral) and new rituals, e.g. scattering the ashes of deceased relatives in beauty spots. Place and spirit are closely bound up with identity for many rural people. Several people turn to science for clues to the meaning and purpose of life.

Related field work found participants often chose spiritual language to describe their experiences of nature. So, reflecting on the values associated with off-shore diving and marine ecosystems, one participant characterised the experience as of 'religious value' 'which is strange really because I am an atheist. I was in one place and visibility opened up and it was like a cathedral, with jewel anemones lighting up everywhere. I felt like I was in the presence of God, if there is such a thing. I was crying when I came out of the water.'

This illustrates how spiritual value cuts across the assumptions of individualistic, welfare economics. Nature puts us humans in our place. It provides us with life, takes it away, and will be there long after the last human. Nature is transcendent, not primarily the provider of warm glows. Alongside the question of what nature can do for us (which conceivably might be priced and included in trade-offs), consideration of the spiritual also challenges us to ask what we are duty-bound to do for nature. As monetary valuation grows in popularity, we need new social structures and mechanisms, as well as new accounts of value, that will incorporate these duties in decision-making.

References

Gottleib, R. S. (ed) (2010). *The Oxford Handbook of Religion and Ecology*. Oxford University Press, Oxford.

International Union for the Conservation of Nature. Several texts available at: http://www.iucn.org/about/work/programmes/gpap_home/gpap_capacity2/gpap_pub/gpap_spiritualpub/ (accessed 4 Feb 2015).

Pungetti, G., Oviedo, G., and Hooke, D. (eds) (2012). *Sacred Species and Sites; advances in biocultural conservation*. Cambridge University Press, Cambridge UK.

UK National Ecosystem Assessment Follow-On (2014). *Work Package Report 6: Shared, Plural and Cultural Values of Ecosystems*. UNEP World Conservation Monitoring Centre, Cambridge UK.

White, L. (1967). The historical roots of our ecologic crisis. *Science*, vol 155, no 3767, pp 1203–1207.

World Bank (2003). *Faith in Conservation: New Approaches to Religions and the Environment*. World Bank, Washington DC.

World Bank (2006). *Faiths and the Environment: World Bank Support 2000–05*. World Bank, Washington DC.

28

TOWARDS EFFECTIVE ECOSYSTEM SERVICES ASSESSMENT IN MARINE AND COASTAL MANAGEMENT

Mahé Charles, Rémi Mongruel, Nicola Beaumont, Tara Hooper, Harold Levrel, Eric Thiébaut and Linwood Pendleton

Introduction

Marine and coastal ecosystems are among the world's most productive (MA, 2005), but they suffer from increasing human pressures and adaptive challenges. Their management is further complicated by the fact that many human uses and activities depend on the ecological health of these areas. While economic valuations of coastal and marine areas have existed for decades, marine managers are now starting to explicitly consider how their management affects the provision of ecosystem services (ES). To do so, they need targeted research to help them understand which ecosystem services may need to be formally measured in order to provide important information for marine management and planning.

Ecosystem services assessments (ESA) have become part of the suite of assessment techniques used in marine policy and marine planning (Börger et al., 2014). Practitioners, therefore, need to determine: a) which of the many ecosystem services should be assessed; b) what measures of ES output or value will best inform decisions; and c) which methodologies should be used to quantify ES in order to ensure that Marine and Coastal Ecosystem Services Assessment (MCESA) will inform policy efficiently.

This chapter outlines a framework for characterizing complex marine and coastal ecological functions and associated ES. Then, it describes possible steps for undertaking MCESA to improve the usefulness of such information in marine management. The 'triage' approach proposed by Pendleton et al. (2015) includes a formalized process for making an initial diagnostic to help natural and social scientists, practitioners and stakeholders narrow the focus of the MCESA in support of management. We illustrate the application of the 'triage' process with one case study.

Marine and coastal ecosystem services

Ecosystem service assessments bring together a range of concepts from the natural and social sciences. Given that they are undertaken by researchers, practitioners, managers and stakeholders from various backgrounds, a shared conceptual and analytical frameworks is needed (Granek et al., 2010). Reaching a shared understanding of the meaning of marine ecosystem services and the impact of decisions on them requires a classification framework, such as that proposed by Liquete et al. (2013) (Table 28.1).

The basic ecosystem service terminology of the MA (2005) creates the foundation from which other detailed typologies and frameworks have emerged (e.g. Beaumont et al. (2007), TEEB (2010); Fletcher et al. (2011); UK NEA (2011); Böhnke-Henrichs et al. (2013) and CICES, see Haines-Young and Potschin, 2016). These more refined frameworks seek to increase clarity and avoid double-counting by defining the distinction between underlying ecological functions and processes, the services they deliver and the benefits ultimately provided to people.

Table 28.1 Marine and coastal ecosystem services using the classification proposed by Liquete et al. (2013).

	Marine Ecosystem Services	*Specific components*
Provisioning services	Food provision	Fishing activities (either commercial or subsistence fishing) and aquaculture
	Water storage and provision	Water use for desalination plants, industrial cooling processes or coastal aquaculture
	Biotic materials and biofuels	Medicinal, ornamental and other industrial resources (oil and fishmeal); biomass to produce energy
Regulation and maintenance services	Water purification	Treatment of human wastes through dilution, sedimentation, trapping or sequestration, etc
	Air quality regulation	Absorption by vegetal or water bodies of air pollutants like particulate matter, ozone or sulphur dioxide
	Coastal protection	Natural defence of the coastal zone against inundation and erosion from waves, storms or sea level rise
	Climate regulation	Sequestration by the ocean of greenhouse and climate active gases
	Weather regulation	Influence of coastal vegetation and wetlands on air moisture or the formation of clouds
	Ocean nourishment	Natural cycling processes leading to the availability of nutrients in seawater for the production of organic matter
	Life cycle maintenance	The maintenance of key habitats that act as nurseries, spawning areas or migratory routes
	Biological regulation	Control of fish pathogens, biological control on the spread of vector borne human diseases
Cultural services	Symbolic and aesthetic values	Contribution to local identity, value of charismatic habitats and species such as coral reefs or marine mammals
	Recreation and tourism	Coastal activities (bathing, snorkelling, scuba diving) and offshore activities (sailing, recreational fishing, whale watching)
	Cognitive effects	Inspiration for arts and applications, material for research and education, information and awareness

Source: Liquete et al. (2013)

These distinctions are important in deciding what can, and should, be quantified and valued during the MCESA.

Core 'ecosystem processes' tend to be defined as the basic physical, chemical or biological processes which occur within ecosystems, while 'ecosystem functions' or 'beneficial ecosystem processes' or 'intermediate services' are referred to as specific ecosystem processes that define the capacity of an ecosystem to provide services (Table 28.2).

Natural habitats, ecological functions and marine and coastal ecosystems

Ecosystems may be characterized by their organization (natural capital stock, structure and pattern), their operation (flows, functioning, processes) and their outcomes for humans (services, goods and benefits, income) (Fisher et al., 2009). One way to capture the organization of an ecosystem is to describe habitats, which can be classified using international multi-tiered nomenclatures such as the European Nature Information System (EUNIS) habitat classification (Galparsoro et al., 2012). Habitats are defined as recognizable spaces that can be distinguished by their abiotic characteristics and associated biological assemblages (ICES, 2005). Habitat classification systems such as EUNIS constitute a well-recognized starting point for describing ecosystems. Understanding where habitats are located and the frequency and scale at which they occur remains fundamental in MCESA. One of the main advantages of the EUNIS system in MCESA is that it is a hierarchical classification system which can then be used at different spatial scales and with different levels of knowledge on habitat distribution.

Table 28.2 List of physio-chemical processes, core ecosystem processes, ecological functions, ecosystem services and benefits linked to marine and coastal ecosystems.

Physico-chemical features	*Core ecosystem processes*	*Ecological functions (beneficial ecosystem processes; intermediate services)*	*Ecosystem services (final services)*	*Benefits for the society of some user groups*
-Hydrology -Nutrients -Oxygen -Temperature -pH and salinity -Depth -Exposure -Density -Turbidity -Light -Wind, wave and tides	-Production -Decomposition -Nutrient cycling -Water cycling -Hydrological processes -Ecological interactions -Evolutionary processes	-Primary production -Competition for food & space -Population control -Biologically mediated habitats -Resilience and resistance -Microbial loops -Carbon fixation -Biomodification of sediments -Delivery and settlement of organisms	-Food provision -Raw materials -Bioremediation of waste -Residential/ industrial water supply -Disturbance prevention	-Fishing harvest and fish consumption -Raw material harvesting and consumption -Damage avoidance for public health -Damage avoidance for private properties -Risk mitigation -Leisure and recreation -Feel good or warm glow -Existence and option use value

Adapted from: Atkins et al. (2011), Beaumont et al. (2007), Boyd and Banzhaf (2007), Fisher et al. (2009), Fletcher et al. (2011), TEEB (2010)

While habitat mapping is a useful first step, it has to be recognized that a single habitat represents a place where many biophysical processes occur, and a system which provides several ecological functions: thus, one habitat may contribute to several ecological functions and ecosystem services, while a single service may depend on several ecological functions and habitats. Relationships between habitats, functions and services are generally multiple, complex, and rarely have simple one-to-one linkages. To reflect these interdependencies, an important next step must be to identify and select both the major ecological functions and the main ecosystem services within the management area.

Links between the biophysical features of ecosystems and the social goods and benefits they deliver are not easy to establish, and require moving beyond the conceptual frameworks mentioned previously. Recent works providing further insight into these issues include that of Potts et al. (2014), who have suggested the extent to which a range of UK species and habitat may deliver 25 key marine and coastal ecosystem services and that of Townsend et al. (2011), who propose a detailed schematic view of the way major marine and coastal ecological functions and ecosystem goods and services are most likely to interact.

Indicators for assessing marine and coastal ecosystem services

Indicators of supply and demand are necessary to characterize ecosystem services. Supply indicators underline the ability or the capacity of the ecosystem to deliver a particular ecosystem service flow. Demand indicators characterize the requirement or concern of human populations, subject to economic and social circumstances. In this case, the service may be direct (e.g. fish consumption) or indirect (e.g. revenues in fisheries industry). Monetary values of ecosystem services are demand indicators, as they are subject to a particular social and economic context.

When conducting an ecosystem service assessment, indicators should be selected collectively by scientists and end-users, who may prefer particular indicators depending on the scope of the assessment and their judgement criteria. Table 28.3 proposes examples of supply and demand indicators for a MCESA.

Table 28.3 Selected examples of supply and demand indicators for marine and coastal ecosystem service assessment.

Marine and coastal ES	*Specific components*	*Supply indicator*	*Demand indicator*
Food provision	Fisheries and aquaculture	Fish landings and production (volume)	Fish consumption Employment and revenue in fishing industry (indirect)
Water storage and provision	Industrial use of seawater	Sea water use settlements	Marine water consumption
Biotic materials and biofuels	Medicinal sector	Production of material used for medicines	Consumption of medicines using marine material
	Ornamental resources	Production of ornamental living material from the sea	Consumption of ornamental living material from the sea
	Energy resources	Production of marine biomass for fuel	Consumption of marine biomass for fuel

Marine and coastal ES	Specific components	Supply indicator	Demand indicator
Water purification	Treatment of human waste		Water quality standard
Air quality regulation	Absorption of pollutants		Air quality standards (for all pollutants except CO_2)
Coastal protection	Natural defence	Mangrove or coral reefs extent	
Climate regulation	Carbon sequestration	Carbon stock exchange Carbon sequestration capacity (potential)	Value of carbon sequestration capacity
Ocean nourishment	Nutrient and organic matters	Primary productivity Algal biomass	Value of organic matter production
Life cycle maintenance	Maintenance of habitats	Biodiversity indicators (Habitats extension or status, diversity of species)	
Symbolic and aesthetic values	Heritage	No of UNESCO heritage sites (potential) No of sites or species used for cultural events (potential)	No of persons placing high values on sea (potential)
Symbolic and aesthetic values	Aesthetic value		Frequentation for Nature based motivation
Recreation and tourism	Recreational activities (non-market activities)		No of tourists
	Recreational fishing		No of recreational fishers Value of recreational fisheries
	Tourism industry (market activities) No of tourists	Protected or preserved area for ecotourism (potential)	Expenses of tourists Employment and revenue in tourism industry (indirect)
Cognitive effects		No of actions for education or research (potential)	

Adapted from: UNEP (2009) and Liquete et al. (2013)

Agreeing on the purpose, scope, methods and tools for the marine and coastal ecosystem services assessment

Marine and coastal ecosystem services assessment can promote understanding of the services and provide estimates of values for the benefits arising from them, given changing levels of pressure and alternative management strategies. However, there are many ecosystem services that could be assessed and a variety of approaches that could be used to manage natural resources.

It is therefore necessary to identify the overarching decision needs for which better ecosystem service information would be useful and then detail how the assessment should be undertaken in terms of effort and focus. Following Lopes and Videira (2013), Pendleton et al. (2015) propose a formalized and transparent, structured decision-making process, called a 'triage' process, that seeks to tailor ecosystem service assessments to the policy actions that would benefit most from MCESA. It helps identify those ecosystem services that could and should be formally assessed in a given situation. Three main steps are proposed (see Figure 28.1).

The first step in this 'triage' process focuses on the identification of the general scope of the MCESA. Initially, it is critically important to consider why and for whom this assessment is being undertaken, and how best to initiate the assessment in a given context. The recent literature fails to find evidence that ecosystem service assessments are used in practice or are influential in policymaking (Laurans et al., 2013). One reason is that such assessments were not designed to be influential. To identify decision points where MCESA would be useful to decision-makers, Pendleton et al. (2015) use a series of questions to identify decision trade-offs that could involve ecosystem services. Further, they ask whether decision-makers are willing to consider ES information, and the geographic and temporal scales of the proposed policy or decision. This helps determine if a MCESA is needed at all and, if so, the realm of possible services that should be considered.

The second step aims to prioritize the specific policy actions most in need of better data as well as which type of ecosystem service change is most likely to be in question as a result of those policies.

In the third and final step of the 'triage', analysts and decision-makers work together to decide exactly how to quantify ecosystem services in terms of, for example, measures of ecological output, economic impact and economic value.

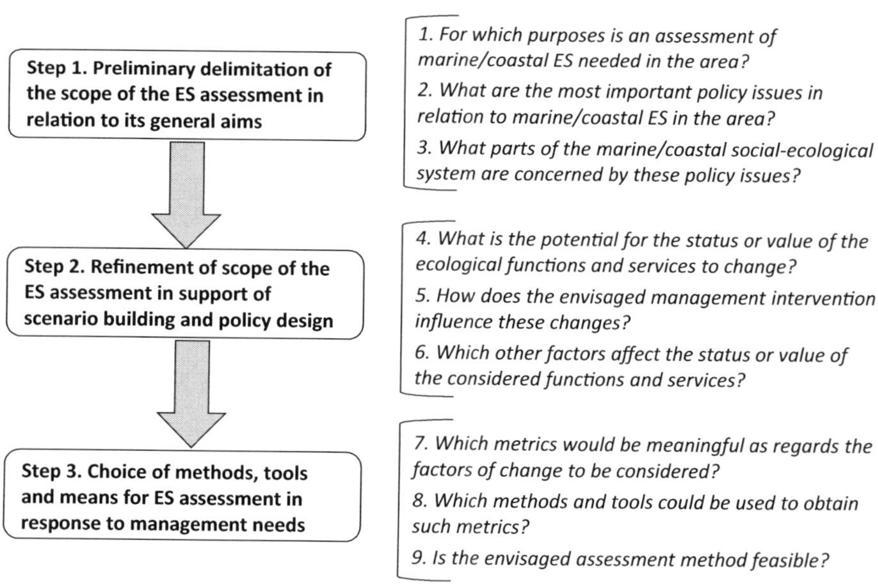

Figure 28.1 Three main stages of a marine and coastal ecosystem service assessment: the 'triage' approach.

Source: Pendleton et al. (2015)

The 'triage' process steps and related key questions

Step1: Preliminary delimitation of the scope of the ecosystem service assessment in relation to its general aims

For which purposes is a MCESA needed in the area?

It is essential to understand the operational needs of the stakeholders who envisage using a MCESA. Examples of possible aims for MCESA are to:

- improve and/or integrate knowledge;
- provide initial diagnosis of issues that may require marine management;
- raise awareness of specific issues or of the values of the marine environment more generally;
- explore possible future changes in the ecosystems or human pressures;
- design a new marine and coastal policy;
- compare operational management options;
- facilitate trade-offs; and
- increase the welfare of relevant populations.

What are the most important policy issues that affect marine ecosystem services in the area?

The policy issue may be linked to the impacts of particular marine uses, the claims of certain stakeholders or the possible change in collective rules. A precise definition of the policy issue is required as well as a collective selection of relevant policy issues.

What parts of the marine social-ecological system are influenced by these policy issues?

It is important to determine the ecosystem components, functions and services that could be influenced by policy issues as well as the identification of the stakeholders and institutions whose actions are concerned by these policy issues.

Step 2: Refinement of scope of the MCESA assessment in support of scenario building and policy design

What is the potential for the status or value of ecological functions and services to change?

If an ecosystem service is not likely to change in quantity or value, there may be little benefit in formally assessing it. Expert knowledge and consensus could help determine which ecological functions or ecosystem services are likely to experience significant changes in the future. Such changes may be the result of on-going natural processes or could result from the evolution of human pressures on the ecosystems, which can be the consequences of emerging uses such as marine renewable energy development. Specific methodologies for assessing the sensitivity of ecosystem services to change may also be implemented when needed (Tillin et al., 2010; MMO, 2012; OSPAR, 2003). Sensitivity assessments aim at gathering existing information on key characteristics of a species or habitat and its response to environmental change, and then communicating this synthesized information in an accessible way to decision-makers (Hiscock and Tyler-Walters, 2006).

How does the envisaged management action influence these changes?

Pendleton et al. (2015) argue that if the specific management action is unlikely to have a significant influence on the production value of ecosystem services there may not be the need to formally assess them. When conducting the MCESA, it is also necessary to weigh the likelihood of a future management action or policy being implemented and whether it will influence the production or the value of an ecosystem service. If, for instance, the proposed policy has little effect on the provision of value of an ecosystem service, then there is very limited justification for performing an assessment of that service.

Which other factors affect the status or value of the considered ecosystem functions and services?

The aim is to assess the role of wider social, economic, environmental and political issues such as climate change or national policies, particularly those beyond the control of local management bodies (a marine protected area management council, for example). If these wider issues have a dominant impact on the value of an ecosystem service, these factors may swamp the impact of the proposed local change, which may mean the expected change in value from local management action cannot be easily determined.

In order to identify the ecosystem services that should be assessed in the context of scenario or policy design, Pendleton et al. (2015) recommend addressing the questions in Step 2 in a simultaneous and integrated way. Each ecosystem service could be given a score (high, moderate, low) in response to each question based on relevant criteria (Table 28.4). The scores for each service, and how it scores for each question, can then inform the decision-making process. For

Table 28.4 Criteria for scoring marine and coastal ES in the second step of the 'triage' process.

Usefulness of ES assessment	*Potential for the ecosystem service value to change*	*Influence of management on ecosystem service change*	*Other factors affecting the ecosystem service*
High	Service is sensitive to impacts and value change will be large	Management will have a large influence on value, a strong probability of coming into effect and is locally driven	Local environmental factors have the strongest influence on value
Moderate	Service is sensitive to impacts and value change will be small OR Service is robust and value change will be large	Management will have a large influence on value and at least a reasonable probability of coming into effect, but is not locally driven OR Management will have a moderate influence on value, at least a reasonable probability of coming into effect and is locally driven	Other factors (social, economic, political, global environmental change) have a similar influence on value to that of local environmental factors

Usefulness of ES assessment	*Potential for the ecosystem service value to change*	*Influence of management on ecosystem service change*	*Other factors affecting the ecosystem service*
Low	Service is robust and value change will be small	Management will have a small influence on value and/or a low probability of coming into effect	Other factors have the strongest influence on value

Source: Pendleton et al. (2015)

example, the scoring can be used to determine whether a MCESA is appropriate, which policies would be best informed by such an assessment and which ecosystem services should be assessed.

Step 3: Choice of methods, tools and means for ecosystem service assessment in response to management needs

Which metrics represent meaningful ways to quantify the factors of change to be considered?

Even for a given ecosystem service, different units of quantification, i.e. types of metrics, could be more or less meaningful to stakeholders and decision-makers depending on the context. Changes related to ecological status would require biophysical metrics. Changes affecting human activities may be expressed in terms of social or monetary values and/or other economic indicators, such as jobs. Changes related to trade-offs may best be quantified in terms of social perception. For the unit of assessment to be meaningful, it is important that the metric clearly expresses the relevant characteristics of an ecosystem service (effective or potential, supply, demand and perception by a given social group).

Which methods and tools could be used to obtain such metrics?

Assessment methods must be selected once the metrics and indicators for estimating the changes in ecosystem services have been chosen. The method chosen should provide data that is useful for the proposed policy decision and the stage of the management process it is intended to support. Broad objectives associated with early management stages like initial diagnosis and policy design may require broad-brush assessment methods (e.g. approximate quantifications) while more operational objectives, like the comparison of management options, could require more focused, refined and precise methods. Valuation methodologies using ecological, economic and social indicators are numerous and can provide single indicators, or multi-criteria or integrated assessments. They are more or less powerful depending on the number of ecosystem services to be included. The precision of these methods also can vary considerably. Thus, it is important to make sure that the metric and the methodologies are appropriate to the decision being made. Finally, stakeholders and decision-makers should understand, trust and accept the results of the proposed method. Precise description of these methodologies is beyond the scope of this chapter, but they are detailed widely in the literature (for example Bateman et al., 2002; EPA, 2009; Huang et al., 2011). A range of available social and economic assessment methods can be organized in different categories: stated preferences vs revealed preferences; deliberative process vs authoritarian process; group-based method vs individual-based method; and multiple attributes vs single attribute.

Is the envisaged valuation method feasible?

Finally, the manpower and cost requirements for assessing different services can vary depending on the methodologies proposed, and must be explicitly considered. Where resources for primary data collection are limited, the availability of supporting data (both ecological and socio-economic) will also have a strong influence on the scope of an assessment.

Case study illustrating the implementation of the first two steps of the 'triage' process

A case study in France, drawn from the European VALMER[1] project, is described to illustrate the implementation of the first two steps of the 'triage' process, namely defining the aim and the scope of the MCESA so that it makes a useful and meaningful contribution to policy, given the available means and knowledge. 'Useful' means that it is linked to the management issue. 'Meaningful' means it is interpretable. The case of the Iroise Sea Marine Nature Park involved possible changes in the way kelp forests are managed. The ecosystem services assessment is therefore applied for a well-defined policy issue.

Created in 2007 off the coast of Finistère (France), the Iroise Marine Nature Park was created as a specific type of Marine Protected Area (MPA) under the French Law dedicated to the conservation of the marine ecosystems, the maritime heritage and the sustainable development of the human activities in the sea. In line with these general goals, the first Park Management Plan for 2010–2015 defined a series of ten specific objectives, including knowledge of marine ecosystems, conservation of habitats, pollution mitigation, material extraction activities, fisheries, kelp exploitation, tourism and recreational activities, and architectural, maritime, archaeological and traditional heritage. The park encompasses 3,500km^2 of natural ecosystems, which are home to large numbers of species of marine mammals, seabirds, seaweed and macro-algae (PNMI, 2010). Application of the first two steps of 'triage' process was necessary to refine the scope of the assessment.

Kelp harvesting (*Laminaria hyperborea*) is a professional activity that takes place in the park area with associated alginate production occurring on land. The sustainable exploitation of kelp fields is a management issue for the PNMI park staff, who as a result has selected it as a major topic that could benefit from an assessment. Implementing the first step of the 'triage' process, it was expected that conducting the MCESA would mainly be used as a tool to compare management options. However, this comparison of management options necessitates improved knowledge on the variety of ecosystem services delivered by kelp. Eventually, a comparison of management options that considers this variety of ecosystem services may help understanding trade-offs between kelp exploitation, recreational uses and conservation of kelp ecosystems.

An expert panel identified 11 provisioning services, six regulation and maintenance services and 10 cultural services, and selected five expected to have high potential for their value to change (Table 28.5). Combining the results of the three questions, the ecosystem services which ranked highest were: provisioning services for alginate and abalone commercial fisheries; the regulating service linked to the maintenance of habitat for abalone; and the cultural service of ecotourism. Other related ecosystem services also scored moderately high. Other ES that were initially considered, such as water quality regulation, climate regulation and coastal

Table 28.5 Scores for each ecosystem service affected by kelp harvest in the Iroise Sea Marine Natural Park (France).

Ecosystem service characterization – Iroise Sea Marine Nature Park (France)				*Triage of ecosystem services*					
Types of marine and coastal ecosystem services		MCES code	Specific ecosystem functions and services	Likely use of value in policy decision	Potential for the ES value to change	Influence of management on ES change	Other factors	Mean	Median
Provisioning services	Food provision	P1	Abalone commercial fisheries	3	2	3	2	2.50	2.50
		P2	Angling commercial fisheries (pollock and European sea bass)	3	1	2	3	2.25	2.50
		P3	Lobster commercial fisheries (fish pots)	3	1	1	3	2.00	2.00
		P4	Alginates for food industry	3	3	3	1	2.50	3.00
		P5	Aquaculture	Potential					
	Biotic materials and biofuels	P6	Biofuel	Potential					
		P7	Crop fertilizer & pest management	1	3	2	1	1.75	1.50
		P8	Alginates for other industries	3	3	3	1	2.50	3.00
		P9	Molecule for medicines (non alginate)	1	2	1	3	1.75	1.50
		P10	Molecule for cosmetics (non alginate)	1	2	1	3	1.75	1.88
		P11	Disservice Bycatch (*Saccorhiza polyschides*)	2	2	2	2	2.00	2.00
Maintenance and regulation services	Coastal protection	R1	Natural coastal defence	1	2	2	1	1.50	1.50
	Biological regulation	R2	Disservice Parasite Anisakis	Potential					

(Continued)

Table 28.5 (Continued)

Ecosystem service characterization – Iroise Sea Marine Nature Park (France)				*Triage of ecosystem services*					
Types of marine and coastal ecosystem services		MCES code	Specific ecosystem functions and services	Likely use of value in policy decision	Potential for the ES value to change	Influence of management on ES change	Other factors	Mean	Median
	Ocean nourishment	M1	Strong primary productivity	3	2	1	1	1.75	1.50
	Life cycle maintenance	M2	Improvement of kelp resilience	2	2	2	2	2.00	2.00
		M3	Key habitats that support a strong biodiversity	3	2	3	1	2.25	2.50
		M4.1	Key habitats for commercial fishes	3	1	2	3	2.25	2.50
		M4.2	Key habitats for abalone	3	2	3	2	2.50	2.50
		M4.3	Key habitats for European lobster	3	1	1	3	2.00	2.00
		M4.4	Key habitats for bottle-nose dolphins	3	2	2	2	2.25	2.00
		M4.5	Key habitats for grey seals	2	2	2	2	2.00	2.00
		M4.6	Key habitats for seabirds	3	2	2	2	2.25	2.00
Cultural services	Symbolic and aesthetic values	C1	Traditional activity	3	3	1	2	2.25	2.50
		C2	Remarkable marine and seascape	3	2	2	1	2.00	2.00
		C3	Remarkable species	3	2	3	1	2.25	2.50
	Recreation and tourism	C4	Recreational fishing (shell, crustacean & fish)	2	3	2	1	2.00	2.00
		C5	Boating	3	1	1	2	1.75	1.50
		C6	Kayaking	3	1	1	2	1.75	1.50
		C7	Seascape & sea life watching (ecotourism)	3	2	3	2	2.50	2.50
	Cognitive effects	C8	Material for research	2	3	2	1	2.00	2.00
		C9	Material for arts	1	1	1	3	1.50	1.00
		C10	School excursion / awareness campaign	2	2	1	3	2.00	2.00

Source: authors

Note: numerical scores (1–3) in order of increasing importance.

protection, either scored low or could not be linked scientifically to expected changes in ecosystem conditions.

Considering the aim of the MCESA in this case study, and the set of ecosystem services to be valued, two methodologies were selected. First, simple indicators were determined based on the available data for the ecosystem services that were selected. In addition to this, the simulation of the impacts of various fisheries management options on the four key services was undertaken to compare scenarios and trade-offs, using a dynamic system model of the kelp ecosystem and a spatial model of kelp distribution. Both biological models together with other information on kelp fishing fleet, processing industry and ecotourism fed a social-ecological simulation model of kelp ecosystem services.

Lessons learned from applying the two first steps of the 'triage' process

Scoping is the first step in any type of analytical assessment. Often, though, the scoping process is conducted in ways that are not transparent or strategic. While careful scientific approaches are often applied to other parts of ecosystem analysis for planning, that same level of rigor is rarely applied to the scoping stage. As a result, the data provided may not be useful, relevant or credible in a decision-making context.

A structured decision-making approach offers a framework for making MCESA scoping more transparent and rigorous. It also helps to ensure that MCESA's are more useful in policy and decision-making.

In applying a formalized process to identify and assess marine and coastal ecosystem services in our case study, we found that the structured decision-making approach of the "triage" process proved useful to marine area managers and stakeholders. The approach helped to identify the marine management issues and also helped to select ecosystem services for which a formal assessment was judged useful by decision-makers. Additionally, the transparency of the process used to choose the services to be assessed helped stakeholders in both study areas to be more confident in the design of future management options. The 'triage' process also helped scientists and MPA managers of this park to share their multidisciplinary knowledge of both the functioning of kelp ecosystems and also the limits of proposed management options.

Structured decision-making, however, comes at a cost. The process requires that analysts and stakeholders work together to narrow the scope of the MCESA. For many stakeholders, this may be their first introduction to ecosystem services terminology and methods. For many analysts, structured decision-making is new ground and may be beyond their area of expertise. These challenges, however, can be minimized if appropriate social scientists are involved in the process of stakeholder engagement in the structured decision-making process. Often, the scope of the MCESA can be narrowed partially by simply interviewing managers and stakeholders – leaving the more contentious and difficult aspects of scoping for the 'triage' phase of the process. Furthermore, if a long term view is taken, the 'triage' process can build knowledge and capacity for both decision-makers and analysts – making it easier to discuss ecosystem services in the future and to propose MCESA and monitoring approaches that will provide useful information for managers.

Acknowledgements

This chapter synthesizes an application developed for the European VALMER project (www.valmer.eu). VALMER (Valuing Ecosystem Services in the Western Channel) aimed to examine in six case studies how improved marine ecosystem services assessment can support effective and informed marine management and planning. It was selected under the European cross-border cooperation programme INTERREG IVA France (Channel)-England, co-funded by the ERDF.

Note

1 www.valmer.eu

References

Atkins, J.P., Burdon, D., Elliott, M., and Gregory, A.J. (2011). Management of the marine environment: integrating ecosystem services and societal benefits with the DPSIR framework in a systems approach. *Marine Pollution Bulletin*, vol 62, no 2, pp 215–226.

Bateman, I. J., Carson, R. T., Day, B., et al. (13 authors) (2002). *Economic Valuation with Stated Preference Techniques. A Manual.* Edward Elgar, Cheltenham.

Beaumont, N. J., Austen, M. C., Atkins, J. P., et al. (14 authors) (2007). Identification, definition and quantification of goods and services provided by marine biodiversity: implications for the ecosystem approach. *Marine Pollution Bulletin*, vol 54, pp 253–265.

Böhnke-Henrichs, A., Baulcomb, C., Koss, R., Hussain, S., and de Groot, R. S. (2013). Typology and indicators of ecosystem services for marine spatial planning and management. *Journal of Environmental Management*, vol 130, pp 135–145.

Börger, T., Beaumont, N., Pendleton, L., et al. (15 authors) (2014). Incorporating ecosystem services in marine planning: the role of valuation. *Marine Policy*, vol 46, pp 161–170.

Boyd, J., and Banzhaf, S. (2007). What are ecosystem services? The need for standardized environmental accounting units. *Ecological Economics*, vol 63, no 2, pp 616–626.

EPA – Environmental Protection Agency (2009). Valuing the protection of ecological systems and services. U.S. Environmental Protection Agency, Office of the Administrator, Science Advisory Board.

Fisher, B., Turner, R.K., and Morling, P. (2009). Defining and classifying ecosystem services for decision making. *Ecological Economics*, vol 68, no 3, pp 643–653.

Fletcher, S., Saunders, J., and Herbert, R.J. (2011). A review of the ecosystem services provided by broad-scale marine habitats in England's MPA network. *Journal of Coastal Research*, vol 64, pp 378–383.

Galparsoro, I., Connor, D. W., Borja, A., et al. (27 authors) (2012). Using EUNIS habitat classification for benthic mapping in European seas: present concerns and future needs. *Marine Pollution Bulletin*, vol 64, pp 2630–2638.

Granek, E.F., Polaski, S., Kappel, C.V., et al. (17 authors) (2010). Ecosystem services as a common language for coastal ecosystem-based management. *Conservation Biology*, vol 24, pp 207–216.

Hiscock, K., and Tyler-Walters H. (2006). Assessing the sensitivity of seabed species and biotopes – The Marine Life Information Network (MarLIN). *Hydrobiologia*, vol 555, pp 309–320.

Huang, I.B., Keisler, J., and Linkov, I. (2011). Multi-criteria decision analysis in environmental sciences: ten years of applications and trends. *Science of the Total Environment*, vol 409, no 19, pp 3578–3594.

ICES (2005). *Report of the Working Group on Habitat Mapping (WGMHM)*. Bremerhaven, ICES CM 2005/E:05. Available at: http://brage.bibsys.no/xmlui/bitstream/handle/11250/102755/1/WGMHM05.pdf

Laurans, Y., Rankovic, A., Billé, R., Pirard, R., and Mermet, L. (2013). Use of ecosystem services economic valuation for decision making: questioning a literature blindspot. *Journal of Environmental Management*, vol 119, pp 208–219.

Liquete, C., Piroddi, C., Drakou, E.G., et al. (7 authors) (2013). Current status and future prospects for the assessment of marine and coastal ecosystem services: a systematic review. *PLoS ONE*, vol 8, no 7.

Lopes, R., and Videira, N. (2013). Valuing marine and coastal ecosystem services: an integrated participatory framework. *Ocean & Coastal Management*, vol 84, pp 153–162.

MA (2005). *Ecosystems and Human Wellbeing: Synthesis.* Island Press, Washington DC.

MMO- Marine Management Organisation (2012). Interactions – between multiple activities and between activities and environment. In: *MMO. 2012. East Inshore and East Offshore Marine Plan Areas Evidence and Issues Report 2012.*

OSPAR (2003). *Criteria for the Identification of Species and Habitats in need of Protection and their Method of Application (The Texel-Faial Criteria).* OSPAR Convention for the Protection of the Marine Environment of the North-East Atlantic. Meeting of the OSPAR Commission, Bremen, 23–27 June 2003. Reference Number: 2003–13. OSPAR 03/17/1-E, Annex 5.

Pendleton, L., Mongruel, R., Beaumont, N., Hooper, T., and Charles, M. (2015). A triage approach to improve the relevance of marine ecosystem services assessments. *Marine Ecology Progress Series*, vol 530, pp 183–193.

PNMI–Parc Naturel Marin de la mer d'Iroise (2010). *Plan de gestion du Parc Marin de la mer d'Iroise 2010–2015, Résumé, Parc Naturel Marin d'Iroise, Le Conquet.* Available at: http://www.parc-marin-iroise.fr/Le-Parc/Objectifs/Plan-de-gestion

Potschin, M. and R. Haines-Young. (2016). Defining and measuring ecosystem services. In Potschin, M., Haines-Young, R., Fish, R. and Turner, R. K. (eds) *Routledge Handbook of Ecosystem Services*. Routledge, London and New York, pp 25–44.

Potts, T., Burdon, D., Jackson, E., et al. (7 authors) (2014). Do marine protected areas deliver flows of ecosystem services to support human welfare? *Marine Policy*, vol 44, pp 139–148.

TEEB (2010). *The Economics of Ecosystems and Biodiversity: Mainstreaming the Economics of Nature: A Synthesis of the Approach, Conclusions and Recommendations of TEEB.* Sukhdev, P., Bishop, J., Gundimeda, H., et al. (10 authors). Bonn/Brussels.

Tillin, H. M., Hull, S. C., and Tyler-Walters, H. (2010). *Development of a Sensitivity Matrix (pressures-MCZ/MPA features).* Report to the Department of Environment, Food and Rural Affairs from ABPMer, Southampton and the Marine Life Information Network (MarLIN)/ Plymouth: Marine Biological Association of the UK. Defra Contract No. MB0102 Task 3A, Report No. 22.

Townsend, M., Thrush, S. F., and Carbines, M.J. (2011). Simplifying the complex: an 'Ecosystem Principles Approach' to goods and services management in marine coastal ecosystems. *Marine Ecology-Progress Series*, vol 434, pp 291–301.

UK National Ecosystem Assessment (2011). *The UK National Ecosystem Assessment Technical Report.* UNEP-WCMC, Cambridge UK.

UNEP (2009). *Report from the Workshop on Ecosystem Service Indicators: Developing and Mainstreaming Ecosystem Service Indicators for Human Wellbeing: Gaps, Opportunities and Next Steps.* UNEP World Conservation Monitoring Centre, Cambridge UK.

29
FRESHWATER

Kate A. Brauman

Introduction

Water is integral to life on Earth. It makes up the majority of our cells and the cells of the biota we depend on for life. We harness the power of water as it moves from the atmosphere, subsurface, and oceans, creating energy, using it for cleaning, and transporting ourselves and our things. As a result, people have long managed landscapes and waterscapes to control water and optimize the benefits it provides (Andreassian, 2004). In the context of Earth's water cycle, however, people affect only a small fraction of the total available water. Liquid freshwater on or near the earth's surface and in the biosphere, the water that we can manage, is less than 0.3% of all water on the planet (Shiklomanov, 2000). Relatively little water has been created or destroyed since water first appeared on earth; the water cycle moves water among forms and locations (Morbidelli et al., 2000).

In the context of the on-going cycling of water, this chapter considers the role of ecosystems and ecosystem services related to freshwater. All ecosystems, both aquatic and terrestrial, depend on water. People intensively manage water, and both managed and unmanaged ecosystems affect water – not by creating water but by mediating its stocks, flows, and constituents. This chapter considers both hydrologic and aquatic services. Hydrologic services describe the way terrestrial landscapes affect water, and include services such as water availability for drinking or irrigation. Freshwater aquatic services encompass the services provided by freshwater ecosystems themselves, such as provision of fish. The services provided by transitional ecosystems such as wetlands are also considered within this context. Economic valuation is an important element of ecosystem services assessment; this chapter outlines basic principles of economic valuation specific to hydrologic and aquatic ecosystem services. Finally, management for hydrologic and aquatic services is considered, particularly Payment for Watershed Services (PWS) projects.

Freshwater related services

The ecosystem services framework organizes biophysical processes by the benefits they impart to people and defines services by their beneficiaries (de Groot et al., 2002; Boyd and Banzhaf, 2006). For water-related services, this requires evaluating how water is used – beneficiaries may be consuming water, using it *in situ*, or taking advantage of the products of a freshwater system

(Brauman et al., 2007). Thus, though water quality and water purification are regularly used as practical shorthand to discuss ecosystem services, in fact they are, respectively, a biophysical attribute of water and an ecosystem function. To be an ecosystem service, a beneficiary must define whether the water is clean enough for boating, swimming, or drinking. It is also crucial to determine if a change in water quality is attributable to ecosystem filtration processes.

As with any ecosystem service, either total production or a change in service delivery can be evaluated and reported for water-related services (Fisher et al., 2008). Total provisioning is generally calculated based on the implicit assumption that service provision in the absence of an ecosystem would be zero. For aquatic systems, this is plausibly the case: a wetland, lake, or river could be drained and no services would be provided. For hydrologic services, however, in which a terrestrial ecosystem affects the flow of water through the landscape, water would cycle in the absence of an ecosystem – though the flow would differ – there is no zero alternative. Thus, for hydrologic services, a change in service provision must be considered.

Hydrologic services

Discussions of freshwater ecosystem services have generally focused on the many ways people use freshwater (e.g. Postel and Carpenter, 1997; Vorosmarty et al., 2005). Ecosystems come into play because landscapes affect water, altering the partitioning of water to the subsurface, along the surface, and into the atmosphere; the speed of water flow; and the constituents within water (Brauman et al., 2007). Examples of final hydrologic services include water removed from water bodies, such as water for drinking or irrigation; water used in place, such as uses for transportation or hydropower; water damages, including changes in the frequency and intensity of damages from flooding; and cultural services, including religious, aesthetic, and recreational uses. Hydrologic services are also instrumental in supporting the aquatic ecosystems that provide services described in the following section.

The connections between ecohydrologic processes and hydrologic services are illustrated in Figure 29.1. To connect ecohydrologic processes in the landscape to hydrologic services, we can evaluate ecosystem effects on water quantity, quality, location and timing. All attributes are important, but in most cases one or two will be most important for a given service. These attributes provide a straightforward way to focus on specific ecohydrologic processes of interest in water management.

Principles of landscape hydrology are well understood, if not always well quantified, and a number of reviews have evaluated the way ecosystems affect the flow of water (see Brauman et al., (2007) and Ponette-González et al., (2014b)). For example, landscape change affects water yield by affecting evapotranspiration, such that more or less water is transferred to the atmosphere instead of becoming available as river flow. Though ecosystems cannot create water, by affecting the amount and location of moisture in the atmosphere, changes in vegetation can also affect precipitation regimes (Swann et al., 2012).

Landscapes may be actively managed specifically for water, such as when forested watersheds are maintained to ensure water quality at drinkable standards. Of note, even in a case of conservation such as this, the ecosystem service is a function of landscape change. Of interest to water users is the difference between water quality given forest conservation and if the forest were to be removed.

In many cases, land management has focused on optimizing landscapes for non-water-related services and landscape effects on water have become apparent via unexpected negative consequences. For example, nutrient and sediment inputs from agriculture or timber harvesting have sometimes rendered streams and lakes undrinkable or unswimmable.

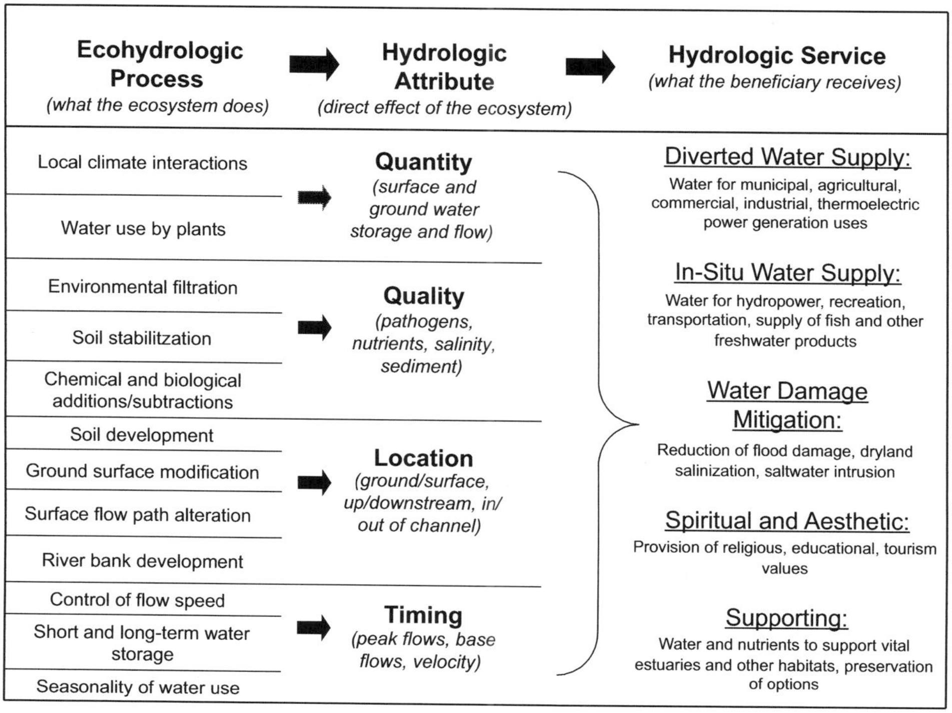

Figure 29.1 Hydrologic Ecosystem Services.

Ecohydrologic processes in terrestrial systems impact attributes of water that affect how the water can be used. Figure adapted from Brauman et al. (2007).

Evaluating landscape effects on hydrologic services can be difficult because the desirable quantity and quality of water depends on its use. More water may or may not be desirable contingent on whether the additional water fulfils demand for dry season flows or overtops stream banks causing flooding and waterlogging. This is quite different from carbon sequestration or pollination, whose value increases monotonically, meaning that more is better.

Ecosystems that bridge terrestrial and aquatic landscapes, including floodplains, wetlands, and estuaries, provide many of the same regulating services described in this section (Zedler and Kercher, 2005; Barbier et al., 2011). Wetlands serve a particularly important role removing nutrients and pollutants from freshwater, a capability that has been harnessed in phytoremediation projects (Susarla et al., 2002)

Aquatic services

Freshwater ecosystems, including lakes, streams, wetlands and estuaries, provide a range of services, illustrated in Figure 29.2. Provisioning services, notably freshwater fish, are particularly important. Freshwater systems also provide regulating services, stemming from ecosystem processes such as in-stream nutrient processing, and cultural services, including those related to fishing. In many of the Native American cultures of the northwestern US, for example, salmon fishing is an integral part of identity (White, 1996). In the Millennium Ecosystem Assessment, freshwater ecosystems and the services they provide are referred to as inland waters (Finlayson and D'Cruz, 2006).

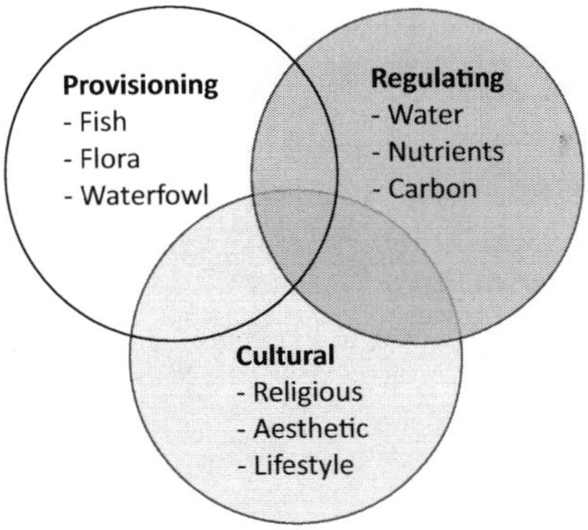

Figure 29.2 Aquatic Ecosystem Services.

Examples of ecosystem services provided by inland waters. These services interact, such that harvest may affect regulating processes within the ecosystem, and cultural value may be related to provisioning or regulating services.

Streams, lakes, and other inland waters support flora and fauna that provide important food and material goods, including fish, algae, rice, waterfowl, clams, mussels, and pelts (Carpenter et al., 2011). Though production of many of these products can be managed or farmed, most aquaculture relies on functioning aquatic systems for associated regulating services such as waste removal and to balance the growth of unwanted organisms.

Inland waters also provide important regulating services, affecting fluxes of water, carbon, nutrients, and other constituents into sediments, the atmosphere, and downstream water bodies. Inland waters, particularly wetlands, serve an important role in retaining and moving water. Biologically, inland waters are important as both sources and sinks of greenhouse gasses (Tranvik et al., 2009) and in the cycling and movement of nutrients. As nitrogen and phosphorus are assimilated in algae blooms, they are sequestered in sediments; denitrificaton subsequently occurs (Bernhardt, 2013; Finlay et al., 2013). Nutrient cycling within inland waters is important locally: for example, it controls algae blooms and the appeal of a location for swimming. Nutrient cycling is also important for controlling nutrient export and the subsequent impact on services such as drinkability of downstream water.

Fish and other animals and plants that are harvested from aquatic systems are themselves integral to aquatic ecosystem functioning. In the process of gathering essential nutrition, for example, fish maintain the trophic structure of lakes and can also affect greenhouse gases, nutrients, and water clarity. Fish also move nutrients through the aquatic system and may affect disease regulation by, for example, helping to control vector-borne disease (Holmlund and Hammer, 1999).

Aquatic ecosystem services are affected by and linked to hydrologic services. Landscape effects on the quantity, quality, and timing of water entering an aquatic system can affect the ability of a freshwater ecosystem to function and thereby provide desired services. Removing water from inland water systems for human use can reduce or eliminate aquatic ecosystem service provision; studies have suggested that 20% to 50% of mean annual flow must remain in-stream to support freshwater ecosystems (Smakhtin et al., 2004). *In situ* uses of water may also affect aquatic ecosystems if flow regimes are altered or fish passage blocked, which can result from dams and reservoir construction (Brismar, 2002). Non-consumptive water uses that affect water quality, such as warm water effluent from thermoelectric cooling, may also affect freshwater fisheries.

Determining the economic value of freshwater services

Human well-being is broadly defined within the ecosystem services framework, including health and happiness as well as income. In much research as well as for management purposes, however, there is considerable interest in assessing the monetary value of freshwater services.

Evaluating the economic value of freshwater-related ecosystem services is in many ways no different than economic valuation of other types of ecosystem services (Farber et al., 2002; Thompson and Segerson, 2009). One might, for example, consider increased operational costs for a water treatment plant if erosion increased after hillsides were deforested (Dearmont et al., 1998) or evaluate the market price of fish provided by a lake. Based on extrapolation from local case studies, the total value of services provided by rivers and lakes was recently estimated at over US$ 4,000 per hectare each year and services from inland wetlands at over US$ 25,000 per ha per year (de Groot et al., 2012). The same review found the value of water flow regulation by terrestrial ecosystems to average nearly US$ 350 per ha per year and by inland wetlands to be more than US $5,500 per ha per year. These numbers provide useful insight into the importance of water-related ecosystem services, though the authors acknowledge that the globally non-representative data set they rely on provides limited insight given wide regional and local variations in ecosystem function and ecosystem service demand.

For freshwater-related services, one is frequently asked the value of "the water." The answer to this is complex for a number of reasons. First, though water is often paid for, the value of water is not reflected by its market price, which generally reflects the cost of treatment and transport, not the cost of water itself. Second, water has immense value when it is in short supply, but as supply increases, value decreases, and value may even become negative if flooding or waterlogging occurs. Third, in many cases, water itself is not the final product of interest. Instead, the value of water stems from its role as a factor of production for some other good or service. Specifics of valuing water have been addressed at length in books such as Young (2005).

Hydrologic ecosystem services are regulating services, so valuation can be complex because ecosystems affect the flow of water but do not create it. As a result, the value of a hydrologic service lies in the way an ecosystem changes the volume, quality, timing and location of water, not in the total amount or quality of water. And changes to water have impact within a larger landscape context. When water quality is of interest, for example, the nutrient sequestration function of a riparian buffer provides a service only in the context of whether nutrients are added in upland areas and must be removed. The value of the riparian buffer stems from the difference in water quality with and without the buffer. For valuation, it is thus important to be explicit about alternate land use or land cover scenarios and the attributes of water delivery under each one (Ponette-González et al., 2014a).

Value is a function of how water will be used: whether there is enough water, and if it is of an appropriate quality, for activities like drinking, fishing, swimming, and transportation. Value of aquatic services is similar, though inland waters provide a range of valuable provisioning as well as regulating services. For both hydrologic and aquatic services, the cultural value of clear water for swimming or a lifestyle supported by fishing may far outweigh the use value of regulating and provisioning services (Chan et al., 2012).

For all freshwater services, valuation hinges on translating changes in a biophysical attribute to changes in a service of interest (Keeler et al., 2012). For example, a change in nutrient concentration may be directly of value for drinking water, but for swimming value is more closely tied to water clarity. While water clarity is a function of nutrient concentration, additional analysis is necessary to make this link.

Management of freshwater services

Ecosystem services provide a compelling framework for management because they directly link the impacts of landscape actions to beneficiaries. This is particularly useful for hydrologic services, which often have impacts that are felt off-site. Management for water services does not require that beneficiaries understand the biophysical connections between land management and hydrologic impact, only that they articulate how they use water and what attributes are important to them. Monetary valuation allows these impacts to be easily integrated into traditional management frameworks such as cost benefit analysis.

Payment for Watershed Services (PWS) projects, in which upstream residents are compensated for managing land to affect water in certain ways, are becoming increasingly popular (Goldman-Benner et al., 2012). In the Latin American Andes, for example, a large number of PWS designed to increase the quality and reliability of drinking water have been implemented (Bennett et al., 2013). PWS may be implemented because regulation is impossible or unenforceable, though these programs require substantial institutional support to function properly (Engel et al., 2008). Ecosystem services may also be adopted for watershed management because they reflect system-wide impacts and can improve communication among stakeholders (Brauman et al., 2014).

As with economic valuation, management of hydrologic services focuses on changes in service delivery. Because precipitation, the primary driver of water availability, is highly variable, quantifying a comparison case generally requires a hydrologic model. A large number of hydrologic models exist, some focused specifically on hydrologic services; differences between these model types are reviewed by Vigerstol and Aukema (2011). Even with the use of models, predicting hydrologic response to land-cover change under changing climate is challenging, and as a result it is not always clear that PWS projects are providing water services of interest (Guswa et al., 2014). This is a particular problem in the tropics, where PWS schemes are in the widest use, because many of the general principles of forest and landscape hydrology are based on research undertaken in temperate locations and may not be directly transferable to the tropics (Ponette-González et al., 2014b). Managing and measuring specific hydrologic fluxes of interest may help ensure that desired services are delivered (Ponette-González et al., 2014a).

Trade-offs among ecosystem services

The value of many water-related ecosystem services is widely recognized and people frequently manage land and waterscapes to enhance these services. However, management actions to enhance one element of freshwater availability or a single freshwater ecosystem service may have negative impacts on ecosystem function and threaten the delivery of other services.

One example of an often overtaxed ecosystem service is the dilution, absorption, and processing of pollutants (Hinga et al., 2005). In Paris, for example, waste was disposed of in the River Seine until the population exceeded the river's capacity to absorb and remove waste. At that point, sewage had to be managed (Delleur, 2003). Another example is the overharvest of fish, which can decimate a population beyond its ability to renew itself (Carpenter et al., 2011).

Dams provide an example of one type of service being enhanced to the detriment of others. When dams are built, freshwater systems are altered to enhance the year-round delivery of freshwater for irrigation or hydropower at the expense of fisheries (Brismar, 2002). In Southeast Asia, on the Mekong river, an ecosystem services study by Ziv et al. (2012) provided recommendations for placing dams to optimize the combined value of stored water for hydropower and the nutrition and income provided by fish harvested in the river.

Each of these examples illustrates the peril of exploiting the services provided by freshwater systems without managing for ecosystem function. Recognizing the benefits of these systems as ecosystem services and managing for them explicitly provides an avenue for more sustainable use.

Conclusions

Research on freshwater ecosystem services is becoming increasingly common (Brauman, 2015). The link from biophysical processes to human well-being, which is central to the theory of ecosystem services, has enormous potential to stimulate novel research questions. This research differs from traditional hydrology, ecology, and economic investigations because it focuses on landscape change and on new variables of interest. The interdisciplinary nature of freshwater services research presents a challenge, however, requiring integration from the beginning stages of problem formation and transparency about disciplinary assumptions (Kaghram et al., 2010).

In management, the appeal of using freshwater ecosystem services is evidenced by their widespread adoption (Bennett et al., 2013). People are increasingly interested in "green infrastructure" projects that deliver desired freshwater services alongside a variety of other ecosystem services. Using the ecosystem services framework to assign economic value to freshwater services also allows them to be integrated into traditional policy-analysis tools such as cost-benefit analysis. However, because freshwater services are a function of both their biophysical setting and their socio-economic context, many elements of a socio-ecological system must be well understood in order to manage them. This complexity can make providing guidelines difficult and may make implementing freshwater services programs challenging in practice.

References

Andreassian, V. (2004). Waters and forests: from historical controversy to scientific debate. *Journal Of Hydrology*, vol 291, no 1–2, pp 1–27.

Barbier, E.B., Hacker, S.D., Kennedy, C., et al. (6 authors) (2011). The value of estuarine and coastal ecosystem services. *Ecological Monographs*, vol 81, no 2, pp 169–193.

Bennett, G., Carroll, N., and Hamilton, K. (2013). *Charting New Waters: State of Watershed Payments 2012*. Forest Trends, Washington, DC.

Bernhardt, E.S. (2013). Cleaner lakes are dirtier lakes. *Science*, vol 342, no 6155, pp 205–206.

Boyd J., and Banzhaf S. (2006). *What Are Ecosystem Services? The Need for Standardized Accounting Units*. Resources for the Future, Washington DC.

Brauman, K.A. (2015). Hydrologic ecosystem services: linking ecohydrologic processes to human well-being in water research and watershed management. *Wiley Interdisciplinary Reviews: Water*. vol 2, no 4, pp 345–358.

Brauman, K.A., Daily, G.C., Duarte, T.K., and Mooney, H.A. (2007). The nature and value of ecosystem services: an overview highlighting hydrologic services. *Annual Review of Environment and Resources*, vol 32, pp 67–98.

Brauman, K.A., Meulen, S., and Brils, J. (2014). Ecosystem services and river basin management. In: Brils, J., Brack, W., Müller-Grabherr, D., Négrel, P. and Vermaat, J. E. (eds) *Risk-Informed Management of European River Basins*. Springer, Berlin.

Brismar, A. (2002). River systems as providers of goods and services: a basis for comparing desired and undesired effects of large dam projects. *Environmental Management*, vol 29, no 5, pp 598–609.

Carpenter, S.R., Stanley, E.H., and Vander Zanden, M.J. (2011). State of the world's freshwater ecosystems: physical, chemical, and biological changes. *Annual Review of Environment and Resources*, vol 36, pp 75–99.

Chan, K.M.A., Guerry, A. D., Balvanera, P., et al. (17 authors) (2012). Where are cultural and social in ecosystem services? A framework for constructive engagement. *BioScience*, vol 62, no 8, pp 744–756.

de Groot, R., Brander, L., Van Der Ploeg, S., et al. (17 authors) (2012). Global estimates of the value of ecosystems and their services in monetary units. *Ecosystem Services*, vol 1, no 1, pp 50–61.

de Groot, R.S., Wilson, M.A., and Boumans, R.M.J. (2002). A typology for the classification, description and valuation of ecosystem functions, goods and services. *Ecological Economics*, vol 41, no 3, pp 393–408.

Dearmont, D., McCarl B. A., and Tolman, D. A. (1998). Costs of water treatment due to diminished water quality: a case study in Texas. *Water Resources Research*, vol 34, no 4, pp 849–853.

Delleur, J. W. (2003). The evolution of urban hydrology: past, present, and future. *Journal of Hydraulic Engineering*, vol 129, no 8, pp 563–573.

Engel, S., Pagiola, S., and Wunder, S. (2008). Designing payments for environmental services in theory and practice: an overview of the issues. *Ecological Economics*, vol 65, no 4, pp 663–674.

Farber, S.C., Costanza, R., and Wilson. M.A. (2002). Economic and ecological concepts for valuing ecosystem services. *Ecological Economics*, vol 41, no 3, pp 375–392.

Finlay, J.C., Small, G.E., and Sterner R.W. (2013). Human influences on nitrogen removal in lakes. *Science*, vol 342, no 6155, pp 247–250.

Finlayson, C. M., and D'Cruz, R. (2006). Inland Water Systems. In: *Millenium Ecosystem Assessment*. Island Press, Washington DC.

Fisher, B., Turner, K., Zylstra, M., et al. (18 authors) (2008). Ecosystem services and economic theory: integration for policy-relevant research. *Ecological Applications*, vol 18, no 8, pp 2050–2067.

Goldman-Benner, R.L., Benitez, S., Boucher, T., et al. (8 authors) (2012). Water funds and payments for ecosystem services: practice learns from theory and theory can learn from practice. *Oryx*, vol 46, no 1, pp 55–63.

Guswa, A.J., Brauman, K.A., Brown, C., et al. (6 authors) (2014). Ecosystem services: challenges and opportunities for hydrologic modeling to support decision making. *Water Resources Research*, vol 50, no 5, pp 4535–4544.

Hinga, K.R., Ahmed, M.T., and Lewis, N. (2005). Waste processing and detoxification. In: *Millenium Ecosystem Assessment*. Island Press, Washington DC.

Holmlund, C.M., and Hammer, M. (1999). Ecosystem services generated by fish populations. *Ecological Economics*, vol 29, no 2, pp 253–268.

Keeler, B.L., Polasky, S., Brauman, K.A., et al. (8 authors) (2012). Linking water quality and well-being for improved assessment and valuation of ecosystem services. *Proceedings of the National Academy of Sciences*, vol 109, no 45, pp 18619–18624.

Khagram, S., Nicholas, K.A., Bever, D., et al. (12 authors) (2010). Thinking about knowing: conceptual foundations for interdisciplinary environmental research. *Environmental Conservation*, vol 37, no 4, pp 388–397.

Morbidelli, A., Chambers, J., Lunine, J. I., et al. (7 authors) (2000). Source regions and timescales for the delivery of water to the Earth. *Meteoritics and Planetary Science*, vol 35, no 6, pp 1309–1320.

Ponette-González, A.G., Brauman, K.A., Marín-Spiotta, E., et al. (7 authors) (2014a). Managing water services in tropical regions: from land cover proxies to hydrologic fluxes. *AMBIO*, vol 44, pp 367–375.

Ponette-González, A.G., Marín-Spiotta, E., Brauman, K.A., et al. (6 authors) (2014b). Hydrologic connectivity in the high-elevation tropics: heterogeneous responses to land change. *BioScience*, vol 64, no 2, pp 92–104.

Postel, S., and Carpenter, S. (1997). Freshwater ecosystem services. In: Daily, G. (ed.) *Nature's Services: Societal Dependence on Natural Ecosystems*. Island Press, Washington DC.

Shiklomanov, I. A. (2000). Appraisal and assessment of world water resources. *Water International*, vol 25, no 1, pp 11–32.

Smakhtin, V., Revenga C., and Döll, P. (2004). A pilot global assessment of environmental water requirements and scarcity. *Water International*, vol 29, no 3, pp 307–317.

Susarla, S., Medina, V. F., and McCutcheon, S. C. (2002). Phytoremediation: an ecological solution to organic chemical contamination. *Ecological Engineering*, vol 18, no 5, pp 647–658.

Swann, A.L.S., Fung, I.Y., and Chiang, J.C.H. (2012). Mid-latitude afforestation shifts general circulation and tropical precipitation. *Proceedings of the National Academy of Sciences*, vol 109, no 3, pp 712–716.

Thompson, B. H., and Segerson, K. (2009). *Valuing the Protection of Ecological Systems and Services*. EPA Science Advisory Board, Washington DC.

Tranvik, L.J., Downing, J.A., Cotner, J.B., et al. (32 authors) (2009). Lakes and reservoirs as regulators of carbon cycling and climate. *Limnology and Oceanography*, vol 54, no 6.2, pp 2298–2314.

Vigerstol, K.L., and Aukema, J.E. (2011). A comparison of tools for modeling freshwater ecosystem services. *Journal of Environmental Management*, vol 92, no 10, pp 2403–2409.

Vorosmarty, C. J., Leveque, C., and Revenga, C. (2005). Fresh water. In: *Millenium Ecosystem Assessment*. Island Press, Washington DC.

White, R. (1996). *The Organic Machine: The Remaking of the Columbia River*. Hill and Wang, New York.

Young, R. A. (2005). *Determining the Economic Value of Water: Concepts and Methods*. Resources for the Future, Washington DC.

Zedler, J.B., and Kercher, S. (2005). Wetland resources: status, trends, ecosystem services, and restorability. *Annual Review of Environment and Resources*, vol 30, pp 39–74.

Ziv, G., Baran, E., Nam, S., Rodríguez-Iturbe, I., and Levin, S.A. (2012). Trading-off fish biodiversity, food security, and hydropower in the Mekong River Basin. *Proceedings of the National Academy of Sciences*, vol 109, no 15, pp 5609–5614.

30

FOREST-RELATED ECOSYSTEM SERVICES

Sandra Luque and Louis Iverson

Introduction

Forests are a crucial element not only of landscapes but also of human living conditions. Covering nearly a third of the earth's land surface (see Box 30.1), they stabilize surface soil, prevent erosion and play an essential role in water resource management at the watershed and local levels. They regulate climate and improve air quality. At the same time they are an important resource for the regional economy (wood production, recreation and tourism) and are an important cultural and social heritage of the local and regional human activities. They provide habitats for a multitude of animal and plant species and are essential for the biological diversity in forest ecosystems over large areas.

Likewise, for centuries, forests have served humans as shelter or a place for natural safety for communities during times of famine or other events that impact agricultural and food production: forests provide fruits, leaves, gum, nuts, timber and wood for fuel. Thus, throughout history, forests supported peoples' livelihoods, especially when crops failed.

Today, the world's forests are in a state of flux due to land-use and climate change, deforestation, afforestation, wildfires, insects and pathogen outbreaks. In the face of both anthropogenic and natural forces there is an increasing need to assess the value of our forests. The incorporation of the ecosystem service (ES) concept into the framework of forest management stems from a need to create a more holistic perception of forests, recognizing not only their economic value, but also their cultural and ecological values, including their regulation capability.

While timber production often dominated the way in which forests were managed in the 20th century, new challenges and increasing pressures in the 21st century have stimulated a more balanced approach, involving the delivery of multiple goods and services. Contemporary sustainable forest management seeks to meet productivity targets while still managing for biodiversity conservation and other ES. Yet integrative forest management practices at the landscape level are complex and require the understanding of patterns at different scales as well as their interrelationships through processes. This chapter sets out the importance of an integrative landscape perspective for managing forests, one which focuses on mosaics of patches and their dynamics in order to integrate ecological values (e.g., the maintenance of ecosystem health and biodiversity conservation) with economic or cultural ones (e.g., timber and recreation).

Box 30.1 Forest worldwide: an important resource

The world has just under 4 billion hectares of forest, or 30.3% of its total land area. The 10 most forest-rich countries account for two-thirds of the total forest area. In descending order of forest area they are: the Russian Federation, Brazil, Canada, the USA, China, Australia, the Democratic Republic of the Congo, Indonesia, Peru and India. The first five of these account for more than half of the world's forest area (SOFO, 2014).

Importance of forests for ecosystem service provision

Forests are important sources of timber, yet they can also provide a wide range of other ES such as habitat quality for a diverse set of species, recreation, non-timber products, water quality, carbon sequestration and landscape character. Likewise, forests, particularly tropical forests, contribute more than other terrestrial biomes to climate-relevant cycles and related biophysical processes. Forest ecosystem services, as with other nature's services, have also been claimed to be of great economic value (Costanza *et al.*, 1997; Pearce and Pearce, 2001; Pearce and Moran, 2001; de Groot *et al.*, 2012). For example, Costanza *et al.* (2014) estimated an ES value of 53,822,007$/ha/yr for tropical forests and 31,372,007$/ha/yr for temperate and boreal forests, for an overall value of over $16 trillion in annual value from ES in forests. In forest valuation studies, service components like carbon storage or hydrological protection frequently bring higher values than forest products. For example, of the bundled ES value estimated for forests mentioned above, only 6% of temperate forest and 1.6% of tropical forest valuation is from the provisioning service 'raw materials' (de Groot *et al.*, 2012).

Hence, the variety of forest landscapes and the successive forms of forest uses observed during different historical periods exemplify the diversity and intensity of multiple needs; they also demonstrate the importance of spiritual values and of social and political realities. In all, forest landscapes provide more than trees; a forested landscape provides a living society with multiple functions.

Historical importance

Forests and their derived products have played a substantial role in the development of civilization, providing humans with building materials and fuel for thousands of years. The long history of wood utilization dates back to 400,000 years ago – the age of the oldest carbon dated wood spear, found in Germany. Other man-made artefacts that have been dated to outstanding ages include a 3,000 year old staircase, a 1,300 year old building in Japan and a 1,200 year old Viking canoe (Grabner and Klein, 2014). Native peoples also relied on forests for subsistence and cultural resources, and they actively managed forests for these values. In the USA, tribal forestry is still very much alive, generating important revenues while the forest remains protected under a sustainable use scheme (*Tribal Forest Protection Act of 2004* – U.S. Public Law 108–278 108th Congress). However, despite increased awareness of the benefits of forests with respect to carbon sequestration and storage, water retention, climate regulation and the provision of habitat, deforestation rates remain disturbingly high, especially in the tropics (Hansen *et al.*, 2013; Costanza *et al.*, 2014). An integrative multifunctional forest management approach could help to maintain sustainable systems (Gramfeld *et al.*, 2013; Schindler *et al.*, 2014). But planning and implementing multifunctional forest management is challenging because of the trade-offs and synergies among ES.

Present threats to forest ecosystem services

Forest loss and its impact on people

Many of the world's remaining forests are threatened by human activities and climate change. Although the pace of deforestation has slowed globally, losses still continue. Estimates vary according to the methods used and there is disagreement on recent net changes in forest area.[1] However, most sources agree that globally, there is a continuing loss of forest cover and a higher rate of loss of forest cover in tropical areas than in temperate ones.

Without harmonized indicators and comparable figures on the impact of forest loss, policy-makers are unlikely to take decisive action to discourage policies that favour the conversion of forests to agriculture and other land uses. The FAO (SOFO, 2014) highlights the critical knowledge gaps that exist in analysing data on the socioeconomic benefits of forests. They suggest that despite international efforts, we are still lacking empirical evidence on the role and contribution of well-managed forests to sustainable development and a green economy. In addition, current data collection, which focuses on forests and trees, needs to be complemented by information about the benefits that people receive. This, they conclude, will be best done by increasing collaboration with public organizations undertaking such surveys.

Land-use change and forest degradation due to human pressures

Alteration of the earth's vegetation is perhaps the most ecologically significant impact that people have had, because of its serious implications for the maintenance of biodiversity. Since vegetation change occurs at a variety of spatial and temporal scales, it is essential that we take cross-scale effects into account. It is vital that we have the ability to measure such changes and to develop predictive models of future change at different scales to be able to plan adaptive management measures.

ES shifts and conflicts

In parts of the Amazon rainforest, rising temperatures and changing rainfall patterns are connected with increased risk of catastrophic dieback, with potentially dangerous local, regional and global consequences. In the Congo Basin, a recent analysis of deforestation trends published by the World Bank highlights the intense pressure that agricultural expansion, mineral exploitation, growing energy needs and an improved transportation network will pose for the integrity of this rainforest (Megevand *et al.*, 2013).

Another example is Ecuador, which is one of the nine most biodiverse countries. Its mega-diverse flora comprises more than 25,000 plant species, which makes it as important as Brazil in terms of species richness per unit area. However, despite Ecuador's significance as a biodiversity hotspot, information about the country is completely lacking in the 2005 Millennium Ecosystem Assessment (MA, 2005), and slash-and-burn practices have fragmented and degraded a significant portion of the original forested landscape. Conversion of natural forests into agricultural land and pastures has affected about 50% of the lower part of the southeastern tropical Andes of Ecuador, in the valley of the Rio San Francisco region (Bendix *et al.*, 2013). These changes also encompass pressures due to the dichotomy and conflicts between forest areas and adjacent pastures. An unintended consequence of this conversion is that pastures are unsustainable and are therefore abandoned after some time. This is a common process in Brazil, as well as in many other tropical and subtropical forest areas. One of the on-going challenges is to restore these degraded areas through reforestation or reconversion to pasture; but any alternative may have negative consequences for the natural system and the local populations. Scenario analysis may help trade-offs among various ES to be understood. Field observations, measurements and experiments, combined with numerical models and calibration, could also provide a foundation for deriving sustainable land use strategies based on a good understanding of the complexity of the ecological systems and the associated services in these very fragile areas.

If the world is to improve livelihoods for the people while simultaneously mitigating and adapting to climate change, it is vital that we find the balance between conserving and regenerating forest areas through economic growth for poverty reduction. In this regard, additional forest research is critical. By bringing relevant and reliable scientific information to national, regional and global policymakers, forest research can provide a positive on-site impact on livelihoods, the environment and sustainable development. To better understand the potential impacts of management on livelihoods and the forest resource base, we need to not only continue current research but also build research where forests are key to sustaining livelihoods.

Mitigation in an era of human-induced climate change

Forests are being affected by climate change, and these effects will likely intensify into the future (Iverson *et al.*, 2014). Evidence is mounting that increasing fires and drought in western North America can be tied to the changing climate (Dennison *et al.*, 2014; Peters *et al.*, 2014). The northward spread and large-area forest die-off due to the mountain pine beetle in the Rocky Mountains and the drought-related tree mortality increases in the US Southwest are well documented (e.g., Vose *et al.*, 2012; Creeden *et al.*, 2014; Williams *et al.*, 2013). In Europe, the extreme drought of 2003 (the European Heat Wave 2003, Ciais *et al.*, 2005), a series of devastating storms (e.g., Central Europe 1990, France 1999, Slovakia 2004, Sweden 2005, Central Europe 2007), and several recent severe fire seasons (e.g., Portugal 2003, Greece 2007), all point towards increasing climate variability due to human-mediated climate change (IPCC, 2014). Shifts in the altitudinal zones affected by bark beetle damages in Austria and in western North America, and latitudinal range shift of biotic disturbance agents across continents (Battisti *et al.*, 2005 (Bentz *et al.*, 2010; Dukes *et al.*, 2009)), provide additional signs of changes that may be considerably more severe in the future (Tebaldi *et al.*, 2006). Numerous other examples of the impacts of increased climate variability on forests are accumulating worldwide (e.g., Allen *et al.*, 2010). Thus, the development of adaptive forest management strategies under the increased frequency and intensity of expected extreme meteorological events is a challenge for the sustainability of forests in the future.

As climate changes, societal demands for goods and services from forests are also changing. The recent decision of European government leaders to increase the share of renewable energy in Europe to 20% by 2020 is expected to result in a much greater demand for forest biomass for bio-energy generation. This higher demand will intensify the competition for resources between forest industry, the energy sector and nature conservation/other protective functions and services (including biodiversity, protection from natural hazards, landscape aesthetics, recreation and tourism).

According to the recent US National Climate Assessment, bioenergy could also emerge as a new market for wood and, aside from some negative competitive potential mentioned above, could aid in the restoration of forests killed by drought, insects and fire (Vose *et al.*,

2012; Joyce *et al.*, 2014). Ironically, much of the restoration needed is in response to direct or indirect climate-related disturbances. Though not yet implemented at the federal level, several regional or state policies in the US also are encouraging a large step-up in energy proportion from renewables over the next decade, including from biomass. However, the rapid expansion of fossil fuel energy via hydrologic fracturing ("fracking") is slowing economic and sociological incentives for bioenergy, but with plenty of environmental negatives, including large amounts of methane leakage to the atmosphere. Methane has a much higher climate warming potential than CO_2 (Miller *et al.*, 2013).

How to adapt landscape systems to climate change is challenging scientifically. In order to support meeting targets established by forest landscape managers, we need more focus on biodiversity conservation as a proxy for the ecological dimensions of a sustainable forest management, while still improving our understanding of ecological processes to set up baselines towards future planning and scenarios. This is particularly challenging because of increasing demands and pressures to intensify wood production and timber exploitation, in addition to agriculture intensification that is increasing often at the expense of treed landscapes. Still, there are demands for improving actions in favour of safeguarding biodiversity, and in a more general way, improving the functioning of forest ecosystems. The need to optimize resource production simultaneously with improving environmental quality represents a challenge and an opportunity for the years to come. Reorganization of forest management systems are needed to find the right balance for successful management adaptation within an ecosystem services approach, while considering bundles and trade-offs at different scales. In particular, we need to consider the valorization of wood resources and production, and their vulnerabilities in relation to more intensive management practices.

A holistic landscape framework could provide a comprehensive and integrative approach from the plot level to the landscape level, considering adaptive management and an analysis of ecological thresholds (Kjellström, 2004; Andersson *et al.*, 2005; Kremer, 2007; Iverson *et al.*, 2014). We need, then, to validate concepts, methodologies and tools based on strong scientific evidence, while at the same time working in tandem with the managers charged to implement policies and actions on the ground. Adaptive management seems then a key component within the set of actions that will help balance multiple objectives under changing environmental conditions, and will improve natural resources management in a wide range of territories.

Forest management within an integrative ecosystem framework

While it has been asserted that greater biodiversity positively influences the delivery of multiple services (Mace *et al.*, 2012; Nelson *et al.*, 2009; Vihervaara *et al.*, 2015), evidence to support this from natural systems at scales relevant to management is still scarce. Sustainability of forest ecosystems affected by the use of and trade in forest-based resources requires an understanding of the links and balance between productivity and soil processes, and their interaction with natural and anthropogenic disturbances. In the past three decades, forest ecosystem models have been developed at different scales within an ecosystem framework, from the plot level to landscapes, to analyse various questions (see, for example, reviews by Chertov *et al.*, 2003; Komarov *et al.*, 2003; and applications from Mäkipää *et al.*, 1999; Morris *et al.*, 1997; Romero-Calcerrada and Luque, 2006; He *et al.*, 2008). Modelling was used to analyse the impacts of different systems of harvesting, forest disturbances, natural development of forests, climate change and carbon balance. Forest ecosystem modelling can effectively extend the classical approach where growth functions and tables are used for the prediction of forest growth and soil nutrition in the changing environment under new silvicultural regimes. The level of the basic forest unit (stand, inventory 'compartment') can now be modelled well in relation to the problems of upscaling the stand's productivity in different climatic and site conditions. Moreover, there are combined models which are able to describe the biological turnover of the elements; first of all, carbon and nitrogen, in the 'soil–vegetation' system (Chertov *et al.*, 2001; Komarov *et al.*, 2003). The models allow an estimation of the forest productivity, carbon and nitrogen dynamics, and water regime in the forest ecosystem. Models are also used to make inventories of carbon sinks and sources under the reporting requirements of the Climate Convention and Kyoto Protocol (see Liski *et al.*, 2006; Peltoniemi *et al.*, 2006; Mäkipää *et al.*, 2008). In all, forest ecosystem models, used within the ecosystem framework, are useful to test and develop our understanding of forest functioning and dynamics. They are also required to meet the demand from policymakers and managers to predict the impacts of different scenarios of use and management of forest resources and its associated services.

Multifunctional sustainable forest management (MSFM)

Management of forest stands has substantially changed in situ forest properties, mostly in terms of tree species composition and the amount of coarse woody debris. In several countries, particularly in Nordic countries that practice intensive use of their forest resources (Hanski, 2000; Luque and Vainikainen, 2008), many forest properties are carefully controlled. Regional characteristics, such as the spatial structure of forest landscapes, are also frequently changed (Luque *et al.*, 2004).

Adequate selection of nature reserves for the maintenance of biodiversity has been under extensive research over the past decades (see Cabeza and Moilanen, 2001; ReVelle *et al.*, 2002, Rodrigues and Gaston, 2002; Kallio *et al.*, 2008; Mönkkönen *et al.*, 2011). Most studies have implicitly assumed that land parcels have equal economic value. This unjustified assumption may severely undermine the efficiency of conservation. For example, using county-level data for the US, Ando *et al.* (1998) showed that accounting for heterogeneity in land prices results in a marked increase in efficiency in terms of either the cost of achieving a fixed coverage of species or the coverage attained from a fixed budget.

Gramfeld *et al.* (2013) provided evidence that diverse, mixed forests, in particular, showed higher levels of multiple ecosystem services. Importantly, the same study found that no single tree species was able to promote all services, and some services were negatively correlated with each other. The pros and cons of species mixtures for productivity and other ecosystem functions have been discussed at length since the early 19th century (see reviews in Naeem *et al.*, 2009; Pretzsch, 2005). Only recently, however, have scientists begun to explicitly investigate how species diversity might be important for the simultaneous provision of multiple functions or services (e.g., Hector and Bagchi, 2007; Paillet *et al.*, 2010; Zavaleta *et al.*, 2010; Gramfeld *et al.*, 2013). Gramfeld *et al.* (2013), working on boreal and temperate production forests, showed that the relationships between tree species richness and multiple ecosystem services were positive and that all services considered attained higher levels with five tree species than with one species.

In all, considering both economic and ecological values for site selection means that, in practice, areas are selected as part of a conservation network according to their benefit-cost ratio (Juutinen *et al.*, 2008; Juutinen *et al.*, 2014). Thus, management decisions should be a compromise among the sites that provide high benefits in terms of biodiversity value and conservation and its associated services (Kallio *et al.*, 2008). This also implies that we accept trade-offs between ecological benefits and economic costs. Another important challenge is to identify the best trades-offs among several services (Schwenk *et al.*, 2012), potentially aided by quantitative methods to evaluate management options (Carpenter *et al.*, 2009; Gramfeld *et al.*, 2013) (See Box 30.2). Multicriteria analyses can help forest owners and forest managers consider the best pathways to potential 'win-win' situations or at least good compromises.

Box 30.2 ***Fostering ES in forests***

To achieve the challenging goals of operationalization of the concepts of ecosystem services within a forest management context, we need:

- *Landscape approaches*, rather than single stand or forest land approaches, in order to account for spatial interactions, bundles and connectivity networks that determine the success or failure of conservation management targets.
- *Regionally and locally tailored management* (adapt at the scale of practice) practices (e.g., lower harvest intensities in areas of greater hydric stress).
- *To avoid the dominance of a single management* strategy (or of the lack of any management) over large areas: diversity of species calls for a diversity of management practices.

The challenge that lies ahead

The challenge that lies ahead demands awareness of the increasing pressures on forests and forest resources, and concern about the continuous changes in climate conditions that will increase forest degradation through such things as soil erosion, desertification, droughts, pests, diseases, storms and fires. Such impacts put at risk the health, vitality and productivity of forests, which all can have adverse impacts on economies, biological diversity and the environment, as well as on the social and cultural benefits of societies. Fortunately, however, many forests are quite resilient and may be able adapt to the altered conditions (Tebaldi *et al.*, 2006; Bentz *et al.*, 2010; Thompson *et al.*, 2009), especially if assisted through active and science-based management (FAO, 2012; Park *et al.*, 2014). But for many other forests, there is an urgent need to address and take action through effective research on and implementation of sustainable forest management. Thus, societies that depend on forests will also need to adjust and adapt to new conditions and transformation adapting to changes (i.e., social-ecological resilient society).

The future will bring both challenges and opportunities. Global challenges include the demands that a growing population will make on global ecosystems, whose resilience is being tested by energy and water scarcity, continuing pollution, and a host of increasing disturbances and human demands. Based on present consumption rates, the supply of ES will fall increasingly short of demand. Forests are at present used very inequitably, and there are many people, particularly in developing countries, who have severely limited access to the benefits that forest ES can deliver. Substantial improvements in resource efficiency and management practices, as aforementioned, is essential to secure a sustainable future for all, while simultaneously tackling climate change through adaptation and mitigation measures.

A vision for the future

Future research should aim at developing and improving methods to measure and value biodiversity and ecosystem resilience (See Box 30.3). Evaluating different habitat types in terms of disturbance frequency and intensity can be imitated in the management and use of such ecosystems. This approach may help detect when ecosystems are approaching the limits of their natural functioning or productive capacity. Future efforts should also aim at improving measures on the

Box 30.3 Goals for better futures with forests

Forests should be conserved for the multiple benefits that they provide, but we still need to:

- Improve information worldwide and communication channels on consumption of forest products for food security and health
- Support forest management, considering multiple tree species to sustain the full range of benefits that the society obtains from forests
- Support interdisciplinary research to provide more evidence that is needed to help re-direct policies, more effectively enhancing the socioeconomic benefits of forests
- Coordinate management across ownership boundaries
- Provide economic and cultural benefits to local communities; e.g., patrimonial values, identity, recreation and entertainment
- Sustain long-term wood and biomass production
- Promote widely non-conventional socioeconomic benefits from forests; e.g., wood products for green buildings; forests for health – medicinal plants, natural organic food; wood quality for musical instruments, boats, toys

importance of forests for society at large (See Box 30.3); we need to improve our understanding of the people who live in and around forests – in many cases depending directly on forests for their livelihoods. In all, well-managed forests have tremendous potential to contribute to sustainable development and promote food security. We need then stronger collaborative efforts to collect data and monitor trends, raise awareness and monitor progress towards an integrative sustainable forest management.

Note

1 See www.fao.org/forestry/fra/remotesensingsurvey/en/).

References

Allen, C., Macaladyb, A., Chenchounic, H., *et al.* (17 authors) (2010). A global overview of drought and heat-induced tree mortality reveals emerging climate change risks for forests. *Forest Ecology and Management*, vol 259, pp 660–684.

Andersson, M., Dahlin, B., and Mossberg, M. (2005). The forest time machine: a multi-purpose forest management decision-support system. *Computers and Electronics in Agriculture*, vol 49, pp 114–128.

Ando, A., Camm, J, Polasky, S., and Solow, A. (1998). Species distributions, land values, and efficient conservation. *Science*, vol 279, pp 2126–2128.

Battisti, A., Stastny, M., Netherer, S., *et al.* (7 authors) (2005). Expansion of geographic range in the pine processionary moth caused by increased winter temperatures. *Ecological Applications*, vol 15, pp 2084–2096.

Bendix, J., Beck, E., Bräuning, A., et al. (6 authors) (eds) (2013). *Ecological Studies, Volume 221: Ecosystem Services, Biodiversity and Environmental Change in a Tropical Mountain Ecosystem of South Ecuador.* Springer Verlag, Berlin and Heidelberg..

Bentz, B.J., Régnière, J., Fettig, C.J. *et al.* (9 authors) (2010). Climate change and bark beetles of the western United States and Canada: direct and indirect effects. *Bioscience*, vol 60, no 8, pp 602–613.

Cabeza, M., and Moilanen, A. (2001). Design of reserve networks and the persistence of biodiversity. *TREE*, vol 16, pp 242–248.

Carpenter, S. R., Mooney, H. A., Agard, J., *et al.* (15 authors) (2009). Science for managing ecosystem services: beyond the Millennium Ecosystem Assessment. *Proceedings of the National Academy of Sciences (PNAS)*, vol 106, pp1305–1312.

Chertov, O., Komarov, A., Kolström, M., *et al.* (2003). Modelling the long-term dynamics of populations and communities of trees in boreal forests based on competition on light and nitrogen. *Forest Ecology and Management*, vol 176, pp 355–369.

Chertov, O.G., Komarov, A.S., Nadporozhskaya, *et al.* (5 authors) (2001). ROMUL – a model of forest soil organic matter dynamics as a substantial tool for forest ecosystem modelling. *Ecological Modelling*, vol 13, pp 289–308.

Ciais, P., Reichstein, M., Viovy, N., *et al.* (33 authors) (2005). Europe-wide reduction in primary productivity caused by the heat and drought in 2003. *Nature*, vol 437, pp 529–533.

Costanza, R., d'Arge, R., de Groot, R., *et al.* (13 authors) (1997). The value of the world's ecosystem services and natural capital. *Nature*, vol 387, pp 253–260.

Costanza, R., de Groot, R., Sutton, P., *et al.* (8 authors) (2014). Changes in the global value of ecosystem services. *Global Environmental Change*, vol 26, pp 152–158.

Creeden, E.P., Hicke, J.A., and Buotte, P.C. (2014). Climate, weather, and recent mountain pine beetle outbreaks in the western United States. *Forest Ecology and Management*, vol 312, pp 239–251.

de Groot, R., Brander, L., van der Ploeg, S. *et al.* (17 authors) (2012). Global estimates of the value of ecosystems and their services in monetary units. *Ecosystem Services*, vol 1, pp 50–61.

Dennison, P. E., Brewer, S. C., Arnold, J. D., and Moritz, M. A. (2014). Large wildfire trends in the western United States, 1984–2011. *Geophys Res Lett*, vol 41, no 8, pp 2928–2933.

Dukes, J. S., Pontius, J., Orwig, D., *et al.* (16 authors) (2009). Responses of insect pests, pathogens, and invasive plant species to climate change in the forests of northeastern North America: what can we predict? *Canadian Journal of Forest Research*, vol 39, no 2, pp 231–248.

FAO (2012). *Forest Management and Climate Change: A Literature Review.* United Nations Food and Agriculture Organizatinon, Rome.

Grabner, M., and Klein, A. (2014). Utilization of different Austrian wood species in past times – knowledge for the future. In: *The International Forestry Review* Vol. 16 (5), IUFRO WC Sustaining Forests, Sustaining People: The Role of Research. Salt Lake City, UT. 5 October 2011.

Gramfeld, L., Sna, T., Bagchi, R., et al. (9 authors) (2013). Higher levels of multiple ecosystem services are found in forests with more tree species. *Nature Communications*, vol 4, p 1340.

Hansen, M.C., Potapov, P. C., Moore, R. *et al.* (15 authors) (2013). High-resolution global maps of 21st-century forest cover change. *Science*, vol 342, no 6160, pp 850–853.

Hanski, I. (2000). Extinction debt and species credit in boreal forests: modelling the consequences of different approaches to biodiversity conservation. *Annales Zoologici Fennici*, vol 37, pp 271–281.

He, H., Keane, R.K., and Iverson, L.R. (2008). Forest landscape models, a tool for understanding the effect of the large-scale and long-term landscape processes. *Forest Ecology and Management*, vol 274, pp 371–374.

Hector, A., and Bagchi, R. (2007). Biodiversity and ecosystem multifunctionality. *Nature*, vol 448, pp 188–190.

Iverson, L., Echeverria, C., Nahuelhual, L., and Luque, S. (2014). Ecosystem services in changing landscapes: an introduction. *Landscape Ecology*, vol 29, no 2, pp 181–186.

Iverson, L., Prasad, A. M., and Matthews, S. (2014). Climate as an agent of change in forest landscapes. In: Azevedo, J., Perera, A. and Pinto, M. (eds) *Forest Landscapes and Global Change: Challenges for Research and Management*. Springer, New York.

IPCC (2014). *Climate Change 2014: Impacts, Adaptation, and Vulnerability.* Working Group II Contribution to the IPCC 5th Assessment Report Intergovernmental Panel on Climate Change, Stanford CA.

Joyce, L.A., Running, S.W., Breshears, D.D., *et al.* (8 authors) (2014). *Forests. Climate Change Impacts in the United States*. U.S. Global Change Research Program. doi:10.7930/J0Z60KZC.

Juutinen, A., Ollikainen, M., Mönkkönen, M., *et al.* (6 authors) (2014). Optimal contract length for biodiversity conservation under conservation budget constraint. *Forest Policy and Economics*, vol 47, pp 14–24.

Juutinen, A., Luque, S., Mönkkönen, *et al.* (5 authors) (2008). Cost-effective forest conservation and criteria for potential conservation targets: a Finnish case study. *Environmental Science & Policy*, vol2, pp 613–626.

Kallio, M., Hänninen, R., Vainikainen, N., and Luque, S. (2008). Biodiversity value and the optimal location of forest conservation sites in Southern Finland. *Ecological Economics*, vol 67, pp 232–243.

Kjellström, E. (2004). Recent and future signatures of climate change in Europe. *Ambio*, vol 33, pp 193–198.

Komarov, A., Chertov, O., Zudin, S., *et al.* (2003). EFIMOD 2 – A model of growth and elements cycling in boreal forest ecosystems. *Ecological Modelling*, vol 170, no 2–3, pp 373–392.

Kremer, A. (2007). How well can existing forests withstand climate change? In: Koskela, J., Buck, A. and Tessier du Cros, E. (eds) *Climate Change and Forest Genetic Diversity: Implications for Sustainable Forest Management in Europe*. Biodiversity International, Rome.

Liski, J., Lehtonen, A., Palosuo, T., *et al.* (7 authors) (2006). Carbon accumulation in Finland's forests 1922–2004: estimate obtained by combination of forest inventory data with modelling of biomass, litter and soil. *Annals of Forest Science*, vol 63, pp 687–697.

Luque, S., Riutta, T., Joensuu, J., *et al.* (4 authors) (2004). Multi-source forest inventory data for biodiversity monitoring and planning at the forest landscape level. In: Marchetti, M. (ed.) *Monitoring and Indicators of Forest Biodiversity in Europe – From Ideas to Operationality*. European Forest Institute Proceedings, no 51, Joensuu.

Luque, S., and Vainikainen, N. (2008). Habitat quality assessment and modelling for biodiversity sustainability at the forest landscape level. In: Lafortezza, R., Chen, J., Sanesi, G. and Crow, T. (eds) *Patterns and Processes in Forest Landscapes: Multiple Use and Sustainable Management, part 3: Landscape-Scale Indicators and Projection Models*. Springer Verlag, Berlin and Heidelberg.

Mace, G. M, Norris, K., and Fitter, A. H. (2012). Biodiversity and ecosystem services: a multilayered relationship. *Trends Ecol Evol*, vol 27, no 1, pp 19–26.

Mäkipää, R., Hakkinen, M., Muukkonen, P., and Peltoniemi, M. (2008). The cost of monitoring changes in forest soil carbon stocks. *Boreal Environment Research*, vol 13, pp120–130.

Mäkipää, R., Karjalainen, T., Pussinen, A., and Kellomäki, S. (1999). Effects of climate change and nitrogen deposition on the carbon sequestration of a forest ecosystem in the boreal zone. *Canadian Journal of Forest Research*, vol 29, pp 1490–1501.

Megevand, C., Mosnier, A., Hourticq, J., *et al.* (5 authors) (2013). Deforestation Trends in the Congo Basin. Directions in Development – Environment and Sustainable Development World Bank eLibrary. http://dx.doi.org/10.1596/978-0-8213-9742-8

MA (2005). *Ecosystems and Human Well-Being, Volume 1: Current State and Trends*. Island Press, Washington DC.

Miller, S.M., Wofsy, S.C., Michalak, A.M., *et al.* (12 authors) (2013). Anthropogenic emissions of methane in the United States. *Proceedings of the National Academy of Sciences*, vol 110, pp 20018–20022.

Mönkkönen, M., Reunanen, P., Kotiaho, J.S., *et al.* (5 authors) (2011). Cost-effective strategies to conserve boreal forest biodiversity and long-term landscape-level maintenance of habitats. *European Journal of Forest Research*, vol 130, pp 717–727.

Morris, D. M., Kimmins, J. P. (Hammish), and Duckert, D. R. (1997). The use of soil organic matter as a criterion of the relative sustainability of forest management alternatives: a modelling approach using FORECAST. *Forest Ecology and Management*, vol 94, pp 61–78.

Naeem, S., Bunker, D.E., Hector, A., *et al.* (6 authors) (2009). *Biodiversity, Ecosystem Functioning, & Human Wellbeing, an Ecological and Economic Perspective.* Oxford University Press, Oxford.

Nelson, E., Mendoza, G., Regetz, J. *et al.* (14 authors) (2009). Modeling multiple ecosystem services, biodiversity conservation, commodity production, and tradeoffs at landscape scales. *Front Ecol Environ*, vol 7, no 1, pp 4–11.

Paillet, Y.L., Bergès, J., Hjältén, P., *et al.* (24 authors) (2010). Biodiversity response to forest management varies widely with taxonomic and ecological groups: a meta-analysis on species richness in Europe. *Conservation Biology*, vol 24, no 1, pp 101–112.

Park, A., Puettmann, K., Wilson, E., *et al.* (4 authors) (2014). Can boreal and temperate forest management be adapted to the uncertainties of 21st century climate change? *Critical Reviews in Plant Sciences*, vol 33, no 4, pp 251–285.

Pearce, D.W., and Moran, M. (2001). *The Value of Biological Diversity: A Handbook*. OECD, Paris.

Pearce, D. W., and Pearce, C. (eds) (2001). *Valuing Environmental Benefits: Case Studies from the Developing World*. Edward Elgar, Cheltenham.

Peltoniemi, M., Palosuo, T., Monni, S., and Mäkipää, R. (2006). The factors affecting the uncertainty of sinks and stocks of carbon in Finnish forest soil and vegetation. *Forest Ecology and Management*, vol 232, pp 75–85.

Peters, M., Iverson, L., and Matthews, S. (2014). Spatio-temporal trends of drought by forest type in the conterminous United States, 1960–2013 [scale 1:12,000,000]. Res. Map NRS-7. U.S. Department of Agriculture, Forest Service, Northern Research Station. Newtown Square, PA.

Pretzsch, H. (2005). Diversity and productivity in forests: evidence from long-term experimental plots. In: Scherer-Lorenzen, M., Körner, C. and Schulze, E. D. (eds) *Ecological Studies, Volume 176: Forest Diversity and Function: Temperate and Boreal Systems*. Springer-Verlag, Berlin.

ReVelle, C. S., Williams, J. C., and Boland, J. J. (2002). Counterpart models in facility location science and reserve selection science. *Environ Model Asses*, vol 7, pp 71–80.

Rodrigues, A.S.L., and Gaston, K. J (2002). Optimization in reserve selection procedures-why not? *Biological Conservation*, vol 107, pp 123–129.

Romero-Calcerrada, R., and Luque, S. (2006). Habitat quality assessment using weights-of-evidence based GIS modelling: the case of *Picoides tridactylus* as keystone species indicator of the biodiversity value of the Finnish forest. *Ecological Modelling*, vol 196, pp 62–76.

Schindler, S., Sebesvari, Z., Damm, C., *et al.* (21 authors) (2014). Multifunctionality of floodplain landscapes: relating management options to ecosystem services. *Landscape Ecol*, vol 29, pp 229–244.

Schwenk, W.C., Donovan, T.M., Keeton, W.S., and Nunery, J.S. (2012). Carbon storage, timber production and biodiversity: comparing ecosystem services with multi-criteria decision analysis. *Ecological Applications*, vol 22, no 5, pp 1612–1627.

SOFO (2014). Enhancing the socioeconomic benefits from forests. In: *State of the World's Forests*. United Nations Food and Agriculture Organizatinon (FAO), Rome. www.fao.org/publications

Tebaldi, C., Hayhoe, K., and Arblaster, J. M. (2006). Going to the extremes: an intercomparison of model-simulated historical and future changes in extreme events. *Climatic Change*, vol 79, pp 185–211.

Thompson, I., Mackey, B., McNulty, S., and Mosseler, A. (2009). *Forest Resilience, Biodiversity, and Climate Change. A Synthesis of the Biodiversity/Resilience/Stability Relationship in Forest Ecosystems*. Secretariat of the Convention on Biological Diversity, Montreal. Technical Series no. 43.

Vihervaara, P., Mononen, L., and Auvinen, A.P. (2015). How to integrate remotely sensed data and biodiversity for ecosystem assessments at landscape scale. *Landscaper Ecology*, vol 30, pp 501–516.

Vose, J.M., Peterson, D.L., and Patel-Weynand, T. (2012). *Effects of Climatic Variability and Change on Forest Ecosystems: A Comprehensive Science Synthesis for the U.S. Forest Sector*. U.S. Department of Agriculture, Pacific Northwest Research Station Forest Service, General Technical Report PNW-GTR-870. Portland, OR.

Williams, A.P., Allen, C. D., Macalady, A. K., *et al.* (15 authors) (2013). Temperature as a potent driver of regional forest drought stress and tree mortality. *Nature Climate Change*, vol 3, pp 292–297.

Zavaleta, E.S., Pasari, J.R., Hulvey, K.B., and Tilman, G.D. (2010). Sustaining multiple ecosystem functions in grassland communities requires higher biodiversity. *PNAS*, vol 107, pp 1443–1446.

31
DRYLANDS

Lindsay C. Stringer and Andrew J. Dougill

Introduction

Drylands are divided into arid, semi-arid and dry sub-humid sub-types, and cover approximately 6 billion hectares, or 40%, of the world's terrestrial surface. Nearly 75% of the world's drylands are located in developing countries. Estimates suggest that globally, more than 2 billion people live in the drylands, while billions of others depend upon goods and services from these systems (UNDP-UNCCD, 2011). Water is the key factor that limits dryland primary productivity. For an area to be classed as a dryland, it must have an aridity index of ≤0.65, where annual mean potential evapotranspiration is at least 1.5 times greater than annual mean precipitation (Middleton and Thomas, 1994). The combination of low rainfall and high evaporation leads to low soil moisture. Together with slow processes of nutrient cycling, this has an important influence on the types of land use that can be sustained, as well the nature and distribution (both temporally and spatially) of the ecosystem services from these environments. Across the arid, semi-arid and dry sub-humid dryland subtypes, land uses include rangelands, dry forests, arable cropland and urban areas. Some uses also encompass areas protected for conservation or used for mining.

Drylands are often considered vulnerable to land degradation and desertification, both of which disrupt the delivery of ecosystem services. Land degradation is a "reduction of or loss in the biological or economic productivity and complexity of rain-fed cropland, irrigated cropland, range, pasture, forest, or woodlands resulting from land uses or from processes arising from human activities and habitation patterns" (UNCCD, 1994, p. 5). Desertification is more explicitly focused on the drylands and defined as "land degradation in arid, semi-arid, and dry sub-humid areas resulting from various factors, including climatic variations and human activities" (UNCCD, 1994, p. 4). Across the various subtypes, semi-arid drylands are most vulnerable to ecosystem services losses (MA, 2005). This is because fewer people live in hyper-arid and arid areas whereas semi-arid drylands have higher population densities while remaining sensitive to degradation.

To conserve drylands, prevent desertification and the loss of ecosystem services and livelihoods they support, the United Nations Convention to Combat Desertification (UNCCD, 1994) was negotiated. The UNCCD recognises that land uses and management practices in dryland ecosystems result from the complex interplay of processes and decisions across scales, from the local to the international. It was considered that an international framework could promote and catalyse global, regional, sub-regional and national actions to prevent ecosystem service loss and reverse existing damage. In doing so, it would improve the development prospects for billions of poor people who inhabit the world's drylands.

This chapter examines the dryland ecosystem services that people draw upon for their subsistence, for their livelihoods and to meet consumer demands from non-dryland areas. It begins by outlining the key dryland ecosystem services and explores the main threats and types of degradation that are being experienced. Important pressures and feedbacks that contribute to dryland ecosystem service degradation are identified and the interactions between them are outlined. The chapter then explores the main challenges in measuring and monitoring dryland ecosystem service losses and unpacks them in relation to scale, methodological approaches and the types of knowledge and stakeholders involved in their assessment. Key policies and approaches used within the context of the UNCCD are then presented drawing on different southern African dryland case studies, illustrating some of the mechanisms that attempt to safeguard ecosystem service provision.

Key dryland ecosystem services

The ecosystem services supported by drylands are often closely intertwined across a range of temporal and spatial scales, and the nonlinearity of the ecological processes involved has to be understood (Reynolds *et al.*, 2007). Drylands can have multiple thresholds and multiple ecological and social states, some of which assist the provision of certain ecosystem services, some of which do not. The broad types of ecosystem service are now discussed.

Provisioning services

Food is provided in drylands from wild game and birds, wild fruits, melons and nuts, as well as commercial game and livestock farming. In Botswana's Kalahari, Sallu *et al.* (2010) suggest that wild foods remain vital sources of subsistence for the rural poor, while livestock production (particularly cattle) underpins the livelihoods of many people. Some dryland plants and animals are used to treat human and animal diseases, with roots, bark, leaves and tubers providing important medicinal services. For example, tribes in dryland northern Mexico use *Prosopis glandulosa*, a shrub or small tree, to treat a wide range of ailments, harnessing the plant's antibiotic, astringent and antiseptic properties. The plant stems can be used to treat fever; the bark is used for bladder infections and measles; sunburn and stings are treated with a decoction of the beans; while gum exuded from the trunk has a range of dermatological uses and is applied to remedy stomach ailments (Davidow, 1999). The value of dryland plants to global consumers can also be seen in the growth of markets for cosmetic products (e.g. *Aloe* spp.), health supplements or gum resins and food additives (e.g. gum Arabic from *Acacia senegal*). Freshwater is another important provisioning service. In drylands, freshwater is stored and retained for domestic, agricultural and industrial use, often being accessed through seasonal rivers and aquifers. Across many of the world's major semi-arid savanna regions, such as Australia, boreholes that exploit groundwater have enabled the expansion of livestock farming into areas that are otherwise economically unviable.

Fibre and fuel from trees, shrubs and grasses may be used in construction, or as fuelwood, or in charcoal production, as is the case in countries such as Malawi, Zambia, Mozambique and Tanzania (Dewees *et al.*, 2010). Grasses provide important fodder for livestock and wild fauna, but in some locations, such as southern Africa's savannas, nutritious species such as *Stipagrostis uniplumis* and *Eragrostis pallens* are being degraded and lost in areas of intensive grazing (Dougill *et al.*, 2014).

Regulating services

Climate regulation is most often considered in relation to tropical biomes that exhibit greater tree density and aboveground biomass, yet dry forests such as Zambia's miombo woodland, and semi-arid rangelands, provide vital global carbon sinks (Ciais *et al.*, 2011; Stringer *et al.*, 2012). Water regulation takes a variety of forms. Domestic stock (cattle, sheep, goats) are highly reliant upon surface water. Shallow aquifers in areas such as South Africa's Molopo River, while mostly saline, provide freshwater pools for livestock. Water regulation services are also provided by wetland areas in drylands. For example, the Okavango Delta, in semi-arid northwestern Botswana, provides important seasonal flooding in the otherwise dry winter months.

Genetic diversity, particularly in dryland wild animal populations, also has an inherent value, offering resistance to diseases, pests and pathogens, as well as to invasive species and climatic shocks and stresses, such as drought and climate variability.

Cultural services and benefits

Drylands have long been inhabited by humans, resulting in strong historical connections to the environment and important spiritual, heritage and inspirational benefits. In some countries, areas of particular spiritual importance are protected by law. In Australia, the 1972 Aboriginal Heritage Act provides protection for both archaeological sites (e.g. rock paintings and remains of modified trees) and anthropological sites (e.g. ceremonial or burial grounds). Dryland cultural benefits include landscapes and views that are aesthetically valued, often for their remoteness and wilderness (e.g. Arizona's Grand Canyon, or Namibia's Namib Desert). Such characteristics attract people from outside the dryland biome for tourism and recreational purposes. The same can be said of activities such as game viewing and the hunting of large native African mammals, especially in countries such as Kenya, Tanzania and South Africa.

The extensive local or indigenous knowledges of dryland populations further provide an important educational benefit in many parts of the world. Local knowledge, built upon experiences in dealing with inherent dryland rainfall variability, has been shown to be especially important in terms of managing future change, including climate change, across a wide range of dryland environments.

Underpinning ecological or supporting services

Ecological processes such as soil formation take place very slowly, especially in drylands where rates of decay are limited by moisture. Insects such as termites, ants and dung beetles play an important role in dryland soil formation. The limited water means that soil nutrient cycles differ significantly from those in more temperate systems (Delgado-Baquerizo *et al.*, 2013). Processes such as desiccation, grazing and fire can promote nutrient cycling under these water-limited conditions, releasing elements and minerals bound up in plants into more available forms that can support primary production. This can be visibly noted from satellite imagery in the form of green pulses of regrowth in fire scars (Scholes and Walker, 1993). Drylands further provide an important supporting safety-net in the form of refuge in times of stress in other, non-dryland areas (e.g. droughts that occur outside the dryland biome), with essential grazing and browse resources delivered by drought-tolerant dryland species that are adapted to variability in water availability.

The biodiversity of drylands requires special attention. Wild species directly contribute to both provisioning and cultural services. However, they also indirectly support many other types of service, and understanding these links is important in assessing and valuing ecosystem services, particularly if double-counting is to be avoided. For example, *Acacia* tree species contribute to soil development through their roots and litter, as well as nutrient cycling (offering a supporting service), forage and fuel (yielding provisioning services, as leaves and pods are eaten by livestock and wood is used as fuel). *Acacia* trees further support other species by providing shade, shelter and nesting sites for birds and other biodiversity, as well as having a strong cultural significance in many settings. Even in commercial ranches in southern Africa, farmers have a cultural connection to *Acacia erioloba* (camel thorn).

Drylands exhibit a high diversity of flowering plants, all of which rely on pollination. Studies of *Acacia tortilis* in arid/semi-arid Kenya, where pastoralist livelihoods dominate, suggest that key pollinators include native bee species (*Apoidea*) and butterflies (*Lepidoptera*), as well as flies (*Diptera*). Research continues to highlight significant concerns on the declining numbers of pollinators, not just in the drylands but globally (Potts *et al.*, 2010), and their likely impact on agricultural crop yields (Aizen *et al.*, 2009).

Soil biodiversity is especially important for a range of services. Clusters of aboveground bush species growing in areas in which biological soil crusts are created by a range of cyanobacteria and micro algae together help to regulate carbon dioxide fluxes and retain water (Thomas, 2012). Biological soil crusts demonstrate complex interactions between species, leading to the provision of ecosystem services which would not otherwise be delivered. Crusts are nevertheless increasingly threatened by ecological changes associated with bush encroachment across semi-arid rangeland systems (Elliott *et al.*, 2014).

In regions such as southern and eastern Africa, large areas of dryland forest and rangeland are demarcated for biodiversity conservation, and receive various different levels of protection. While this can help to sustain cultural as well as other types of ecosystem services, it often results in trade-offs, and reduces the ability of the poor to be able to access particular areas (Hulme and Murphree, 2001).

Threats and types of ecosystem service degradation

Dryland degradation encompasses multiple interacting processes, driven by human and environmental (including climate) factors operating across different temporal and spatial scales. Climate and hydrological processes and patterns operate at macro- and meso-scales, interacting with local topography, to set the dryland biophysical context. This is overlain with socio-political and economic structures and processes such as markets, property rights, population and demographic changes, human migration patterns and technological changes that span a range of levels. Together, these factors combine with land use and management practices both past and present, shaping decision-making about ecosystem management at smaller scales. As such, the state of the world's drylands and the ecosystem services they provide results from a complex interplay of factors, stemming from both within and without the dryland biome. As ecosystem services are degraded and lost, the effects are experienced both directly by land users and communities dependent upon the degraded area for their livelihoods, and indirectly in other places, through e.g. water quality decline, food insecurity and increased food prices (Table 31.1).

Degradation can cross-cut the range of ecosystem service categories; for example, overgrazing of rangelands causes shifts towards less palatable species, including bushes. This can cause reduced livestock productivity and a decline in provisioning services (especially food). It also increases the bare ground left open to wind erosion, and breaks up biological soil crusts, which can increase the rate of carbon turnover and reduce climate regulation functions. Fire and drought can also play an important role, interacting with grazing patterns to cause soil erosion and loss of species and genetic diversity.

Table 31.1 Pressures and ecosystem service degradation under different dryland land uses.

Land use	*Key proximal drivers of degradation*	*Key forms of degradation*
Rangeland	• Overgrazing • Over-extraction of groundwater • Fire • Drought	• Bush encroachment • Soil erosion (including disturbance of biological soil crusts and climate regulation function, and reactivation of otherwise stable dune fields) • Loss of genetic diversity and species richness
Dry forests	• Land clearance for agriculture • Increasing energy demand for charcoal • Mining	• Deforestation • Loss of cultural services and aesthetic value • Loss of soil organic carbon • Loss of genetic diversity and species richness
Cropland	• Lack of soil inputs/fallowing • Population growth • Demand for food production	• Soil nutrient loss • Soil erosion • Soil sealing (affecting climate and water regulation)

In many drylands, particularly those in the developing world, ecosystem services underpin billions of livelihoods (UNDP-UNCCD, 2011). Those who are most heavily dependent on the natural resource based for their survival, often the rural poor, are the most directly threatened by ecosystem service losses. While dryland agricultural households and economies have received a lot of attention in research literature, other sectors, such as mining and energy, can play a key role in driving degradation, particularly through the destruction of dry forest. In Africa, miombo woodlands are the most extensive dryland forest formation, covering approximately 27 million km^2 (Kalaba *et al.*, 2013). Vast areas of miombo are being destroyed to meet increasing local demands for charcoal, while the extractive industries are clearing huge forest swathes in pursuit of mining activities that supply other countries with metals such as copper and cobalt (Figure 31.1).

In light of the various drivers of dryland ecosystem service degradation, it becomes important to identify a baseline from which measurements can take place, such that changes can be monitored. While various studies have been undertaken to assess the condition of drylands, each has employed different methods, measuring different variables (using different indicators) over different temporal and spatial scales. This means it is impossible to compare the findings. The most widely cited report on the state of the world's drylands is still the MA, which estimates that 10–20% of the drylands are degraded (with 65–85% probability and medium certainty; MA, 2005). Measurement challenges are further exacerbated by the need to focus on slow variables (e.g. soil formation) and fast variables (e.g. crop yields) and recognition that problems and solutions that are manifest at one scale both influence, and are influenced by, those at other scales. These challenges are described below.

Challenges in measuring and monitoring dryland ecosystem services

The context and scale of analysis often determines whether ecosystem service losses are viewed as problematic. Some understandings of degradation focus on biophysical changes associated with ecosystem functioning, while others link more directly to the land's potential to be productive and deliver benefits to humans. For households dependent upon the resource base for their livelihoods, local losses of dryland ecosystem services can increase food insecurity, particularly if they disrupt one of their core income streams.

Figure 31.1 Evidence of drivers of miombo woodland loss in Copperbelt District, Zambia industrial removal for copper smelter use (top) and local charcoal sales (bottom).

Understanding the context of the local land use system is vital in determining when interventions to reduce and avoid ecosystem service losses are needed. However, different groups can bring divergent perspectives about the same environmental changes. For example, predominantly pastoral livelihood strategies in dryland Botswana are underpinned by a diverse range of grass and forb species. Changes to grass community composition as a result of woody species encroachment can have a negative economic and livelihood impact for those dependent upon livestock. Simultaneously, while local provisioning services are declining, this change can result in an increased potential for global climate regulation, via enhanced carbon sequestration in the aboveground biomass and roots of the bushes. In this scenario, pastoralists could receive carbon payments if credits for increased carbon storage are traded on the voluntary carbon market. However, critiques abound, noting at best that carbon markets offer a supplementary income source (e.g. Mathur *et al.*, 2014), and that these kinds of responses to ecosystem service changes often overlook the losses of local cultural benefits (amongst others).

Similarly, across the world, protected areas have been established to conserve the biodiversity that underpins a range of different dryland ecosystem services. Sometimes this has involved the resettlement of local populations and the restriction of access to particular areas. The impact has not been to reduce ecosystem services per se, but to reduce the access of some groups to certain services (particularly provisioning services upon which livelihoods often directly depend). Measuring ecosystem services and their losses is therefore a complex task, requiring consideration of multiple perspectives and demanding mixed methods approaches to capture the costs, benefits and trade-offs, both within and between different ecosystem service categories.

Dryland monitoring has typically relied on ecological and soil-based approaches at local scales, alongside remote sensing analyses over larger areas. However, these tend to emphasise ecosystem service stocks, rather than the flows of services upon which dryland land users depend. Increasingly, prominence has been given to multi-spectral remote sensing to calculate the greenness of live canopy vegetation, through use of the Normalized Difference Vegetation Index (NDVI) (see also Liquete *et al.*, 2016). However, this faces limitations in that NDVI is unable to distinguish between grass-dominated rangeland systems, bush-encroached areas and land with natural tree cover, due to an inability to determine vegetation height. This presents a need for ground checking, as well as consideration of rainfall data. Recognising the importance of rainfall is particularly important given its high variability across even small dryland areas. Without triangulation of data sources, NDVI analyses become meaningless (Stringer *et al.*, 2012).

The scientific literature increasingly acknowledges the importance of not just drawing on multiple methods but also spanning disciplines and using natural and social science methods. Approaches that draw upon local knowledge and integrate it with scientific findings are becoming more widespread (Stringer and Reed, 2007). These allow scientific rigour and accuracy to be combined with enhanced relevance and sensitivity to local context, and can help to identify ways in which populations have traditionally been able to sustain ecosystem services. Both conceptually and methodologically, integrated research has required new frameworks that are capable of connecting different disciplinary perspectives with other forms of knowledge. Some frameworks have been tested in a range of dryland settings and found to provide useful information to policymakers (Reed *et al.*, 2011).

Policies and approaches that safeguard ecosystem service provision

Reynolds *et al.* (2007) note that the interconnection between ecological and social issues in drylands means that the options to support livelihoods and ecological management, including the maintenance of ecosystem services, need also to be integrated. This requires a variety of policies and approaches that can respond to different needs over varying temporal and spatial scales, and also highlights the close relationship between policies and mechanisms that aim to sustain ecosystem service provision and those that aim to enhance human development.

The UNCCD entered into force in 1996 and provides a global framework for catalysing policy action to ensure the sustainability of dryland ecosystems and the livelihoods and global markets they support. Through its multi-scale structure it encourages the development of regional, sub-regional and national policies and actions to combat desertification. At the national level, parties to the UNCCD affected by desertification are obliged to create National Action Programmes (NAPs). NAPs identify the main drivers and impacts of desertification within a country and ways to address them (Stringer *et al.*, 2007). However, UNCCD implementation remains dependent on donor aid, and its influence on sustainable development has been weaker than anticipated because economic imperatives have caused other issues, such as health and education, to be prioritised. Further, there is widespread international disagreement about the main goal of the UNCCD and whether environmental or developmental aspirations should dominate (Bauer and Stringer, 2009).

Despite calls from the international community for countries to look across economic, environmental and social dimensions in decision-making, policy decisions have largely focused on the "economic bottom line". In drylands, decisions tend to result from consideration of whether the benefits of taking steps to reduce particular ecosystem service losses outweigh the costs of action over a given timeframe. This leads to reliance on assessment methods that draw on valuation approaches and tools, and which point to the use of economic incentives or disincentives to shape behaviours. Whether these approaches take into account different stakeholder perspectives depends largely upon the political context. Some initiatives, such as community-based natural resource management, aim to harness local priorities and build alternative livelihood options through the engagement of communities in the conservation of particular ecosystem services (Stringer *et al.*, 2012). Others, such as fencing and subsidies to private cattle ranchers in Botswana's livestock sector, place less emphasis on social equity and are more top-down in their operation.

While the social, economic and political context is not uniform across all drylands, the challenges in sub-Saharan Africa are indicative of a range of policy options that are found in dryland locations across the world. This can be illustrated by case studies from the region.

Community-based natural resource management in Zimbabwe

The Communal Area Management Programme for Indigenous Resources (CAMPFIRE) programme began in Zimbabwe in 1989 to provide alternative livelihoods for dryland communities through the sustainable use of natural resources and ecosystem services (Logan and Moseley, 2002). It sought to "give wildlife and other resources a tangible cash value; and to link this benefit as closely as possible to the landowner" (Child, 1996, p. 372). Tourists would pay an operator to go on either a hunting or photography safari, or stay in a tourist lodge, with a share of the revenue being channelled to the local communities, so communities received financial compensation for not using the ecosystem (Frost and Bond, 2008). It was thought that this would restrict the conversion of natural habitat to agricultural use and thus conserve a range of ecosystem services. The model that was implemented combined a payment for ecosystem service (PES) approach with community-based natural resource management (see also Brouwer, 2016). Prior to the political and economic crisis in Zimbabwe in the early 2000s, the programme had been widely heralded as a success, with attention drawn to flagship projects such as that in Mahenye.

Key difficulties tended to relate more to social and development issues rather than environmental problems. For example, projects were criticised for failing to recognise the complexity and diversity of local communities, thus inhibiting their empowerment (Ribot, 2003); and for oversimplifying local governance, in particular, failing to note that Rural District Councils are not considered representative by local communities (Measham and Lumbasi, 2013). Together, the failings of institutional structures and external economic and political pressures mean that, despite some successes, CAMPFIRE projects have now collapsed in many parts of Zimbabwe, and so the ecosystem service and other community gains have been small.

Policy and economic mechanisms in Botswana's Kalahari rangelands

Kalahari rangeland systems support pastoral livelihoods through cattle, smallstock and game production and include extensive areas assigned for wildlife conservation. Concerns are growing over dual threats of rangeland degradation, through bush encroachment and loss of nutritious perennial grass species, and increasing rural poverty levels (Chanda *et al.*, 2003). Botswana's government oriented its policies and economic incentives towards private land ownership and the fencing of ranches with a view to implementing rotational grazing approaches. However, recent ecological studies (Dougill *et al.*, 2014) have shown bush encroachment to be equally prevalent in both communal grazing areas and private ranches. The shift in land tenure, combined with the introduction of fencing, has also had major impacts on regional patterns of key wildlife species (Verlinden *et al.*, 1998). The expansion of cattleposts and fenced ranches is associated with large areas of empty savanna in the southern Kalahari, with low numbers of key ungulate species. Policies that favour cattle production and private land tenure and fencing therefore lead to critical trade-offs with land use decisions that place more emphasis on wildlife management, biodiversity, the provision of veld products and cultural services (Sallu *et al.*, 2010).

Payment for ecosystem services schemes in Zambia's miombo forests

Zambia's miombo woodland systems provide foods such as mushrooms, insects, fruits, seeds, wild vegetables, honey and oils, as well traditional medicinal plants, building materials and fuelwood. To the local residents, "miombo woodlands are a pharmacy, a supermarket, a building supply store and a grazing resource" (Dewees *et al.*, 2010, p. 61). Around 60% of Zambia is forested, and there are active plans that aim to conserve the forest and reduce deforestation, through e.g. global initiatives such as REDD+ (Reducing Emissions from Deforestation and [forest] Degradation). Projects are increasingly being developed around PES schemes that value services such as climate regulation (carbon sequestration) and biodiversity. Some of these projects follow a Joint Forest Management (JFM) approach, where communities are empowered to manage forest resources, with plans for any carbon payments to be used to fund wider community activities. However, JFM has been beset with implementation challenges, including disagreements over how benefits can be fairly distributed (Leventon *et al.*, 2014).

While forest and energy policies align and support moves to reduce deforestation, agricultural policy's objective is to increase the area of farmland under production. Approaches such as conservation farming that can promote synergy between agriculture and forestry are poorly developed (Kalaba *et al.*, 2014). This lack of inter-sectoral cooperation and coordination is a common problem in dryland nations' policies (Akhtar-Schuster *et al.*, 2011), indicating the need for a more holistic approach to bridge sectoral divides if ecosystem services are to be conserved.

Synthesis and conclusion

This chapter has explored the key dryland ecosystem services and the threats they face, particularly in sub-Saharan Africa. It has demonstrated that a wide range of factors interact to degrade dryland ecosystems and that these have both human and environmental drivers. Populations living in dryland biomes often feel the more immediate impact of degradation and ecosystem service loss due to their reliance on the ecosystem to make a living. However, drylands supply important services to consumers from other parts of the world, too, so the indirect impacts of their degradation are widely felt. Measuring and monitoring the condition of drylands remains challenging and necessitates involvement of a range of stakeholders – particularly land users who have important local knowledge. Integrated monitoring can help to improve policy formulation and move towards more sustainable land use trajectories. Our case studies have shown that policies and economic incentives can sometimes encourage and reward ecosystem management that damages ecosystem services, and that it is vital to have adequate governance mechanisms in place to ensure that initiatives that seek to conserve ecosystem services operate in an equitable way. As dryland environments experience the impacts of climate change, it is imperative that the ecosystem services they provide are sustained.

References

Aizen, M.A., Garibaldi, L.A., Cunningham, S.A., and Klein, A.M. (2009). How much does agriculture depend on pollinators? Lessons from long-term trends in crop production. *Annals of Botany*, vol 103, pp 1579–1588.

Akhtar-Schuster, M., Thomas, R.J., Stringer, L.C., Chasek, P., and Seely, M.K. (2011). Improving the enabling environment to combat land degradation: institutional, financial, legal and science-policy challenges and solutions. *Land Degradation and Development*, vol 22, no 2, pp 299–312.

Bauer, S., and Stringer, L. C. (2009). The role of science in the global governance of desertification. *Journal of Environment and Development*, vol 18, no 3, pp 248–267.

Brouwer, R. (2016). Payments for ecosystem services. In Potschin, M., Haines-Young, R., Fish, R. and Turner, R.K. (eds) *Routledge Handbook of Ecosystem Services*. Routledge, London and New York, pp 548–553.

Chanda, R., Totolo, O., Moleele, N. M., Setshogo, M., and Mosweu, S. (2003). Prospects for subsistence livelihood and environmental sustainability along the Kalahari Transect: the case of Matsheng in Botswana's Kalahari rangelands, *Journal of Arid Environments*, vol 54, no 2, pp 425–445.

Child, B. (1996). The practice and principles of community-based wildlife management in Zimbabwe: the CAMPFIRE programme. *Biodiversity and Conservation*, vol 5, no 3, pp 369–398.

Ciais, P., Bombelli, A., Williams, M., *et al.* (2011). The carbon balance for Africa: synthesis of recent research studies. *Phil Trans. R. Soc. A.*, vol 369, no 1943, pp 2038–2057.

Davidow, J. (1999). *Infusions of Healing: A Treasury of Mexican-American Herbal Remedies*, Simon and Schuster, New York.

Delgado-Baquerizo, M., Fernando T. M., Gallardo, A., *et al.* (2013). Decoupling of soil nutrient cycles as a function of aridity in global drylands. *Nature*, vol 502, no 7473, pp 672–678.

Dewees, P.A., Campbell, B.M., Katerere, Y., *et al.* (2010). Managing the miombo woodlands of southern Africa: policies, incentives and options for the rural poor. *Natural Resources Policy Research*, vol 2, pp 57–73.

Dougill, A.J., Perkins, J. S., Akanyang, L., *et al.* (2014). *Land Use, Rangeland Degradation and Ecosystem Service Provision: New Analyses from Southern Kalahari, Botswana*. Report for the Economics of Land Degradation Initiative, Leeds, UK. Available at: http://www.see.leeds.ac.uk/research/sri/eld/

Elliott, D. R., Thomas, A. D., Hoon, S. R., and Sen, R. (2014). Niche partitioning of bacterial communities in biological crusts and soils under grasses, shrubs and trees in the Kalahari. *Biodiversity Conservation*, vol 23, no 7, pp 1709–1733.

Frost, P.G.H., and Bond, I. (2008). The CAMPFIRE programme in Zimbabwe: payments for wildlife services. *Ecological Economics*, vol 65, no 4, pp 776–787.

Hulme, D., and Murphree, M. W. (eds) (2001). *African Wildlife and Livelihoods – The Promise and Performance of Community Conservation*. James Currey Ltd, Oxford.
Kalaba, F. K., Quinn, C. H., and Dougill, A. J. (2013). Contribution of forest provisioning ecosystem services to rural livelihoods in the Miombo woodlands of Zambia, *Population and Environment*, vol 35, no 2, pp 159–182.
Kalaba, F. K., Quinn, C. H., and Dougill, A. J. (2014). Policy coherence and interplay between Zambia's forestry, energy, agricultural and climate change policies and multilateral environmental agreements. *International Environmental Agreements: Politics, Law and Economics*, vol 14, no 2, pp 181–198.
Leventon, J., Kalaba, F. K., Dyer, J. C., Stringer, L. C., and Dougill, A. J. (2014). Delivering community benefits through REDD: the case of Joint Forest Management in Zambia. *Forest Policy and Economics*, vol 44, pp 10–17.
Liquete, C., Haas, E., Bondo, T., Hirzinger, C., Schnelle, M., Reisinger, D., Lyon, D., Finisdore, J., and Ledwith, M. (2016). A practical approach to mapping of ecosystems and ecosystem services using remote sensing. In Potschin, M., Haines-Young, R., Fish, R., and Turner, R. K. (eds) *Routledge Handbook of Ecosystem Services*. Routledge, London and New York, pp 205–212.
Logan, B. I., and Moseley, W.G. (2002). The political ecology of poverty alleviation in Zimbabwe's Communal Areas Management Programme for Indigenous Resources (CAMPFIRE), *Geoforum*, vol 33, no 1, pp 1–14.
MA (2005). *Ecosystems and Human Well-Being: Desertification Synthesis*. World Resources Institute, Washington DC.
Mathur, V. N., Afionis, S., Paavola, J., Dougill, A. J., and Stringer, L. C. (2014). Experiences of host communities with carbon market projects: towards multi-level climate justice. *Climate Policy*, vol 14, pp 42–62.
Measham, T. G., and Lumbasi, J. A. (2013). Success factors for Community-Based Natural Resource Management (CBNRM): lessons from Kenya and Australia. *Environmental Management*, vol 52, no 3, pp 649–659.
Middleton, N. J., and Thomas, D.S.G. (1994). *World Atlas of Desertification*. United Nations Environment Programme. Edward Arnold, Nairobi.
Potts, S. G., Biesmeijer, J. C., Kremen, C., *et al.* (2010). Global pollinator declines: trends, impacts and drivers. *Trends in Ecology & Evolution*, vol 25, no 6, pp 345–353.
Reed, M. S., Buenemann, M., Atlhopheng, J., *et al.* (2011). Cross-scale monitoring and assessment of land degradation and sustainable land management: a methodological framework for knowledge management. *Land Degradation & Development*, vol 22, no 2, pp 261–271.
Reynolds, J. F., Stafford Smith, M., Lambin, E. F., *et al.* (2007). Global desertification: building a science for dryland development. *Science*, vol 316, pp 847–851.
Ribot, J. C. (2003). Democratic decentralisation of natural resources: institutional choice and discretionary power transfers in sub-Saharan Africa. *Public Administration and Development*, vol 23, no 1, pp 53–65.
Sallu, S.M., Twyman, C., and Stringer, L.C. (2010). Resilient or vulnerable livelihoods? Assessing livelihood dynamics and trajectories in rural Botswana. *Ecology and Society*, vol 15, no 4, art 3.
Scholes, R. J., and Walker, B. H. (1993). *An African Savanna: Synthesis of the Nylsvley Study*. Cambridge University Press, Cambridge UK.
Stringer, L. C., Dougill, A. J., Thomas, A. D., *et al.* (2012). Challenges and opportunities in linking carbon sequestration, livelihoods and ecosystem service provision in drylands. *Environmental Science and Policy*, vol 19–20, pp 121–135.
Stringer, L.C., and Reed, M. S. (2007). Land degradation assessment in southern Africa: integrating local and scientific knowledge bases. *Land Degradation and Development*, vol 18, no 1, pp 99–116.
Stringer, L. C., Thomas, D.S.G., and Twyman, C. (2007). From global politics to local land users: applying the United Nations Convention to Combat Desertification in Swaziland. *The Geographical Journal*, vol 173, no 2, pp 129–142.
Thomas, A. D. (2012). Impact of grazing intensity on seasonal variations in soil organic carbon and soil CO_2 efflux in two semiarid grasslands in southern Botswana. *Phil Trans. R Soc. B.*, vol 367, no 1606, pp 3076–3086.
UNCCD (United Nations Convention to Combat Desertification) (1994). *United Nations Convention to Combat Desertification in Those Countries Experiencing Serious Drought and/or Desertification Particularly in Africa: Text with Annexes*. UNEP, Nairobi.
UNDP-UNCCD (2011). *The Forgotten Billion: MDG Achievement in the Drylands*. UNCCD-UNDP, New York and Nairobi.
Verlinden, A., Perkins, J. S., and Murray, M. (1998). How are people affecting the distribution of less migratory wildlife in the southern Kalahari of Botswana? A spatial analysis. *Journal of Arid Environments*, vol 38, no 1, pp 129–141.

32

ECOSYSTEM SERVICES SUPPLIED BY MEDITERRANEAN BASIN ECOSYSTEMS

Berta Martín-López, Elisa Oteros-Rozas, Emmanuelle Cohen-Shacham, Fernando Santos-Martín, Marta Nieto-Romero, Claudia Carvalho-Santos, José A. González, Marina García-Llorente, Keren Klass, Ilse Geijzendorffer, Carlos Montes and Wolfgang Cramer

Importance of the Mediterranean ecosystems for biodiversity and ecosystem services

The Mediterranean Basin is located across the south of Europe and the north of Africa. A key defining feature is the climate variability, with contrasting temperatures and rainfall conditions between winter and summer. These characteristics, together with the relatively irregular nature of precipitation, determine the high spatial-temporal variability of biophysical conditions that create a complex mosaic of ecosystems (Le Houérou, 2005). The transformation of landscapes by people, over more than three millennia, has promoted a high diversity of ecosystems as well as a high level of endemism and species richness (Myers *et al.*, 2000). The anthropogenic transformation of ecosystems was the result of people adapting to the unpredictable and changing environment (Blondel *et al.*, 2010). Thus, landscapes of the Mediterranean are characterized by the co-evolution of social and ecological system where original forests and scrublands have been transformed into a shifting mosaic of patches containing different states of maturity of forests and more or less intensively used croplands. Many of the cultural landscapes resulting from this co-evolutionary process are highly multifunctional (Blondel *et al.*, 2010). Overall, the Mediterranean Basin is widely recognized as one of the world's most important biodiversity hotspots, being home to almost 10% of the vascular plants and 3% of the planet's terrestrial vertebrates (Myers *et al.*, 2000). This high level of endemism and species diversity both characterizes its uniqueness and makes it vulnerable to drivers of change, such as the effects of climate (e.g., the extent, intensity and frequency of fire events and droughts), land-use change (e.g., urbanization, agriculture intensification or land-abandonment) or the introduction of alien species (Cuttelod *et al.*, 2008).

The multifunctional landscapes of the Mediterranean provide many examples of social-ecological systems with a diversity of disturbance levels, where ecosystems with extensive disturbance and human management can reach a peak of species and ecosystem services diversity (García-Llorente *et al.*, 2012). We can therefore think of Mediterranean systems as complex adaptive social-ecological systems in which the relationships between people and nature have created the socio-cultural and ecological conditions to deliver a diverse flow of ecosystem services.

For example, multifunctional areas, such as the cork oak savannahs, *montados* or *dehesas* in the Iberian Peninsula provide a range of ecosystem services (Bugalho *et al.*, 2011).

The traditional extensive management of Mediterranean landscapes is currently being replaced, either by more intensive land-use management practices, such as intensive agriculture, or by land-abandonment. In many areas, this transformation causes a deterioration of most services, especially those involved in the regulation of ecosystem processes or those associated with spiritual enrichment, heritage, recreation and aesthetic experiences (Bugalho *et al.*, 2011; Martín-López *et al.*, 2012).

This chapter synthesizes the main results from the sub-global Millennium Assessments that have been undertaken in the Mediterranean Basin, and analyses the relationships between the provision of ecosystem services and the effect of different drivers of change in different Mediterranean ecosystems. It concludes by looking at the challenges facing the management of ecosystem services in the region.

Ecosystem services assessment initiatives in the Mediterranean Basin

The Millennium Ecosystem Assessment has inspired several 'sub-global' assessments in the Mediterranean Basin, namely the Spanish National Ecosystem Assessment (S-NEA) and the Israel National Ecosystem Assessment (I-NEA) (Table 32.1).

The S-NEA started in 2009, supported by the Biodiversity Foundation of the Ministry of Environment. It involved around 60 biophysical and social scientists and assessed the implications of the degradation of ecosystems and loss of biodiversity for human well-being in Spain (S-NEA, 2014). The ambition was to develop the basis for a new generation of policy instruments that focussed on the relationships between ecosystems, biodiversity and human well-being, and which also comply with various regulations, agreements and international initiatives that link the biodiversity conservation policies of Spain and Europe. Using the conceptual framework shown in Figure 32.1, the S-NEA sought to evaluate and provide interdisciplinary information on the consequences of changes in terrestrial and marine ecosystems and the loss of biodiversity for human well-being over the last five decades. The goal was to move the debate on the conservation of ecosystems and biodiversity beyond the academic world and to show its relevance to society. The main message of the S-NEA was that '*the present and future economic, social and cultural life of the inhabitants of the ecosystems of Spain is closely linked to the conservation of their ability to generate provisioning, regulating and cultural services that determine different components of our well-being*'. The assessment

Table 32.1 Main characteristics of the sub-global Millennium Assessments of ecosystem services in the Mediterranean Basin: the case of the Spanish (S-NEA) and Israeli National Ecosystem Assessments (I-NEA).

	S-NEA	*I-NEA*
Timeline	2009- on-going	2012–2015
Overall objective	Evaluate and provide interdisciplinary information on the consequences of changes in aquatic and terrestrial ecosystems and the loss of biodiversity for human well-being.	Increase awareness of the multifaceted value of nature and human dependence on functioning ecosystems; to assist decision-makers in incorporating the value of ecosystem services into landscape planning processes.

	S-NEA	I-NEA
Timeline	2009- on-going	2012–2015
Conceptual framework	The conceptual framework is modified from the MA (2005). In addition, the Driver-Pressure-State-Impact-Response (DPSIR) was used to analyse the complex relationships established between ecosystems and human systems from (Santos-Martín *et al.*, 2013).	The conceptual framework is an adaptation of the UK NEA (2011), CICES (Haines-Young and Potschin, 2013) and the MA (2005).
Assessment users	Multiple stakeholders, such as government, academics, expert staff, NGOs and the private sector.	Managers, planners and decision-makers.
Ecosystem typology	14 ecosystem types were evaluated based on a set of general operational issues appropriate for articulating the assessment at a national scale.	6 broad ecosystem types: Mediterranean, desert, inland water, marine (Mediterranean and Red Seas), agricultural, and urban ecosystems that are further sub-divided into specific ecosystems.
Ecosystem services assessed	22 services (7 provisioning, 8 regulating and 7 cultural).	20 services (8 provisioning, 9 regulating and 3 cultural).
Methodological procedures	Creation of a typology of ecosystem services based on MA (2005). Selection of ecosystem services indicators being temporally explicit, scalable and quantifiable. Assessing ecosystem services through biophysical, socio-cultural and economic methods.	Collection and synthesis of existing data and information from scientific and grey literature. Quantification and valuation of services and benefits using multiple indicators found in the data collected to present biophysical, economic, health and social measures.
Organizational structure	Approximately 60 researchers from different disciplines in the ecological and social sciences and from more than 20 universities and research centres.	Assessment board (approximately 40 representatives; stakeholders and clients), assessment team (approximately 120 experts; leads and contributing authors), two assessment co-chairs, project management and support provided by Hamaarag.
Scale	Multi-scale approach, including national and local scale (five case studies have been included).	Multi-scale approach, including national and local scale (different case studies will be included based on existing data).
Impacts	The results are designed for managers, the business sector, association networks, NGOs and civil society in general.	Assessment outputs are designed to serve the needs of policy- and decision-makers, as well as planners and practitioners, and serve as a broad knowledge base on the state of ecosystem services in Israel.
Future challenges	Multidimensional framework for assessing ecosystem services, including methods ranging from biophysical (supply-side) to socio-cultural and economic approaches (demand-side).	As the assessment progresses it faces challenges in integrating data from different sources collected at different scales with different methods and shaping the content to meet the practical needs of potential users.

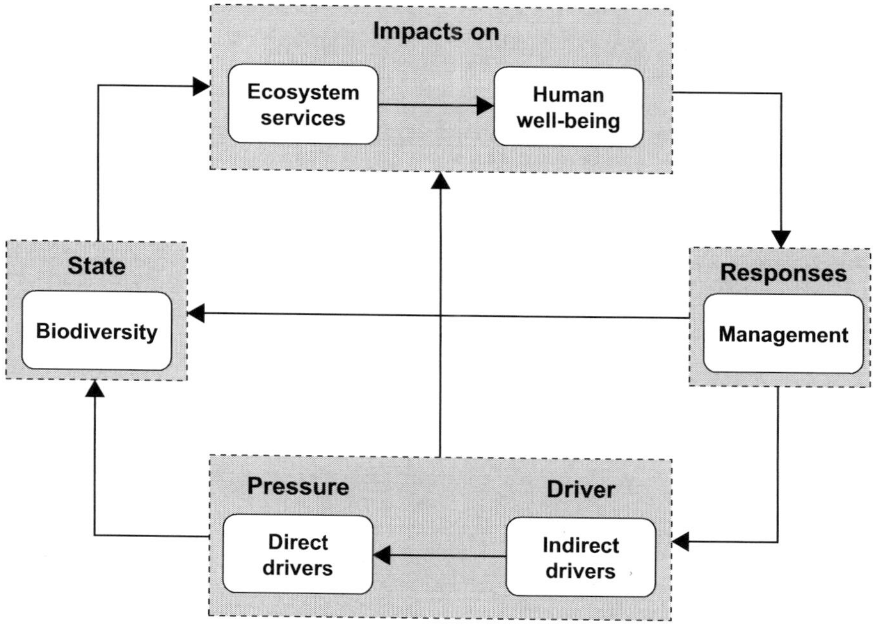

Figure 32.1 Conceptual framework of the Spanish ecosystem assessment (S-NEA).
Source: Santos-Martin et al. (2013)

found that around 45% of Spain's ecosystem services have been degraded, with regulating services being the most negatively affected; coastal and inland aquatic ecosystems have suffered the greatest deterioration, while forests and mountains have been the best-conserved ecosystems. It also established that the synergistic interactions between economic and demographic drivers have promoted dramatic land-use changes, leading to biodiversity loss (Santos-Martín *et al.*, 2013). Finally, it concluded that while there is still sufficient natural capital to provide ecosystem services to present and future generations, steps need to be taken to deal with drivers of change to ensure the sustainability of Spanish social-ecological systems (S-NEA, 2014).

The I-NEA[1] which was conducted by Hamaarag, was modelled after the UK National Ecosystem Assessment (UK NEA, 2011). It was initiated in September 2012 to undertake an assessment by integrating existing data and information from different sources (Table 32.1), and to increase the awareness of the public and decision-makers about the multifaceted value of nature and societal dependence on the functioning ecosystems. Additionally, it has sought to help managers and decision- and policy-makers to incorporate the value of ecosystem services and biodiversity into planning, land management and policy development. The I-NEA was guided by a Board of over 40 representatives of local and national government ministries and agencies, NGOs and the business sector. These represented the future users of the I-NEA and so have been actively involved in the assessment process. The I-NEA's conceptual framework (Figure 32.2) captures the project's scope, which is to present a comprehensive picture of the current state and trends of the country's ecosystem services across all ecosystem types, to establish their value, to evaluate the effect of drivers of change affecting services and to design management and policy response options. The I-NEA has also sought to identify significant and pressing knowledge gaps.

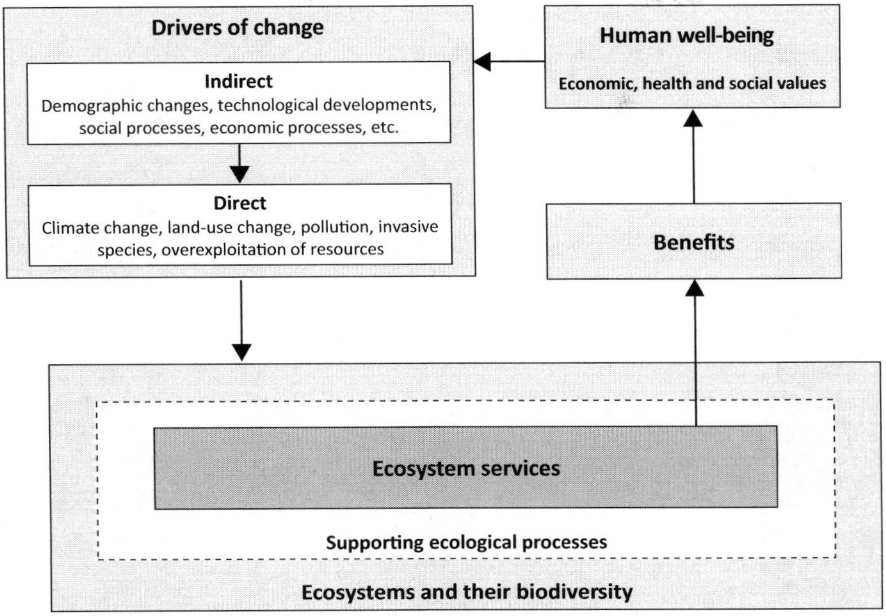

Figure 32.2 Conceptual framework of the Israel ecosystem assessment (I-NEA), adapted from Orenstein and Izhaki (2014). Please note that the I-NEA refers to supporting services (MA, 2005) as 'supporting ecological processes'.

Ecosystem services delivered by Mediterranean ecosystems

Agroecosystems

Agroecosystems are those ecosystems managed with the intention of producing, distributing and consuming food, fuel and fibre (Power, 2010). Thus, while they are significant sources of provisioning services, to a lesser extent they also provide important regulating and cultural services (Nieto-Romero *et al.*, 2014). Provisioning services include food such as olives, cereals and wine grapes, or meat and dairy products, as well as, timber and fibres (e.g., Oteros-Rozas *et al.*, 2014; Willaarts *et al.*, 2012). The regulating services that are frequently overlooked, which include fire prevention by grazing activities, which decrease woody vegetation density (Ruiz-Mirazo *et al.*, 2011), water regulation through specific agrarian and forestry practices as in the *dehesas* and *montados* (Willaarts *et al.*, 2012), or pollination through the maintenance of higher floral diversity communities (Potts *et al.*, 2006). Similarly, cultural services are very important in the context of traditional Mediterranean agroecosystems (Blondel *et al.*, 2010). The cultural benefits include existence and aesthetic values of landscapes (Sayadi *et al.*, 2009; García-Llorente *et al.*, 2012) and recreational activities (Fleischer and Tsur, 2000).

Trade-offs between ecosystem services occur in agroecosystems (Power, 2010). Although regulating and cultural services are necessary inputs for certain provisioning services, such as food or fibre, the maximization of provisioning services have impacted certain regulating and cultural ecosystem services and associated benefits in agroecosystems. In large parts of Mediterranean agroecosystems, land-use simplification, either by intensification or land abandonment, has hindered the provision of regulating services such as the control of soil erosion or water flow, cultural benefits such as sense of place and identity, and specific provisioning services such as those related to livestock (Oteros-Rozas *et al.*, 2014).

Forests and scrublands

Mediterranean forests and scrublands are important ecological infrastructure for biodiversity and the delivery of ecosystem services to people (FAO, 2013). Although scrublands are natural in the region, the majority have resulted from human disturbance of forest ecosystems through fires and cutting, or from the abandonment of agricultural land (Calvo *et al.*, 2012). Forests and scrublands are important for the regulation of water quantity and quality, climate and air quality, as well as erosion control, pollination and pest regulation. As a result of the vegetation structure, composition and associated management practices, forest and scrubland can provide provisioning services including firewood, timber and pulp, biomass for energy production and non-timber products such as cork, charcoal, resin, fodder for livestock, mushrooms, pine kernels, chestnuts, honey, fruits for liquors and aromatic plants used to cook, as medicines or for cosmetics (e.g., Calvo *et al.*, 2012). However, in some parts of the Mediterranean Basin, scrublands have been considered as marginal lands due to their low productivity, and have been replaced by forests in order to better protect soils from erosion and increase land profitability (Calvo *et al.*, 2012).

Forests and scrublands are also important providers of cultural benefits such as: recreational activities in the form of tourism, hunting and relaxation; local ecological knowledge related to traditional practices such as charcoal production; scientific knowledge; environmental education; scenic beauty and cultural identity (López-Santiago *et al.*, 2014).

Many human factors are responsible for the degradation of Mediterranean forests. In the east and the south of the Mediterranean Basin, the drivers underpinning ecosystem service depletion are habitat homogenization due to the expansion of cultivated areas and urbanization, desertification and effects of climate change, and increasing use of forests for fuelwood, grazing, and mass tourism (García-Ruíz *et al.*, 2011). In the north of the Mediterranean Basin, the abandonment of farmland and land-use intensification, as well as the growth of the tourist sector, have the highest impact (Kuvan, 2012). However, overall, Mediterranean forests and shrubs are expanding, mainly as a result of farmland abandonment (García-Ruíz *et al.*, 2011). This can impact the ecosystem services supply in several ways, such as a decrease of water yield in drier regions and increased fire risk (Scarascia-Mugnozza *et al.*, 2000).

Wetlands

The Ramsar Convention on Wetlands defines these ecosystems as areas of "marsh, fen, peatland or water, whether natural or artificial, permanent or temporary, with water that is static or flowing, fresh, brackish or salt, including areas of marine water the depth of which at low tide does not exceed six meters" (Article 1.1). The main ecological characteristic of Mediterranean wetlands is their negative water balance, with a higher evapotranspiration than precipitation. This has significant consequences for the hydrological cycle and, therefore, for specific regulating services (Mediterranean Wetlands Outlook, 2012).

Wetlands contribute to supply provisioning services such as crops, fish or meat, and freshwater (Martín-López *et al.*, 2011; Mediterranean Wetlands Outlook, 2012); regulating services such as flood attenuation, erosion control, hydrological regulation, water purification and pollination (García-Llorente *et al.*, 2011); and cultural benefits, such as ecotourism (Beltrame *et al.*, 2013), scientific knowledge, inspiration for art and cultural heritage. In addition, wetlands are important for endemic species and for aquatic and terrestrial migratory species (García-Llorente *et al.*, 2011).

Temporal and spatial trade-offs between ecosystem services are apparent in Mediterranean wetlands (e.g., Cohen-Shacham *et al.*, 2011; Martín-López *et al.*, 2011). For example, the drainage of the Hula Wetland in Israel for intensive crop production, critically affected regulating services, by triggering the release of large amounts of nutrients downstream that impacted the Sea of Galilee a few decades later. The change in land-use also reduced the output of traditional foods from activities such as buffalo farming, fishing and hunting, and raw material for the construction of huts, mats, ropes, baskets and small fishing rafts (Cohen-Shacham *et al.*, 2011).

In the past, wetlands have often been used for their grazing resources or considered non-usable and unhealthy areas. As a result, they were often ignored in biodiversity conservation strategies and have been lost by conversion to agriculture (Mediterranean Wetlands Outlook 2012). The intensification of land-uses outside wetland areas can also have negative impacts on the provision of regulating services and can lead to social conflicts (Gómez-Baggethun *et al.*, 2013). For example, the broad-scale hydrological flows responsible for the existence of wetlands makes them especially sensitive to pressures generated across multiple scales. Within the Mediterranean Basin, there are more than 3,500 dams that affect water and sediment discharge and hence wetland function (Cuttelod et al., 2008). In addition, Mediterranean coastal wetlands are among the most threatened in the world due to global sea-level rise (Cuttelod *et al.*, 2008).

As a result of the combined effect of land-use changes, water overexploitation and climate change, the social perception of wetland values has been fundamentally transformed. This has prompted several recent international efforts to protect and sustainably use them. Since 1991, the efforts to recognize the value of wetlands have been coordinated by the MEDWET initiative, a partnership between the European Commission, the Ramsar Secretariat, 26 governments of Mediterranean and peri-Mediterranean countries and several non-governmental organizations. Today, more than 360 wetland sites in the Mediterranean Basin have been listed as being of international importance for conservation under the Ramsar Convention.

Sustainable management in the Mediterranean

Learning from the past

The co-evolution of ecosystems and human societies in the Mediterranean Basin since Neolithic times has produced a unique and characteristic landscape. Indeed, so close are the links between society and ecology that some authors suggest that Mediterranean landscapes were 'designed' by different cultures and that 'natural' ecosystems no longer really exist (Blondel *et al.*, 2010). This notion has important implications for our understanding of ecosystem dynamics and integrity. It poses a particular challenge for ecosystem services assessments because services have been co-produced through the interaction of historically transformed ecosystems and particular traditional management practices. Thus the supply of many ecosystem services often now depends on the maintenance of traditional management practices (Box 32.1). However, the output of other ecosystem services may also depend on modern land management strategies, such as organic farming (Nieto-Romero *et al.*, 2014). Specific regulating services related to soil and water management are highly associated with the maintenance of traditional practices and their embedded local ecological knowledge (Martín-López *et al.*, 2012). Such knowledge, which is also referred to as experiential, traditional or ecoliteracy, is a cumulative body of understandings and beliefs, developed by stakeholders at a local scale through the interaction with their environment (Berkes *et al.*, 2000). The importance and relevance of this knowledge system in the Mediterranean Basin lies in its local and integrative nature and the support it gives to the sustainable use and management of ecosystem services. However, the effect of indirect drivers

Box 32.1 The case of transhumant landscapes: how human traditional practices can promote the delivery of ecosystem services

Transhumance is a customary management practice to move livestock from one grazing ground to another, as from lowlands to highlands, with the change of seasons or with changing weather. Seasonal movements of flocks allow herders to match grazing pressure with seasonal peaks in pasture availability, favouring an optimal exploitation of existing resources. Transhumant pastoralism still persists in many Mediterranean countries, although on a smaller scale than in the past.

A recent assessment of ecosystem services has shown that many important ecosystem services are highly dependent on the maintenance of transhumant pastoralist practices in Spain (Oteros-Rozas *et al.*, 2014). In fact, some regulating services, such as fire prevention, maintenance of soil fertility provided by sheep manure, holm-oak regeneration, habitat connectivity and seed dispersal, are dependent on the maintenance of this traditional practice and associated local ecological knowledge (Oteros-Rozas *et al.*, 2013).

of change in rural systems of the Mediterranean Basin, through such processes as land abandonment or land-use intensification, is eroding these local knowledge systems (Oteros-Rozas *et al.*, 2013; Iniesta-Arandia *et al.*, 2015) and related ecosystem services (Martín-López *et al.*, 2012).

On the basis of the review of current knowledge of ecosystem services in the Mediterranean Basin, some important insights emerge for improved ecosystem services management. Assuming that traditional mosaic landscapes can supply both a more diverse bundle of ecosystem services as well as greater quantities of services (Bugalho *et al.*, 2011; García-Llorente *et al.*, 2012), an appropriate objective would be to restore their structural characteristics and avoid further landscape homogenization. The sustainable supply and use of ecosystem services across urban-rural gradients should be considered in planning policy in the Mediterranean Basin (García-Nieto *et al.*, 2013).

Insights for landscape management

To ensure the maintenance of ecosystem services relevant for human well-being in the Mediterranean, a reform of policy instruments used in landscape planning is probably required. For example, the wider public and decision-makers need to be aware of the importance of counteracting the deterioration of regulating and cultural services caused by the direct drivers of change, such as land-use, climate or overexploitation of water. It is also likely that a diversity of land-uses, including both traditional and modern practices, increases the ability of the social-ecological system to deliver ecosystem services to people. While more research is required to substantiate this, the conclusion should be considered in developing regional and local land-use policy. Further, it might be valuable to recognize the role of traditional management practices in keeping and shaping multifunctional landscapes and the need to empower local communities to manage them by drawing on their local ecological knowledge-systems in decision-making. Finally, we should design institutional systems able to include multiple stakeholder groups and organizations, and their values across different spatial and administrative scales.

Note

1 http://www.hamaarag.org.il/en/content/inner/ecosystem-services

References

Beltrame, C., Cohen-Shacham, E., Trouillet, M., and Guillet, F. (2013). Exploring the links between local management and conservation applying the Ecosystem Services concept: conservation and tourism service in Camargue (France). *International Journal of Ecosystem Services, Biodiversity Science and Management*, vol 9, pp 166–177.

Berkes, F., Colding, J., and Folke, C. (2000). Rediscovery of traditional ecological knowledge as adaptive management. *Ecological Applications*, vol 10, pp 1251–1262.

Blondel, J., Aronson, J., Bodiou, J.Y., and Boeuf, G. (2010). *The Mediterranean Region Biological Diversity in Space and Time*. Oxford University Press, Oxford.

Bugalho, M. N., Caldeira, M. C., Pereira, J. S., Aronson, J., and Pausas, J. G. (2011). 'Mediterranean cork oak savannas require human use to sustain biodiversity and ecosystem services'. *Frontiers in Ecology and Environment*, vol 9, pp 278–286.

Calvo, L., Baeza, J., Marcos, E., Santana, V., and Papanastasi, V. P. (2012). Post-fire management of shrublands. In: Moreira, F., Arianoutsou, M., Corona, P. and de las Heras, J. (eds) *Post-Fire Management and Restoration of Southern European Forests*. Springer, Dordrecht, Heidelberg, London, and New York.

Cohen-Shacham, E., Dayan, T., Feitelson, E., and de Groot, R. S. (2011). Ecosystem service trade-offs in wetland management: drainage and rehabilitation of the Hula, Israel. *Hydrological Sciences Journal*, vol 56, pp 1582–1601.

Cuttelod, A., García, N., Abdul Malak, D., Temple, H., and Katariya, V. (2008) The Mediterranean: a biodiversity hotspot under threat. In: Vié, J-C., Hilton-Taylor, C. and Stuart, S. N. (eds) *The 2008 Review of the IUCN Red List of Threatened Species*. IUCN, Gland.

FAO (2013). *State of Mediterranean Forests 2013*. FAO, Rome.

Fleischer, A., and Tsur, Y. (2000). Measuring the recreational value of agricultural landscape. *European Review of Agricultural Economics*, vol 27, pp 385–398.

García-Llorente, M., Martín-López, B., Díaz, S., Montes, C. (2011). Can ecosystem properties be fully translated into service values? An economic valuation of aquatic plants services. *Ecological Applications*, vol 21, pp 3083–3103.

García-Llorente, M., Martín-López, B., Iniesta-Arandia, I., López- Santiago, C. A., Aguilera, P. A., and Montes, C. (2012). The role of multi-functionality in social preferences toward semi-arid rural landscapes: an ecosystem service approach. *Environmental Science and Policy*, vol 19–20, pp 136–146.

García-Nieto, A. P., García-Llorente, M., Iniesta-Arandia, I., and Martín-López, B. (2013). Mapping forest ecosystem services: from providing units to beneficiaries. *Ecosystem Services*, vol 4, pp 126–138.

García-Ruiz, J. M., Lopez-Moreno, J. I., Vicente-Serrano, S. J., Lasanta-Martinez, T., and Beguería S. (2011). Mediterranean water resources in a global change scenario. *Earth Science Reviews*, vol 105, pp 121–139.

Gómez-Baggethun, E., Kelemen, E., Martín-López, B., Palomo, I., and Montes, C. (2013). Scale misfit in ecosystem service governance as a source of environmental conflict. *Society & Natural Resources*, vol 26, pp 1202–1216.

Haines-Young, R., and Potschin, M. (2013). *Common International Classification of Ecosystem Services (CICES)*: Consultation on version 4. EEA Framework Contract No EEA/IEA/09/003.

Iniesta-Arandia, I., García del Amo, D., García-Nieto, A. P., Piñeiro, C., Montes, C., and Martín-López, B. (2015). Factors influencing Local Ecological Knowledge maintenance in Mediterranean semi-arid watersheds. *AMBIO*, vol 44, pp 285–296.

Kuvan, Y. (2012). Assessing the impacts of tourism on forests: mass tourism and policy in Turkey. *Environmental Engineering & Management Journal*, vol 11, pp 1415–1424.

Le Houérou, H. N. (2005). *Atlas of Climatic Diagrams for the Isoclimatic Mediterranean Zones*. Montpellier, France.

López-Santiago, C.A., Oteros-Rozas, E., Martín-López, B., Plieninger, T., González, E., and González, J. A. (2014). Using visual stimuli to explore the social perceptions of ecosystem services in cultural landscapes: the case of transhumance in Mediterranean Spain. *Ecology & Society*, vol 19, p 27. Available at: http://www.ecologyandsociety.org/vol19/iss2/art27/

MA (2005). *Ecosystems and Human Well-Being: Wetlands and Water Synthesis*. Island Press, World Resources Institute, Washington DC.

Martín-López, B., García-Llorente, M., Palomo, I., and Montes, C. (2011). The conservation against development paradigm in protected areas: valuation of ecosystem services in the Doñana social-ecological system (south-western Spain). *Ecological Economics*, vol 70, pp 1481–1491.

Martín-López, B., Iniesta-Arandia, I., García-Llorente, M., Palomo, I., Casado-Arzuaga, I., García del Amo, D., Gómez-Baggethun, E., Oteros-Rozas, E., Palacios-Agundez, I., Willaarts, B., González, J. A., Santos-Martín, F., Onaindia, M., López-Santiago, C., and Montes, C. (2012). Uncovering ecosystem service bundles through social preferences. *PLoS ONE*, vol 7, pp e38970.

Mediterranean Wetlands: Outlook (2012). *First Mediterranean Wetlands Observatory Report – Technical Report.* Tour du Valat, France.

Myers, N., Mittermeier, R. A., Mittermeier, C. G., da Fonseca, G.A.B., and Kent, J. (2000). Biodiversity hotspots for conservation priorities. *Nature*, vol 403, pp 853–858.

Nieto-Romero, M., Oteros-Rozas, E., González, J. A., and Martín-López, B. (2014). Exploring the knowledge landscape of ecosystem services assessments in Mediterranean agroecosystems: insights for future research. *Environmental Science & Policy*, vol 37, pp 121–133.

Orenstein, D., and Izhaki, I. (2014). From planning to implementation: Israel's National Ecosystem Assessment. In: Klass, K. (ed.) *Ecosystems and Human Wellbeing – A National Assessment.* Jerusalem), Hamaarag (in Hebrew).

Oteros-Rozas, E., Martín-López, B., González, J. A., Plieninger, T., López, C. A., and Montes C. (2014). Socio-cultural valuation of ecosystem services in a transhumance social-ecological network. *Regional Environmental Change*, vol 14, pp 1269–1289.

Oteros-Rozas, E., Ontillera-Sánchez, R., Sanosa, P., Gómez-Baggethun, E., Reyes-García, V., and González, J. A. (2013). Traditional ecological knowledge among transhumant pastoralists in Mediterranean Spain. *Ecology and Society*, vol 18, p 33.

Palomo, I., Montes, C., Martín-López, B., González, J. A., García-Llorente, M., Alcorlo, P., and García-Mora, M. R. (2014). Incorporating the social-ecological approach to protected areas in the Anthropocene. *BioScience*, vol 64, pp 181–191.

Potts, S. G., Petanidou, T., Roberts, S., O'Toole, C., Hulbert, A., and Willmer, P. (2006). Plant-pollinator biodiversity and pollination services in a complex Mediterranean landscape. *Biological Conservation*, vol 129, pp 519–529.

Poulos, S. E., and Collins, M. B. (2002). Fluviatile sediment fluxes to the Mediterranean Sea: a quantitative approach and the influence of dams. In: Jones, S. J. and Frostick, L.-E. (eds) *Sediment Flux to Basins: Causes, Controls and Consequences.* Geological Society (Special Publications), London.

Power, A. G. (2010). Ecosystem services and agriculture: tradeoffs and synergies. *Philosophical Transactions of the Royal Society of London. Series B, Biological sciences*, vol 365, pp 2959–2971.

Ruiz-Mirazo, J., Robles, A. B., and González-Rebollar, J. L. (2011). Two-year evaluation of fuelbreaks grazed by livestock in the wildfire prevention program in Andalusia (Spain). *Agriculture, Ecosystems & Environment*, vol 141, pp 13–22.

Santos-Martín, F., Martín-López, B., García-Llorente, M., Aguado, M., Benayas, J., and Montes, C. (2013). Unraveling the relationships between ecosystems and human wellbeing in Spain. *PLoS ONE*, vol 8, pp e73249.

Sayadi, S., González-Roa, M. C., and Calatrava-Requena, J. (2009). Public preferences for landscape features: the case of agricultural landscape in mountainous Mediterranean areas. *Land Use Policy*, vol 26, pp 334–344.

Scarascia-Mugnozza, G., Oswald, H., Piussi, P., and Radoglou, K. (2000). Forests of the Mediterranean region: gaps in knowledge and research needs. *Forest Ecology and Management*, vol 132, pp 97–109.

S-NEA (Spanish National Ecosystem Assessment) (2014). *Ecosystems and Biodiversity for Human Wellbeing. Synthesis of the Key Findings.* Spanish Ministry of Agriculture, Food and Environment, Madrid. Available at: http://www.ecomilenio.es

UK-NEA (UK National Ecosystem Assessment) (2011). *The UK National Ecosystem Assessment: Synthesis of the Key Findings.* UNEP-WCMC, Cambridge UK.

Willaarts, B., Volk, M., and Aguilera, P. (2012) Assessing the ecosystem services supplied by freshwater flows in Mediterranean agroecosystems. *Agricultural Water Management*, vol 105, pp 21–31.

33

ECOSYSTEM SERVICES PROVIDED BY SOIL LIFE

Wim H. van der Putten and Diana H. Wall

Introduction

Soils harbour an enormous diversity of microbes and fauna that play a major role in delivering ecosystem services (Wall et al., 2012). This diversity can be assessed by a large array of microscope-based and molecular identification techniques and varies largely from 100 to more than 9,000 microbial taxa per g soil and from 10 to several thousand taxa of fauna present under one square metre of vegetation (Bardgett and van der Putten, 2014). The services produced by soil life are strongly dependent on the abiotic environment: the global location determines climate (minimum, maximum and average temperature, number of frost days, plant cover, etc.), whereas within climatic regions, soil type – and related to that, vegetation composition – determines which soil organisms may be present, due to habitat filtering activities. Finally, within ecosystems, effects of humans predominantly influence soil ecosystem services by land management, soil tillage, crop species used, water table management, fertilization, pesticide use, and soil sealing, amongst others.

The massive diversity of soil life has given rise to studies on the importance of belowground biodiversity for ecosystem functioning. It only needs a hand full of belowground species to provide an ecosystem function (Setälä et al., 2005), but that does not mean that the other thousands of species are not useful for soil functioning and for the provisioning of ecosystem services. Explanations for high species diversity in soil include the heterogeneity of soil as a habitat, the role of plant diversity, the inactive state of many soil biota, and the well-conserved presence of many soil biota in microsites that protect (microbial) species from being predated, or outcompeted (Bardgett and van der Putten, 2014). In any case, it cannot be simply concluded that the large diversity implies considerable functional redundancy. Some groups, such as those involved in an ecosystem function, such as decomposition, might be relatively species-rich, whereas other groups, such as those in nitrogen transformation by nitrification, can be relatively species-poor.

A well-established concept in soil ecology is that there are basically two groups of soil biota, one group feeding on live plant roots and another group feeding on dead plant materials, such as aboveground and belowground litter originating from decaying leaves and roots, and exudates from living roots. Predators of both groups of soil biota are usually categorized in the group of species feeding on dead plant remains, but in practice they feed on many species of soil organisms (Wardle et al., 2004). Some soil ecosystem services are based on activities of soil biota in the compartment that is directly interacting with plant roots, such as mutualistic symbioses with plant roots, plant growth promotion, or defence against soil-borne root pathogens by bacteria that are feeding on root exudates (Mendes et al., 2011). Other ecosystem services, such as those related to decomposition of organic matter and nutrient provision to growing plants, are provided by soil organisms that feed on dead plant tissues (Wardle et al., 2004). These examples show the concept of two groups of soil biota that are not distinct, however species interactions occur between them.

It is well known that soil biota can influence soil formation, soil structure, and the transformation of soil organic matter into humus (Wardle, 2002). This role of soil biota is pivotal for a number of provisioning, supporting, and regulating ecosystem services that are related to soil structure, soil fertility, water infiltration, and water storage capacity of soils. Although all soil organisms may contribute to these processes, a large-scale field study in European agricultural fields has shown that not all soil life correlates equally well with individual soil ecosystem services (de Vries et al., 2013), suggesting that there is no single relationship between the diversity and complexity of belowground biodiversity and the processes that underpin ecosystem services.

A factor that has been often overlooked is how soil life contributes to dynamics of natural vegetation, which can also be translated into ecosystem services. For example, soil pathogens that are present in many soils can control the abundance of individual plant species, which promotes plant species diversity (Packer and Clay, 2000). Plant species that benefit disproportionally from symbiotic mutualists in the soil, such as mycorrhizal fungi or nitrogen-fixing plant species, could become dominant, as their abundance is not being controlled, but promoted (Bever, 2002). What will happen when plants are no longer controlled by soil pathogens becomes clear when observing the performance of introduced exotic plant species in their new ranges. Introduction in new ranges most likely coincides with release from native soil-borne pathogens, which promotes the invasiveness of the introduced exotics. This has happened with some tree species that have been introduced into other continents (Reinhart et al., 2003; Gundale et al., 2014). Therefore, soil-borne pathogens provide important regulating services resulting in abundance control, thus preventing species from breaking out in an uncontrolled way. The relatively limited amount of cases where introduced species escape from such abundance control shows how important soil-borne pathogens are in unmanaged ecosystems.

Interestingly, over time, introduced exotic plant species may become less abundant (Speek et al., 2013), and it has been assumed that increased exposure to local pathogens may play a role in this process (Mitchell and Power, 2003; Hawkes, 2007). For example, it has been shown that exotic plant species in New Zealand introduced longest ago are more susceptible to soil-borne pathogen effects than recently introduced plant species (Diez et al., 2010). Therefore, control of introduced exotic plant and possibly also animal species, such as introduced earthworms in New England and introduced New Zealand flatworms in the United Kingdom, may be an important ecosystem service provided by soil organisms. The natural abundance decline of introduced exotic plant species may take decades to centuries (Speek et al., 2013), but in some cases soil pathogen incidence may already occur within a three-decade period (Dostál et al., 2013). Therefore, it may be possible that such soil-borne biotic resistance against invasive exotic plant species can develop at much faster rates than currently expected.

During the last one or two decades, awareness that soil organisms also can influence aboveground species and species interactions has increased (Wardle, 2002). For example, root-feeding insects can change the attractiveness and suitability of plants to aboveground plant-feeding insects and parasitoid wasps, which are the natural enemies of plant-feeding insects (Soler et al., 2005). These so-called aboveground-belowground interactions have become increasingly studied, revealing a number of examples where soil life influences aboveground interactions between plants, enemies and their enemies (Bezemer and Van Dam, 2005). Soil life can also influence aboveground interactions between plants and pollinating insects (Poveda et al., 2003). The examples show that soils can also influence aboveground regulating ecosystem services varying from pest control and pollination to lifecycle maintenance and habitat protection.

Soils also influence human health, directly or indirectly. Soils contain a number of human pathogens, which may influence health and even death incidences, such as Q fever (Jeffery and van der Putten, 2011); in other words, they may provide a disservice in relation to human well-being. In contrast, soils possibly control soil-borne pathogen outbreaks by competition or antagonistic interactions, but very little is known about this subject. Microbes in soils are also the basis for antibiotics that cure human diseases, such as penicillin. These antibiotics are naturally produced in soil during microbial interactions and can be cultured by growing microbes under specific conditions. The wealth of belowground biodiversity for curing new diseases, such as those due to multi-resistant bacteria, is still to be explored (Hol et al., 2014).

The services and benefits provided by soils

In a previous soil biodiversity summary report by (Turbé et al., 2010), ecosystem services and benefits provided by soil life can be grouped into six thematic categories (Table 33.1):

Soil organic matter recycling and fertility. This includes soil formation, soil structure, water infiltration, nutrient cycling, and, consequently, primary production. Without this capacity, life would be impossible, as all dead plant materials would accumulate on top of the soil surface and no new nutrients would become available for any further plant growth.

Regulation of carbon flux and climate control. This service operates directly via the carbon storage in the soil, and indirectly by carbon storage in vegetation. Following the Kyoto protocol, there was some concern that soils would not be able to take up enough carbon dioxide. However, during the past one and a half decade it has been shown that in most countries soil organic matter declines rather than increases.

Water cycle regulation. This ecosystem service includes infiltration, storage, purification, transfer to aquifers and surface effluents, erosion prevention, and regulation of flows in effluents. The current flooding or drying catastrophes of rivers might be better counteracted when soil life is in the right state. Reduced soil biodiversity will lead to loss of stress resistance, or resilience.

Decontamination and bioremediation. This ecosystem service concerns neutralization of contaminants, which can be due to binding to soil particles, chemical degradation, or biological degradation. Many sites, especially in heavily populated regions, are under high pressure from toxic elements. Work in this research area may help to determine if, how, and where soil life may contribute to the growth and development of plant species and other aboveground biodiversity.

Pest control. Soil life may neutralize pests and pathogens of plants, animals, and humans by biological control. Increased pesticide use has made us become more dependent on chemical pest control activities. Still, many species are controlled by soil life, and perhaps a nice demonstration of this power could be provided by the invasiveness of introduced exotic plant species, which tend to be released from their natural enemies.

Human health. As explained above, soils can be sources of novel anti-microbial compounds, as well as being able to break down or neutralize human or animal pathogens. These anti-microbial compounds are the result of interactions between microbes that occur naturally in the soil, following excessive testing.

The ecosystem services may be compared in different ways, depending on whether or not the naming of the Millennium Ecosystem Assessment (MA) or CICES is adopted (Table 33.1).

Table 33.1 Comparison of the services classification of this report with MEA nomenclature.

Theme	*MA nomenclature and, where appropriate, the corresponding final service at the CICES Group level ()*	*Type of service*
Soil organic matter recycling and fertility, including soil formation	Decomposition (Mediation by biota), nutrient cycling, soil formation (Soil formation and composition), primary production, erosion regulation (mass flow regulation)	Supporting and provisioning
Regulation of carbon flux and climate control	Climate regulation (Atmospheric composition and climate regulation)	Regulating
Water cycle regulation	Water regulation and water purification (Water condition)	Regulating
Decontamination and bioremediation	(Mediation by biota)	Regulating
Pest control	Diseases regulation (Pest and disease control)	Regulating
Human health	Diseases regulation (Pest and disease control)	Regulating

Source: adapted from Turbé et al., 2010

Forward look and conclusions

Currently, soil is often used for maximizing single ecosystem services, such as primary production. In order to obtain more services from the same piece of land, it will be essential to explore multi-functionality in ecosystem services. For example, instead of sealing all urban soil with pavement, opening up a fraction of the surface will enable some water infiltration, which may prevent sewer systems over-flooding during heavy rain events (Setälä et al., 2014). Such multi-functionality could also be aimed for in agricultural environments, for example by growing plant species-rich vegetation on arable margins, which may serve as habitat for overwintering natural enemies of pest species (Tscharntke et al., 2005). In some cases, providing one ecosystem service may have such a negative trade-off with others that it results in a disservice. For example, intensified production of first generation bioenergy crops may reduce the use of fossil fuels and thereby save on greenhouse gas production. However, high input-output crop production enhances the risk that soils produce enhanced amounts of greenhouse gases, and primary production used for biofuels cannot be used for producing food and feed. In this case, biofuels trade-off analyses need to include soil-related ecosystem services in order to obtain a more complete overview of all costs and benefits. Therefore, in the future it will be essential to analyse the net effects of all ecosystem services, aiming at the net effect of all services will be positive, rather than neutral or negative. For example, in a recent study Brady et al. (2015) developed a method in order to value changes in supporting soil ecosystem services and associated soil natural capital in agriculture. This method included changes in income streams of future farming (Brady et al., 2015). These types of analyses have the potential to enhance economic underpinning of the value of soil biodiversity-related ecosystem services.

In conclusion, soil-borne ecosystem services are crucial for supporting life on earth. Besides primary production, soil ecosystem services may also be considered for other effects, such as climate regulation, pest control, provision of novel antibiotics, and other services. Such multi-functional use of soils is still in its infancy and several world-wide programmes, including the Global Soil Biodiversity Initiative (http://www.globalsoilbiodiversity.org/), have been launched in order to enhance awareness of soil ecosystem services. The next step is to enable stakeholders and end users to include soil ecosystem services in their decision-making. For example, excessive soil sealing in urban environments may cause disservices due to loss of capacity of bare ground to uptake the peak of rain water, which requires increased investments in sewer systems. Awareness that unsealed soil provides ecosystem services in urban areas may be used in urban decision-making to save on costs of infrastructure (Setälä et al., 2014). Ten years after completion of the Millennium Ecosystem Assessment, soils and soil ecosystem services have achieved a prominent place on the international research agenda. Now, it is time for the next step towards a more integrated approach of soil ecosystem services, focussing on optimizing multi-functionality, which may contribute to sustainability of society.

References

Bardgett, R. D., and van der Putten, W. H. (2014). Belowground biodiversity and ecosystem functioning. *Nature*, vol515, pp 505–511.

Bever, J. D. (2002). Negative feedback within a mutualism: host-specific growth of mycorrhizal fungi reduces plant benefit. *Proceedings of the Royal Society B-Biological Sciences*, vol 269, no 1509, pp 2595–2601.

Bezemer, T. M., and Van Dam, N. M. (2005). Linking aboveground and belowground interactions via induced plant defenses. *Trends in Ecology & Evolution*, vol 20, no 11, pp 617–624.

Brady, M. V., Hedlund, K., Cong, R-G., Hemerik, L., Hotes, S., Machado, S., Mattsson, L., Schulz, E., and Thomsen, I. K. (2015). Valuing supporting soil ecosystem services in agriculture: a natural capital approach. *Agronomy Journal* (in press).

de Vries, F. T., Thebault, E., Liiri, M., Birkhofer, K., Tsiafouli, M. A., Bjornlund, L., Jorgensen, H. B., Brady, M.V., Christensen, S., de Ruiter, P. C., d'Hertefeldt, T., Frouz, J., Hedlund, K., Hemerik, L., Hol, W.H.G., Hotes, S., Mortimer, S. R., Setala, H., Sgardelis, S. P., Uteseny, K., van der Putten, W. H., Wolters, V., and Bardgett, R. D. (2013). Soil food web properties explain ecosystem services across European land use systems. *Proceedings of the National Academy of Sciences of the United States of America*, vol 110, no 35, pp 14296–14301.

Diez, J. M., Dickie, I., Edwards, G., Hulme, P. E., Sullivan, J. J., and Duncan, R. P. (2010). Negative soil feedbacks accumulate over time for non-native plant species. *Ecology Letters*, vol 13, no 7, pp 803–809.

Dostál, P., Müllerová, J., Pyšek, P., Pergl, J., and Klinerová, T. (2013). The impact of an invasive plant changes over time. *Ecology Letters*, vol 16, no 10, pp 1277–1284.

Gundale, M. J., Kardol, P., Nilsson, M.-C., Nilsson, U., Lucas, R. W., and Wardle, D. A. (2014). Interactions with soil biota shift from negative to positive when a tree species is moved outside its native range. *New Phytologist*, vol 202, no 2, pp 415–421.

Hawkes, C. V. (2007). Are invaders moving targets? The generality and persistence of advantages in size, reproduction, and enemy release in invasive plant species with time since introduction. *American Naturalist*, vol 170, no 6, pp 832–843.

Hol, W.H.G., de Boer, W., and Medina A. (2014). Beneficial interactions in the rhizosphere. In: Dighton, K. and Krumins, J.A.K. (eds) *Interactions in soil: promoting plant growth*. Springer Verlag, Berlin and Heidelberg.

Jeffery, S., and van der Putten, W. H. (2011). *Soil Borne Human Diseases*. Publications Office of the European Union, Luxembourg

Mendes, R., Kruijt, M., de Bruijn, I., Dekkers, E., van der Voort, M., Schneider, J.H.M., Piceno, Y. M., DeSantis, T. Z., Andersen, G. L., Bakker, P., and Raaijmakers, J. M. (2011). Deciphering the rhizosphere microbiome for disease-suppressive bacteria. *Science*, vol 332, no 6033, pp 1097–1100.

Mitchell, C. E. and Power, A. G. (2003). Release of invasive plants from fungal and viral pathogens, *Nature*, vol 421, no 6923, pp 625–627.

Packer, A., and Clay, K. (2000). Soil pathogens and spatial patterns of seedling mortality in a temperate tree. *Nature*, vol 404, no 6775, pp 278–281.

Poveda, K., Steffan-Dewenter, I., Scheu, S., and Tscharntke, T. (2003). Effects of below- and above-ground herbivores on plant growth, flower visitation and seed set. *Oecologia*, vol 135, no 4, pp 601–605.

Reinhart, K. O., Packer, A., van der Putten, W. H., and Clay, K. (2003). Plant-soil biota interactions and spatial distribution of black cherry in its native and invasive ranges. *Ecology Letters*, vol 6, no 12, pp 1046–1050.

Setälä, H., Bardgett, R. D., Birkhofer, K., Brady, M., Byrne, L., de Ruiter, P. C., de Vries, F. T., Gardi, C., Hedlund, K., Hemerik, L., Hotes, S., Liiri, M., Mortimer, S. R., Pavao-Zuckerman, M., Pouyat, R.,

Tsiafouli, M., and van der Putten, W. H. (2014). Urban and agricultural soils: conflicts and trade-offs in the optimization of ecosystem services. *Urban Ecosystems*, vol 17, no 1, pp 239–253.

Setälä, H., Berg, M. P., and Jones, T. H. (2005). Trophic structure and functional redundancy in soil communities. In: Bardgett, R. D., Usher, M. B. and Hopkins, D.W. (eds), *Biological Diversity and Function in Soils*. Cambridge University Press, Cambridge UK.

Soler, R., Bezemer, T. M., van der Putten, W. H., Vet, L.E.M., and Harvey, J. A. (2005). Root herbivore effects on above-ground herbivore, parasitoid and hyperparasitoid performance via changes in plant quality. *Journal of Animal Ecology*, vol 74, no 6, pp 1121–1130.

Speek, T.A.A., Davies, J.A.R., Lotz, L.A.P., and van der Putten, W. H. (2013). Testing the Australian Weed Risk Assessment with different estimates for invasiveness. *Biological Invasions*, vol 15, no 6, pp 1319–1330.

Tscharntke, T., Klein, A. M., Kruess, A., Steffan-Dewenter, I., and Thies, C. (2005). Landscape perspectives on agricultural intensification and biodiversity – ecosystem service management. *Ecology Letters*, vol 8, no 8, pp 857–874.

Turbé, A., De Toni, A., Benito, P., Lavelle, P., Lavelle, P., Ruiz, N., van der Putten, W. H., Labouze, E., and Mudgal, S. (2010). *Soil Biodiversity: Functions, Threats and Tools for Policy Makers*. Bio Intelligence Service, IRD and NIOO, Report for European Commission (DG Environment).

Wall, D. H., Bardgett, R. D., Behan-Pelletier, V., Herrick, J. E., Jones, T. H., Ritz, K., Six, J., Strong, D. R., and van der Putten, W. H. (2012). *Soil Ecology and Ecosystem Services*. Oxford University Press, Oxford.

Wardle, D. A. (2002). *Communities and Ecosystems. Linking the Aboveground and Belowground Components*. Princeton University Press, Princeton NJ and Oxford.

Wardle, D. A., Bardgett, R. D., Klironomos, J. N., Setälä, H., van der Putten, W. H., and Wall, D. H. (2004). Ecological linkages between aboveground and belowground biota. *Science*, vol 304, no 5677, pp 1629–1633.

34

THE IMPORTANCE OF GRASSLANDS IN PROVIDING ECOSYSTEM SERVICES

Opportunities for poverty alleviation

Benis N. Egoh, Janne Bengtsson, Regina Lindborg, James M. Bullock, Adam P. Dixon and Mathieu Rouget

Introduction

The importance of ecosystems in providing ecosystem services and supporting human well-being has been largely recognised by the scientific community and policymakers, but grasslands have been mostly neglected by these communities. The focus has primarily been on ecosystems like forests due to their importance in providing ecosystem services such as carbon storage and the importance of carbon storage in mitigating threats from climate change (Canadell and Raupach, 2008). However, grasslands are one of the most widespread ecosystems in the world, covering about 26% of the terrestrial area (Boval and Dixon, 2012). Although their importance is not recognised as much as that of the forest, grasslands are important providers of ecosystem services to humans and society. Indeed, grasslands are also large carbon sinks and, according to Minahi et al. (1993), almost as important as forests in the recycling of greenhouse gases. Permanent grasslands store large amounts of carbon in the soil (Lal, 2004; Soussana et al., 2010; Bullock et al. 2011; Lemaire et al. 2011), much more than agricultural soils, and as much as forest soils (i.e. if trees are not included) (Farley et al. 2013). However, grasslands are better known around the world for their provision of ecosystem services related to grazing, such as fodder provision for the delivery of meat and dairy (Boval and Dixon, 2012). This service plays a key role not only in developing (e.g. most southern African countries and Mexico) countries, where poverty is prominent and small-scale farmers depend on the wider landscape for their livelihoods, but also in developed countries such as the USA and Australia (Franzluebbers and Steiner, 2016, and MacLeod and McIvor, 2016).

Several definitions exist for grasslands, some also including areas with woody vegetation (Allen et al., 2011). According to FAO (http://www.fao.org/agriculture), grassland ecosystems may be loosely defined as areas dominated by grasses (members of the family *Poaceae*, excluding bamboos) or grass-like plants with few woody plants, maintained by fire, grazing, and/or climate (dry or freezing conditions which restrict woody plant growth) (Vidrih et al., 2010). Savannahs, although not typically classified as grasslands, are the most extensive C_4 grassy biome, being a mixture of continuous grass cover with discontinuous tree cover, with their distinctiveness as a biome dependent on the intensity of woody vegetation (see Bond, 2008). The extent of woody vegetation cover is dependent on climatic factors (e.g. presence or absence of precipitation, frequency of fire) and management aspects (e.g. grazing). Grasslands can be classified into natural, semi-natural, and improved grasslands, the difference between the three being linked to management. While natural grasslands are not actively managed by humans but driven by ecological processes such as fires, drought, or grazing, semi-natural (grazed by livestock or mown) and improved grasslands (often ploughed and fertilised) are managed by humans and often referred to as anthropogenic grasslands (Lemaire et al., 2011). The term 'grassland' is also sometimes used synonymously with 'pastureland' when referring to an improved grazing-land ecosystem (Allen et al., 2011).

Most grasslands around the world are therefore either cultivated for crops or grazed by livestock for meat production, which in the latter case has often involved intensification through some combination of tillage, re-sowing, and nutrient addition (see Franzluebbers and Steiner [2016] for an American example and MacLeod and McIvor [2016] for an Australian example). According to Suttie et al. (2005), no grassland is entirely natural, i.e. never utilised by humans. In the European Union (EU) most grasslands are considered to be anthropogenic, having developed from former forested land by the removal of trees and shrubs through burning, felling, and controlled grazing, or from the draining of wetlands, and maintained by on-going management, which prevents succession (Vidrih et al., 2010; Bullock et al., 2011).

The value of natural and semi-natural grasslands to humans is different depending on where you are in the world. The majority of grasslands are located in tropical developing countries, where they are particularly important to the livelihoods of some one billion poor people (Boval and Dixon, 2012). In Europe, natural and semi-natural grasslands used to be important in the agricultural economy (e.g. Emanuelsson, 2009), but are nowadays part of the cultural landscape and not necessarily viewed as more important in securing livelihoods than other ecosystem types (Lindborg et al., 2008; Navarro and Pereira, 2012). While the primary management of natural and semi-natural grasslands in the developed world (e.g. Europe) is grazing or hay cutting, in developing countries grasslands are used not only for fodder production, but also for collecting wild foods, fencing and weaving materials such as grasses, and even fish from the

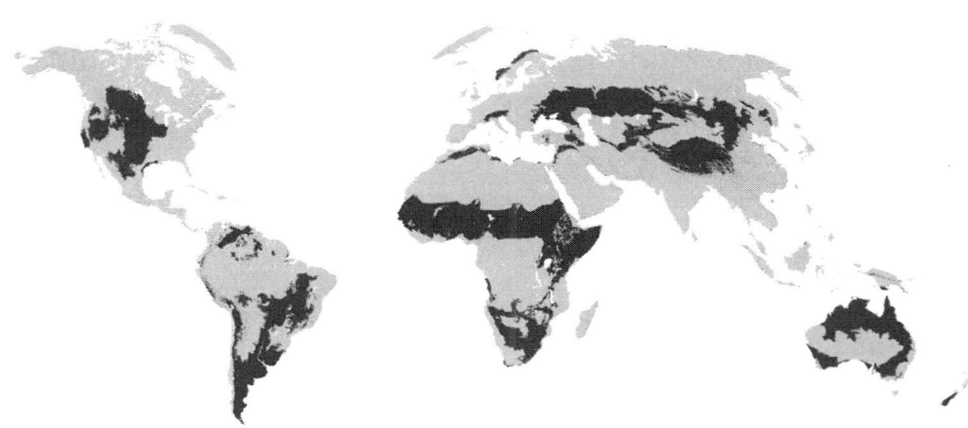

Figure 34.1 Map of global grassland distribution based on broad classification that includes some woodlands.

Source: adapted from Nixon et al., 2014

Note: map is not comprehensive and excludes some grassland types (e.g. European semi-natural grasslands) due to lack of fine scale data

temporal water bodies (e.g. ponds) found in grasslands (Morris, 2011). Although there has been concern in developed countries about the adverse effects on health of consumption of saturated animal fats, along with other environmental concerns, the demand for ruminant products associated with grasslands (dairy and meat) has remained more or less stable (Kearney, 2010). For example, there is some evidence that livestock systems in semi-natural grasslands in the UK produce meat products which have both better food value and lower amounts of saturated fats (Bullock et al., 2011). In developing countries, high rates of population growth, coupled with aspirations for a better diet, have increased the overall requirement for food (Wilkins, 2001; Kearney, 2010). These different dynamics result not only in different ways in which grasslands are appreciated and valued around the world, but also in different types of threats facing them.

Grasslands (semi-natural and natural grasslands) are facing significant threats, and the need to better manage them cannot be overemphasised. Due to their potential in accumulating litter and generating fertile soils, grasslands are constantly faced with threats ranging from conversion to crop production to intensive fodder provision. For example, at least 80% of North America's central grasslands have been converted to cropland, mostly for animal feed or food production for humans (Foley et al., 2005; Ceballos et al., 2010), while over 90% of the UK's semi-natural grasslands have been lost since the 1930s, mostly to cropland or improved grassland (Bullock et al., 2011). Other threats include overgrazing, urbanisation, afforestation, alien plant invasions, and climate change (Neke and Du Plessis, 2004). An important emerging threat that may further add to the decline of grasslands is biofuels. Fargione et al. (2008) outlined the potential emerging threats from biofuels on grasslands in the US. According to the authors, the corn ethanol boom was associated with a 4.9 million-ha increase in corn cropland used for ethanol between 2005 and 2008, resulting in the loss of more than 850,000 ha of set-aside grassland in the United States. This shows that biofuels could become one of the biggest threats to grasslands in the coming years, at least in the US. Conversion of grasslands to croplands (including biofuels) has significant implications for biodiversity and ecosystem services such as water provisioning.

Much cultivated land in Europe is being abandoned or taken out of agricultural production and converted into grasslands (Navarro and Pereira, 2012). For example, in the Northern Alps and Jura mountain of France, farmers specialised in milk production in the 1960s and converted ploughed fields into improved grasslands (Giraudoux et al., 1997). Such practices have led to an increase in grasslands in some parts of Europe, with an increase in grassland biodiversity. However, the projection is that semi-natural grasslands will continue to be under threat, not due to conversion (as much is now protected in Europe) but rather to lack of management and the subsequent encroachment of trees (Bullock et al., 2011).

Grasslands are well known for their biodiversity. Several biodiversity hotspots contain grasslands species which are endemic. Examples include the Maputaland-Pondoland-Albany of South Africa, Cerrado in Brazil, and locations in the horn of Africa (Phoenix et al., 2006). Grasslands have also been shown to be hotspots for small arthropods in Canada (Scudder, 2010). Wilson et al. (2012) showed that grasslands are a hotspot for global plant species, containing in some areas as many as 89 species in a square metre. While conservation of grasslands has been promoted mainly due to their high biodiversity, conservation objectives in the grassland have not focused on ecosystem services provision. This is partly because the full range of ecosystem services provided by grasslands is not understood due to lack of studies on ecosystems services in the grasslands (but see Bullock et al., 2011 for the UK). This chapter identifies and discusses the range of ecosystem services provided by grasslands based on an expert review of the current literature. It then looks at the potential of grasslands to contribute to poverty alleviation through a South African case study.

Ecosystem services provided by grasslands: an expert review

In 2014, a workshop with experts from Europe and South Africa explored the variety of ecosystem services provided by grasslands in general. First, a search of the ISI web of Knowledge (by February 2014) using the terms "grassland" AND "ecosystem services" was carried out, and 433 articles were found. The abstract of these articles were read and ecosystem services provided by grasslands were listed and later reviewed by experts based on their experience especially in Europe and Southern Africa. Second, specific search terms based on different ecosystem service were used together with the term "grassland" to examine other ecosystem services provided by grasslands that were not identified by the experts or listed in the first search. The experts then discussed these ecosystem services based on their importance linked to demand. The Millennium Ecosystem Assessment (MA) framework was used as a guide to establish which ecosystem services were delivered by the grassland ecosystems and to understand gaps in knowledge about ecosystem service provision for grasslands. The MA was chosen because it is used globally and easy to understand. Table 34.1 shows a list of ecosystem services provided by grasslands around the world and their relative importance as identified by expert judgement from the workshop. Franzluebbers and Steiner (2016) and MacLeod and McIvor (2016) provide additional examples of issues drawn from America and Australia.

Provisioning services

As discussed by experts during the workshop and ranked as very high in importance (Table 34.1), the most obvious ecosystem service provided by grasslands is fodder for animals, with the benefit of meat and dairy production to fulfil the growing demand for food associated with a growing human population around the world (Wilkins, 2001), and the production of other materials from animals, such as wool from sheep or leather from a variety of livestock (see Franzluebbers and Steiner (2016) and MacLeod and McIvor (2016)). However, there are many other services provided by grasslands that are not as well recognised. For example, in many parts of the world, like South Africa, grasslands are sources of medicinal plants used by rural communities to cure various ailments (Dzerefos and Witkowski, 2001). These communities not only depend on grasslands for medicinal plants, but part of their livelihood is also secured through the sale of grass for fencing or brooms. According to Dold and Cocks (2002), medicinal plants generate about $2.5 million annually in the eastern province of South Africa. In an arid country like South Africa, grasslands are particularly important for water provisioning services, and large areas of the grassland have been set aside as protected areas and are conserved under the water act. Grasslands provide almost all of the provisioning ecosystems services listed in the MA (2005) framework, except for timber and fuel wood production. This is understandable, as grasslands predominantly consist of grasses and hardly contain large trees that can be used for timber production. Grasslands containing woody vegetation, however, can be used as a source of fuel wood.

Regulating services

Grasslands are important for water regulating services. Plants influence the hydrological regulation from plant to plot scale by altering the rate of water infiltration into soils, the storage capacity of surface soils, and the surface runoff, through their varying water use efficiencies and surface roughness (Macleod and Ferrier, 2011, pp. 229–238). Grasslands are also important for water supply where they act as sponges, holding water and slowly releasing it during the dry season, making it available when it is most needed (Macleod and Ferrier, 2011). It is important to note that water regulation also involves the contribution of grasslands to the provision of clean and timely water. In Europe, a positive relationship between riparian grasslands and water quality has been established, indicating the role of grasslands in water purification services (see

Table 34.1 Ecosystem services provided by grasslands from literature review and importance based on expert knowledge.

Service	*Sub-category*	*Explanation*	*Source/example*	*Importance in South Africa and/or Europe*
Provisioning services				
Food	Livestock	Grasslands are mostly known for grazing of animals on natural grasslands or production of fodder through cultivation of grasslands.	Lemaire et al., 2011	Very high
	Captured fish	Fishing occurs mostly in seasonally inundated grasslands.	McGrath et al., 2007	Low
	Aquaculture	Areas within grasslands in proximity to rivers get flooded with water during rainy season generating ponds that are used for aquaculture	Hossain et al., 2007, 2009	Low
	Wild plants and animals	Wild game is hunted from grasslands as bush-meat (e.g. birds).	Magige et al., 2008	Medium
Fibre	Wool and leather, building material	Grasslands are used for wool and leather production. Grasslands also provide building material (e.g. roofing).	Potter et al., 2000; Palmer and Ainslie, 2006	High
Genetic resources		Grasslands have a unique contribution to the maintenance of global genetic library. Humans depend on a limited number of grassland species for nutrition, medicine, fibre, and shelter.	Sala and Paruelo, 1997	High
Biochemical, natural medicines, pharmaceuticals		Wild plants and animal are harvested from grasslands for food and medicinal purposes.	Dzerefos and Witkowski, 2001; Morris, 2011	Medium
Ornamental resources		Many grasses and grassland forbs are used for ornamental purpose and grasslands serve as a source for these grasses.	Gao et al., 2008	Medium

(Continued)

Table 34.1 (Continued)

Service	*Sub-category*	*Explanation*	*Source/example*	*Importance in South Africa and/or Europe*
Fresh water		Grasslands facilitate water infiltration hence enhancing base flow. They serve as sponges that soak up water during the rainy season and gradually release it during the dry season.	Macleod and Ferrier, 2011	High
Regulating services				
Air quality regulation		Grasslands provide air purification services.	Bullock et al., 2011	Low
Climate regulation	Carbon sequestration and storage	New grasslands have been shown to sequester as much as 0.6 Mg C $ha^{-1}year^{-1}$. Root turnover creates the largest organic carbon input to grassland soils and favours soil carbon storage.	Schuman et al., 2002; Soussana et al., 2004	Medium
	Other greenhouse gases	Grasslands are important in GHG balances, both as sinks for greenhouse gasses (e.g. CO2) but also sources of N2O and CH4 from cattle grazing.	Soussana et al., 2004	Low
Water regulation		Grasslands facilitate water infiltration, hence reducing runoff. In some places (e.g. grasslands of South Africa), they serve as sponges that soak up water during the rainy season and gradually release it during the dry season.	Bhark and Small, 2003; Turpie et al., 2008	Very high
Erosion regulation		Grasslands can decrease soil erosion in agricultural landscapes.	Brazier et al., 2007; Macleod and Ferrier, 2011	Medium
Water purification and waste treatment		European scientists have found positive relationships between riparian grasslands and water quality.	Lemaire et al., 2011	High

Disease regulation	Grassland species can influence disease spread both positively and negatively through predator-prey interactions where these species are intermediary hosts for certain parasites which cause human or crop diseases.	Ostfeld et al., 2008; Thies et al., 2005	Medium
Pest regulation	Beetles found in natural grasslands have been shown to provide the service of pest control in nearby croplands, where they prey on pests such as aphids.	Gardiner et al., 2009	Medium
Pollination	Grasslands serve as population sources for pollinating insects in agricultural landscapes.	Öckinger and Smith, 2007	Medium
Natural hazard regulation	Coastal grasslands are important in buffering storms and other extreme natural events.	Everard et al., 2010	Medium
Cultural services			
Spiritual and religious values	Grasslands are used as sacred sites.	Gokhale et al., 1997	Medium
Knowledge systems	Scientists use grasslands for experimental purposes (e.g. understanding species distribution or life history traits).	Lindborg, 2007	High
Educational values	Grassland plants are used for educational purposes in schools.	Bebbington, 2005	Low
Inspiration	Grasslands have been used as an inspiration for art (e.g. grassland portraits).	Essig, 2008	Low
Aesthetic values	Humans appreciate the beauty of grasslands, especially when they are species-rich.	Lindemann-Matthies et al., 2010	High

(Continued)

Service	*Sub-category*	*Explanation*	*Source/example*	*Importance in South Africa and/or Europe*
Social relations		Grasslands can form the basis of social relations (e.g. grazing associations).	Cleaver, 2002	Medium
Sense of place		Sense of place has been identified as an ecosystem service provided by grasslands to humans in Europe.	Lamarque et al., 2011	Low
Cultural heritage values		Many grasslands have great cultural value (e.g. in Mongolia and Europe).	Buckley et al., 2008	Very High
Recreation and ecotourism		Grasslands are important for recreational activities such as hiking and bird watching, which, if not well-managed, can have negative impact on biodiversity.	Fernandez-Juricic et al., 2005	High
Supporting Services (underpinning ecological processes)				
Soil formation		Grassland soils have high organic matter content due to the high surface cover of the grass sward important for soil formation.	Brazier et al., 2007	High
Photosynthesis		Grassland plants produce plant biomass and oxygen necessary for other organisms.	Igamberdiev and Lea, 2006	High
Primary production		Photosynthesis (above). Ecological processes such as grazing may also promote aboveground productivity in grasslands.	Frank and McNaughton, 1993	High
Nutrient cycling		Grasslands are important in cycling nutrients such C, N, P, and K.	Dubeux et al., 2007	Very High

Table 34.1). Pollination services are important in food production, and grasslands can be important in providing pollination services both to agricultural and natural areas. Grasslands also play a role in pest regulation by enhancing predator (natural enemy) populations that prey on pests (e.g., Gardiner et al., 2009). Grasslands are important for preventing soil erosion (Fullen, 1998) and in regulating greenhouse gases. For example, grassland systems have higher resident times for soil organic matter than cultivated systems and even some forest systems, thereby regulating carbon dioxide release from the system (Six and Jastrow, 2002; Farley et al., 2013). Essentially, grasslands provide all regulating services in the MA framework, as listed in Table 34.1.

Cultural services

Grasslands provide all cultural services listed in the MA framework. An important benefit from the grassland is recreation. In Italy, 76% of visitors in two parks (Monte Sole and Lakes of Suviana) containing grasslands gave a high rating (from 8 to 10 on a scale of 1 to 10) to these parks as places for recreational activities, and about 56% of the visitors were willing to pay for a grassland conservation programme (Marzetti et al., 2011). Grasslands can be protected as nature reserves or national parks and are often advertised as hotspots for local and national tourism (Fischer et al., 2008) (see also MacLeod and McIvor (2016)). The cultural landscape can also be a tourist attraction in itself, e.g. the Swiss Alps and the Mongolian grassland landscape, potentially playing a significant role in the World Heritage context (Buckley et al., 2008).

The social dynamics of herders are also important ecosystem services provided by grasslands through herders associations, although this role is declining in many developed countries (Poschlod and WallisDeVries, 2002). In Southern Africa (at least in Caprivi in Namibia), grass-cutting is controlled and, during cutting season, all villagers are given the opportunity to cut the grass together over a short period with the opportunity to improve social relations as villagers share equipment (e.g. machetes or even transport) (Egoh, 2002). Such activities improve social relations amongst rural communities. Historically, grassland management practices, such as hay cutting, were socially important in many areas of Europe (Emanuelsson, 2009), a phenomenon that can still be found today in some areas. In addition, products from grasslands (e.g. grass brooms made from grass or twigs) have immense cultural value in some areas. In South Africa, for example, urban residents still prefer using traditional grass brooms over industrially manufactured brooms because of the cultural significance they attached to the use of these brooms, e.g. as wedding presents, and for their ability to offer households protection from lightning attributed to sorcery (Cocks and Dold, 2004). Livestock keeping *per se* plays an important role in SA's cultural-economic history and is still an important part of present-day ritual/supernatural dimensions of livestock ownership (Ainslie, 2013).

Underlying ecosystem structures and processes, or supporting services

Grasslands provide all supporting services listed in the MA (Table 34.1; also see Franzluebbers and Steiner (2016) and MacLeod and McIvor (2016)). They have high productivity turnover, which translates to many of the supporting ecosystem services. Litter accumulation in grasslands not only contributes to accumulation of soil organic matter and other nutrients but also to the formation of fertile soils (Six and Jastrow, 2002). Grassland soils have high organic matter content due to the high surface cover of the grass sward and large root production. The high productivity in grasslands as a result of photosynthesis also produces oxygen used by other organisms. In a nutshell, grasslands contribute to nutrient cycling and distribution in the landscape, which results in high productivity, which is one of the reasons for grasslands being preferred habitats for cultivation. Hence grasslands with high productivity potential as a result of rich soils are often the ones being transformed to cultivated land (Stoate et al., 2009) in many areas of the world, like the North American prairies, the steppes in eastern Europe (e.g. Ukraine), or South Africa.

Disservices

Despite the range of ecosystem services provided by grasslands, some disservices or negative impacts on humans have been recorded. One example is grassland vegetation in proximity to urban areas, which may cause substantial damage to property and loss of human lives due to the frequent fires that are part of the ecological processes in the ecosystem in certain parts of the world. Grassland fires generally have low intensity, but the frequency makes the fire dangerous when in proximity to urban areas. On average, about 85 homes have been lost to bush fires from forest or grasslands in Australia per year, and this is predicted to increase with climate change (McAneney, 2005; Pitman et al., 2007). Grasslands can also harbour some crop pests and diseases (D'Arcy, 1995). The conversion of forested land into grasslands and shrublands has been shown to provide the optimal habitats for *M. limnophilus* and *C. longicaudatus*, which are intermediate hosts of *E. multilocularis*, responsible for the human alveolar echinococcocosis (AE), a highly pathogenic zoonotic disease caused by the larval stage of the cestode (Giraudoux et al., 2003). Despite the negative effects of grasslands to humans, the positive contributions of this ecosystem appear to outnumber the negative ones.

Case study: the potential for creating markets for ecosystem services in the grasslands of South Africa and implications for poverty alleviation

After nearly 20 years of democracy with the abolition of apartheid in 1994, close to half the population of South Africa could reasonably be said to be living 'in poverty', and income inequality is now wider than before (Marais, 2011). A new government has just been elected and job creation is top on their agenda, as it is for most countries around the world, especially developing countries. Several initiatives are being put forward for job creation and poverty alleviation in South Africa. One of these initiatives is the creation of a 'green economy' to shape development strategies and economic growth in South Africa and several other developing countries (Musango et al., 2014). In this regards, more and more governments see payments for ecosystem services (PES) as one of the mechanisms for greening the economy and alleviating poverty.

An example of PES is Reducing Emissions from Deforestation and Forest Degradation (REDD+), which is being implemented across the globe, but is only valid in countries with large forested areas. In South Africa, an initiative that has gained momentum is the Working for Water program (WfW), whereby invasive trees are being removed to improve both biodiversity and water supply while creating jobs. While the WfW programme is one of the most successful, several other programs have been initiated and include: Working for Land (WfL), Working on Fire (WoF), and Working for Wetlands (WfW) (see www.environment.gov.za/projectsprogrammes). These initiatives have mostly been put in place to improve the environment. For example, the key objective of the Working for Land program is to ensure that degraded ecosystems are restored to their former or original state so that they are able to maintain or support ecosystem diversity and functioning and deliver ecosystem services such as soil erosion control. The South African government is looking to expand job creation through these types of initiatives, with the goal of achieving biodiversity conservation, ecosystem service provision (e.g. soil erosion control, water regulation), and job creation simultaneously. South African grasslands present an opportunity to meet these goals and these objectives are embedded in the goals of the grassland program of South Africa (www.grasslands.org.za).

The grassland program in South Africa was established because grasslands are seen as the food basket of the country, as most cultivated areas are surrounded by grasslands (approximately 24%

of transformed areas covered with agriculture; Neke and Du Plessis, 2004). Other economic activities within the grasslands include forestry and mining. Unsustainable economic activities in the biome are seen as threats to biodiversity and the ecosystem services it provides to humans, and therefore a grassland program has been established to encourage sustainable use of grassland resources. The 5-year program was funded by the Global Environmental Facility (GEF) through the United Nations Development Programme (UNDP) to complement and facilitate synergies between existing and proposed grassland biodiversity conservation initiatives, which are also relevant to ecosystem services (see www.grasslands.org.za). Since grasslands are highly productive ecosystems, there are opportunities to establish several PES programs to alleviate poverty (e.g. compensation of farmers for reducing livestock stocking rates or other improved land management or payments for restoration of grasslands by the mining sector).

The need for sustainable use of grassland resources and the potential role of PES for poverty alleviation has been recognised recently in some countries, such as China, where locals could earn up to a third of their household income by reducing herd size (Wilkes and Mandula, 2010). Another similar initiative to promote sustainable land management in the US, called the Conservation Effects Assessment project, is explained in Franzluebbers and Steiner (2016) Box 34.1. In South Africa, most of the regulating ecosystem services mapped by Egoh et al. (2008) are concentrated in the grasslands biome (Figure 34.2) and a map of poverty in the country also shows that the poorest communities are found mostly in the grasslands (Figure 34.3). This scenario presents an opportunity for PES initiatives within the grassland biome, since regulating services are mostly delivered by intact, well-managed ecosystems. Blignaut et al. (2008) identified areas with potential for poverty alleviation in South Africa by overlaying maps of ecosystem service from Egoh et al. (2008) with poverty maps to highlight areas where restoration would improve ecosystem services while creating jobs. More than 60% of the priority areas identified are located in the grassland biomes, highlighting the potential of restoration interventions in this biome to alleviate poverty in the country. According to Blignaut and his colleagues, natural grazing for livestock in the grasslands biome in South Africa had the potential to have generated R8,172/km^2 ($770) in 2008. They argue that the creation of incentives for sound land management practices, such as PES schemes and the creation of markets for ecosystem goods and services, is required if the agricultural sector is expected to remain a major employer and generator of income.

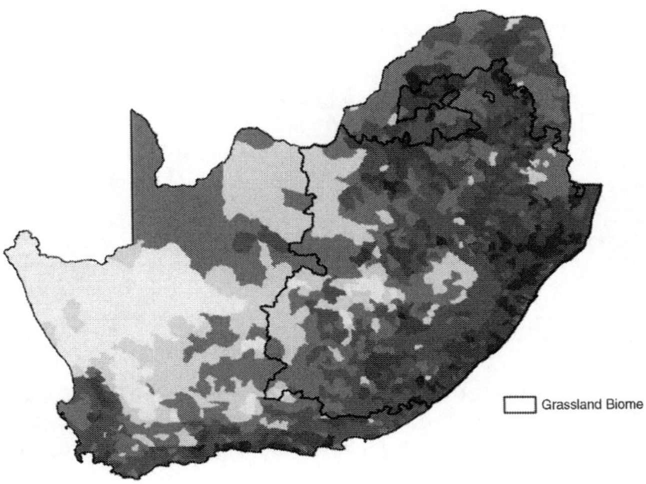

Figure 34.2 Map of South Africa showing distribution of four ecosystem services: water supply, water regulation, carbon storage, and soil retention.

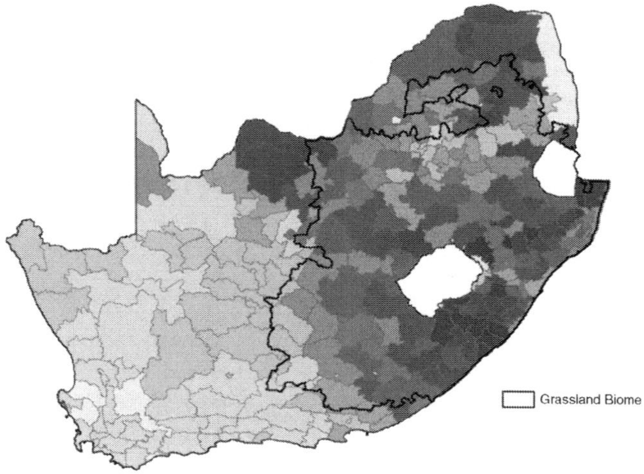

Figure 34.3 Map of South Africa showing distribution of poverty. The darker areas are the high areas. The grassland biome is concentrated in the middle belt, indicated by a black line.

Source: Blignaut et al., 2008; Egoh et al., 2008

Conclusions

Grasslands are recognised as important ecosystem for food production around the world. The principal ecosystem service they are known for is fodder provision, but they provide a large range of other important services. These ecosystem services need to be highlighted and taken account of in assessing the value of the world's grasslands. Many grassland types are threatened by cropping, intensification, afforestation, and urbanisation, but an emerging threat is the expansion of biofuel production. In the past decade, conservation of grasslands and other biomes was mainly based on their high biodiversity, at least in South Africa. In addition, the range of ecosystem services provided by grasslands is a valid reason for sustainable use and conservation of the grassland biome. The South African case study also shows that this type of approach could be useful in poverty alleviation not only in South Africa but also around the world, improving livelihoods while curbing the threats facing grasslands.

Acknowledgements

We thank all the experts who attended the workshop on grassland ecosystem services (O'Connor, T, Everson, C, Everson, T and O'Farrell, P) for their contribution during discussions on ecosystem services provided by grasslands. STIAS is thanked for funding the workshop on grassland ecosystem services. This research was supported by the South African Research Chair initiative of the Department of Science and Technology and National Research Foundation of South Africa.

References

Ainslie, A. (2013). The sociocultural contexts and meanings associated with livestock keeping in rural South Africa. *African Journal of Range and Forage Science*, vol 30, pp 35–38.

Allen, V.G., Batello, C., Berretta, E.J., et al. (11 authors) (2011). An international terminology for grazing lands and grazing animals. *Grassland and Forage Science*, vol 66, pp 2–28.

Bebbington, A. (2005). The ability of A-level students to name plants. *Journal of Biological Education*, vol 39, pp 62–67.

Bhark, E.W., and Small, E.E. (2003). Association between plant canopies and the spatial patterns of infiltration in shrubland and grassland of the Chihuahuan desert, New Mexico. *Ecosystems*, vol 6, pp 185–196.
Blignaut, J., Marais, C., Rouget, M., et al. (7 authors) (2008). *Making Markets work for People and the Environment: Employment Creation from Payment for Eco-Systems Services*. Second Economy Strategy project report. Trade and Industrial Policy Strategy. South Africa.
Bond, W.J. (2008). What limits trees in C4 grasslands and savannas? *Annual Review of Ecology, Evolution and Systematics*, vol 39, pp 641–659.
Boval, M., and Dixon, R.M. (2012). The importance of grasslands for animal production and other functions: a review on management and methodological progress in the tropics. *Animal*, vol 6, pp 748–762.
Brazier, R.E., Bilotta, G.S., and Haygarth, P. M. (2007). A perspective on the role of lowland, agricultural grasslands in contributing to erosion and water quality problems in the UK. *Earth Surface Processes and Landforms*, vol 32, pp 964–967.
Buckley, R., Ollenburg, C., and Zhong, L. (2008). Cultural landscape in Mongolian tourism. *Annals of Tourism Research*, vol 35, no 1, pp 47–61.
Bullock, J.M., Jefferson, R.G., Blackstock, T.H., et al. (8 authors) (2011). Semi-natural grasslands. In: *The UK National Ecosystem Assessment Technical Report*. UNEP-WCMC, Cambridge UK.
Canadell, J.G., and Raupach, M.R. (2008). Managing forests for climate change mitigation. *Science*, vol 320, pp 1456–1457.
Ceballos, G., Davidson, A., List, R., Pacheco, J., Manzano-Fischer, P., et al. (7 authors) (2010). Rapid decline of a grassland system and its ecological and conservation implications. *PLoS ONE*, vol 5, pp 1–12.
Cleaver, F. (2002). Reinventing institutions: bricolage and the social embeddedness of natural resource management. *European Journal of Development Research*, vol 14, no 2, pp 11–30.
Cocks, M.L., and Dold, A.P. (2004). A new broom sweeps clean. The economic and cultural value of grass brooms in the Eastern Cape Province, South Africa. *Forests, Trees and Livelihoods*, vol 14, no 1, pp 33–42.
D'Arcy, C.J. (1995). Symptomatology and host range of barley yellow dwarf. In: D'Arcy, C. J. and Burnett, P. A. (eds) *Barley Yellow Dwarf: 40 Years of Progress*. Am. Phytopathol. Soc. Press, St. Paul MN.
Dixon, A. P., Faber-Langendoen, D., Josse, C., Morrison, J., & Loucks, C. J. (2014). Distribution mapping of world grassland types. *Journal of Biogeography*, vol 4, no 11, 2003–2019.
Dold, A.P., and Cocks, M.L. (2002). The trade in medicinal plants in Eastern Cape Province, South Africa. *South African Journal of Science*, vol 98, pp 589–597.
Dubeux, J.C.B., Sollenberger, L.E., Mathews, B.W., Scholberg, J.M., and Santos, H.Q. (2007). Nutrient cycling in warm-climate grasslands. *Crop Science*, vol 47, pp 915–928.
Dzerefos, C.M., and Witkowski, E.T.F. (2001). Density and potential utilization of medicinal grassland plants from abe bailey nature reserve, South Africa. *Biodiversity Conservation*, vol 10, pp 1875–1896.
Egoh, B., Reyers, B., Rouget, M., et al. (6 authors) (2008). Mapping ecosystem services for planning and management. *Agriculture Ecosystems & Environment*, vol 127, pp 135–140.
Egoh, B.N. (2002). *Spatial Pattern of Natural Resource Use by Local Community in Eastern Caprivi, Namibia*. M.Sc. Thesis, University of Cape Town.
Emanuelsson, U. (2009). *The Rural Landscapes of Europe: How Man Has Shaped European nature*. Formas, Stockholm.
Essig, M. (2008). Portraits of grasses: a story of art and restoration from the Apache Highlands ecoregion. *Ecological Restoration*, vol 26, pp 206–212.
Everard, M., Jones, L., and Watts, B. (2010). Have we neglected the societal importance of sand dunes? An ecosystem services perspective. *Aquatic Conservation: Marine and Freshwater Ecosystems*, vol 20, pp 476–487.
Fargione, J., Hill, J., Tilman, D., Polasky, S., and Hawthorne, P. (2008). Land clearing and the biofuel carbon debt. *Science*, vol 319, pp 1235–1238.
Farley, K.A., Bremer, L.L., Harden, C.P., and Hartsig, J. (2013). Changes in carbon storage under alternative land uses in biodiverse Andean grasslands: implications for payment for ecosystem services. *Conservation Letters*, vol 6, pp 21–25.
Fernández-Juricic, E., Venier, P., Renison, D., and Blumstein, D.T. (2005). Sensitivity of wildlife to spatial patterns of recreationist behavior: a critical assessment of minimum approaching distances and buffer areas for grassland birds. *Biological Conservation*, vol 125, pp 225–235.
Fischer, M., Rudmann-Maurer, K., Weyand, A., and Stöcklin, J. (2008). Agricultural land use and biodiversity in the Alps – How cultural tradition and socioeconomically motivated changes are shaping grassland biodiversity in the Swiss Alps. *Mountain Research and Development*, vol 28, pp 148–155.
Foley, J.A., DeFries, R., Asner, G.P., et al. (19 authors) (2005). Global consequences of land use. *Science*, vol 309, pp 570–574.

Frank, D.A., and McNaughton, S.J. (1993). Evidence for the promotion of aboveground grassland production by native herbivores in Yellowstone National Park. *Oecologia*, vol 96, pp. 157–161.

Fullen, M. A. (1998). Effects of grass ley set-aside on runoff, erosion and organic matter levels in sandy soils in east Shropshire, UK. *Soil and Tillage Research*, vol 46, pp 41–49.

Gao, H., Liu, J., and Guo, A. (2008). Preliminary evaluation of adaptability and utilization value of ornamental grasses in Nanjing. *Prataculturai Science*, vol 8, 033 (online).

Gardiner, M.M., Landis, D.A., Gratton, C., et al. (10 authors) (2009). Landscape diversity enhances biological control of an introduced crop pest in the north-central USA. *Ecological Applications*, vol 19, pp 143–154.

Giraudoux, P., Craig, P.S., Delattre, P. et al. (10 authors) (2003). Interactions between landscape changes and host communities can regulate *Echinococcus multilocularis* transmission. *Parasitol*, vol 127, pp S121–31.

Giraudoux, P., Delattre, P., Habert, M., et al. (10 authors) (1997). Population dynamics of fossorial water vole (*Arvicola terrestris scherman*): a land usage and landscape perpective. *Agriculture Ecosystems and Environment*, vol 66, pp 47–60.

Gokhale, Y., Velankar, R., Subhash Chandran, M.D., and Gadgil, M. (1997). Sacred woods, grasslands and water bodies as self-organized systems of conservation in India. Paper presented at the Regional Workshop on the Role of Sacred Groves in Conservation and Management of Biological Diversity, Kerala Forest Research Institute, Peechi.

Hossain, M.S., Chowdhury, S.R., Das, N.G., and Rahman, M. M. (2007). Multi-criteria evaluation approach to GIS-based land-suitability classification for tilapia farming in Bangladesh. *Aquaculture International*, vol 15, pp 425–443.

Hossain, M.S., Chowdhury, S.R., Das, N.G., Sharifuzzaman, S.M., and Sultana, A. (2009). Integration of GIS and multicriteria decision analysis for urban aquaculture development in Bangladesh. *Landscape Urban Plan*, vol 90, pp 119–133.

Igamberdiev, A.U., and Lea, P.J. (2006). Land plants equilibrate O_2 and CO_2 concentrations in the atmosphere. *Photosynthesis Research*, vol 878, no 1, pp 177–194.

Jonsson, M., Bommarco, R., Ekbom, B., et al. (8 authors) (2014). Ecological production functions for biological control services in agricultural landscapes. *Methods in Ecology and Evolution*, vol 5, pp 243–252.

Kearney, J. (2010). Food consumption trends and drivers. *Philosophical Transactions of the Royal Society B: Biological Sciences*, vol 365, pp 2793–2807.

Lal, R. (2004). Soil carbon sequestration to mitigate climate change. *Geoderma*, vol 123, pp 1–22.

Lamarque, P., Tappeiner, U., Turner, C., et al. (8 authors) (2011). Stakeholder perceptions of grassland ecosystem services in relation to knowledge on soil fertility and biodiversity. *Regional Environmental Change*, vol 11, pp 791–804.

Lemaire, G., Hodgson J., and Chabbi. A. (eds) (2011). Grassland Productivity and Ecosystem Services. CABI, Wallingford.

Lindborg, R. (2007). Evaluating the distribution of plant life-history traits in relation to current and historical landscape configurations. *Journal of Ecology*, vol 95, pp 555–564.

Lindborg, R., Bengtsson, J., Berg, A., et al. (11 authors) (2008). A landscape perspective on conservation of semi-natural grasslands. *Agriculture Ecosystems & Environment*, vol 125, no 1, pp 213–22.

Lindemann-Matthies, P., Junge, X., and Matthies, D. (2010). The influence of plant diversity on people's perception and aesthetic appreciation of grassland vegetation. *Biological Conservation*, vol 143, pp 195–202.

MA (2005). *Ecosystems and Human Well-Being: Synthesis*. Island Press, Washington DC.

Macleod, C.J.A., and Ferrier, R.C. (2011). Temperate grasslands in catchment systems: the role of scale, connectivity and thresholds in the provision and regulation of water quality and quantity. In: Lemaire, G., Hodgson J. and Chabbi. A. (eds) *Grassland Productivity and Ecosystem Services*. CABI, Wallingford.

MacLeod, N. D., and McIvor, J. G. (2006). Reconciling economic and ecological conflicts for sustained management of grazing lands. *Ecological Economics*, vol 56, no 3, pp 386–401.

Magige, F.J., Holmern, T., Stokke, S., Mlingwa, C., and Roskaft, E. (2008). Does illegal hunting affect density and behaviour of African grassland birds? A case study on ostrich (*Struthio camelus*). *Biodiversity and Conservation*, vol 18, pp 1361–1373.

Marais, H. (2011). *South Africa Pushed to the Limit. The Political Economy of Change*. UCT Press, Cape Town.

Marzetti, S., Disegna, M., Villani, G., and Speranza, M. (2011). Conservation and recreational values from semi-natural grasslands for visitors to two Italian parks. *Journal of Environmental Planning and Management*, vol 54, no 2, pp 169–191.

McAneney, J. (2005) Australian bushfire: quantifying and pricing the risk to residential properties. In: Morrison, R. J., Quin, S. and Bryant, E. A. (eds) *Proceedings of a Symposium of the Planning for Natural Hazards – How Can We Mitigate the Impacts?* University of Wollongong, Wollongong.

McGrath, D.G., Almeida, O.T., and Merry, F.D. (2007). The influence of community management agreements on household economic strategies: cattle grazing and fishing agreements on the lower Amazon floodplain. *International Journal of the Commons*, vol 1, pp 67–87.

Minahi, K., Goudriaan, J., Lantinga, E.A., and Kimura, T. (1993). Significance of grasslands in emission and absorption of greenhouse grasses. In: Barker, M.J. (ed.) *Grasslands for Our World*. SIR Publishing, Wellington.

Morris, C.D. (2011). Rangeland management for sustainable conservation of natural resources. In: Lemaire, G., Hodgson, J. and Chabbi, A. (eds) *Grassland Productivity and Ecosystem Services*. CABI Publishing, Wallingford.

Musango, J.K., Brent, A.C., and Bassi, A. M. (2014). Modelling the transition towards a green economy in South Africa. *Technol. Forecast. Soc. Chang.*, vol 87, pp 257–273.

Navarro, L., and Pereira, H. (2012). Rewilding abandoned landscapes in Europe. *Ecosystems*, vol 15, pp 900–912.

Neke, K.S., and du Plessis, M.A. (2004). The threat of transformation: quantifying the vulnerability of grasslands in South Africa. *Conservation Biology*, vol 18, pp 466–477.

Öckinger, E., and Smith, H.G. (2007). Semi-natural grasslands as population sources for pollinating insects in agricultural landscapes. *Journal of Applied Ecology*, vol 44, pp 50–59.

Ostfeld, R.S., Keesing, F., and Eviner, V. (2008). *Infectious Disease Ecology: Effects of Ecosystems on Disease and of Disease on Ecosystems*. Princeton University Press, Princeton NJ.

Palmer, A.R., and Ainslie, A. (2006). Arid rangeland production systems of Southern Africa. *Science et changements planétaires*, vol 17, pp 98–104.

Phoenix, G.K., Hicks, W.K., Cinderby, S., et al. (12 authors) (2006). Atmospheric nitrogen deposition in world biodiversity hotspots: the need for a greater global perspective in assessing N deposition impacts. *Global Change Biology*, vol 12, pp 470–476.

Pitman, A.J., Narisma, G.T., and McAneney, J. (2007). The impact of climate change on the risk of forest and grassland fires in Australia. *Climatic Change*, vol 84, pp 383–401.

Poschlod, P., and WallisDeVries, M.F. (2002). The historical and socioeconomic perspective of calcareous grasslands – lessons from the distant and recent past. *Biological Conservation*, vol 104, pp 361–376.

Potter, L., Lee, J., and Thorburn, K. (2000). Reinventing Imperata: revaluing Alang-Alang grasslands in Indonesia. *Development and Change*, vol 31, pp 1037–1053.

Sala, O.E., and Paruelo, J.M. (1997). Ecosystem services in grasslands. In Daily, G. C. (ed.) *Nature's Services: Societal Dependence on Natural Ecosystems*, Island Press, Washington, DC.

Schuman, G. E., Janzen, H. H., and Herrick, J. E. (2002). Soil carbon dynamics and potential carbon sequestration by rangelands. *Environmental Pollution*, vol 116, pp 391–396.

Scudder, G.G.E. (2010). Grasslands: biodiversity hotspots for some arthropods in British Columbia. In: Shorthouse, J. D. and Floate, K. D. (eds) *Arthropods of Canadian Grasslands, Vol. 1: Ecology and Interactions in Grassland Habitats*. Biological Survey of Canada, Ottawa.

Six, J., and Jastrow, J. (2002). *Organic Matter Turnover*. Marcel Dekker, New York.

Soussana, J.F., Loiseau, P., Vuichard, N., et al. (7 authors) (2004). Carbon cycling and sequestration opportunities in temperate grasslands. *Soil Use and Management*, vol 20, pp 219–230.

Soussana, J.F., Taller, T., and Blanfort, V. (2010). Mitigating the greenhouse gas balance of ruminant production systems through carbon sequestration in grasslands. *Animal*, vol 4, no 3, pp 334–350.

Stoate, C., Báldi, A., Beja, P., et al. (7 authors) (2009). Ecological impacts of early 21st century agricultural change in Europe–a review. *Journal of Environmental Management*, vol 91, pp 22–46.

Suttie, J.M., Reynolds, S. G, and Batello, C. (2005). *Grasslands of the World*. Plant Production and Protection Series (FAO), no 34. Food and Agriculture Organization of the United Nations, Rome.

Thies, C., Roschewitz, I., and Tscharntke, T. (2005). The landscape context of cereal aphid-parasitoid interactions. *Proc. R. Soc. B.*, vol 272, pp 203–210.

Turpie, J.K., Marais, C., and Blignaut, J.N. (2008). Evolution of a payments for ecosystem services mechanism that addresses both poverty and ecosystem service delivery in South Africa. *Ecological Economics*, vol 65, pp 788–798.

Vidrih, M., Čop, J., Trdan, S., Vidrih, A., and Eler, K. (2010). Grassland use in European Union and Slovenia. *Biotechnology in Animal Husbandry*, vol 26, pp 191–203.

Wilkes, A., and Mandula, J.T. (2010). The myth of community and sustainable grassland management in China. *Frontiers of Earth Science in China*, vol 4, pp 59–66.

Wilkins, J.R. (2001). Facing the growing needs of mankind – grassland and rangelands In: Nösberger, J., Geiger, H. H. and Struik, P. C. (eds) *Crop Science: Progress and Prospects*. CABI, Wallingford.

Wilson, J.B., Peet, R.K., Dengler, J., and Pärtel, M. (2012). Plant species richness: the world records. *Journal of Vegetation Science*, vol 23, pp 796–802.

Briefing Note 34.1 Ecosystem services and grasslands in America

Alan J. Franzluebbers and Jean L. Steiner

Historically, grasslands occupied a large portion of Canada and the US, throughout the Great Plains region east of the Rocky Mountains. Today, significant expanses of grassland and chaparral are found in the more arid western and intermountain regions of the Rocky Mountains. Other significant grasslands occur in the eastern part of the US as a result of land conversion from forest and cropland to introduced forages on private farms, as well as the natural rangelands of the Florida peninsula and the southern and eastern coastal regions (USDA-NRCS, 2014).

Plate 34.1 Tallgrass Prairie National Preserve in Kansas.

Source: National Park Service

Grasslands in the US are valued for a wide range of ecosystem services, and provide a primary source of forage for grazing livestock (Steiner and Franzluebbers, 2009). Importantly, they also serve to control hydrologic functioning, balance the dynamics of CO_2 to and from the atmosphere, and provide open ground for a wide diversity of wildlife and plant communities (Glaser, 2014).

Native prairies are a historically sustainable and resilient land cover, rooted by vigorous grasses and forbs that provide vital ecosystem services such as water cycling, nutrient cycling, gas exchange with the atmosphere, climate regulation, food and feed production, and aesthetic experience – essentially the plethora of services. Even shelter was made available to early European settlers of the Great Plains from sod cut from the ground. The American experience certainly suggests that grasslands must be considered a vital ecosystem component, keeping soil from eroding and allowing the underlying earth to function to its fullest capacity across individual fields as well as expansive ecoregions (Franzluebbers, 2013). In a radio address on "The Strength and Quietness of Grass" to the American public in 1940, USDA Secretary Henry A. Wallace stressed the importance of grass-based agriculture in the context of the challenges facing the future of the country: "The strength and quietness of grass should be, must be, permanently a part of our agriculture if it is to have the strength it will need in the future" (Kirschenmann et al., 2009).

To better understand the impacts of land management on grassland resources and how they contribute to ecosystem services, the USDA-Natural Resources Conservation Service (NRCS) partnered with the Agricultural Research Service, the National Institute of Food and Agriculture, and others, in a program called the Conservation Effects Assessment Project. The rangeland component of the project began in 2006 to assess and quantify the impacts and effects of conservation practices on environmental quality at national, regional, and watershed scales; strengthen the scientific basis associated with conservation programs and enhanced environmental quality of managed lands; and provide a solid scientific foundation for NRCS conservation practices (Briske, 2011).

Plate 34.2 Hayland in the Piedmont region of Virginia.

Photo credit: Alan Franzluebbers

The pasture/hayland component of the project began in 2008 and led to a thorough literature synthesis to assess the intended purposes of forages, i.e. improving forage yield and quality; maintaining species, vigor, and regrowth; providing feedstock for biofuel; controlling insects, plant diseases, and weeds; improving livestock nutrition and health; optimizing nutrient management and uptake; reducing soil erosion; improving quality of soil and water; improving riparian and watershed function; protecting air quality; enhancing carbon sequestration; and providing fish and wildlife benefits (Nelson, 2012).

Plate 34.3 Brangus cattle grazing native prairie in Oklahoma.

Photo credit: Mike Brown, USDA-ARS, El Reno, Oklahoma

Grasslands in America continue to provide an abundance of forage for livestock production and livelihoods for people that value open space. Unfortunately, grasslands are also being exploited through excessive cultivation of the deep, rich soils that developed over the millennia in the Great Plains, resulting in enormous loss of soil organic matter, native fertility, and soil sediment; or through excessive stocking on semi-arid and arid rangelands located in brittle environments of the southwestern US, resulting in loss of vegetative cover, low resilience, excessive soil erosion, and poor rural livelihoods. Some key environmental, social, and economic issues in grassland environments include the need to make social and policy changes before disaster occurs, such as that of the Dust Bowl of the 1930s, and the excessive pollution of ground and surface waters during the post-World War II agricultural revolution. Others involve balancing land preservation with land utilization for economic and ecological stability; effective technology transfer mechanisms for ecologically based agricultural business models; and defining the extent of land use changes in response to biofuel production systems, as well as defining the economic, environmental, and social trade-offs of such changes (Steiner and Franzluebbers, 2009).

Preservation of grasslands in America has involved an ongoing effort by private and government organizations to understand their intrinsic qualities, but also to simply enjoy the solitude and serenity of an expansive and undulating sea of grass, the harmonizing sounds of soft blades of grass whispering in the wind, or the quietness of nature undisturbed yet awaiting the coming storm thundering in the distance. The US government has protected some of these areas from development by establishing National Parks and Preserves, including Theodore Roosevelt National Park in North Dakota (http://www.nps.gov/thro/index.htm) and Tallgrass Prairie National Preserve in Kansas (http://www.nps.gov/tapr/index.htm). The US Forest Service administers other grassland areas, including Caddo National Grasslands Wildlife Management Area in Texas (https://www.tpwd.state.tx.us/huntwild/hunt/wma/find_a_wma/list/?id=4) and Buffalo Gap National Grassland in South Dakota (http://www.fs.usda.gov/recarea/nebraska/recarea/?recid=30329). Some states also have taken the initiative to preserve the heritage of grassland areas, including Great Valley Grasslands State Park in California (http://www.parks.ca.gov/?page_id=559).

References

Briske, D. D. (ed.) (2011). Conservation benefits of rangeland practices: assessment, recommendations, and knowledge gaps. USDA-Natural Resources Conservation Service, Washington, DC. Available at: www.nrcs.usda.gov/wps/portal/nrcs/detail/national/technical/nra/ceap/?&cid=stelprdb1045811.

Franzluebbers, A. J. (2013). Ecosystem services from forages. In: Bittman, S. and Hunt, D. (eds) *Cool Forages: Advanced Management of Temperate Forages*. Pacific Field Corn Association, Agassiz.

Glaser, A. (ed.) (2014). *America's Grasslands Conference: The Future of Grasslands in a Changing Landscape*. Proc. 2nd Biennial Conf. Conserv. America's Grasslands, 12–14 August, Manhattan, KS. National Wildlife Federation and Kansas State University, Washington DC and Manhattan KS.

Kirschenmann, F., Gompert, T., and Williams, A. (2009). Grass in the timeline of agriculture. In: Wedin, W. F. and Fales, S. L. (eds) *Grassland: Quietness and Strength for a New American Agriculture*. Am. Soc. Agron., Crop Sci. Soc. Am., Soil Sci. Soc. Am., Madison, WI.

Nelson, C. J. (ed.) (2012). *Conservation Outcomes from Pastureland and Hayland Practices: Assessment, Recommendations, and Knowledge Gaps*. Allen Press, Lawrence, KS.

Steiner, J. L., and Franzluebbers, A. J. (2009). Farming with grass – for people, for profit, for production, for protection. *Journal of Soil and Water Conservation*, vol 64, pp 75A-80A.
USDA-NRCS (2014). Non-federal grazing land, 2007. United States Department of Agriculture, Natural Resources Conservation Service. Available at: http://www.nrcs.usda.gov/Internet/FSE_MEDIA/stelprdb1041691.png

Briefing Note 34.2 Ecosystem services and grasslands in Australia

Neil MacLeod and John McIvor

Australian grasslands

Australia's size (~7.68 million km^2), latitudinal range (11°S to 44°S), and annual rainfall (mean ~100mm to 4000mm) have created a diversity of grassland environments, covering 70% of the continent (McIvor, 2005). Prior to European colonisation in the 18th century, these grasslands were subject to low-intensity occupation by native animals and indigenous peoples, whose populations and movements followed seasonal cycles of rainfall and resource availability and did not include cultivation or livestock rearing. European settlement triggered significant landscape modification through agricultural development and introduction of exotic plants and animals (McIvor, 2005). Grazing and cropping remain the dominant use of the grasslands.

Agriculture occurs within three zones (Plate 34.4) classified by vegetation and climate (McIvor, 2005). The *pastoral* zone spans the arid and semi-arid regions of northern and central Australia and is dominated by livestock grazing native pastures. The intermediate rainfall *wheat-sheep* zone extends through southeastern Queensland, central New South Wales, northern Victoria, southern South Australia, and southwestern Western Australia, and is principally mixed cereal and livestock production, with grain legumes and pastures sown in rotation with crops. The *high rainfall* zone is confined to the coastal belt and tableland areas of eastern Australia, including Tasmania, and is dominated by beef, dairy, and sheep production, primarily on sown pastures.

Ecosystem services from the grasslands

The recognition that humankind benefits from the functioning of natural ecosystems is growing (MA, 2005). Australian grasslands encompass ecological units of considerable geographical and biological diversity with the capacity to provide high levels of ecosystem services (MacLeod and Brown, 2014).

The most dominant is *provisioning* services based on pasture and crop growth – the annual value of livestock and crop production is ~USD39 billion (ABS, 2014). Increasing emphasis is being placed on alternative uses with less direct linkages to resource productivity and increasing amenity and lifestyle dimensions (MacLeod and Brown, 2014). Examples include crop pollination and honey production (~USD1 billion per annum – Gibbs and Muirhead, 1998), indigenous foodstuffs and medicinal produce ("bush tucker"), genetic material of use for recovering degraded pastures, and sites for energy extraction (oil, gas, solar, wind) or infrastructure. While the economic value of these services is not known, networks supporting their capture are developing (MacLeod and Brown, 2014).

The remaining service categories are difficult to define, measure, and value, but are important components of the total services value of grasslands. *Regulating* services include carbon sequestration

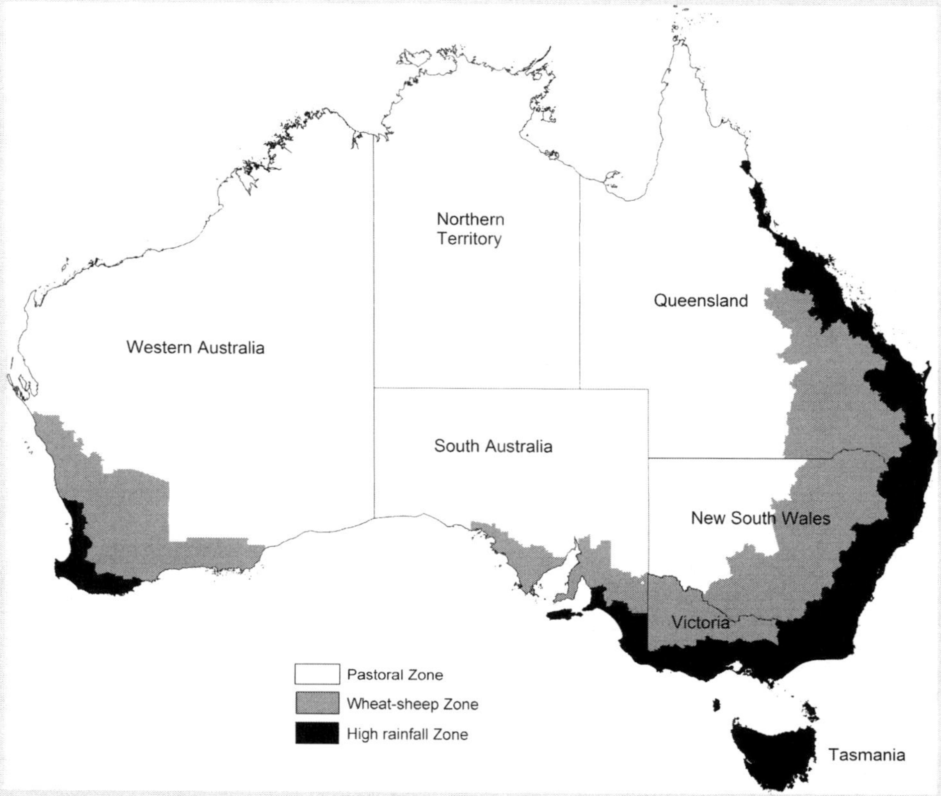

Plate 34.4 Agricultural production zones – Australia.
Source: authors

by plants and soil contributing to climate stabilisation and pollution control. Payment schemes for carbon regulation through management of vegetation and soils are the most developed application of a market for a regulating service. *Cultural* services include benefits from spiritual enrichment, cognitive development, and aesthetics; and are the most difficult to quantify due to their extreme subjectivity and scope for controversy and conflict. *Supporting* services include provision by native vegetation of critical habitat and water cycling, and have attributes that are amenable to the creation of markets for securing financial returns (MacLeod and Brown, 2014).

Impairment of ecosystem services

The ongoing supply of valuable ecosystem services from grasslands cannot be taken for granted, and while the production benefits have expanded since European settlement, the impact on landscape resources has been massive. Landscape dysfunction includes soil structure decline and accelerated erosion, native and exotic pasture decline, exotic weed ingress, shrub encroachment, tree decline and habitat loss, wildlife decline and extinction, salinity, loss of access to genetic materials, and diminished "amenity" options (MacLeod and Brown, 2014). Despite increasing adherence to conservation farming practices, the magnitude of impairment is high, as illustrated by one assessment (Tothill and Gillies, 1993) of the condition of the northern grazing lands (comprising much of the *pastoral* zone): of ~75 million hectares, 32% were moderately degraded (i.e. slight soil deterioration, increased presence of undesirable species) and 12% were severely degraded (i.e. severe soil deterioration, predominance of undesirable species). Degradation on this scale prompted a national assessment of all land and water resources, which confirmed an alarming level of impairment of ecosystems services (NLWRA, 2002). Research programs have explored sustainable utilisation of pastoral resources, including quantifying tradeoffs between conservation and production (MacLeod and McIvor, 2006) and ecological limits to exploitative use (McIntyre *et al.*, 2000). The National, State, and Territory governments support a range of resource conservation initiatives to restore and protect grassland ecosystems services, including the Landcare, Bushcare, and Reef Rescue programs, and conservation organisations such as Greening Australia, the Nature Conservancy, and Conservation Australia provide support for community-based conservation projects.

References

ABS (2014). *Value of Agricultural Commodities Produced, Australia*, year ended 30 June 2013 (Series 7503.0), Australian Bureau of Statistics, Canberra.

Gibbs, D.M.H., and Muirhead, I.F. (1998). *The Economic Value and Environmental Impact of the Australian Beekeeping Industry*. Australian Honey Bee Industry Council, Raceview.

MA (2005). *Ecosystems and Human Well-Being: Synthesis*. Island Press, Washington DC.

MacLeod, N. D., and Brown, J. R. (2014). 'Valuing and rewarding ecosystem services from rangelands. *Rangelands*, vol 36, no 2, pp 12–19.

MacLeod, N. D., and McIvor, J. G. (2002). Reconciling economic and ecological conflicts for sustained management of grazing lands. *Ecological Economics*, vol 56, pp 386–401.

McIntyre, S., McIvor, J.G., and MacLeod, N.D. (2000). Principles for sustainable grazing in eucalypt woodlands: landscape-scale indicators and the search for thresholds. In: Hale, P., Maloney, D. and Sattler, P. (eds) *Management for Sustainable Ecosystems*, University of Queensland, Brisbane.

McIvor, J. G. (2005). Australian Grasslands. In: Suttie, J. M., Reynolds, S. G. and Batello, C. (eds) *Grasslands of the World*. Food and Agriculture Organisation, Rome.

NLWRA (2002). *Landscape Health in Australia: A Rapid Assessment of the Relative Condition of Australia's Bioregions and Sub-Regions*. Environment Australia and the National Land and Water Resources Audit, Canberra.

Tothill, J. C., and Gillies, C. (1993). *The Pasture Lands of Northern Australia*. Tropical Grasslands Society of Australia, Brisbane.

35

CULTIVATED LANDS

Tobias Plieninger, Christopher M. Raymond and Elisa Oteros-Rozas

Introduction

Cultivated landscapes are – together with urban lands – those biomes that have undergone the strongest degree of human impact. They cover approximately 38% of the global land area (Foley *et al.*, 2011). By definition, cultivated lands are managed for a small number of agricultural crops that can be used for different purposes, such as food, feed, fibre or energy. Many cultivated lands provide important market and non-market ecosystem services. These benefits have been captured under the umbrella of "multifunctionality", a normative term that underpins many current agricultural support policies of OECD member countries. Multifunctional agriculture acknowledges that agriculture has several functions beyond the production of agricultural commodities, such as landscape aesthetics, biodiversity conservation and contribution to the socio-economic viability of rural areas (Renting *et al.*, 2009). Central to multifunctionality is the assumption that non-marketed ecosystem services can be produced jointly with agricultural goods, being in a complementary rather than competitive relationship.

However, most intensive agricultural systems are more harmful than supportive to biodiversity and ecosystem services. For example, the Millennium Ecosystem Assessment (2005) concluded that intensified agriculture has led to diminished water availability and quality, reduced carbon sequestration and increased eutrophication. The Global Environmental Outlook 3 (Secretariat of the Convention on Biological Diversity, 2010) identified agriculture as a key pressure on biodiversity, causing habitat loss, degradation and fragmentation as well as excessive nutrient loads. Consequently, the global expansion of croplands and agriculture-related features such as changes in the nitrogen and phosphorous cycles is now considered as a process in which mankind has transgressed or is close to irreversibly transgressing planetary boundaries (Steffen et al., 2015). The unsustainability of many agricultural systems is likely to increase, given the continuously rising demand for agricultural commodities. The hotspots of the resulting global competition for agricultural land occur at the interfaces between: a) forests and agriculture; b) urban land use and intensive agriculture; c) bioenergy, feed crops and food crops; and d) extensive cropland and grazing lands (Lambin and Meyfroidt, 2014). Competition for land results in a global trend towards greater agricultural intensification in richer nations and land clearing extensification in poorer nations. If this trend continues, approximately 1 billion ha of land would be cleared globally by 2050, with greenhouse gas emissions reaching −3 Gt of CO_2 equivalents y^{-1} and

−250 Mt of N use y^{-1} (Tilman *et al.*, 2011). To address the combined effects of degradation of ecosystems and their services, food insecurity, population growth and increasing affluence of societies (leading to more varied and resource-intensive diets), some researchers are calling for "sustainable intensification" (Garnett *et al.*, 2013). According to this approach (whose basic assumptions are under debate), food production would be increased from existing farmland in ways that place far less pressure on ecosystems and their services and that do not undermine the capacity to continue producing food in the future.

In this chapter we provide an overview of the relevance of cultivated lands for the provision of ecosystem services. First, we review the ecosystem services that are provided and consumed in these lands. We then present three cultivation systems that have the potential for "sustainable intensification" by maintaining biodiversity and ecosystem services while producing agricultural and forestry commodities. We end with some considerations of policy mechanisms to safeguard ecosystem services provision on cultivated lands.

Ecosystem services provided and consumed in cultivated lands

To understand the relationship between agriculture and ecosystem services, it is important to acknowledge that agricultural cultivation plays a key role in the delivery of ecosystem services (*ecosystem services from agriculture*), but at the same time makes use of a wide range of ecosystem services (*ecosystem services to agriculture*) (Swinton *et al.*, 2007; Zhang *et al.*, 2007). Furthermore, agriculture is, as outlined earlier, an important source of dis-services, but can at the same time suffer from ecosystem dis-services itself (e.g., pests). This conceptual relationship is outlined in Figure 35.1.

Services from cultivated land

Cultivated lands are the result of anthropogenic transformations of ecosystems for the provisioning of crops. In fact, they are the main suppliers of food in the world, which justifies the attraction of growing interest in their sustainability, particularly in the face of the expected increase of food demand by 70% by 2050 (Burney *et al.*, 2010). Moreover, approximately 2.5 billion people make their living out of crop cultivation and livestock raising, most of which do so through small-scale farming (FAO, 2012). *Provisioning services* are therefore the most evident ecosystem services from cultivated lands, comprising not only food for direct human consumption, but also feed and fodder for animals, fibres, biofuels, medicines, pharmaceutical products, dyes, chemicals and other raw materials (Millennium Ecosystem Assessment, 2005).

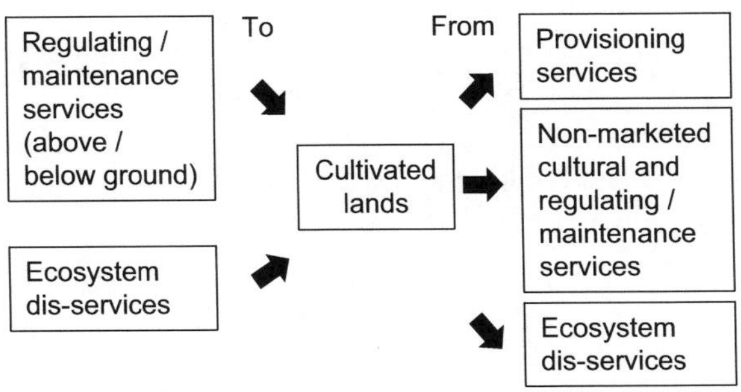

Figure 35.1 Conceptual model of ecosystem services and dis-services to and from cultivated lands
Source: adapted from Zhang et al., 2007.

Cultivated lands are also important providers of *regulating services*, such as pollination, nutrient cycling, soil formation, reduction/control of soil erosion, pest control, carbon storage, water and soil quality control through detoxification of noxious chemicals, water regulation, climate regulation, flood and fire prevention and habitat conservation (e.g., Rositano and Ferraro, 2014).

Further, *cultural services* are important in the context of cultivated lands, in particular traditional/local and scientific knowledge, education, existence and spiritual values linked to biodiversity, recreational activities and scenic beauty (e.g., Oteros-Rozas et al., 2014). Conservation of biodiversity may also be considered a cultural ecosystem service influenced by agriculture, since most cultures appreciate nature as an explicit human value (Power, 2010).

The supply of many of these ecosystem services (e.g., those related to soils and climates regulation) is linked to low-input, small-scale and traditional farming practices, in which energy and nutrient cycles tend to be closed locally. However, the past 50 years were marked by an intensification of agriculture, with a decoupling of agriculture from local ecosystem services and a growing application of nitrogen and phosphorus fertilizers as well as oil-dependent machinery (Stoate *et al.*, 2009). While world average yields of major food crops increased by a factor two in the last 50 years, the total amount of external nitrogen brought in through fertilizers increased seven times in the same period, the amount of phosphorus increased three times and the amount of water used for irrigation doubled (Foley *et al.*, 2005), leading to a number of ecosystem dis-services generated by agriculture.

Services to cultivated land

Agricultural production also depends on ecosystem services (mainly *regulating services*) provided by natural ecosystems, but these have been often neglected (Zhang *et al.*, 2007; Power, 2010). One of the most explored ecosystem services to agriculture is pollination, which contributes to the yield and quality of insect-pollinated crops, through the structural and plant species diversity of hedgerows that offer habitats for pollinating insect populations, for example. The production of more than 75% of the world's most important crops that feed humanity and 35% of the food produced are dependent upon animal pollination (Klein *et al.*, 2007). Pest control through natural enemies such as generalist and specialist predators and parasitoids has also been acknowledged as an ecosystem service to agriculture (Shackelford *et al.*, 2013). Another way in which insects provide ecosystem services is through dung burial, recycling wastes generated by large animals while improving forage palatability. Non-crop areas of cultivated landscapes can provide habitats where the abovementioned beneficial animals obtain their food resources, mate, reproduce and overwinter (Zhang *et al.*, 2007). Further, soil structure and fertility are determined by earthworms, macro- and micro-invertebrates, microorganisms and non-crop plants that can affect the quality and quantity of agricultural output (Zhang *et al.*, 2007). Genetic diversity is key to provide raw material for primary production, in particular through offering crop and animal breeds with variations that are the basis for the selection of desirable traits. In addition, at the species level, genetic diversity can also enhance biomass output per unit of land through better utilization of nutrients and reduced losses to pests and diseases. Other regulating ecosystem services to agriculture are hydrological and atmospheric regulation (Zhang *et al.*, 2007).

Trade-offs

Ecosystem services to and from agriculture are in close interaction with each other, giving rise to the notion of ecosystem services trade-offs. Empirically, a frequent pattern is that intensive forms of crop cultivation cause trade-offs with many regulating and cultural services (Chan and

Satterfield, 2016). For example, in Europe there has been a trade-off between agricultural production and the supply of cultural services from cultivated lands, particularly between 1950 and 1990, as a result of the simplification of cultivated landscapes. However, in the last two decades, in many areas of Europe, rural recreation activities are spreading hand in hand with the appreciation of scenic beauty and cultural heritage related with rural livelihoods, giving place to diverse trends (Harrison *et al.*, 2010). All in all, the question of whether and how intensive or extensive agriculture influences the various ecosystem services trade-offs is an important issue, requiring targeted research (Power, 2010).

Land use systems that foster ecosystem services

Several land use systems hold the promise to integrate agricultural production and ecosystem services. Prominent examples are agroforestry systems, organic agriculture and high nature value farming systems approaches. In this section, we briefly characterize these approaches and explore the major types of ecosystem services that can be fostered through these approaches. In a broader sense, comprehensive safeguarding of ecosystem services may require a "landscape approach" that integrates agricultural production, biodiversity conservation and rural livelihoods and links agricultural land, "natural" areas and appropriate institutional mechanisms to exploit synergies and to manage trade-offs between ecosystem services (Tscharntke *et al.*, 2005; Sayer *et al.*, 2013).

Agroforestry systems

Agroforestry comprises land-use systems and technologies in which trees and agricultural crops are grown together on the same land management unit, either in a temporal sequence (e.g., improved fallow systems) or in a spatial arrangement (e.g., alley cropping systems). A general attribute of agroforestry systems is that they aim to integrate commodity production with sustainability issues (in particular related to poverty alleviation, food security, and soil and biodiversity conservation), while striving to be compatible with local farming practices (Nair, 2007). According to estimates of the World Agroforestry Centre, agroforestry is found on approximately 46% of the global agricultural land area (10.1 billion ha). It is particularly prevalent in Southeast Asia, Central America and South America (Zomer *et al.*, 2009). There is evidence that many agroforestry systems are able to foster multiple ecosystem services on the same land area. One central strength of agroforestry systems (both trees and crops) is that they complementary acquire resources such as solar radiation, water and nutrients. Through complementary acquisition, they also increase provisioning services, with land resource efficiency being up to 40% higher than in conventional agricultural systems (Dupraz *et al.*, 2007). Other important ecosystem services include carbon sequestration in vegetation and soils, soil protection, microclimate moderation and reduction of water runoff (Nair, 2007).

Organic farming systems

Organic agriculture is an agricultural production system that is based on practices such as crop rotation, green manure, compost and biological pest control. In 2009, the global extent of organic agricultural land was 37.2 million hectares (including in-conversion areas), corresponding to 0.9% of agricultural lands (Willer, 2011). An important feature of organic farming is capitalizing on local ecosystem services for enhancement of food production by replacing major external inputs such as pesticides and fertilizers (Sandhu *et al.*, 2008). The most relevant ecosystem services are pollination, biological control, nutrient cycling and soil fertility, i.e., both below- and aboveground ecosystem services. Organic farming also includes activation of local ecological knowledge, promotion of place-based agricultural research and relatively higher degrees of labour required per land unit. Using an experimental approach to economic valuation, Sandhu *et al.* (2008) evaluated the contributions of organic farming to safeguarding ecosystem services in comparison to conventional agriculture. Considering a set of 12 ecosystem services in an agricultural landscape in New Zealand, they concluded that the total economic value of all but one (aesthetic services) ecosystem services was consistently higher in organic agriculture.

High nature value farming systems

High Nature Value (HNV) farming systems, a largely European concept, are "predominantly low-intensity systems which often involve a relatively complex interrelationship with the natural environment" (Baldock *et al.*, 1993). These systems occur in a variety of environments, climatic conditions, economic contexts and production systems, covering about 32% (75 million ha) of the farmland within the European Union (Paracchini *et al.*, 2008). HNV systems comprise livestock systems, arable systems and permanent cropping systems. Key determinants of biodiversity and ecosystem services around HNV farming systems are low-intensity practices, in terms of fertilizer and pesticide inputs, machinery and livestock stocking levels used; the presence of semi-natural vegetation; and diversity of land cover (Plieninger and Bieling, 2013). HNV farming systems have been primarily studied for the biodiversity values that they support. However, HNV farming also results in multiple ecosystem services, comprising provisioning, e.g., high-quality food and maintenance of genetic resources; regulating, e.g., soil quality regulation, pollination and water purification; and cultural services, e.g., heritage, recreation and ecotourism (Oppermann *et al.*, 2012).

Policy mechanisms for fostering ecosystem services

Despite growing awareness of the full range of ecosystem services, most agricultural lands are managed for the short-term production of food, fibre and fuel – often at the expense of other ecosystem services. Most proposed solutions to the underprovision of ecosystem services in agricultural ecosystems involve government or market approaches (Swinton *et al.*, 2007). However, a third major approach – cooperative solutions where landowners work together to provide ecosystem services – is increasingly promoted in the agricultural sector (Reed *et al.*, 2014). Cooperative approaches are particularly needed in cases where the spatial scale of management in agricultural ecosystems (e.g., a 500 hectare farm) does not match the spatial scale of ecosystem processes necessary to provide ecosystem services (e.g., a tri-county watershed) (Cumming *et al.*, 2006). As market, government and cooperative solutions all involve costs and benefits, they must be carefully examined to determine the most efficient and politically as well as socially feasible way to achieve a desired level of ecosystem services provision.

Instruments for enhancing ecosystem services include control measures, economic instruments, market facilitation, facilitation of public-private partnerships and conservation trusts, and research and extension services (Oskam *et al.*, 2011). Perhaps the most typical instruments are agri-environmental schemes, which governments around the world have supported since the early 1990s. They are payments to farmers aimed at encouraging or enforcing the production of environmental goods, such as soil protection or the restoration of native vegetation. Agri-environmental schemes typically link public funding for agriculture to the provision of societal benefits, such as climate regulation and water quality (Burton and Schwarz, 2013).

Box 35.1 Capitalizing on farm policies to foster ecosystem services in cultivated landscapes of Europe

With an estimated US$253 billion worth of subsidies for agricultural producers in OECD countries in 2009 (OECD, 2010), public agricultural policies have decisive influence on the management of cultivated landscapes. In June 2013, an agreement was reached on the future design of the European Union's Common Agricultural Policy for the 2014–2020 period, aiming for a stronger consideration of aspects of environmental sustainability. At the core of this agreement was the introduction of a "greening" component to which 30% of direct payments are now dedicated, aiming "that all EU farmers in receipt of support [. . .] deliver environmental and climate benefits as part of their everyday activities" (European Commission, 2011). For this, farmers have to comply with three mandatory principles:

- Crop diversification: to improve the resilience of agroecosystems, farms will need to cultivate at least two crops when their arable land exceeds 10 hectares and at least 3 crops when their arable land exceeds 30 hectares.
- Maintenance of permanent pasture: to strengthen retention of soil carbon and grassland habitats, land claimed as having been permanent pasture must be preserved.
- Establishment of ecological focus areas: to deliver water and habitat protection, farms will have to dedicate 5% and later 7% of farmland as "ecological focus areas", such as field margins, hedges, trees, fallow land, landscape features, biotopes, buffer strips and afforested areas.

The "greening" of the Common Agricultural Policy is a landmark decision and involves an estimated increase from US$3.9 billion to US$11.8 billion on environment-related expenditures. To promote crop diversity, maintain permanent grassland, and establish focus areas can contribute to increased flows of some ecosystem services in agricultural landscapes. However, the proposed underlying policy mechanisms are simplistic in their design and ignore much of the science of ecosystem services. In particular, they do not account for key features such as ecosystem services bundles, site-specificity and regionalization, appropriate spatial scales or funding permanence. Consequently, the "greening" mechanisms result in only limited improvements in ecosystem services provision from cultivated lands.

Source: Plieninger et al., 2012

Box 35.2 Using tender schemes to protect and restore ecosystem services on private farmland in Australia

Unlike farmers in the United States and Europe, Australian farmers have not received ongoing funding through agricultural support schemes to protect, manage or restore ecosystem services. Rather, funding for conservation actions on-farm is ad hoc and is subject to the funding objectives of short-term conservation programs or phases (generally 3–4 years in duration), which creates challenges for the long-term protection of biodiversity and ecosystem services on private land (Fitzsimons *et al.*, 2013). Between 1994 and 2008, the Australian Government-funded Natural Heritage Trust (in its various phases) injected $3 billion into natural resources management programs, policy instruments and incentives which in the early years were managed principally by formal institutions as well as community-based groups. The Natural Heritage Trust (NHT) was set up by the Australian Government in 1997 to help restore and conserve Australia's environment. Funding for the trust was largely derived from the sale of the then-publically owned telecommunications company Telstra. Under various phases of NHT and partner programs, landholders could apply for grants to undertake activities such as native vegetation planting and fencing. NHT also provided funding for the National Landcare Program, which was established by the Australian Government in 1989 in response to a joint submission by the National Farmers' Federation and the Australian Conservation Foundation. Between 2009 and 2012, the Australian Government supported the Environmental Stewardship Program, which provided incentive payments for land managers for long-term protection of ecosystem services on private land.

A variety of initiatives have been supported under the Environmental Stewardship Program, including the Bush Tender in Victoria and Land for Wildlife scheme in New South Wales. Here we focus on a successful scheme managed by the South Australian Murray-Darling Basin Natural Resources Management Board between 2009 and 2012, named the Conservation Stewardship Program (CSP). The CSP was a tender scheme which assisted land managers with undertaking management activities that protected and restored the threatened ecological communities of iron grass natural temperate grasslands and peppermint box grassy woodlands over a 15-year period in parts of the Adelaide and Mount Lofty Ranges, mid-north of South Australia and the South Australian Murray Darling Basin (SA Murray-Darling Basin NRM Board, 2014). Land managers who owned or managed land containing iron grass natural temperate grassland or peppermint box grassy woodland were invited to submit a bid to carry out works to protect or restore native vegetation on their property and determine the cost of carrying out the works. Proposed activities included fencing, grazing pressure reduction and pest animal and plant control or buffering. Sixty-seven funding agreements, from 10 to 15 years, were offered to land managers whose tender was successful (i.e., offered the best biodiversity value for money in their tender). Over 9600 hectares of iron grass grasslands and peppermint box woodlands have been protected and managed on private land as a result of the CSP (SA Murray-Darling Basin NRM Board, 2014).

Box 35.3 Defending family and peasant farming for food security, food sovereignty and the delivery of ecosystem services in Latin America and the Caribbean

Family farming represents today more than 80% of farms in Latin America and Caribbean, providing at country level between 27% and 67% of food production, covering between 12% and 67% of agricultural land and generating between 57% and 77% of farming employment in the region (Baquero *et al.*, 2007). Family farming, in addition, contributes to rural development by favouring generational turnover, creating or reinforcing local support networks, preserving cultural aspects and local identity, promoting the conservation of local landraces and breeds and applying practices that help preserving and improving soil while demanding less fuel and other inputs (FAO, 2012), for example.

In Latin America and the Caribbean, the unequal modernization of agriculture as a consequence of the different effects of macroeconomic policies – which have most frequently privileged business agriculture – has triggered a bipolar agriculture: business vs. family (FAO, 2012). Different dimensions, rationales, management and therefore capacity for the delivery of ecosystem services define these two models. In the face of the food crises of the last 10 years, family farming is increasingly acknowledged as a provider of multiple ecosystem services, as a poverty and climate change alleviator and therefore as the most adequate option toward social-ecological resilience. In this sense, several programs have been developed in the last 10 years to promote family farming in the region. At a country level some examples are: the *Programa Nacional de Fortalecimiento de la Agricultura Familiar (PRONAF)* in Brazil and Chile, the *Secretaría de Desarrollo Rural y Agricultura Familiar* in Argentina, the program of *Apoyo a la Agricultura Familiar* in El Salvador, the *Programa de Fomento a la Producción de Alimentos para la Agricultura Familiar* in Paraguay, the *Estrategia del Buen Vivir* in Ecuador and the *Estrategia de Agricultura Familiar* in Costa Rica.

Within policy instruments to support cultivated lands there is another framing, not explicitly linked to the ecosystem services conceptualization, but to the improvement of human well-being through food sovereignty and the democratization of food production. This discourse is mostly predominant in Latina America and the Caribbean, where the inclusion of the ecosystem services concept is being increasingly adopted by formal institutions, particularly in relation to schemes of Payments for Ecosystem Services.

Boxes 35.1–35.3 illustrate three cases from Europe, Australia and Latin America in which instruments to safeguard ecosystem services from cultivated land are being implemented (including the difficulties of doing so) at various spatial scales, involving both public and private funding sources. These cases, as well as many other examples, show that the implementation of policy mechanisms for ecosystem services faces a number of issues in relation to the performance of management practices, quality assurance (standards and certification), cost of information, bringing together the supply and demand side, coordinating conservation programs with markets, free-riding, large transaction costs and few buyers and sellers (Ribaudo *et al.*, 2010).

Conclusions

Cultivated lands are typically managed for small numbers of agricultural crops, not for a broader suite of ecosystem services. On these, agriculture both provides and depends on ecosystem services. Depending on the particular land use practices and environmental contexts, agriculture can generate both ecosystem services and dis-services (also called negative externalities), with the latter being dominant in most intensive agricultural systems. Landscape approaches that comprise land-use systems such as agroforestry, organic agriculture or HNV farming can potentially integrate agricultural production with the provision of a wider set of ecosystem services. Many different policy mechanisms have been applied in the agricultural sectors, but few have received a financial scale and a suitable design that would effectively safeguard ecosystem services on cultivated lands.

References

Baldock, D., Beaufoy, G., Bennet, G., and Clark, J. (1993). *Nature Conservation and New Directions in the EC Common Agricultural Policy*. Institute for European Environmental Policy, London.

Baquero, F.S., Fazzone, M.R., and Falconi, C. (2007). *Políticas para la agricultura familiar en América Latina y El Caribe*. Oficina Regional de la FAO para América Latina y el Caribe, Santiago.

Burney, J.A., Davis, S.J., and Lobell, D.B. (2010). Greenhouse gas mitigation by agricultural intensification. *Proceedings of the National Academy of Sciences of the United States of America*, vol 107, pp 12052–12057.

Burton, R.J.F., and Schwarz, G. (2013). Result-oriented agri-environmental schemes in Europe and their potential for promoting behavioural change. *Land Use Policy*, vol 30, pp 628–641.

Chan, K.M.A. and Satterfield, T. (2016). Managing cultural ecosystem services for sustainability. In Potschin, M., Haines-Young, R., Fish, R. and Turner, R.K. (eds) *Routledge Handbook of Ecosystem Services*. Routledge, London and New York, pp 343–351.

Cumming, G.S., Cumming, D.H., and Redman, C.L. (2006). Scale mismatches in social-ecological systems: causes, consequences, and solutions. *Ecology and Society*, vol 11, art 14.

Dupraz, C., Burgess, P., Gavaland, A., *et al.* (10 authors) (2007). *SAFE (Silvoarable Agroforestry for Europe) Synthesis Report*. SAFE Project (August 2001–January 2005).

European Commission (2011). *The CAP Towards 2020: Meeting the Food, Natural Resources and Territorial Challenges of the Future*. COM (2010) 672 final, Brussels.

FAO (2012). *Marco estratégico de mediano plazo de cooperación de la FAO en Agricultura Familiar en América Latina y el Caribe 2012–2015*. Available at: http://www.fao.org/docrep/019/as169s/as169s.pdf

Fitzsimons, J., Pulsford, I., and Wescott, G. (2013). Challenges and opportunities for linking Australia's landscapes: a synthesis. In: Fitzsimons, J., Pulsford, I. and Wescott, G. (eds) *Linking Australia's Landscapes: Lessons and Opportunities from Large-scale Conservation Networks*. CSIRO Publishing, Melbourne.

Foley, J.A., DeFries, R., Asner, G.P., *et al.* (19 authors) (2005). Global consequences of land use. *Science*, vol 309, pp 570–574.

Foley, J.A., Ramankutty, N., Brauman, K.A., *et al.* (21 authors) (2011). Solutions for a cultivated planet. *Science*, vol 478, pp 337–342.

Garnett, T., Appleby, M., Balmford, A., *et al.* (10 authors) (2013). Sustainable intensification in agriculture: premises and policies. *Science*, vol 341, pp 33–34.

Harrison, P.A., Vandewalle, M., Sykes, M.T., *et al.* (17 authors) (2010). Identifying and prioritising services in European terrestrial and freshwater ecosystems. *Biodiversity and Conservation*, vol 19, pp 2791–2821.

Klein, A. M., Vaissiere, B.E., Cane, J.H., *et al.* (7 authors) (2007). Importance of pollinators in changing landscapes for world crops. *Proceedings of the Royal Society B-Biological Sciences*, vol 274, pp 303–313.

Lambin, E., and Meyfroidt, P. (2014). Trends in global land-use competition. In: Seto, K. C. and Reenberg, A. (eds) *Rethinking Global Land Use in an Urban Era*. MIT Press, Cambridge MA.

Millennium Ecosystem Assessment (2005). *Ecosystems and Human Well-Being: Synthesis*. Island Press, Washington DC.

Nair, P.K.R. (2007). The coming of age of agroforestry. *Journal of the Science of Food and Agriculture*, vol 87, pp 1613–1619.

OECD (2010). *Agricultural Policies in OECD Countries at a Glance*. Organisation for Economic Co-operation and Development, Paris.

Oppermann, R., Beaufoy, G., and Jones, G. (2012). *High Nature Value Farming in Europe – 35 European Countries, Experiences and Perspectives*. Verlag Regionalkultur, Ubstadt-Weiher.

Oskam, A., Meester, G., and Silvis, H. (2011). *EU Policy for Agriculture, Food and Rural Areas*. Wageningen Academic Publishers, Wageningen.

Oteros-Rozas, E., Martín-López, B., González, J., *et al.* (6 authors) (2014). Socio-cultural valuation of ecosystem services in a transhumance social-ecological network. *Regional Environmental Change*, vol 14, pp 1269–1289.

Paracchini, M.L., Petersen, J.-E., Hoogeveen, Y., *et al.* (6 authors) (2008) *High nature value farmland in Europe. An Estimate of the Distribution Patterns on the Basis of Land Cover and Biodiversity Data*. Office for Official Publications of the European Communities, Luxembourg.

Plieninger, T., and Bieling, C. (2013). Resilience-based perspectives to guiding high nature value farmland through socio-economic change. *Ecology and Society*, vol 18, art 20.

Plieninger, T., Schleyer, C., Schaich, H., *et al.* (7 authors) (2012). Mainstreaming ecosystem services through reformed European agricultural policies. *Conservation Letters*, vol 5, pp 281–288.

Power, A.G. (2010). Ecosystem services and agriculture: tradeoffs and synergies. *Philosophical Transactions of the Royal Society B: Biological Sciences*, vol 365, pp 2959–2971.

Reed, M.S., Moxey, A., Prager, K., *et al.* (9 authors) (2014). Improving the link between payments and the provision of ecosystem services in agri-environment schemes. *Ecosystem Services*, vol 9, pp 44–53.

Renting, H., Rossing, W., Groot, J., *et al.* (8 authors) (2009). Exploring multifunctional agriculture. A review of conceptual approaches and prospects for an integrative transitional framework. *Journal of Environmental Management*, vol 90, pp S112-S123.

Ribaudo, M., Greene, C., Hansen, L., and Hellerstein, D. (2010). Ecosystem services from agriculture: steps for expanding markets. *Ecological Economics*, vol 69, pp 2085–2092.

Rositano, F., and Ferraro, D. (2014). Ecosystem services provided by agroecosystems: a qualitative and quantitative assessment of this relationship in the Pampa region, Argentina. *Environmental Management*, vol 53, pp 606–619.

SA Murray-Darling Basin NRM Board (2014). *Environmental Stewardship in Iron Grass Grasslands and Peppermint Box Woodlands*. Available at: http://www.naturalresources.sa.gov.au/samurraydarlingbasin/projects/environmental-stewardship

Sandhu, H.S., Wratten, S.D., Cullen, R., and Case, B. (2008). The future of farming: the value of ecosystem services in conventional and organic arable land. An experimental approach. *Ecological Economics*, vol 64, pp 835–848.

Sayer, J., Sunderland, T., Ghazoul, J., *et al.* (12 authors) (2013). Ten principles for a landscape approach to reconciling agriculture, conservation, and other competing land uses. *Proceedings of the National Academy of Sciences of the United States of America*, vol 110, pp 8349–8356.

Secretariat of the Convention on Biological Diversity (2010). *Global Biodiversity Outlook 3*. CBD, Montréal.

Shackelford, G., Steward, P.R., Benton, T.G., *et al.* (7 authors) (2013). Comparison of pollinators and natural enemies: a meta-analysis of landscape and local effects on abundance and richness in crops. *Biological Reviews*, vol 88, pp 1002–1021.

Steffen, W., Richardson, K., Rockström, J., *et al.* (18 authors) (2015). Planetary boundaries: guiding human development on a changing planet, *Science*, vol 346, no 6223.

Stoate, C., Báldi, A., Beja, P., *et al.* (9 authors) (2009). Ecological impacts of early 21st century agricultural change in Europe – A review. *Journal of Environmental Management*, vol 91, pp 22–46.

Swinton, S.M., Lupi, F., Robertson, G.P., and Hamilton, S.K. (2007). Ecosystem services and agriculture: cultivating agricultural ecosystems for diverse benefits. *Ecological Economics*, vol 64, pp 245–252.

Tilman, D., Balzer, C., Hill, J., and Befort, B.L. (2011). Global food demand and the sustainable intensification of agriculture, *Proceedings of the National Academy of Sciences of the United States of America*, vol 108, pp 20260–20264.

Tscharntke, T., Klein, A. M., Kruess, A., Steffan-Dewenter, I., and Thies, C. (2005). Landscape perspectives on agricultural intensification and biodiversity – Ecosystem service management, *Ecology Letters*, vol 8, pp 857–874.

Willer, H. (2011). *The World of Organic Agriculture 2012: Statistics and Emerging Trend*. IFOAM, FIBL, Bonn, Frick.

Zhang, W., Ricketts, T.H., Kremen, C., Carney, K., and Swinton, S.M. (2007). Ecosystem services and dis-services to agriculture. *Ecological Economics*, vol 64, pp 253–260.

Zomer, R.J., Trabucco, A., Coe, R., and Place, F. (2009) Trees on farm: analysis of global extent and geographical patterns of agroforestry. ICRAF Working Paper no 89. World Agroforestry Centre, Nairobi.

36

ECOSYSTEM SERVICES PROVIDED BY URBAN GREEN INFRASTRUCTURE

Thomas Elmqvist, Erik Gómez-Baggethun and Johannes Langemeyer

Introduction

In a rapidly urbanizing world, cities have become main centres of demand for ecosystem services, and sources of environmental impacts. The pressure on ecosystems from cities will further accelerate in the future, with approximately 60% of the urban land present in 2030 forecast to be built in the period 2000–2030 (see Elmqvist et al. 2013). Urbanization therefore presents fundamental challenges for sustainbility, but it also presents unprecedented opportunities for novel forms of designing and managing urban ecosystems so as to secure their capacity to sustain ecosystem services and their resilience to cope with change. For example, as IPCC predicts higher frequency and intensity of climate extremes, urban ecosystems, i.e. "green and blue infrastructure", are often viewed to have a crucial role in building adaptive capacity to cope with climate and other global change (Depietri et al. 2012, Farrugia et al. 2013). Strategies based on investments in urban green infrastructure and ecosystem-based adaptation to climate change are gaining considerable interest across the globe, particularly since such investments simultaneously generate multiple ecosystem services, enhancing human well-being (e.g. Elmqvist et al. 2013). Indeed, as Box 36.1 shows, there has been considerable convergence in the literature dealing with ecosystem services and green infrastructure, with the latter now seen as a necessary structural element for the former in urban areas.

Despite growing attention to cities in the sustainability agenda, ecosystem services provided in urban landscapes are still poorly understood, particularly when it comes to designing, creating and restoring ecological structures and processes in order to maintain and enhance ecosystems benefits to humans and related values in urban areas (e.g. Pataki et al., 2011; Gómez-Baggethun et al. 2013). Analyses of urban ecosystem services also present some unique challenges and options when compared to rural areas. Specifically, since the number of beneficiaries on a surface basis usually is very high, even small units (e.g. individual trees) may contribute to generating high values, and small green space patches can provide manifold benefits with regards to multiple citizens' demands. Also, in urban areas it is very clear that ecosystem services are generated by a tightly coupled social-ecological system, where the feedbacks between flows of services, values, institutions, management and practices are very evident (Figure 36.1).

Ecosystem services in urban areas

In the following we provide a description of ecosystem services provided by urban ecosystems, drawing on previous research (e.g. Bolund and Hunhammar 1999; Gómez-Baggethun and Barton 2013, Gómez-Baggethun et al. 2013). To do so, we rely on well-established classification frameworks, such as those from the Millennium Ecosystem Assessment (MA), The Economics of Ecosystems and Biodiversity (TEEB) and the Common International Classification of Ecosystem Services (CICES). A classification of services, building on the typologies developed in the above-mentioned initiatives, is given in Table 36.1.

Provisioning services

Food supply

Food supply from urban agricultural land and gardens have played a critical role in the development of cities throughout history (e.g. Keshavarz, 2015). Although data on food production yields in cities is scattered and fragmented (Gittleman et al. 2012), it is apparent that most cities only produce a small share of the food they consume, relying heavily on the (global) hinterland to cover their demands for food. However, urban food production has been shown to play an important role in food security during particular periods, especially economic and political crises (Barthel and Isendahl, 2013). Barthel et al. (2010) estimate that during food shortages caused by World War II in Sweden about 10% of the country's food production took place within cities. Local food provision plays also an important role in many African cities; for example, 90% of all vegetables consumed in Dar es Salaam (Tanzania) (Jacobi et al. 2000) and 60% of vegetables consumed in Dakar (Senegal) originate from urban agriculture (Mbaye and Moustier, 2000). In Asia, the share of rice supplied by the city to urban residents ranges from 7% in Phnom Penh to 58% in Hanoi to about 100% in cities like Vientiane (Anh, 2004; Ali et al. 2005).

Water supply

Securing water supply in cities in the face of climate change is currently a major challenge (Fitzhugh and Richter, 2004). Ecosystems, such as forests, lakes and wetlands, capture, filter and store water used in cities for drinking, irrigation and other human uses, and the watersheds from which cities get water thus provide important services to the urban population. Two well-known examples are the Omerli Watershed and the Catskill Watershed, which provide drinking water to the megacities Istanbul and New York City, respectively. While the Omerli Watershed is currently under severe threat from urban development (Wagner et al. 2007), the Catskill Watershed is among the best-known examples of an active protection and management of ecosystem services (Chichilnisky and Heal, 1998). Aware of their importance for sustained water supply, various cities in the Andean region are taking measures to protect the mountain ecosystems (e.g. *paramos*) around them. An example is current efforts by authorities of Bogotá, Colombia, to protect and restore the Andean forest ecosystems on the city's eastern border (*Cerros Orientales*).

Regulating services

Urban temperature regulation

Heat islands in cities generated by the use of fossil fuels and heat waves (Meehl and Tebaldi, 2004) can create severe hazards to health, especially among children and elderly people (Depietri

Table 36.1 Classification of ecosystem services in urban areas and underlying ecosystem functions and components.

Ecosystem functions	*Ecosystem service type*	*Examples*	*Key references*
Provisioning services			
Energy conversion into edible plants through photosynthesis	Food supply	Vegetables and other food produced in urban allotments and peri-urban areas agricultural belts	Altieri et al. (1999); Orsini et al. (2014)
Water capture, filtration and storage by urban and peri-urban ecosystems	Water supply	Forests and wetlands capture, filter and store water used in cities for drinking, irrigation and other human uses	Chichilnisky and Heal (1998)
Regulating services			
Photosynthesis, shading and evapotranspiration	Urban temperature regulation	Trees and other urban vegetation reflect the sun, provide shade, create humidity and block wind	Bowler et al. (2010)Bolund and Hunhammar (1999) Baró et al. (2015)
Dry deposition of gases and particulate matter	Air purification	Absorption of pollutants by urban vegetation in leaves, stems and roots	Escobedo et al. (2011); Baró et al. (2014)
Carbon sequestration and storage by fixation in photosynthesis	Global climate regulation	Carbon sequestration and storage by soils and the biomass of urban shrubs and trees	Nowak (1994) McPherson (1998)
Mechanical insurance by physical barriers and absorption of kinetic energy	Moderation of extreme events	Storm, flood and wave buffering by reefs, mangroves and wetlands; heat absorption during heat waves	Danielsen et al. (2005); Costanza et al. (2006)
Percolation and regulation of runoff and river discharge	Runoff mitigation	Soil and vegetation percolate water during precipitation events	Shuster et al. (2008) Farrugia et al. (2013)
Movement of floral gametes by biota	Pollination and seed dispersal	Urban ecosystem provides habitat for birds, insects and pollinators	Andersson et al. (2007); Jansson and Polasky (2010)
Removal or breakdown of xenic nutrients	Waste treatment	Effluent filtering and nutrient fixation by urban soils and wetlands	Vauramo and Setälä (2010)

Ecosystem functions	*Ecosystem service type*	*Examples*	*Key references*
Absorption, refraction and dispersion of sound waves by vegetation and water	Noise reduction	Absorption, refraction and dispersion of sound waves by vegetation barriers, especially thick vegetation	Fang and Ling (2003)
Supporting/Habitat services			
Presence of ecological niches	Habitat provision	Urban green spaces provide habitat for many rare and endangered species	Blair (1996)
Cultural services			
Ecosystems with recreational values	Recreation and relaxation	Urban green areas provide opportunities for recreation, meditation and relaxation	Chiesura (2004); van den Berg (2010)
Ecosystems with aesthetic values	Aesthetic benefits	Urban parks and other green and blue spaces in sight from houses	Tyrväinen (2001); Cho et al. (2008)
Human experience of ecosystems	Cognitive development	Urban green spaces offer opportunity for cognitive coupling to seasonality and ecosystem dynamics	Barthel et al. (2010)
Human perception of ecosystems	Sense of place and social cohesion	Affectively charged attachments to urban ecosystems that play a role in defining identity, sense of place and community	Gotham and Bromley (2002); Tidball et al. (2014)

Source: modified from Gómez-Baggethun and Barton 2013 and Gómez-Baggethun et al. 2013

et al. 2012). The European heat wave of 2003, for example, accounted for more than 70,000 excess deaths (EEA, 2010). Urban green and blue infrastructure has the capacity to locally buffer heat extremes (Hardin and Jensen, 2007). For example, water areas buffer temperature extremes by absorbing heat in summertime and releasing it in wintertime. Likewise, urban vegetation, especially trees, reduces temperature by reflecting solar radiation, and through functions like shading and evapotranspiration (Bowler et al. 2010; McPhearson, 2011). Water released from plants absorbs heat as it evaporates, and reduces air temperature (Nowak and Crane, 2000).

Air purification

Air pollution constitutes a severe risk to human health in cities worldwide. Urban green infrastructure, especially trees, removes pollutants from the air and filters out gases and airborne particulates (Nowak, 1996). For most pollutants, such as ozone (O_3), sulfur dioxide (SO_2), nitrogen dioxide (NO_2) and carbon monoxide (CO), air purification capacity by green infrastructure has been found to be modest when compared with overall city emission. However, recent research indicates that green infrastructure can make substantial contribution in the removal of other important pollutants, such as particulate matter smaller than 10μm (PM10) (Baró et al. 2014).

Global climate regulation

Cities account for approximately 80% of global emissions of greenhouse gases. Greenhouse gases in cities include carbon dioxide (CO_2), methane (CH_4), nitrous oxide (NO_2), chlorofluorocarbons (CFCs) and tropospheric ozone (O_3). Urban trees act as sinks of CO_2 by storing excess carbon as biomass during photosynthesis (Nowak et al. 2007). Because the amount of CO_2 stored is proportional to the biomass of the trees, increasing the number of trees can potentially slow the accumulation of atmospheric carbon in urban areas. For example, research has estimated a carbon sequestration capacity of 6187 t/y in Barcelona (Baró et al. 2014) and 16,000 t/y in Philadelphia (Nowak et al. 2007). It should be noted, however, that the amount of carbon a city can offset locally through green infrastructure is modest when compared to overall city emissions (Baró et al. 2014, Pataki et al. 2011).

Moderation of extreme events

The increased frequency and intensity of environmental extremes expected with climate change poses increasing adaptation challenges for cities (Meehl and Tebaldi, 2004; Zahran et al. 2008). Many megacities around the world are located in coastal areas, which put their dwellers at special risk from extreme hazards like storms, floods, hurricanes and tsunamis. The devastating effects of events like the Indian Ocean Tsunami in 2004, Hurricane Katrina in 2005 and the earthquake and tsunami in Japan in 2011 have resulted in new visions for risk management and vulnerability reduction in cities, based on innovative combinations of grey and green infrastructure, including nature-based solutions benefitting from the protective capacities of natural ecosystems (Costanza et al. 2006; Finlay, 2010; Depietri et al. 2012). Coastal ecosystems such as wetlands, mangroves, deltas and coral reefs act as natural barriers, buffering flooding and storms, and reducing human and material damage from extreme events (e.g. Danielsen et al. 2005).

Runoff mitigation

Recent flood disasters in the United States (2005, 2008, 2012); the Philippines (2012, 2013); and Britain (2014) illustrate how vulnerable coastal cities are to storm surge flooding (Aerts et al. 2014). Urban landscapes characterized by 50–90% impervious surface can lose 40–83% of rainfall to surface runoff; in comparison, only 13% surface runoff can occur in forests (Bonan, 2002). Increasing vegetation cover and reducing impermeable surface area through ecological restoration can decrease volumes of surface storm water runoff, and hence the vulnerability to water flooding, in cities. Vegetation stabilizes the soil and reduces surface runoff following precipitation events, reducing the pressure on drainage systems. Urban vegetation slows down flooding effects by intercepting rain water through tree canopies, plant leaves and stems (Bolund and Hunhammar, 1999; Pataki et al. 2011). Similar effects are observed through establishment of bioswales, green roofs, rain gardens, bioretention filters and catch basins (Shuster et al. 2008).

Pollination and seed dispersal

Urban ecosystems may provide habitat for birds, insects and other fauna critical to pollination and seed dispersal to urban ecosystems, and even adjacent agricultural areas (Andersson et al. 2007). Urban gardens, for example, may have a crucial role in the network of habitats supporting pollinators and seed dispersers, such as bumblebees and birds. However, habitat fragmentation and loss from urban development threaten these species. Jansson and Polasky (2010) found that while the impact of development on pollination might be modest, the erosion of the resilience of the service (measured through change in response diversity) could be high. In turn, specific management practices for urban gardens, cemeteries, parks and other green spaces may promote functional groups of insects and bird communities, which enhance pollination and seed dispersal (Andersson et al. 2007).

Waste treatment

Ecosystems filter out and decompose organic waste from urban effluents by storing and recycling waste through dilution, assimilation and chemical re-composition (TEEB, 2010). Plant communities in urban soils can play an important role in the decomposition of many labile and recalcitrant litter types (Vauramo and Setälä 2010). Wetlands and aquatic ecosystems are crucial to filtering sewage from cities, reducing pollutants and nutrient loads before effluent waters return into the water cycle (Karathanasis et al. 2003). Green infrastructure can enhance nutrient retention by adding coarse woody debris, constructing in-channel gravel beds and increasing the width of vegetation buffer zones and tree cover (Booth, 2005).

Noise reduction

Noise pollution is a major problem for human well-being and health in cities. Plant barriers, such as belt trees, can attenuate noise pollution from traffic and other human activities through absorption, deviation, reflection and refraction of sound waves by branches and leaves (Ishii 1994). Different plant species mitigate noise differently. Important factors for noise reduction include density, width, height and length of tree belts as well as leaf size and branching characteristics. For example, the wider the vegetation belt, the higher the density, and the more foliage and branches, the greater the noise reduction effect (Fang and Ling, 2003).

Habitat services

Biodiversity conservation

Urban areas may be very rich in species (Melles et al. 2003; Müller et al. 2010, Elmqvist et al. 2013). Species diversity often peaks at intermediate levels of urbanization, at which many native and non-native species thrive, but typically declines as urbanization intensifies (Blair, 1996). Well-designed green roofs can provide habitat for species affected by urban land-use changes (Brenneisen, 2003), green lanes can support wetland fauna (Colding and Folke, 2009), hardwood deciduous trees are an important source for species with high dispersal capacity (Zetterberg, 2011), and old trees provide habitat for species of birds, insects and fungi, including many rare and endangered species. Maintaining agro-biodiversity and functional response diversity in urban gardens can enhance resilience in urban food production systems in the face of climate change (Barthel et al. 2010).

Cultural services and benefits

Recreation and relaxation

The recreational possibilities of urban parks, forests and lakes are manifold and among the highest-valued ecosystem services in cities (e.g. Chiesura, 2004; Konijnendijk et al. 2013; Langemeyer et al. 2014). Recreational benefits result from physical exercises or mental recreation in the urban environment, reducing stress and providing a sense of peacefulness and tranquility. Studies have shown that benefits from urban green spaces have been associated with reduced stress (White et al. 2013) and increased physical and mental health (Alcock et al. 2014). For example, urban gardeners show lower levels of stress and stress release compared to other citizens (Hawkins et al. 2011; van den Berg et al. 2011). Ulrich (1984) documented several decades ago that a view of green spaces through a window could accelerate recovery from surgery. In addition, van den Berg et al. (2010) found that living in proximity to green space was correlated with fewer stress-related health problems and a higher general health perception. The recreational value of green spaces depends not only on ecological characteristics such as biological and structural diversity, but also on built features such as benches and sport facilities. Recreational quality of urban ecosystems also varies with aspects such as accessibility, penetrability, safety, privacy and comfort, as well as with factors that may cause sensory disturbance in the form of visual or aural impacts such as garbage or noise (Rall and Haase, 2011).

Environmental learning

Technology and urbanization tend to reduce people's awareness of ecosystem and seasonal dynamics. Krasny and Tidball (2009) have documented that exposure to nature and green space may provide multiple opportunities for cognitive development, which may result in increased potential for environmental stewardship. For example, urban forests and urban gardens serve environmental education purposes in urban areas where people are often physically and mentally disconnected from the physical environment. Likewise, gardeners and other users of urban gardens, cemeteries and other green spaces may develop and retain important ecological knowledge useful in local management (Barthel et al. 2010). The benefits of preserving local ecological knowledge have been highlighted in terms of increased resilience and adaptive capacity in urban systems (Buchmann, 2009).

Sense of place and social cohesion

The creation of more inclusive cities is a major challenge for cities in the 21st century (United Nations, 2014b, p. 11). Experiencing green spaces and participating in their management, such as by tending urban gardens, has been shown to foster a stronger identification with locations and the development of a positive sense of place and place attachment (Tidball et al. 2014). Attachment and positive sense of place may further enable community building, social cohesion and integration, based on shared interests and identity as well as social connections (Gotham and Brumley, 2002; Glover, 2004; Rosol, 2006).

Conclusions

Urbanization and technological progress have fostered the misleading conception that urban people no longer depend on nature in any direct form. Empirical data shows, however, that decoupling of cities from ecological systems can only occur locally and partially, thanks to the appropriation of vast areas of ecosystem services provision beyond the city boundaries (Gómez-Baggethun and Barton, 2013). Protecting and restoring urban ecosystem services can enhance a city's capacity to secure long-term condition for heath, security, social relations and other components of human well-being. Ecosystem services that are relevant in urban contexts include urban temperature regulation, moderation of climate extremes, recreation and relaxation, environmental learning, and social cohesion. Besides their direct contribution to quality of life, urban green spaces can be a major source of resilience for cities through, for example, buffering extreme events and being potential areas for food supply; thereby enhancing capacity to deal with environmental and socio-economic shocks. In contrast, loss of green spaces may increase vulnerability to shocks and negative health impacts on urban inhabitants, which may both carry economic costs.

Worldwide, city authorities are looking for innovative ways to maintain and increase urban green infrastructure and ecosystem services through urban planning (Rosenzweig et al. 2009). Yet many studies have suggested a lacking integration of ecosystem services into urban planning (Primmer and Furman, 2012; Kabisch, 2015). Trade-offs may arise between ecosystem services under different land-uses. For instance, clearing a patch of forest to create a park enhances recreational values but generally reduces biodiversity (Gómez-Baggethun et al. 2013). Trade-offs arise not only between ecosystem services but also across the values different persons attach to those services. Ecosystem outputs perceived as services by some may be perceived as disservices by others; for instance, the shade of a tree might be welcomed by some people for its cooling effect, while reducing daylight in other people's houses. There is a growing literature on "ecosystem disservices" (Baró et al. 2014, von Döhren and Haase, 2015, Kronenberg, 2015), which are important to include in the future analyses, but so far there are limited quantifications of these due to methodological challenges. Values may also be perceived and expressed in different dimensions and languages, with distinctions between ecological, socio-cultural and economic values most commonly proposed. Trade-offs may also emerge across different valuation languages (Martín-López et al. 2014); for example, Langemeyer et al. (2014) observed strong deviations between the socio-cultural and economic expression of values related to cultural ecosystem services from an urban park. Reaching a comprehensive picture of the multiple potential ecosystem services and nuisances of restoring or losing urban ecosystems therefore involves endorsing integrated valuation approaches, capable of combining multiple value dimensions, stakeholder perspectives, knowledge systems and fields of expertise (Gómez-Baggethun and Martín-López, 2015; Langemeyer, 2015).

An important contribution of the ecosystem service approach has been to provide a consistent framework to integrate information from various fields of knowledge concerned with the urban environment, emphasize the potential multi-functionality of urban ecosystems and facilitate an arena for interdisciplinary dialogue based on a holistic perspective of the multiple ecosystem services urban green spaces provide (Langemeyer, 2015). However, there is still limited empirical evidence for many commonly cited benefits (Pataki et al. 2011). For example, little evidence supports that urban green space can reduce urban greenhouse gas emissions or reduce air and water pollutant concentrations. On the other hand, there is evidence supporting substantial reductions in urban runoff and effects on local temperature regulation (Elmqvist et al. 2015) and for manifold cultural ecosystem services provided by urban green spaces (Langemeyer et al. 2014; Camps-Calvet et al. 2015). The effectiveness of nature-based solutions to urban problems using investments in ecological infrastructure should also be compared with other strategies, such as those based on "civic engineering", and often considered as a complement to them. For example, whereas restoration of urban woodlands is likely to be an effective measure to enhance biodiversity and opportunities for recreation, caps on car use or taxes on fuels may be a more effective measure to reduce urban greenhouse gas emissions and to improve air quality in cities.

While studies on single ecosystem services in cities are increasing, analyses of multiple services are still rare, and there is still a need for "integrated valuations" of urban ecosystem services (Gómez-Baggethun and Martín-López, 2015). To accomplish this, there is a need for increased synthesizing, interpreting and communicating knowledge about the multiple values generated by urban green infrastructure. This constitutes a major challenge for urban green infrastructure governance, and new incentives for an operationalization of the ecosystem service framework in urban policy-making, planning and management are needed (Haase et al. 2014; Kabisch, 2015).

The future of ecosystem service research in urban areas is likely to focus on the integration of green infrastructure, built infrastructure and institutional arrangements designed to create innovative urban design of functions that combine living systems with built systems to improve well-being and the resilience of the urban region. This may be facilitated by the creation of urban learning labs with broad stakeholder representation, where new knowledge is generated and new solutions implemented in a co-production mode.

References

Aerts, J.C.J.H., Botzen, W.J., Emanuel, K. et al. (8 authors) (2014). Evaluating flood resilience strategies for coastal megacities, *Science*, vol 344, no 6183, pp 473–475.

Alcock, I., White, M. P., Wheeler, B. W., Fleming, L. E., and Depledge, M. H. (2014). Longitudinal effects on mental health of moving to greener and less green urban areas. *Environmental Science and Technolology*, vol 48, pp 1247–1255.

Ali, M., de Bon, H., and Moustier, P. (2005). Promoting the multifunctionality of urban and peri-urban agriculture. *Urban Agriculture Magazine*, vol 15, pp 9–11.

Altieri, M. A., Companioni, N., Cañizares, K., et al. (7 authors) (1999). The greening of the 'barrios': urban agriculture for food security in Cuba. *Agriculture and Human Values*, vol 16, no 2, pp 131–140.

Andersson, E., Barthel, S., and Ahrné, K. (2007). Measuring social-ecological dynamics behind the generation of ecosystem services. *Ecological Applications*, vol 17, no 5, pp 1267–1278.

Andersson, E., McPherson, T., Kremer, P., et al. (8 authors) (2015). Scale and context dependence of ecosystem service providing units. *Ecosystem Services*, in press, DOI: http://dx.doi.org/10.1016/j.ecoser.2014.08.001

Anh, M.T.P. (2004). Urban and peri-urban agriculture in Hanoi: opportunities and constraints for safe and sustainable food production. In *Technical Bulletin / World Vegetable Center*, no 32. AVRDC, Shankua, Taiwan.

Baró, F., Chaparro, L., Gómez-Baggethun, E., Langemeyer, J., Nowak, D. J., and Terradas, J. (2014). Assessing ecosystem services provided by urban forests in relation to air quality and climate change mitigation policies in Barcelona, Spain. *AMBIO*, vol 43, pp 466–479.

Baró, F., Haase, D., Gómez-Baggethun, E., and Frantceskaki, N. (2015). Mismatches between ecosystem services supply and demand in urban areas: a quantitative assessment in five European cities. *Ecological Indicators*, vol 55, pp 146–158.

Barthel, S., Folke, C., and Colding, J. (2010). Social–ecological memory in urban gardens: retaining the capacity for management of ecosystem services. *Global Environmental Change*, vol 20, no 2, pp 255–265.

Barthel, S., and Isendahl, C. (2013). Urban gardens, agriculture, and water management: sources of resilience for long-term food security in cities. *Ecological Economics*, vol 86, pp 224–234.

Blair, R. B. (1996). Land use and avian species diversity along an urban gradient. *Ecological Applications*, vol 6, no 2, pp 506–519.

Bolund, P., and Hunhammar, S. (1999). Ecosystem services in urban areas. *Ecological Economics*, vol 29, no 2, pp 293–301.

Bonan, G. B. (2002). *Ecological Climatology: Concepts and Applications*. Cambridge University Press, Cambridge UK.

Booth, D. B. (2005). Challenges and prospects for restoring urban streams: a perspective from the Pacific Northwest of North America. *Journal of the North American Benthological Society*, vol 24, no 3, pp 724–737.

Bowler, D. E., Buyung-Ali, L., Knight, T. M., and Pulllin, A.S. (2010). Urban greening to cool towns and cities: a systematic review of the empirical evidence. *Landscape and Urban Planning*, vol 97, no 3, pp 147–155.

Brenneisen, S. (2003). *Ökologisches ausgleichspotenzial von extensiven dachbegrünungen: bedeutung für den arten- und naturschutz und die stadtentwicklungsplanung*. Doctoral Dissertation, Institute of Geography, University of Basel.

Buchmann, C. (2009). Cuban home gardens and their role in social–ecological resilience. *Human Ecology*, vol 37, no 6, pp 705–721.

Camps-Calvet, M., Langemeyer, J., Calvet-Mir, L., and Gómez-Baggethun, E. (2015). Socio-cultural valuation of ecosystem services from urban gardens: case study from Barcelona, Spain. *Environmental Science and Policy*. Accepted.

Chichilnisky, G., and Heal, G. (1998). Economic returns from the biosphere. *Nature*, vol 391, pp 629–630.

Chiesura, A. (2004). The role of urban parks for the sustainable city. *Landscape and Urban Planning*, vol 68, no 1, pp 129–138.

Cho, S.-H., Poudyal, N. C., and Roberts, R. K. (2008). Spatial analysis of the amenity value of green open space. *Ecological Economics*, vol 66, no 2–3, pp 403–416.

Colding, J., and Folke, C. (2009). The role of golf courses in biodiversity conservation and ecosystem management. *Ecosystems*, vol 12, no 2, pp 191–206.

Costanza, R., Mitsch, W. J., and Day, J. W. (2006). A new vision for New Orleans and the Mississippi delta: applying ecological economics and ecological engineering. *Frontiers in Ecology and the Environment*, vol 4, no 9, pp 465–472.

Danielsen, F., Sørensen, M. K., Olwig, M. F. et al. (2005). The Asian tsunami: a protective role for coastal vegetation. *Science*, vol 310, no 5748, pp 643–643.

Depietri, Y., Renaud, F. G., and Kallis, G. (2012). Heat waves and floods in urban areas: a policy-oriented review of ecosystem services. *Sustainability Science*, vol 7, pp 95–107.

Elmqvist, T. (2011). Ecosystems, ecosystem services, and social dystems in urban landscapes. In Niemelä, J. et al. (eds.) *Urban Ecology: Patterns, Processes, and Applications*. Oxford University Press, Oxford.

Elmqvist, T., Fragkias, M., Goodness, J., Güneralp, B., Marcotullio, P. J., McDonald, R. I., Parnell, S., Schewenius, M., Sendstad, M., Seto, K. C., and Wilkinson, C. (eds.) (2013). *Urbanization, Biodiversity and Ecosystem Services: Challenges and Opportunities. A Global Assessment*. Springer, Dordrecht, Heidelberg, New York, London.

Elmqvist, T., Setälä, H., Handel, S. N., van der Ploeg, S., Aronson, J., Blignaut, J. N., Gómez-Baggethun, E., Nowak, D. J., Kronenberg, J., and de Groot, R. (2015). Benefits of restoring ecosystem services in urban areas. *Current Opinion in Environmental Sustainability*, vol 14, pp 101–108.

Escobedo, F. J., Kroeger, T., and Wagner, J. E. (2011). Urban forests and pollution mitigation: analyzing ecosystem services and disservices. *Environmental Pollution*, vol 159, no 8–9, pp 2078–2087.

Fang, C-F., and Ling, D-L. (2003). Investigation of the noise reduction provided by tree belts. *Landscape and Urban Planning*, vol 63, no 4, pp 187–195.

Farrugia, S., Hudson, M. D., and McCulloch, L. (2013). An evaluation of flood control and urban cooling ecosystem services delivered by urban green infrastructure. *International Journal of Biodiversity Science, Ecosystem Services and Management*, vol 9, no 2, pp 136–145.

Finlay, E. (2010). Green infrastructure to combat climate change. *Northwest Climate Change Action Plan and GRaBS project*. Community Forest Northwest, Edinburgh UK.

Fitzhugh, T. W., and Richter, B.. D (2004). Quenching urban thirst: growing cities and their impacts on freshwater ecosystems. *Bioscience*, vol 54, pp 741–754.

Gittleman, M., Jordan, K., and Brelsford, E. (2012). Using citizen science to quantify community garden crop yields. *Cities and the Environment (CATE)*, vol 5, no 1, p 4.

Glover, T. D. (2004). Social capital in the lived experiences of community gardeners. *Leisure Sciences*, vol 26, pp 143–162.

Gómez-Baggethun, E., and Barton, D. N. (2013). Classifying and valuing ecosystem services for urban planning. *Ecological Economics*, vol 86, pp 235–245.

Gómez-Baggethun, E., Gren, Å., Barton, D. N., Langemeyer, J., McPhearson, T., O'Farrell, P., Andersson, E., Hamstead, Z., and Kremer, P. (2013). Urban ecosystem services. In Elmqvist, T. (ed.) *Urbanization, Biodiversity and Ecosystem Services*. Springer, Dordrecht, Heidelberg, New York, London.

Gómez-Baggethun, E., and Martín-López, B. (2015). Ecological economics perspectives on ecosystem services valuation. In Martinez-Alier, J., and Muradian, R. (eds.) *Handbook on Ecological Economics*. Edward Elgar, Cheltenham, UK, Northampton, MA, USA.

Gotham, K., and Brumley, K. (2002). Using space: agency and identity in a public-housing development. *City and Community*, vol 1, pp 267–289.

Haase, D., Larondelle, N., McPhearson, T., Schwarz, N., Hamstead, Z., Kremer, P., Langemeyer, J., et al. (21 authors) (2014). Quantitative review of urban ecosystem services assessment: concepts, models and implementation. *AMBIO*, vol 43, pp 413–433.

Hanski, I., and Mononen, T. (2011). Eco-evolutionary dynamics of dispersal in spatially heterogeneous environments. *Ecology Letters*, vol 14, pp 1025–1034.

Hardin, P. J, and Jensen, R. R. (2007). The effect of urban leaf area on summertime urban surface kinetic temperatures: a Terre Haute case study. *Urban Forestry & Urban Greening*, vol 6, pp 63–72.

Hawkins, J. L., Thirlaway, K. J., Backx, K., Clayton, D. A. (2011). Allotment gardening and other leisure activities for stress reduction and healthy ageing. *HortTechnology*, vol 21, no 5, pp 577–585.

Ishii, M. (1994). Measurement of road traffic noise reduced by the employment of low 848 physical barriers and potted vegetation. *Inter-noise*, vol 29–31, pp 595–597.

Jacobi, P., Amend, J., and Kiango, S. (2000). Urban agriculture in Dar es Salaam: providing for an indispensable part of the diet. In Bakker, N., Dubbeling, M., Gündel, S., et al. (eds.) *Growing Cities, Growing Food, Urban Agriculture on the Policy Agenda. A Reader on Urban Agriculture*. ZEL, Felding, Germany.

Jansson, Å., and Polasky, S. (2010). Quantifying biodiversity for building resilience for food security in urban landscapes: getting down to business. *Ecology and Society*, vol 15, no 3, pp 20.

Kabisch, N. (2015). Ecosystem service implementation and governance challenges in urban green space planning – The case of Berlin, Germany. *Land Use Policy*, vol 42, pp 557–567.

Karathanasis, A. D., Potter, C. L., and Coyne, M. S. (2003). Vegetation effects on fecal bacteria, BOD, and suspended solid removal in constructed wetlands treating domestic wastewater. *Ecological Engineering*, vol 20, pp 157–169.

Keshavarz, N. (2015). History of urban gardens in Europe. In Bell, S., et al. (eds.) *Urban Allotment Gardens in Europe*. Routledge, London. In review.

Konijnendijk, C. C., Annerstedt, M., Busse Nielsen, A., and Maruthaveeran, S. (2013). *Benefits of Urban Parks: A Systematic Review*. International Federation of Parks and Recreation Administration (IFPRA), Copenhagen and Alnarp.

Krasny, M., and Tidball, K. (2009). Community gardens as contexts for science, stewardship, and civic action learning. *Cities and the Environment (CATE)*, vol 2, no 1, pp 1–18.

Kronenberg, J. (2015). Why not to green a city? Institutional barriers to preserving urban ecosystem services. *Ecosystem Services*, vol 12, pp 218–227.

Kruuse, A. (2011). GRaBS Expert Paper 6: The Green Space Factor and the Green Points System. In GRaBS (ed.) *The GRaBS Project*. Town and Country Planning Association & GRaBS, London.

Langemeyer, J. (2015). *Urban Ecosystem Services: The Value of Green Spaces in Cities*. Doctoral dissertation. Universitat Autònoma de Barcelona & Stockholm University.

Langemeyer, J., Baró, F., Roebeling, P., and Gómez-Baggethun, E. (2014). Valuing Cultural Ecosystem Services to Assess Urban Green Infrastrucutre Strategies: The Case of Park Montjuïc, Barcelona, Spain. *Ecosystem Services*. In press. 10.1016/j.ecoser.2014.11.016.

MA (2005). *Millennium Ecosystem Assessment: Ecosystems and Human Well-Being: Synthesis*. Island Press, Washington DC.

Martín-López, B., Gómez-Baggethun, E., García-Llorente, M., and Montes, C. (2014). Trade-offs across value-domains in ecosystem service assessment. *Ecological Indicators*, vol 37, pp 220–228.

Mbaye, A., and Moustier, P. (2000). Market-oriented urban agricultural production in Dakar. In N. Bakker, M. Dubbeling, S. Gündel, et al. (eds.) *Growing Cities, Growing Food, Urban Agriculture on the Policy Agenda. A Reader on Urban Agriculture*. Zentralstelle für Ernährung und Landwirtschaft (ZEL) Food and Agriculture Development Centre, Felding, Germany.

McPhearson, T. (2011). Toward a sustainable New York City: Greening through urban forest restoration. In Slavin, E. (ed.) *Sustainability in America's Cities: Creating the Green Metropolis*. Island Press, Washington DC.

McPherson, E. G. (1998). Atmospheric carbon dioxide reduction by Sacramento's urban forest. *Journal of Arboriculture*, vol 24, pp 215–223.

Meehl, G. A., and Tebaldi, C. (2004). More intense, more frequent, and longer lasting heat waves in the 21st century. *Science*, vol 305, no 5686, pp 994–997.

Melles, S.U.O., Glenn, S.M.G., and Martin, K.O.B. (2003). Urban bird diversity and landscape complexity: species–environment associations along a multiscale habitat gradient. *Conservation Ecology*, vol 7, no 1, 5: online http://www.conseco.org/vol7/iss1/art5.

Müller, N., Werner, P., and Kelcey, J. G. (2010). *Urban Biodiversity and Design*. John Wiley & Sons.

Newman, O. (1981). *Community of Interest*. Anchor Books, New York.

Nowak, D. J. (1994). Atmospheric carbon dioxide reduction by Chicago's urban forest. In McPherson, E. G., Nowak, D. J., and Rowntree, R. A. (eds.) *Chicago's Urban Forest Ecosystem: Results of the Chicago Urban Forest Climate Project*. US Dept. of Agriculture, Forest Service, Northeastern Forest, Radnor, PA, US.

Nowak, D. J. (1996). Notes: estimating leaf area and leaf biomass of open-grown deciduous urban trees. *Forest Science*, vol 42, no 4, pp 504–507.

Nowak, D. J., and Crane, D. E. (2000). The urban forest effects (UFORE) model: quantifying urban forest structure and functions. In Hansen, M., and Burk, T. (eds.) *Integrated Tools for Natural Resource Inventories in the 21st Century*. North Central Research Station, St. Paul.

Nowak, D. J., Hoehn III, R. E., Crane, D. E., et al. (2007). *Assessing Urban Forest Effects and Values: Philadelphia's Urban Forest*. U.S. Department of Agriculture, Newtown Square PA.

Orsini, F., Gasperi, D., Marchetti, L., Piovene, C., Draghetti, S., Ramazzotti, S., Bazzocchi, G., and Gianquinto, G. (2014). Exploring the production capacity of rooftop gardens (RTGs) in urban agriculture: the potential impact on food and nutrition security, biodiversity and other ecosystem services in the city of Bologna. *Food Security*, vol 6, pp 781–792.

Pataki, D. E., Carreiro, M. M., Cherrier, J., et al. (10 authors) (2011). Coupling biogeochemical cycles in urban environments: ecosystem services, green solutions, and misconceptions. *Frontiers in Ecology and the Environment*, vol 9, no 1, pp 27–36.

Potschin, M., and Haines-Young, R. (2011). Ecosystem services: exploring a geographical perspective. *Progress in Physical Geography*, vol 35, no 5, pp 575–594.

Pourias, J., Duchemin, E., and Aubry, C. (2015). Products from urban collective gardens: food for thought or for consumption? Insights from Paris and Montreal. *Journal of Agriculture, Food Systems and Community Development*, vol 5, no 2, pp 175–199.

Primmer, E., and Furman, E. (2012). Operationalising ecosystem service approaches for governance: do measuring, mapping and valuing integrate sector-specific knowledge systems? *Ecosystem Services*, vol 1, no 1, pp 85–92.

Rall, E. L., and Haase, D. (2011). Creative intervention in a dynamic city: a sustainability assessment of an interim use strategy for brownfields in Leipzig, Germany. *Landscape and Urban Planning*, vol 100, pp 189–201.

Rosenzweig, C., Solecki, W. D., Cox, J., et al. (12 authors) (2009). Mitigating New York City's heat island: integrating stakeholder perspectives and scientific evaluation. *Bulletin of the American Meteorological Society*, vol 90, no 9, pp 1297–1312.

Rosol, M. (2006). *Gemeinschaftsgärten in Berlin. Eine qualitative Untersuchung zu Potenzialen und Risiken bürgerschaftlichen Engagements im Grünflächenbereich vor dem Hintergrund des Wandels von Staat und Planung*. [Community gardens in Berlin. A qualitative study on potentials and risk of civic engagement in the field of urban green spaces in the context of landscape planning politics.] Doctoral Dissertation, Humboldt-Universität zu Berlin.

Shuster, W., Morrison, M., and Webb, R. (2008). Front-loading urban stormwater management for success: a perspective incorporating current studies on the implementation of retrofit low-impact development. *Cities and the Environment (CATE)*, vol 1, no 2, article 8, pp 1–15.

TEEB (2010). *The Economics of Ecosystems and Biodiversity. Ecological and Economic Foundations*. UNEP/Earthprint, London, Washington.

Tidball, K. G., Weinstein, E. D., and Krasny, M. E. (2014). Synthesis and conclusion: applying greening in red zones. In Tidball, K. G., and Krasny, M. E. (eds.) *Greening in the Red Zone*. Springer, Dordrecht, Heidelberg, New York and London.

Tyrväinen, L. (2001). Economic valuation of urban forest benefits in Finland. *Journal of Environmental Management*, vol 62, pp 75–92.
Ulrich, R. S. (1984). View through a window may influence recovery from surgery. *Science*, vol 224, pp 420–421.
United Nations (2014). Open working group proposal. Sustainable Development Goals. A/68/970. Available at: http://undocs.org/A/68/970
van den Berg, A. E., and Custers, M.H.G. (2011). Gardening promotes neuroendocrine and affective restoration from stress. *Journal of Health Psychology*, vol 16, pp 3–11.
van den Berg, A. E., Maas, J., Verheij, R. A., et al. (5 authors) (2010). Green space as a buffer between stressful life events and health. *Social Science & Medicine*, vol 70, no 8, pp 1203–1210.
Vauramo, S., and Setälä, H. (2010). Decomposition of labile and recalcitrant litter types under different plant communities in urban soils. *Urban Ecosystems*, vol 14, no 1, pp 59–70.
von Döhren, P., and Haase, D. (2015). Ecosystem disservices research: a review of the state of the art with a focus on cities. *Ecological Indicators*, vol 52, pp 490–497.
Wagner, I., Marshalek, J., and Breil, P. (2007). *Aquatic Habitats in Sustainable Urban Water Management: Science, Policy and Practice*. Taylor and Francis/Balkema, Leiden.
White, M. P., Alcock, I., Wheeler, B. W., and Depledge, M. H. (2013). Would you be happier living in a greener urban area? A fixed-effects analysis of panel data. *Psychological Science*, vol 24, pp 920–928.
Zahran, S., Brody, S. D., Vedlitz, A., et al. (5 authors) (2008). Vulnerability and capacity: explaining local commitment to climate-change policy. *Environment and Planning C: Government and Policy*, vol 26, no 3, pp 544–562.
Zetterberg, A. (2011). *Connecting the Dots: Network Analysis, Landscape Ecology, and Practical Application*. Doctoral dissertation, KTH Royal Institute of Technology.

Briefing Note 36.1 Green infrastructure and ecosystem services

Susannah Gill

The terms 'green infrastructure' and 'ecosystem services' are increasingly used in understanding and planning natural and semi-natural environments in both rural and urban areas. In practice, the two terms are sometimes used interchangeably; at other times they are taken to be competing approaches, with policymakers and practitioners, in the UK at least, aligning themselves with one or the other. In this short discussion piece, the two terms are considered in order to establish a better understanding and alignment of the links between them.

Both green infrastructure and ecosystem services have many definitions. One definition of green infrastructure is that it is 'an interconnected network of green space that conserves natural ecosystem values and functions and provides associated benefits to human populations' (Benedict and McMahon, 2002, p.12), whilst ecosystem services have been defined as 'the benefits people obtain from ecosystems' (Millennium Ecosystem Assessment, 2003, p.3). These definitions suggest that it can be understood that green infrastructure can provide ecosystem services. Indeed, a review of both academic (e.g. Cilliers et al., 2013; Farrugia et al., 2013; Lovell and Taylor, 2013; Gill et al., 2007) and more policy or practitioner-oriented (e.g. European Commission, 2013; Natural Economy Northwest, 2010; Natural England, no date) literature suggests that this is the most common understanding of the relationship between the two.

Both green infrastructure and ecosystem services are systems approaches, and a similar logic has emerged for each (Plate 36.1). For ecosystem services, a 'cascade model' has been used to summarise the logic (Potschin and Haines-Young, 2011), employing terminology such as *biophysical structure or process, function, service, benefit*, and *value*. Similarly, for green infrastructure, practitioners have developed a model which links *type* to *function, benefit*, and *value* (e.g. North West Green Infrastructure Think Tank, 2008; The Mersey Forest, 2010). This can be seen, to some extent, as mirroring the

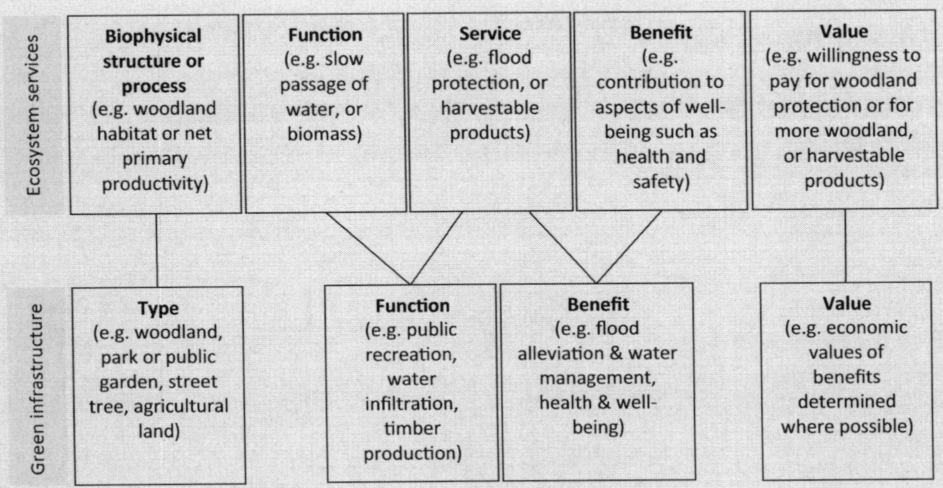

Plate 36.1 The connections between the terminology employed in the ecosystem services 'cascade model' (Potschin and Haines-Young, 2011) and a green infrastructure model (North West Green Infrastructure Think Tank, 2008; The Mersey Forest, 2010). The connecting lines show the links between the terms.

'cascade model' (Plate 36.1). Whilst there may be slight differences of interpretation in the precise use of terms such as *function* and *benefit*, this muddling of terminology is also apparent within each field (e.g. De Groot et al, 2002). According to these models, green infrastructure types can therefore be understood to be a complex indicator for specific ecosystem services, based on the accumulated knowledge about their biophysical structures, processes, and functions. As such, it makes sense to see green infrastructure and ecosystem services as different terms relating to the same approach, rather than as distinctive approaches.

Perhaps one of the main differences relates to their relative and historical prominence within the academic and policy worlds. Ecosystem services historically have a much stronger academic underpinning, as evidenced by the occurrence of the term in the academic literature, whereas it could be argued that green infrastructure is the more policy-facing of the two terms. The term ecosystem services first appears in the academic literature in 1983 (Ehrlich and Mooney, 1983) and now has 8,608 references[1] (Plate 36.2) and, as of 2012, a dedicated journal entitled *Ecosystem Services*. Gómez-Baggethun et al. (2010) suggest that it is with the Millennium Ecosystem Assessment, in 2003, that ecosystem services started to become more firmly embedded in policy agendas. By contrast, it seems that the concept of green infrastructure has largely been developed and applied in policy and practice (e.g. Mell, 2010; Gómez-Baggethun and Barton, 2013). According to Pankhurst (2012), the term green infrastructure was first applied in water management and land use planning contexts in the United States in the 1980s-1990s, which suggests a more practice-oriented origin. Green infrastructure first appears in the academic literature in 1995 (Hauserman, 1995; Walmsley, 1995), which is 12 years later than the appearance of ecosystem services. It is also much less referenced, with 468 academic references[1] (Plate 36.2), or a ratio of 1:18 in comparison to those for ecosystem services.

The two also appear to have had slightly different audiences in terms of their policy influence, at least in the UK. Ecosystem services tend to be largely rural in focus and relate to environmental

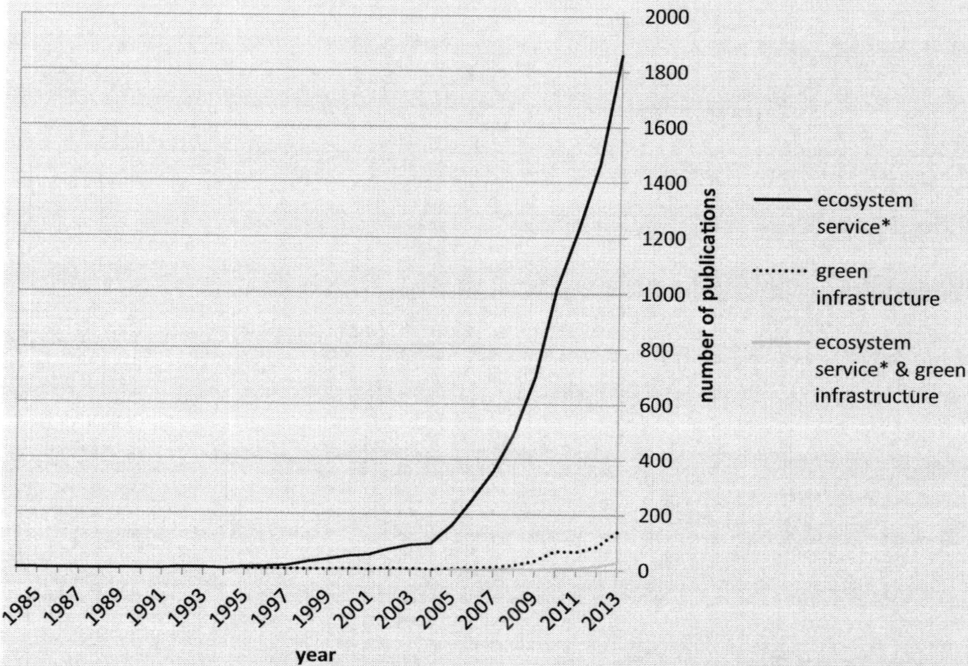

Plate 36.2 References to ecosystem services and green infrastructure within the academic literature[1].

management and biodiversity conservation; whilst green infrastructure tends to be largely urban or peri-urban in focus and relate to planning and regeneration. This is reflected in the language used by each. For example, the ecosystem service language refers to ecosystems and habitats, which are used as the basis for assessments. The assessments are often undertaken for natural or semi-natural ecosystems, but can also include urban ecosystems (e.g. UK National Ecosystem Assessment, 2011; Gómez-Baggethun and Barton, 2013). On the other hand, the green infrastructure language refers to a typology which is the basis of mapping assessments, rather than to ecosystems or habitats. The green infrastructure typologies are sometimes derived from urban planning classifications for green space (e.g. The Mersey Forest, 2010; the types used were developed from the now superseded Planning Policy Guidance 17, Department of Communities and Local Government, 2002) in order to make them more applicable in practice. In addition, the word 'infrastructure' is used specifically to appeal to an urban planning audience who are already familiar with the provision of grey infrastructure such as roads.

Despite their relative and historical differences in academic and policy worlds, there is an increasing alignment between green infrastructure and ecosystems services. Within the academic literature the terms were first brought together in 2005 (Moll, 2005), and 63 publications now refer to both (Plate 36.2). In public policy discourse, green infrastructure is often the term that is used to apply to urban ecosystems, which then provide a range of ecosystem services (Gómez-Baggethun and Barton, 2013). As such, it is argued here that green infrastructure and ecosystem services should be seen as different terms relating to the same approach, complementing and supporting each other, rather than as distinctive and conflicting approaches. In many respects, ecosystem services can be seen as the science that underpins green infrastructure planning, whilst green infrastructure may be viewed as the public policy discourse relating to ecosystem services, especially, but not exclusively, within an urban context. It can be understood that green infrastructure can provide ecosystem services.

Acknowledgements

Thank you to Renate Bürger-Arndt, University of Göttingen, and Julie Marsaud, France Nature Environnement for reviewing and discussing this paper as part of transnational cooperation for the EU Interreg IVC ForeStClim project.

Note

1 The figures used here were derived by searching the article title, abstract, keywords within the Scopus database for "ecosystem service*", "green infrastructure", and the two combined, after Potschin and Haines-Young (2011). Using different databases, such as Web of Science, yields different figures (e.g. Costanza and Kubiszewski, 2012).

References

Benedict, M.A., and McMahon, E.T. (2002). Green infrastructure: smart conservation for the 21st century. *Renewable Resources Journal*, vol 20, no 3, pp 12–17.

Cilliers, S., Cilliers, J., Lubbe, R., and Siebert, S. (2013). Ecosystem services of urban green spaces in African countries-perspectives and challenges. *Urban Ecosystems*, vol 16, no 4, pp 681–702.

Costanza, R., and Kubiszewski, I. (2012). The authorship structure of "ecosystem services" as a transdisciplinary field of scholarship. *Ecosystem Services*, vol 1, no 1, pp 16–25.

De Groot, R.S., Wilson, M.A., and Boumans, R.M.J. (2002). A typology for the classification, description and valuation of ecosystem functions, goods and services. *Ecological Economics*, vol 41, pp 393–408.

Department for Communities and Local Government (2002). *Planning Policy Guidance 17: Planning for Open Space, Sport and Recreation*. Available at: http://webarchive.nationalarchives.gov.uk/20120919132719/www.communities.gov.uk/archived/publications/planningandbuilding/planningpolicyguidance17 (accessed 30 April 2014).

Ehrlich, P.R., and Mooney, H.A. (1983). Extinction, substitution, and ecosystem services. *Bioscience*, vol 33, no 4, pp 248–254.

European Commission (2013). *Building a Green Infrastructure for Europe*. Available at: http://ec.europa.eu/environment/nature/ecosystems/docs/green_infrastructure_broc.pdf (accessed 30 April 2014).

Farrugia, S., Hudson, M.D., and McCulloch, L. (2013). An evaluation of flood control and urban cooling ecosystem services delivered by urban green infrastructure. *International Journal of Biodiversity Science, Ecosystems Services and Management*, vol 9, no 2, pp 136–145.

Gill, S.E., Handley, J.F., Ennos, A.R., and Pauleit, S. (2007). Adapting cities for climate change: the role of the green infrastructure. *Built Environment*, vol 33, no 1, pp 115–133.

Gómez-Baggethun, E. and Barton, D.N. (2013). Classifying and valuing ecosystem services for urban planning. *Ecological Economics*, vol 86, pp 235–245.

Gómez-Baggethun, E., de Groot, R., Lomas, P.L., and Montes, C. (2010). The history of ecosystem services in economic theory and practice: from early notions to markets and payment schemes. *Ecological Economics*, vol 69, no 6, pp 1209–1218.

Hauserman, J. (1995). Green infrastructure and landscape architects help design a statewide greenway network in Florida. *Landscape Architecture*, vol 85, no 7, pp 58–61.

Lovell, S.T., and Taylor, J.R. (2013). Supplying urban ecosystem services through multifunctional green infrastructure in the United States. *Landscape Ecology*, vol 28, pp 1447–1463.

MA (2003). *Ecosystems and Human Well-Being: A Framework for Assessment*. Island Press, London. Available at: www.maweb.org/en/Framework.aspx (accessed 30 April 2014).

Mell, I.C. (2010). *Green Infrastructure: Concepts, Perceptions and Its Use in Spatial Planning*. PhD thesis, Newcastle University. Available at: https://theses.ncl.ac.uk/dspace/bitstream/10443/914/1/Mell10.pdf (accessed 30 April 2014).

Moll, G. (2005). Repairing ecosystems at home. *American Forests*, vol 111, no 2, pp 41–44.

Natural Economy Northwest (2010) *Green Infrastructure Prospectus: A Prospectus for Green Infrastructure – Underpinning the Sustainable Development of Northwest England*. Available at: www.greeninfrastructurenw.co.uk/resources/Prospectus_V6.pdf (accessed 30 April 2014).

Natural England (n.d.) Green infrastructure. Available at: www.naturalengland.org.uk/ourwork/planningdevelopment/greeninfrastructure (accessed 30 April 2014).

North West Green Infrastructure Think Tank (2008). *North West Green Infrastructure Guide*. Available at: www.ginw.co.uk/resources/GIguide.pdf (accessed 30 April 2014).

Pankhurst, H.J. (2012). *Green Infrastructure: Mainstreaming the Concept. Understanding and Applying the PRINCIPLEs of Green Infrastructure in South Worcestershire*. Natural England Commissioned Reports, Number 079, Available at: http://publications.naturalengland.org.uk/publication/46011 (accessed 30 April 2014).

Potschin, M.B., and Haines-Young, R.H. (2011). Ecosystem services: exploring a geographical perspective. *Progress in Physical Geography*, vol 35, no 5, pp 575–594.

The Mersey Forest (2010). *Liverpool Green Infrastructure Strategy – Technical Document*. Prepared for Liverpool City Council and Liverpool Primary Care Trust. Available at: www.ginw.co.uk/liverpool/Technical_Document.pdf (accessed 30 April 2014).

UK National Ecosystem Assessment (2011). *The UK National Ecosystem Assessment: Synthesis of the Key Findings*, UNEP-WCMC, Cambridge UK. Available at: http://uknea.unep-wcmc.org/ (accessed 30 April 2014).

Walmsley, A. (1995). Greenways and the making of urban form. *Lands-cape and Urban Planning*, vol 33, no 1–3, pp 81–127.

PART IV

Ecosystem services

Linking and informing agendas – introduction

All of the preceding contributions to this *Handbook* point to a field of research that is developing a durable set of concepts and methodologies for understanding the environmental basis of human well-being. How these concepts and methodologies might serve to link and inform wider agendas across science and policy is the focus of the following chapters.

From a general starting point, we know that the framework of ecosystem services is holistic in outlook and that, in principle, the breadth of concerns falling within its ambit is very wide indeed. We also know that, to a significant extent, the field's origins are trans-disciplinary. It arises from a community much larger than academic research alone, and so the boundaries between the theory and practice of an ecosystem services perspective, or indeed between producers and users of ecosystem services knowledge, cannot be sharply delineated. A commitment to application and animating knowledge in a practical context is thus part of the perspective's 'DNA'.

And yet, the integrating, synthesising and applied ambitions of the perspective need to be tempered with the challenge of 'mainstreaming'. This is a maturing field, but the added value of taking an ecosystem services perspective, or applying the ecosystems approach, remains tentative and fragile with respect to many salient areas of science and policy. Many of the chapters that make up this part are designed to elaborate emerging thinking at the interface of ecosystems services and these salient areas: matters of health, food and water security, climate change, planning and green infrastructure and so forth. How does an ecosystem services perspective inform these areas, and how might these areas inform the future of ecosystem services research? These questions are also given context by opening and closing contributions that tackle the wider policy, institutional and practical arrangements that foster, govern and sometimes constrain the advancement of the field and the way ecosystem service-based knowledge is utilised and applied.

In the opening chapter to this part, ten Brink and Kettunen set out the landscape within which the policy mainstreaming of ecosystem services is occurring, cautioning that there remain major needs and opportunities for the integration and mainstreaming of the ecosystem services concept across diverse policy areas. Asserting influence in a world of multiple sectoral and horizontal policy objectives and incentives, stakeholder interests and institutional roles is a daunting challenge, and for these authors progress will be dictated partly by being transparent in purpose, open to alternative approaches, and building the evidence base across scales as part of a multi-level governance process.

Many of the mainstreaming contexts that are highlighted by ten Brink and Kettunen are elaborated in detail in the subsequent contributions. We begin by exploring how an ecosystem service perspective maps on to strategies now pervading issues of climate change. The contribution by Locatelli shows how ecosystem services play an important role in strategies for both climate change mitigation and adaptation, and how these strategies are being adopted to varying extents within policy. While ecosystem services are considered by the author as part of the solution to climate change, they are also affected by changing climatic conditions. Thus ecosystem-based approaches to climate change should also consider adaptation for ecosystems as a further important dimension of policies and local initiatives.

In the contributions by Power, and Hendry and Gooch, we consider how ecosystem services might be framed within debates about food and water security. Food security has represented a paradigmatic global challenge since the 1970s, and surfaced and re-surfaced in a range of definitional guises. In her chapter Power sets out the pre-occupying concerns of the current debate, showing how food security for small holder farmers and rural populations is presaged on the provision of a whole range of other (non-production) ecosystem services. An ecosystem service perspective may help animate and expand thinking in a field generally pre-occupied with the development and roll-out of new agricultural technologies. In turn, Hendry and Gooch consider water as providing a range of ecosystem services and examine how the concept of ecosystem services is both framing and re-conceptualising the water policy agenda within an emerging water security paradigm.

In their chapter, exploring the links between poverty and ecosystem services, Schaafsma and Fisher offer an overview of the contribution of ecosystem services to well-being in developing countries, arguing that different ecosystem services are necessary to both provide a stable flow of goods and benefits and the means to deal with one-off shocks and disturbances. They discuss different discourses on the relationship between changes in poverty and ecosystem services and emphasise the importance of access and power in the distribution of ecosystem goods and benefits when considering the trade-offs in ecosystem services management and poverty issues.

The importance of addressing connections between biodiversity, ecosystem services and human health are addressed in the chapter by Kretsch, who notes several important barriers to effective practical co-operation between relevant sectors on these issues. Although policy interest is growing, these challenges include *inter alia* a general lack of mutual understanding of relevant issues, differences in technical language, differing geographic and temporal scales of operation and narrow perspectives of individual remits. The author explores some of these ecosystem-based approaches to population health, discussing linkages to current international policies on health and biodiversity and wider relevance to other sectors.

The idea of 'greening' business is a persistent concern of policies for sustainability, and arguably a context where the field of ecosystem services can assert significant influence on thinking. The contribution by Duke explains how the concepts of natural capital and ecosystem services are increasingly having purchase on businesses, and indeed are helping to transform and refine the way businesses think about their relationship with the natural world. This includes the uptake of natural capital accounting and the development of new standards and metrics which businesses can apply to guide and measure performance. The author sees innovations in the finance sector in support of green growth, and increasing consumer awareness of the environmental impacts of businesses, as helping to drive these developments.

In the chapter by Brouwer, we are offered insight into the related development of market-based mechanisms for delivering ecosystem services, specifically 'Payments for Ecosystem Service' (PES) schemes. This is an important area, in which the practical ambitions and controversies of the ecosystem services perspective are being played out. An important message of this chapter is the considerable variability that defines approaches to PES and that, despite these schemes having a number of virtues, the factors that contribute to their functioning remain poorly understood.

The challenge of maximizing biodiversity and ecosystem service benefits in conservation decision-making provides the focus of a chapter by Grantham and colleagues. Drawing out the potential tension between a focus on ecosystem services and outcomes for biodiversity, these authors set out the types of steps that would need to be taken to align thinking and ensure mutually reinforcing outcomes.

The contributions by Opdam and von Haaren and colleagues offer companion pieces on a major context for innovation in the field of ecosystem services: planning processes. Opdam's contribution explores the challenge of incorporating consideration of services into planning processes from a landscape perspective. The need for tools and methods that foster collaboration, creativity and ownership are key to this author's vision, as is the need to be flexible in the way in which core concepts and languages of the field are used. Opdam suggests, for instance, that the term 'landscape services' may be more appropriate to bridging and connecting terms. The contribution of von Haaren et al. also considers the landscape scale, but links discussion more directly to the concerns of integrated spatial planning. They present a strong picture of the potential of an ecosystem services perspective to make up for shortcomings of conventional approaches, but point as well to contexts and sensitivities by which planners may be reluctant to adopt an ecosystem services logic. These chapters also include a Briefing Note authored by Geneletti on the themes of Impact Assessment to support understanding of some core related concepts.

In our final two chapters we draw back again to the wider policy and institutional context in which ecosystem services concepts and knowledge are adopted and applied. In the contribution by Primmer, the case for an institutional perspective is made. There is a need to understand the formal and informal institutions that shape how ecosystems are used and managed, and build institutions that embed the analysis of ecosystem services in their governance functions. This Part closes with a review by Russel and colleagues of the theory and practice of knowledge use in decision-making and an analysis of why certain forms of ecosystem service knowledge are used in particular ways. The authors explore the possible strategies and activities to enhance ecosystem services knowledge utilisation in decision-making, and argue that the way in which this knowledge is used in practice often substantially varies from the direct, linear form which many natural scientists expect to find. In a fitting close to the part, these authors argue that expectations of when knowledge can affect policy need to be adjusted to arrive at more realistic understandings of use.

Key discussion and debating points

- Ecosystem services is a synthesising framework, but its uptake and utilisation across relevant sectors and policy areas is uneven. What institutional and practical arrangements would be necessary to further mainstream and strengthen this perspective in research and policy areas with related, if quite different, traditions and models of working?
- The greening of business provides one salient context in which ideas of natural capital and ecosystem services are being asserted. What factors foster or impede recognition of these concerns in business context? How can the business case be better made and approaches developed that better incorporate sustainability considerations into the concerns and practices of business?
- Market-based mechanisms, such as 'Payments for Ecosystem Services' schemes, are an emerging feature of ecosystem service management. What are the risks and opportunities

associated with the development of market-based mechanisms for ecosystem service management? How should market-based mechanisms be governed and their outcomes monitored?

- The generation of knowledge is an important facet of any research field, of which ecosystem services is no exception. Yet linear models of knowledge creation and utilisation have been widely challenged and discredited within the social sciences. What alternative models of knowledge production and exchange might be used to advance an ecosystem services perspective in decision-making, and how might these be realised in ways that challenge and extend how we think about valid and useful knowledge?

37

A POLICY PERSPECTIVE ON MAINSTREAMING ECOSYSTEM SERVICES

Opportunities and risks

Patrick ten Brink and Marianne Kettunen

Ecosystem services – an integral part of 21st-century biodiversity policy

The ecosystem service (ES) concept is a major focus for discussion within biodiversity policy. Globally, ES are an integral element of the global Strategic Plan for Biodiversity 2011–2020 under the UN Convention for Biological Diversity (CBD), to which over 190 parties committed at the CBD COP 10 in Nagoya,[1] Japan in 2010 (CBD, 2010). Several global biodiversity targets for 2020 (so-called Aichi Targets) are dedicated to the conservation and restoration of ecosystem services. In particular, Aichi Target 14 addresses 'essential ecosystem services'[2] and Target 15[3] is focused on carbon sequestration and natural hazards management.[4] Aichi Targets 1 and 2[5] focus on the values of biodiversity, with ecosystem services values being implicitly integrated. These targets provide a commitment from signatories to recognise and integrate ESS into decision-making at all appropriate levels. The establishment of the Intergovernmental Science-Policy Platform on Biodiversity and Ecosystem Services (IPBES) in 2012 is another example of a global commitment to the ES concept, complementing the global targets. The platform is intended to improve the Science Policy Interface with specific focus on ecosystem services,[6] which should support the implementation of the Aichi Targets.

National Biodiversity Strategies and Action Plans (NBSAPs) under the CBD need to integrate the targets of the Strategic Plan for Biodiversity 2011–2020, and ecosystem services are becoming integrated in national policies, strategies and implementation around the world (UNEP-WCMC and IEEP, 2013). At a European level, for example, a dedicated target of the EU Biodiversity Strategy to 2020[7] on ecosystem services states that '*by 2020, ecosystems and their services are maintained and enhanced by establishing green infrastructure and restoring at least 15% of degraded ecosystems' (Target 2).*

Although the Millennium Ecosystem Assessment (MA) was only published in 2005, and global commitments can take years to develop and agree, the relatively rapid uptake of ES in global biodiversity policy can be explained by the following:

- The evidence of on-going ecosystem degradation and loss of biodiversity, and the growing appreciation that this degradation and loss is leading to a loss of ecosystem services, many of which are public goods (of public benefits and open access to the public, without markets or prices) important for human well-being.

- A recognition that the coherence, effectiveness and efficiency of public policies and private decisions can be improved through the integration of nature and ecosystem services.

In addition, the uptake of ES builds on an existing foundation of the appreciation of the values of nature in academia (e.g. seminal works by Daily, 1997 and Costanza et al., 1997) and in policy circles, even if the term ES was often not used explicitly. The concept of nature's value to mankind already gained global attention when the CBD was agreed in Rio in 1992, which explicitly and implicitly recognised the importance of nature, as seen by the definition of biological resources as including *'genetic resources, organisms or parts thereof, populations, or any other biotic component of ecosystems with actual or potential use or value for humanity'* (Article 2, CBD, 1992).

While the concept of ES has been effectively mainstreamed in biodiversity policy, it still needs to be integrated more widely across policy areas.

Implementing ecosystem services in practice: mainstreaming the concept into sectoral and horizontal policies

A major challenge and opportunity for the ecosystem service concept is that of its integration beyond biodiversity policy into other environmental, sectoral and horizontal policies.[8] This is needed to avoid inappropriate trade-offs between ES and between ES and biodiversity conservation across policies. It is also needed to help identify opportunities where the appreciation and understanding of ES can create win-win solutions for delivering different policy objectives. The integration of ES into sectoral and horizontal policies will improve the policy coherence and added value of individual policies. Examples across policy areas are illustrated below. Which of these policy areas and related mechanisms will offer the greatest potential will be country-specific, depending on national circumstances and dedicated commitments.

One area of major potential for enhancing policy coherence is that of climate change. Nature-based solutions, i.e. policy and management responses that build on the maintenance or restoration of nature's functioning and processes, can be used to support the achievement of climate-related policy goals. These solutions build on the understanding of how ES can contribute to such things as the cost-effective climate change mitigation (e.g. peatland protection and restoration, protection of old growth forests), adaptation to the adverse impacts of climate change and natural hazards management (e.g. forests reducing risk of flooding, avalanches or mudslides). On climate mitigation, key policy processes include the global United Nations Framework Convention on Climate Change (UNFCCC) and related national and local climate strategies implementing global commitments. Improving the recognition of nature-based means for mitigation within these processes through the REDD+ instrument (a financial mechanism, Reducing Emissions from Deforestation and Degradation, to internalise carbon related ecosystem services provided by forests; see Luque and Iverson, 2016), for example, can support both climate and biodiversity objectives. On climate adaptation and disaster mitigation, improving the links between the UNFCCC, the UNCCD[9] and UNCBD and associated implementation of commitments would be important, helping climate mitigation and adaptation while simultaneously addressing land desertification, degradation and biodiversity conservation. Climate policy is, however, also a potential source of policy dissonance where it runs counter other policy objectives. For example, monoculture carbon plantation forests can be low in biodiversity and also lead to a smaller set of ecosystem service provision to the detriment of others (e.g. cultural services) and attention is needed in designing REDD+ schemes and carbon plantations more generally to ensure that the potential multiple ES benefits (or losses) are not overlooked (ten Brink et al., 2011).

Another fundamental challenge is that of water security and the provision of clean water. (see also Hendry and Gooch, 2016) Both can benefit from nature's role in water retention, water and waste regulation and water provision (Russi et al., 2013). Here integration can be facilitated by making nature-based solutions for water management an integral part of implementing Millenium Development Goals (MDGs) and Sustainable Development Goals (SDGs) through, for example, international development cooperation and national water security measures. Multi-country river basin management processes are also venues for integrating ES and potentially support international diplomacy where water is a source of international conflict. A dialogue around the evidence for ecosystem-based management approaches in multi-country river basins could potentially contribute to diplomacy initiatives, conflict resolution and solution identification. Where ecosystems are fundamental to the provision of clean water and/or water use decisions (e.g. dam construction, over-abstraction) risk undermining biodiversity and ecosystem functions, services and benefits to society, there are particular merits in taking account of the roles of nature in discussions, decisions and cooperation. Examples of where this could arguably be useful include: the Volta River Basin in Ghana and Burkina Faso, the Okavango Delta in southern Africa, the Aral Sea in central Asia and the Mekong, which runs from the Tibetan plateau down to Vietnam in East Asia. While there are synergies between biodiversity and water security over the short- to long-term, as biodiversity often is fundamentally important for clean water provision, there can also be conflicts as people, industry and nature 'compete' for water. When short-term water security objectives lead to over-extraction, such that less is available for ecosystems to function (i.e. if the water tables fall), it can lead to ecosystem tipping points with loss of biodiversity and, in some cases, ecosystem changes – i.e. from woodland to grassland, leading to often quite different ecosystem service provision.

The long term viability of livelihoods and food security can be improved through ecosystem-based management of fisheries, helping ensure that viable fish stock are maintained by, for example, protecting the nursery functions of coastal ecosystems (TEEB, 2011). On land, sustainable agricultural management practices can help avoid soil erosion, maintaining the agro-ecosystem's natural ability to maintain both soil fertility and carbon stocks. Similarly, investment in protecting the genetic diversity of crops can help crop resilience, while the protection of wild pollinators and natural predators can support food production and lead to cost savings from avoided pollination costs, avoided losses from reduced yields and lesser costs of pesticides and herbicides (Hajjar et al., 2008). At the same time, agricultural policies and practice can lead to policy conflict as policies and incentives focusing on short-term outputs may lead to a loss of soil fertility, carbon storage and water retention potential from soils. Furthermore, the conversion of biodiverse areas to agricultural land can involve important trade-offs between biodiversity, ecosystem services and food provisioning (TEEB, 2011). The use of certain pesticides and the loss of natural areas can have negative effects on wild pollinators and thus result in loss of biodiversity, pollinating services and, subsequently, reduced farm output (Vanbergen et al., 2013, Power, 2016).

For the energy and transport sectors, the conflict with biodiversity is well known: habitat fragmentation through infrastructure, pollution and water abstraction each affect biodiversity and ecosystem services provision. Large hydroelectric dams can lead to major inundation upstream and falling water tables downstream, which can lead to ecosystem tipping points and biodiversity losses as well as social impacts when upstream relocation of human settlements is necessary (Koenig, 2002). This provides a fundamental challenge and constant source of policy conflict. Environmental impact assessments (EIA) and other assessment tools can be useful in informing decision-making and minimising trade-offs, but have often proved weak tools in the face of calls for economic development. Evidence of the values of ecosystem service losses could help strengthen policymakers' resolve to make better use of EIAs. As regards synergies, there are cost-effective nature-based solutions for the energy sector that merit more attention, such as providing shading of rivers, which could reduce possible constraints on use of cooling waters for power stations, or reduced soil erosion, which could lead to losses of output from hydropower plants (TEEB, 2011; Russi et al., 2013).

The potential contribution of nature to poverty alleviation, regional development and the wider transition to a green economy[10] is also being understood and emphasised (ten Brink et al., 2012; IEEP and Milieu, 2013). On poverty alleviation, there are opportunities for policy synergies through development cooperation and national poverty alleviation programmes that recognise the social benefits of ecosystems and integrate them in investment decisions. Similarly, the UNCSD[11] (Rio +20 process) and associated MDGs and SDGs promise to be important policy processes to integrate ecosystem services for poverty alleviation. As regards national and regional development, national plans and regional strategies can be used to encourage win-wins through building on national and local nature-related branded products, creating cost savings from the provision of clean water, increasing opportunities for recreation and (nature-based) tourism and attracting investment as a result of better environmental quality. Again, it is important to understand the risks of trade-offs (e.g. tourism activities on the environment), and put in place measures to safeguard assets and avoid unsustainable losses. Finally, innovations for green economy can usefully build in benefits created by bio-innovations such as biomimicry and pharmaceuticals. Improving access and sharing benefits under the CBD Access and Benefit Sharing regimes could prove a key driver both for innovation in pharmaceuticals and innovation and eco-efficiency more widely. Green economy strategies, national development plans and NBSAPs could also each be used as tools to help take account of and enable potential win-wins and hence be a source of policy coherence.

Ecosystem services can also support health objectives through, for example, benefits from improved air quality and from mitigated heat island effects in cities, access to green areas, and the range of medicines linked to genetic materials and traditional knowledge (ten Brink et al., 2012, see also Fish et al., 2016). National health strategies are useful governance tools here, as are municipal climate and health policies, plans and investments.

Finally, there is also a need for a reappraisal as to how the financial sector, notably insurance, banking and pension-investment funds, take account of nature, nature-based solutions and costs of damages in their policies, premia charged and investment choices.

Approaches and means for mainstreaming ES into policy

Progress with the mainstreaming of ES in sectoral and horizontal policies varies across policy areas and governance levels (global, national, regional and local as well as private sector policies). In general, mainstreaming and integration needs to take place at three different levels – conceptual integration (where policy documents explicitly or implicitly take ES into account), operational integration (where specific measures or instruments that address ES-related objectives are identified and committed to) and, finally, integration through implementation (where measures achieve integration on the ground in investment decisions) (Kettunen et al., 2014). For example, within the European Union there has been a range of policy developments and reforms that have provided opportunities for taking up the ecosystem service concept across different sectoral policies, potentially directly affecting the Union's 28 member countries – there is good conceptual integration for a range of EU policy areas and generally weaker operational integration (Kettunen et al., 2014).

The integration of the ES concept into policy development and implementation needs a good evidence base, a range of tools and instruments (e.g. impact assessments and strategies), engagement by different stakeholders and mobilization of resources to facilitate the uptake. Integrating scientific knowledge on multiple benefits of ES in policymaking throughout the policy cycle – from policy framing to policy formulation, negotiation, implementation and review – is essential. Different environmental impact assessment tools can have these benefits, for example, if applied early enough, thoroughly enough and if the results are taken into account. Strategies and action plans can also be useful processes for coherence and good governance – e.g. the NBSAPs, green economy strategies and national development plans each have the potential to support the integration of ecosystem services and nature based solutions. Resource mobilization is equally essential, and demonstrating nature-based solutions to objectives behind bilateral aid and development assistance can offer a way forward. Similarly, the reform of incentives harmful to biodiversity (and wider reform of Environmentally Harmful Subsidies, EHS) will also potentially reduce pressures on the environment while liberating funds (Oosterhuis and ten Brink, 2014) which can help meet policy objectives more efficiently.

More generally, strengthening the implementation of the Polluter Pays Principle (PPP) via regulation and economic incentives will be essential to halt biodiversity loss, protect the flow of ecosystem services and encourage investment in natural capital. Strengthening the beneficiary (or user) pays principle and beneficiary-provider gets paid principles (i.e. by having more Payments For Ecosystem Services (PES, see also Brouwer, 2016) or other incentives, such as tax relief or reverse auctions), will also help, though there are limits to how much PES can fund due to resource availability and ethical and cultural resistance to the ideas of payments for services, where this is seen simply as 'responsible behaviour' (ten Brink et al., 2011). There is the further challenge of implementation at practical levels such as permitting (e.g. land-use change permits, which can impact ES), inspection and non-compliance enforcement, which require a range of other stakeholders (e.g. permitting agents, inspectors, judges), institutions (e.g. courts and the rule of law), governance processes and tools (e.g. fees, fines and criminal sentences). This requires institution-building, capacity-building, political will and individuals to champion change.

The uptake and integration of the ecosystem service concept into different policy areas can take place through a range of processes, requiring a range of institutional roles and tools. Which areas are ripe for progress and which tools will prove helpful is country-dependent. The main opportunities do lie with national-level sectoral policies and decision-making processes in countries across the world, supported or encouraged by global agreements, conventions and protocols. Furthermore, in certain countries, regional (i.e. state) and local-level policies can also play a crucial role in the uptake through planning processes, investment and other instruments (e.g. as has been seen in the EU, Australia, the US, Brazil and Japan).

While, in principle, there is therefore promise for improved policy coherence and reduced policy dissonance, there remain major challenges – political will, vested interests, institutional roles, time, geography and economics itself. Decisions are taken within certain jurisdictions (e.g. by institutional and geographic reach), represent certain interests (e.g. sectoral, geographic) and are usually short-term in nature. Impacts abroad, impacts in the medium- to long-term and impacts on groups that are under-represented or not represented at all are often given less attention in decision-making. Similarly, decisions often favour the economic, arguably biasing decision-making by not fully factoring in other values. Given the global, long-term and public goods nature of the biodiversity and ecosystem services, this would suggest that decision-makers need to broaden their horizons – becoming 'ombudsmen' for future generations and more responsible statesmen and stateswomen in a world run by national self-interest, where short-term private gain is more important than (longer-term) public goods. The challenge is daunting. An improved and transparent evidence base on the multiple values of nature and the flow of ecosystem services will be a key tool for progress, though one not without risks.

Ensuring the appropriate use of ecosystem service concept in policy- and decision-making

Across all policy sectors and all levels of governance, there can be both opportunities and risks associated with the use of the ecosystem services concept – and particularly any associated monetary estimates – in supporting the policy- and decision-making processes. Ecosystem services are an anthropocentric concept that may focus attention too narrowly on mankind's benefits and hence overlook the wider ethical argument. Increased focus on monetary valuation of services may lead to perceptions that prices should be put on nature, and hence a commoditization of nature. This has led to major discussions within communities (e.g. within the biodiversity community in the European Union) and across the world (e.g. with the ALBA group[12] strongly opposed to the concept of valuing nature). Some see that, globally, decision-making models are generally economic and hence to have a chance of influence one needs to 'talk' economics (while also noting other arguments using non-economic terms); others think that the dominant economic model itself needs changing, and talking economics will only exacerbate the problem. A third way would be to use economic arguments to support biodiversity in the short to medium term, use them to support awareness of the values of nature by those who think primarily in economic terms and then broaden the awareness to other dimensions necessary to have a full picture of nature and its benefits to society and the economy. While this wider debate remains open and unclear as to which path is the most beneficial – or indeed risky – for biodiversity, it is useful to look at the specific, concrete issues to illustrate the opportunities and risks, and look at measures to minimise the risk.

For example, there can be both opportunities and risks in the use of the ecosystem services concept linked to protected areas (Kettunen and ten Brink, 2013). While the argument of economic benefits is being made to encourage protection and financing, bridging the gap between conservation objectives and poverty alleviation, some may use the argument to say that in the future only protected areas that offer ecosystem services should be funded. This would overlook the biodiversity's intrinsic value and important conservation objectives, undermining one of the fundamental purposes of protected areas. The level of risk of using the ES concept, for protected areas and more widely, depends on how the concept is interpreted and used, which in turn depends on personality, perception and politics. Arguments of anthropocentric benefits and economic value can be used both positively and negatively in policy- and decision-making, depending on understanding and intention. Good information and governance (due to transparent processes and participation) is essential to use the tools and terms constructively. The level of risk of the ES concept can also be reduced by focusing efforts on integration – i.e. where there are nature-based solutions to other policies – and making progress there, rather than staying within the biodiversity community.

In summary, the way forward will require a complex, multiple-level governance response to develop and integrate the growing evidence base into policymaking and implementation at all levels, as well as integration of ES into business decisions and those of communities and citizens (see also Fish et al., 2016). There is an overall need for increased policy coherence to take account of synergies and manage trade-offs for effective and efficient policymaking and implementation. The progress of policy coherence and its impact on the ground will vary widely across the globe, and it can only be hoped that examples of success will contribute to progress with integration and implementation so that there is less policy dissonance and greater improved prospects for biodiversity.

Acknowledgements

With thanks to Elizabeth van Dijl and to Jean-Pierre Schweitzer for comments. This chapter builds on on-going research in the context of the EU 7FP research project, OPERAs (Operational Potential of Ecosystem Research Applications; Grant agreement no: 308393).

Notes

1 Tenth Conference of the Parties (COP10) of the Convention of Biological Diversity (CBD).
2 Target 14 states that '*By 2020, ecosystems that provide essential services, including services related to water, and contribute to health, livelihoods and well-being, are restored and safeguarded, taking into account the needs of women, indigenous and local communities, and the poor and vulnerable*'.
3 Both within *Strategic Goal D: Enhance the benefits to all from biodiversity and ecosystem services.*
4 Target 15 states that '*By 2020, ecosystem resilience and the contribution of biodiversity to carbon stocks has been enhanced, through conservation and restoration, including restoration of at least 15 per cent of degraded ecosystems, thereby contributing to climate change mitigation and adaptation and to combating desertification*'.
5 under *Strategic Goal A: Address the underlying causes of biodiversity loss by mainstreaming biodiversity across government and society.*
6 http://www.ipbes.net/
7 COM (2011). *Our Life Insurance, our Natural Capital: An EU Biodiversity Strategy to 2020.*
8 Horizontal policies refer to policy frameworks that aim to address cross-cutting issues across different sectoral policies. They include, for example, policy frameworks for impact assessments.
9 UN Convention to Combat Desertification.
10 UNEP defines a green economy as '*one that results in improved human well-being and social equity, while significantly reducing environmental risks and ecological scarcities. In its simplest expression, a green economy can be thought of as one which is low carbon, resource efficient and socially inclusive*' (UNEP, 2011).
11 United Nations Conference on Sustainable Development.
12 States of the Bolivarian Alliance for the Americas (ALBA) – Bolivia, Cuba, Ecuador, Nicaragua and Venezuela.

References

Brouwer, R. (2016). Payments for ecosystem services. In: Potschin, M., Haines-Young, R., Fish, R. and Turner, R. K. (eds) *Routledge Handbook of Ecosystem Services*. Routledge, London and New York, pp 548–553.

CBD (1992). *Convention on Biological Diversity*. Available at: http://www.cbd.int/convention/text/default.shtml (accessed 19 September 2015)

CBD (2010). *COP 10 Decision X/2: Strategic Plan for Biodiversity 2011–2020*, Convention on Biological Diversity, Available at: http://www.cbd.int/decision/cop/?id512268 (accessed 16 April 2014).

Costanza, R., D'Arge, R., de Groot, R., et al. (11 authors) (1997). The value of the world's ecosystem services, *Nature*, vol 387, pp 253–260.

Daily, G. (ed.) (1997). *Nature's Services: Societal Dependence on Natural Ecosystems*. Island Press, Washington DC.

Fish, R., Saratsi, E., Reed, M., and Keune, H. (2016). Stakeholder participation in ecosystem service decision making. In: Potschin, M., Haines-Young, R., Fish, R., and Turner, R. K. (eds) *Routledge Handbook of Ecosystem Services*. Routledge, London and New York, pp 256–270.

Hajjar, R., Jarvis, D. I., and Gemmill-Herren, B. (2008). The utility of crop genetic diversity in maintaining ecosystem services. *Agriculture, Ecosystems and Environment*, vol 123, pp 261–270.

Hendry, S., and Gooch, G. (2016). Ecosystem services and water security. In: Potschin, M., Haines-Young, R., Fish, R. and Turner, R. K. (eds) *Routledge Handbook of Ecosystem Services*. Routledge, London and New York, pp 501–508.

IEEP and Milieu (2013). *The Guide to Multi-Benefit Cohesion Policy Investments in Nature and Green Infrastructure*. European Commission, Brussels. Available at: http://ec.europa.eu/regional_policy/sources/docgener/studies/pdf/guide_multi_benefit_nature.pdf. (accessed 16 April 2014).

Kettunen, M. and ten Brink, P. (eds) (2013). *The Social and Economic Benefits of Protected Areas: An Assessment Guide*. Earthscan / Taylor & Francis, London.

Kettunen, M., ten Brink, P., Underwood, E., and Salomaa, A. (2014). *Policy Needs and Opportunities for Operationalising the Concept of Ecosystem Services*, unpublished report in the context of EU FP7 OPERAs project.

Koenig, D. (2002). Toward local development and mitigating impoverishment in development-induced displacement and resettlement. RSC Working Paper No. 8, International Development Centre, University of Oxford. Available at: http://www.rsc.ox.ac.uk/files/publications/working-paper-series/wp8-development-induced-displacement-resettlement-2002.pdf (accessed 17 December 2014).

Luque, S. and Iverson, L. (2016). Forest-related ecosystem services. In: Potschin, M., Haines-Young, R., Fish, R. and Turner, R. K. (eds) *Routledge Handbook of Ecosystem Services*. Routledge, London and New York, pp 383–393.

MA (2005). *Ecosystems and Human Well-Being, Summary for Decision Makers*. Island Press, Washington DC.

Oosterhuis, F., and ten Brink, P. (eds) (2014). *Paying the Polluter. Environmentally Harmful Subsidies and their Reform*. Edward Elgar, Cheltenham and Northampton MA.

Power, A. (2016). Can ecosystem services contribute to food security? In: Potschin, M., Haines-Young, R., Fish, R. and Turner, R. K. (eds) *Routledge Handbook of Ecosystem Services*. Routledge, London and New York, pp 491–500.

Russi, D., ten Brink, P., Farmer, A., et al (8 authors) (2013). *The Economics of Ecosystems and Biodiversity for Water and Wetlands*, United Nations Environment Programme, Geneva. Available at: http://www.teebweb.org/wp-content/uploads/2013/04/TEEB_WaterWetlands_Report_2013.pdf (accessed 16 April 2014).

TEEB (2011). *The Economics of Ecosystems and Biodiversity (TEEB) in National and International Policy Making*. ten Brink, P. (ed.) Earthscan, London and Washington DC.

ten Brink, P., Bassi, S., Bishop, J., Harvey, C. A., Karousakis, K., Markandya, A., Nunes, P.A.L.D., McConville, A. J., Ring, I., Ruhweza, A., Sukhdev, P., Vakrou, A., van der Esch, S., Verma, M., and Wertz-Kanounnikoff S. (2011). *Rewarding Benefits through Payments and Markets*. In: TEEB (2011).

ten Brink, P., Mazza, L., Badura, T., Kettunen, M., and Withana, S. (2012). *Nature and its Role in the Transition to a Green Economy*, United Nations Environment Programme, Geneva. Available at: http://www.teebweb.org/wp-content/uploads/2013/04/Nature-Green-Economy-Full-Report.pdf (accessed 16 April 2014).

UNEP (2011). *Towards a Green Economy – Pathways to Sustainable Development and Poverty Eradication, A Synthesis for Policy Makers*, United Nations Environment Programme, Geneva. Available at: http://www.unep.org/greeneconomy/Portals/88/documents/ger/GER_synthesis_en.pdf (accessed 16 April 2014).

UNEP-WCMC and IEEP (2013). *Incorporating Biodiversity and Ecosystem Service Values into NBSAPs*. UNEP-WCMC and IEEP, Cambridge, London, and Brussels. Available at: http://www.ieep.eu/assets/1200/Guidance_doc_A4_FINAL.pdf (accessed 16 April 2014).

Vanbergen, A. J., and the Insect Pollinators Initiative (2013). Threats to an ecosystem service: pressures on pollinators. *Frontiers in Ecology and the Environment*, vol 11, pp 251–259.

38

ECOSYSTEM SERVICES AND CLIMATE CHANGE

Bruno Locatelli

Introduction

Ecosystem services play an important role in strategies for tackling climate change: mitigation and adaptation (Turner et al., 2009). Mitigation aims at reducing emissions sources or enhancing sinks of greenhouse gases, and adaptation aims at adjusting natural or human systems to moderate harm or exploit beneficial opportunities from climate variations (Figure 38.1). Because of their different rationales, these strategies have different priority sectors and locations: mitigation prioritizes larger emission sources or stronger potential sinks, whereas adaptation prioritizes vulnerable people, ecosystems and activities. While some sectors are mostly concerned by one of the two strategies (e.g., energy by mitigation or health by adaptation), ecosystems and their services are clearly relevant to both.

Ecosystems contribute to mitigation because of their capacity to remove carbon from the atmosphere and to store it. Ecosystems contribute also to adaptation because they provide services that can help people adapt to both current climate hazards and future climate change (Figure 38.2). While ecosystem services are part of the solution to climate change, they are also affected by changing climatic conditions. Ecosystem-based approaches to climate change should recognize the multiple links between ecosystem services and climate change: management can enhance the contribution of ecosystem services to adaptation and mitigation ('ecosystem-based adaptation and mitigation') and, as climate change will affect ecosystems and their services, adaptation measures are needed to reduce negative impacts and maintain ecosystem functions ('adaptation for ecosystem services').

This chapter explores the links between ecosystem services and climate change. It first describes the ecosystem services that contribute to mitigation and adaptation, as well as the threat of climate change to ecosystem services. Here the focus is on provisioning services (e.g., food and timber) and regulating services (e.g., water regulation and pest control), as there is little evidence on how adaptation benefits from cultural services (e.g., recreation, aesthetic and spiritual benefits). In the section on adaptation services, only services that contribute directly to human well-being and resilience are considered, and so supporting services (e.g., primary production and nutrient cycling) are excluded. However, because they are important for ecological resilience, they will be considered in the section on climatic threats. The chapter will end with an overview of policy instruments related to ecosystem-based adaptation and mitigation, and the trade-offs that arise when pursuing the strategies jointly.

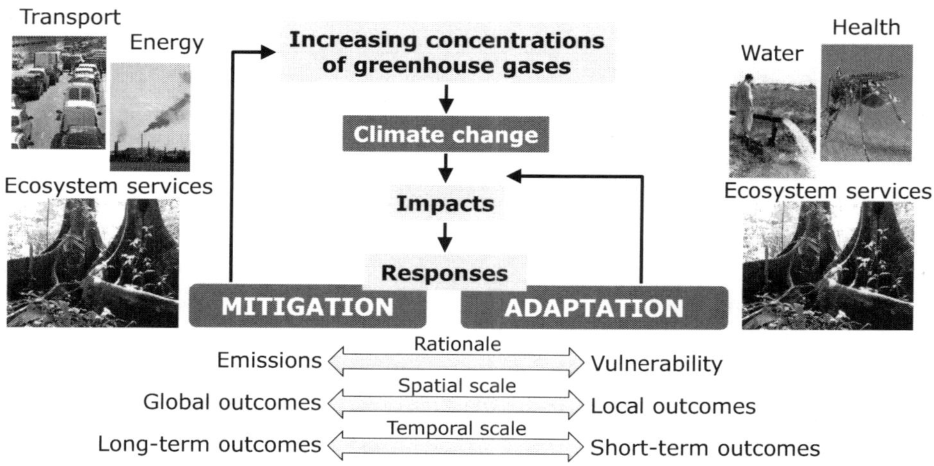

Figure 38.1 Differences between climate change adaptation and mitigation.

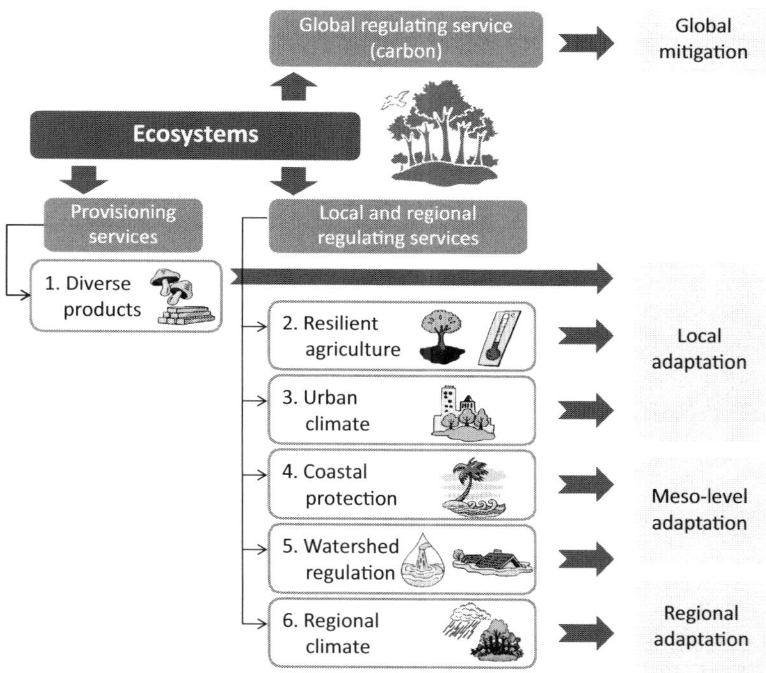

Figure 38.2 Contribution of ecosystem services to climate change adaptation and mitigation.

Mitigation services

Ecosystems contribute to mitigation because of their capacity to remove carbon from the atmosphere and to store it. Terrestrial ecosystems absorb around 3 billion tons of atmospheric carbon per year (Pg/yr) through net growth, which accounts for 30% of anthropogenic CO_2 emissions (Canadell and Raupach, 2008). Forest ecosystems play a crucial role in carbon sequestration, particularly tropical forests, but tropical deforestation causes carbon emissions, estimated between 0.8 to 2.8 Pg/yr (Baccini et al., 2012; Harris et al., 2012), equivalent to 6–17% of global anthropogenic CO_2 emissions to the atmosphere (Van der Werf et al., 2009).

Thus, ecosystem management can contribute to climate change mitigation. Afforestation (converting long-time non-forested land to forest) and reforestation (converting recently non-forested land to forest), for example, increase carbon in the vegetation, whereas forest conservation contributes to reducing carbon emissions from deforestation. Agricultural management can also enhance carbon sequestration through soil conservation and the introduction of trees in agroforestry systems (Uprety et al., 2012).

Adaptation services

Well-managed ecosystems can help societies to adapt to current climate hazards and future climate change by providing a range of ecosystem services (Doswald et al., 2014; Pramova et al., 2012b). Six key areas are reviewed here.

Products and local communities

Provisioning services play an important role in the coping and adapting strategies of rural communities, particularly in developing countries (Innes and Hickey, 2006). Many rural communities use forest products as safety nets for coping with stresses, as when agriculture production fails due to drought. During floods in Peru (Takasaki et al., 2004) and droughts in Tanzania (Enfors and Gordon, 2008), coping strategies include collecting forest wild products. After a hurricane in Honduras, the collection and trade of forest products helped households recover (McSweeney, 2005).

Timber and non-timber forest products (such as firewood, wild fruits, mushrooms and fodder) also contribute to livelihood diversification, an adapting and anticipatory strategy that reduces the sensitivity of households and communities to climate variations. Numerous studies have demonstrated the importance of ecosystem provisioning services for livelihood diversification and resilience in Bolivia (Robledo et al., 2004) or in Cameroon (Bele et al., 2011), for example. In Morogoro (Tanzania), the main strategy of rural communities for dealing with climate variability is livelihood diversification, partly through firewood or fruits (Paavola, 2008). Complex cropping systems, with associated multiple species of crops, fodder and trees, provide a continuous harvest of products despite climate variations in Mali (Djoudi et al., 2013) and Bangladesh (Rahman et al., 2012), for example.

The coping and adapting strategies of the poorest or most vulnerable households often rely heavily on ecosystem products, because of the lack of alternatives and the limited requirements in financial, physical or human assets for collecting these products. This reliance has been observed among young and poor households with limited land access during a flood in Peru (Takasaki et al., 2004), households with low income or headed by older and less-educated individuals during droughts in Malawi (Fisher et al., 2010) and the poorest and least-educated people after a flood in Indonesia (Liswanti et al., 2011). As a result, farmers and pastoralists with limited access to forest products are more vulnerable to rainfall variations than others in Kenya (Owuor et al., 2005) and the lack of access of mangrove resources increases the vulnerability of poor coastal communities in the Philippines (Walton et al., 2006). As the use of provisioning services for coping with stresses often results from a lack of alternative strategies, it can be a symptom of poverty rather than a solution for adaptation (Pattanayak and Sills, 2001). Ecosystem services as a safety net can be a poverty trap, particularly when resource availability is low, the population in need is large and alternatives are lacking (Levang et al., 2005).

Local climate regulation in agriculture

As trees in or near agricultural fields provide regulating services that reduce the vulnerability of cropping systems to climate variations, the introduction of trees in agriculture, such as in agroforestry and silvopastoralist systems, is considered as an effective adaptation strategy. Tree roots explore soil deeply for water and nutrients, which benefits crops during droughts. Trees improve fertility and protect soils from erosion by increasing soil organic matter, porosity, infiltration and soil cover (Verchot et al., 2007). Nitrogen-fixing trees contribute to the resilience of crops to droughts due to improvements in soil nutrients and water infiltration, as research has shown, for example, in Malawi and Zambia (Garrity et al., 2010). In Niger, cereal production was less affected by recent droughts in areas with tree regeneration (Sendzimir et al., 2011). As shade trees control temperature and humidity and protect from winds and storms, they can also improve the resilience of coffee and cacao production in, for example, Mexico (Lin, 2010).

Studies on agroforestry systems highlight trade-offs. For example, high tree cover increases soil protection but reduces the light available to crops in the understorey, which requires the identification of the context-specific tree cover that maximizes the benefits of agroforestry. Other trade-offs occur between average yields and resilience: tree cover buffers crops against climate stress but decreases average yields in the absence of stress. In agroforestry, tree ecosystem services may contribute differently to crop adaptation to climate change depending on climate scenarios and production systems (Verchot et al., 2007).

Despite the benefits of agroforestry, its expansion has been constrained by policies promoting intensive agriculture systems that exclude trees or, in some cases, induce deforestation (Morton et al., 2008). Other approaches are possible, in which agricultural intensification occurs in association with trees, so that ecosystem services and incomes are secured (Steffan-Dewenter et al., 2007). The social and biophysical context determines how land-use and agricultural policies balance land sparing (maximization of agricultural production in some areas and conservation of natural ecosystems in others) and land sharing (integration of conservation and production in heterogeneous landscapes) (Fischer et al., 2008).

Local climate regulation in cities

As urban forests and trees regulate temperatures (through shade and evaporative cooling) and water (through rain interception and infiltration), they play a role in urban adaptation to climate variability and change. Because of their impermeable surfaces, cities are vulnerable to flooding, but urban parks or trees can reduce runoff through infiltration. The urban heat island effects, which increase the health impacts of heat waves, are moderated by green cover, as observed, for example, in Manchester, UK (Gill et al., 2007).

In cities, ecosystem-based adaptation requires a good understanding of landscape ecology and the potential of green infrastructure to improve the well-being of vulnerable communities, as in the case of Durban, South Africa (Roberts et al., 2012). Adaptation needs also to be designed at multiple scales, including ecosystem management outside the urban areas and for upper watershed protection. For example, three scales are proposed in Beijing, China, for a green infrastructure at the regional scale (forest belts), in the city (urban parks and green corridors) and in neighbourhoods (road and vertical greening) (Li et al., 2005). However, urban ecosystem-based adaptation raises concerns about high opportunity costs of land and possible management constraints: for example, during droughts, scarce water consumed to maintain trees may be needed for other uses.

Protection of coastal areas

By stabilizing land and absorbing and dissipating wave energy, coastal ecosystems such as mangroves can protect coastal areas from climate-related threats: tropical storms, sea-level rise, floods and erosion. The protection services of mangroves against storms were evident after a cyclone in Orissa, India (Das and Vincent, 2009) and are recognized by coastal communities in Bangladesh (Iftekhar and Takama, 2008). Coastal forest management has been suggested to control beach erosion from future sea level rise in Zanzibar, Tanzania (Mustelin et al., 2010), for example. In Vietnam, planting mangroves reduce the costs of maintaining sea dykes built for protecting coastal settlements (Adger, 1999).

It is unclear how much mangrove is needed to reduce the vulnerability of a coastal area to different threats and how the protective role is influenced by topography, bathymetry or mangrove extent and species. Mangrove width is an important factor, but the minimal width for a given area also depends on mangrove structure and species. Plans should be based on a good understanding of coastal dynamics and mangrove role (Feagin et al., 2010). In addition, as coastal ecosystems cannot guarantee complete protection from extreme events, they should be part of a broader disaster risk reduction and adaptation strategy (Baird et al., 2009).

Protection of watersheds

Ecosystems influence the hydrological functioning of watersheds through their contribution to rainfall interception, evapotranspiration, water infiltration, and groundwater recharge. This influence can reduce the impacts of climate variations on downstream population. For example, ecosystems can preserve base flows during dry seasons if they facilitate groundwater recharge; they can also reduce peak flows or floods during rainfall events if they contribute to rainfall interception and infiltration. In addition, ecosystems can reduce soil erosion and landslide hazards, which are partially climate-related. Higher base flow in forested watersheds reduced the impacts of droughts on downstream farming communities in Flores, Indonesia (Pattanayak and Kramer, 2001). Natural forest regeneration improved water provision to agriculture during extended dry periods, stabilized hillsides and reduced the impacts to soil erosion and landslides on communities in Bolivia (Robledo et al., 2004).

Even though hydrological studies on forests and water could inform decisions on adaptation, few deal with extreme events and social vulnerability to water and soil-related hazards. In addition, the influence of forests on floods is debated. Even if forests can reduce storm flow because of their higher infiltration, this effect is questioned in the case of large rainfall events once soils are saturated with water (Bruijnzeel, 2004). However, in spite of such controversies, the role of forests in the most frequent medium-scale floods should not be overlooked (Locatelli and Vignola, 2009). Similarly, controversies exist on the effect of forests on base flow, because it results from two competing ecosystem processes: in forests, high transpiration reduces base flow whereas high infiltration increases soil water recharge and base flow. Regarding soil erosion and landslides, hydrological literature confirms that surface erosion is generally low in forests; however, uncertainties remain about the role of forests in landslide prevention, especially when high rainfall intensity overwhelms the role of roots in stabilizing soils (Sidle et al., 2006).

Climate regulation at regional and continental scale

At the regional and continental scale, ecosystems play a role in recycling rainfall and generating flows of atmospheric water vapour. While evapotranspiration by forests reduces total water flows in a watershed, it also pumps water back into the atmosphere, which can increase rainfall in the region (Ellison et al., 2012). Forests may also act as a pump of atmospheric moisture, attracting moist air from oceans to inland regions (Makarieva and Gorshkov, 2007; Sheil and Murdiyarso, 2009), but this role of forests in hydrological processes at the regional scale is debated (Meesters et al., 2009). The role of forests and trees in regulating atmospheric water and regional rainfall has been overlooked by scientific assessments on ecosystems and climate change, despite its place, for example, in moderating droughts effects due to global climate change.

Climate threats on ecosystem services

Most ecosystems are vulnerable to climate change even under low- and medium-range scenarios of global warming (Scholes and Settele, 2014). They are likely to be affected by gradual changes in temperature or precipitation and climate-related disturbances (e.g., flooding, drought and wildfire), in association with other threats (e.g., land use change, pollution, overexploitation of resources). These changes and disturbances will affect ecosystem structure and function, the ecological interactions among species and their geographical ranges, which will result in changes in biodiversity and ecosystem services (Locatelli et al., 2008). Ecosystem vulnerability has consequences for the global climate: if changes and disturbances release carbon into the atmosphere, vegetation-climate feedback will amplify global warming (Canadell et al., 2004). Local and regional ecosystem services may also be affected by climate change, such as water regulation or timber production, with direct implications for dependent societies (Shaw et al., 2011).

The resilience of ecosystems in a context of climate change depends on multiple factors, such as other non-climatic pressures, landscape configuration and species richness and diversity (Locatelli et al., 2008). Nutrient cycling and primary production are important components of the functioning, resistance and resilience of the ecosystem and we need to understand more the ecological mechanisms that facilitate the maintenance and adaptation of ecosystem services during periods of change (Lavorel et al., 2015).

Where short-term or non-climatic threats to ecosystems are minimized, specific measures for climate change adaptation can be incorporated into management. Management can reduce the risks linked to climate change and increase the capacity of ecosystems and species to adapt (Scholes and Settele, 2014). Actions can buffer ecosystems from perturbations, such as through fire or pest management, or facilitate ecological adjustments to changing climates, such as by reducing landscape fragmentation to facilitate species migration (Guariguata et al., 2008). Adaptation must, however, be an on-going process rather than seeking to maintain existing conditions or targeting a new equilibrium (Stein et al., 2013).

Existing policy instruments

Ecosystem-based mitigation of climate change is now recognized by international agreements and policy instruments. For example, the contribution of tropical afforestation and reforestation is acknowledged in the Clean Development Mechanism (CDM) of the Kyoto Protocol, and several plantation projects are rewarded through this mechanism or voluntary carbon agreements. Another initiative is REDD+ (Reducing Emissions from Deforestation and forest Degradation (Angelsen et al., 2012)). This aims to maintain carbon stocks based on the provision of financial incentives to protect forests from deforestation and degradation, and enhance carbon stocks through sustainable forest management.

The place of ecosystem-based approaches in the international discussions is not as clear for adaptation as it is for mitigation, but some initiatives have been developed at national and local scales (Locatelli et al., 2011). Among the 44 National Adaptation Programmes of Action (NAPAs) submitted by least developed countries to the UN Framework Convention on Climate Change (UNFCCC) by mid-2010, more than half recognized the importance of ecosystem services (Pramova et al., 2012a). Around 25% of the adaptation projects proposed in the NAPAs included ecosystem management activities for improving human well-being and adaptation through such measures as soil rehabilitation, erosion control and water regulation.

Ecosystem-based approaches to climate change: the way forward

Many projects and programs are contributing to effective mitigation and adaptation strategies through the conservation of biodiversity and ecosystem services (World Bank, 2009), though they rarely consider both adaptation and mitigation (Locatelli et al., 2011). A comprehensive approach must encompass three dimensions: ecosystem-based mitigation, ecosystem-based adaptation, and adaptation for ecosystems (Figure 38.3). To ensure that ecosystems mitigate climate change and help people adapt, management must reduce current threats to ecosystem services (e.g., deforestation and forest degradation) as a first step. It should also address future threats by developing adaptation measures. In ecosystem-based approaches to climate change, 'adaptation for ecosystems' is thus needed to ensure that ecosystem-based adaptation and mitigation work in the long term.

The management of ecosystem services can provide joint benefits for both mitigation and adaptation where, for example, the spatial distributions of carbon, hydrological services or biodiversity are positively correlated (Locatelli et al., 2014). For example, mangrove conservation and restoration simultaneously contribute to protecting coastal areas and to storing large amounts of carbon (Donato et al., 2011). Forest conservation projects for mitigation, such as REDD+

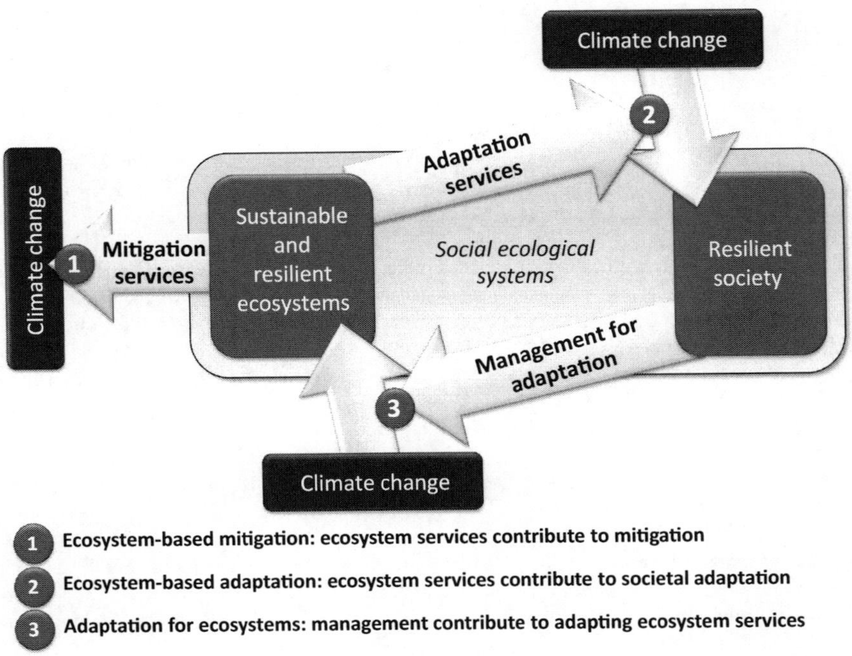

Figure 38.3 The three pillars of ecosystem-based approaches to climate change.

projects, can improve the adaptation of local livelihoods by increasing the provision of local regulation ecosystem services to communities, protecting them from hydrological variations. They can also contribute to diversifying incomes and economic activities through the use of provisioning services such as non-timber forest products. REDD+ projects can also facilitate ecological adaptation to climate change by reducing anthropogenic pressures on forests, enhancing connectivity between forest areas and conserving biodiversity hotspots (Locatelli et al., 2015).

But trade-offs between adaptation and mitigation can occur. Adaptation can lead to increased emissions: for example, if ecosystem management aims at improving water balance for adapting water users to climate change, the best outcomes may in some cases be achieved through ecosystems with low carbon content, such as grasslands, rather than forests (Locatelli and Vignola, 2009). Conversely, mitigation can increase vulnerability. For example, a monoculture using species with fast growth and high water consumption can perform well in terms of carbon storage and mitigation, but cause downstream water shortages and biodiversity losses, which can then increase social and ecological vulnerability to climate change. The IPCC have warned that the widespread transformation of ecosystems for mitigation, such as planting fast-growing tree species or bioenergy plantations, will negatively impact ecosystems and biodiversity (Scholes and Settele, 2014). A REDD+ project may increase livelihood's vulnerability if it restricts the rights and access of local people to forest provisioning services.

Although adaptation and mitigation present notable differences, particularly in their objectives, spatial and temporal scales, there is an increasing need to pursue them jointly (Warren, 2011). Given that ecosystems can provide mitigation and adaptation services at the same time, policies and local initiatives related to ecosystem management can integrate both climate change strategies and avoid trade-offs between them. Beyond the adaptation-mitigation integration, there is a need to mainstream climate change in the policy domains of ecosystem management and rural development (Kok and de Coninck, 2007).

References

Adger, W. N. (1999). Social vulnerability to climate change and extremes in coastal Vietnam. *World Development*, vol 27, pp 249–269.

Angelsen, A., Brockhaus, M., Sunderlin, W., and Verchot, L. (2012). *Analysing REDD+: Challenges and Choices*. CIFOR, Center for International Forestry Research, Bogor.

Baccini, A., Goetz, S. J., Walker, W. S., et al. (12 authors) (2012). Estimated carbon dioxide emissions from tropical deforestation improved by carbon-density maps. *Nature Clim. Change*, vol 2, pp 182–185.

Baird, A. H., Bhalla, R. S., Kerr, A. M., Pelkey, N. W., and Srinivas, V. (2009). Do mangroves provide an effective barrier to storm surges? *Proceedings of the National Academy of Sciences*, vol 106, p E111.

Bele, M., Somorin, O., Sonwa, D., Nkem, J., and Locatelli, B. (2011). Forests and climate change adaptation policies in Cameroon. *Mitigation and Adaptation Strategies for Global Change*, vol 16, pp 369–385.

Bruijnzeel, L. A. (2004). Hydrological functions of tropical forests: not seeing the soil for the trees? *Agriculture, Ecosystems and Environment*, vol 104, pp 185–228.

Canadell, J. G., Ciais, P., Cox, P., and Heimann, M. (2004). Quantifying, understanding and managing the carbon cycle in the next decades. *Climatic Change*, vol 67, pp 147–160.

Canadell, J. G., and Raupach, M. R. (2008). Managing forests for climate change mitigation. *Science*, vol 320, pp 1456–1457.

Das, S., and Vincent, J. R. (2009). Mangroves protected villages and reduced death toll during Indian super cyclone. *Proceedings of the National Academy of Sciences of the United States of America*, vol 106, pp 7357–7360.

Djoudi, H., Brockhaus, M., and Locatelli, B. (2013). Once there was a lake: vulnerability to environmental changes in northern Mali. *Regional Environmental Change*, vol 13, pp 493–508.

Donato, D. C., Kauffman, J. B., Murdiyarso, D., et al. (6 authors) (2011). Mangroves among the most carbon-rich forests in the tropics. *Nature Geoscience*, vol 4, pp 293–297.

Doswald, N., Munroe, R., Roe, D., et al. (9 authors) (2014). Effectiveness of ecosystem-based approaches for adaptation: review of the evidence-base. *Climate and Development*, vol 6, no 2, pp 185–201.

Ellison, D., Futter, M. N., and Bishop, K. (2012). On the forest cover–water yield debate: from demand- to supply-side thinking. *Global Change Biology*, vol 18, pp 806–820.

Enfors, E. I., and Gordon, L. J. (2008). Dealing with drought: the challenge of using water system technologies to break dryland poverty traps. *Global Environmental Change*, vol 18, pp 607–616.

Feagin, R. A., Mukherjee, N., Shanker, K., et al. (15 authors) (2010). Shelter from the storm? Use and misuse of coastal vegetation bioshields for managing natural disasters. *Conservation Letters*, vol 3, pp 1–11.

Fischer, J., Brosi, B., Daily, G.C., et al. (10 authors) (2008). Should agricultural policies encourage land sparing or wildlife-friendly farming? *Frontiers in Ecology and the Environment*, vol 6, pp 380–385.

Fisher, M., Chaudhury, M., and McCusker, B. (2010). Do forests help rural households adapt to climate variability? Evidence from southern Malawi. *World Development*, vol 38, pp 1241–1250.

Garrity, D. P., Akinnifesi, F. K., Ajayi, O. C., et al. (8 authors) (2010). Evergreen agriculture: a robust approach to sustainable food security in Africa. *Food Security*, vol 2, pp 197–214.

Gill, S. E., Handley, J. F., Ennos, A. R., and Pauleit, S. (2007). Adapting cities for climate change: the role of the green infrastructure. *Built Environment*, vol 33, pp 115–133.

Guariguata, M. R., Cornelius, J. P., Locatelli, B., Forner, C., and Sánchez-Azofeifa, G. A. (2008). Mitigation needs adaptation: tropical forestry and climate change. *Mitigation and Adaptation Strategies for Global Change*, vol 13, no 8, pp 793–808.

Harris, N. L., Brown, S., Hagen, S. C., et al. (9 authors) (2012). Baseline map of carbon emissions from deforestation in tropical regions. *Science*, vol 336, pp 1573–1576.

Iftekhar, M. S., and Takama, T. (2008). Perceptions of biodiversity, environmental services, and conservation of planted mangroves: a case study on Nijhum Dwip Island, Bangladesh. *Wetlands Ecology and Management*, vol 16, pp 119–137.

Innes, J. L., and Hickey, G. M. (2006). The importance of climate change when considering the role of forests in the alleviation of poverty. *International Forestry Review*, vol 8, pp 406–416.

Kok, M.T.J., and de Coninck, H.C. (2007). Widening the scope of policies to address climate change: directions for mainstreaming. *Environmental Science & Policy*, vol 10, pp 587–599.

Lavorel, S., Colloff, M. J., McIntyre, S., et al. (10 authors) (2015). Ecological mechanisms underpinning climate adaptation services. *Global Change Biology*, vol 21, pp 12–31.

Levang, P., Dounias, E., and Sitorus, S. (2005). Out of the forests, out of poverty? *Forests, Trees and Livelihoods*, vol 15, pp 211–235.

Li, F., Wang, R., Paulussen, J., and Liu, X. (2005). Comprehensive concept planning of urban greening based on ecological principles: a case study in Beijing, China. *Landscape and Urban Planning*, vol 72, pp 325–336.

Lin, B. B. (2010). The role of agroforestry in reducing water loss through soil evaporation and crop transpiration in coffee agroecosystems. *Agricultural and Forest Meteorology*, vol 150, pp 510–518.

Liswanti, N., Sheil, D., Basuki, I., Padmanaba, M., and Mulcahy, G. (2011). Falling back on forests: how forest-dwelling people cope with catastrophe in a changing landscape. *International Forestry Review*, vol 13, pp 442–455.

Locatelli, B., Evans, V., Wardell, A., Andrade, A., and Vignola, R. (2011). Forests and climate change in Latin America: linking adaptation and mitigation. *Forests*, vol 2, pp 431–450.

Locatelli, B., Imbach, P., and Wunder, S. (2014). Synergies and trade-offs between ecosystem services in Costa Rica. *Environmental Conservation*, vol 41, pp 27–36.

Locatelli, B., Kanninen, M., Brockhaus, M., et al. (6 authors) (2008). *Facing an Uncertain Future: How Forest and People Can Adapt to Climate Change*. CIFOR, Bogor.

Locatelli, B., Pavageau, C., Pramova, E., and Di Gregorio, M. (2015). Integrating climate change mitigation and adaptation in agriculture and forestry: opportunities and trade-offs. *Wiley Interdisciplinary Reviews: Climate Change*, vol 6, pp 585–598.

Locatelli, B., and Vignola, R. (2009). Managing watershed services of tropical forests and plantations: can meta-analyses help? *Forest Ecology and Management*, vol 258, pp 1864–1870.

Makarieva, A., and Gorshkov, V. (2007). Biotic pump of atmospheric moisture as driver of the hydrological cycle on land. *Hydrology and Earth System Sciences*, vol 11, pp 1013–1033.

McSweeney, K. (2005). Natural insurance, forest access, and compounded misfortune: forest resources in smallholder coping strategies before and after Hurricane Mitch, eastern Honduras. *World Development*, vol 33, pp 1453–1471.

Meesters, A., Dolman, A., and Bruijnzeel, L. (2009). Comment on "Biotic pump of atmospheric moisture as driver of the hydrological cycle on land" by A. M. Makarieva and V. G. Gorshkov. *Hydrology and Earth System Sciences*, vol 13, no 7, pp 1299–1305.

Morton, D., Defries, R., Randerson, J., et al. (6 authors) (2008). Agricultural intensification increases deforestation fire activity in Amazonia. *Global Change Biology*, vol 14, pp 2262–2275.

Mustelin, J., Klein, R. G., Assaid, B., et al. (7 authors) (2010). Understanding current and future vulnerability in coastal settings: community perceptions and preferences for adaptation in Zanzibar, Tanzania. *Population and Environment*, vol 31, pp 371–398.
Owuor, B., Mauta, W., and Eriksen, S. (2005). Adapting to climate change in a dryland mountain environment in Kenya. *Mountain Research and Development*, vol 25, pp 310–315.
Paavola, J. (2008). Livelihoods, vulnerability and adaptation to climate change in Morogoro, Tanzania. *Environmental Science & Policy*, vol 11, pp 642–654.
Pattanayak, S. K., and Kramer, R. (2001). Worth of watersheds: a producer surplus approach for valuing drought mitigation in Eastern Indonesia. *Environment and Development Economics*, vol 6, pp 123–146.
Pattanayak, S. K., and Sills, E. O. (2001). Do tropical forests provide natural insurance? The microeconomics of non-timber forest product collection in the Brazilian Amazon. *Land Economics*, vol 77, pp 595–612.
Pramova, E., Locatelli, B., Brockhaus, M., and Fohlmeister, S. (2012a). Ecosystem services in the National Adaptation Programmes of Action. *Climate Policy*, vol 12, pp 393–409.
Pramova, E., Locatelli, B., Djoudi, H., and Somorin, O. A. (2012b). Forests and trees for social adaptation to climate variability and change. *Wiley Interdisciplinary Reviews: Climate Change*, vol 3, pp 581–596.
Rahman, S. A., Imam, M. H., Snelder, D. J., and Sunderland, T. (2012). Agroforestry for livelihood security in agrarian landscapes of the Padma floodplain in Bangladesh. *Small-Scale Forestry*, vol 11, pp 529–538.
Roberts, D., Boon, R., Diederichs, N., et al. (9 authors) (2012). Exploring ecosystem-based adaptation in Durban, South Africa: "learning-by-doing" at the local government coal face. *Environment and Urbanization*, vol 24, pp 167–195.
Robledo, C., Fischler, M., and Patino, A. (2004). Increasing the resilience of hillside communities in Bolivia: has vulnerability to climate change been reduced as a result of previous sustainable development cooperation? *Mountain Research and Development*, vol 24, pp 14–18.
Scholes, R., and Settele, J. (2014). *Terrestrial and Inland Water Systems.* Working Group 2 Contribution to the Fifth Assessment Report. IPCC, International Panel on Climate Change, Geneva.
Sendzimir, J., Reij, C. P., and Magnuszewski, P. (2011). Rebuilding resilience in the Sahel: regreening in the Maradi and Zinder regions of Niger. *Ecology and Society*, vol 16, 10.5751/ES-04198–160301.
Shaw, M. R., Pendleton, L., Cameron, D. R., et al. (10 authors) (2011). The impact of climate change on California's ecosystem services. *Climatic Change*, vol 109, pp 465–484.
Sheil, D., and Murdiyarso, D. (2009). How forests attract rain: an examination of a new hypothesis. *Bioscience*, vol 59, pp 341–347.
Sidle, R. C., Ziegler, A. D., Negishi, J. N., et al. (6 authors) (2006). Erosion processes in steep terrain–Truths, myths, and uncertainties related to forest management in Southeast Asia. *Forest Ecology and Management*, vol 224, pp 199–225.
Steffan-Dewenter, I., Kessler, M., Barkmann, J., et al. (10 authors) (2007). Tradeoffs between income, biodiversity, and ecosystem functioning during tropical rainforest conversion and agroforestry intensification. *Proceedings of the National Academy of Sciences*, vol 104, p 4973.
Stein, B. A., Staudt, A., Cross, M. S., et al. (10 authors) (2013). Preparing for and managing change: climate adaptation for biodiversity and ecosystems. *Frontiers in Ecology and the Environment*, vol 11, pp 502–510.
Takasaki, Y., Barham, B. L., and Coomes, O.T. (2004). Risk coping strategies in tropical forests: floods, illnesses, and resource extraction. *Environment and Development Economics*, vol 9, pp 203–224.
Turner, W. R., Oppenheimer, M., and Wilcove, D. S. (2009). A force to fight global warming. *Nature*, vol 428, pp 278–279.
Uprety, D. C., Dhar, S., Hongmin, D., et al. (6 authors) (2012). *Technologies for Climate Change Mitigation: Agriculture Sector.* UNEP Risø Centre on Energy, Climate and Sustainable Development, Copenhagen.
Van der Werf, G. R., Morton, D. C., DeFries, R. S., et al. (8 authors) (2009). CO2 emissions from forest loss. *Nature Geoscience*, vol 2, pp 737–738.
Verchot, L. V., Van Noordwijk, M., Kandji, S., et al. (10 authors) (2007). Climate change: linking adaptation and mitigation through agroforestry. *Mitigation and Adaptation Strategies for Global Change*, vol 12, pp 901–918.
Walton, M.E.M., Samonte-Tan, G.P.B., Primavera, J. H., Edwards-Jones, G., and Le Vay, L. (2006). Are mangroves worth replanting? The direct economic benefits of a community-based reforestation project. *Environmental Conservation*, vol 33, pp 335–343.
Warren, R. (2011). The role of interactions in a world implementing adaptation and mitigation solutions to climate change. *Philosophical Transactions of the Royal Society A: Mathematical, Physical and Engineering Sciences*, vol 369, pp 217–241.
World Bank (2009). *Convenient Solutions to an Inconvenient Truth: Ecosystem-Based Approaches to Climate Change.* The World Bank, Washington DC.

39

CAN ECOSYSTEM SERVICES CONTRIBUTE TO FOOD SECURITY?

Alison G. Power

Dimensions of food security

The concept of food security encompasses many dimensions, including availability of food, access to food, adequacy in terms of nutrition and food safety, and acceptability according to cultural norms. An even more comprehensive definition includes food sovereignty and sustainability as essential elements of food security (De Schutter, 2010; Chappell and LaValle, 2011). Ericksen (2008) notes that the concept of availability itself includes at least three components: production, distribution, and exchange. Despite these multiple dimensions, most discourse about food security focuses on availability or, even more narrowly, agricultural production. Yet more than three decades ago, Amartya Sen noted that, "starvation is the characteristic of some people not *having* enough food to eat. It is not the characteristic of there *being* not enough food to eat. While the latter can be a cause of the former, it is but one of many *possible* causes" (Sen, 1981).

While food production has risen dramatically over the past 50 years (Hazell and Wood, 2008), widespread poverty and food insecurity have persisted (Figure 39.1, Barrett, 2010). Measures of food security that focus on agricultural productivity and per capita food production tell a more optimistic tale than measures that focus on poverty, nutritional status, or well-being. Although public attention on food security is often precipitated by disasters that affect productivity, such as hurricanes, earthquakes, or prolonged drought, most food insecurity results from poverty (Barrett, 2010). Moreover, although agricultural productivity increased significantly during the 1960s, 1970s, and 1980s, there is mounting evidence that productivity gains are leveling out, despite the introduction and adoption of new agricultural technologies (Peng et al., 1999; Tilman et al., 2001). Hence, advancing food security will require improving access to food, food safety, nutritional adequacy, and sustainability, as well as increasing productivity (i.e., availability). This chapter will address the potential contribution of ecosystem services to several of these dimensions of food security (see also Plieninger et al., 2016).

Increasing food availability through ecosystem services

Ecosystem services contribute to food availability via two pathways. The dominant pathway is through support to agricultural production of domesticated plants and animals. A secondary, but sometimes important, pathway is through harvest of wild plants and animals from

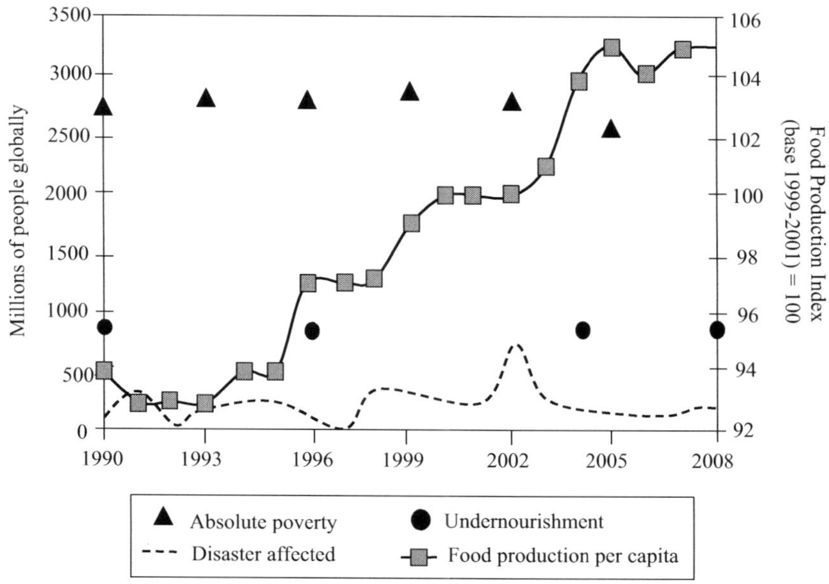

Figure 39.1 Increase in food production versus other indicators of trends in food security 1990–2008
Source: adapted from Barrett, 2010.

natural ecosystems. The majority of agricultural products are derived from terrestrial ecosystems, whereas many important wild foods are harvested from aquatic systems. However, both agriculture and wild harvest contribute food via terrestrial and aquatic systems.

Agriculture is both a consumer and producer of ecosystem services (Millennium Ecosystem Assessment, 2005; Zhang et al., 2007; Power, 2010). In general, agriculture is defined and shaped by management for provisioning services, which contribute food, forage, fiber, and bioenergy for human use. Ecosystem services important to agricultural productivity are generated at a variety of scales, both internal to the agroecosystem and within the broader landscape. Productivity depends on ecosystem services from natural and semi-natural ecosystems in the landscape, such as regulation of water flow, water purification, soil regeneration and fertility, pollination, biological pest control, and genetic biodiversity for crop and livestock breeding. But many of these ecosystem services can also be supplied locally, within the agroecosystem itself. In addition to supplying provisioning services in the form of agricultural products, agroecosystems can produce ecosystem services that support provisioning (e.g., pollination, pest control, genetic diversity for future agricultural use, soil retention, regulation of soil fertility, and nutrient cycling), as well as regulating services important to human well-being (e.g., flood control, water quality control, carbon storage, and climate regulation through greenhouse gas emissions).

Ecosystem services important to agriculture thus flow to agroecosystems from natural habitats within the landscape and are also provided within the agroecosystem itself. These services serve as the foundation for ecological intensification, an approach to agricultural intensification that can lead to increased productivity, environmental protection, and preservation of the natural resource base (Dore et al., 2011; Bommarco et al., 2013). Ecological intensification can provide benefits in both developed and developing countries. Where lack of resources prevents farmers from achieving yields equivalent to the maximum attainable yield in their agroecological zone, these yield gaps can be reduced by taking advantage of ecosystem services. In developed countries where farmer yields are close to attainable yields, employing ecosystem services can replace some portion of purchased inputs and reduce environmental damage resulting from intensive agricultural practices. Hence ecosystem services can play an important role in a variety of agroecosystems. In this discussion, two common ecosystem services, pest control and pollination, will be used to illustrate the contribution of ecosystem services to food production.

Pest control services and agricultural productivity

Yield loss due to pests is an important constraint on global agricultural production (Oerke, 2006), and is one of the major problems in smallholder agriculture in both developed and developing countries. Natural enemies are an important, affordable mechanism for controlling pests at multiple scales. Evidence is accumulating that heterogeneous landscapes that include natural and semi-natural areas support higher biodiversity of natural enemies of agricultural pests, and that enemy diversity is often enhanced by proximity to natural ecosystems. Increased enemy diversity can lead to better delivery of biological pest control services in agroecosystems, because enemy diversity is associated with stronger pest suppression in both tropical and temperate agriculture (e.g., Snyder et al., 2008; Letourneau et al., 2009). A recent meta-analysis of 38 studies demonstrated a positive relationship between landscape diversity and natural enemy diversity and abundance, as well as predation and parasitism rates (Chaplin-Kramer et al., 2011). Several recent studies have demonstrated that landscapes with higher habitat heterogeneity have lower pest populations (e.g., O'Rourke et al., 2011) and higher efficacy of biological pest control (Thies and Tscharntke, 1999; Geiger et al., 2010; Meehan et al., 2011). However, in their meta-analysis Chaplin-Kramer et al. (2011) found that the influence of landscape complexity on pest abundance and crop damage was highly variable among the limited subset of studies that directly examined pest response, suggesting that more research on pest control efficacy is needed.

A particularly effective approach to measuring the impact of natural enemies on pest damage to crops is to employ experimental exclosures that exclude specific types of enemies. For example, Chaplin-Kramer and Kremen (2012) experimentally excluded natural enemies from broccoli infested with the cabbage aphid in landscapes with and without significant natural area. As predicted, they found that pest suppression increased with landscape complexity. Several studies have excluded vertebrate natural enemies (birds and/or bats) within tropical agroforestry systems, such as shaded coffee and cacao, and found similar influences of landscape diversification on pest suppression (e.g., Karp et al., 2013; Classen et al., 2014).

Despite abundant data on the positive association between pest suppression and both local and landscape scale diversification, the impacts of this suppression on yield is variable (Letourneau et al., 2011; Kremen and Miles, 2012). Some studies have demonstrated an effect of pest control services on yield. For example, Kellermann et al. (2008) estimated significant yield increases due to bird predation on coffee berry borer, one of the most important coffee pests worldwide. Excluding vertebrate predators (birds and bats) resulted in lower yields of cacao in Indonesia (Maas et al., 2013) and lower fruit set in coffee systems in Tanzania (Classen et al., 2014). However, other landscape-level studies have not detected a strong effect of pest suppression on yields.

Increased diversity of natural enemies and stronger pest suppression is also associated with local (within-farm) plant diversity (Letourneau et al., 2011), suggesting that diversification can increase pest control services at both local and landscape scales. Polycultures and other species mixtures are commonly practiced by smallholders for a variety of reasons, including pest control. At the farm scale, a recent meta-analysis found that both yield and biological pest control were higher in polycultures than monocultures, but there was no evidence that biocontrol had any effect on yield (Iverson et al., 2014). One potential reason for this variability in farm-level outcomes is that the effects of pest suppression on yield are often masked by the use of chemical pest control in conventional, non-diversified systems.

Several studies have estimated the value of natural pest control services to conventional, large-scale agriculture. Monetary value estimates are typically based on avoided crop losses due to insect pests, as well as the avoided cost of pesticides when relevant. Pest suppression provided by Brazilian free-tailed bats to cotton production in the Winter Garden region of Texas has been estimated as 2–29% of the value of the cotton crop, depending on the pest pressure in any given year (Boyles et al., 2011). Meehan et al. (2011) estimated that the loss of pest control services due to landscape simplification in the Midwestern US cost farmers US$34–103 million in 2007. Natural enemy control of soybean aphid in soybean was estimated at a minimum of US$239 million in four US states in 2007–08 (Landis et al., 2008). Losey and Vaughn (2006) calculated that natural pest control services are worth at least US$13.6 billion annually in agricultural crops in the US. Though clearly inexact, these estimates indicate the potential importance of pest control services to global food production.

Most estimates of economic value are more relevant to agricultural systems in developed countries than in developing countries. In developing country smallholder agriculture, crop losses to pests may affect household consumption and nutritional status directly, as well as indirectly through loss of income from crops sales. Several studies have estimated the value of pest control services for smallholder cash crops such as coffee (Kellermann et al., 2008) and cacao (Maas et al., 2013), which can be critically important to household cash incomes and rural livelihoods.

Pollination services and agricultural productivity

Global crop production also relies on pollination services. The yields of more than 70% of the leading global crops are dependent on, or enhanced by, pollination (Klein et al., 2007). Despite the dominance in the human diet of wind-pollinated grass crops like rice, wheat, and maize, approximately 35–40% of the total global volume of food crop production comes from animal-pollinated crops. Both local and global food supply is negatively affected by the loss of pollinators (Aizen et al., 2009). In both developed and developing countries, crop production by smallholders is particularly likely to benefit from natural pollination services, due to proximity to natural areas in the landscape. However, agricultural production in developing countries tends to rely more heavily on pollination than agriculture in developed countries, and it relies more heavily on wild pollinators than managed honey bees. Therefore, smallholders in developing countries are especially vulnerable to the loss of natural pollination services.

In addition to the contribution of pollination services to yields, crop quality is often improved by pollination, particularly in fruits such as strawberries and raspberries, which produce misshapen fruits if not adequately pollinated. Recent experiments with strawberries demonstrate that pollination by bees increases fruit weight, firmness, and redness, while reducing malformations and sugar–acid ratios (Klatt et al., 2014). These qualities increase the commercial grade, shelf life, and market value of the product, and would be expected to lead to higher returns to farmers.

Maintaining natural and semi-natural ecosystems in the landscape can enhance pollination of agricultural crops, and the proximity of natural ecosystems can also influence the delivery of pollination services. For example, Ricketts (2004) found that pollination rates, yield, and quality of coffee decreased with increasing distance from natural forest remnants. A meta-analysis of data from 29 studies on 21 different crops indicated that species richness of pollinators decreased with greater distance from natural areas (Garibaldi et al., 2011). In general, analyses of a range of crops indicate that species richness of pollinators and pollinator visitation rates decline with distance from natural or semi-natural vegetation (Ricketts et al., 2008). In addition, pollination services can also be enhanced by on-farm diversification (Kremen and Miles, 2012).

Stability and sustainability of agricultural production

In addition to the contribution ecosystem services can make to agricultural productivity, ecosystem services flowing from natural ecosystems may have important influences on the temporal stability and sustainability of production. Not only do diverse communities of natural enemies provide better overall pest suppression, but they can also provide a buffer against population variability of any particular species. Several studies have shown that a more diverse community of parasitoids leads to lower temporal variability in parasitism rates of agricultural pests (e.g., Macfadyen et al., 2011; Veddeler et al., 2010).

Biological pest control services can maintain pest populations at consistent, low levels that contrast with the boom and bust pest cycles typically engendered by the use of pesticides. Because natural enemies such as predatory beetles often modify their search patterns to focus on incipient pest outbreaks, but rarely eradicate pest populations, low levels of both pest and predators can persist. In contrast, pesticides are likely to eliminate both pest and predator. When pest populations rebound, as they usually do, predators are not locally available to exert control on these rapidly growing populations.

Pollination services in agroecosystems may also be stabilized by proximity to natural ecosystems in heterogeneous landscapes. In their meta-analysis of pollinator communities, Garibaldi et al. (2011) found that spatial and temporal variability in pollination services increased with greater distance from natural areas, despite pollination by domesticated honey bees. Diverse communities of wild pollinators can provide a buffer against population variability of any particular species, including honey bees (Winfree and Kremen, 2009).

Ecosystem services internal to agricultural systems can strongly influence yield stability through the accumulation of natural capital (stocks of natural assets that yield "a flow of valuable goods or services into the future," Costanza et al., 1997). In smallholder systems, careful management of crop residues and manure can build or maintain soil organic matter, thereby accumulating natural capital that can prevent soil degradation and yield declines. The ability of crops to utilize added nutrients effectively depends on soil organic matter and overall soil health. As described by Tittonell and Giller (2013) for tropical systems, smallholders practice a type of precision agriculture by recognizing soil heterogeneity and adjusting management practices according to soil fertility gradients. These practices rely on natural capital to stabilize yields.

Ecosystem services flowing to agricultural systems thus play an essential role in buffering stochastic variation in productivity through enhanced abundance and diversity of beneficial species. In addition, ecosystem services may be instrumental in buffering low-income rural households against seasonal food availability. One of the key challenges of smallholder households is the seasonality of agricultural production, due to annual cycles of temperature and rainfall. In developing countries, smallholders characteristically depend on agriculture for both subsistence and income, and this reliance on a seasonal, natural-resource based livelihood often leads to cyclical food insecurity. Due to the inherent seasonality of crop production, harvest of wild foods can play an especially important role in poor households, providing a bridge between periods of agricultural harvests. Wild harvests also provide alternative food resources to buffer food security in the event of system shocks due to unusual weather events, extreme market fluctuations, or even civil conflicts.

Increasing food access through ecosystem services: income generation

Lack of access to food is a fundamental impediment to food security that often results from poverty. Disaggregating the beneficiaries of ecosystem services can aid in evaluating the potential contributions of ecosystem services to food security (Daw et al., 2011, Poppy et al., 2014). In general, low-income households are more likely to depend on the benefits of ecosystem services than are higher income households. Lack of economic resources often leads to dependence on natural resources, along with vulnerability to natural fluctuations of these resources (TEEB, 2010). Livelihoods of the rural poor rely on the combination of natural resources and family labour, and the returns to labour are determined by the quantity and quality of those resources (Barrett et al., 2011). Compared to other rural households, the poor are liable to depend on the provisioning services provided by natural ecosystems, and the marginal value of those benefits is typically higher. Moreover, the poor benefit in multiple ways – through subsistence, enhanced nutrition, and generation of cash income. Daw et al. (2011) make a persuasive argument that the valuation of the benefits of ecosystem services should be disaggregated, based on their greater impact on the well-being of the poor.

Smallholder households in developing countries often rely on agriculture for both consumption and income. To the extent that ecosystem services contribute to greater productivity, as well as greater stability in the face of natural perturbations, they may also contribute to poverty alleviation. Although many smallholders remain in chronic poverty, other households experience transient, periodic, and/or seasonal poverty (Figure 39.2). Ecosystem services can increase the stability of production, buffer against extreme weather events and pest outbreaks, and supply alternative foodstuffs through wild harvests. All these contributions can translate into market goods that increase family income at critical moments of scarcity, allowing households to avoid slipping into extreme poverty. In addition, where ecosystem services replace purchased inputs, such as the substitution of biological pest control for pesticides, household resources are conserved. Thus provisioning and other services supplied by ecosystems may support both subsistence and income, buffering households against food insecurity.

Ecosystem services internal to agroecosystems can influence income generation through the accumulation of natural capital. Stores of natural capital have direct and indirect effects on income generation. For example, natural capital in the form of soil fertility and soil organic matter can result in higher yields that may translate into marketable yield surpluses. Marenya and Barrett (2009) found that the marginal returns to fertilizer use varied significantly among smallholder farmers in Western Kenya, depending on soil carbon content. While it is commonly assumed that mineral fertilizers will increase yields in nutrient poor soils, lack of organic matter can limit crop response to nutrient amendments, as noted above. If yield response is weak, the marginal value of fertilizers is low, and many farmers forgo them. Soil organic matter can

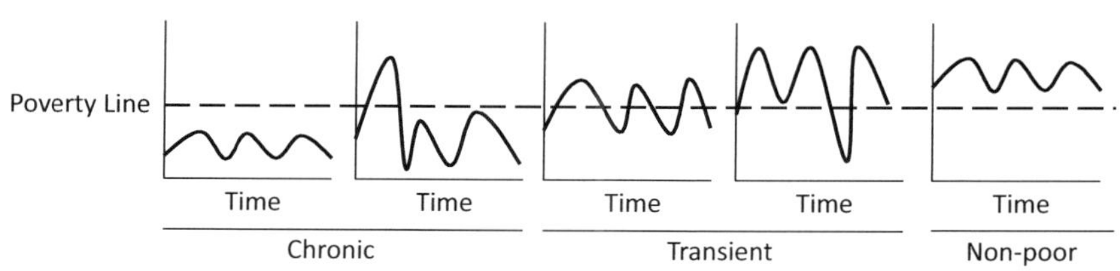

Figure 39.2 Categorization of the chronic poor, transient poor, and the non-poor.

Source: adapted from Hulme et al., 2001

strongly mediate agronomic response to purchased fertilizers, thereby indirectly influencing yields and cost-benefit ratios of purchased inputs.

While ecosystem services can contribute to household income generation, access to ecosystem services may be as important as any other access issue for the poor. Understanding access mechanisms for particular ecosystem services is essential for evaluating the potential contribution to food security, whether considering the provision of wild foods or food provisioning from agriculture (Ribot and Peluso, 2003).

Increasing food adequacy through ecosystem services

Diverse landscapes expand the provisioning services provided by both natural habitats and agroecosystems and can lead to a greater diversity of foods. At the household level, this has the potential to lead to increased dietary diversity, as well as greater potential for income generation through the sale of these foods. Household-level dietary diversity appears to be a useful indicator of food security. Dietary diversity is consistently associated with measures of nutrient adequacy, as well as child nutritional status and growth, in developing countries (Ruel, 2003).

Ecosystem services can also play a significant role in the nutritional quality of human diets through the provision of micronutrient-rich foods. In an assessment of global agricultural vulnerability to the loss of pollination services, Gallai et al. (2009) conclude that global agricultural production would not be drastically affected by losses in pollination. Of course, this does not mean that local production would not suffer substantially in some regions. More importantly, the production of fruits and vegetables would be severely disrupted by pollinator losses (Gallai et al., 2009). Since fruits and vegetables are the primary dietary source of many micronutrients, this could result in reductions in the nutritional adequacy of human diets, particularly for the poor.

Pollinator-dependent crops contribute disproportionately to the production of foods rich in vitamin A and other micronutrients. Along with iron deficiency, vitamin A deficiency is one of the most important micronutrient deficiencies in developing countries (Kennedy et al., 2003). In a novel analysis of the spatial overlap between pollinator-dependent crops and micronutrient deficiencies, Chaplin-Kramer et al. (2014) report that the severity of iron, vitamin A, and folate deficiencies often corresponds with plant-based micronutrient dependence on pollination. For example, nearly 50% of the plant-derived sources of vitamin A in South and Southeast Asia depend on pollination. In general, micronutrient dependence on pollination coincides with areas of poverty and food insecurity, suggesting that any declines in pollinator-dependent crops are likely to have direct impacts on the nutritional status of the poor.

Summary

Sustaining the provision of ecosystem services holds great promise for improving global food security, while simultaneously providing significant environmental advantages. In particular, ecosystem services can contribute to improvements in: the availability of food through increased productivity; access to food through increased income generation; the stability and sustainability of agricultural production through buffering of both environmental and socioeconomic variability; and food adequacy through increased dietary diversity and nutritional quality. In particular, ecosystem services can enhance the livelihoods and well-being of smallholders through the provisioning of wild foods from natural habitats and the ecological intensification of agricultural production.

References

Aizen, M. A., Garibaldi, L. A., Cunningham, S. A., and Klein, A. M. (2009). How much does agriculture depend on pollinators? Lessons from long-term trends in crop production. *Annals of Botany*, mcp076.

Barrett, C.B. (2010). Measuring food insecurity. *Science*, vol 327, no 5967, pp 825–828.

Barrett, C.B., Travis, A.J., and Dasgupta, P. (2011). On biodiversity conservation and poverty traps. *Proceedings of the National Academy of Science of the United State of America*, vol 108, no 34, pp 13907–13912.

Bommarco, R., Kleijn, D., and Potts, S. G. (2013). Ecological intensification: harnessing ecosystem services for food security. *Trends in Ecology & Evolution*, vol 28, pp 230–238.

Boyles, J. G., Cryan, P. M., McCracken, G.F., and Kunz, T.H. (2011). Economic importance of bats in agriculture. *Science*, vol 332, pp 41–42.

Chaplin-Kramer, R., Dombeck, E., Gerber, J., et al. (8 authors) (2014). Global malnutrition overlaps with pollinator-dependent micronutrient production. *Proceedings of the Royal Society B*, vol 281, pp 1794.

Chaplin-Kramer, R., and Kremen, C. (2012). Pest control experiments show benefits of complexity at landscape and local scales. *Ecological Applications*, vol 22, no 7, pp 01936–1948.

Chaplin-Kramer, R., O'Rourke, M.E., Blitzer, E.J., and Kremen, C. (2011). A meta-analysis of crop pest and natural enemy response to landscape complexity. *Ecology Letters*, vol 14, pp 922–932.

Chappell, M.J., and LaValle, L.A. (2011). Food security and biodiversity: can we have both? An agroecological analysis. *Agriculture and Human Values*, vol 28, no 1, pp 3–26.

Chen, H., and Lin, Y. (2013). Promise and issues of genetically modified crops. *Current Opinion in Plant Biology*, vol 16, pp 255–260.

Classen, A., Peters, M.K., Ferger, S.W., et al. (10 authors) (2014). Complementary ecosystem services provided by pest predators and pollinators increase quantity and quality of coffee yields. *Proceedings of the Royal Society B*, vol 281, no 1779, p 2013348.

Costanza, R., d'Arge, R., de Groot, R., et al. (13 authors) (1997). The value of the world's ecosystem services and natural capital. *Nature*, vol 387, pp 253–260.

Daw, T., Brown, K., Rosendo, S., and Pomeroy, R. (2011). Applying the ecosystem services concept to poverty alleviation: the need to disaggregate human well-being. *Environmental Conservation*, vol 38, pp 370–379.

De Schutter, O. (2010). *UNGA HRC Sixteenth Session Committee Report on the Right to Food*, UN Doc A/ HRC/16/49.

Dore, T., Makowski, D., Malezieux, E., et al. (6 authors) (2011). Facing up to the paradigm of ecological intensification in agronomy: revisiting methods, concepts and knowledge. *European Journal of Agronomy*, vol 34, pp 197–210.

Ericksen, P.J. (2008). Conceptualizing food systems for global environmental change research. *Global Environmental Change*, vol 18, pp 234–245.

Gallai, N., Salles, J. M., Settele, J., and Vaissière, B. E. (2009). Economic valuation of the vulnerability of world agriculture confronted with pollinator decline. *Ecological Economics*, vol 68, no 3, pp 810–821.

Garibaldi, L.A., Steffan-Dewenter, I., Kremen, C., et al. (23 authors) (2011). Stability of pollination services decreases with isolation from natural areas despite honey bee visits. *Ecology Letters*, vol 14, pp 1062–1072.

Gebbers, R., and Adamchuk, V.I. (2010). Precision agriculture and food security. *Science*, vol 327, pp 828–831.

Geiger, F., de Snoo, G. R., Berendse, F., et al. (20 authors) (2010). Landscape composition influences farm management effects on farmland birds in winter: a pan-European approach. *Agriculture, Ecosystems and Environment*, vol 139, pp 571–577.

Godfray, H.C., Beddington, J. R., Crute, I. R., et al. (10 authors) (2010). Food security: the challenge of feeding 9 billion people. *Science*, vol 327, pp 812–818.

Hazell, P., and Wood, S. (2008). Drivers of change in global agriculture. *Proceedings of the Royal Society B*, vol 363, pp 495–515.

Iverson, A. L., Marín, L. E., Ennis, K. K., et al. (8 authors) (2014). Do polycultures promote win-wins or trade-offs in agricultural ecosystem services? A meta-analysis. *Journal of Applied Ecology*, vol 51, no 6, pp 1593–1602.

Karp, D.S., Mendenhall, D.S., Sandi, R.F., et al. (7 authors) (2013). Forest bolsters bird abundance, pest control and coffee yield. *Ecology Letters*, vol 16, pp 1339–1347.

Kellermann, J. L., Johnson, M. D., Stercho, A. M., and Hackett, S. C. (2008). Ecological and economic services provided by birds on Jamaican Blue Mountain coffee farms. *Conservation Biology*, vol 22, no 5, pp 1177–1185.

Kennedy, G., Nantel, G., and Shetty, P. (2003). The scourge of "hidden hunger": global dimensions of micronutrient deficiencies. *Food, Nutrition and Agriculture*, vol 32, pp 8–16.

Klatt, B. K., Holzschuh, A., Westphal, C., Clough, Y., Smit, I., Pawelzik, E., and Tscharntke, T. (2014). Bee pollination improves crop quality, shelf life and commercial value. *Proceedings of the Royal Society of London B: Biological Sciences*, vol 281, no 1775, p 20132440.

Klein, A. M., Vaissiere, B. E., Cane, J. H., Steffan-Dewenter, I., Cunningham, S. A., Kremen, C., and Tscharntke, T. (2007). Importance of pollinators in changing landscapes for world crops. *Proceedings of the Royal Society of London B: Biological Sciences*, vol 274, no 1608, pp 303–313.

Kremen, C., and Miles, A. (2012). Ecosystem services in biologically diversified versus conventional farming systems: benefits, externalities, and trade-offs. *Ecology and Society*, vol 17, no 4, pp 40–65.

Landis, D. A., Gardiner, M. M., van der Werf, W., and Swinton, S. M. (2008). Increasing corn for biofuel production reduces biocontrol services in agricultural landscapes. *Proceedings of the National Academy of Sciences*, vol 105, no 51, pp 20552–20557.

Letourneau, D.K., Armbrecht, I., Rivera, B.S., et al. (17 authors) (2011). Does plant diversity benefit agroecosystems? A synthetic review. *Ecological Applications*, vol 21, no 1, pp 9–21.

Letourneau, D.K., Jedlicka, J.A., Bothwell, S.G., and Moreno, C.R. (2009). Effects of natural enemy biodiversity on the suppression of arthropod herbivores in terrestrial ecosystems. *Annual Review of Ecology, Evolution, and Systematics*, vol 40, pp 573–592.

Losey, J. E., and Vaughan, M. (2006). The economic value of ecological services provided by insects. *Bioscience*, vol 56, no 4, pp 311–323.

MA (2005). *Ecosystems and Human Well-Being: General Synthesis*. Island Press, Washington DC.

Maas, B., Clough, Y., and Tscharntke, T. (2013). Bats and birds increase crop yield in tropical agroforestry landscapes. *Ecology Letters*, vol 16, pp 1480–1487.

Macfadyen, S., Craze, P.G., Polaszek, A., et al. (6 authors) (2011). Parasitoid diversity reduces the variability in pest control services across time on farms. *Proceedings of the Royal Society B*, vol 278, pp 3387–3394.

Marenya, P.P., and Barrett, C.B.B. (2009). State-conditional fertilizer yield response on Western Kenyan farms. *American Journal of Agricultural Economics*, vol 91, no 4, pp 991–1006.

Meehan, T.D., Werling, B.P., Landis, D.A., and Gratton, C. (2011). Agriculture landscape simplification and insecticide use in the Midwestern U.S. *Proceedings of the National Academy of Science of the United State of America*, vol 108, no 28, pp 11500–11505.

Oerke, E. C. (2006). Crop losses to pests. *Journal of Agricultural Science*, vol 144, pp 31–43.

O'Rourke, M.E., Rienzo-Stack, K., and Power, A.G. (2011). A multi-scale, landscape approach to predicting insect populations in agroecosystems. *Ecological Applications*, vol 21, no 5, pp 1782–1791.

Peng, S., Cassman, K.G., Virmani, S.S., Sheehy, J., and Khush, G.S. (1999). Yield potential trends of tropical rice since the release of IR8 and the challenge of increasing rice yield potential. *Crop Science*, vol 39, pp 1552–1559.

Plieninger, T., Raymond, C. and Oteros-Rozas, E. (2016). Cultivated lands. In: Potschin, M., Haines-Young, R., Fish, R. and Turner, R.K. (eds) *Routledge Handbook of Ecosystem Services*. Routledge, London and New York, pp 442–451.

Poppy, G.M., Chiotha, S., Eigenbrod, F., et al. (12 authors) (2014). Food security in a perfect storm: using the ecosystem services framework to increase understanding. *Philosophical Transactions of the Royal Society B*, vol 369, p 20120288.

Power, A.G. (2010). Ecosystem services and agriculture: tradeoffs and synergies. *Philosophical Transactions of the Royal Society B*, vol 365, pp 2959–2971.

Ribot, J. C., and Peluso, N. L. (2003). A theory of access. *Rural Sociology*, vol 68, pp 153–181.

Ricketts, T.H. (2004). Tropical forest fragments enhance pollinator activity in nearby coffee crops. *Conservation Biology*, vol 18, pp 1262–1271.

Ricketts, T.H., Regetz, J., Steffan-Dewenter, I., et al. (13 authors) (2008). Landscape effects on crop pollination services: are there general patterns? *Ecology Letters*, vol 11, pp 499–515.

Ruel, M.T. (2003). Operationalizing dietary diversity: a review of measurement issues and research priorities. *Journal of Nutrition*, vol 133, pp S3911–3926.

Sen, A. (1981). *Poverty and Famines: An Essay on Entitlement and Deprivation*, Clarendon Press, Oxford; emphasis in original.

Snyder, G.B., Finke, D.L., and Snyder, W.E. (2008). Predator biodiversity strengthens aphid suppression across single and multiple-species prey communities. *Biological Control*, vol 44, pp 52–60.

TEEB (2010). *The Economics of Ecosystems and Biodiversity: Mainstreaming the Economics of Nature: A Synthesis of the Approach, Conclusions and Recommendations of TEEB*. Progress Press, Malta.

Thies, C., and Tscharntke, T. (1999). Landscape structure and biological control in agroecosystems. *Science*, vol 285, no 5429, pp 893–895.

Tilman, D., Fargione, J., Wolff, B., et al. (10 authors) (2001). Forecasting agriculturally driven global environmental change. *Science*, vol 292, pp 281–284.

Tittonell, P., and Giller, K.E. (2013). When yield gaps are poverty traps: the paradigm of ecological intensification in African smallholder agriculture. *Field Crops Research*, vol 143, pp 76–90.

Veddeler, D., Tylianakis, J., Tscharntke, T., and Klein, A. M. (2010). Natural enemy diversity reduces temporal variability in wasp but not bee parasitism. *Oecologia*, vol 162, pp 755–762.

Winfree, R., and Kremen, C. (2009). Are ecosystems stabilized by differences among species? A test using crop pollination. *Proceedings of the Royal Society B*, vol 276, pp 229–237.

Zhang, W., Ricketts, T.H., Kremen, C., Carney, K., and Swinton, S.M. (2007). Ecosystem services and dis-services to agriculture. *Ecological Economics*, vol 64, pp 253–260.

40

ECOSYSTEM SERVICES AND WATER SECURITY

Sarah Hendry and Geoffrey Gooch

Introduction

Water is a cross-cutting issue that is interconnected with many different ecosystems services. For example, within the Millennium Development Goals (MDGs, UN General Assembly, 2000), those goals related to water, such as improving access to drinking water and sanitation, the only ones that affected the achievement of every other MDG. The interconnectivity of the 'water / food / energy nexus' (see, for example, Bonn Nexus, 2011) reminds us of the threat of the 'perfect storm' (Beddington, 2009); water is required to support the supply of food and energy whilst remaining indispensable to human and other life in itself, and the demand for all three is rising inexorably. Although the volume of water on earth remains constant, and is theoretically self-renewing and self-cleansing through the hydrological cycle at a global level, the proportion of freshwater is only 2.5% of the total water available (UNEP, 2008). Within that percentage, the tiny fraction of freshwater available for human use is threatened locally by pollution and over-abstraction, and by climate change, urbanisation and population growth. Water provides ecosystem services in all categories. Using the CICES system (Haines-Young and Potschin, 2013; Potschin and Haines-Young, 2016) it is a provisioning service in its own right, and also essential to provision of food, some fuels and fibre. Regulating services include flood regulation and water purification, and again water plays a role in the regulation of disease and of climate. And in cultural services, water provides aesthetic, spiritual, educational and recreational benefits. It is also part of the mechanism for the underpinning 'supporting' services in classifications that use these (MA, 2005). All life as we understand it depends on water.

Water security

The water security debate is an emerging paradigm, sharpened by the various demands and policy contexts noted above, and especially by the multiple and growing pressures on the resource. Water security can be defined in many ways, reflecting different political perspectives and academic disciplines (Magsig, 2015). Often it is seen as part of a 'national security' concept, with possible military implications linked to fears of water wars. There is also an emerging discourse around environmental or ecological security, perhaps most relevant to this book, but also, there is the concept of human security and the provision of water services of different types, for health, well-being and food production. The 'water wars' issue is a good example of lazy conceptualisation and theoretical trends. There is far more evidence of cooperation over water, and there has only ever been one water 'war' as such. In the 1980s, Boutros Boutros-Ghali, then Secretary-General of the UN, suggested that the next wars in the Middle East would be over water, not politics or oil; and that has not yet (thankfully) been the case. But there are many examples of smaller-scale conflicts over water locally, and in many regions, and water services of many different types are targets in those conflicts. The ecological implications of human impacts on the water resource, direct and indirect, have been extensively noted in other parts of this book, and will be returned to further below. The human security dimension is perhaps the most interesting if seen from the perspective of water services of different types. Linked to equally difficult questions around development, sustainable or otherwise, the concept of water services recognises that water is essential to life, and the provision of basic needs, but also that water services are vital for economic development. It is therefore necessary to acknowledge the tensions between economic development, and the protection of the resource base on which development itself also depends.

Specific to water, a number of definitions have emerged from the domain of public policy. Some years ago, the Global Water Partnership stated that '[w]ater security, at any level from the household to the global, means that every person has access to enough safe water at affordable cost to lead a clean, healthy and productive life, while ensuring that the natural environment is protected and enhanced' (Rogers and Hall, 2000, p. 12). The United Nations Development Programme considers that 'water security is about ensuring that every person has reliable access to enough safe water at an affordable price to lead a healthy, dignified and productive life, while maintaining the ecological systems that provide water and also depend on water' (UNDP, 2006, p. 3). Most recently, UN-Water defined water security as '[t]he capacity of a population to safeguard sustainable access to adequate quantities of acceptable quality water for sustaining livelihoods, human well-being, and socio-economic development, for ensuring protection against water-borne pollution and water-related disasters, and for preserving ecosystems in a climate of peace and political stability' (UN-Water, 2013, p. 1). Thus human needs are included, as are developmental needs, together with the protection of the environment and ecosystem services. The next two sections of this chapter will consider first, the meanings and relevance of water services, and second, the meanings and role of regulation in managing water supply. The concluding section will consider how the concept of ecosystem services is both framing and reconceptualising the water policy agenda.

Water supply and water services

In the context of this book, water services might be understood as water-related ecosystem services, and that is certainly one way that the phrase can be interpreted. However, more generally, the term 'water services' is often used to designate the provision of (urban) water services, that is, the supply of drinking water, water for commercial use and wastewater services. It can also be understood as the provision of basic sanitation, likely not to be waterborne but still with consequences for the water environment. These services may be provided by public or private providers, or through self-supply at individual or community levels. They are relevant to basic human needs (drinking, cooking and basic hygiene) and linked to the debate around the human right to water, which is also helping to frame and re-conceptualise debates over water services.

In a seminal article, Gleick argued that 50 litres / person / day (LPD) was sufficient to provide for health and hygiene (Gleick, 1996), but not for waterborne sanitation. A range of 20–40 LPD is also often cited; for example, South Africa provides a Basic Water supply of 25 LPD (DWAF,

2003), which should be provided free by municipalities to indigent households and supplied to all at a lifeline rate. This is also incorporated into the South African concept of 'the Reserve', an innovative mechanism for prioritising and recognising both basic human needs and ecological requirements in terms of environmental flow. The core ecosystem service of providing drinking water has been estimated at somewhere around 10% of global withdrawals (UN-Water, 2009).

Water services may, however, be defined more broadly than this. The European Water Framework Directive[1] (2000/60/EC, WFD) defines water services as:

> all services which provide, for households, public institutions or any economic activity:
>
> (a) abstraction, impoundment, storage, treatment and distribution of surface water or groundwater,
>
> (b) waste-water collection and treatment facilities which subsequently discharge into surface water.
>
> *(WFD Art. 2)*

This understanding of water services goes beyond the provision of domestic water and indeed beyond the provision of piped urban water supply to industry or commercial users. As it specifies 'any economic activity', it is at least arguable that it includes water for agriculture; and the extent of the definition has been referred to the European Court of Justice (European Commission v Germany C-525/12). Under the WFD, states must have pricing policies to incentivise sustainable water use, and 'take account of the costs of water services, including environmental and resource costs' (WFD Art. 9). When EU Member States submitted their River Basin Management Plans, within which all these requirements should be reported, the analysis conducted by the European Commission suggested that most states had addressed cost recovery and pricing policy in terms of urban supply but had not applied it to agricultural water. Globally, agricultural use amounts to some 70% of all water withdrawals, and in many countries farmers are the group who are least likely to be paying the full cost of the water they use. This then affects the supply of other provisioning services depending on water, especially food and fibre.

Agriculture is of course not a homogenous sector; it ranges from vast grain monocultures, pastoral ranges and hothouse horticulture, to the subsistence farmers eking a living in many parts of the world. It is difficult to obtain good data on sectoral use, and certainly not within sectors; it is likely that subsistence farmers are less efficient and have less advanced irrigation technology, though this may also mean that substantial return flows are returned to the basin. A relevant question is whether this use – subsistence farming – should come within the concept of basic human needs, found for example in the UN Watercourses Convention (where 'special regard' should be given to 'vital human needs', UN 1997 Art. 10) but also in the human rights discourse. The UN Economic and Social Council, in its General Comment 15, has suggested that small-scale agricultural use of water should be included in the human right to water (UN, 2002, para.7) as part of the right to food. But acceptance of this inclusion would significantly expand the overall requirement for water for basic human needs, and may also cause some tensions with another water paradigm, Integrated Water Resources Management (IWRM). IWRM has been much discussed in policy and academic literature, and this is not a place to revisit that debate at any length. It is, however, generally taken to include an integrated approach to the resource (especially managing surface water and groundwater together); an integrated approach to catchments (recognising catchment boundaries, and recognising the interrelationship between land

and water); and a participative approach, engaging stakeholders in the management process. Notably, one of the few texts dealing specifically with law and ecosystem services (Ruhl et al., 2007) recommends a multi-layered planning system that is highly compatible with many versions of IWRM. Often, the introduction of IWRM will involve some form of legislative reform in order to introduce the specified processes and mechanisms, and perhaps also to provide for the reallocation of water. Water allocation may be carried out within a prioritisation of uses within a state (or a basin), in which case basic human needs, or water for food production or industry, may be a priority. If a state explicitly recognises a constitutional or human right to water, then that will inevitably be prioritised, but it may not be the only priority. If there is a human right to water, then human rights discourse this will 'trump' any other right, at which point the extent of that right in volumetric terms, and specifically whether it includes subsistence farming, may make a significant difference to the volumes of water left for other purposes. Nonetheless, both IWRM and the human right to water theoretically remain high on global, regional and national policy agendas. Whether or not these policy aims are implemented is another matter. In addition to potential tensions between the different uses of water, there may be tensions with other ecosystem services provided by water.

Water services beyond basic human needs

A broader analysis of the ecosystem services provided by water takes us beyond water supply for basic human needs. Some water is used for industrial purposes, perhaps 20% of withdrawals, but again sectoral figures are variable and may be calculated in different ways. Water may be used as an ingredient, as part of a cooling process, or as a means of exiting industrial waste. Significant amounts of water are needed for the production of soft and alcoholic drinks, much of which is now pumped from aquifers in what is a form of water mining. Water is also used in many production techniques, and may in this way conflict with other uses such as agriculture, human consumption and spiritual values. The leather industry on the banks of the Ganges, for example, is a major polluter and competes in this way with other uses of water.

Some domestic supply in developed countries is well in excess of the basic requirements suggested by Gleick or the WHO, and if there is a wastewater system this will at least double the minimum supply required. Where human wastes are disposed of via waterborne systems, this could be seen as a regulating, purification service, mediating wastes through water flows, but given the scale and density of human habitation, this will not function without some type of further treatment. In the near future, there is likely to be much more emphasis on treating wastewater to make it fit for reuse, and on different treatment options for different forms of reuse. Both the provision of a networked supply and almost all treatment options for wastewater (domestic and industrial) will require energy, which in turn may use more water, reminding us again of the water-food-energy nexus. Similarly, most non-subsistence agricultural operations will require energy in different forms, whether for machinery, transport or the production of fertiliser and pesticides. Whilst provision of basic human needs is a clear imperative in terms of domestic water security, so too is food, and water for food production. Again, this is not the place to explore in detail the widely debated concept of virtual water (Allan, 2011), or water 'footprints' (Hoekstra and Chapagain, 2008), but these concepts can be helpful in understanding the flows and exchanges of benefits based on water that take place through international trade. Ironically, water-rich, developed countries often import large amounts of irrigated foodstuffs from countries with fewer water resources. This is also true for timber, another vital provisioning service. Forest management is linked closely to water management in terms of competition for water, and in terms of sustainable management of indigenous forestry for climate regulation. It is unsurprising that, in the developing world, much of the ecosystem services debate is focused around forest protection (Bennett and Carroll, 2014), partly as a result of the funding mechanisms in the UN Convention on Climate Change.[2]

As noted above, agricultural production is the major consumer of fresh water. Countries may choose to restrict (by regulation or incentive or both) the crops grown by farmers, or the total extent of agricultural activity, and import water-intensive foodstuffs from elsewhere, to increase the water available for other purposes. They may choose to encourage large-scale cash crop agriculture at the expense of small subsistence farmers, or to maximise the latter. All of these policy choices may be influenced or determined at the national level by perceptions of food security, water security or broader notions of security in its military or civilian senses. Fisheries are also a vital provisioning service and a major source of protein in developing countries. Other uses of the water environment may impact natural fisheries negatively, whilst aquaculture brings its own local environmental impacts.

Hydropower is also a major user of water and contributes to human security in different ways. Although non-consumptive, it affects flow regulation and, hence, availability for other sectors. Hydroelectric power is one of the most important energy sources today, especially for developing countries. The industrialised countries have developed almost all of their hydropower potential, but in other parts of the world there is still a large potential for development. As rivers are often trans-boundary, there may be conflicting interests involved. For example, in the Mekong, Laos is in the process of building dams on the main Mekong River. Cambodia and Vietnam have objections, as the dam will impact agricultural production, especially in the Mekong Delta, which is the major rice producing area of Vietnam, and one of the most important in the world. In the meantime, hydro also has significant, usually negative, effects on ecosystems.

Finally, in terms of direct use by humans, the cultural (spiritual, recreational) services provided by water are usually related to water *in situ*, in its 'natural' state at any given time. In many parts of the world, water is seen as inherently spiritual and an important cultural signifier. For example, the Ganges River in India is associated with the cycle of life and death, and in parts of Africa water bodies such as lakes and waterfalls are seen as the habitat of spirits. In developed countries, the value of water is still seen as an important contributor to quality of life in a social as well as an economic sense. To access these services, therefore, it is necessary to maintain or protect some waterbodies and their surrounding environments.

Water for the environment

Many ecosystem services involving water are not directly 'water services' in anthropocentric terms, but rather services provided to the environment, on which human and other life nonetheless depends. Protection or preservation of ecosystems, as might be required under the UN Watercourses Convention[3] (UN 1997 Art. 20), the Convention on Biological Diversity[4] (CBD) or the Ramsar Convention[5] on the protection of wetlands for migratory birds will assist in this aim, as will many regional, national and local legal instruments, whether in accordance with these treaties or independent from them.

The need for environmental flows can be built into the regulatory and management systems for water in a variety of ways. Water can be retained from other uses for the environment, with specific temporal and volume levels of water flows mandated before any abstractions can be made for other purposes (with the likely exception of basic human needs). Specific environmental allocations of water can be made to provide for ecosystem needs. Both of these techniques can operate in a planned system, such as one based on IWRM, and many countries are aiming at adopting these approaches. Thus, under the WFD, environmental flows should be maintained or increased to achieve 'good ecological status'. In South Africa, the Reserve includes an environmental reserve, which has proved harder to calculate than the 25 LPD mandated for human use. In the Murray-Darling in Australia, there is an environmental watering plan, which includes buy-backs of water for the environment (Government of Australia Water Act, 2007). Where there is a water market in place, it may also be possible for private interests such as fisheries trusts to purchase water for the environment. In all of these systems, trade-offs between the environment and human uses for development operate in different ways, but all give some recognition of ecological needs. In the EU, the objective of the WFD is good ecological status, but with grounds for exemptions and extensions. In South Africa and some Australian states, there is an ecological classification system for waterbodies that does not have an overall goal of a specific class, but recognises that human needs and development activities will mean that some waters will achieve a lower class. Under Australian federal law, perhaps reflecting its more recent adoption, protection of ecosystem services and functions are an explicit driver. Maintaining environmental flows will increase the availability of many provisioning and regulating services, both those used directly by humans and those that indirectly support the resource base. Only by protecting all those services will there be water security in the fullest sense.

Water services, water security and water regulation

Just as the concept of water services can hold several different meanings, so too can the concept of water regulation. Again, it is possible that the ecosystems debate can bring fresh conceptualisation and integration of different perspectives to bear on regulation, and a more human-centred approach is likely to lead to a narrower understanding of the term. Regulation in a non-legal sense is used in relation to the water environment in several different ways, in the form of flow regulation for irrigation and protection of instream and riparian zone ecosystems; or regulating services, in the ecosystem paradigm. Perhaps amongst these different understandings, the lawyers have the narrowest understanding of all, but law nonetheless relates to all aspects of water management.

Water law applies both to the water resource and water services. In terms of managing the resource base, legal frameworks can and should establish the systems and processes for water management, including stakeholder engagement and IWRM. Law will also be used to allocate water, and where changes are being made to water rights' regimes these changes may be contentious. They are likely to require a high-level Act or Code, which in turn should provide opportunity for wider debate, including the engagement of stakeholders over a period of time (see, for example Hodgson, 2006). If, for example, there is to be provision for environmental flow, this may be addressed in these rules. Finally, water resources law (or environmental law) will also regulate water quality, controlling inputs into the system from human sources and linking to broader environmental law frameworks at different levels of government. Controls over water quality and quantity are linked, which can be seen clearly in the ecosystem services approach; there must be enough water in the system, and of sufficient quality, to continue to maintain the desired services.

In terms of the (legal) regulation of water services, this is first concerned with drinking water quality and other service standards for water supply. The latter include pressure and access / availability, for example, as well as: distance to the source for rural areas and collection time in urban areas; number of households served by a standpipe; and hours / days for which piped supply is available. Drinking water quality as such is likely to be subject to standards based on the WHO guidelines or a subset of these (WHO, 2004). All of these might be applied and enforced by public health departments. Secondly, for sanitation and wastewater, there may be standards detailing the type of facilities provided and access to these, any maintenance obligations and perhaps customer standards for sewer flooding. Increasingly, regulatory and policy attention is turning to the correct treatment and effective reuse of wastewater (WHO, 2006; European Commission *Water Blueprint*, 2012; and UN-Water, 2014).

There are also links between water services and water resources law. Service providers at bulk level will be abstractors of raw water and will also be responsible for discharges from wastewater systems. Further, upstream catchment protection and water safety planning (which is suggested to be mandatory under the current WHO guidelines) bring water supply issues firmly into the ambit of land use and catchment management. The more upstream protection, the less downstream treatment will be required. The regulation of both water services delivery and the raw water resource are critical to the maintenance of water security at a domestic level; if the water supplied is unsafe, and / or wastewater is allowed to contaminate surface or groundwater, then the starting point of all the definitions of water security noted above, the life and health of the individual, will fail. If the sources are protected and the wastewater managed for reuse, then the wider set of ecosystem services is better protected.

One further aspect of legal regulation may be of interest in the ESS debate. In a trans-boundary context, the principal international legal instrument, the UN Watercourses Convention, enables the concept of benefit-sharing, within the core principle of equitable and reasonable utilisation, as an alternative (or more likely in addition) to allocating water (see, for example, in the context of ecosystem services, Rieu-Clarke and Spray, 2013). In theory, at least, if all the states in a trans-boundary basin are cooperating, benefit-sharing is a preferable way to maximise the returns on every cubic metre of water available. Alternatively, if cooperation is not forthcoming, as is currently evident in a number of trans-boundary basins worldwide, benefit-sharing will be seen as a threat that detracts from security of different types, including water security. The debate outlined above over food (and water) security and the policy choices available also plays out internationally, but only a cooperative approach will maximise the protection of the resource and the services it provides.

Conclusions

In conclusion, water plays a unique role in the provision of many ecosystem services, and is also vital from the perspective of different forms of security. This chapter has sought to identify the ecosystem services delivered by water, and the many different uses to which water is put, in the context of the emerging debate around water security. Whilst the security debate is often framed in narrow terms of state (or individual) interest, there is some recognition, internationally and by national policymakers, that a broader and more cooperative approach to managing water for all its human uses is essential to cope with current global changes, especially population growth, climate change, urbanisation and environmental degradation. Whilst the ecosystem services paradigm is primarily anthropocentric, it nonetheless recognises the fundamental biophysical imperatives on which all life depends – the deepest and most fundamental form of security. Drawing that paradigm into the water security debate, and exposing different disciplines, sectors and policymakers to the ecosystem imperative, could help to foster a wider and more cooperative approach.

Notes

1 http://ec.europa.eu/environment/water/water-framework/index_en.html
2 http://unfccc.int/2860.php
3 http://www.unwatercoursesconvention.org/
4 http://www.cbd.int/
5 http://www.ramsar.org/

References

Allan, T. (2011). *Virtual Water: Tackling the Threat to our Planet's Most Precious Resource*. I.B.Tauris, London.

Beddington, J. (2009). Keynote Speech, Sustainable Development UK Conference. 19 March 2009. Available at: http://www.govnet.co.uk/news/govnet/professor-sir-john-beddingtons-speech-at-sduk-09

Bennet, G., and Carroll, N. (2014). *Gaining Depth: State of Watershed Investment 2014*. A Report by Forest Trends' Ecosystem Marketplace. Available at: http://www.forest-trends.org/documents/files/SOWI2014.pdf

Bonn Nexus (2011). *The Water, Energy and Food Security Nexus – Solutions for a Green Economy*. Policy Recommendations. Available at: http://www.water-energy-food.org/en/whats_the_nexus/messages_policy_recommendations.html

DWAF (2003). *Strategic Framework for Water Services*. Available at: http://www.dwaf.gov.za/Documents/

European Commission (2012). *A Water Blueprint for Europe*. Publications Office of the European Union, Luxembourg.

Gleick, P. (1996). *Basic Water Requirements for Human Activities Meeting Basic Needs*. Originally published in *Water International*, vol 21, pp 83–92.

Government of Australia Water (Cwlth) Act 2007 No.137. Available at: https://www.comlaw.gov.au/Details/C2007A00137

Haines-Young, R., and Potschin, M. (2013). *Common International Classification of Ecosystem Services (CICES)*. Report to the European Environment Agency. Available at: www.cices.eu

Hodgson, S. (2006). *Modern Water Rights: Theory and Practice*. FAO Legislative Series 92 Rome.

Hoekstra, A.Y., and Chapagain, A. K. (2008). *Globalization of Water*. Blackwell, Malden MA, Oxford, and Victoria.

MA (2005). *Ecosystems and Human Well-Being: Synthesis*. Island Press, Washington DC.

Magsig, B-O. (2015). *International Water Law and the Quest for Common Security*. Routledge, London and New York.

Potschin, M., and Haines-Young, R.(2016). Defining and measuring ecosystem services. In: Potschin, M., Haines-Young, R., Fish, R. and Turner, R. K. (eds) *Routledge Handbook of Ecosystem Services*. Routledge, London and ä New York, pp 25–44.

Rieu-Clarke, A., and Spray, C. (2013). *Ecosystem Services and International Water Law: Towards a More Effective Determination and Implementation of Equity?* (16)2 PER/PELJ 12/212.

Rogers, P., and Hall, A. (2000). *Integrated Water Resources Management* GWP TAC Background Paper No.4. Available at: http://www.gwp.org/en/The-Challenge/IWRM-Resources/

Ruhl, J. B., Kraft, S., and Lant, C. (2007). *The Law and Policy of Ecosystem Services*. Island Press, Washington DC.

UN (1997). *UN Watercourses Convention*. Available at: http://www.unwatercoursesconvention.org/the-convention/

UN (2002). Committee on Economic, Social and Cultural Rights. General Comment No. 15 on the Right to Water E/C.12/2002/11.

UN General Assembly (2000). *Millennium Declaration* A/RES/55/2.

UNDP (2006). *Human Development Report. Beyond Scarcity: Power, Poverty and the Global Water Crisis*. Available at: http://hdr.undp.org/sites/default/files/reports/267/hdr06-complete.pdf

UNEP (2008). *Vital Water Graphics*. Available at: http://www.unep.org/dewa/vitalwater/index.html

UN-Water (2009). *World Water Development Report 3: Water in a Changing World*. Available at: http://www.unesco.org/water/wwap/wwdr/

UN-Water (2013). *Analytical Brief on Water Security and the Global Water Agenda*. Available at: http://www.unwater.org/topics/water-security/en/

UN-Water (2014). *The United Nations World Water Development Report 2014: Water and Energy*. Paris, UNESCO. Available at: http://unesdoc.unesco.org/images/0022/002257/225741e.pdf

WHO (2004). *Guidelines for Drinking Water Quality, 3rd Edition*. Available at: http://www.who.int/water_sanitation_health/dwq/gdwq3/en/

WHO (2006). *Guidelines for the Safe Use of Wastewater, Excreta and Greywater*. Available at: http://www.who.int/water_sanitation_health/wastewater/gsuww/en/index.html

41

WHAT ARE THE LINKS BETWEEN POVERTY AND ECOSYSTEM SERVICES?

Marije Schaafsma and Brendan Fisher

Introduction

Whilst it is increasingly recognised that ecosystems contribute to the well-being of the poor, we are now only beginning to unpick the myriad relationships between changes in ecosystem services and poverty across various contexts. Much more work is needed, with a core question being how we can manage ecosystems for the benefit of contemporary poor people as well as future generations.

Ecosystem services and poverty have mostly been studied separately. Where natural scientists often engage in ecosystem services assessments and quantification of services outputs (such as tonnes of carbon stored), social scientists focus mainly on assessment of poverty and well-being and look at causes in the political economy. For example, the recent report on chronic poverty (ODI, 2014) does not mention sustainable ecosystem management but contains a chapter on the increased vulnerability resulting from natural hazards. Ecosystem assessments, on the other hand, tend to focus on the positive well-being impacts of ecosystem services and often pay less attention to disservices.

This chapter gives an overview of the evidence and study of the relationship between poverty and ecosystem services in the academic literature. After discussing different approaches to defining and measuring poverty, the chapter gives an overview of the contribution of ecosystem services to well-being in developing countries, the different discourses on the relationship between poverty and ecosystem services, and the trade-offs in ecosystem services management and poverty issues. The chapter ends with an outlook for future research linking poverty and ecosystem services.

What is poverty?

People derive 'well-being' from, among other things, ecosystem 'goods and benefits' that they can derive from ecosystem services. Well-being has been described and defined in multiple ways and has been discussed for millennia. Two broad approaches have emerged: the eudaimonic and the hedonic approach (Ryan and Deci, 2001). Aristotle's concept of eudaimonia is the actualisation of human potential: fulfilling one's true nature is possible when people have the freedom to live the life that is most congruent with their deeply held values. Fulfilling needs or desires includes not only physical needs, but also psychological needs of autonomy, competence and relatedness: factors that allow people to achieve subjective well-being. In the hedonic approach, human well-being builds on constructs of happiness and satisfaction.

Poverty is broadly defined as deprivation of well-being. But the wide range in concepts of well-being implies that there is no unanimously supported definition of poverty and therefore no single measure of poverty. Nowadays, poverty is generally considered to be a multidimensional state, with both subjective and objective aspects. The former aspects include personal, emotional and psychological interpretations of well-being, whilst the latter use indicators such as income, housing and food security.

Traditional economic approaches have tended to focus on income and have defined poverty as a lack of purchasing power, which builds on the concept of utility or welfare. For example, the chronic poverty report defines extreme poverty based on a statistical measure of $1.25 per person per day, and severe poverty as $0.70 per person per day; or in some cases below national poverty lines or national food poverty lines (ODI, 2014). Asset-based approaches tend to look at material assets, such as the quality of shelter, clothing and furniture, as well as access to services such as health and education. These focus mostly on basic needs.

Alternatives to the income and assets approaches include the capabilities approach and the concept of social exclusion. Sen (1985) defined poverty as the deprivation of capabilities that enable people to be and do things of worth to them, or to have the freedom to promote or achieve functionings they intrinsically value. This includes resources and commodities, but also community life and social aspects of shame and self-respect. In this definition, freedom has to be real and effective, allowing people to pursue valuable acts (agency) and creating real opportunities to achieve functionings. Sen's framework of capabilities provides moral guidance, but does not provide an easy way of measuring poverty. This is partly because poverty indicators are context-specific, and therefore people should be allowed to define the criteria to assess their own quality of life. The capabilities approach also highlights the importance of non-material dimensions, such as emotional well-being, democracy and family life (Robeyns, 2005).

The Millennium Development Goals (MDGs) build on Sen's capabilities approach. The UNDP now views poverty as a denial of choices and opportunities for living a tolerable life. The MDGs were the result of a global agreement to reduce absolute poverty and mark a move away from economic growth and income-based development strategies. The MDGs are based on an objective poverty definition and tend to measure a subset of functionings and resources that relate to basic needs, rather than the full set of freedoms. As such, the MDGs only deliver well-being as long as people value its outcomes, which may be compromised by the global rather than cultural and context-specific definition. The Multidimensional Poverty Index (MPI), which also builds on the capabilities approach, includes three different dimensions of poverty: health, education and living standards, measured by ten indicators (Alkire and Santos, 2014). The MPI focuses on realised functionings (ex post) as proxies for capabilities (*ex ante*). Critical in poverty assessment using the MPI is the definition of the minimum threshold below which the level of functioning is unacceptable, i.e. poor. The thresholds vary across cultures and contexts, and in applications of the MPI these thresholds are adjusted to local preferences.

The concept of social exclusion (SE) focuses on deprivation and marginalisation whereby certain groups or individuals in society are systematically deprived of resources, opportunities and rights. This includes material deprivation, but also social, political or economic participation in society. The SE approach considers how people become poor and how that affects their lives, i.e. the dynamics (Laderchi et al., 2003). The Environmental Entitlements framework by Leach et al. (1999) explicitly incorporates rights and resources (endowments) and the means to exert those rights and use the resources (entitlements) that allow people to convert ecosystem services into capabilities. This helps to consider issues of access to resources and social differentiation of that access (Fisher et al., 2014).

The material, relational or social, and subjective aspects of well-being are brought together in a 3-dimensional human well-being concept by the Well-being in Development Countries research Group (WeD) (McGregor and Sumner, 2010). This concept and its operationalisation have been developed with respect to developing countries that rely on natural resources. It is centred on the idea that human well-being is produced in relationships and engagement with others, in which social meaning is constructed. To this end, Opschoor (2007) emphasises the importance of social relationships between individuals, groups and the state, all competing for wealth and power, which leads him to define poverty as 'a social condition of chronic insecurity resulting from a malfunctioning of economic, ecological, cultural and social systems, causing groups of people to lose the capacity to adapt and survive and to live beyond minimal levels of satisfaction of their needs and aspirations.' With this definition of poverty we can see a move towards closer linkages between the state of the natural environment and human well-being.

In several conceptualisations of well-being, natural resources are regarded as inputs to functionings that allow people to generate well-being, not as a vital component of well-being in itself. For example, Reardon and Vosti (1995) define poverty as being poor in natural and other assets, where natural assets are considered as inputs to production. Income approaches ignore (environmental) externalities. In the capabilities approach, environmental factors play a role by allowing people to convert resources (goods and services) into capabilities and realise functionings (Robeyns, 2005). However, the importance of the natural environment to most, if not all, human capabilities is not adequately addressed in theories relating to justice like the capabilities approach (Holland, 2007). Despite the crucial role of natural resources for multiple dimensions of well-being, from basic needs to sense of place and cultural identity, only one of the MDGs relates to environmental sustainability, including a mix of targets related to sustainable policy development, biodiversity loss, access to safe drinking water and quality of life of slum dwellers. This list, and the position of environmental targets within development criteria, may change with the formulation of the Sustainable Development Goals (SDGs): the proposed 17 SDGs contain targets related to climate change, marine and terrestrial ecosystems. However, the role of ecological sustainability is unlikely to become as central to the SDGs as proposed by Holland (2008) as a meta-capability, or in ecological economics frameworks.

Contribution of ecosystem services to well-being in developing countries

Different ecosystem services contribute to different dimensions of human well-being, from income to livelihood support and protection against (natural) shocks. The literature on the importance of ecosystem services for poor people has tended to focus on services provided by more intact ecosystems as well as on provisioning services, such as food, firewood and construction material. For example, Cavendish (2000) shows how environmental goods contribute significantly to rural income and key economic activities, such as agricultural production.

This is not necessarily a critique: poverty can be measured by looking at a limited set of dimensions, and nutrition, energy and housing are probably one of, if not, the most important ones. If we simply look at the amount of the globe that humans have transformed for agricultural and pastoral activities, food production benefits seem to be favoured over other (non-provisioning) ecosystem services on global scale. However, other services may be more important at local levels (Raudsepp-Hearne et al., 2010), such as the protection against cyclones provided by mangroves in India, which significantly reduced the loss of human lives (Das and

Vincent, 2009). The contribution of cultural and regulatory services, including those that regulate water flows throughout the year in support of agriculture, soil quality or settings of spiritual meaning, also form important contributions to well-being. Land use decisions about a natural landscape can include full conversion, sustainable management, or full protection: each of these options will have different potential impacts on poverty. The governance of the distribution of the ecosystem goods and benefits, e.g. through access regulations, participatory management or use agreements, and the extent to which the needs of the poor are considered in governance, is an important factor in the poverty outcomes of land use decisions.

The use of ecosystem services is not homogenous across the poor: absolute and relative dependence on natural resources tends to differ across income classes (e.g. Cavendish, 2000). Rich members of communities tend to benefit more from ES in absolute terms, but relative to income poorer people depend more on ES, reflecting in some cases up to a quarter of their cash income (e.g. Schaafsma et al., 2014a).

The human well-being and ecosystems service relationship is innately dynamic. A given ecosystem may or may not provide a stable flow of goods and benefits, but not protect (or protect greatly) people against one-off shocks and disturbances. Ecosystem functioning can also confer costs to people at some constant background rate (e.g. disease vectors) or bigger shocks at more stochastic intervals (e.g. random crop raiding by elephants). Such fluctuations are key processes in the persistence of poverty. The magnitude, spatial and temporal aspects of poverty are important determinants of people's ability to resist the negative impacts of such fluctuations, as well as the role ecosystems can play. Their vulnerability to these impacts depends on their asset levels, as well as their social networks (Adger, 2006). More diverse bundles of ecosystem services are more likely to provide community resilience against a range of (environmental) pressures and threats (Roe et al., 2011), such as climate change, political unrest or abrupt changes in commodity markets. Bundles of ecosystem services provide multiple options to adapt or react to these threats. For example, forest-adjacent communities tend to rely mostly on ecosystem services supporting agriculture for their livelihoods, but may rely on other forest products out of farming seasons or after crop failures (a 'safety net,' e.g. wild fruit).

The relationship between ecosystem services and poverty is hence far less linear and more complicated than a focus on provisioning services as inputs to well-being generation would suggest. We need a better understanding of how ecosystem services, in particular regulating and cultural services, contribute to human well-being dimensions such as health and security.

Poverty and ecosystem degradation: correlation and causation

Although society may share the objective to achieve a sustainable future with high biodiversity and low poverty, suggested policy interventions to achieve this ideal depend on the beliefs about the structural cause-effect relationship between poverty and ecosystem degradation. Whilst some see these as two independent issues, others argue that conservation efforts can only be successful if poverty is addressed, that conservation is only legitimate where poverty is not compromised, or that natural resources are necessary for poverty reduction (Adams et al., 2004).

At global scales, aggregate indices of well-being, such as the Human Development Index and Gross Domestic Product, and environmental degradation are positively correlated, which provides no evidence for the claim (e.g. in MA, 2005) that declining ecosystems will result in loss of human well-being (Raudseppe-Hearne et al., 2010), and undermines arguments against further exploitation of natural resources. However, these well-being indices ignore various dimensions of well-being that are not (easily) substitutable, such as security and freedom, and exclude natural capital assets as part of the inclusive wealth of societies (UNU-IHDP and UNEP, 2012). At the same time, macro-level studies (although mostly for abiotic natural capital) suggest that countries rich in natural resources tend not to be able to turn this wealth into economic growth or more equality when institutions are weak.

At the subnational scale, Fisher and Christopher (2007) and Turner et al. (2012) show that areas of high biodiversity and ecosystem services spatially overlap with areas of high poverty. Another common pattern is that the poor, especially in urban areas, tend to be more exposed to environmental degradation, such as air pollution, toxins and wastes (Brulle and Pellow, 2006). Despite this co-location, no consensus exists about the causal mechanism or the connection between the status of natural resources and poverty, and little empirical evidence exists on the synergies and trade-offs between conservation and poverty at local to regional scales (Opschoor, 2007; Barrett et al., 2011). For a few middle income countries like Costa Rica, recent work has shown that forest conservation has (at the mean) net positive benefits of reduced poverty and increased carbon storage (Ferraro et al., 2011). However, nature-based tourism revenue (Ferraro and Hanauer, 2014) appears to be one of the key mechanisms here for poverty impacts and it is an open question if low-income and least-developed countries can take advantage of such an approach.

There is a growing recognition that the poverty-resource use relationship is context-specific. Evidence suggests that economic growth does not necessarily lead to environmental degradation, and much depends on society's choices when reaching higher wealth levels (Roe et al., 2011). Another common platitude is that environmental degradation is caused by rural poverty or population growth. Evidence shows that degradation has also been driven by rich or urban population segments. Around Dar es Salaam, the largest city of Tanzania, waves of deforestation around the city have taken place as a result of the growing demand for charcoal for urban households (Ahrends et al., 2010). At the local scale, wealthier farmers have been found to engage in more intensive farming and put more pressure on soil and water resources than poorer farmers. Actors and drivers at different scales interact: for example, deforestation by local actors may be driven by international trade and its institutional organisation, whilst the feedback occurs at local scale and those negatively impacted by forest loss are the same local actors.

In the academic literature, static and partial approaches are gradually making way for system-based approaches, which accept uncertainty about the functioning of social-ecological systems as a standard characteristic. This system-based thinking may help to understand the mutual interactions over time and space between environmental and societal change and impacts on well-being, risks and resilience, as well as system characteristics that create 'poverty traps.' These traps exist when poverty reinforces itself and persists (Barrett et al., 2011): once a household becomes poor, it becomes very difficult for its members as well as its descendants to emerge from poverty. A social-ecological trap is a situation where feedbacks between social and ecological systems lead to undesirable and sometimes irreversible situations (Cinner, 2011). For example, asset-less people in low-income and highly unequal countries are especially prone to being trapped (Dasgupta, 2007). Sustained correlation between poverty and environmental degradation have been attributed to high reliance on natural resources of low value by the poor (Angelsen and Wunder, 2003), shared vulnerability to shocks such as climate change, institutional failure (to control unsustainable resource use by purely self-interested individuals) and policy failure. For example, in coastal communities, poverty and weak institutions can lead to overfishing and the use of destructive fishing technologies, which destroy the coral reef habitat, reduce the ability of fish populations to recover and take away the control over macro-algae populations so that remaining corals are overgrown (Cinner, 2011).

Overall, understanding of the cultural and political processes that may create, maintain and change these situations is limited (Bourguignon et al., 2007). It requires in-depth exploration of the mechanisms that link ecosystem services and well-being across time and space, such as access to and control over ecosystem services, institutions (e.g. land tenure, property rights), human mobility and vulnerability, but also the type of resources and ecosystems and their seasonality. These mechanisms have locally specific effects. For example, property rights may provide secure and steady access to natural resources or motivation to invest in sustainable resource use, but the poor may also invest in sustainable practices without secure property rights (Cinner et al., 2012) as long as benefits of doing so are clear. The importance of power to well-being, and power asymmetries to differentiated access as a cause of poverty, should be emphasised to make ecosystem services assessments relevant to poverty issues.

Trade-offs in ecosystem services management and poverty alleviation

Given that poverty is multidimensional and ecological systems deliver a suite of goods and services (and potential disservices) that are relevant to different poverty dimensions, ecosystem management and poverty alleviation policies are not independent. Stakeholders may have different and conflicting visions and perspectives on how to manage, govern and interact with ecological systems. Such different visions make decision-making, i.e. choices and trade-offs, extremely complicated (McShane et al., 2011), but the process can be made more open and transparent by making benefits and losses explicit and showing which stakeholders win and lose. In turn, the distribution of these outputs determine the societal responses that affect the ecological system and hence ecosystem service production (Reyers et al., 2013) and the success of ecosystem management (Halpern et al., 2013). Very few empirical assessments in the academic literature show the distributional outcomes of ecosystem policy options. Van Beukering et al. (2003) show how local communities may benefit most from conservation of the Leuser Ecosystem of Sumatra, Indonesia; the logging industry will benefit most in absolute and relative terms from deforestation in the short run, but the wealthier stakeholders will suffer from reduced ecological services in the long run. Disaggregation of ecosystem services benefits using the Kaldor Hicks tableau or value chain analysis may help to understand the distribution of costs and benefits. An analysis of the timber value chain in Tanzania, for example, reveals how a considerable proportion of the benefits of illegal timber harvesting falls to a small group of timber dealers who use this money to collude and bribe officials at various level, thereby maintaining the corrupted system that is so beneficial to them (Schaafsma et al., 2014b). Equity weights may also help to differentiate the benefits that fall to poorer stakeholders.

Balancing the needs of development and resource use with poverty alleviation involves trade-offs between different policy objectives or criteria, such as equity and efficiency, welfare of current and future generations (Glotzbach and Baumgartner, 2012), biodiversity and ecosystems targets (Cimon-Morin et al., 2013), or between different ecosystem services targets (Rodriguez et al., 2006). One of the important critiques of Payments for Ecosystem Services (PES) is that they do not deliver win-win outcomes for both the poor and the environment, and the focus on efficient resource use shifts attention away from the unequitable outcomes for the most vulnerable in society (Muradian et al., 2010). Similarly, conservation efforts in practice cannot focus solely on biodiversity impacts, but have to balance conservation goals with societal impacts (Halpern et al., 2013). Of course, societal support is necessary for effective interventions (Simpson and Vira, 2010). Poor people may be unable to oppose choices that lead to detrimental environmental degradation because of their lack of voice in policy processes. This is not limited to the poor living in the here and now; negative impacts that occur far away in space or time are often given less importance.

In practice, gains for both well-being and conservation can be found in situations where the initial levels of both are low: conservation may in such cases help people to increase their natural capital (see Box 41.1). Win-lose situations can occur when the ecosystem quality or quantity is high but people are poor. In such cases, conservation efforts tend to reduce access to important ecosystem services. The opportunity costs of conservation tend to fall disproportionally on local people, who are often (relatively) poor (Fisher et al., 2011). For example, protecting biodiversity and mitigating climate change in Tanzania's Eastern Arc Mountains prohibits the use of land for agriculture, and trees for timber or charcoal. Despite the large gains to the international community, conversion and forest degradation remains most valuable to local actors.

Box 41.1 Linking poverty alleviation and ecosystem management in coastal Mozambique

"The land is tired. The weather is changing, the soil is changing, and it is hard to live as a farmer. So, many people think that they will be able to support their families by fishing, but the fish are disappearing." – Former fisherman, Angoche, Mozambique, 2012

The area of Mozambique where this quote comes from is now called the Primeiras e Segundas Environmental Protection Area (P&S), after being gazetted by the Mozambican Government in 2012. It is an area where one-third of all households face chronic food insecurity. That rate is doubled for female-headed households. Over three-quarters of all people live on less than $1.25 per day. Less than 30% of household heads have finished primary school and over half of the children are stunted due to chronic malnutrition.

In this region close to 80% of households farm, with the majority farming less than one hectare. Yields are low and soils degraded. A third of households fish, but the evidence and local perception speaks to a major decline in this resource. People rely almost entirely on fuelwood and charcoal for cooking and heating, despite its growing scarcity. There are only small remnants of coastal forests left, and the mangroves, while extensive, are in some places being lost at a rate of 2.5% per year.

In P&S poverty is food insecurity, nutrient deficiency, paltry incomes, minimal access to schooling and dwindling livelihood opportunities. The mixed-livelihood strategies (farming, fishing, forest utilisation) here are completely reliant on a declining natural resource base. This is the context in which the CARE-WWF Alliance along with several government agencies are attempting to link natural resource management and the provision of ecosystem services with poverty alleviation. Through farmer field schools teaching climate-smart agricultural, and through establishing no-take fish sanctuaries protecting juveniles of important marine species, the Alliance, their collaborators and the local communities are seeking proof that in places where ecosystem degradation and poverty rates are high, managing land and seascapes for the provision of ecosystem services reveals double dividends for nature and people.

Preliminary results suggest that the farm and fisheries interventions can improve the biophysical conditions (fish diversity and biomass, soil quality) on the coast and that these are beginning to translate into human benefits – improved diet. However, the challenges are many. Mineral and gas exploration, cyclones, drought and a changing climate all pose threats to the functioning of the local ecosystems and livelihood strategies. Having a clearer understanding of the tradeoffs across livelihood decisions and the resilience local ecosystems need to continue to provide ecosystem services should help the communities and their partners make more sustainable decisions for the management of the region's natural systems.

In designing more equitable policy interventions, different perspectives exist on what constitutes fair and just: equity in outcomes such as risks, costs, benefits (distributional justice), equity in access (contextual justice), or procedural justice (McDermott et al., 2013). Whilst poverty alleviation achievements are measured in terms of poverty outcomes, procedural justice – policies in which the poor can participate and have power in governance – can help to improve distributional justice. Access is an important lever: for example, prohibiting forest access for conservation reasons could considerably reduce the (material) wellbeing of members of forest-adjacent communities, whilst giving locals privileged access might be able to compensate for protection-related restrictions (Wunder et al., 2014).

However, how to best regulate access and management for poverty alleviation purposes is still under debate. Persha et al. (2011) argue that community forest management or other locally run regimes are beneficial to livelihoods as well as forests where forests are larger. However, this depends on the quality of the forest and the level of enforcement of access rules. Community-based strategies are not necessarily equitable: they do not prevail benefit allocation to the poor, and may result in elite-capture. Yet, top-down command-and-control strategies are often too expensive to enforce and implement and therefore ineffective (Simpson and Vira, 2010). A global study has found that rural smallholders tend to benefit more from government forests than community-managed forests, where government forests are of better quality and the enforcement of regulations is lower compared to well-run community-managed forests (Jagger et al., 2014). However, such outcomes depend on context.

Development of research on poverty-ecosystem services nexus

Research on the link between poverty alleviation and ecosystem services has a long list of questions to address, and this chapter has only scratched the surface of some of the topics of debate and emerging consensus.

In the development of poverty concepts, it is now widely acknowledged that poverty is a multidimensional state, with objective and subjective aspects, but much less clarity exists about the measurement of (lack of) wellbeing. The dynamic nature of individual's poverty makes monitoring, and therefore consistent measuring, crucial for understanding how to eradicate extreme poverty in the long run, and how ecosystem management could contribute to that cause. Some ecosystem services may be important for poverty alleviation, whilst others may contribute to poverty reduction. Alternatively, they might not have a reduction or alleviation function, but simply/only prevent people from falling deeper into poverty. Some ecosystem goods and benefits may be crucial to the chronically poor, others to the transient poor.

Although the importance of ecosystem services to the poor has been acknowledged widely, this has not led to the widespread inclusion of ecological indicators in poverty assessments and well-being concepts explicitly, despite their role in supporting almost all aspects of human life. The root of this issue may be in the lack of integration between development and natural science disciplines, but recent interdisciplinary conceptual frameworks may help to address this. For example, Fisher et al. (2014) developed a new framework for ecosystem services and poverty alleviation, aiming to bring together topics of: ecosystems, their services, goods and benefits; human well-being, its multiple dimensions and link to poverty; social differentiation and issues of scale; individual characteristics, including endowments, entitlements, capitals, and preferences; access and control; drivers of ecosystem change; and human responses of adaptation and mitigation.

Most of the work on the role of ecosystem services in poverty has focused on provisioning services, whilst intermediate (supporting) services, regulating services and cultural services have received less attention. As many of the ecosystem service assessments take a more quantitative approach, and cultural services, in particular, do not lend themselves easily to quantification, they have often remained beyond the scope of such assessments (Rodriguez et al., 2006). The value of landscape aesthetics, recreation and cultural and spiritual significance has already been demonstrated in social and behavioural sciences (Daniel et al., 2012). A further extension would be to incorporate not only individual values, but also the shared values that societies attribute to the natural environment, including senses of justice and fairness, human responsibility and ownership towards the environment (see Jax, 2016, Kenter 2016). This has already required the expansion of the ecosystem service toolbox to include non-monetary quantification and qualitative methods to reflect those values.

The bulk of the literature has focused on services provided by natural habitats. This focus needs to be expanded, and quickly, to include semi-natural, peri-urban and novel ecological systems. Further integration of knowledge on agricultural ecosystems and landscapes, and the link between those systems and (semi-)natural systems, is also necessary in order to draw attention to the links between ecosystem services and agricultural production. The importance of agricultural systems to human well-being was a core part of the UK National Ecosystem Assessment. As most of the poor in African and Asian nations live in rural areas, and global food security is increasingly recognised as a threat to global political and economic stability, TEEB (The Economics of Ecosystems and Biodiversity) has recently launched a new programme on agriculture and food. Research on ecosystem services and poverty should therefore also consider the impacts of agricultural developments on well-being ecosystems local and global scales in a wider social cost-benefit analysis.

Finally, a gradual shift towards social-ecological systems thinking seems to be influencing the research in this ecosystem service-poverty nexus. In this there is a growing recognition of the importance of understanding resilience and vulnerability in this space. It may help change objectives from efficient resource use towards managing for the resilience of socio-ecological systems with the dual end goals of improved human well-being and ecological sustainability.

Acknowledgements

MS was funded by an ESPA Early Career Fellowship Grant (FELL-2014–104) with support from the Ecosystem Services for Poverty Alleviation (ESPA) programme. The ESPA programme is funded by the Department for International Development (DFID), the Economic and Social Research Council (ESRC) and the Natural Environment Research Council (NERC). BF was supported by a grant to the CARE-WWF Alliance from the Sall Family Foundation.

References

Adams, W.M., Aveling, R., Brockington, D., et al. (9 authors) (2004). Biodiversity conservation and the eradication of poverty. *Science*, vol 306, no 5699, pp 1146–1149.

Adger, W.N. (2006). Vulnerability. *Global Environmental Change*, vol 16, no 3, pp 268–281.

Ahrends, A., Burgess, N.D., Milledge, S.A., et al. (9 authors) (2010). Predictable waves of sequential forest degradation and biodiversity loss spreading from an African city. *Proceedings of the National Academy of Sciences*, vol 107, no 33, pp 14556–14561.

Alkire, S., and Santos, M.E. (2014). Measuring acute poverty in the developing world: robustness and scope of the multidimensional poverty index. *World Development*, vol 59, pp 251–274.

Angelsen, A., and Wunder, S. (2003). Exploring the forest – Poverty link. *CIFOR Occasional Paper*, vol 40, pp 1–20.

Barrett, C.B., Travis, A.J., and Dasgupta, P. (2011). On biodiversity conservation and poverty traps. *Proceedings of the National Academy of Sciences*, vol 108, no 34, pp 13907–13912.

Bourguignon, F., Ferreira, F.H., and Walton, M. (2007). Equity, efficiency and inequality traps: a research agenda. *The Journal of Economic Inequality*, vol 5, no 2, pp 235–256.
Brulle, R.J., and Pellow, D.N. (2006). Environmental justice: human health and environmental inequalities. *Annual Review of Public Health*, vol 27, pp 103–124.
Cavendish, W. (2000). Empirical regularities in the poverty-environment relationship of rural households: evidence from Zimbabwe. *World Development*, vol 28, no 11, pp 1979–2003.
Cimon-Morin, J., Darveau, M., and Poulin, M. (2013). Fostering synergies between ecosystem services and biodiversity in conservation planning: a review. *Biological Conservation*, vol 166, pp 144–154.
Cinner, J.E. (2011). Social-ecological traps in reef fisheries. *Global Environmental Change*, vol 21, no 3, pp 835–839.
Cinner, J.E., McClanahan, T.R., MacNeil, M.A., et al. (17 authors) (2012). Comanagement of coral reef social-ecological systems. *Proceedings of the National Academy of Sciences*, vol 109, no 14, pp 5219–5222.
Daniel, T.C., Muhar, A., Arnberger, A., et al. (21 authors) (2012). Contributions of cultural services to the ecosystem services agenda. *Proceedings of the National Academy of Sciences*, vol 109, no 23, pp 8812–8819.
Das, S., and Vincent, J.R. (2009). Mangroves protected villages and reduced death toll during Indian super cyclone. *Proceedings of the National Academy of Sciences*, vol 106, no 18, pp 7357–7360.
Dasgupta, P. (2007). *Poverty Traps: Exploring the Complexity of Causation, 2020 Focus Brief on the World's Poor and Hungry People*. IFPRI, Washington DC.
Ferraro, P.J., and Hanauer, M.M. (2014). Quantifying causal mechanisms to determine how protected areas affect poverty through changes in ecosystem services and infrastructure. *Proceedings of the National Academy of Sciences*, vol 111, no 11, pp 4332–4337.
Ferraro, P.J., Hanauer, M.M., and Sims, K.R.E. (2011). Conditions associated with protected area success in conservation and poverty reduction. *Proceedings of the National Academy of Sciences*, vol 108, no 34, pp 13913–13918.
Fisher, B., and Christopher, T. (2007). Poverty and biodiversity: measuring the overlap of human poverty and the biodiversity hotspots. *Ecological Economics*, vol 62, no 1, pp 93–101.
Fisher, B., Lewis, S.L., Burgess, N.D., et al. (9 authors) (2011). Implementation and opportunity costs of reducing deforestation and forest degradation in Tanzania. *Nature Climate Change*, vol 1, no 3, pp 161–164.
Fisher, J.A., Patenaude, G., Giri, K., et al. (8 authors) (2014). Understanding the relationships between ecosystem services and poverty alleviation: a conceptual framework. *Ecosystem Services*, vol 7, pp 34–45.
Glotzbach, S., and Baumgartner, S. (2012). The relationship between intragenerational and intergenerational ecological justice. *Environmental Values*, vol 21, no 3, pp 331–355.
Halpern, B.S., Klein, C.J., Brown, C.J., et al. (11 authors) (2013). Achieving the triple bottom line in the face of inherent trade-offs among social equity, economic return, and conservation. *Proceedings of the National Academy of Sciences*, vol 110, no 15, pp 6229–6234.
Holland, B. (2007). Justice and the environment in Nussbaum's "Capabilities Approach": Why sustainable ecological capacity is a meta-capability. *Political Research Quarterly*, vol 61, no 2, pp 379–332.
Jagger, P., Luckert, M.K., Duchelle, A.E., Lund, J., and Sunderlin, W.D. (2014). Tenure and forest income: observations from a global study on forests and poverty. *World Development*, vol 64, pp 53–65.
Jax, K. (2016). Ecosystem services and ethics. In Potschin, M., Haines-Young, R., Fish, R. and Turner, R.K. (eds) *Routledge Handbook of Ecosystem Services*. Routledge, London and New York, pp 301–303.
Kenter, J.O. (2016). Deliberative and non-monetary valuation. In Potschin, M., Haines-Young, R., Fish, R. and Turner, R.K. (eds) *Routledge Handbook of Ecosystem Services*. Routledge, London and New York, pp 271–288.
Laderchi, C.R., Saith, R., and Stewart, F. (2003). Does it matter that we do not agree on the definition of poverty? A comparison of four approaches. *Oxford Development Studies*, vol 31, no 3, pp 243–274.
Leach, M., Mearns, R., and Scoones, I. (1999). Environmental entitlements: dynamics and institutions in community-based natural resource management. *World Development*, vol 27, no 2, pp 225–247.
MA (2005). *Ecosystems and Human Well-being: Synthesis*. Island Press, Washington, DC.
McDermott, M., Mahanty, S., and Schreckenberg, K. (2013). Examining equity: a multidimensional framework for assessing equity in payments for ecosystem services. *Environmental Science & Policy*, vol 33, pp 416–427.
McGregor, A., and Sumner, A. (2010). Beyond business as usual: what might 3-D wellbeing contribute to MDG momentum? *IDS Bulletin*, vol 41, no 1, pp 104–112.
McShane, T.O., Hirsch, P.D., Trung, T.C., et al. (14 authors) (2011). Hard choices: making trade-offs between biodiversity conservation and human well-being. *Biological Conservation*, vol 144, no 3, pp 966–972.

Muradian, R., Corbera, E., Pascual, U., Kosoy, N., and May, P.H. (2010). Reconciling theory and practice: An alternative conceptual framework for understanding payments for environmental services. *Ecological Economics*, vol 69, no 6, pp 1202–1208.

ODI–Overseas Development Institute (2014). *The Chronic Poverty Report 2014–2015: The Road to Zero Extreme Poverty*. ODI, London.

Opschoor, J.B. (2007). Environment and poverty: perspectives, propositions, policies. *ISS Working Paper Series/General Series*, vol 437, pp 1–38.

Persha, L., Agrawal, A., and Chhatre, A. (2011). Social and ecological synergy: local rulemaking, forest livelihoods, and biodiversity conservation. *Science*, vol 331, no 6024, pp 1606–1608.

Raudsepp-Hearne, C., Peterson, G.D., Tengö, M., et al. (8 authors) (2010). Untangling the environmentalist's paradox: why is human well-being increasing as ecosystem services degrade? *BioScience*, vol 60, no 8, pp 576–589.

Reardon, T., and Vosti, S.A. (1995). Links between rural poverty and the environment in developing countries: asset categories and investment poverty. *World Development*, vol 23, no 9, pp 1495–1506.

Reyers, B., Biggs, R., Cumming, G.S., Elmqvist, T., Hejnowicz, A.P., and Polasky, S. (2013). Getting the measure of ecosystem services: a social-ecological approach. *Frontiers in Ecology and the Environment*, vol 11, no 5, pp 268–273.

Robeyns, I. (2005). Assessing global poverty and inequality: income, resources, and capabilities. *Metaphilosophy*, vol 36, no 1–2, pp 30–49.

Rodríguez, J.P., Beard, T.D., Bennett, E.M., et al. (8 authors) (2006). Trade-offs across space, time, and ecosystem services. *Ecology and Society*, vol 11, no 1, p 28.

Roe, D., Thomas, D., Smith, J., Walpole, M., and Elliott, J. (2011). *Biodiversity and Poverty: Ten Frequently Asked Questions–Ten Policy Implications*. International Institute for Environment and Development, London.

Ryan, R.M., and Deci, E.L. (2001). On happiness and human potentials: a review of research on hedonic and eudaimonic well-being. *Annual Review of Psychology*, vol 52, no 1, pp 141–166.

Schaafsma, M., Burgess, N.D., Swetnam, R.D., et al. (6 authors) (2014b). Market signals of unsustainable and inequitable forest extraction: assessing the value of illegal timber trade in the Eastern Arc Mountains of Tanzania. *World Development*, vol 62, pp 155–168.

Schaafsma, M., Morse-Jones, S., Posen, P., et al. (29 authors) (2014a). The importance of local forest benefits: economic valuation of non-timber forest products in the Eastern Arc Mountains in Tanzania. *Global Environmental Change*, vol 24, pp 295–305.

Sen, A. (1985). Well-being, agency and freedom: the Dewey lectures 1984. *The Journal of Philosophy*, vol 82, no 4, pp 169–221.

Simpson, R. D., and Vira, B. (2010). Assessing intervention strategies. In Ash, N., Blanco, H., Brown, C., Garcia, K., Henrichs, T., Lucas, N., Raudsepp-Hearne, C., Scholes, R., Simpson, R.D., Tomich, T.P., Vira, B. and Zurek, M. (eds.) *Ecosystems and Human Well-Being: A Manual for Assessment Practitioners*, Island Press, pp 221–253.

Turner, W.R., Brandon, K., Brooks, T.M., et al. (8 authors) (2012). Global biodiversity conservation and the alleviation of poverty. *BioScience*, vol 62, no 1, pp 85–92.

UNU-IHDP and UNEP (2012). *Inclusive Wealth Report 2012. Measuring Progress toward Sustainability*. Cambridge University Press, Cambridge UK.

van Beukering, P.J., Cesar, H.S., and Janssen, M.A. (2003). Economic valuation of the Leuser National Park on Sumatra, Indonesia. *Ecological Economics*, vol 44, no 1, pp 43–62.

Wunder, S., Angelsen, A., and Belcher, B. (2014). Forests, livelihoods, and conservation: broadening the empirical base. *World Development*, vol 64, pp 1–11.

42
ECOSYSTEM SERVICES AND HEALTH

Conor E. Kretsch

Introduction

Although academic interest in the relationship between the environment and human health dates back well over two thousand years, the intimate connections between the state of the natural environment and human health have largely been forgotten across much of the modern health sciences. Public discussions on environmental health tend to focus on a core set of issues dealing with environmental exposures and related risks – air and water quality, health and safety at work, waste management, food safety, and so on. It is only relatively recently that this dialogue has broadened to include considerations of biodiversity, ecosystems, and the benefits we derive from them. When one considers the changing nature of global public health and related policy, it is clear that a strong basis for reconnecting the spheres of biodiversity and human health exists.

The role of the environment in this complex mix is now far more widely acknowledged. The interactions between people and biodiversity can, for better or worse, determine the baseline health status of a community, provide the foundations for good health and secure livelihoods, or create the conditions responsible for significant morbidity or mortality. Conversely, it is increasingly clear that biodiversity policies and programmes for environmental management must account for the potential knock-on impacts (positive or negative) on human health and well-being (e.g. Bagnoli et al., 2008).

This chapter provides an overview of the main linkages between ecosystems and human health, and discusses emerging strategies for integrating ecosystem services and biodiversity into public health science, policy, and practice.

Biodiversity, ecosystems, or ecosystem services?

When addressing the health and other benefits we receive from the natural environment, there is often confusion as to whether or to what extent we should distinguish biodiversity from ecosystems and / or ecosystem services. For example, one perspective considers that ecosystems, as complex systems, have emergent properties that are not held by their individual components; in other words, ecosystems are more than the sum of their parts, and therefore a primary focus on units of biodiversity may not produce a complete understanding of the benefits that living systems provide to human health and well-being. An alternative perspective is that individual elements of biodiversity can have specific impacts on, or benefits to, human health regardless of whether they are viewed in isolation or as part of a complex system (for example, a plant which provides medicinal benefits). Furthermore, since many elements of biodiversity – and any benefits to society they may hold – remain to be examined or even discovered by science, the wider systems view can overlook health benefits not yet fully understood or realised.

This variety of perspectives may, to some extent, result from debate over the role that biodiversity plays in sustaining ecosystem function, and from different views as to whether biodiversity itself should be seen as an ecosystem service (Lele et al., 2013). It also frequently stems from different uses of the term "biodiversity" – which is often taken to simply refer to the number or diversity of species. The definition of biodiversity adopted by the UN Convention on Biological Diversity ("the variability among living organisms from all sources including, *inter alia*, terrestrial, marine and other aquatic ecosystems and the ecological complexes of which they are apart; this includes diversity within species, between species, and of ecosystems")[1] is widely accepted in policy contexts and in the practical implementation of nature conservation strategies, as well as informing much in applied ecology and conservation science.

Although the relationships between biodiversity and ecosystem services are not fully understood, there is a consensus that biodiversity underpins many ecosystem functions, and the loss of biodiversity represents a significant threat to many ecosystem services important to human well-being (Díaz et al., 2006), including those supporting human health (CBD and WHO, 2015).

Bearing in mind the overlap in definitions and differences in perspectives, for the purposes of this chapter ecosystems are considered to be a component of biodiversity, and the services we derive from ecosystems for our health are considered to be variously dependent upon a diversity of genes, species, and habitats, and / or on the range of complex interactions (biotic and abiotic) between them. The terms *ecosystems* and *biodiversity* are therefore used interchangeably in the text.

Why health?

It is perhaps reasonable to enquire as to why health should be singled out from other constituents of well-being as a topic for specific focus from an ecosystem services perspective. The Millennium Ecosystem Assessment identified five main aspects of well-being, of which one was health (MA, 2005), but demonstrated a significant degree of interaction with other aspects (security, basic material for good life, and freedom of choice), so that focus on health alone might seem redundant.

One clear reason for considering health on its own is that different components of ecosystems and different ecosystem services influence human well-being in different ways (see Table 42.1). Whilst some ecosystem services benefit our well-being in several ways (for example, food production supports livelihoods and food security, and also has important social and cultural dimensions) this is not universally the case. Health also interacts with and affects many other aspects of well-being – such as one's ability to work and socialise, to participate in family life, and be an active part of the community. As such, in many cases health may be singled out as the most crucial element of well-being. Health is also an important component in self-reported assessments of well-being, and an important consideration in determinations of quality of life (Skevington, 2009).

At the community level, health is a universal and persistent concern. Directly and indirectly, health considerations form a major part of household budgets, through health insurance, primary care costs, purchase of medicines or medical devices, health-related taxes, and, to some

Table 42.1 Types of ecosystem services and their support for health.

Category of ecosystem services	*Examples of services important to human health*
Supporting services	Nutrient management
	Assimilation and detoxification of wastes
Provisioning services	Provision of foods and medicinal compounds and related genetic resources
	Support for livelihoods
Regulating services	Regulation of pests and diseases
	Regulation of flooding and other risks
	Purification of water and air
	Regulation of climate
Cultural services	Recreational options
	Cognitive development and restorative support
	Inspiration and spiritual outlets
	Scientific (incl. health science) research and education
	Social cohesion and community support

degree, food expenses. Health is an important aspect of government services at the local level, and health services appropriate a significant portion of national governmental expenditure.[2] Health is also increasingly recognised as an important indicator of sustainable development (McMichael, 2006). Today it is well understood that sustainable health care systems are a prerequisite for socially and environmentally sound development, not just a result of it.

Another reason to consider health as distinct from (but closely connected to) other elements of well-being arises from recent discussions on how "health" should be defined in the 21st century. Perhaps the most widely used definition for health comes from the Constitution of the WHO (1948),[3] which considers health to be "a state of complete physical, mental and social well-being, and not merely an absence of disease or infirmity". This definition was ground-breaking and of great significance at the time in raising the profile of mental health issues, and in recognising that health was not just determined by the biophysical status of the individual but was also influenced by society. However, it is worth noting that the continuing relevance of the WHO definition of health has been questioned in recent years – is this definition still adequate in an increasingly globalised world and in the face of pervasive human impacts on the global environment? Some commentators have indicated that the definition should also account for the ability to adapt to health challenges (e.g. Lancet, 2009; Huber et al., 2011) – suggesting that truly healthy individuals are not necessarily those who remain free from disease, but those that can recover from illness and return to active life. Whether one agrees with this perspective – and there are many who do not – this focus on adaptation has useful correlations with the concept of ecosystem health, with its similar notions of resistance and resilience (Costanza, 1992; Rapport et al., 1998), and therefore it may be particularly useful in terms of understanding the importance of biodiversity to sustaining health and well-being in the context of global change and the resilience of social-ecological systems (see for example Bunch et al., 2011).

From another environmental perspective, efforts to mainstream biodiversity across public and private sectors require targeting of relevant information to individual stakeholders, which means that issues of specific relevance to the health sector should be identified and explained. Conversely, noting the relevance that environmental policies may have for public health – both positive and negative – means that the relationships must likewise be properly understood and

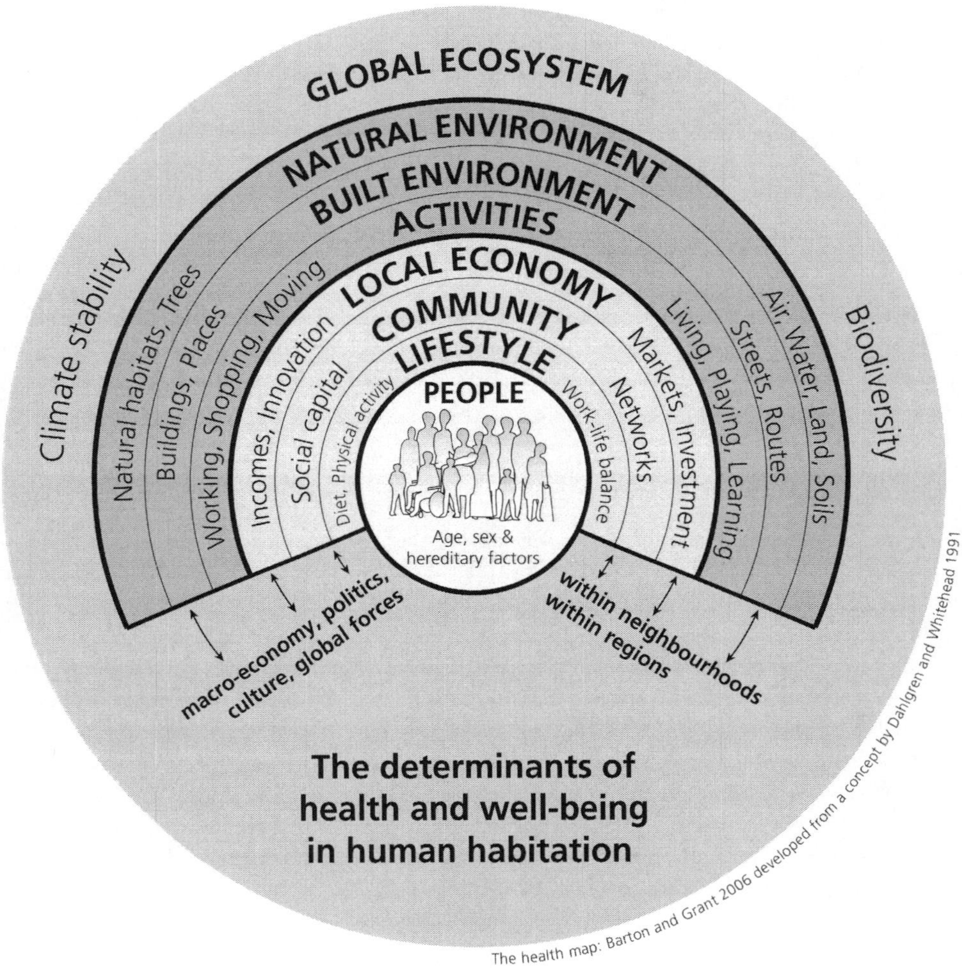

Figure 42.1 Health map for the local human habitat.

Source: Barton and Grant, 2006, after Dahlgren and Whitehead, 1991[4]

accounted for by conservation groups and other environmental agencies. Finally, health also offers an additional means of valuing the contribution of ecosystems to human well-being, and an alternative to the current focus on economic valuation, which may help to address some of the current ethical issues associated with ecosystem service accounting (see for example Luck et al., 2012).

How ecosystems and health are connected

Our health is supported by biodiversity and ecosystem services in many ways – through direct provision of essential materials for health (water, food, shelter, medicines) and by assimilation of pollutants, buffering against natural disaster, and regulating cycles of disease and infection. Ecosystems are now recognised as having particular importance for a number of health issues across the three core areas of public health intervention – infectious disease, non-communicable disease (including mental health and toxicology), and trauma. In this way ecosystems may be viewed as the settings within and from which health outcomes are determined (see Figure 42.1). Some of the key areas for consideration are discussed briefly below (for a more detailed overview, see also: CBD, 2008; Corvalan et al., 2005; Chivian and Bernstein, 2008; Sala et al., 2009; CBD and WHO, 2015).

Support for food production and food security, dietary health, and livelihood sustainability

Biodiversity provides the genetic resources that are the foundation of modern and traditional agriculture and are essential to future food security (Ford-Lloyd et al. 2011). Research has shown that diverse diets based on a broad variety of food species promote health, and can help to protect against disease by providing essential nutrients and enhancing livelihood security (Fanzo et al., 2013). This is particularly important in the context of nutrition transitions, where increasingly urbanised or developing societies move from diverse diets rich in nutrients to energy-dense, nutrient-poor diets (Frison et al., 2006). Movements away from traditional biodiversity-based food production systems towards intensive monocultures or industrial agriculture can increase vulnerability of crops and livestock to environmental extremes, and increase risks of food emergencies following natural disasters. Loss of agricultural biodiversity can therefore threaten health, livelihood sustainability, and economic security (Frison et al., 2011). Wild biodiversity is also important for human nutrition, with many people, particularly in poorer communities, being heavily dependent upon wild species for food and related traditional uses. For example, the importance of wild fisheries for human nutrition is well documented (Kawarazuka and Béné, 2010), and plants and animals harvested from the wild have been shown to contribute significantly to nutritional requirements for many local and indigenous communities (e.g. Burlingame, 2000; Hoffman and Cawthorn, 2012). Some of these food resources are under increasing threat from over-exploitation, and from global contaminant transport associated with industrial activities (Srinivasan et al., 2010; Kuhnlein and Chan, 2000).

Ecology of pathogens and vectors, and the regulation and control of infectious diseases

Pathogenic (disease-causing) organisms have important roles to play in ecology and evolution, and in the maintenance of ecosystem services. Bacteria, viruses, fungi, protozoa, and other parasites that cause harm to one certain species generate a range of impacts on the ecology and health of ecosystems. Many such pathogens have natural hosts which are not susceptible to disease, but act as carriers which facilitate the persistence and circulation of disease organisms in their environment. Ecosystem change can alter the ecology of carriers, hosts, and the diseases they carry. This can alter disease ecology, potentially increasing the risk of the emergence or spread of infectious diseases in animals, plants, and humans, including novel zoonotic outbreaks (disease spread to humans from animals) and global pandemics (Gottdenker et al., 2014). In recent years, the emergence and / or spread of many novel and severe diseases including HIV/AIDS, Ebola, Marburg, Hantavirus pulmonary syndrome, SARS, MERS, and highly pathogenic avian influenza has been attributed to human impacts on the environment or interactions with wildlife, particularly through agricultural intensification and related land-use change, hunting, and the global wildlife trade. Other socially and economically important diseases which are known to be affected by ecosystem change include toxoplasmosis, cholera, schistosomiasis, and malaria. There is also a risk that programmes to tackle certain diseases may impact negatively on biodiversity, through land clearance, wildlife culls, or use of chemicals, potentially affecting the sustainability of ecosystem goods and services upon which communities depend. Another group of microbes that are important from a health perspective are the commensal organisms that normally live on or within other organisms, and which have positive effects on health – including, for example the many microbes that are symbiotic with plants, and human gut and skin microbes that offer protection from infection and support healthy development. There is growing interest in how human behaviour and ecosystem change affects these organisms, and how this can have unexpected impacts on health (Hanski et al., 2012; Lehman et al., 2015).

Biodiversity provides vital natural products and therapeutic compounds, with medicinal, economic, social and cultural value

Many species provide culturally important traditional medicines and raw materials. Loss of species or habitats can impact cultural traditions and livelihoods, particularly for indigenous peoples and marginalised communities. In 1995, the WHO estimated that up to 80% of people in Africa and Asia depend on traditional medicines for their primary health care (WHO, 2002). Many modern drug products are also derived from wild species. Examples include pain killers (e.g. Zinconitide from cone snail toxin), cardiac drugs (e.g. Lanoxin from *Digitalis* plants), anti-cancer drugs (e.g. Taxol from *Taxus* trees and Hycamtin from *Camptotheca* trees), diabetic treatments (e.g. exanitide from the Gila monster *Heloderma suspectum*) and anti-parasitics (e.g. quinine from *Cinchona* trees). Many other potentially important species are yet to be investigated or discovered (Leal et al., 2012).

Social, cultural, and spiritual values and community cohesion

Biodiversity and cultural diversity are inherently linked, with habitats and species often forming integral elements of the social and bio-cultural landscape upon which communities are founded (CBD and WHO, 2015). The "sense of place" a person holds for a particular location, such as their home town, and which they may share with their community, is often strongly influenced or determined by their awareness and use of local biodiversity. Where access to greenspace is important for physical recreation, it can play a role in addressing non-communicable disease and "diseases of affluence" (e.g. diabetes, obesity, cardio-pulmonary illness), particularly in urban areas. Promoting access to, and stewardship of, local biodiversity and natural habitats can also support programmes to address other ills such as depression and anti-social behaviour. Ecosystem change and landscape degradation can result in loss of public goods, and feelings of disconnection from open spaces or the wider countryside. This can have negative implications for physical and mental well-being and social cohesion.

Resources for medical science

Studies in wildlife physiology can yield important clues to human physiology and disease mechanisms, and have helped lead to many important developments in human and veterinary medicine. Examples of species of interest to medical science include bears (for insights into osteoporosis, cardio-vascular disorders, and diabetes), sharks (osmoregulation and immunology), marine worms (tissue repair and wound treatment), bats (audiology and ENT), and horse-shoe crabs (ophthalmology and molecular cell biology) (Chivian and Bernstein, 2008). Wildlife physiology and biochemistry, and the diversity of life strategies, are also important for veterinary medicine, diagnostics, and related biotechnology (e.g. Wildt et al., 2010).

Ecosystem services for climate change adaptation and disaster risk reduction

Climate change is anticipated to have a significant impact on human health, and many of the changes are directly associated with climate impacts on biodiversity and ecosystem services (Patz et al., 2014). For example, changes in the populations or distribution of disease vectors (such as mosquitoes which carry malaria) or hosts (such as wildfowl which carry avian influenza) could lead to changes in disease patterns or increase the risk of outbreaks. Loss of ecosystem services also places communities at greater risk from other climate impacts, such as flooding, drought, wild fires, and crop failure.

Healthy ecosystems can provide important natural buffers and defences against severe weather events and other natural disasters. Biodiversity also provides essential food resources to people displaced by crises or conflict. Ecosystem disturbance can result in the loss of these important assets and leave whole communities exposed to increased risk. Habitat loss is also a contributory factor in dryland salinity and desertification, impacting livelihoods and exacerbating drought and food insecurity. People who have been displaced by disaster or conflict may be more susceptible to illness, and more dependent on ecosystem services for food, shelter, disease regulation, and medicine. However, disaster relief and response efforts can cause further environmental damage and dislocate people from the ecosystems that they depend upon (Sudmeier-Rieux and Ash, 2009). Biodiversity conservation and management of ecosystem services can be an important element in emergency preparedness, response, and recovery efforts (Sudmeier-Roux et al., 2006).

Mental health

There has been much work exploring the benefits which exposure to nature and natural spaces can have for mental health and well-being. There is strong evidence for the beneficial use of nature exposure – including wild animals in wild settings, and domesticated animals – in the treatment of depression, anxiety, and personality disorders. In particular, it has been suggested that contact with nature is important for childhood development, and children who grow up with an awareness of nature and a sense of environmental values are more likely to conserve nature as adults, and may be less likely to engage in certain anti-social behaviours (Roe and Aspinall, 2011).

Experience and knowledge of local biodiversity and natural landscapes, including local important or endemic features, also contribute to a person's sense of place and identity, influencing human attachments to landscape, culture, and community. Loss of (or loss of access to) these features can impact psychological well-being, leading to feelings of dislocation and impacting feelings of identity for individuals and communities (Horwitz et al., 2001). Connectedness with nature has also been associated with general life satisfaction and happiness (Zelenski and Nisbet, 2014) and with improved outcomes for patients in healthcare settings (Velarde et al., 2007). However, much of the current evidence for these connections arise from studies in urban and western contexts (Lovell et al., 2014), and the role of cultural and socio-economic factors may be far more significant (Clark et al., 2014); there is much additional research required before any lessons from current evidence can be universally applied in policy contexts (Hartig et al., 2014).

Certain negative associations between biodiversity and health status are also important to recognise. For example, research has shown that the geographic occurrence of biodiversity hotspots may correlate with focal points for disease organisms (e.g. Jones et al., 2008). In certain specific areas or situations, disease risk within a community may increase with proximity to wild areas or intact natural habitats, though the underlying causes are rarely known. However, disease organisms are also considered to be an essential element of healthy ecosystems, and a high diversity of parasites or pathogens may actually be important for maintaining a favourable conservation status for some species of wildlife (e.g. Hudson et al., 2006). Therefore, it may often be the case that the status of local biodiversity is less important as a risk factor for disease outbreaks than the integrity of the ecosystems involved, or the interactions that take place between biodiversity and human communities, as they may alter disease ecology (Jones et al., 2013). It is increasingly clear that each situation must be assessed at the landscape level, accounting for the role of people and ecosystems in the wider ecological landscape, and for the root causes of biodiversity change or health impacts associated with the political and economic activities of a growing human population (CBD and WHO, 2015).

These themes are frequently interlinked by specific aspects of biodiversity, or by interdependence on ecosystem services. For example, the susceptibility of a community or an individual to a specific disease may be related to nutrition security, which in many regions is dependent upon agricultural biodiversity. Similarly, the ability of a community to treat diseases, and therefore the potential socio-economic impacts of that disease, may be dependent upon the availability of particular species with medicinal value. In addition, the animal origins of many important human diseases (zoonotic infections, transmitted to humans from wildlife and livestock) creates additional concerns for the wider human health implications of the emergence of infectious diseases in wildlife.

Barriers to co-operation

The health and biodiversity sciences have much to gain from one another, and yet at national and regional levels the policies and practices of achieving primary health care and conserving biodiversity are still largely distinct from one another and rarely converge (Langlois et al., 2012). Bridging these gaps requires new thinking in both the health and biodiversity sciences and in other areas such as development, agriculture and land use planning. This means enhanced co-operation and a critical shift in how these disciplines analyse risks and opportunities, and how they communicate.

In biodiversity policy, global agreements such as the Convention on Biological Diversity (CBD) seek to integrate or mainstream biodiversity and its conservation across relevant spheres of government and in the private sector.[5] However, environmental concerns have often been considered to be an "add on" to programmes of work in other sectors, including the health sector, despite the rhetoric of global policy forums. Similarly, the health sector has traditionally not been involved in environmental decision-making relevant to biodiversity and ecosystem services. Campbell-Lendrum (2005) identified three core difficulties that prevent health sector participation in environmental matters: a lack of awareness of the relevance of environmental matters to health; a methodological approach focused on discrete cause-effect relationships rather than systemic issues; and a lack of health sector input to processes addressing the environmental root causes of health problems. The core issues identified by Campbell-Lendrum may be further broken down to include cultural, resource, and political barriers, knowledge gaps, and differing temporal scales of operation (see Table 42.2).

These issues can also be looked at from the other direction – with the biodiversity and environmental community also lacking in awareness of the linkages, and in understanding and acceptance, of other sectors' issues (see also e.g. Keune et al., 2013). Furthermore, the environmental community as a whole has had difficulty in the effective communication of key issues in environmental science to the public and decision-makers (e.g. Bickford et al., 2012). Chivian

Table 42.2 Barriers to co-operation between health and biodiversity communities.

Issue preventing co-operation	*Practical implications*
Lack of awareness of relevance of interlinkages	Poor communication between health and biodiversity communities Lack of understanding of core issues Limited evidence, particularly in local contexts or relating to specific health issues or ecosystems Perceptions of inherent conflict between sectoral "agendas"
Narrow methodological focus	Limited resources / focus of investment Difference in technical language between sectors Different scales of operation (geographic, temporal, biological) Limited view of functional remits Lack of awareness of cultural factors (including human cultures generally, and institutional cultures that influence decision-making)
Lack of input to processes on environmental root causes	Lack of understanding of core issues Lack of political will Limited resources / focus of investment Different approaches to risk

and Bernstein (2008) have noted that various efforts to highlight economic, aesthetic, scientific, and moral arguments have not been sufficient to address the current biodiversity crisis, and propose health as a critical alternative cross-cutting issue. Complicating this further are issues of trade-offs – biodiversity conservation having negative impacts on human well-being, (see for example McShane et al., 2011) and perceptions (right or wrong) of biodiversity as being bad for health, sometimes phrased as "ecosystem disservices" (see for example Lyytimäki and Sipilä, 2009), though that phrase is also used to describe harm to ecosystems caused by human activities.

Furthermore, when one considers the range of linkages between health and biodiversity, then it is clear that several other sectors must be brought to the table – including those involved with food production, travel and transport, trade, poverty reduction, economic development, infrastructure planning, and education.

Generally, one of the major barriers to cross-sector engagement in biodiversity issues is a concern that a significant additional burden of human and financial resources is required if biodiversity is to be included in decision-making or practices in non-environment sectors. When the value of such integration and the concept of ecosystem services is poorly understood, there is little incentive to adopt more systemic approaches, and important opportunities for improving health can be missed (Langlois et al., 2012).

Similarly, the biodiversity community may be reluctant to broaden its focus or to divert limited resources if health and related social issues are to be incorporated into biodiversity policy and practice. Therefore, ecosystem approaches to health should ideally (as far as is practicable) aim to work within existing systems – similar to how corporate environmental management plans seek to work within existing quality management systems (e.g. ISO 14001 builds on ISO 9001) – including integrating transdisciplinary approaches into educational curricula, including those for professional development (Gómez et al., 2013).

In order to effectively make the case for co-operation, the evidence base linking health with biodiversity and ecosystem services must be clear and effectively communicated to stakeholders. The complexity of the linkages between biodiversity and health has been touched upon above.

Many of the links between biodiversity and human health are well understood and supported by direct and empirical evidence – for example, the importance of biodiversity to drug discovery, medical research, and nutrition. However, many of the other proposed connections – such as the importance of biodiversity to reducing human infectious disease risk, its links with mental health, or its potential to support interventions for physical fitness – are generally weak, are still contested by scientists, and / or are only understood in a wide or general sense; i.e. the relevance of biodiversity to health outcomes in particular local communities at small geographic scales is often highly uncertain and very difficult to ascertain. As a result, the information needed to inform and support evidence-based decision-making for integrating biodiversity into specific human health initiatives or policies is often not available (Myers et al., 2013). In these cases, at the very least, applying the precautionary principle based on shared goals – such as maximising or securing human well-being – utilising existing knowledge about health-environment linkages can inform more collaborative decision-making processes.

In response to these challenges, various approaches to health based on ecosystems-thinking have emerged, each of which aims to understand the intimate links between the environment, development, and human and animal health, which are described below.

Ecosystem approaches to health

The ecosystem approach to health may be defined as a strategy for securing and enhancing health that recognises the fundamental importance of biodiversity and the goods and services it provides; it aims to understand the intimate links between the environment, human, and animal health and human development, and recognises the need for integrated public health policies and development programmes that view the protection of ecosystems as an important part of achieving their objectives. Whilst the relationships between biodiversity and human health are enormously varied and often complex, the basic concept of the ecosystem approach to health is perhaps relatively simple. In practical terms, implementing the ecosystem approach requires the development of mutual understanding and co-operation across disciplines, and should be seen as a method for maximising resource efficiency and achieving better conservation, health, and development outcomes (Bunch et al., 2008).

New ecosystem approaches to all aspects of health, including research and monitoring, disease prevention and control, education and reporting, and delivery of primary care, are being explored, and many new disciplines and sub-discipline, such as the science of conservation medicine and the "One Health" approach have emerged and developed rapidly, exploring the linkages between the health of ecosystems, wildlife, and livestock as key elements among the social, economic, and environmental determinants of human health. Veterinary medicine has increasingly embraced the ecosystem approach, particularly in the context of emerging diseases, animal health monitoring, and protected areas management (CBD and WHO, 2015).

Figure 42.2 illustrates some of the key determinants of health outcomes – including linkages between policy, social, environmental, economic and personal factors – and highlights several areas where ecosystem approaches to public health can be implemented. Building capacity for the ecosystem approach to health within relevant sectors and disciplines and enabling those groups to collaborate requires clear and accurate information on the importance of biodiversity to health, and its relevance to specific health goals. At higher strategic levels of governance, it can be useful to address these within the context of the broader range of national and global sustainability issues which influence health outcomes. It is also important that the experiences of scientists, decision-makers, and stakeholders working in specific fields (conservation, disease control, poverty alleviation etc) are taken into account.

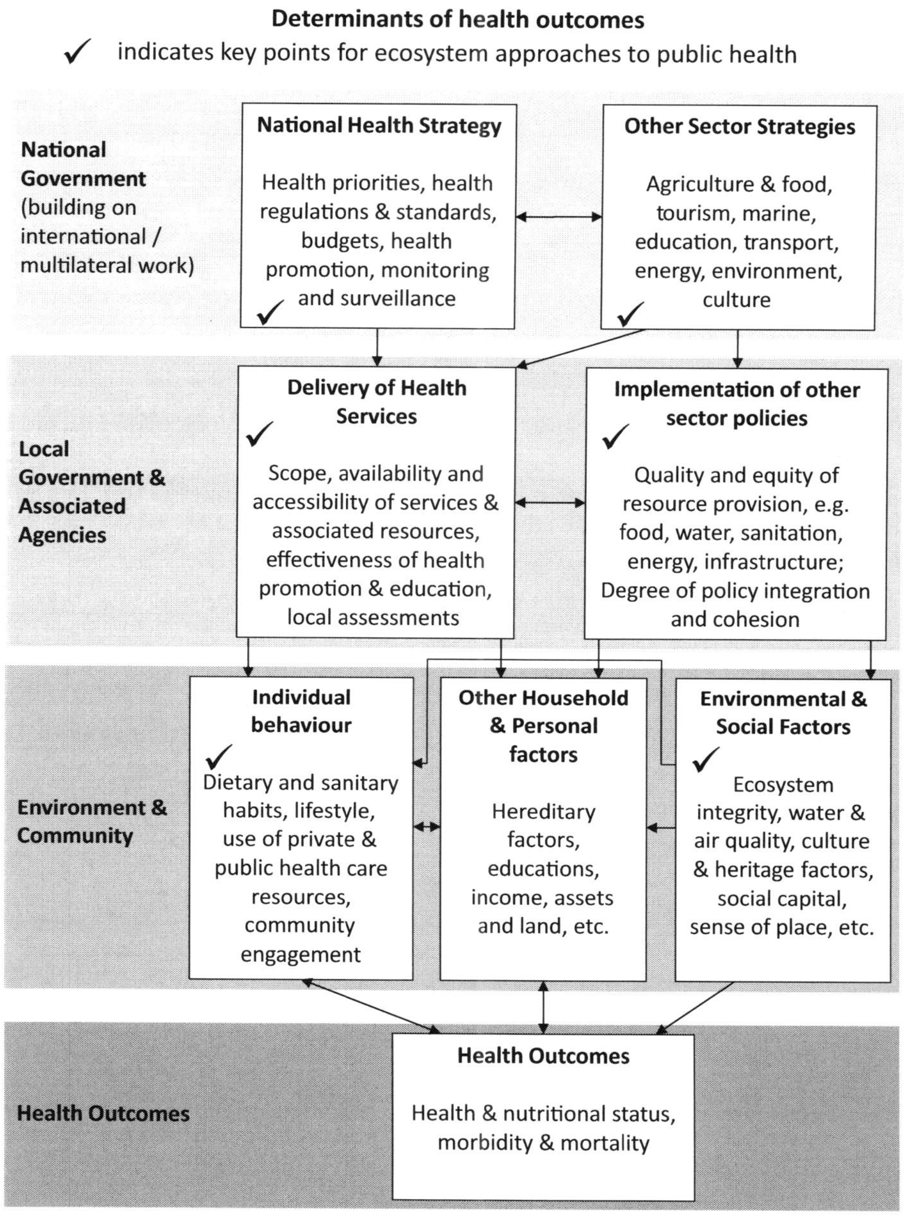

Figure 42.2 Opportunities for mainstreaming ecosystem approaches in health.

Source: COHAB Initiative Secretariat; reproduced with permission

Building the knowledge base should seek to cover the range of resources, from literature to practical experience and community knowledge. This is particularly important when dealing with issues of cultural heritage, spiritual well-being, and social cohesion, where the lack of scientific research into particular practices, experiences, or belief systems does not necessarily reflect a lack of legitimacy. It is important to note also that stores of local and indigenous knowledge of ecosystems and the wider environment can provide important information on past, present, and potential future problems at the interface of environment and health and provide opportunities for novel integrated approaches (Theobald et al., 2015).

To facilitate this, an appropriate mechanism or framework should be put in place to: (1) support collection and analysis of information from sources varying from scientific research to community inputs; (2) support knowledge transfer; and (3) enable communication of salient points to decision makers and practitioners. It can be useful to centre this type of framework on a virtual (e.g. internet-based) forum, or on a conference or workshop setting that promotes the sharing of experience and "lessons learned".

Conclusions

This chapter has provided an overview of the key linkages between biodiversity, ecosystem services, and human health. Whilst many such linkages are now well-established, others are less certain, or are anticipated based on current knowledge, and considerable additional research is required. For example, little is known about how specific health outcomes may be affected by particular changes in ecosystem structure or function, or how multiple ecosystem changes may interact to affect health status or risks. Furthermore, it is clear that the relationship between human health and the health of ecosystems is frequently mediated by a complex set of social, cultural, and economic factors, affecting not only the degree to which ecosystems influence health outcomes, but also how various groups of people within a given population access, utilise, and value local biodiversity and the ecosystem services associated with health.

To ensure that these linkages can be effectively assessed and understood, and to enable science to be translated into effective policy, continued effort is required to facilitate co-operation between the health and biodiversity / ecosystem services sectors, as well as other relevant sectors such as agriculture, tourism, trade, and economics, and with businesses and civil society. New approaches and initiatives continue to be developed to help foster these partnerships, based upon ecosystem approaches to health that view the conservation and sustainable use of biodiversity as being critical to protecting and enhancing the health of human communities.

In as much as the loss of biodiversity can threaten the health of human populations, its conservation, restoration, and sustainable use can help to strengthen modern and traditional healthcare systems, protect against natural disasters, combat poverty and hunger, prevent disease outbreaks, and promote resilience, stability and security for millions of people worldwide. In this context, biodiversity and ecosystem services may be particularly important to reducing, or adapting to, the impacts of climate change. In many regions, critical health services provided by biodiversity – the provision of fresh water and clean air, the regulation of the global climate, the provision of food resources, and the regulation of pests and diseases – are under particular threat. Thus it becomes increasingly important not only to protect remaining ecosystems, but also to restore and enhance degraded ecosystems throughout the world.

Notes

1 Convention on Biological Diversity (CBD) Article 2. Available at https://www.cbd.int/convention/text
2 The WHO maintains a database of national health accounts, including public and private expenditures, updated annually at http://www.who.int/health-accounts/en/ see also the OECD's "Health at a Glance" reports and related data, available at http://www.oecd.org/health
3 Available at http://www.who.int/governance/eb/who_constitution_en.pdf
4 Licenced under Creative Commons Attribution-NonCommercial-ShareAlike 3.0 Unported (CC BY-NC-SA 3.0) (http://creativecommons.org/licenses/by-nc-sa/3.0/deed.en)
5 CBD Articles 6 and 10

References

Bagnoli, P., Goeschl, T., and Kovács, E. (2008). *People and Biodiversity Policies Impacts, Issues and Strategies for Policy Action: Impacts, Issues and Strategies for Policy Action.* OECD Publishing, Paris.

Barton, H., and Grant, M. (2006). A health map for the local human habitat. *The Journal for the Royal Society for the Promotion of Health*, vol 126, no 6, pp 252–253. Developed from the model by Dahlgren and Whitehead, 1991.

Bickford, D., Posa, M.R.C., Qie, L., Campos-Arceiz, A., and Kudavidanage, E. P. (2012). Science communication for biodiversity conservation. *Biological Conservation*, vol 151, no 1, pp 74–76.

Bunch, M., McCarthy, D., and Waltner-Toews, D. (2008). A family of origin for an ecosystem approach to managing for sustainability. In: *The Ecosystem Approach: Complexity, Uncertainty, and Managing for Sustainability.* Columbia University Press, New York.

Bunch, M. J., Morrison, K. E., Parkes, M. W., and Venema, H. D. (2011). Promoting health and well-being by managing for social–ecological resilience: the potential of integrating ecohealth and water resources management approaches. *Ecology and Society*, vol 16, no 1, p 6.

Burlingame, B. (2000). Wild nutrition. *Journal of Food Composition and Analysis*, vol 13, pp 99–100.

Campbell-Lendrum, D., (2005). How much does the health community care about global environmental change? *Global Environ. Change*, vol 15, pp 296–298.

Chivian, E., and. Bernstein, A. (eds) (2008). *Sustaining Life – How Human Health Depends on Biodiversity.* Oxford University Press, New York.

Clark, N.E., Lovell, R., Wheeler, B.W., et al. (6 authors) (2014). Biodiversity, cultural pathways, and human health: a framework. *Trends in Ecology & Evolution*, vol 29, no 4, pp 198–204.

Convention on Biological Diversity (CBD) (2008). Summary Report from the Second International Conference on Health and Biodiversity. Galway, Ireland, 25–28 February 2008. CBD Secretariat, Montreal.

CBD and WHO (2015). *Connecting Global Priorities – Biodiversity and Human Health, a State of Knowledge Review.* WHO, Geneva.

Corvalan, C., Hales, S., and McMichael, A. J. (2005). *Ecosystems and Human Well-Being: Health Synthesis. A Report of the Millennium Ecosystem Assessment.* WHO, Geneva.

Costanza, R. (1992). Toward an operational definition of ecosystem health. In: Costanza, R., Norton, B.G. and Haskell, B.D. (eds) *Ecosystem Health: New Goals for Environmental Management.* Island Press, Washington, DC.

Dahlgren, G., and Whitehead, M. (1991). "The main determinants of health" model. In: Dahlgren G. and Whitehead M. (eds) *European Strategies for Tackling Social Inequities in Health: Levelling Up Part 2.* Copenhagen: WHO Regional Office for Europe.

Díaz, S., Fargione, J., Chapin III, F. S., and Tilman, D. (2006). Biodiversity loss threatens human well-being. *PLoS Biol*, vol 4, p e277.

Fanzo, J., Hunter, D., Borelli, T., and Mattei, F. (2013). *Diversifying Food and Diets – Using Agricultural Biodiversity to Improve Nutrition and Health.* Earthscan, Abingdon.

Ford-Lloyd, B.V., Schmidt, M., Armstrong, S.J., et al. (20 authors) (2011). Crop wild relatives – Undervalued, underutilized and under threat? *BioScience*, vol 61, no 7, pp 559–565.

Frison, E.A., Cherfas, J., and Hodgkin, T. (2011). Agricultural biodiversity is essential for a sustainable improvement in food and nutrition security. *Sustainability*, vol 3, no 1, pp 238–253.

Frison, E.A., Smith, I.F., Johns, T., Cherfas, J., and Eyzaguirre, P.B. (2006). Agricultural biodiversity, nutrition, and health: making a difference to hunger and nutrition in the developing world. *Food & Nutrition Bulletin*, vol 27, no 2, pp 167–179.

Gómez, A., Balsari, S., Nusbaum, J., Heerboth, A., and Lemery, J. (2013). Perspective: environment, biodiversity, and the education of the physician of the future. *Academic Medicine*, vol 88, no 2, pp 168–172.

Gottdenker, N. L., Streicker, D. G., Faust, C. L., and Carroll, C. R. (2014). Anthropogenic land use change and infectious diseases: a review of the evidence. *EcoHealth*, vol 11, no 4, pp 619–632.

Hanski, I., von Hertzen, L., Fyhrquist, N., et al. (14 authors) (2012). Environmental biodiversity, human microbiota, and allergy are interrelated. *Proceedings of the National Academy of Sciences*, vol 109, no 21, pp 8334–8339.

Hartig, T., Mitchell, R., De Vries, S., and Frumkin, H. (2014). Nature and health. *Annual Review of Public Health*, vol 35, pp 207–228.

Hoffman, L. C., and Cawthorn, D. M. (2012). What is the role and contribution of meat from wildlife in providing high quality protein for consumption? *Animal Frontiers*, vol 2, no 4, pp 40–53.

Horwitz, P., Lindsay, M., and O'Connor, M. (2001). Biodiversity, endemism, sense of place, and public health: inter-relationships for Australian inland aquatic systems. *Ecosystem Health*, vol 7, no 4, pp 253–265.
Huber, M., Knottnerus, J.A., et al (14 authors) (2011). How should we define health? *British Medical Journal*, vol 343, no 6, p d4163.
Hudson, P.J., Dobson, A.P., and Lafferty, K.D. (2006). Is a healthy ecosystem one that is rich in parasites? *Trends in Ecology & Evolution*, vol 21, no 7, pp 381–385.
Jones, B.A., Grace, D., Kock, R., et al. (11 authors) (2013). Zoonosis emergence linked to agricultural intensification and environmental change. *Proceedings of the National Academy of Sciences*, vol 110, no 21, pp 8399–8404.
Jones, K.E., Patel, N.G., Levy, M.A., et al. (7 authors) (2008). Global trends in emerging infectious diseases. *Nature*, vol 451, no 7181, pp 990–993.
Kawarazuka, N., and Béné, C. (2010). Linking small-scale fisheries and aquaculture to household nutritional security: an overview. *Food Security*, vol 2, no 4, pp 343–357.
Keune, H., Kretsch, C., De Blust, G., et al. (24 authors) (2013). Science–policy challenges for biodiversity, public health and urbanization: examples from Belgium. *Environmental Research Letters*, vol 8, no 2, p 025015.
Kremen, C. (2005). Managing ecosystem services: what do we need to know about their ecology? *Ecology Letters*, vol 8, no 5, pp 468–479.
Kuhnlein, H.V., and Chan, H.M. (2000). Environment and contaminants in traditional food systems of northern indigenous peoples. *Annual Review of Nutrition*, vol 20, no 1, pp 595–626.
Lancet (editorial) (2009). What is health? The ability to adapt. *Lancet*, vol 373, p 78.
Langlois, E., Campbell, K., Prieur-Richard, A.-H., Karesh, W., and Daszak, P. (2012). Towards a better integration of global health and biodiversity in the new sustainable development goals beyond Rio + 20. *EcoHealth*, vol 9, pp 381–385.
Leal, M.C., Puga, J., Serôdio, J., Gomes, N.C., and Calado, R. (2012). Trends in the discovery of new marine natural products from invertebrates over the last two decades–where and what are we bioprospecting? *PLoS One*, vol 7, no 1, e30580.
Lehman, R. M., Acosta-Martinez, V., Buyer, J. S., et al. (15 authors) (2015). Soil biology for resilient, healthy soil. *Journal of Soil and Water Conservation*, vol 70, no 1, pp 12A-18A.
Lele, S., Springate-Baginski, O., Lakerveld, R., Deb, D., and Dash, P. (2013). Ecosystem services: origins, contributions, pitfalls, and alternatives. *Conservation and Society*, vol 11, no 4, p 343.
Lovell, R., Wheeler, B.W., Higgins, S.L., Irvine, K.N., and Depledge, M.H. (2014). A systematic review of the health and well-being benefits of biodiverse environments. *Journal of Toxicology and Environmental Health, Part B*, vol 17, no 1, pp 1–20.
Luck, G.W., Chan, K.M., Eser, U., et al (7 authors) (2012). Ethical considerations in on-ground applications of the ecosystem services concept. *BioScience*, vol 62, no 12, pp 1020–1029.
Lyytimäki, J., and Sipilä, M. (2009). Hopping on one leg – the challenge of ecosystem disservices for urban green management. *Urban Forestry & Urban Greening*, vol 8, no 4, pp 309–315.
MA (2005). *Ecosystems and Human Well-Being: Synthesis*. Island Press, Washington DC.
McMichael, A. J. (2006). Population health as the "bottom line" of sustainability: a contemporary challenge for public health researchers. *Eur. J. Public Health*, vol 16, no 5, pp 79–82.
McMichael, A.J., and Beaglehole, R. (2000). The changing global context of public health. *Lancet*, vol 356, no 9228, pp 495–499.
McShane, T.O., Hirsch, P.D., Trung, T.C., et al. (14 authors) (2011). Hard choices: making trade-offs between biodiversity conservation and human well-being. *Biological Conservation*, vol 144, no 3, pp 966–972.
Myers, S. S., Gaffikin, L., Golden, C. D., et al. (8 authors) (2013). Human health impacts of ecosystem alteration. *Proceedings of the National Academy of Sciences*, vol 110, no 47, pp 18753–18760.
Patz, J.A., Frumkin, H., Holloway, T., Vimont, D.J., and Haines, A. (2014). Climate change: challenges and opportunities for global health. *JAMA*, vol 312, no 15, pp 1565–1580.
Rapport, D. J., Costanza, R., and McMichael, A. J. (1998). Assessing ecosystem health. *Trends in Ecology and Evolution*, vol 13, pp 397–402.
Roe, J., and Aspinall, P. (2011). The restorative outcomes of forest school and conventional school in young people with good and poor behaviour. *Urban Forestry & Urban Greening*, vol 10, no 3, pp 205–212.
Sala, O. E., Meyerson, L. A., and Parmesan, C. (eds) (2009). *Biodiversity Change and Human Health: From Ecosystem Services to Spread of Disease*. Island Press, Washington DC.

Skevington, S. (2009). Quality of life, biodiversity and health: observations and applications. In: Sala, O.E., Meyerson, L. A. and Parmesan, C. (eds) *Biodiversity Change and Human Health: From Ecosystem Services to the Spread of Disease*. Island Press, Washington DC.

Srinivasan, U.T., Cheung, W.W., Watson, R., and Sumaila, U.R. (2010). Food security implications of global marine catch losses due to overfishing. *Journal of Bioeconomics*, vol 12, no 3, pp 183–200.

Sudmeier-Rieux, K. and Ash, N. (2009). *Environmental Guidance Note for Disaster Risk Reduction: Healthy Ecosystems for Human Security*. IUCN, Gland.

Sudmeier-Rieux, K., Masundire, H. M., and Rizvi, A. H. (2006). *Ecosystems, Livelihoods and Disasters: An Integrated Approach to Disaster Risk Management, Number 4*. IUCN, Gland.

Theobald, E.J., Ettinger, A.K., Burgess, H.K., et al. (11 authors) (2015). Global change and local solutions: tapping the unrealized potential of citizen science for biodiversity research. *Biological Conservation*, vol 181, pp 236–244.

Velarde, M.D., Fry, G., and Tveit, M. (2007). Health effects of viewing landscapes–Landscape types in environmental psychology. *Urban Forestry & Urban Greening*, vol 6, no 4, pp 199–212.

Wildt, D.E., Comizzoli, P., Pukazhenthi, B., and Songsasen, N. (2010). Lessons from biodiversity – The value of nontraditional species to advance reproductive science, conservation, and human health. *Molecular Reproduction and Development*, vol 77, no 5, pp 397–409.

World Health Organisation (WHO) (1948). Preamble to the Constitution of the World Health Organisation as adopted by the International Health Conference, New York, 19–22 June, 1946; signed on 22 July 1946 by the representatives of 61 States (Official Records of the World Health Organisation, no. 2, p. 100) and entered into force on 7 April 1948.

World Health Organisation (WHO) (2002). *Traditional Medicines Strategy 2002–2005*. WHO, Geneva.

Zelenski, J. M., and Nisbet, E. K. (2014). Happiness and feeling connected: the distinct role of nature relatedness. *Environment and Behavior*, vol 46, no 1, pp 3–23.

43

ECOSYSTEM SERVICES AND THE GREENING OF BUSINESS

Guy Duke

Introduction – business impacts and dependencies on natural capital and ecosystem services

Businesses are linked to natural capital and ecosystem services through impacts and dependencies. All businesses have an impact on natural capital and ecosystem services, either directly (e.g. through consumption of biotic products, or clearance or conversion or disturbance of natural ecosystems) or indirectly (e.g. through energy use, which contributes to climate change, which, in turn, affects ecosystems). Conversely, all businesses are dependent, either directly (e.g. for raw materials) or indirectly (e.g. for clean water, or an attractive living environment for their employees), on ecosystem services (Figure 43.1).

For example agricultural businesses depend on numerous species and ecosystem services, including genetic diversity, pollination, freshwater supplies and nutrient cycling. They impact natural capital and ecosystem services by clearance and conversion of land, through soil and water pollution and through greenhouse gas emissions. Forestry businesses depend on ecosystem services, including freshwater supply, climate regulation and nutrient cycling, and have an impact on natural capital and ecosystem services through commercial logging. Mining and quarrying can lead to large-scale destruction of habitats and have indirect impacts through road-building and pollution. The oil and gas industries depend on supplies of freshwater, and have impacts through upstream operations (drilling, construction, etc.) and downstream combustion and greenhouse gas emissions. The personal care and cosmetics industry depends on numerous natural ingredients. The water supply and sanitation sector is highly dependent on a range of ecosystem services for sustainable and cost efficient operations. The transport industry has large impacts on natural capital and ecosystem services. Many tourism businesses depend on ecosystem services, including the amenity value of natural areas. Many manufacturing industries depend on a range of ecosystem services, and impact through supply of raw materials, footprint of facilities and pollution from production processes (TEEB, 2012).

Natural capital accounting

One of the main ways in which businesses are engaging with the concepts of natural capital and ecosystem services is through natural capital accounting (NCA; see also Houdet et al.,

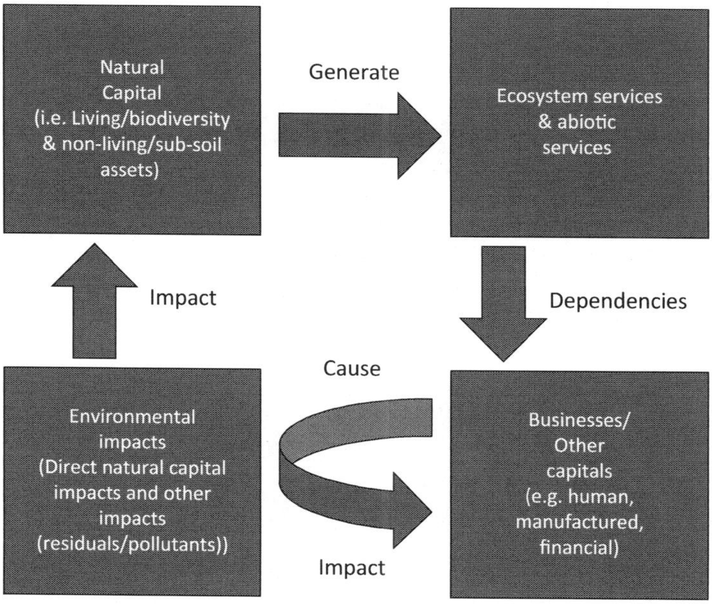

Figure 43.1 Inter-relationships between natural capital, ecosystem and abiotic services, business and other capitals, and environmental impacts.

Source: modified according to Spurgeon, 2014

2016). The Natural Capital Coalition (www.naturalcapitalcoalition.org) is among the key initiatives seeking to strengthen corporate natural capital accounting and reporting. Launched in 2012 as the TEEB for Business Coalition follow-on to TEEB, it was rebranded as the Natural Capital Coalition in 2014. It is a global, multi-stakeholder open-source platform for supporting the development of methods for natural and social capital valuation in business. Its aim is to achieve a shift in behaviour to preserve and enhance, rather than deplete, the earth's natural capital. To achieve this, the Coalition brings together global leaders on natural capital in a convening platform to scale action. Key aims are to: build the business case to demonstrate that integrating natural capital in business decision-making leads to better business decisions (with the benefits of greater resilience, reduced costs, improved security of supply and, ultimately, a more sustainable business model); and support the development and testing of a harmonized framework and methodologies for valuing natural capital that can be applied in business (in strategic planning, management, disclosure in corporate reporting, and investor criteria and benchmarking).

Business risks related to NC and ES

Businesses face a wide range of risks related to biodiversity, ecosystem services and natural capital, including coastal flooding, desertification and food security (PricewaterhouseCoopers, 2010). Figure 43.2 sets out some of the interconnections between biodiversity loss and other global risks as identified by the Global Risks Network of the World Economic Forum. As the figure shows, biodiversity loss (and by extension, decline in ecosystem services) is

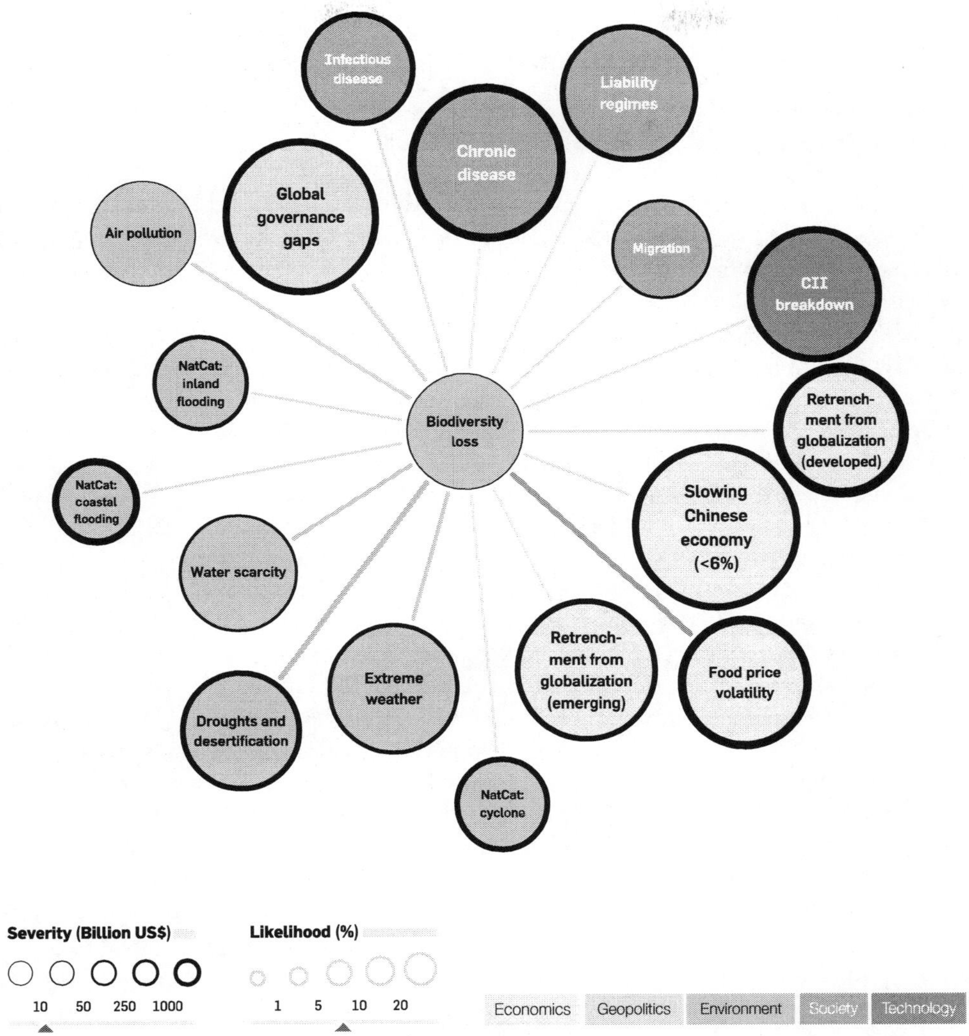

Figure 43.2 Biodiversity loss at the nexus of many risks.

Source: World Economic Forum, 2010

connected to a number of other risks, with estimated severities in dollar terms ranging from tens of billions for inland flooding and infectious disease, to many hundreds of billions for food price volatility and chronic disease (PricewaterhouseCoopers, 2010). These risks will affect not only companies with direct reliance on natural resources but also the supply chains and growth objectives of most industry sectors in the developed and developing world. The 13th Annual Global CEO Survey 2010 of 1200 CEOs conducted by PricewaterhouseCoopers found that 27% of CEOs (but 53% in Latin America and 45% in Africa) were either "extremely" or "somewhat" concerned about "biodiversity loss" as one of a range of threats to their business growth. However, overall business concern was low, relative to other risks (PricewaterhouseCoopers, 2010).

Business opportunities and innovations related to natural capital and ecosystem services

Consideration of natural capital and ecosystem services can help businesses to recognize and pursue a wide range of new business opportunities. This is illustrated by the work of the UK Ecosystem Market Task Force, which reported in early 2013. The establishment of the business-led Task Force was a key commitment in the UK Government's Natural Environment White Paper, *The Natural Choice*. The White Paper's ambition is to create "a green economy, in which economic growth and the health of our natural resources sustain each other, and markets, business and Government better reflect the value of nature". The Task Force also responded to the recent *Industrial Strategy: UK Sector Analysis from the Department for Business, Innovation and Skills* which identified "increasing demand for environmental products, processes and standards" as one of the main four drivers for growth over the next decade.

The Task Force was appointed in late 2011 and made up of 10 business leaders under the Chairmanship of Sir Ian Cheshire, Group Chief Executive of Kingfisher plc. Other businesses represented included Jaguar Land Rover, Unilever and United Utilities. The government asked the Task Force to "review the opportunities for UK business from expanding green goods, services, products, investment vehicles and markets which value and protect our natural environment" (Ecosystem Markets Task Force, 2013, p. 2). The approach adopted by the Task Force was explicitly built on the ecosystem services framework of the Millennium Ecosystem Assessment (MA). Specifically, the Task Force commissioned research to underpin its thinking, which took as a starting point the UK National Ecosystem Assessment (NEA). The research first reviewed systematically almost 1500 pages of the 2011 NEA Report[1] to extract any business opportunities identified, either explicitly or implicitly. The research team then used innovative thinking and stakeholder consultation to elaborate and assess the full range of business opportunities that had been extracted (Duke et al., 2012). This research identified eight "types" of business opportunity, many of which are likely to be applicable in other national settings (Box 43.1).

The evidence gathered enabled the Task Force to home in on a subset of "more promising opportunities" which was then subjected to deeper analysis. This involved exploring the economic case (costs, benefits, impacts) and making practical recommendations for the further development of each opportunity (Duke et al., 2013). The Task Force focused on identifying win-win business opportunities that deliver substantial benefits for both nature and business. In its final report, *Realising Nature's Value* (EMTF, 2013) the Task Force presented 22 opportunities for the UK, grouped under four themes (Box 43.2; those considered by the task force to present the most significant win-win opportunities are highlighted in italics).

Box 43.3 provides an example of an innovative business model relating to biodiversity offsetting.

Box 43.1 Types of business opportunity linked to valuing and/or protecting ecosystem services

- ***Product markets*** – products derived from and/or sustaining ecosystem services, and related certification services.
- ***Offsetting*** – business opportunities linked to offsetting impacts on biodiversity, carbon or other natural assets or ecosystem services.

- ***Payment for ecosystem services (PES)*** **–** a variety of schemes through which the beneficiaries, or users, of ecosystem services provide payment to the stewards, or providers, of ecosystem services (see also Brouwer, 2016).
- ***Environmental technologies*** **–** these prevent or treat pollution, enhance management of ecosystems, and enable more efficient resource use.
- ***Markets for cultural services*** **–** e.g. for tourism, recreation and preventive or curative health treatments, based on nature's services.
- ***Financial and legal services*** **–** e.g. financial services for investment in nature-based businesses, legal services to secure property rights which underpin PES or offsetting.
- ***Ecosystem knowledge economy*** **–** services that deliver knowledge about ecosystems and ecosystem services; the UK could emerge as an international leader in this respect.
- ***Corporate ecosystem initiatives*** **–** measures taken by companies to reduce negative impacts and enhance positive impacts on nature, in order to enhance brands, meet consumer demand, manage supply chain issues or simply "do the right thing".

Source: Duke et al., 2012

Box 43.2 Business opportunities that protect and/or value nature

Carbon and markets for nature

- *Biodiversity offsetting: securing net gain for nature through planning and development*
- *Sustainable local woodfuel: active sustainable management supporting local economies*
- Carbon reduction through investing in nature
- Environmental bonds

Food cycle

- *Bio-energy and anaerobic digestion on farms: closing the loop using farm waste to generate energy*
- *Nature-based certification and labelling: connecting consumers with nature*
- Common Agricultural Policy
- Food waste

Water cycle

- *Water cycle catchment management (including water and wastewater catchment management, Sustainable Urban Drainage Systems and soft flood defences)*
- Water trading
- Water supply pipe ownership
- Water metering
- Very long term planning
- Privatization of flood defences

Natural capital: cross-cutting themes

- Managing natural resource security
- Using nature to enhance resilience
- Business accounting for nature: mainstreaming standards and metrics
- Knowledge Economy: UK expertise enabling business opportunities to enhance nature

Source: EMTF, 2013

Box 43.3 The Environment Bank Ltd – an innovative business model to compensate for development impacts on biodiversity and other natural capital assets[2]

The Environment Bank (EB) is a UK company operating in emerging markets for environmental assets, in particular relating to providing compensation for developmental impacts on biodiversity. EB has developed an innovative business model as a broker, involving metrics, trading and delivery systems. This model can be adapted to trade in other natural capital assets (e.g. nature-based carbon, water quality). EB focuses on compensating for development impacts on non-protected habitats for which habitat restoration or creation is known to be feasible. EB takes a third-party, market-based approach to ensure efficient and effective delivery of habitat compensation schemes.

There are significant potential benefits to developers using EB as a broker, including time-savings, reduced long-term liabilities and reputational benefits. EB's business model is highly scalable and can deliver considerable company growth as the market expands in England. The model could be applied elsewhere in the EU wherever governments encourage a market-based, brokered approach for habitat compensation schemes.

Research has shown that compensation of residual development impacts could, over 20 years, deliver 3,000 km^2 of ecological creation/restoration in England, and 10–50,000 km^2 across the EU. The majority of this would arise from impacts that are currently not compensated for at all, and therefore represents a significant biodiversity gain. These compensation schemes would also deliver other ecosystem services and enhance ecosystem resilience to climate change.

The UK Ecosystem Markets Task Force identified the potential annual market value for such offsetting as £90–470m in England, and £750m-£7.5bn in the EU.

Business engagement with the concepts of natural capital and ecosystem services

Businesses are becoming increasingly aware of the need to extend their engagement with a new set of market-based sustainability drivers. These drivers relate to the sustainability of their supply chains, changing customer preferences for more sustainable products, and emerging opportunities in relation to new markets and investment opportunities base on ecosystems. A number of companies have taken the lead, for example, in declaring support for valuing and accounting for natural capital and ecosystem services, such as those companies involved in the Natural Capital Leaders Platform of the Cambridge Institute for Sustainability Leadership.[3]

Many companies remain unaware of their dependencies and impacts on nature and unengaged or uncertain as to the risks and opportunities related to natural capital and ecosystem services. Many remain reluctant to make commitments, perceiving more risks and costs of engaging than clear opportunities. This is underlined by the fact that less than one in five companies see biodiversity as an important business issue, while only two of the world's largest 100 companies manage it as a strategic risk (PricewaterhouseCoopers, 2010).

However, several business-facing organizations, such as Trucost, have elevated the discussion among some leading companies, such as PUMA and its parent company Kering, through quantifying the impacts of companies on ecosystem values and through highlighting related business opportunities. This has led to increasing interest from businesses in approaches to natural capital accounting to support decision-making and reporting (see below). Despite this activity, there is still low awareness among most companies as to both the need and the opportunity for action.

New business models

The nature of ecosystem markets means they often involve a combination of public and private goods and complex temporal and spatial relationships between sources of supply and demand. Consequently, diverse business models are needed to develop the wide range of market opportunities related to natural capital and ecosystem services. These can involve:

- Collectives (e.g. groups of farmers collaborating to deliver catchment management in a "payment for ecosystem services" deal);
- The public sector entering into or supporting transactions in different ways, such as public-private partnership agreements for green infrastructure, or joint public and private sector purchasing of ecosystem services; and,
- Social enterprises, for which delivering public (environmental) interests, rather than profits, is part of the return achieved through business activity.

Regulation for new ecosystem markets

Many business opportunities related to natural capital and ecosystem services can benefit from government action in terms of policy or regulation. Beyond helping to develop specific business opportunities, macro-scale policy or regulatory provisions can have significant influence over the use of natural resources. For example, the landfill tax escalator introduced in UK during the 1990s and accelerated in the 2000s has been linked to improvements in recycling rates and waste management practices and has possibly also enhanced the waste sector's reputation (Duke et al., 2013).

Regulation can also establish compliance markets (e.g. for carbon under the EU Emissions Trading Scheme, or for biodiversity and ecosystems under the US Wetland Mitigation and endangered species offsetting schemes) and influence the way ecosystems are classified as assets. Indeed, the long-term prospects for ecosystem markets are heavily dependent on macro-scale policy decisions. The general idea of environmental tax reform applies specifically to the development of environmental markets; shifting the burden of taxation away from employment to resource use can improve the commercial returns on investments in ecosystem markets relative to other economic opportunities.

The adjustment of approaches to development planning, including proper accounting for impacts on biodiversity, natural capital and ecosystem services, and the use of tools such as offsetting and habitat banking, can also help to deliver "no net loss" of biodiversity and ecosystem services when businesses implement development projects.

Thus, while many of the business opportunities for protecting and valuing nature involve only micro-level changes that encourage markets to take better account of the value of nature, more fundamental changes in the way that we take account of the value of ecosystems in regulation, planning, economic development and wider decision-making could have more profound impacts on the working of markets and the role of business.

Finance

The finance industry, which includes credit unions, banks, insurers, accountancy companies and investment funds, is giving increasing attention to natural capital and ecosystem services, both in terms of risk and in terms of opportunity.

The degradation of natural capital and the decline in ecosystem services is increasingly recognized as a significant risk for lenders, investors and insurers. It can reduce the ability of a borrower to repay a loan, or of an investor to realize the expected return on investment, and can reveal insurance premiums to be miscalculated.

Furthermore, a number of risk factors related to the degradation of natural capital and decline in ecosystem services can result in companies facing "stranded assets", that is, assets that have suffered from unanticipated or premature write-downs, devaluations or conversion to liabilities. These risk factors include changing resource landscapes (e.g. phosphate scarcity), government regulations (e.g. requiring carbon or biodiversity offsets), evolving social norms (e.g. consumer preference for certified products) and litigation and statutory interpretations (e.g. disclosure on environmental impacts). Such risk factors are poorly understood and regularly mispriced, resulting in an over-exposure to such risks throughout our financial and economic systems.

The concept of stranded assets is often used in relation to the so-called carbon bubble. This refers to a significant portion of remaining fossil fuel reserves, which may become stranded assets if the need to avoid dangerous climate change means that these reserves cannot be burned. Similarly, the degradation of natural capital and decline of ecosystem services could result in significant areas of land becoming stranded assets. The Bank of England Governor, Mark Carney, has called the failure of businesses to build in such environmental risk factors a "tragedy of horizons" that could lead to market failure.[4]

In managing risk related to natural capital and ecosystem services, the Equator Principles (2013) provide useful guidance for the finance industry for cases in which a single asset is being financed. However, most financing is not of a single asset, but is at the corporate level. Here, it is more difficult for the finance industry to judge a company's impact on natural capital and ecosystem services. There is a need for much better disclosure, including both natural capital accounting, and rules on disclosure (of natural capital accounts), to inform lending and investment decisions.

Accounting and reporting for business impacts on natural capital will be critical to the more appropriate allocation of (financial) capital and credit.

The Natural Capital Declaration[5] (NCD) is an initiative focused on the financial sector, which seeks to include natural capital-related risks in the cost of capital. The NCSD notes that these risks result in a range of challenges to businesses, including legal liability, credit risk, volatility, unexpected falls in cash flows, and reputational, regulatory and portfolio risks, each presenting different financial pressures requiring additional mitigating measures. The NCD seeks to clarify how financial institutions are exposed to material natural capital risks through companies, and to encourage financial institutions to allocate capital to "natural capital positive business opportunities".

Box 43.4 NatureVest

NatureVest is a division of The Nature Conservancy (a US-based not-for-profit) supported by JP Morgan Chase. It was launched to transform the way we protect natural capital by capitalising on the growing impact investment sector and by fostering ways to advance investment in conservation. As part of that effort, NatureVest: convenes investors; develops and executes innovative financial transactions; and continues to build an investment pipeline across multiple sectors, including agriculture, fisheries and environmental markets. These include offsets, multi-use forestry and agriculture, and green bonds. There is a need for increased investor demand for such opportunities. The Norwegian and Dutch government pension funds are leading the way in this. Typically, credit is the most informed capital, while bond holders are less well informed, so relatively more effort is required with bond holders to green their investments.

While the finance industry tends to consider natural capital and ecosystem services in terms of risk, it is also beginning to recognize that there are opportunities. This involves, notably, reallocating capital from the exploitation of fossil fuels ("dead natural capital") to living natural capital (e.g. Box 43.4).

Standards and metrics

In recent years, the discussion on what and how to measure in determining the environmental impact of companies has deepened and broadened. There are now a range of voluntary frameworks and a few mandatory ones, including, in the UK, a future requirement for FTSE-listed companies to disclose their carbon dioxide emissions.[6] But while there has been some progress internationally in determining standards and metrics in relation to greenhouse gases and water, there has been less progress in agreeing to and implementing those relating to biodiversity. For many in business it is clear that what cannot be measured cannot be managed, making this aspect of how businesses engage with ecosystem markets quite fundamental.

For the economic value of ecosystems to be properly realized through market-based and other approaches will require reliable and comparable data to be collected and presented. This is vital not only for businesses themselves in understanding the value of their activities, but also for the effective engagement of stakeholders, ranging from investors to clients and consumers. UN negotiations, leading to the agreement of Sustainable Development Goals by 2015, present an additional new context within which new metrics for business would ideally be embedded.

The International Integrated Reporting Committee (IIRC) is, for example, leading a process that will elaborate and address some of the key challenges embedded in this complex question.[7] While the process takes a view beyond ecosystems, there are important cross-cutting themes being identified by it. Business for Social Responsibility[8] is a coalition that has taken a more focused approach in relation to ecosystems, producing a review of available tools.

UNEP-WCMC works with the Global Reporting Initiative (GRI) on the development of ecosystem services indicators for business reporting and has produced various materials that are helping to determine simple but meaningful standards.[9] Others have reviewed the case for strong metrics in relation to conservation action, irrespective of market or business involvement.

Consumer Awareness

For ecosystem markets to work on a larger scale there will need to be broad backing from consumers. This needs to be seen not only in "green" product choices, but also in a more broadly positive public view of the wider involvement of companies (through, for example, exemplary offsetting schemes and payment for ecosystem services). This is by no means guaranteed. Awareness as to the role of markets and other economic tools being harnessed to promote sustainability goals in relation to ecosystems is still in its early stages and awareness could develop in negative as well as positive directions.

For example, some campaigners point to ecosystem markets as having the potential to worsen the impacts of business on nature.[10] If this becomes a widespread view (accurate or not) there will be reluctance among companies to risk their reputations in the use of tools and approaches that do not yet enjoy widespread public backing. Others are working hard to raise awareness through the provision of information about the value of ecosystems to society.[11]

Aside from the conceptual merits of placing values on nature, and amid attempts to raise public awareness, there is also an increasingly intense debate among practitioners as to the best way to engage consumers, such as through "meeting people where they are at", rather than changing their underlying values. This question is far from simple and goes way beyond the provision of information. Campaignstrategy.org[12] presents a wealth of information on the dimensions of this debate. For example, a paper on the challenges of how to communicate and raise awareness around climate change provides a case in point.

With this in mind, it is vital to identify positive narratives that actually connect with people and through which it will be possible to frame the emerging process of realizing the economic value of natural systems. This involves seeking routes through which to respond to the overly negative and polarizing commentaries that are emerging in relation to ecosystem markets.

Resources for Businesses Wanting to Engage with Ecosystem Services

An increasing number of web-based platforms have been established to help businesses to integrate the concepts of natural capital and ecosystem services. These include platforms established by governments at national and supra-national levels, and business-led initiatives. Examples are provided in Box 43.5.

Box 43.5 Examples of business and biodiversity initiatives

INTERNATIONAL

- **The Economics of Ecosystems and Biodiversity (TEEB)** (www.teebweb.org): features the TEEB study report on business and enterprise, which examines business risks and opportunities, offers practical tools and explores new business models and the alignment of business actions in relation to biodiversity and ecosystem services with other corporate responsibility initiatives.
- **European Business and Biodiversity Campaign** (business-biodiversity.eu): aims to promote the business case for biodiversity in the EU member states through workshops, seminars and a cross-media communication strategy.

- **EU Business @ Biodiversity Platform** (ec.europa.eu/environment/biodiversity/business): addresses action at a European level, building national platforms in Europe; and focuses on three workstreams for the period 2014–2016, on natural capital accounting, innovation and financing.
- **Global Partnership for Business and Biodiversity** (www.cbd.int/business): arises from engagement between business and the Convention on Biological Diversity (CBD).
- **Natural Capital Business Hub** (www.naturalcapitalhub.org): a collaborative platform of more than 40 global companies.
- **Natural Capital Coalition** (www.naturalcapitalcoalition.org) (formerly the TEEB for Business Coalition): is developing a Natural Capital Protocol seeking to harmonize natural capital accounting in business.
- **Natural Capital Declaration (NCD)** (www.naturalcapitaldeclaration.org): a finance sector initiative to integrate natural capital considerations into loans, equity, fixed income and insurance products, as well as in accounting, disclosure and reporting frameworks, through the development of metrics and tools.
- **Proteus** (www.proteuspartners.org): a partnership between businesses and UNEP World Conservation Monitoring Centre (UNEP-WCMC) to make available global information on biodiversity to enable better-informed decisions.
- **UN Global Compact** (www.unglobalcompact.org): an initiative for businesses committed to principles of human rights, labour, environment and anti-corruption.
- **UNEP Finance Initiative (UNEP FI)** (www.unepfi.org): a partnership between UNEP and over 180 financial sector institutions, working to understand the impacts of environmental and social considerations on financial performance.
- **World Business Council for Sustainable Development (WBCSD)** (www.wbcsd.org): brings together some 200 companies dealing with sustainable development and business.

NATIONAL AND OTHER

- **Brazil: Brazilian Business and Biodiversity Initiative** (ibnbio.org): a platform for information sharing and the development of strategies on biodiversity protection and sustainable use of natural resources.
- **Canada: Canadian Business and Biodiversity Council (CBBC)** (www.businessbiodiversity.ca): assists businesses to operate in an environmentally responsible and sustainable manner.
- **Central and Eastern Europe: CEEWeb Cooperation with Business** (www.ceeweb.org/about-us/cooperation-with-business): provides guidelines for cooperation of business and NGOs in biodiversity-related initiatives, with a focus on the Visegrad countries (Czech Republic, Hungary, Poland, Slovakia).
- **Finland: Corporate Responsibility Network (FIBS)** (www.fibsry.fi): raises awareness of biodiversity, provides tools to assess dependencies on ecosystem services, helps businesses manage impacts on nature.
- **France, Plateforme Entreprises et Biodiversité (Orée)** (www.entreprises-biodiversite.fr): brings together public and private stakeholders to share best practices and foster implementation at landscape level.

- **Germany: Biodiversity in Good Company** (www.business-and-biodiversity.de): brings together private businesses committed to including protection of biodiversity in their environmental management systems.
- **India: India Business & Biodiversity Initiative (IBBI)** (www.businessbiodiversity.in): aims to guide and mentor business organizations in India on biodiversity conservation and sustainable use.
- **Japan: Japanese Business Initiative for Biodiversity (JBIB)** (www.jbib.org): supports Japanese companies interested in promoting their contribution to the environment and society.
- **Netherlands: Platform Biodiversiteit, Ecosystemen en Economie** (platformbee.nl): this national platform works on encouraging companies to integrate natural capital considerations in their strategies and operations.
- **Portugal: Initiative on Business and Biodiversity** (www.icnf.pt/portal/naturaclas/gest-biodiv1/business-biodiversity): promotes the introduction of biodiversity strategies in businesses through voluntary arrangements.
- **Spain: Spanish Business and Biodiversity Initiative** (www.fundacion-biodiversidad.es/empresaybiodiversidad): a national initiative which aims to increase the engagement of the business sector in order to achieve the international Aichi Biodiversity Targets.
- **UK: Natural Capital Leaders Platform** (www.cisl.cam.ac.uk/Business-Platforms/Natural-Capital-Leaders-Platform): run by the University of Cambridge Institute for Sustainability Leadership (CiSL), convenes companies with significant environmental impacts and dependencies that are willing to take action to review, value, re-design strategies, set targets and report on natural capital use.

Future directions

The UK Ecosystem Markets Task Force wrote in their final report (EMTF, 2013, p. 4): "The business reality is that change will happen: as resource prices face upward pressure, or as governments act to protect nature through regulation or taxes, to try to reflect the true costs of what were once seen as free natural products and services .... There are both risks and opportunities for business here. The companies that can lead on this issue and innovate will see both opportunity and reputational wins. The followers will benefit as workable solutions are confirmed and can be implemented, but the laggards are likely to lose out from price pressures and loss of share to nimbler rivals".

In the future (TEEB, 2012), we are likely to see more and more companies taking action to identify their impacts and dependencies on natural capital and ecosystem services, to assess the business risks and opportunities associated with these impacts and dependencies, taking action to mitigate risks and to grasp emerging opportunities. This is likely to be accompanied by the development, harmonization and widespread uptake of natural capital accounting tools and information systems, the integration of business actions for natural capital and ecosystem services into wider corporate responsibility strategies, and increasing engagement and partnership by business with business peers and stakeholders in government, NGOs and civil society to improve guidance and policy.

Acknowledgment

This chapter has drawn extensively from Duke et al. (2012) and Duke et al. (2013), which were funded by the UK Department for Environment, Food and Rural Affairs and the UK Natural

Environment Research Council through the Valuing Nature Network. The author would like to thank the co-authors of that report, and in particular Tony Juniper, for material used here.

Notes

1 http://uknea.unep-wcmc.org/Resources/tabid/82/Default.aspx
2 Adapted from: http://ec.europa.eu/environment/biodiversity/business/workstreams/Workstream2-Innovation-for-Biodiversity-and-Business/Outputs-to-date.html
3 http://www.cisl.cam.ac.uk/business-action/natural-resource-security/natural-capital-leaders-platform
4 www.theguardian.com/environment/2014/oct/13/mark-carney-fssil-fuel-reserves-burned-carbon-bubble
5 www.naturalcapitaldeclaration.org
6 See for example http://www.businessgreen.com/bg/news/2185657/coalition-confirms-introduction-mandatory-carbon-reporting
7 See http://www.theiirc.org/
8 www.bsr.org
9 See http://www.unep-wcmc.org/developing-standardised-indicators_587.html
10 See for example http://www.guardian.co.uk/commentisfree/2012/aug/06/price-rivers-rain-greatest-privatisation
11 See for example http://www.ucsusa.org/ssi/biodiversity/communicating-ecosystem-services/
12 See http://www.campaignstrategy.org/

References

Brouwer, R. (2016). Payments for ecosystem services. In: Potschin, M., Haines-Young, R., Fish, R. and Turner, R.K. (eds) *Routledge Handbook of Ecosystem Services*. Routledge, London and New York, pp 548–553.

Duke, G., Conway, M., Dickie, I., et al. (7 authors) (2013). EMTF Second Phase Research: Opportunities for UK Business that Protect and/or Value Nature. Final Report. ICF GHK, London.

Duke, G., Dickie, I., Juniper, T., et al. (9 authors) (2012). *Opportunities for UK Business that Value and/or Protect Nature's Services*. Final Report to the Ecosystem Markets Task Force and Valuing Nature Network. GHK, London.

EMTF (2013). *Realising Nature's Value: The Final Report of the Ecosystem Markets Task Force*. Department for Environment, Food and Rural Affairs, London.

Hanson, C., Ranganathan, J., Iceland, C,. and Finisdore, J. (2012). *The Corporate Ecosystem Services Review. Guidelines for Identifying Business Risks & Opportunities Arising from Ecosystem Change*. World Resources Institute, Washington DC.

Houdet, J.R.A., Finisdore, J., Martin-Ortega, et al. (8 authors)(2016). Accounting for ecosystem services in business. In: Potschin, M., Haines-Young, R., Fish, R. and Turner, R.K. (eds) *Routledge Handbook of Ecosystem Services*. Routledge, London and New York, pp 220–227.

PricewaterhouseCoopers (2010). *Biodiversity and Business Risk*. A Global Risks Network briefing. World Economic Forum, Geneva.

Spurgeon, J. (2014). *Natural Capital Accounting for Business. Guide to Selecting an Approach*. Final report to the EU Business and Biodiversity Platform.

TEEB (2012). The Economics of Ecosystems and Biodiversity in Business and Enterprise. Bishop, J. (ed.). Earthscan, London and New York.

The Equator Principles June 2013. Available at: http://www.equator-principles.com/resources/equator_principles_III.pdf

UN-SEEA (2013). *United Nations System of Environmental-Economic Accounting – Experimental Ecosystem Accounting*. United Nations Statistics Division, New York.

WBCSD (2011). *The Guide to Corporate Ecosystem Valuation*. Available at: http://www.wbcsd.org/work-program/ecosystems/cev/downloads.aspx

World Economic Forum (2010). *Global Risks 2010. A Global Risks Network Report*. World Economic Forum, Geneva.

44

PAYMENTS FOR ECOSYSTEM SERVICES

Roy Brouwer

Introduction

Payments for ecosystem services (PES) aim to translate the often non-market value of environmental goods and services into financial incentives to preserve the ecosystems that provide these services (Salzman, 2005; Wunder et al., 2008). The basic principle behind PES is that resource owners and communities who are in a position to provide ecosystem services should be compensated for the cost of that provision, and that those who benefit from these services should pay for them, thereby internalizing the benefits.

Wunder (2005) outlines five criteria to describe PES: a voluntary transaction, where well-defined ecosystem services (ES) are bought by an ES user from an ES provider under the agreed ES quantity and quality conditions in the transaction (conditionality requirement). In practice, PES is used as a more generic term for a variety of arrangements where local communities, farmers and other water and land managers are paid for conservation activities that deliver ES (Vatn, 2010), of which biodiversity and landscape preservation, carbon sequestration and water protection are most common (Duncan, 2006). The economic market principle behind PES is illustrated in Figure 44.1. An ecosystem supplies a number of goods and services for which there exists societal demand. Using PES, market conditions are approximated and a price is derived. The question typically is to what extent the price paid actually reflects scarcity conditions, given demand and supply. In practice, many 'market distortions' exist, including government interventions ranging from fixed price levels to mandatory participation, which make

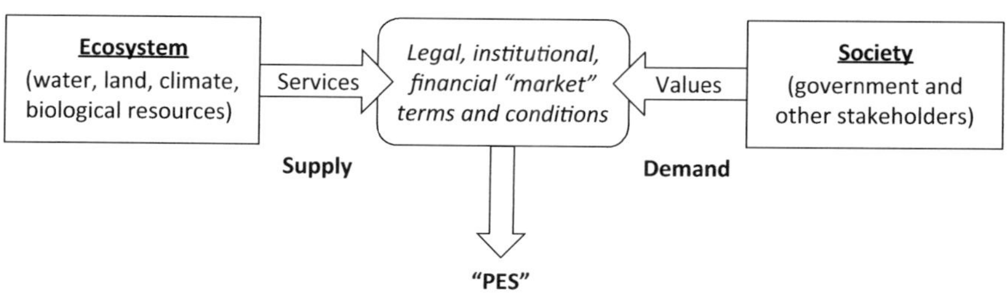

Figure 44.1 Payments for ecosystem services (PES) as the result of demand and supply.

PES a less than ideal economics textbook case for the internalization of externalities through the creation of a new market and market functioning.

The advantages of PES

Hybrid forms of PES have many attractive characteristics relative to other conservation approaches and are often carried out with the participation of local stakeholders as a form of co-management. Ascertaining their advantages requires measuring the additional effects of actual programmes in the field, also referred to as the 'additionality requirement' (Daniels et al., 2010). Such impact evaluation can also help in identifying opportunities for further improvements in efficiency of these programmes (Kerr and Jindal, 2007). The effectiveness of PES schemes may depend on several factors (Farley and Costanza, 2010). Some of these relate to the clarity of the definition of the ES and a careful assessment of ES demand and supply, including who are the beneficiaries who are willing to pay for the ES provision (Mayrand and Paquin, 2004). This may be different from the question of who finances the PES scheme. Here, options include donations and grants from national and international organizations, government payments and subsidies, payments from beneficiaries and market development for related goods and services at the national and international level (Mayrand and Paquin, 2004). PES schemes also need clear and enforceable rules and transaction mechanisms. Rules specify the rights and obligations of parties and the responsibilities and powers of institutions. Key legal issues in PES rule-making in this context are clarification of rights and tenure (Greiber, 2009), and establishing effective compliance and enforcement mechanisms (Smith et al., 2006). Porras et al. (2008) found in their review of PES schemes that land security played an important role in their successful implementation.

PES efficiency is not only determined by the extent to which incremental ES are provided, but also by the cost at which this is achieved. These costs include the opportunity cost of alternative land use activities, the implementation and maintenance costs of land use changes and the transaction costs of programme management and monitoring (Wunder et al., 2008). The latter are usually hard to quantify. The importance of having adequate insight in the opportunity costs of alternative land use to assess the incentive compatibility of a PES design is illustrated in Figure 44.2.

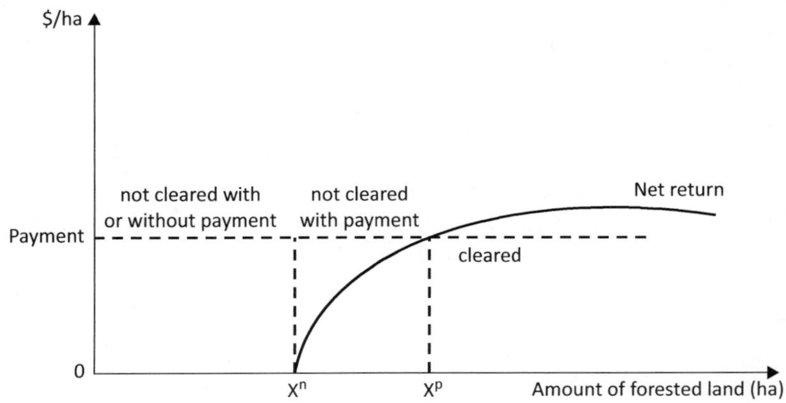

Figure 44.2 Illustration of the impact of compensation payments like PES on alternative land uses.

Source: modified from Pfaff et al., 2008

Landowners or users have unknown opportunity costs of alternative land uses. This is represented in Figure 44.2 by the net revenues curve. A rationally behaving landowner or user will only start exploiting the land if the net revenues are positive (at point X^n). Land will remain forested as long as the net revenues are not positive. Investments and maintenance costs are needed to clear the land. Once the net revenues are positive, a land owner or user will start clearing (more) land. The land owner or user will only stop doing this from an economic point of view when the compensation involved for conservation is equal to or higher than the net revenues of the alternative land use. It is only here when a PES-compensation scheme becomes effective and changes the land owner or user's behaviour. In Figure 44.2 this is the case between X^n and X^p. Paying a landowner or user for not clearing the land of forest between points 0 and X^n is not effective, as the land would not have been cleared without a payment. The compensation payment is too low from point X^p onwards to convince the farmer not to clear unless non-economic motives and considerations also play a role. In order to inform decision-making with regards to incentive compatible contract designs, a number of approaches have been suggested in the literature, such as screening contracts and procurement auctions (Ferraro, 2008).

It has furthermore been argued that it is easier to convince beneficiaries to participate in a PES scheme when the costs and benefits of ES provision are visible and quantifiable (Rojahn and Engel, 2005). Generally speaking, beneficiaries will be more inclined to pay for specific services than general conservation services. In identifying beneficiaries, it is also important to identify potential free riders who could benefit from the provision of services without contributing in the PES system. This too may affect contributors' support for the PES scheme or lead to their withdrawal from the scheme (Mayrand and Paquin, 2004). PES schemes must furthermore generate a sufficient and sustainable flow of revenues to land users to make sure that they implement and maintain land use changes that will generate the required ES (Pfaff et al., 2008). Payments under PES schemes must therefore be ongoing, as opposed to one-time payments, and be open-ended to allow them to last over time (Pagiola and Platais, 2002). Payment methods also matter for PES efficiency. An ES buyer may be indifferent about the mode of payment as long as the provider signs the contract. But the contract's sustainability may eventually depend on the unforeseen development effect of payments on household incomes, changes in consumption and demand for land and labour. These changes may have side-effects on conservation beyond what is stipulated in the contract. Wunder (2005) therefore argued to carefully think about or experiment with different payment modes, including cash versus non-cash and the periodicity of payment.

Judging the success of PES

Assessments of success and failure factors have also focused on PES as a mechanism for both environmental protection and poverty reduction. Bulte et al. (2008) show that tying together PES and poverty reduction may result in lower efficiency in meeting either objective. Thus, it may be better to focus programmes that concentrate on one or the other objective separately. This is closely related to the so-called Tinbergen (1952) rule, which says that ideally the number of policy instruments (in this case PES) should be equal to the number of policy targets. When deviating from this, the number of policy instruments should at least exceed the number of policy goals in order to be effective. If there are more policy targets than policy instruments, this may undermine reaching the different targets. Wunder et al. (2008) also hint at this in their comparative analysis of PES in developed and developing countries between user-financed and government-financed schemes using different criteria, including design, costs, environmental effectiveness and livelihood outcomes. The user-financed programmes were found to be better targeted, more closely adapted to local conditions and needs, with better monitoring and a greater willingness to enforce conditionality and fewer confounding side-objectives than government-financed programmes.

Although the literature on PES has grown exponentially, the factors that contribute to the functioning of the schemes are still not well understood. In an attempt to assess the institutional-economic driving forces behind the environmental performance of PES in a more quantitative manner, Brouwer et al. (2011) conducted a meta-analysis of close to 50 payment for watershed services (PWS) schemes, and found a number of systematic patterns despite the limited number of observations. Contrary to expectations, the price paid for ES delivery or the scale of implementation did not show up as significant factors. Instead, direct payments by downstream hydropower companies to upstream land owners for reduced sediment loads are identified as a successful PWS example, confirming findings in the literature about the important role of user-financed schemes. The careful selection of ES providers also played a significant role for this was expected to improve environmental outcomes. However, a wide variety of selection criteria appear to be applied in PWS schemes in practice, related to the location, quality and quantity of land, and the socio-economic status of potential participants. Moreover, having more intermediaries proved to have a negative impact on the environmental performance of PWS schemes, possibly due to an associated increase in transaction costs and delayed decision-making. Similarly, a negative effect on environmental performance was found if a scheme was voluntary. Although voluntary agreements have been propagated in the PES literature (Wunder, 2005), 20 percent of the PWS schemes analysed by Brouwer et al. (2011) appear to be mandatory in practice. If land users or land owners do not have full autonomy in participation, this may result in a lack of commitment from ES providers. However, the results from the meta-analysis suggest that mandatory participation is significantly more likely to be successful in reaching the environmental objectives involved than voluntary participation. Finally, community commitment and participation, i.e. if the contract is concluded with the entire community instead of individual service providers, had a positive effect on environmental outcome. A possible explanation for this may be that the community plays an important role in compliance and enforcement.

An important message from the study is the need for establishing quantifiable environmental objectives and monitoring progress towards reaching these objectives. Less than half of the schemes in the analysis used quantifiable indicators and monitored the impact of the schemes on environmental performance. In the vast majority of these cases, the indicators furthermore referred to the efforts put into scheme implementation, such as area with forest cover, instead of the actual impacts and outcomes of the scheme. Due to the absence of reliable longer-term scientific data and adequate cross-evaluation of the additional effects of PES on ES provision, the empirical evidence base of the impacts of these schemes remains very limited (Tognetti et al., 2010; Naeem et al. 2015).

Conclusion

Despite the global increase in PES schemes, the factors that contribute to the functioning of PES are still poorly understood. With respect to their environmental performance, there is a clear lack of international monitoring guidelines and time series data, including baseline level information, to assess the incremental impact of PES in a specific area or region. Similarly, from a behavioural economics point of view, how land owners or users respond to the provided financial incentives and the incentive compatibility of PES schemes has so far hardly ever been properly investigated. In some cases, this too is due to a lack of data or information asymmetry between the buyer of the ecosystem service and the supplier of the ecosystem service, in particular about the opportunity costs of land and labour. There is some evidence of the existence of a link between motivation and behaviour in the context of PES (e.g. Van Hecken and Bastiaensen, 2010; García-Amado et al., 2013), implying that non-monetary incentives such as improved extension and other social services may also play a role, but the evidence base for this is scarce. As a result, it is very difficult, if not impossible, to assess the costeffectiveness and economic efficiency of PES schemes worldwide.

Besides additionality, conditionality requires careful identification and definition of the provided ES involved. From an economic point of view, this means assessing the extent to which ES and biodiversity are suitable for 'marketing' based on their degree of public or private good characteristics. The causal link between a payment and the provision level of an ecosystem service plays a crucial role here in justifying and legitimizing payments. User-financed PES schemes seem to hold most promise and may financially be most viable in the long term as a market-based instrument, because they are less dependent on continued government funding as long as the conditionality criterion is met. Due to the often missing link between the efforts to preserve, for example, forests and the services these forests provide, it is hardly ever the case that payments relate directly to the provision level of a particular ES. Payments are more often based on the effort involved in providing ecosystem services than the provision level of a specific ecosystem service, due to a lack of scientific evidence and long-term monitoring of environmental change. Moreover, multiple services can often be provided at the same time, even though payments only relate to one (e.g. Phan et al., 2014).

In practice, PES schemes are developed and implemented in a wide variety of shapes and forms, often based on more than one single ecosystem objective. Both the ecological and socio-economic impacts of PES will be directly linked to the institutional design of the schemes. More research is needed in this area. Current and future PES schemes may benefit from improvements in their institutional design and embedding. This may include applying appropriate screening procedures for participation and varying financial compensation levels to enhance the effectiveness of the necessary land use changes and impacts on the provision level of the ES involved. However, these additional benefits will have to be compared to the transaction costs related to selection, monitoring and enforcement of the contractual agreements commonly concluded in PES schemes.

References

Brouwer, R., Tesfaye, A., and Pauw, P. (2011). Meta-analysis of institutional-economic factors explaining the environmental performance of payments for watershed services. *Environmental Conservation*, vol 38, no 4, pp 1–13.

Bulte, E. H., Lipper, L., Stringer, R., and Zilberman, D. (2008). Payments for ecosystem services and poverty reduction: concepts, issues and empirical perspective. *Environment and Development Economics*, vol 13, pp 245–254.

Daniels, A. E., Bagstad, K., Esposito, V., Moulaert, A., and Rodriguez, C. M. (2010). Understanding the impacts of Costa Rica's PES: are we asking the right questions? *Ecological Economics*, vol 69, pp 2116–2126.

Duncan, E. (2006). *Payments for Environmental Services. An Equitable Approach for Reducing Poverty and Conserving Nature*. Report, World Wide Fund For Nature. Available at: http://wwf.panda.org/about_our_earth/about_freshwater/freshwater_resources/?73340/Payments-for-Environmental-Services-An-equitable-approach-for-reducing-poverty-and-conserving-nature

Engel, S., Pagiola, S., and Wunder, S. (2008). Designing payments for environmental services in theory and practice: an overview of the issues. *Ecological Economics*, vol 65, no 4, pp 663–674.

Farley, J., and Costanza, R. (2010). Payments for ecosystem services: from local to global. *Ecological Economics*, vol 69, no 11, pp 2060–2068.

Ferraro, P. J. (2008). Asymmetric information and contract design for payments for environmental services. *Ecological Economics*, vol 65, no 4, pp 810–821.

García-Amado, L. R., Pérez, M. R., and García, S. B. (2013). Motivation for conservation: assessing integrated conservation and development projects and payments for environmental services in La Sepultura Biosphere Reserve, Mexico, Chiapas. *Ecological Economics*, vol 89, pp 92–100.

Greiber, T. (2009). *Payments for Ecosystem Services. Legal and Institutional Frameworks*. IUCN, Gland.

Kerr, J., and Jindal, R. (2007). USAID PES Brief 4: Impact evaluation of PES programs. In: USAID (ed.) *Lessons and Best Practices for Pro-poor Payment for Ecosystem Services*. USAID PES Source Book. Blacksburg, VA. SANREM CRSP. Available at: http://moderncms.ecosystemmarketplace.com/repository/moderncms_documents/PES_Sourcebook.1.pdf.

Mayrand, K., and Paquin, M. (2004). Payments for environmental services: a survey and assessment of current schemes. Commission for Environmental Cooperation of North America, Unisfera International Center, Montreal. Available at: http://www.cec.org/Storage/56/4894_PES-Unisfera_en.pdf

Muradian, R., Corbera, E., Pascual, U., Kosoy, N., and May, P. H. (2010). Reconciling theory and practice: An alternative conceptual framework for understanding payments for environmental services. *Ecological Economics*, vol 69, pp 1202–1208.

Naeem, S., Ingram, J. C., Varga, A., et al. (45 authors) (2015). Get the science right when paying for nature's services. *Science*, vol 13, pp 1206–1207.

Pagiola, S., and Platais, G. (2002). Payments for environmental services. Environment Strategy Notes no. 3. The World Bank Environmental Department, Washington, DC. Available at: http://siteresources.worldbank.org/INTEEI/Resources/EnvStrategyNote32002.pdf

Pfaff, A., Robalino, J. A., and Sanchez-Azofeifa, G. A. (2008). Payments for environmental services: empirical analysis for Costa Rica. Working Paper series SAN08-5, Terry Sanford Institute of Public Policy, Duke University, Durham, NC. Available at: http://sanford.duke.edu/research/papers/SAN08–05.pdf

Phan, T. D., Brouwer, R., and Davidson, M. (2014). The economic costs of avoided deforestation in the developing world: a meta-analysis. *Journal of Forest Economics*, vol 20, no 1, pp 1–16.

Porras, I., Grieg-Gran, M., and Neves, N. (2008). All that glitters. A review of payments for watershed services in developing countries. Natural Resource Issues, no 11. International Institute for Environment and Development, London.

Rojahn, A., and Engel, S. (2005). Direct payment for biodiversity conservation, watershed protection and carbon sequestration: contract theory and empirical evidence. Working paper, Institute for Environmental Decision, Chair of Environmental Policy and Economics. ETH, Zurich.

Salzman, J. (2005). Creating markets for ecosystem services: notes from the field. *NYUL Rev.*, vol 80, p 870.

Smith, M., de Groot, D., Perrot-Maîte, D., and Bergkamp, G. (2006). Establishing payments for watershed services. IUCN, Gland. Available at: http://data.iucn.org/dbtw-wpd/edocs/2006–054.pdf

Tinbergen, J. (1952). *On the Theory of Economic Policy*. North-Holland Publishing Company, Amsterdam.

Tognetti, S. S., Aylward, B., and Bruinzeel, L. A. (2010). Assessment needs to support the development of arrangements for payments for ecosystem services from tropical montane cloud forests. In: Bruijnzeel, L. A., Scatena, F. N. and Hamilton, L. S. (eds) *Tropical Montane Cloud Forests: Science for Conservation and Management*. Cambridge University Press, Cambridge UK.

Van Hecken, G., and Bastiaensen, J. (2010). Payments for ecosystem services in Nicaragua: do market-based approaches work? *Development and Change*, vol 41, no 3, pp 421–444.

Vatn, A. (2010). An institutional analysis of payments for environmental services. *Ecological Economics*, vol 69, pp 1245–1252.

Wunder, S. (2005). Payments for environmental services: some nuts and bolts. Occasional Paper no. 42. Center for International Forestry Research (CIFOR), Bogor, Indonesia.

Wunder, S., Engel, S., and Pagiola, S. (2008). Taking stock: a comparative analysis of payments for environmental services programs in developed and developing countries. *Ecological Economics*, vol 65, pp 843–852.

45

MAXIMIZING BIODIVERSITY AND ECOSYSTEM SERVICE BENEFITS IN CONSERVATION DECISION-MAKING

Hedley S. Grantham, Rosimeiry Portela, Mahbubul Alam, Daniel Juhn and Lawrence Connell

Introduction

Despite important conservation efforts over the last few decades, such as a growth of protected areas globally (Watson et al., 2014), biodiversity has been subject to an unprecedented decline, with an estimated 39 percent decline of terrestrial and marine species, and 76 percent of freshwater species, from 1970 to 2008 (WWF, 2014). Habitat loss, fragmentation, and degradation continue to be the main drivers of biodiversity loss, and these remain largely unabated (CBD, 2011). Traditionally, this conservation problem has been framed as trying to reduce the loss of biodiversity as much as possible (Myers et al., 2000; Mace, 2014). The solution to this problem has predominantly been through site-based nature protection strategies (particularly protected areas), designed in a way that aims to cost-effectively reduce threats and maximize biodiversity persistence (Wilson et al., 2009). Principally, conserving biodiversity has been justified based on its intrinsic value (Mace, 2014), but with the recognition that biodiversity loss might mean a loss of potential option value for people in the future (e.g. new medicines, resilience to climate change) (Callicott, 2006).

Departing from an exclusive focus on biodiversity, the conservation movement over the last decade or so has expanded nature conservation to incorporate the concept of human welfare (Naughton-Treves et al., 2005; Mace, 2014). Indeed, the broad recognition of the concept of ecosystem services and its application for conservation and sustainable management of natural resources is nothing short of extraordinary. Important advances in the recognition of the relevance of ecosystem services to human well-being have been achieved with global efforts such as the Millennium Ecosystem Assessment (MA, 2003), which expanded the concept beyond the scientific community, reaching policy- and decision-makers (Carpenter et al., 2009). Subsequently, The Economics of Ecosystems and Biodiversity (TEEB) made one of the most compelling cases for global action on biodiversity protection: economic growth depends on it (Kumar, 2010). The effort called for society to make nature's values visible, and for decision-makers and the business community to assess, communicate and take actions that incorporate the role of biodiversity and ecosystem services, from freshwater provision to climate regulation, in economic activities. That built momentum for the Intergovernmental Platform on Biodiversity and

Ecosystem Services (IPBES), which seeks to gather scientific information and advance methods and analytical approaches to measuring biodiversity and ecosystem services as a means to strengthen dialogue between the scientific community and stakeholders.

A human-centric approach to conservation, focusing on ecosystem services, has created a debate on how this might affect outcomes for biodiversity (Soule, 2013; Kareiva, 2014). At the core of the debate is the issue of whether scarce resources for biodiversity conservation, such as threatened species with little economic value, will be reallocated to ecosystem services in the future, and particularly, toward ecosystem services with high utilitarian value for people in human-dominated landscapes and seascapes. Currently, there is little evidence to suggest this one way or another.

Relatedly, the concern around "new conservation" has been whether ecosystem service-focused conservation leads to co-benefits for biodiversity. At the global scale, Turner et al. (2007) found that priority biodiversity conservation maps had a disproportionate share of terrestrial ecosystem services values, though overlap varied among regions. Larsen et al. (2011) asserted that the biodiversity and ecosystem service relationships are complex, with both synergies and trade-offs. However, even for individual ecosystem services, there has been research at sub-national scales to suggest that the spatial values and benefits of protecting individual ecosystem services align neither with each other nor biodiversity (Chan et al., 2006). Understanding the benefits of biodiversity conservation and various types of ecosystem services thus can help understand the co-benefits and conflicts of different conservation interventions (Chan et al., 2011).

To maximize conservation objectives, there is a need to assess multiple possible interventions

Protected areas have long been the primary strategy for biodiversity conservation, and the basis for most conservation priority-setting exercises that seek to direct conservation investment (Watson et al., 2014). Increasingly, other types of land uses are being recognized for their contribution to conservation objectives, such as forestry concessions and agricultural areas (e.g. Wilson et al., 2010). Many ecosystem services have high economic value (Costanza et al., 2014) and our approach here encourages the consideration of other types of interventions beyond protected areas, including market-based instruments (e.g. payments for ecosystem services, conservation concessions, subsidies, taxes, trading schemes). This reflects societies' recognition that market-based interventions, such as payment for ecosystem services, might particularly suit some ecosystem services, like those based on water (e.g. Goldman-Benner et al., 2012) and carbon (e.g. Wendland et al., 2010), when there are favourable market, cultural, and governance preconditions (Wunder, 2013), (see also Brouwer, 2016). Interventions like payments might generate financially sustainable strategies that ideally free up limited resources toward biodiversity conservation (although may instead crowd out direct investment in conservation), and reflects a scenario that the benefits of conservation can exceed the economic costs (Balmford and Whitten, 2003). Some market-based approaches might also work well for biodiversity objectives (Pirard et al., 2012). We encourage here a holistic, integrated approach which recognizes that, for conservation interventions to be planned effectively, they often need to be planned while considering development trade-offs and economic efficiency.

To maximize benefits, it is important to prioritize the set of interventions being considered. A core principle of conservation prioritization is to measure potential interventions based on the performance measures, and not on the underlying set of values (Game et al., 2013). For example, it is not recommended to prioritize the placement of protected areas for biodiversity solely based on biodiversity values, but to include factors like financial costs and being realistic where opportunities lie (Game et al., 2013). Busch and Grantham (2013) explored the biodiversity and climate mitigation benefits of three conservation strategies in Indonesia: 1) protected areas based on achieving targets for species representation, 2) carbon payments based on maximizing emission reductions from forests, and considering constraints such as land use profitability, and 3) biodiversity payments based on maximizing biodiversity benefits cost-effectively. Payments were based on similar payment constraints for each conservation strategy. They found that each strategy leads to different outcomes for biodiversity and climate mitigation benefits. This is because there are likely to be other constraints to consider when applying various types of conservation strategies (e.g. financial costs, opportunity costs), and the spatial benefits of biodiversity and carbon are not necessarily spatially aligned (see Busch and Grantham, 2013). This example suggests that there is indeed a risk of biodiversity benefits being reduced if conservation funds were to be allocated to carbon payments. Given that carbon payments might generate funds independently of biodiversity budgets (e.g. Norway-Indonesia letter of intent on REDD, 2010), these expected carbon payments should therefore be considered when biodiversity planning based on potential new protected areas and/or biodiversity payments. A coordinated, comprehensive, and complementary set of conservation interventions that considers multiple potential management options and full set of values will likely lead to more effective conservation than designing them individually or relying on one or the other. The next section discusses broadly how to do that.

A science-based approach for maximizing biodiversity and ecosystem services benefits in conservation decision-making

Here we describe a broad, generic, science-based approach towards the design of complementary sets of conservation interventions, one that aims to understand multiple sets of societal objectives around nature conservation outcomes, including both biodiversity and ecosystem services benefits; a range of management interventions to be considered to achieve objectives; and science to support good decision-making that maximizes objectives and looks for synergies and co-benefits, but also recognizes trade-offs. This approach allows the concept of ecosystem services and "new conservation" to provide support to biodiversity conservation, and vice versa, through clear understanding of co-benefits. We use a generic set of steps well-established for identifying conservation priorities (Wilson et al., 2009; Gregory et al., 2012; Marcot et al., 2012; Martinez-Harms et al., 2015). We discuss each step in relation to maximizing objectives for biodiversity and ecosystem services, and highlight key research areas that could increase our ability for identifying the best set of interventions.

Problem formulation

For any type of conservation decision analysis, it is important to start with an understanding of the reasons why conservation interventions might be necessary, contextual information for implementing future interventions, and identifying potential solutions. In many situations this might start with a scoping study or similar effort. Here we do not attempt to comprehensively guide what might be included in such a study, but key information likely includes: understanding the sociocultural setting; important economic sectors; decision-making authority and process by which decisions are made and enforced; conservation, natural resource management, and development strategies and plans; human development issues; tenure systems, and other governance opportunities and constraints. As part of scoping, an assessment of the institutional, legal, and policy context relevant to biodiversity and ecosystem services is particularly important, as it provides a basis for understanding potential interventions. Understanding this might also help recognize what interventions are currently being implemented, and any monitoring and evaluation that might have been applied to understand effectiveness.

Ideally, the key outcomes from proper problem formulation will include an understanding of problems and goals to address, and identifying potential interventions for helping solve these problems and achieve these goals. Also likely important is identifying: scale of assessment; important biodiversity and ecosystem services values; drivers and actors of change, and trends in these values; resources available for implementation; and other contextual information. The process of prioritization is likely as important as the science. The importance of planning processes and stakeholder involvement during any conservation planning exercise is well-established and discussed extensively elsewhere (e.g. Reyers et al., 2010; Schlossberg and Shuford, 2005).

Objectives and performance measures

Objectives are concise statements of the fundamental interests that could be affected by a decision (Gregory et al., 2012). Performance measures are quantitative expressions of the objectives that are used to evaluate the performance of an intervention in relation to the objectives (Marcot et al., 2012). They are based on the predicted outcome of an intervention. Predicting the performance outcomes of different interventions for biodiversity and ecosystem services values allows a measure of their potential co-benefits. It also provides an insight into what values might be missed if taking a purely biodiversity or ecosystem services approach to conservation.

For biodiversity, objectives and performance measures are fairly well-established, at least for biodiversity distribution (e.g. species and ecosystems). People often place the greatest value on maintaining a species' population or habitat size at a level that is sufficient to ensure the continued persistence of the species, with diminishing returns to increases above this threshold (Gaston and Fuller, 2008). This is similar for ecosystem concern and protection (Keith et al., 2015). Other than biodiversity distribution, there are also other types of values that can indicate biodiversity importance, particularly the location of important biological processes (e.g. bird migration sites), which are more difficult to identify and measure (Senior et al., 2014). Several guidelines have been developed that standardize the identification of valued biodiversity (e.g. IUCN Species and Ecosystems Red lists, and Key Biodiversity Areas). Performance is typically measured in the area of a species or ecosystem, important place, or in some cases as species population size.

Objectives for ecosystem services are complex, and there has not been much previous success in defining these. In fact, Martinez-Harms et al. (2015) found in their literature review of past ecosystem service assessments that less than 10% even stated their objectives. Properly defined objectives considering both biophysical and socioeconomic components of ecosystem services are important, and more research is needed on this. One might consider supply-side (e.g. restore or maintain habitat), and/or use-side (e.g. reducing pressures and abatement of threats) objectives for an ecosystem service. Objectives can also be based on the types of beneficiaries of ecosystem services (e.g. focus on financially poor people or farmers), and on ecosystem service use now and/or in the future. There is also hundreds of years of work on natural resource management objectives that might be considered.

Performance measures for ecosystem services have also not been well defined in the past. Martinez-Harms et al. (2015) found in their review that only 8% of ecosystem services assessments included performance measures for ecosystem services, and those that did primarily based them on biophysical values independent of their use. Biophysical estimates of ecosystem services are a matter of underlying physical parameters, done by empirical, as well as modelling approaches, and can be a prerequisite step for the monetary valuation. Measurements of the biophysical flows of ecosystem services can be modelled using spatially explicit quantitative tools (e.g. InVEST, ARIES, MIMES, etc.) (Bagstad et al., 2013). The use of these models to assess ecosystems and their ability to provide ecosystem services has evolved tremendously in the last few years. However, there still remain numerous challenges, particularly for how to include the relationship of supply and use of an ecosystem service, and identify where benefits of conservation outweigh costs. Very rarely have performance measures for ecosystem services been transformed into non-linear values or utility functions. In other words, assuming, for example, there are diminishing returns to the benefits of an intervention e.g. restoration of a forest for firewood for a local community.

One measure of the economic value or utility of ecosystem services protection is monetary valuation, as this has direct relevance for decision-making, particularly for understanding potential market-based instruments such as carbon trade (Hepburn, 2012) and payments for ecosystem services (Farley and Costanza, 2010). Valuation is not only used to provide an estimate of the monetary value of some goods and services, but those stocks of assets (i.e. the aggregate of future flows of services given a certain discount rate). Economic valuation generates information that is beyond quantifying the benefits provided by nature. Studies have shown how values generated by conventional economic accounting change drastically when ecosystem services are taken into account (Costanza et al., 2014). Biodiversity and habitats around the world are being degraded and converted to other land uses for economic reasons. Conversion of forests, for example, into non-forest use comes with opportunity costs (i.e. forgone opportunities based on the forest). It is therefore important to measure what these opportunity costs are. The application of economics to conservation can be used to generate information on opportunity costs and helps make more informed policy and land use decisions.

Despite huge interests from both policy and science, little work has been done to improve valuation methods. Generally speaking, market-based approaches are normally preferred and are better suited to provisioning services. However, many ecosystem services are typically public goods, and are not transacted in markets, particularly many regulating and cultural services. But there are real connections between these ecosystem services and economy. For example, pollination benefits crop fields but is not usually considered as an input cost to production. Absent a valuation exercise, the value of pollination is only visible in the market if pollinator habitat is lost and farmers must rent, for example, managed honeybees to replace pollination services from nature. Similarly, many other regulating and cultural services (and some provisioning services such as medicinal plants) have indirect links to market and/or non-use values (e.g. bequest value). To generate credible information through economic valuation, these links have to be identified and valued.

The application of bioeconomic models can be used to explore the potential outcomes and performance of some interventions. These types of models allow the integration of important biophysical information and ecological processes with economic decision behaviour. They are particularly useful for understanding the potential outcomes of alternative market-based approaches like payments for ecosystem services. For example, Busch et al. (2011) explored the outcomes of different REDD+ (i.e. carbon payment mechanisms) scenarios across 85 countries and assessed the co-benefits of policies to biodiversity using bioeconomic models. The model considered outcomes of potential payments based on land use dynamics and opportunity costs to foregone production.

Decision analysis

The aim of a decision analysis is to help prioritize between interventions that best achieve objectives. There are numerous approaches to decision analysis, from simple (e.g. consequence tables, described in Gregory et al., 2012) to more complex ones (e.g. multi-objective portfolio optimization, see Bryan (2010) described below).

Before exploring any priority setting analysis, it can be useful to first determine the overall feasibility of any potential intervention within an area. If the enabling conditions, such as governance, institutional capacity, regulatory frameworks, and socioeconomic characteristics of the region are not sufficient (e.g. CATIE and Global Mechanism of the United Nations Convention to Combat Desertification, 2011), it might rule out options to be included in the evaluation a priori.

Also, before choosing between interventions, it might be useful to tailor the design of the intervention. Spatial prioritizations can be used to identify where objectives can be achieved most effectively for an intervention. This can be for biodiversity and/or ecosystem services, or both and include other factors that help with prioritization (e.g. opportunities and constraints). For example, Wendland et al. (2010) prioritized areas in Madagascar for payments for ecosystem services, integrating biodiversity and ecosystem services benefits using a multi-criteria approach. As part of the prioritization, they incorporated opportunity costs of agricultural values, and identified where the most impact will be by including avoided deforestation. Often spatial prioritizations are applied without considering the specific intervention (e.g. national park, conservation concession), but on a general outcome of preserving the values within that location. These types of analyses aim to identify solutions that maximize objectives and assume that specific types of interventions can be tailored later to that site. A good example is Egoh et al. (2010), who applied several spatial prioritizations approaches in an area of South Africa using an optimization approach that either: 1) achieved specific targets around the protection of biodiversity and ecosystem services cost-effectively; or 2) explored trade-offs in maximizing objectives for a fixed budget (see also Egoh et al., 2016).

There are several approaches to prioritize between alternative interventions. A method gaining popularity in the conservation literature is cost-effectiveness analysis, which can be a useful tool to prioritize between interventions in different places. For each intervention, using the performance measures discussed in the section above, benefit-cost ratios can be calculated to determine which interventions where provide the highest returns. Measurement of co-benefits can be included in this analysis (e.g. Budiharta et al., 2014). All these approaches are amenable to incorporate various types of objectives including social preferences, which can be varied depending on stakeholder interest, and these help explore trade-offs between competing objectives.

Multi-intervention optimization can prioritize between alternative sets of interventions, but the methods can be complex. Within a region of Australia, Bryant (2010) used a cost-effectiveness measure across an impressive 46 types of interventions considering 23 ecosystem services and biodiversity objectives. The author applied a portfolio algorithm to select the best set of interventions that maximized benefits and minimized financial costs. Results have informed AU$46 million of investment. Wilson et al. (2010) optimized across different land use zones for the achievement of biodiversity objectives while minimizing the opportunity costs to timber harvesting by estimating the contribution of different types of land uses to biodiversity objectives. Grantham et al. (2013) optimized across different types of zones within a marine protected area network for achieving targets for biodiversity, fisheries, and tourism objectives.

The use of scenario planning is a complementary approach to optimization. Developing a manageable set of alternative scenarios allows a more detailed assessment of each scenario's performance, such as the application of ecosystem service models with complex spatial and/or temporal dynamics that would otherwise be difficult to incorporate into an optimization framework. The outputs of prioritizations and optimizations discussed above can form the basis for some of the scenarios. Scenarios are also very amendable to stakeholder participation in their identification. Palomo et al. (2011) demonstrate how to achieve participatory scenario planning that factors in broad sets of interventions as potential futures for multiple objectives. There are numerous studies in the literature of spatial planning process that compare alternative scenarios of land uses or sea uses and their predicted impacts on biodiversity, ecosystem services and other values (e.g. agricultural or fisheries production) (e.g. Nelson et al., 2009, Law et al., 2015).

Embedding science into decision-making

Integrating policy-makers, managers, scientists, and stakeholders into a scientific process is an important, but complex, process, particularly when there are conflicting values, different opinions, and perceptions. It is beyond the scope of this chapter to discuss how to effectively embed science into management and policy, and the importance of social learning processes and adaptive management (but see Cowling et al., 2008 for a good introduction). We do want to highlight that this can be time- and effort-consuming. Ideally, the process involves participation of different parties to help define the problems, goals, and objectives, investigate scenarios and measures, feasibility of possibilities, etc (see also Fish et al., 2016). Given the uncertainties associated with responses of human and ecological systems, it is also important to apply precautionary and adaptive mechanisms to respond to surprises/unexpected events resulting from the implementation of different interventions.

Discussion

The four steps discussed above will likely reveal important justifications for a portfolio approach to conservation for achieving biodiversity and ecosystem services objectives. We do not suggest that our approach will stop funds and efforts being re-allocated from biodiversity objectives towards ecosystem services ones, and vice versa. Naturally, not all conservation goals are likely to be achieved in any given region. However, what we have described here allows transparent choices to be made with the full understanding of the range of potential interventions and their benefits, costs, and trade-offs.

Nature conservation requires the implementation of a broad set of policies and management actions, ranging from direct regulations, such as the establishment of protected areas, to the creation of markets to transact ecosystem services. Market-based approaches such as payments for ecosystem services, if properly designed and implemented, could create opportunities for the generation of financial resources to support conservation. This is particularly true within human-dominated landscapes, where ecosystems generate goods and services for people's livelihoods, or for global public goods like carbon and climate mitigation. Although it is likely that valued biodiversity and ecosystem services will not always coincide in space, careful consideration of opportunities can be key for a more efficient use of scarce financial resources toward achieving multiple policy objectives.

The steps proposed here require a multidisciplinary team and sometimes high-level expertise, which can often be obstructed by a lack of budget and capacity. There can also be institutional constraints: relevant government and non-government agencies may not be adequately prepared to address these challenges and the breadth of strategies considered here. Existing policies and regulatory mechanisms might not be well-aligned to adopt these cutting-edge frameworks.

These challenges notwithstanding, what we propose is a more holistic, integrated approach to conservation planning, one that relies on science to support good decision-making, considering objectives and important ecological and socioeconomic constraints and trade-offs. This is a transformative process, and one that better promises outcomes than relying on a handful of traditional conservation approaches.

Acknowledgements

We thank Aaron Bruner, James Watson, and Jonah Busch for helpful reviews on an earlier version of this chapter, and the Gordon and Betty Moore Foundation for support.

References

Bagstad, K.J., Semmens, D., Waage, S., and Winthrop, R. (2013). A comparative assessment of tools for ecosystem services quantification and valuation. *Ecosystem Services*, vol 5, pp 27–39.

Balmford, A., and Whitten, T. (2003). Who should pay for tropical conservation, and how could the costs be met? *Oryx*, vol 37, no 2, pp 238–250.

Brooks, T. M., Mittermeier, R. A., da Fonseca, G.A.B., et al. (8 authors) (2006). Global biodiversity conservation priorities. *Science*, vol 313, pp 58–61.

Bryan, B.A. (2010). Development and application of a model for robust, cost-effective investment in natural capital and ecosystem services. *Biological Conservation*, vol 143, no 7, pp 1737–1750.

Brouwer, R. (2016). Payments for ecosystem services. In: Potschin, M., Haines-Young, R., Fish, R. and Turner, R.K. (eds) *Routledge Handbook of Ecosystem Services*. Routledge, London and New York, pp 548–553.

Budiharta, S., Meijaard, E., Erskine, P.D., et al. (6 authors) (2014). Restoring degraded tropical forests for carbon and biodiversity. *Environmental Research Letters*, vol, 9, no 11, p 114020.

Busch, J., Godoy, F., Turner, W.R., and Harvey, C. A. (2011). Biodiversity co-benefits of reducing emissions from deforestation under alternative reference levels and levels of finance. *Conservation Letters*, vol 4, no 2, pp 101–115.

Busch, J., and Grantham, H. S. (2013). Parks versus payments: reconciling divergent policy responses to biodiversity loss and climate change from tropical deforestation. *Environ. Res. Lett.*, vol 8, p 034028.

Callicott, J. B. (2006). Conservation values and ethics. In: Groom, M., Meefe, G. and Carroll, C.R. (eds) *Principles of Conservation Biology, 3rd edition*. Sinauer, Sunderland MA.

Carpenter, R. S., Mooney, H. A., Agard, J., et al. (15 authors) (2009). Science for managing ecosystem services: beyond the Millennium Ecosystem Assessment. *PNAS*, vol 106, no 5, pp 1305–1312.

CATIE and Global Mechanism of the United Nations Convention to Combat Desertification (2011). *Scorecard Tool to Assess Applicability of Incentive and Market Based Mechanisms*. CBD, Montreal.

Chan, K.M., Hoshizaki, L., and Klinkenberg, B. (2011). Ecosystem services in conservation planning: targeted benefits vs. co-benefits or costs? *PLoS One*, vol 69, e24378.

Chan, K.M., Shaw, M.R., et al. (5 authors) (2006). Conservation planning for ecosystem services. *PLoS Biology*, vol 4, no 11, e379.

Convention on Biological Diversity (CBD) (2011). Quick guide to the Aichi Biodiversity Targets. Target 5, habitat loss halved or reduced. Available at: http://www.cbd.int/doc/strategic-plan/targets/T5-quick-guide-en.pdf

Costanza, R., de Groot, R., Sutton, P., et al. (8 authors) (2014). Changes in the global value of ecosystem services. *Global Environmental Change*, vol 26, pp 152–158.

Cowling, R. M., Egoh, B., Knight, A. T., et al. (9 authors) (2008). An operational model for mainstreaming ecosystem services for implementation. *PNAS*, vol 105, no 28, pp 9483–9488.

Egoh, B., Bengtsson, J., Lindborg, R., et al. (7 authors) (2016). The importance of grasslands in providing ecosystem services: opportunities for poverty alleviation. In: Potschin, M., Haines-Young, R., Fish, R., and Turner, R. K. (eds) *Routledge Handbook of Ecosystem Services*. Routledge, London and New York, pp 421–441.

Egoh, B.N., Reyers, B., Carwardine, J., et al. (11 authors) (2010). Safeguarding biodiversity and ecosystem services in the Little Karoo, South Africa. *Conservation Biology*, vol 24, no 4, pp 1021–1030.

Farley, J., and Costanza, R. (2010). Payments for ecosystem services: from local to global. *Ecological Economics*, vol 69, no 11, pp 2060–2068.

Fish, R., Saratsi, E., Reed. M., and Keune, H. (2016). Stakeholder participation in ecosystem service decision-making. In: Potschin, M., Haines-Young, R., Fish, R. and Turner, R.K. (eds) *Routledge Handbook of Ecosystem Services*. Routledge, London and New York, pp 256–270.

Game, E.T., Kareiva, P. and Possingham, H.P. (2013). Six common mistakes in conservation priority setting. *Conservation Biology*, vol 27, pp 480–485.

Gaston, K. J., and Fuller, R. A. (2008). Commonness, population depletion and conservation biology. *Trends Ecol. Evolut.*, vol 23, pp 14–19.

Goldman-Benner, R.L., Benitez, S., Boucher, T. et al. (8 authors) (2012). Water funds and payments for ecosystem services: practice learns from theory and theory can learn from practice. *Oryx*, vol 46, no 1, pp 55–63.

Grantham, H.S., Agostini, V.N., Wilson, et al. (10 authors) (2013). A comparison of zoning analyses to inform the planning of a marine protected area network in Raja Ampat, Indonesia. *Marine Policy*, vol 38, pp 184–194.

Gregory, R., Failing, L., Harstone, M., et al. (6 authors) (2012). *Structured Decision Making: A Practical Guide to Environmental Management Choices*. John Wiley & Sons, Chichester.

Hepburn, C. (2012). Carbon trading: a review of the Kyoto mechanisms. *Annual Review of Environment and Resources*, vol 32, pp 375–393.

Kareiva, P. (2014). New conservation: setting the record straight and finding common ground. *Conservation Biology*, vol 8, no 3, pp 627–888.

Keith, D.A., Rodríguez, J.P., Brooks, T.M., (16 authors) (2015). The IUCN red list of ecosystems: motivations, challenges and applications. *Conservation Letters* (in press).

Kumar, P. (2010). *The Economics of Ecosystems and Biodiversity (TEEB)*. Earthscan, London.

Larsen, F.W., Londoño-Murcia, M.C., and Turner, W.R. (2011). Global priorities for conservation of threatened species, carbon storage, and freshwater services: scope for synergy? *Conservation Letters*, vol 4, no 5, pp 355–363.

Law, E. A., Meijaard, E., Bryan, B. A., et al. (6 authors) (2015). Better land-use allocation outperforms land sparing and land sharing approaches to conservation in Central Kalimantan, Indonesia. *Biol Conserv*, vol 186, pp 276–286.

MA (2003). *Ecosystems and Human Well-Being: A Framework for Assessment*. Inland Press, Washington DC, Covelo CA, and London.

Mace, G.M. (2014). Whose conservation. *Science*, vol 345, no 6204, pp 1558–1560.

Marcot, B.G., Thompson, M.P., Runge, M. C., et al. (9 authors) (2012). Recent advances in applying decision science to managing national forests. *Forest Ecology and Management*, vol 285, pp 123–132.

Martinez-Harms, M.J., Bryan, B.A., Balvanera, P., et al. (7 authors) (2015). Making decisions for managing ecosystem services. *Biological Conservation*, vol 184, pp 229–238.

Myers, N., Mittermeier, R.A., Mittermeier, C. G., et al. (5 authors) (2000). Biodiversity hotspots for conservation priorities. *Nature*, vol 403, no 6772, pp 853–858.

Naughton-Treves, L., Holland, M. B., and Brandon, K. (2005). The role of protected areas in conserving biodiversity and sustaining local livelihoods. *Annual Review of Environment and Resources*, vol 30, pp 219–252.

Nelson, E., Mendoza, G., Regetz, J., et al. (14 authors) (2009). Modeling multiple ecosystem services, biodiversity conservation, commodity production, and tradeoffs at landscape scales. *Frontiers in Ecology and the Environment*, vol 7, no 1, pp 4–11.

Norway-Indonesia letter of intent on REDD (2010). Letter of Intent between the Government of the Kingdom of Norway and the Government of the Republic of Indonesia on "Cooperation on reducing greenhouse gas emissions from deforestation and forest degradation". Available at: http://www.forestsclimatechange.org/fileadmin/photos/Norway-Indonesia-LoI.pdf

Palomo, I., Martín-López, B., López-Santiago, C., et al. (4 authors) (2011). Participatory scenario planning for protected areas management under the ecosystem services framework: the Doñana social-ecological system in southwestern Spain. *Ecology and Society*, vol 16, no 1, p 23.

Pirard, R. (2012). Market-based instruments for biodiversity and ecosystem services: a lexicon. *Environmental Science & Policy*, vol 19, pp 59–68.

Reyers, B., Roux, D. J., Cowling, R. M. et al. (6 authors) (2010). Conservation planning as a transdisciplinary process. *Conservation Biology*, vol 24, no 4, pp 957–965.

Schlossberg, M., and Shuford, E. (2005). Delineating "public" and "participation". *URISA Journal*, vol 16, no 2, pp 15–26.

Senior, M.J., Brown, E., Villalpando, P., and Hill, J. K. (2014). Increasing the scientific evidence base in the "High Conservation Value" (HCV) approach for biodiversity conservation in managed tropical landscapes. *Conservation Letters*, online Dec/2014, DOI: 10.1111/conl.12148

Soule, M. (2013). The "new conservation". *Conservation Biology*, vol 27, pp 895–897.

Turner, W., Brandon, K., Brooks, T. M., et al. (6 authors) (2007). Global conservation of biodiversity and ecosystem services. *BioScience*, vol 57, no 10, pp 868–873.

UNSD (2012). *System of Environmental Accounting 2012 – Experimental Ecosystem Accounting*. The United Nations Statistical Division (UNSD). Available at: http://unstats.un.org/unsd/envaccounting/seea.asp, accessed 16.09.2015

Watson, J.E., Dudley, N., Segan, D.B., et al. (4 authors) (2014). The performance and potential of protected areas. *Nature*, vol 515, no 7525, pp 67–73.

Wendland, K.J., Honzák, M., Portela, R., et al. (6 authors) (2010). Targeting and implementing payments for ecosystem services: opportunities for bundling biodiversity conservation with carbon and water services in Madagascar. *Ecological Economics*, vol 69, no 11, pp 2093–2107.

Wilson, K.A., Carwardine, J., and Possingham, H.P. (2009). Setting conservation priorities. *Annals of the New York Academy of Sciences*, vol 1162, no 1, pp 237–264.

Wilson, K.A., Meijaard, E., Drummond, S., et al. (16 authors) (2010). Conserving biodiversity in production landscapes. *Ecological Applications*, vol 20, no 6, pp 1721–1732.

World Wildlife Fund (2014). *Living Planet Report 2014: Species and Spaces, People and Places*. McLellan, R., Iyengar, L., Jeffries, B. and Oerlemans, N. (eds.).

Wunder, S. (2013). When payments for environmental services will work for conservation. *Conservation Letters*, vol 6, no 4, pp 230–237.

46

BRIDGING THE GAP BETWEEN ECOSYSTEM SERVICES AND LANDSCAPE PLANNING

Paul Opdam

The concept of ecosystem services has great potential in the planning of landscape change. For example, it frames the relation between humans and nature according to a utilitarian vision. Nature is no longer exclusively seen as an object of value that is threatened by economic and demographic growth and therefore should be protected from further damage, but as a life-support system providing benefits that can be utilized by people. This opens a new perspective on nature and biodiversity in the spatial planning of landscapes.

Consider landscapes as geographically and economically cohesive areas which are adapted by humans to meet new challenges or desires. In such an interpretation of landscape, ecosystem services provided by the landscape add benefits and value to those who live and work there, visit the area for leisure, or extract resources from it. Optimizing such benefits may then become a goal for adapting landscapes. The term 'ecosystem services' also provides a binding concept that brings all the values and benefits that humans attribute to their landscape under the same umbrella (Fürst et al., 2014). Furthermore, the application of ecosystem services moves the human-nature relationship beyond a discourse about protection into one about sustainability. Sustainability then means that the landscape is managed in such a way that it retains the potential to provide these benefits to current and future societies. Therefore, the ecosystem services concept contributes to developing sustainable landscape planning (Potschin and Haines-Young, 2013). Yet, in spite of these positive perspectives, the worlds of ecosystem services and environmental planning are far apart (Termorshuizen and Opdam, 2009; von Haaren and Albert, 2011, see also von Haaren et al., 2016). This chapter seeks to bridge this gap.

Why is it that the concept of ecosystem services has not yet been discovered in landscape planning? And how should the research field of ecosystem services develop to become relevant to applications in landscape planning? The current mainstream of ecosystem services research seems to consider the concept as the basis for an alternative strategy to conserve nature, biodiversity, and natural resources (Chan et al., 2006). Most of this work is organized around two lines of research (Martinez-Harms and Balvanera, 2012). The first is about developing diagnostic methods to identify which services are provided where in the landscape. The second pursues methods to value these services in monetary terms. These two lines of work generate academic discussions about the right classification of ecosystem services, the problem of double counting of values, and estimates of economic values of services which are not treated in the real market

(Schröter et al., 2014). Applications of such methods in the context of land use planning and conservation planning are numerous. These applications have often the character of a policy assessment or a comparative assessment of scenarios designed by scientists. This work is, however, not so relevant to landscape planning.

Before understanding why work on ecosystem services is not always relevant to landscape planning, the nature of the landscape planning process first needs to be specified. I will focus on a type of landscape planning which is becoming popular in many countries around the world. In this type of landscape planning, the local community has an important role in making decisions by deliberating about problems and aims, negotiating solutions and organizing collective action. Such a process can be completely independent of the government, in which case we talk about 'self-governance' (Ostrom, 2009). More often it will be some form of cooperation with the government ('co-governance') such as when the government pays for flower strips in field margins). The main characteristic of these forms of landscape governance is that values are negotiated at the community level and measures in the landscape are driven by motives of individuals and groups in the area, instead of being determined by the government (see also Kenter, 2016). Moreover, such values are connected to multiple dimensions of the human-nature relationship, and not limited to a sectorial interest (such as biodiversity or storm water protection). In this context, valuation is not so much an economic as a social activity (Liu and Opdam, 2014): if awareness is growing that a higher value of landscape benefits can be gained by changing the landscape, actors may decide that adaptation is 'profitable'. In this way, valuation elicits adaptation. Taking action may happen at the level of the individual property, but for sustainable landscape management, collaboration and collective action are mandatory (Levebvre et al., 2014).

Thus it is easy to see why current work on ecosystem services, which is mainly developed within the domains of land use planning and conservation planning, has not captured the attention of landscape planners. I give a few reasons:

1 Landscape planning is about locally perceived values, including economic but also social, cultural, and sustainability values. These values are created and negotiated in a collaborative process within the social network of an area (Liu and Opdam, 2014). Values determined by scientists, on a national or regional level, are not recognized as relevant and legitimate by local communities. Current valuation methods do not provide information that supports collaborative decision-making: they are based on rational rules set in a linear line of reasoning (cf. Haines-Young and Potschin, 2014).
2 Landscape planning is about change: the focus of a planning group is on benefits that can be created or improved rather than on services actually provided. Instead of mapping the actual level of provisioning, there is a need for spatially explicit information about where the ecological conditions of the landscape allow effective measures to increase landscape benefits.
3 The collaborative landscape planning process is better stimulated by focussing on opportunities that can be created in the area than by an analysis of world-level problems.
4 Landscape planning is a creative process: instead of rigid assessment approaches there is a need for flexible methods by which desired values can be transformed into a landscape design that fits the views and preferences of the local actors.
5 Ecosystem services are associated with nature protection, not with multifunctional rural and urban and peri-urban landscape where most people live.

To foster research in ecosystem services that is more relevant to planning science and practice, the science of ecosystem services has to proceed according to a few of the principles of collaborative landscape planning. First, new approaches and tools are required that foster *ownership*. Planning groups need to be able to develop their own feeling that using ecosystem services is in their interests, that it solves their problem, and that it is their selection of services that contributes to their solution. Therefore, knowledge tools should be open and flexible enough as to allow actors to make normative choices (such as choosing preferred services or determining the ambition level of the preferred service) and bring in their local knowledge (about places where changing the landscape can best create synergy with other developments, for example). Second, tools need to support a creative process; for example, a search for the best spatial solution in the local context. This requires that actors are able to create spatial alternatives based on thorough knowledge about the relationship between spatial structure, functioning of the landscape, and benefits aimed for. Instead of focussing too much on assessment, which frames the relation between nature and humans as a problem, it is much more inspiring to focus on how local actors envision the future of their landscape. Techniques based on appreciative enquiry are known to create creativity and ownership (Cristina and Fernando, 2010). Third, considering the multiple values and the diversity of worldviews among stakeholders, methods and tools should facilitate collaboration and collective action. This may be done by, for example, organizing collaborative design workshops where actors apply design rules to create a preferred green infrastructure (Steingröver et al., 2010).

Apart from developing methods and tools that create ownership, creativity, and collaboration, a further contribution to bridging the gap to landscape planning may be that we start using a different conceptual frame to broadcast that we are extending the conventional work on ecosystem services towards the domain of collaborative planning. Actors in a planning group may not recognize ecosystem services as relevant to their planning process because the concept is associated with nature conservation and protected areas. In the Netherlands, for example, ecosystems are understood as protected areas where natural processes reign. In context of landscape planning, the term landscape services (Termorshuizen and Opdam, 2009) was suggested as more appropriate, based on the assumption that the term landscape is associated with a place to live and work which is adjusted to gain benefits. In my experience, actors in local planning groups tend to reject the words 'ecosystem services' as a planning concept because they associate it with nature conservation, protected areas, and restrictive legislation. Another obvious point of attention is the link between ecosystem services and a landscape structure that can be focussed on, and changed by, local actors. Ecosystems services are not created; it is the landscape that is transformed to create added value. We therefore need concepts that connect benefits to particular structures in the landscape. In multifunctional landscapes, ecological networks or green infrastructure is the landscape structure where most landscape services have their functional base. We have proposed that linking functional and structural landscape concepts is a helpful way of aligning actors with different backgrounds and interests through a planning process that evolves from identifying values and opportunities towards deciding about physical adaptation (Opdam et al., 2015). If we solve the gap between the conservation-oriented approach of most current ecosystem services work and the world of landscape planning, with respect to both terminology and tools, the ecosystem services concept has the potential to play an important role in binding actors and fostering collaboration and collective action for landscape adaptation.

Acknowledgments

This chapter is financially supported by the EU-funded Interreg project GIFT-T! (Green Infrastructure for Tomorrow – Together).

References

Chan, K.M.A., Shaw, M. R., Cameron, D. R., Underwood, E. C., and Daily, G. C. (2006). Conservation planning for ecosystem services. *PLoS Biology*, vol 4, pp 2138–2152.

Cristina, M. S., and Fernando, L. (2010). Appreciative inquiry: a positive approach to organizational planning and learning. *Social Research Reports*, vol 10, pp 3–105.

Fürst, C., Opdam, P., Inostroza, L., and Luque, S. (2014). Evaluating the role of ecosystem services in participatory land use planning: proposing a balanced score card. Special issue: integrating ecosystem services in land use planning and decision-making practice. *Landscape Ecology*, vol 29, pp 1435–1446.

Haines-Young, R., and Potschin, M. (2014). The ecosystem approach as a framework for understanding knowledge utilisation. *Environment and Planning C: Government and Policy*, vol 32, pp 301–319.

Kenter, J. O. (2016). Deliberative and non-monetary valuation. In: Potschin, M., Haines-Young, R., Fish, R. and Turner, R. K. (eds) *Routledge Handbook of Ecosystem Services*. Routledge, London and New York, pp 271–288.

Levebvre, M., Espinosa, M., Gomez y Paloma, S., et al. (6 authors) (2014). Agricultural landscapes as multi-scale public good and the role of the Common Agricultural Policy. *Journal of Environmental Planning and Management*. Available at: http://dx.doi.org./10.1080/09640568.2014.891975

Liu, J., and Opdam, P. (2014). Valuing ecosystem services in community-based landscape planning: introducing a wellbeing-based approach. Special issue: integrating ecosystem services in land use planning and decision-making practice. *Landscape Ecology*, vol 29, pp 1347–1360.

Martinez-Harms, M. J., and Balvanera, P. (2012). Methods for mapping ecosystem service supply: a review. *International Journal of Biodiversity Science, Ecosystem Services & Management*, vol 8, pp 17–25.

Opdam, P., Nassauer, J., Wang, Z., et al. (10 authors) (2013). Science for action at the local landscape scale. *Landscape Ecology*, vol 28, pp 1439–1445.

Opdam, P., Westerik, J., Vos, C., and De Vries, B. (2015). The role and evolution of boundary concepts in transdisciplinary landscape planning. *Planning Theory and Practice*, vol 16, pp 63–78.

Ostrom, E. (2009). A general framework for analyzing sustainability of social-ecological systems. *Science*, vol 325, pp 419–422.

Potschin, M., and Haines-Young, R. (2013). Landscapes, sustainability and the place based analysis of ecosystem services. *Landscape Ecology*, vol 28, pp 1053–1065.

Schröter, M., Van der Zanden, E., Van Oudenhoven, A., et al. (7 authors) (2014). Ecosystem services as a contested concept: a reflection on the critique and counter-arguments. *Conservation Letters*, vol 7, no 6, pp 514–523.

Steingröver, E. G., Geertsema, W., and Van Wingerden, W.K.R.E. (2010). Designing agricultural landscapes for natural pest control: a transdisciplinary approach in the Hoeksche Waard (The Netherlands). *Landscape Ecology*, vol 25, pp 825–838.

Termorshuizen, J., and Opdam, P. (2009). Landscape services as a bridge between landscape ecology and sustainable development. *Landscape Ecology*, vol 24, pp 1037–1052.

von Haaren, C., and Albert, C. (2011). Integrating ecosystem services and environmental planning: limitations and synergies. *International Journal of Biodiversity Science, Ecosystem Services & Management*, vol 7, pp 150–167.

von Haaren, C., Albert, C., and Galler, C. (2016). Spatial and landscape planning: a place for ecosystem services. In: Potschin, M., Haines-Young, R., Fish, R., and Turner, R. K. (eds) *Routledge Handbook of Ecosystem Services*. Routledge, London and New York, pp 568–578.

47

SPATIAL AND LANDSCAPE PLANNING

A place for ecosystem services

Christina von Haaren, Christian Albert and Carolin Galler

Introduction

The recognition of the ecosystem services (ES) concept is rising in spatial and landscape planning (Maes et al., 2012; Hauck et al., 2013; see also Odam, 2016). Furthermore, policy changes in Europe, such as those currently discussed for the Environmental Impact Assessment, the Water Framework Directive and the Habitats Directive, may trigger the introduction of the ES concept in practice. In countries with a long tradition of spatial and landscape planning, certain aspects of the ES approach often already exist in evaluation and impact assessment methodologies (e.g. Albert et al., 2014a). Additionally, the use of European or national legislation to derive response options and routines for implementation already incorporates a specific ES perspective. However, the existing achievements of land use and landscape planning do not mean that the ES approach is simply a new name for an old approach for spatial environmental planning. On the contrary, considerable opportunities and challenges could arise by integrating the ES concept into the theoretical framework of spatial and landscape planning (von Haaren and Albert, 2011). Significantly, the long-standing methodological practice, together with the implementation routines in spatial and landscape planning, may lead the way for conceptual adaptations (or variations) of the ES concept in response to implementation conditions (von Haaren et al., 2014). Thus, added value from cross-fertilisation may be expected by bringing the two spheres together. The following sections will outline: the general features of spatial and landscape planning that define the application of the ES concept as well as present deficits; the potential of the ES approach to compensate for shortcomings of spatial and landscape planning; examples of how this could be methodologically achieved; and the added value of the ES approach for implementation.

Introducing ES into spatial and landscape planning

Introducing ES into spatial and landscape planning is a complex endeavour which needs to take into account the varying planning systems and cultures in different countries (Albert et al., 2014b). In the following section, we will refer to the general features of spatial planning in democratic, constitutional states, which use spatial planning and or landscape planning for regulating spatial environmental development.

General features of spatial and landscape planning which frame the integration of ES

Spatial planning serves (in Europe) as an umbrella term encompassing local planning traditions such as "l'aménagement du territoire, Raumordnung, town and country or land use planning, ruimtelijke ordening or urbanismo" (Othengrafen, 2012: 24). Spatial planning refers to "the methods used by the public sector to influence the distribution of people and activities in spaces at various scales as well as the location of the various infrastructures, recreation and nature areas" (Spatial planning act, CEMAT in COMMIN 2007:6). Spatial planning is carried out by, or in the name of, public authorities responding to the demands of society as a whole or specific stakeholder interests. It coordinates the spatial impacts of other sectorial policies, formulates objectives to control future activities and implies an intervention in the physical environment, often via statutory spatial plans (see compilation in Othengrafen, 2012: 24–25). Recently, its scope has expanded beyond controlling land uses and developments to "seeking to influence the spatial (non-physical) processes [. . .] such as economic, social and political forces that determine the spatial (physical) end product . . ." (Greed, 2000 in Othengrafen, 2012: 25).

As to landscape planning, we refer here to the general definition of the European Landscape Convention (ELC), (Chapter I Art. 1), which defines it as a " . . . strong forward-looking action to enhance, restore or create landscapes". A landscape is understood as "an area, as perceived by people, whose character is the result of the action and interaction of natural and/or human factors" (ELC, Chapter I Art. 1f). The term landscape planning is used in this chapter/section synonymously with both environmental planning and the component of spatial planning which deals with spatial environmental issues (in many states, landscape planning is performed as an integral part of spatial planning). Landscape planning may be legally binding. More often it contributes to spatial planning and subsequently leads to other ways of implementation of environmental goals. For instance, it can lead to area protection, impact mitigation, in the case of potentially harmful development, or targeted funding of environmental measures. Each of these implementation options requires spatial specifications of the general environmental goals and standards of directives and laws. They also require area-specific assessment of the sensitivity of ecosystem services against existing and foreseeable pressures and impacts. Additionally, responses which are derived to react to environmental deterioration or to improve the environmental situation should be place-based, in order to ensure optimal relevance for their implementation.

An important feature of both spatial and landscape planning is public and stakeholder participation; an obligation often following national defined obligations to the Aarhus convention (UNECE, 1998) (see also Fish et al., 2016). In Europe, public participation is also a feature of strategic impact assessment, which is a mandatory part of spatial planning. For landscape planning, the ELC also requires the inclusion of "the public". However, in contrast to situations where decisions regarding objectives and valuations are completely left to participatory processes (Daily et al., 2000), public participation in governmental spatial and landscape planning is framed by two considerations. First, in a representative democracy, decisions which affect public interests beyond those of the landowners are usually made by an elected body. Public participation is thus restricted either to (i) informing those political decisions made by the municipal /regional council or (ii) decisions which will only affect the group of those who participate in the participation process. Landscape planning can prepare both kinds of participation situations by providing proposals for environmental change, which are then consulted on. Second, planning processes are opened to participation on issues which can politically be decided at their respective political decision scale (multilevel governance). For example, participation at a regional level cannot include decisions about the impact on a protected species whose presence in a region is not a matter of regional discretion. In a participatory process of landscape planning, information about these protected species may be conveyed. However, it should not be pretended that participation could change the assessment or the need to protect these species, as this may produce a feeling of frustration if citizens feel that their arguments are not considered.

The focus on legally codified values and norms, public participation and the implementation focus of spatial and landscape planning, has resulted both in a typical structure of planning components (Figure 47.1) and particular ways of handling ES.

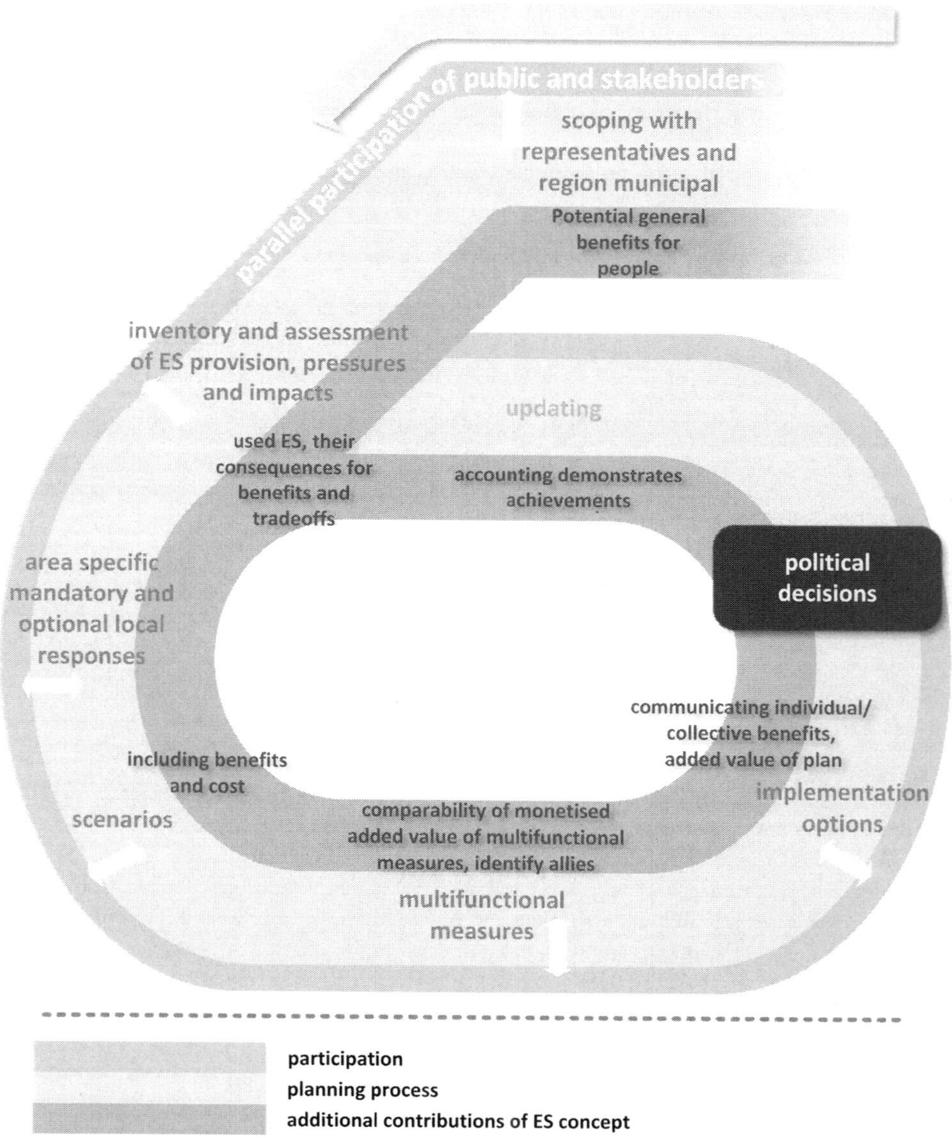

Figure 47.1 The landscape planning process and potential additional contributions of the ES concept.

Source: von Haaren et al., 2008, modified

The landscape planning process can be interpreted as an iterative process (Steinitz et al., 2003; Kato and Ahern, 2008) with several small loops due to stakeholder and public participation input. This input through participation is not only relevant in the phase of determining discretionary planning objectives, scenarios and alternative futures. Instead, nowadays, participation is, or should be, part of the whole planning process, including the inventory stage, in which local knowledge can be included. Even after a plan has been enacted by the city or county council, citizen knowledge can contribute to updating the inventory and evaluation of the landscape. GIS technology and online participation now enable updating to be a continuous process instead of a periodic exercise (e.g. Region Hannover, 2013; Galler et al., 2014).

The role ecosystem services may play in this process is restricted to mapping and evaluating the provision of ES by the landscape. Up to now, this supply was evaluated in landscape planning primarily without consideration of the actual use of ES and the resulting final benefits for individuals or society as a whole. The value bases used as criteria, standards or benchmarks in evaluation, are taken from legislation and experts' substantiation of general goals (for example enshrined in European directives such as the EU Nitrate or Habitat Directive). Individual preferences are integrated only in the context of public participation framed by multilevel governance. Inventories and accountings of goods produced by a landscape (such as wood and farm products) and their economic valuations, understood here in the narrow sense of monetization, are only rarely implemented at the landscape scale.

The scope of what is mapped and evaluated in landscape planning is a consequence of its presumed public character and the assumption that commercial goods, as well as individual interests, need not be represented in governmental planning. Instead, governmental planning favours a more general societal welfare approach expressed by democratically legitimized norms. In principle, this is still plausible and valid. Nevertheless, shortcomings of this limited perspective highlight the need for expanding landscape planning to include additional options and new perspectives offered by the ES approach.

Existing shortcomings of landscape planning include the fact that the advantages of landscape conservation and development are often not clearly illustrated for the users of public goods. Respective potentials waiting to be activated involve communicating better how landscape changes would affect the concrete benefits which individuals or groups draw from the landscape and, subsequently, how their well-being would be affected. Because this is not sufficiently substantiated, participation processes are dominated by groups that have an obvious strong economic interest in the land, such as land owners and farmers or developers. Additionally, tourism organisations, nature conservation NGOs and riding and hunting clubs usually take part in participation, whereas the interest and engagement of individual citizens is often restricted to those who fear individual impairments by the planning objectives. Furthermore, the discussion of landscape or spatial plans in local or regional councils is characterized by the juxtaposition of environmental improvements with the plans versus the costs of implementation and resulting trade-offs. For instance, negative trade-offs for agriculture or housing development may include restrictions and loss of revenue (Wende et al., 2012). The positive effects of the avoided impairments are less obvious. For the involved politicians, such intangible successes are often not visible or attractive enough to encourage their presentation to the public prior to an election. Finally, yet importantly, planning focuses on what can be influenced at respective decision levels. Neither regional/local landscape nor spatial planning explicitly address the driving forces of harmful or advantageous trends in the landscape, trends such as land consumption by housing development or intensification of agriculture. Not knowing that certain features of agricultural policy or building legislation are drivers of such undesirable local trends, this lack of knowledge will stimulate people to blame local decision-makers and undermines understanding of political interconnections.

Potential of the ES approach to equalize shortcomings of spatial and landscape planning

The ES approach holds considerable potential for compensating for these deficits (see Figure 47.1 inner circle) of spatial and landscape planning. To begin with, it may be stated that mapping ecosystem service supply is usually identical with mapping landscape functions in landscape planning (von Haaren et al., 2014). Nevertheless, the ES approach holds additional perspectives and new ways of evaluating ES that are not yet represented in landscape planning (von Haaren and Albert, 2011). In particular, methodological additions such as creating an inventory of the actual utilization of the ES supply (by new indicators and social science methods), as well as monetization and accounting of ES, or exploring the effects on societal, collective and individual well-being, may supplement and inform the future development of landscape planning. Such new components can stimulate discussions about beneficial impacts on individuals, ES trade-offs and, additionally, the state and trends of ES conservation on national and state levels. The latter fosters acknowledgement of the driving forces of ecological problems and, through this, a better acceptance of the local decision space and its relationship to processes taking place at higher spatial scales. Cardinally scaled accounts of ES provisions and economic valuations would highlight to politicians the budget consequences of belated action for ES conservation. Furthermore, the ES approach allows communication of nature conservation objectives in a way that is better aligned with societal and individual human well-being. By integrating the new perspectives of "utilized ES" as well as respective benefits, additional information is provided, which shows people where and how they are personally affected. Finally, the added value of defining the economic value of goods and services may stimulate the identification of new allies among the group of land users by demonstrating future synergies.

Examples for the support of green and blue infrastructures by ES

Landscape planning incorporating the above ES approach will be a promising tool for supporting green and blue infrastructure. It brings together mapping methodologies and is orientated towards participation and implementation of landscape planning with an ES approach focus on individual interests assessing the actual utilization, the economic value and highlighting the consequences of impacts on human welfare. Both green and blue infrastructure is here incorporated under the term "blue-green" infrastructure to emphasize their interdependencies in an ES context. ES like water retention for flood protection, for example, stem from both the properties of the "green" catchment areas and the flood plains or the "blue" river morphology. Assessing actual use of ES may support planning of collective and individual benefits of blue-green infrastructure.

Assessing cultural ecosystem services based on available data

The additional information gained by applying the ES concept in assessment and valuation in landscape planning can be illustrated by a case study for assessing cultural ecosystem services based on available data in the region of Hannover, Germany. Conventional landscape planning, and thus also spatial planning, considers the landscape function of "visual landscape" by implementing an expert-based mapping and evaluation of different landscape character areas, which in this spatial context constitutes a green recreation infrastructure. The evaluation method includes the criteria of diversity, uniqueness and beauty, or naturalness, which have proved to be key determining factors in peoples' perceptions of landscape quality. The evaluation results are summarized on a five-level Likert scale. The analysis results are illustrated as cultural ecosystem services (CES) which are offered by the landscape and are featured in the left map of Figure 47.2. Applying the ES concept in assessment and valuation allows consideration of two further aspects. First, infrastructure for using cultural ecosystem services can be taken into account (the middle map in Figure 47.2). Relevant infrastructure includes bike and hiking paths, information stands and lookout towers. Second, information on the actual utilization of CES can be compiled. Such information could be derived from diverse sources, including visitor statistics. In the example illustrated, the right map of Figure 47.2, geo-tagged pictures from an online database were used as a proxy for mapping actual use of CES. Prior research has shown that this proxy can be used to estimate actual visitation frequencies (Wood et al., 2013).

The added value of this additional information on ES for planning and management is threefold. It can help emphasize the importance of ES for human well-being through illustrating utilized ES and goods. It can provide an enhanced basis for monitoring changes and setting benchmark values and objectives to measure progress. Finally, and most importantly, the differentiated information could potentially help identify provisioning capacities that remain unused and which could be further exploited, if additional infrastructure were provided. Vice versa, areas where there is much utilization because of proximity to settlements but where landscape aesthetic quality is low can be identified and used as an incentive for enhancing the landscape. Additionally, the information could identify areas where public spending on infrastructure is ineffective as it is not reflected by the actual usage.

Assessing multifunctionality of green infrastructure in the common dimension of objective fulfillment and using the results for accounting on higher scales

This case study demonstrates how quantification and accounting of ES (including ES trade-offs) contribute to defining synergies and multifunctionality. For the county of Verden, in the northwest of Germany, an area-specific quantification of multiple ES was performed on the basis of data taken from the regional landscape plan. Four core ES were considered: biodiversity, water supply with respect to water quality (nitrogen input), natural yield/energy crop production with respect to erosion prevention and climate change mitigation with respect to CO_2 sequestration of soils.

The effects of environmental measures on the respective ES were quantified. In this way, trade-offs were identified and different implementation measures were compared with respect to their multifunctional effects on different ES. The multifunctional effects of environmental measures, even of the same type of measure, vary depending on where the measures are located in the case study area (see Figure 47.3).

Accounting for ES in this way allows for the comparison of different implementation strategies on a regional level. In a planned integrative baseline scenario, agri-environmental measures can be combined and allocated for best fulfillment of (all considered) ES provision. The proportional degree to which an objective is fulfilled (for each ES) is used as a reference level, or common dimension, for comparing trade-offs. Furthermore, it can be used for cost-benefit analysis and, thus, estimating the efficiency of measures.

The added value of integrating ES concept into landscape planning has three main dimensions:

- On the basis of an area-specific ES assessment, the areas where different ES overlap can be defined. By referring to the proportional objective fulfillment, a common reference level is introduced that allows for comparative quantification of different ES provisions. In this way, the values of a site for different ES can be compared and multifunctionality of green infrastructure can be assessed.

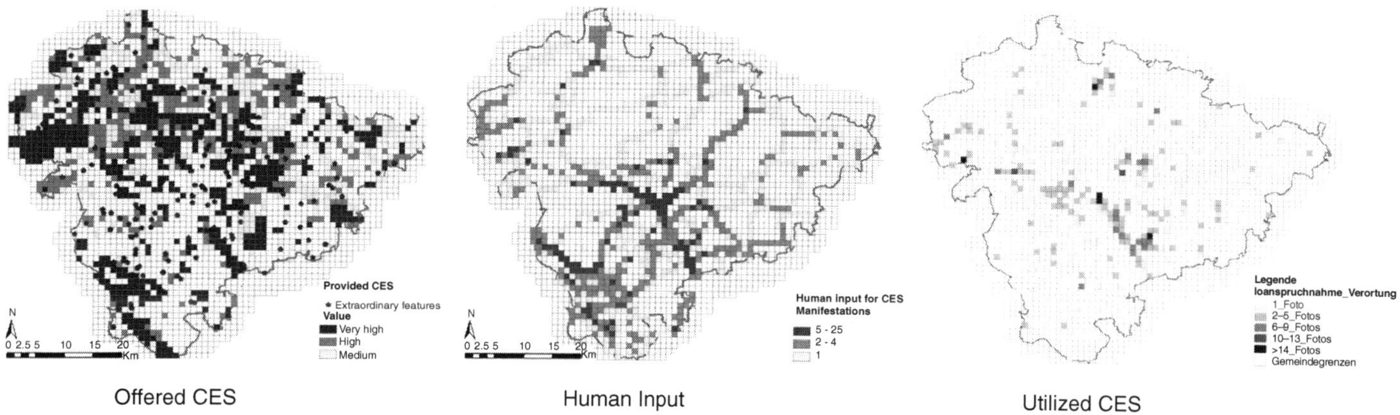

Figure 47.2 Juxtaposition of offered cultural ecosystem services (CES), additional human input, such as recreational routes and other infrastructures, and used CES as derived from geo-tagged pictures uploaded to panoramio.

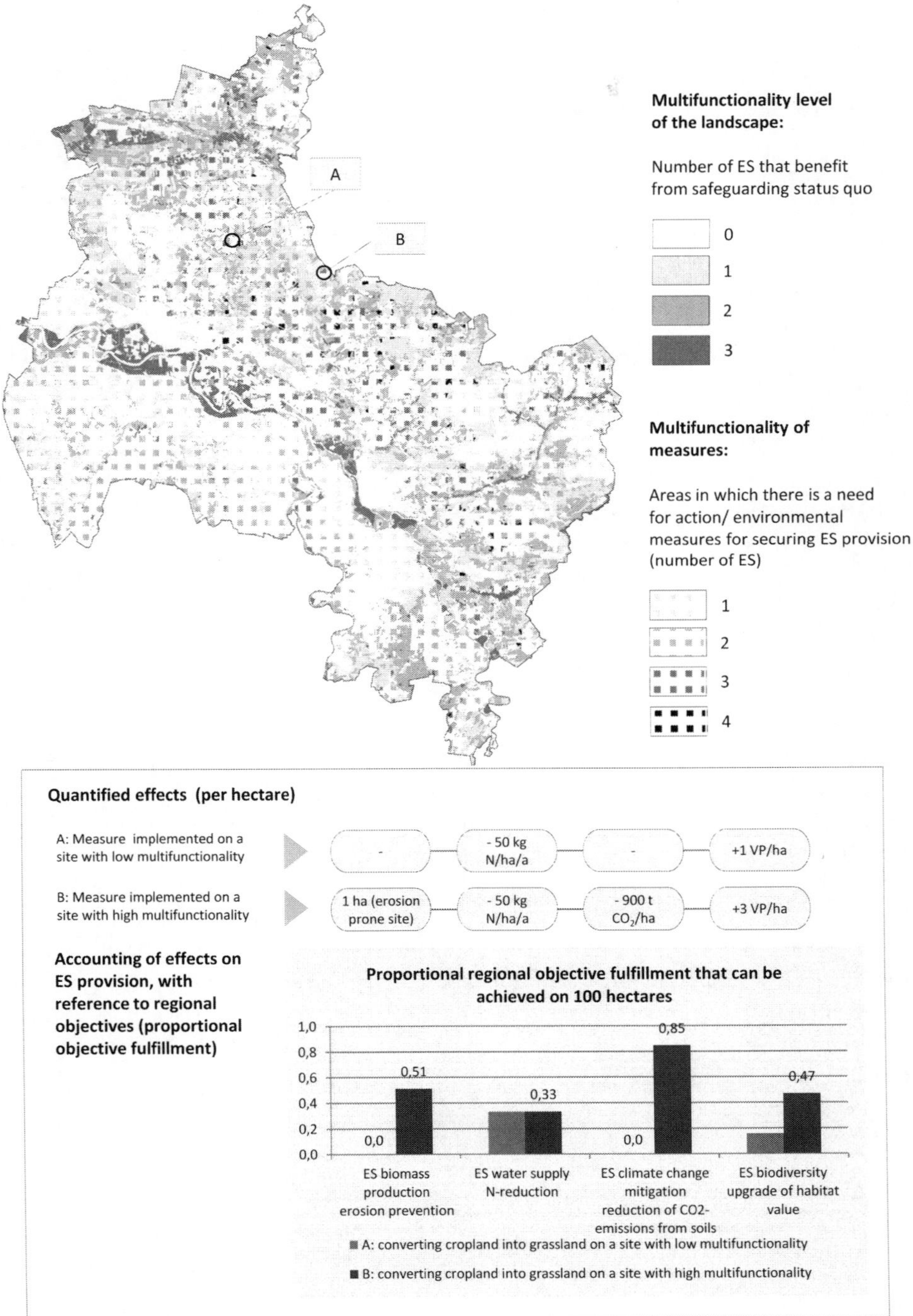

Figure 47.3 Quantification and accounting of multifunctionality and trade-offs in ES provision on the example of agri-environmental measures (converting cropland into extensive grassland).

- Area-specific quantification of the effects on different ES enables the estimation of environmental effectiveness of measures (including trade-offs) when implemented on different sites. This may serve as a basis for a spatially explicit estimation of measures' efficiency that includes a bundle of environmental effects.
- These (spatial) multifunctionality analyses provide valuable information (for planners and actors) for identifying synergies/complementarities among actors within the planning/governance and implementation process. For example, at sites where certain environmental measures contribute to both water quality and biodiversity, cooperation/collaboration between water and nature conservation governance leads to efficient public spending and induces win-win solutions.

The accounting assists in estimating environmental effects that can be achieved on regional scales. In this way, conclusions can be made concerning to what proportion an objective can be achieved by a certain type of measure or by development alternatives. Furthermore, this enables comparisons between effectiveness and efficiency of environmental measures/programs in different regions. Overall, such information about cost-efficient spending of public money may pave the way for better implementation and communication of ES related management measures. Through integration of the ES concept into landscape planning, politicians are able to see monetary effects and different sectors of the administration are encouraged to co-operate.

Added value of ES for implementation

The potential of the ES approach for communicating environmental objectives and strengthening their political support will be demonstrated and realized by the integration of the ES information into the instruments used for implementing spatial and landscape planning objectives. Such implementation instruments range from a direct, legally binding effect of the planning designations (which is often the case with spatial planning designations) to mere recommendations or the provision of options for action. As a rule of thumb, if private interests do not favour safeguarding an ecosystem service but this service has the characteristics of a public good, there is a case for the government to take over responsibilities for conservation. The instruments for realizing these responsibilities can be manifold. Their selection, and the location of allies who might share these responsibilities, can be aided by the new perspectives and evaluations provided by the ES approach. In more concrete terms, the different implementation instruments of spatial and landscape planning could profit from the communicative strength in the following ways:

- Legally binding area designations of spatial planning, restrictions of zoning, nature reserves and water protection areas, can be better communicated by pointing out the individual or collective benefits for locals or another specified group of beneficiaries. Likewise, the real price (including external costs) of some ecosystem goods, such as drinking water or ES destruction, can be compared to the costs of governmental measures, such as buying land for protected areas or compensating land owners for land use restrictions. Such information may greatly support political decisions.
- Restrictions on farming practice for safeguarding public-offered ES need to be implemented in a targeted way, with due regard for sensitive or valuable sites. For instance, landscape planning indicates the relevant sites for preventing soil erosion and grassland conversion, or safeguarding hedgerows of a certain size and function. Implementation

and acceptability could be improved by quantifying the costs of non-compliance to restrictions, such as monetising CO_2 emissions from ploughing organic soils. Upscaling of such calculations, including direct payments to agriculture and comparing this to the private revenue of the farmers, helps to recognize the driving forces of land use intensification.
- Incentives for land users to conserve ES (such as agri-environment measures), which may be proposed by landscape planning in sensitive or particularly valuable areas, can be supported by arguments of actual use (e.g. by local recreation). An additional use value is demonstrated instead of simply pointing out the aesthetic value of the visual landscape. Studies about the willingness of the population to pay for certain services may support the justification of respective budget decisions.
- Marketing of a nature conservation product, such as sheep meat from landscape management (e.g. Albert et al., 2009), can also be supported by ES information. For instance, if a public environmental good is linked to a commercial product, then the added ES value can be illustrated in a more concrete way for the customer if the benefits are quantified, designated to a defined group of customers (in case of local marketing) or monetized or if the added recreational and health benefit can be described.
- Activation of citizens for voluntary ES protection and development initiatives is facilitated if the personal or group benefit of using ES can be defined. For example, the development of green infrastructure in an area important for local recreation could be carried out by local citizens, cycling commuters and hunters, who would also profit from the additional landscape elements. The percentage of green space in the proximity of people's homes has a positive effect on their perceived health (e.g. Maas et al., 2006). If the benefits of daily cycling and walking for health could be monetized by a health cost avoidance calculation, even health insurance could become interested in contributing financially to the development and management of green spaces for a healthy population.

In sum, the broad scope of the ES approach and the additional information created by supplementary methodological approaches can substantially support implementation in spatial and landscape planning.

Discussion and conclusion

At present, spatial and landscape planning already are "a place for ecosystem services". Both types of planning represent a crucial interface for the implementation of spatial sustainable development. Important components of this interface include stakeholder and public participation, as well as support of political decisions. Both these components call for the inclusion of the communicative strength of the entire evaluation spectrum of the ES approach. The integration of evaluations of ES, beyond that of just presenting an area's specific evaluations of offered ES, could lead to a better case for environmental conservation on local and regional political levels. Its principal strength is the likelihood that it may improve communication of public goals by breaking them down into individual, collective and societal benefits and aspects of well-being, including monetized effects. Such effects on citizens can then be understood clearly and unambiguously by decision-makers, stakeholders and other citizens.

The examples presented in this chapter define and illustrate the potentials of the ES approach in the context of spatial and landscape planning. In all examples, the new information garnered from the approach provided arguments to better illustrate public benefits and human welfare. It has been demonstrated that methodologies exist for measuring the additional aspects of the

ES approach which have not been previously present in spatial and landscape planning. Either existing methods could be applied or new methods developed, using existing data. The methods refer in particular to:

- Identifying the actual use and beneficiaries of ES; this information is expected to communicate the added value of landscape planning objectives by targeting certain groups of beneficiaries (e.g. residents using the landscape for recreation) or by referring to the expected gains or losses in well-being in more concrete terms.
- Juxtaposing offered and utilized ES, as well as infrastructure; this allows the deduction of planning proposals from discrepancies between the provision of nature and the societal demand if, for example, a low landscape aesthetic quality coincides with high recreational use in the same area.
- Measuring multifunctionality of ES and management measures as well as the cost-effectiveness of a multifunctional management.

The results can be integrated into the instruments used for implementing spatial and landscape planning.

Further ES characteristic evaluations include cost benefit comparisons over different spatial scales, accounting for change of ES provision over time, and identifying and potentially quantifying and monetizing the effects on the wellbeing of individuals, collectives or the whole society. In addition to the arguments for safeguarding and developing ES which are used in present day planning and which often do not appeal very much to politicians, decision-makers can also be addressed by presenting to them the values and trade-offs in monetary categories or at least quantified accounting.

Nevertheless, it should be considered that the methods and results of quantification and monetization of ES are well-developed when it comes to market goods and services, private benefits and individual preferences. ES which are based on bequest values and existence values are much harder to quantify and express in monetary dimensions. A commodification of these ES may fail to protect them (Kosoy and Corbera, 2010; Hoppichler, 2013, von Haaren et al., 2014). Many planners fear that introducing the ES concept into planning will result in such a commodification (Albert et al., 2014a) and that in many cases the monetized benefits of impairing nature will be much higher than the insufficiently valued monetized loss of ES and biodiversity. For example, the benefits of a river with natural riverine vegetation can be monetized by a survey of the local population's willingness to pay. However, the revenue produced by shipping traffic, which depends on a technically well-developed waterway, is much higher – a fact that will be emphasized by the industry stakeholders. Furthermore, people may be willing to pay but there are no political or administrative mechanisms in place to realize this potential.

The fears of planners regarding commodification of nature have to be taken seriously. These fears may be considered by separately handling non-monetized offered ES and monetized utilised ES. This separation of arguments should be applied in communication with politicians. Furthermore, the distinction between offered and monetized utilized ES is also relevant for implementation. The offered ES are needed as information for existing implementation channels. Monetized results would be introduced only as additional arguments or in the context of additional implementation strategies such as the marketing of conservation products or for influencing political framework conditions. If applied in this way, the ES approach holds great potential for inspiring people's appreciation of the value of nature and hence supports its implementation in spatial and landscape planning.

References

Albert, C., Aronson, J., Fürst, C., and Opdam, P. (2014b). Integrating ecosystem services in landscape planning: requirements, approaches, and impacts. *Landscape Ecology*, vol 29, pp 1277–1285.

Albert, C., Aurbacher, J., von Haaren, C., Mahnkopf, B., and Petermann, C. (2009). Ökonomische Auswirkungen zukünftiger Agrarentwicklungen auf die Landschaftspflege und mögliche Beiträge der Aufpreisvermarktung von Naturschutzprodukten im Landkreis Diepholz. *Berichte über Landwirtschaft*, vol 87, no 3, pp 357–379.

Albert, C., Hauck, J., Buhr, N., and von Haaren, C. (2014a). What ecosystem services information do users want? Investigating interests and requirements among landscape and regional planners in Germany. *Landscape Ecology*, vol 29, pp 1301–1313.

COMMIN (2007). BSR INTERREG III B Project "Promoting Spatial Development by Creating COMmon MINdscapes". European Glossary. Last updated April 18th 2007. Available at: commin.org/upload/Glossaries/European_Glossary/COMMIN_European

Daily, G.C., Söderqvist, T., Aniyar, S., et al. (6 authors) (2000). Ecology – the value of nature and the nature of value. *Science*, vol 289, no 5478, pp 395–396.

Fish, R., Saratsi, E., Reed. M., and Keune, H. (2016). Stakeholder participation in ecosystem service decision-making. In: Potschin, M., Haines-Young, R., Fish, R. and Turner, R.K. (eds) *Routledge Handbook of Ecosystem Services*. Routledge, London and New York, pp 256–270.

Galler, C., Krätzig, S., Warren-Kretzschmar, B., and von Haaren, C. (2014). Integrated approaches in digital/interactive landscape planning. In: Wissen Hayek, U., Fricker, P. and Buhmann, E. (eds) *Peer Reviewed Proceedings of Digital Landscape Architecture 2014 at ETH Zürich*, Herbert Wichmann.

Hauck, J., Schweppe-Kraft, B., Albert, C., et al. (20 authors) (2013). The promise of the ecosystem services concept for planning and decision-making. *Gaia*, vol 22, no 4, pp 232–236.

Hoppichler, J. (2013). Vom Wert der Biodiversität. Wirtschaftliche Bewertungen und Konzepte fu¨r das Berggebiet. Bundesanstalt fu¨r Bergbauernfragen. Forschungsbericht Bundesanstalt für Bergbauernfragen.

Kato, S., and Ahern, J. (2008). "Learning by doing": adaptive planning as a strategy to address uncertainty in planning. *Journal of Environmental Planning and Management*, vol 51, no 4, pp 543–559.

Kosoy, N., and Corbera, E. (2010). Payments for ecosystem services as commodity fetishism. *Ecol Econ*, vol 69, no 6, pp 1228–1236.

Maas, J., Verheij, R.A., Groenewegen, P.P., de Vries, S., and Spreeuwenberg, P. (2006). Green space, urbanity, and health: how strong is the relation? *Journal Epidemiol Community Health*, vol 60, no 7, pp 587–592.

Maes, J., Egoh, B., Willemen, L., et al. (14 authors) (2012). Mapping ecosystem services for policy support and decision making in the European Union. *Ecosystem Services*, vol 1, no 1, pp 31–39.

Opdam, P. (2016). Bridging the gap between ecosystem services ad landscape planning. In: Potschin, M., Haines-Young, R., Fish, R. and Turner, R.K. (eds) *Routledge Handbook of Ecosystem Services*. Routledge, London and New York, pp 564–567.

Othengrafen, F. (2012). *Uncovering the Unconscious Dimensions of Planning. Using Culture as a Tool to Analyse Spatial Planning Practice*. Ashgate, Farnham.

Region Hannover (2013). Landschaftsrahmenplan der Region Hannover. Available at: http://www.hannover.de/Leben-in-der-Region-Hannover/Umwelt/Naturschutz/Landschaftsrahmenplan-der-Region-Hannover

Steinitz, C., Arias, H., Bassett, S., et al. (9 authors) (2003). *Alternative Futures for Changing Landscapes: The Upper San Pedro River Basin in Arizona and Sonora*, Island Press, Washington DC.

UNECE (1998). Aarhus Convention. Convention on access to information, public participation in decision-making and access to justice in environmental matters. Aarhus, Denmark, 25 June 1998. Available at: www.unece.org/fileadmin/DAM/env/pp/documents/cep43e.pdf

von Haaren, C. and Albert, C. (2011) Integrating ecosystem services and environmental planning: limitations and synergies. *International Journal of Biodiversity Science, Ecosystem Services & Management*, vol 7, no 3, pp 150–167.

von Haaren, C., Albert, C., Barkmann, J., et al. (7 authors) (2014). From explanation to application: introducing a practice-oriented ecosystem services evaluation (PRESET) model adapted to the context of landscape planning and management. *Landscape Ecology*, vol 29, no 8, pp 1335–1346.

von Haaren, C., Galler, C., and Ott, S. (2008). *Landscape Planning. The Basis of Sustainable Landscape Development*. Bundesamt für Naturschutz, Bonn.

Wende, W., Wojtkiewicz, W., Marschall, I., et al. (8 authors) (2012). Putting the plan into practice: implementation of proposals for measures of local landscape plans. *Landscape Research*, vol 37, no 4, pp 483–500.

Wood, S.A., Guerry, A. D., Silver, J.M., and Lacaya, M. (2013). Using social media to quantify nature-based tourism and recreation. *Scientific Reports*, vol 3, no 2976, DOI: 10.1038/srep02976

Briefing Note 47.1 Including ecosystem services in impact assessment: challenges and opportunities

Davide Geneletti

Impact assessment (IA) is the process of identifying the future consequences of a current or proposed action (IAIA, 2012). It provides information that can support decision-making. Currently, impact assessment is one of the main existing tools to verify that sustainability strategies are implemented at all levels of decision-making (Sheate, 2009). The emergence of ecosystem services approaches has been leading to a paradigm shift in IA. This has already produced tangible results, both in legislation and in practice. For example, the text of the European Commission proposal for a revised Environmental Impact Assessment (EIA) Directive includes "biodiversity and the ecosystem services it provides" among the environmental aspects to be addressed (EC, 2012). Recent practical guidance reflects the need to mainstream ecosystem services in a more systematic and comprehensive way into IA tools, such as Strategic Environmental Assessment (UNEP, 2014; Geneletti, 2016). This paradigm shift offers a number of opportunities and poses some challenges to the development of IA.

Integration: Ecosystem services approaches promote the integration of different IA tools by blurring the boundaries between them. For example, EIA is traditionally centred on the biophysical aspects of the environment, but when ecosystem services are considered it necessarily has to pay more attention to issues related to human well-being. Analogously, Social IA will have to consider the biophysical environment to give justice, for example, to the treatment of cultural ecosystem services. As argued by Morrison-Saunders et al. (2014), the proliferation of IA types and their increasing specialization can produce silo-based approaches, which undermine the effectiveness of IA in promoting sustainability.

Impact framing: Consideration of ecosystem services helps to move beyond the traditional, and largely "reactive", framing of impacts that characterize IA. Especially in the environmental sector, IA is still largely based on the analysis of the activities that an action (e.g. a project or a plan) requires, and on the prediction of their effects on a number of issues of concern (soil, water, health, etc.). Hence, the environment is essentially seen as a backdrop to absorb impacts, rather than a provider of services that can contribute to the action's objectives (Baker et al., 2013). Ecosystem service approaches can help to explore and assess development options that may safeguard ecosystems and their capability to provide services (Partidartio and Gomes, 2013).

Assessment methods: The lack of widely agreed-upon indicators and assessment methods may undermine the inclusion of ecosystem services in IA (Geneletti, 2011). The practice of IA has developed in some fields, and detailed guidelines exist on how different environmental and socio-economic components should be analysed, and how impacts on them should be modelled. However, the field of ecosystem services is still characterized by debate about the suitability of different mapping and assessing methods, or even about the same definition of services. For example, economic valuation is contested, and proposed methods differ widely in terms of information requirements and assumptions. It might be difficult for IA practitioners to justify the use of specific methods, given that some forms of IA are subject to legal challenge (Baker et al., 2013).

Ecosystem service categories: There is a risk that the scarcity of data and well-established assessment protocols may produce a bias in the type of information of ecosystem services that is included in IA; studies may consider ecosystem services that are easier to identify (and communicate) and for which more information is available, as opposed to services that are actually relevant but difficult to measure. For example, the review conducted by Honrado et al. (2013) concludes that provisioning (and to a lesser extent cultural) ecosystem services are much more frequently addressed by EIAs than regulating services. The problem of bias can be addressed by developing guidance for scoping the most important ecosystem services, as recently done by UNEP (2014). This scoping stage should provide for stakeholder consultation, and consider the time and resource constraints of real-life IA processes.

The integration of ecosystem services in IA has potential benefits in terms of the better design of policies, plans, and projects, but there are also critical issues that need to be recognized. These can be addressed by learning from pilot applications and case studies, as well as by taking advantage of the data, tools, and methods for ecosystem service representation and modelling that are becoming increasingly available.

References

Baker, J., Sheate, W.R., Phillips, P., and Eales, R. (2013). Ecosystem services in environmental assessment – Help or hindrance? *Environmental Impact Assessment Review*, vol 40, pp 3–13.

EC (2012). *Proposal for a Directive of the European Parliament and of the Council amending Directive 2011/92/EU on the Assessment of the Effects of Certain Public and Private Projects on the Environment.* /* COM/2012/0628 final – 2012/0297 (COD).

Geneletti, D. (2011). Reasons and options for integrating ecosystem services in strategic environmental assessment of spatial planning. *International Journal of Biodiversity Science, Ecosystem Services and Management*, vol 7, no 3, pp 143–149.

Geneletti, D. (2016). A conceptual approach to promote the integration of ecosystem services in strategic environmental assessment. *Journal of Environmental Assessment Policy and Management*, in press.

Honrado, J.P., Vieira, C., Soares, C., et al. (7 authors) (2013). Can we infer about ecosystem services from EIA and SEA practice? A framework for analysis and examples from Portugal. *Environmental Impact Assessment Review*, vol 40, pp 14–24.

IAIA (2012). *Fastips No. 1: Impact Assessment.* International Association for Impact Assessment. Available at: www.iaia.org/publications-resources/fastips.aspx (accessed 10 May 2014).

Morrison-Saunders, A., Pope, J., Gunn, J.A.E., Bond, A., and Retief, F. (2014). Strengthening impact assessment: a call for integration and focus. *Impact Assessment and Project Appraisal*, vol 32, no 1, pp 2–8.

Partidario, M.R., and Gomes, R.C. (2013). Ecosystem services inclusive strategic environmental assessment. *Environmental Impact Assessment Review*, vol 40, pp 36–46.

Sheate, W. R. (2009). The evolving nature of environmental assessment and management: linking tools to help deliver sustainability. In: Sheate, W. R. (ed.) *Tools, Techniques and Approaches for Sustainability: Collected Writings in Environmental Assessment Policy and Management.* World Scientific, Singapore.

UNEP (2014). *Integrating Ecosystem Services in Strategic Environmental Assessment: A Guide for Practitioners.* A report of Proecoserv.

48
AN INSTITUTIONAL PERSPECTIVE

Eeva Primmer

Introduction

The mis-led assumption that new and newly organized knowledge will automatically feed to decision-making and improve ecosystem service governance can be corrected by taking an institutional perspective. Ecosystem services are defined, enjoyed and governed by people. The behavior of people follows patterns that have been prescribed by formal rules, such as laws and policies – or mere shared norms and practices. These written and unwritten rules and norms can be conceptualized as formal and informal institutions.

Without an understanding of institutions, we cannot systematically advance sustainable governance of ecosystem services. An institutional perspective unravels the rules that generate and maintain regularities in the behavior of people, groups or organizations. The perspective merges insights from law, economics and organizational studies (Scott, 2001). Sociology, anthropology and philosophy also apply the terms institutional and institutions to address rules and regularities in society. The institutional perspective is important for ecosystem service analyses because it draws attention to the existing rights and responsibilities regarding ecosystem service use and management (Primmer and Furman, 2012). Identifying existing rights and responsibilities is a prerequisite for designing new policies and institutions that fit the existing institutional context.

Formal institutions

Formal institutions are explicitly stated norms and rules about what can be done, what must be done and what must not be done (Ostrom, 1990). As such, formal institutions set the framework for the use of ecosystem services and the management of ecosystems. For example, natural resource legislation or land use zoning might define how a watershed or a forest area can be managed, what can be extracted from the ecosystem, who can make decisions about the use of the resource, who can access the area or who should be engaged when new plans for the ecosystem are developed.

Property rights are an important institution often identified in institutional analysis. The analysis of property rights can reveal differences in ecosystem service management practices between different landowner groups, or it can help identify changes in ecosystem service use induced by alterations in the rights. Property rights extend beyond land-ownership and can be conceptualized broadly. They can cover, for example, formally defined rights to extract water from a stream or a right to hunt certain species in a specific area. However, the rights and responsibilities to ecosystem services can be viewed in an even more comprehensive sense than what is covered by the broadest conceptualizations of property rights.

Formalized rights to benefit from an ecosystem service or to govern an ecosystem are also important institutions. These kinds of formal rights and responsibilities are held by, for example, local inhabitants or state agencies. The units that these rights and responsibilities attach to can be defined in numerous ways; they can relate to activities (e.g., recreation), spatial units (e.g., protected areas), economic sectors (e.g., agriculture), processes (e.g., watershed planning) or policy instruments (e.g., a payment for maintaining an area in natural condition). All these rights can be labeled formal institutions if they are explicitly formally defined.

Informal institutions

Informal institutions are less explicit than formal regulations because they are embedded in customs and traditions or in professional and administrative practices (North, 1990). For example, local people tend to follow traditional, customary rules about how much water they can extract from a water stream (Gómez-Baggethun et al., 2013). In private ownership contexts, the land-owners' normative perceptions influence their decisions to use ecosystem services or, for example, adopt ecosystem service payments (Lockie, 2013; Primmer et al., 2014). In organizational contexts, informal institutions are embedded in administrative and professional norms and practices (March and Olsen, 1984). For example, forestry planners' shared professional norms have been shown to shape the ways in which ecosystems are managed in non-industrial private forests (Primmer, 2011). Correspondingly, the customs of farmers, fishermen and hunters as well as the relevant advisors and administrators condition the practice of ecosystem conservation and use.

The behaviors of different groups of actors also tend to follow particular rules regarding what benefits they can derive from ecosystems and what they should harness and preserve (Gómez-Baggethun et al., 2013; Norgaard, 2010; Lockie, 2013). The benefits that actors have enjoyed for a long time are institutionalized, and their alteration will change the rights. It is important to identify existing rights when new policies are designed. For example, payments for ecosystem services can be designed to secure or enhance public benefits. By doing so, the payments strengthen a collective right to ecosystem services, healthy ecosystems or biodiversity. At the same time, the payment is a compensation for the ecosystem service producer, for giving up a right to use the ecosystem in some other way. It is possible that the popularity of ecosystem service payments rests on this idea: generally, the achieved rights, representing the status quo, are maintained or their loss is compensated (Vatn, 2010). Furthermore, the identification of previously overlooked or ignored ecosystem services and benefits should be accompanied by an analysis of who already holds the right to the benefit, such as unpolluted water, beautiful scenery or pollination. Only after having identified the right to benefit can we analyze how rights will change with potential new policies or institutions.

Institutional change

Institutional change alters rights and responsibilities. An attempt to purposely generate institutional change and reformulate institutional arrangements often takes place through policy design and changes to law. Once these changes have taken place and start to genuinely influence behavior, we can call the resulting rules new formal institutions. However, institutions also evolve outside the purposeful institutional design process, through informal rules becoming such strong norms that they condition people's behavior (Goodin, 1996). In some cases, the formalization of the institution is a mere final step in this kind of an institutionalization process.

Institutional change can take form through specific policy changes redirecting different actors' behavior and their rights and roles for a limited period or in a limited setting. In this sense, institutions overlap with policies. The delineation between the concepts of institutions and policies should be based on the stability and legitimacy of the arrangement. Institutions tend to endure beyond policy changes and enjoy credibility among those who are subject to the institutions.

Institutions evolve over time, whether they are targeted with purposeful design or not. Institutional design might address ecosystem functions or services that have not been regulated before. Informal norms also change over time, sometimes very rapidly. There are examples of traditional, sustainable ecosystem management institutions having eroded fast after the establishment of new policies on natural resource use or conservation (Gómez-Baggethun et al., 2013). On the other hand, local actors can spontaneously devise new rules that fit their operational context and enhance sustainability (Ostrom, 1990).

Institutions can ease or constrain governance

Institutions contribute to the predictability of human behavior and therefore allow coordinating action. In this sense, institutions are a precondition for governing ecosystems sustainably (Ostrom, 1990; Agrawal, 2001; Young, 2002; Paavola, 2007). The recognition of ecosystem services and the new knowledge produced about them is likely to prompt a need to develop new policies and governance mechanisms. The new policies can have high transaction costs for those providing the service and for the society that enjoys ecosystem services (Coggan et al., 2010). By allowing coordination through easing interactions between actors in general, institutions reduce transaction costs (North, 1990).

Professional and administrative actors play an important role in carrying and shaping institutions; intermediaries can help in communicating new ecosystem service policy. At the same time, professional norms and administrative practices can constrain new ecosystem service policy, or at least slow down the institutional change that is envisioned to take place as soon as a new policy has been launched. It has been shown that norms among professionals and administration, together with operational level rules more generally, condition the ways in which new conservation and management policies are implemented (Hardy and Koontz, 2009; Primmer et al., 2013). As new ecosystem service policies are often designed at different governance levels than where they are expected to generate impacts, the design process should consider how the policies fit the institutions of the implementing level.

Applying the institutional perspective

Understanding institutions is important for the analysis of ecosystem services. It is necessary for the analysis of policies governing ecosystem functions and for an in-depth analysis of ecosystem benefits and costs. Identifying the rules regarding rights and responsibilities should be the first step of any ecosystem service policy study. The analysis can be descriptive and brief in situations where an understanding of the governance context is needed. Alternatively, the analysis might go further to test hypotheses derived from different institutional theories on rights and responsibilities, different types of rules, power, institutional constraints or transaction costs, or institutional fit, policy coherence and multilevel governance. Finally, with the strong emphasis of ecosystem service research on increased understanding and accessible information, the institutional context in which this knowledge would be used should not be ignored. It is paramount that new knowledge meets the institutionalized norms and practices that condition ecosystem service use and conservation.

References

Agrawal, A. (2001). Common property institutions and sustainable governance of resources. *World Development*, vol 29, no 10, pp 1649–1672.

Coggan, A., Whitten, S. M., and Bennett, J. (2010). Influences of transaction costs in environmental policy. *Ecological Economics*, vol 69, no 9, pp 1777–1784.

Gómez-Baggethun, E., de Groot, R., Lomas, P. L., and Montes, C. (2010). The history of ecosystem services in economic theory and practice: from early notions to markets and payment schemes. *Ecological Economics*, vol 69, pp 1209–1218.

Gómez-Baggethun, E., Kelemen, E., Martín-López, B., Palomo, I., and Montes, C. (2013). Scale misfit in ecosystem service governance as a source of environmental conflict. *Society & Natural Resources: An International Journal*, vol 26, no 10, pp 1202–1216.

Goodin, R. E. (1996). Institutions and their design. In: Goodin , R. E. (ed.) *The Theory of Institutional Design*. Cambridge University Press, Cambridge UK.

Hardy, S.D., and Koontz, T.M. (2009). Rules for collaboration: institutional analysis of group membership and levels of action in watershed partnerships. *Policy Studies Journal*, vol 37, no 3, pp 393–414.

Lockie, S. (2013). Market instruments, ecosystem services, and property rights: assumptions and conditions for sustained social and ecological benefits. *Land Use Policy*, vol 31, pp 90–98.

March, J. G., and Olsen, J. P. (1984). The new institutionalism: organizational factors in political life. *The American Political Science Review*, vol 78, no 3, pp 734–749.

Norgaard, R.B. (2010). Ecosystem services: from eye-opening metaphor to complexity blinder. *Ecological Economics*, vol 69, pp 1219–1227.

North, D. C. (1990). *Institutions, Institutional Change and Economic Performance*. Cambridge University Press, Cambridge UK.

Ostrom, E. (1990). *Governing the Commons: The Evolution of Institutions for Collective Action*. Cambridge University Press. Cambridge UK.

Paavola, J. (2007). Institutions and environmental governance: a reconceptualization. *Ecological Economics*, vol 63, pp 93–103.

Primmer, E. (2011). Analysis of institutional adaptation: integration of biodiversity conservation into forestry. *Journal of Cleaner Production*, vol 19, no 16, pp 1822–1832.

Primmer, E., and Furman, E. (2012). Operationalising ecosystem service approaches for governance: do measuring, mapping and valuing integrate sector-specific knowledge systems? *Ecosystem Services*, vol 1, pp 85–92.

Primmer, E., Paloniemi, R., Similä, J., and Barton, D. N. (2013). Evolution in Finland's forest biodiversity conservation payments and the institutional constraints on establishing new policy. *Society & Natural Resources*, vol 26, no 10, pp 1137–1154.

Primmer, E., Paloniemi, R., Similä, J., and Tainio, A. (2014). Forest owner perceptions of institutions and voluntary contracting for biodiversity conservation: not crowding out but staying out. *Ecological Economics*, vol 103, pp 1–10.

Scott, R. W. (2001). *Institutions and Organizations, 2nd Edition*. Sage Publications, Thousand Oaks CA.

Vatn, A. (2010). An institutional analysis of payments for environmental services. *Ecological Economics*, vol 69, pp 1245–1252.

Young, O. R. (2002). *The Institutional Dimensions of Environmental Change: Fit, Interplay, and Scale*. The MIT Press, Cambridge MA.

49

THE USE OF ECOSYSTEM SERVICES KNOWLEDGE IN POLICY-MAKING

Drawing lessons and adjusting expectations

Duncan Russel, Andrew Jordan and John Turnpenny

Introduction

The 12 principles of the Convention on Biological Diversity[1] set out an approach for managing ecosystems and the services they provide in a sustainable manner. Many of the principles imply that a strong knowledge base is required to successfully manage ecosystems (e.g. Principle 6: Ecosystems must be managed within the limits of their functioning). In this respect, the Millennium Ecosystem Assessment (MA) and similar national based assessments have generated much new knowledge about the functioning of ecosystems, the impacts of human activities on them and the link between ecosystem health and human well-being (Potschin and Haines-Young, 2011). However, simply possessing 'more knowledge' is no guarantee that it will be embedded into policymaking to facilitate the greater protection of ecosystems. Indeed, as many existing studies have shown (for an overview see: Owens, 2012, Juntti et al., 2009, Jordan and Russel, 2014), generating more knowledge does not necessarily lead to better environmental or ecological outcomes; knowledge is only ever at most a necessary but insufficient condition for protecting important ecosystems. Thus, an important challenge for scientists and practitioners concerned with the protection of global ecosystems is to better understand and adjust expectations of how knowledge on ecosystem services is used, by whom, and in which context to inform decision-making – see Box 49.1 for key definitions and terms around knowledge use used in this chapter.

The majority of studies of knowledge utilisation suggest that in many decision-making contexts, a more nuanced understanding of policymaking is required (e.g. Sabatier, 1998; Sanderson, 2002; Owens, 2005; Russel and Jordan, 2007; Hertin et al., 2009). In this literature, policymaking is seen as a much more multifaceted and recursive process involving political power games, bargaining and negotiation. A case in point in this respect is the Intergovernmental Platform on Biodiversity and Ecosystem Services, which tends to be 'dominated by bargaining processes about the allocation of power and authority' (Beck et al., 2014, p. 84), despite its vast evidence-collecting effort. The knowledge-power relationship is further complicated by the often wholly different understandings and expectations about evidence use amongst politicians, scientists, NGOs and industry groups. See for example Ingold and Gschwend's (2014) study on how the role of science in Swiss environmental policy was influenced by alliances of actors with competing policy visions.

How ecosystem services knowledge is used in decision-making, by whom and in which contexts to make decisions affecting ecosystems forms the heart of this chapter. First, it provides the ecosystem services policy context and related expectations behind knowledge use. Then it draws on the existing literature to provide an overview of the theory and practice of knowledge use in policymaking. It then reflects on how ecosystem service knowledge use in policymaking can be enhanced, how knowledge has (or has not) been used to support the management of ecosystems and draws on insights from the knowledge utilisation literature to explain why ecosystem services knowledge is used the way it is. Finally, it concludes by highlighting the adjustments that need to be made around expectations of ecosystem service knowledge use, and research agendas for enhancing understandings of knowledge use for ecological protection. Ultimately, this chapter argues that knowledge use is often far from the textbook ideal, where knowledge shapes policy in a direct linear fashion, i.e. knowledge is produced and fed into the policymaking process to produce concrete evidence-based policy outputs. However, non-linear forms of knowledge use can have positive impacts on the management of ecosystems if channelled in the right manner, through, for example, enhancing stakeholder dialogue or cumulatively shaping policy discourses over longer time periods.

Box 49.1 Key terms in this chapter

Knowledge: knowledge comes in many different forms. The links between knowledge, evidence and research are complex and contested by different actors. For instance, research can be seen as only one type of evidence, with other forms including, for example, ideological evidence from think tanks, opinion-based evidence from public polling, etc. Evidence is also argued to be only one source of knowledge, working alongside knowledge from the experiences of stakeholders/citizen/policymakers, and political knowledge, among others.

Instrumental knowledge use: where knowledge is directly used to shape a decision in a manner that leads to a concrete policy output over a relatively short time frame

Conceptual knowledge use: where a body of knowledge builds up cumulatively to influence a broader policy agenda over longer periods of time.

Strategic knowledge use: where actors in decision-making contexts selectively use knowledge as ammunition in broader political battles.

Co-production of knowledge: where processes are established for the interaction between knowledge generators, knowledge uses and other stakeholders to enhance use through building up a shared and collective body of knowledge.

Ecosystem services policy: expectations of knowledge use

The 2005 MA was one of a long line of global environmental assessments sponsored by the United Nations to highlight the impact that human activities have had (and appear likely to have) on environmental and ecological systems. Like the 12 principles of the Convention on Biological Diversity and other global assessment processes, the MA also highlighted how long-term human well-being is underpinned by the healthy functioning of environmental and related ecological systems. Crucially, though the MA generated new knowledge on global ecosystems and the services they provide, it did not fully explore the conditions in which it was – or was not – likely to be utilised (MA, 2005, p. 20). Indeed, the MA followed a very similar approach to other global assessment processes (e.g. the IPCC for climate change) in that it sought to systematise knowledge and promote awareness amongst policymakers, but not specific end users or decisions. The MA was built around a model of 'instrumental' use, which assumes – to put it crudely – that once the 'right' type of information has been supplied, rational decision-makers will act upon this evidence base to produce more robust environmental policy. As we show below, however, such a model of knowledge use tends to be the exception rather than the norm and, even where it does exist, the process of knowledge use may be more complex than first meets the eye.

While there are clear examples of where knowledge has been used in an instrumental manner in which there is a direct and linear relationship between knowledge inputs and policy outputs (Haas, 2004; Owens, 2005), the majority of studies suggest the pattern of uptake is in fact more complex and recursive in environmental (e.g. Juntti et al., 2009; Owens, 2012; Jordan and Russel, 2014) and other policy fields (e.g. Sabatier, 1998; Sanderson, 2002; Owens, 2005; Nutley et al., 2007; Hertin et al., 2009).

A body of literature which suggests a more differentiated and nuanced understanding and set of expectations of use is necessary has evolved. This literature is sensitive to the processes of power, bargaining and negotiation in different contexts. Often there are wholly different, and conflicting, understandings and expectations about knowledge use amongst politicians, scientists, NGOs and industry groups (Pielke, 2007). There are also varied pathways through which knowledge can be transmitted; so many, in fact, that the impact of knowledge may only be discernible over many decades (Sabatier, 1998; Rich, 1997). Added to this complexity is the mediating effect of the many knowledge brokers, intermediaries and go-betweens that often operate at the interface between science and policy (Sanderson, 2002). Moreover, knowledge and power have always been very deeply intertwined, further complicating the relationship between knowledge and decision-making (Radaelli, 1995, Juntti et al., 2009). Despite this critique, it may not be surprising that assessments like the MA have been built around an instrumental model of knowledge use, since it still has an enduring appeal as an ideal type, particularly amongst economists, natural scientists and policymakers.

Compared to the richness of the discussion in other policy areas, the debate within the ecosystem services community (both researchers and practitioners) about the conditions in which new knowledge is or is not used, by whom and for what purpose, has barely begun (for exceptions, see Ferraro et al., 2011; Laurans et al., 2013; Jordan and Russel, 2014; Russel et al., 2014). The studies which do exist suggest that when knowledge on ecosystem services has been used, it has had no – or only a rather limited – policy impact. However, such findings could well be an artefact of the approach adopted by researchers: many of their studies more or less implicitly seek – or expect – a highly instrumental form of knowledge utilisation, and when this is not found, 'a common conclusion is that [ecosystem services knowledge] is not used at all' (McKenzie et al., 2014, p. 322). So how can we move understandings forward to paint a more nuanced picture, and to adjust expectations and understandings of ecosystem services knowledge use in policymaking?

Understanding knowledge utilisation

Theories and analytical frameworks

In many respects, the ecosystem service concept is just the latest in a very long line of attempts to ensure that the 'true value' of the environment is reflected in decision-making (Cowell and Lennon, 2014), which can be traced back to the Brundtland report on Sustainable Development (WCED, 1987) and even earlier international policy discourses (e.g. the Club of Rome's Limits to Growth (1972) report). The existing literatures on the fate of these attempts suggest that the impact of knowledge use has been patchy at best (see Juntti et al. (2009) and Owens (2005) for an overview). Indeed, some of the earlier studies on knowledge use in environmental policy making draw out the complex knowledge-policy dynamics in issues such as pesticide policy (Jasanoff, 1987) and air pollution (Sabatier, 1988). In many respects, efforts to implement the concept of ecosystem services has followed similar international initiatives, such as Brundtland, without fully considering the vexed problem of knowledge use. There is thus little to suggest that current attempts to promote the concept of ecosystem services will not encounter problems similar to those past initiatives concerned with giving environmental considerations greater footing in decision-making. Policy learning around ecosystem services, though, may be limited by the characteristics of ecosystem services knowledge, which is complicated and high in epistemic uncertainty (see Dunlop, 2014; Laurans et al., 2013; Schröter et al., 2014). Indeed, knowledge is still rapidly developing and still contested (Schröter et al., 2014), as suggested by the rapid increase in research in this area (e.g. Laurens et al., 2013), and concepts have not yet fully formed (Dunlop, 2014), e.g. the continued refinements around what constitutes cultural ecosystem services (e.g. Church et al., 2014). Thus, knowledge on ecosystem services may not necessarily lend itself to linear and technical forms of knowledge transfer and utilisation (Hockley, 2014).

Recent studies show that actual practice of implementing the concept of ecosystem services consistently falls short of the high-level political ambition to embed ecosystem knowledge in decision-making (see Russel et al., 2014; Jordan and Russel, 2014). Moreover, many studies clearly show a failure to fully learn lessons from the past. For example, Hockley (2014) notes that CBA has long been considered 'the best game in town' for embedding environmental knowledge into decisions on the basis of its transparency and perceived objectivity. It is not, therefore, surprising that it has been strongly promoted by advocates of ecosystem services (TEEB, 2010; HMT, 2012), many of whom are environmental and ecological economists. But the actual use of environmental knowledge in this type of appraisal process has at best been 'sporadic' (Hockley, 2014). Even in the environmental sector, CBAs have been used strategically by environmental ministries and agencies to win political battles with 'non' environmental actors, rather than to directly inform and/or improve policy making (ibid). In fact, the environment ministry in the UK rarely practiced CBA in its textbook form (Russel and Jordan, 2007) and in the US – the country which originally pioneered it – CBAs are 'seriously deficient' (Harrington et al., 2009), suggesting a gap between the expert and user communities (Caplan, 1979), as well as between what environment ministries preach and what they practice. In a similar vein, others have found knowledge on ecosystem services in Europe and North America has been used but often in unexpected and political ways to support political agendas (e.g. see Haines-Young and Potschin, 2014; McKenzie et al., 2014; Ingold and Gschwend, 2014). Such use is far from the instrumental model outlined above.

We now turn to the literature on knowledge utilisation to help us explain and understand this puzzle. Knowledge utilisation is a rapidly maturing academic field (e.g. Weiss, 1979; Sabatier, 1998; Radaelli, 1995; Rich, 1991; Haas, 2004; Owens, 2005, 2012), dating back at least 40 years (Dunlop, 2014). However, as Dunlop (2014) demonstrates, at the core of knowledge utilisation research are a couple of hundred articles, many of them directly informed by the work of one woman – Carol Weiss (for details, see Dunlop, 2014). Indeed, as far back as the 1970s Weiss (1979) referred 'to the problem of little effect' to help conceptualise the limited use of scientific evidence to inform decision-making. There are many broad themes that emerge from across this body of work that show the difficulties associated with conceptualisations around instrumental knowledge use. For instance, the knowledge utilisation literature broadly accepts that 'context matters'; where, when and how utilisation occurs shapes the extent of use (Nutley et al., 2007, p. 303).

It also recognises that knowledge comes in many different forms (Rich, 1991, p. 14; Juntti et al., 2009) with the links between knowledge and evidence and research often being complex and contested by different actors. Nutley et al. (2007, p. 23), for example, argue that research is one type of evidence, and that evidence is one source of knowledge. Thus, knowledge is not viewed as a neutral package of 'facts'; on the contrary, what counts as knowledge and how it is presented (for example in terms of the 'services provided') is an inescapably political act (Radaelli, 1995; Juntti et al., 2009; Jordan and Russel, 2014). Moreover as Rich (1991, p. 15) notes, 'use' is not an 'all-encompassing concept'. Consequently, mapping the extent and pathways of knowledge utilisation ultimately depends on what types of knowledge are being investigated, what the analyst means by 'use' (see for example the typology of different uses below) and whether they see knowledge utilisation as an 'outcome' a 'processes' or some combination of the two (ibid., p. 12).

In addition to these broader themes, the existing literature on knowledge utilisation breaks down the complexities surrounding knowledge use in different ways. For instance, the literature has sought to unpack the different understandings of the term 'use'. Rich (1997, p. 15) usefully distinguishes between four main types: 1) 'use', where knowledge has been received and read; 2) 'utility', when a user judges knowledge as having potential value but the purpose for which has yet to be identified; 3) 'influence', which indicates that knowledge has contributed to a decision; and 4) 'impact', where information has been 'received and understood', leading to clear and concrete action which is more akin to instrumental use. Thus, even if instrumental uses of knowledge on ecosystem services are far and few between, this does not mean that use has not occurred; it may simply be that a user has yet to find a value for this knowledge in their given decision-making context.

Probably the most recognised contributions of the literature have been to order knowledge use around particular 'models' rather than 'types' of knowledge use. These include:

- the classic *rational linear instrumental model* of knowledge use (Weiss, 1979; Owens, 2005, Dunlop, 2014) mentioned above;
- the *conceptual or enlightenment model*, whereby a body of knowledge builds up to shape a broader policy agenda over longer time periods of decades or more (e.g. Weiss, 1979; Radaelli, 1995);
- the *strategic model*, where knowledge is used tactically by different actors in particular decision-making contexts (often selectively) as ammunition in broader political battles (Owens, 2005);
- the *co-production or social model*, whereby knowledge use and generation results from two-way processes of interaction between knowledge generators, users and other stakeholders (Owens, 2012; Haines-Young and Potschin, 2014; McKenzie et al., 2014); and
- the possibility of non-use, but, as Dunlop (2014) notes, while this may well be empirically possible it is in fact very unlikely, with an active rejection being more likely.

Knowledge utilisation: when theory meets practice

What do these models imply for our understanding of ecosystem service knowledge use? Despite the instrumental logic underpinning assessments like the MA, other uses may well be in play, especially as instrumental uses tend to be relatively rare (Dunlop, 2014). Overall, then, when it comes to embedding ecosystem knowledge, little thought has been given to other types of knowledge use, creating many blind spots. For example, how should decision-makers weigh CBA-derived knowledge against that derived locally from key stakeholders, as would arise in the co-production model (Hockley, 2014), and what roles can tools like multi-criteria decision analysis play in facilitating this weighting processes? How does one account for longer-term (conceptual) uses of ecosystem services knowledge, given that the ecosystem services debate is still relatively new in decision-making contexts (Waylen and Young, 2014)? Currently, debates on ecosystem services approaches have provided few clear answers to such questions.

A key theme emerging from the knowledge utilisation literature focuses on the processes that shape 'why' particular forms of knowledge are used in the way that they are. From this perspective, there are a number of different types of knowledge that decision-makers can draw on to help formulate policy, ranging from scientific research through to expert advice and lay knowledge. However, not all knowledge is given equal weight by decision-makers, with some being seen as more legitimate, salient or credible than other knowledge (Rich, 1997, Juntti *et al.*, 2009), e.g. peer reviewed scientific knowledge versus a think tank/ consultancy report. Key underlying factors said to be shaping the perceived legitimacy of a given type of knowledge include: the way power relationships within a policy sector shape debates over what counts as legitimate knowledge and how it is used (e.g. Radaelli, 1995; Juntti et al., 2009); the ways in which knowledge is collected, processed, agreed upon and presented (or framed) to users (Cowell and Lennon, 2014); the role played by scientists as policy advocates in a given policy sector; and how agenda-setting by specific actors within decision-making cycles influences the types of knowledge that decision-makers seek out (Sabatier, 1998; Radaelli, 1995). Ultimately, such factors can influence whether knowledge is used within existing norms, values and problem framings. Such a situation can, in turn, greatly influence the way in which approaches are adopted to implement relatively new concepts like ecosystem services, and the use that is derived from them. Indeed, in much of the discourse around ecosystem services, more economic venues of analysis such as cost-benefit analysis are often perceived to provide more credible and 'usable' knowledge opportunities to link suppliers and users of knowledge (Pearce, 1998; Haas, 2004, pp. 573, 574; Cowell and Lennon, 2014), than those occurring when more scientific framing and/or deliberative/participatory approaches dominate (Hanley, 2001; Fish, 2011). This is an area of ongoing debate in the ecosystem services community.

Another relevant, ongoing debate in the ecosystem services community is the role that local knowledge (i.e. that of practitioners and stakeholders on the ground) plays in decision-making. Indeed, the Intergovernmental Platform on Biodiversity and Ecosystem Services (IPBES) deliberately seeks to include local knowledge alongside more standard forms of scientific knowledge as part of its evidence base (Beck et al., 2014), whereas assessments such as the MA tend to be more exclusively focused on only scientific knowledge.

The existing literature also seeks explanations for why, in certain contexts, '[k]nowledge can speak volumes to power', whereas in others it is ignored (Haas 2004, p. 587). Much of the research on this issue draws upon the 'two culture' approach comprising of groups of knowledge generators and users (Caplan, 1979). Improvements to knowledge utilisation are characterised by a metaphor of 'bridging the gap' between the generator and user cultures (Caplan, 1979). The two cultures approach provides a parsimonious conceptualisation, but has been criticised on the basis that it may 'camouflage . . . the complex interactions that produce a . . . decision' (Rich, 1991, p. 325). Another area of research has sought to provide a more nuanced account through 'boundary work' exploring the boundary between knowledge production and policy (use). This body of work explores how such boundaries are 'managed' (or not) through organisations (e.g. environment ministries), individuals (e.g. policy entrepreneurs) or institutionalising processes and procedures (e.g. the MA and similar national-level assessments) (Owens et al., 2006) that seek to engender knowledge utilisation processes. Others, such as Nutley et al. (2007) and Haas (2004), have focused on more generic factors around expectations for enhancing knowledge use on ecosystem services in decision-making. We discuss these in the next section.

Can the use of ecosystem services knowledge be enhanced?

As we have observed above, instrumental knowledge use tends to be the exception rather than the norm. This does not mean that other forms of knowledge use are necessarily bad, or that strategies for enhancing knowledge use on ecosystems and the services they provide cannot be followed. However, it is useful to consider some of the factors that could be retarding and enabling the use of ecosystem services knowledge in decision-making. Some of these factors have already been discussed above, such as the role that power politics can play in shaping what counts as legitimate knowledge (Radaelli, 1995; Juntti et al., 2009) and in the strategic deployment of knowledge as ammunition in a political battle (Owens, 2005). Other scholars, though, have focused more specifically on the role played by institutions (see for example Juntti et al., 2009; Russel et al., 2014). Institutions are the places in which knowledge is generated, used and mediated and policy is made (Peters, 2005). As a result, there is arguably a need to carefully account for the interaction between internal (i.e. knowledge quality factors) and external (i.e. institutional) factors (Cowell and Lennon, 2014). For example, drawing on the example of marine conservation, Cassarini et al. (2014) highlight the importance of the interaction of knowledge quality factors (e.g. the perceived credibility of published findings) with institutional factors related to how knowledge is spread among decision-makers (e.g. informal distribution between institutional actors) in creating conditions that encourage or restrict knowledge use amongst decision-makers.

Building on concepts from institutional analysis (Peters, 2005), Turnpenny et al. (2014) seek to better conceptualise the role played by institutions in shaping knowledge use. Specifically, they argue that understanding the use of knowledge on ecosystem services in decision-making requires an understanding of the barriers and enabling factors operating at different levels within institutions. The first level stems from those factors that influence individual behaviour (e.g. skills, educational background, resources); the second level relates to organisational culture (e.g. institutional configurations, leadership, procedures for quality control); and the third level concerns wider societal values (e.g. how the environment is weighed or prioritised against economic activity). Turnpenny et al. (2014) argue that through identifying the levels at which barriers to knowledge generation exist, strategies to enhance the uptake of knowledge on ecosystem services can be more targeted. Barriers operating at the societal level, though, are seemingly more difficult to overcome, as they are ingrained in wider societal values and processes (Russel et al., 2014). Ultimately, though, knowledge use can be enhanced by focusing on a mix of measures focusing on the organisational culture and individual behaviour levels (for examples see Russel et al., 2014).

Importantly, lessons for enhancing ecosystem services knowledge use chime with lessons from other policy areas. For example, Nutley et al. (2007) draw on insights from social policy to highlight the importance of: translating and adapting research to specific policy and institutional contexts; the presence of enthusiastic champions; wider institutional ownership and buy-in; the perceived credibility of the knowledge; ongoing financial and technical support; integrating new concepts with existing processes and systems; and the provision of sustained high-level leadership. Perhaps the issue that the knowledge utilisation literature has most focused on is that of institutional champions in promoting knowledge uptake around concepts such as ecosystem services (e.g. Pielke, 2007; Cossarini, et al., 2014; Cowell and Lennon, 2014; Dunlop, 2014). Given what we know about the various ways in which knowledge can be used in the policy process, we need a clearer understanding of the boundaries of what is possible, especially in terms of the skills that advocates of ecosystem services should ideally have (Dunlop, 2014). Such skills must be able to exploit opportunities to push the concept of ecosystem services (Cowell and Lennon, 2014; Russel et al., 2014) and utilise the 'persuasive' power of reports like the MA or similar national-level assessments to 'open doors' for new collaborations and partnerships (Waylen and Young, 2014). What qualities should the expert have in order to facilitate learning? According to Dunlop (2014), the answer is context-specific (also see Nutley et al., 2007) and dependent on the level of governance and also the sector (for example, enhancing knowledge use at more local levels requires an ability to engage hearts as well as minds). Another consistent theme to emerge is that public participation in the vein of a co-production model a potentially important enabler of conceptual enlightenment across stakeholder groups (Fazey et al., 2013). Moreover, stakeholder involvement may enhance a policy's chances of surviving the early stages of consultation, cross-examination and formal approval, so as to enhance the legitimacy of knowledge (Cowell and Lennon, 2014). Building on the local knowledge of key stakeholders may be also important in such processes, as shown by McKenzie et al. (2014) in their study of ecosystem knowledge use in planning processes in North and Central America.

Moving on from discussing ways of enhancing knowledge use, it is important to also remember that non-instrumental uses do not mean non-use of knowledge, and that non-instrumental uses can have positive benefits for the management of ecosystems and the services they provide. Indeed, examples from the emerging literature on knowledge use and ecosystem services provide some interesting examples of the benefits of non-instrumental use. For example, Haines-Young and Potschin (2014) found that non-instrumental uses can have a powerful role in promoting consideration of ecosystem services. They found knowledge was being iteratively co-produced by producers and users in local contexts in England. As new knowledge was produced through this two-way process, debates evolved in a manner that in some instances mediated conflict between different stakeholders. They also found that conceptual enlightenment through co-production between scientists and other stakeholders was actually the catalyst for more instrumental types of use. Similar processes are reported by McKenzie et al. (2014) in their case studies of the integration of ecosystem services into planning processes in Central and North America. In particular, non-instrumental uses of knowledge empowered less powerful indigenous groups, through processes of co-production, giving them a far stronger voice in the planning process. Similar messages also emerge from studies on other policy areas beyond the environment and ecosystems. For instance, while immediate and instrumental utilisation may be limited in the area of social policy, the process of interaction between producers and users may nonetheless have significant effects in terms of fostering greater trust, promoting longer term policy learning and even encouraging the emergence of new problem conceptualisations (Nutley et al., 2007, p. 302).

Conclusions

The whole issue of how knowledge on ecosystem services is used to inform decision-making is massively under-researched (see, though, Fazey et al. (2012), Russel et al. (2014); and a special issue of Environmental and Planning C (Jordan and Russel, 2014)). A number of key areas stand out as requiring more attention.

First, it would be useful to focus on more in-depth case studies employing mixed methods approaches (Jordan and Russel, 2014; Russel et al., 2014) and large 'n' studies (Haines-Young and Potschin, 2014) to better understand the patterns of use and associated enablers and barriers in specific institutional, sectoral and operational contexts, including cross-national comparisons (Russel et al., 2014). Such work may entail examining variation of knowledge in different policy venues, include the planning system, the work of expert committees such as the Intergovernmental Platform on Biodiversity and Ecosystem Services, ex post policy evaluation exercises within governments, and the work of parliamentary committees, among others, to provide more rigorous evaluations of the actual impacts of knowledge on policy outcomes. Such work could also explore different ways of enhancing use, from both knowledge supply and demand perspectives (Lemos et al., 2012, pp. 790–792; Cossarini, 2014).

Second, to support such empirical work, there is a need for more theoretical development to provide better understandings of how ecosystem services knowledge use can be more sensitive to issues of power and control (see Radaelli, 1995; Juntti et al., 2009, Jordan and Russel, 2014); and to synthesise perspectives from the knowledge utilisation and public participation literatures to better conceptualise processes of co-production (McKenzie et al., 2014).

Third, it is important to place ecosystem services research in its historical context. For many knowledge producers and users, in fact, the seemingly 'new' concept of ecosystem services is but the latest candidate in a much longer-running story in which environmentalists have sought to achieve greater policy purchase for their ideas. Perhaps their constant methodological and conceptual refinement is actually a sign of weakness (Cowell and Lennon, 2014), suggesting that not much learning among 'non' environmental actors has occurred in the past. To the extent that this is true, more up-front thinking about the users and uses of ecosystem services knowledge would not go amiss.

To conclude, producing 'more knowledge' is only ever, at most, a necessary but insufficient condition for greater policy success. There is thus a strong case to suggest that expectations need to be adjusted in terms of how knowledge of ecosystem services is used in policymaking contexts. Ecosystem knowledge has to navigate many obstacles before it can influence decisions. Even where knowledge of ecosystem services does influence decision-making, it may be far from the instrumental uses often perceived by knowledge producers. Indeed, non-instrumental uses and processes can have positive impacts on decisions around the management of ecosystems, facilitating communication between stakeholders to better anticipate and mediate potential conflicts. The challenge is how to ensure that these deliver benefits to a sufficient number of those involved so that the whole endeavour is politically sustainable. Politicians are often under intense pressure to deliver policy outputs and outcomes as quickly and as efficiently as possible, even more so in times of economic austerity. But focusing exclusively on instrumental uses can, as noted above, easily lead to frustration and eventually disillusionment in the policy process, including crucial elements such as the MA that were created with the explicit aim of generating knowledge use. This makes the uncomfortable possibility that 'ecosystems may have already been damaged by the time knowledge about the threats feeds through into decision making' (Jordan and Russel, 2014, p. 204) even more likely. Unfortunately, if scientific predictions of future ecosystem loss are to be believed, time is a commodity which is in especially short supply.

Note

1 http://www.cbd.int/ecosystem/principles.shtml

References

Beck, S., Borie, M., Chilvers, J., et al. (16 authors) (2014). Towards a reflexive turn in the governance of global environmental expertise. The cases of the IPCC and IPBES. *GAIA*, vol 23, no 2, pp 80–87.

Caplan, N. (1979). The two-communities: theory and knowledge utilization. *American Behavioural Scientist*, vol 22, pp 459–470.

Church, A., Fish, R., Haines-Young, R., et al. (15 authors) (2014). *Cultural ecosystem services and indicators*. Report from Work Package Five of UK National Ecosystem Assessment Follow-on. UNEP-WCMC, LWEC.

Cossarini, D.M., MacDonald, B. H., and Wells, P. G. (2014). Communicating marine environmental information to decision makers: enablers and barriers to use of publications (grey literature) of the Gulf of Maine Council on the Marine Environment. *Ocean and Coastal Management*, vol 96, pp 163–172.

Cowell, R., and Lennon, M. (2014). 'The utilisation of environmental knowledge in land-use planning: drawing lessons for an ecosystem services approach. *Environment and Planning C: Government and Policy*, vol 32, pp 263–282.

Dunlop, C. A. (2014). The possible experts: how epistemic communities negotiate barriers to knowledge use in ecosystems services policy. *Environment and Planning C: Government and Policy*, vol 32, pp 208–228.

Fazey, I., Evely, A. C., Reed, M. S., et al. (21 authors) (2013). Knowledge exchange: a review and research agenda for environmental management. *Environmental Conservation*, vol 40, pp 19–36.

Ferraro, P., Lawlor, K., Millan, K., and Pattanayak, S. (2011). Forest figures: ecosystems services valuation and policy evaluation in developing countries. *Review of Environmental Economics and Policy*, vol 6, pp 20–44.

Fish, R. D. (2011). Environmental decision making and an ecosystems approach. *Progress in Physical Geography*, vol 35, pp 671–680.

Haas, P. M. (2004). When does truth listen to power? *Journal of European Public Policy*, vol 11, pp 569–592.

Haines-Young, R., and Potschin, M. (2014). The ecosystems approach as a framework for knowledge utilisation. *Environment and Planning C: Government and Policy*, vol 32, pp 301–319.

Hanley, N. (2001). Cost–benefit analysis and environmental policymaking. *Environment and Planning C: Government and Policy*, vol 19, pp 103–118.

Harrington, W., Heinzerling, L., and Morgenstern, R. D. (eds) (2009). *Reforming Regulatory Impact Analysis*. Resources for the Future, Washington DC.

Hertin, J., Turnpenny, J., Jordan, A., et al. (6 authors) (2009). Rationalising the policy mess? Ex ante assessment and the utilisation of knowledge in the policy process. *Environment and Planning A*, vol 41, pp 1185–1200.

HM Treasury (2012). *Accounting for Environmental Impacts: Supplementary Green Book Guidance*. The Stationary Office, London.

Hockley, N. (2014). The use and influence of cost–benefit analysis: a venue for integrating ecosystem knowledge in decision making. *Environment and Planning C: Government and Policy*, vol 32, pp 283–300.

Ingold, K., and Gschwend, M. (2014). 'Science in policy making: neutral experts or strategic policy makers. *West European Politics*, vol 37, no, 5, pp 993–1018.

Jasanoff, S. (1987). EPA's regulation of daminozide: unscrambling the messages of risk. *Science, Technology, and Human Values*, vol 12, nos 3–4, pp 116–124.

Jordan, A., and Russel, D. (2014). Embedding the concept of ecosystem services? The utilisation of ecological knowledge in different policy venues. *Environment and Planning C: Government and Policy*, vol 32, pp 192–207.

Juntti, M., Russel, D., and Turnpenny, J. (2009). Evidence, politics and power in public policy for the environment. *Environmental Science and Policy*, vol 12, pp 207–215.

Laurans, Y., Rankovic, A., Bille, R., Pirard, R., and Mermet, L. (2013). Use of ecosystem services economic valuation for decision making. *Journal of Environmental Management*, vol 119, pp 208–219.

Lemos, M-C., Kirchhoff, C., and Ramprasad, V. (2012). Narrowing the climate information usability gap. *Nature Climate Change*, vol 2, pp 789–794.

MA (2005). *Synthesis Report Millennium Ecosystem Assessment*, United Nations Environment Programme, New York.

Meadows, D., Meadows, D., Randers, J., and Behrens III, W. (1972). *Limits to Growth*. New American Library, New York.

McKenzie, E., Posner, S., Tillmann, P., et al. (6 authors) (2014). Understanding the use of ecosystem service knowledge in decision making: lessons from international experiences of spatial planning. *Environment and Planning C: Government and Policy*, vol 32, pp 320–340.

Nutley, S. M., Walter, I., and Davis, H.T.O. (2007). *Using Evidence*. Policy Press, Bristo.l

Owens, S. (2005). Making a difference? Some perspectives on environmental research and policy. *Transactions of the Institute of British Geographers, New Series*, vol 30, pp 287–292.

Owens, S. (2012). Experts and the environment: the UK Royal Commission on Environmental Pollution 1970–2011. *Journal of Environmental Law*, vol 24, pp 1–22.

Owens, S., Petts, J., and Bulkeley, H. (2006). Boundary work: knowledge, policy, and the urban environment. *Environment and Planning C: Government and Policy*, vol 24, pp 633–643.

Pearce, D.W. (1998). Environmental appraisal and environment policy in the European Union. *Environmental and Resource Economics*, vol 11, pp 489–501.

Peters, G. (2005). *The New Institutionalism in Political Science, 2nd edition.* Bloomsbury, London and New York.

Pielke, R. (2007). *The Honest Broker.* Cambridge University Press, Cambridge UK.

Potschin, M., and Haines-Young, R. (2011). Ecosystem services: exploring a geographical perspective. *Progress in Physical Geography*, vol 35, pp 575–594.

Radaelli, C. M. (1995). The role of knowledge in the policy process. *Journal of European Public Policy*, vol 2, pp 159–183.

Rich, R. F. (1991). Knowledge creation, diffusion and utilization. *Science Communication*, vol 12, pp 319–337.

Rich, R. F. (1997). Measuring knowledge utilization: processes and outcomes. *Knowledge and Policy*, vol 10, pp 11–24.

Russel, D., and Jordan, A. (2007). Gearing up governance for sustainable development. *Journal of Environmental Planning and Management*, vol 50, pp 1–22.

Russel, D., Turnpenny, J., Jordan, A., et al (6 authors) (2014). *Embedding an Ecosystem Services Framework in Appraisal*, report from Work package Eight of the UK National Ecosystem Assessment Follow-on Project, UNEP-WCMC, LWEC.

Sabatier, P. A. (1988). An advocacy coalition framework of policy change and the role of policy orientated learning therein. *Policy Sciences*, vol 21, pp 129–168.

Sabatier, P. A. (1998). The advocacy coalition framework. *Journal of European Public Policy*, vol 5, pp 98–130.

Sanderson, I. (2002). Making sense of what works. *Public Policy and Public Administration*, vol 17, no 3, pp 61–75.

Schröter, M., Zanden, E.H., Oudenhoven, A.P., et al. (7 authors) (2014). Ecosystem services as a contested concept: a synthesis of critique and counter-arguments. *Conservation Letters*, vol 6, pp 3802–3824.

TEEB (2010). *Mainstreaming the Economics of Nature: A Synthesis of the Approach, Conclusions and Recommendations of the TEEB.* Progress Press, Mriehel.

Turnpenny, J., Russel, D., and Jordan, A. (2014). The challenge of embedding an ecosystem services approach: patterns of knowledge utilisation in public policy appraisal. *Environment and Planning C: Government and Policy*, vol 32, pp 247–262.

Waylen, K. and Young, J. (2014). Expectations and experiences of diverse forms of knowledge use: the case of the UK National Ecosystem Assessment. *Environment and Planning C: Government and Policy*, vol 32, pp 229–246.

WCED (World Commission on Environment and Development) (1987). *Our Common Future.* UN, New York.

Weiss, C. (1979). The many meanings of research utilization. *Public Administration Review*, vol 39, pp 426–431.

PART V

Conclusion

50

ON THE CHANGING RELATIONSHIP BETWEEN ECOSYSTEM SERVICES CONTINUANCE AND SUSTAINABILITY

Tim O'Riordan

In the parable of the talents (Matthew 25:14–30), the Master (representing wealth creation) gives three of his servants 10, two, and a single talent (a measure of money) to do what they like with their Master's bounty. The first two invest in banks and fields and double their "talentage." The third makes an interesting observation before placing his talent in the ground.

"Master, I knew you to be a hard man, reaping where you did not sow, and gathering where you scattered no seed, so I was afraid, and I went and hid your talent in the ground. Here you have what is yours."

The usual interpretation is that the servant is slothful and was cast into the "outer darkness" where there was "weeping and the gnashing of teeth."

Ponder a while. The servant recognised that this was, in modern parlance, wealth from "non-sustainable" activity. He was subtly scolding his Master by refusing to invest "dirty wealth" into more "dirty wealth." Though this is not a popular interpretation of one of the most famous parables in the New Testament, it is an allegory for sustainability in the present day. The best overall exposition is by Rockström and Klum (2012).

The modern sustainability parable is based on the evidence that nearly all of human wealth is currently acquired on the back of diminishing and disabling ecosystem services, and the usurpation of labour and the ill-being of the disadvantaged and discriminated-against. And all invested new wealth is similarly acquired on an unjust divestment of life support and social well-being.

The drive to afford coherence and credibility to the role of ecosystem services is a drive to challenge the biases of valuation and power in the determination of market-based exchange. The default position of such exchange is non-sustainability. To shift the default position to sustainability will require global recognition of the essentialness of ecosystem services for viable continuation of wealth creation. Any talent on offer should be a sustainably buttressed talent.

The difficulty here lies in the ways in which markets work, investors think, and politicians act. Investment geared towards future returns is nearly always based on models of investment where there is a presumption of growth, innovation, and overall betterment of income. Indeed, these are important preconditions. In a world of unstable financial markets and diminishing ecosystem functioning, as well as growing social tension and increasing political distrust, it may no longer be possible to guarantee long-term payback from a given investment.

The task for those promoting the full incorporation of ecosystem values into market-based exchange is to prove that the outcome of current transactions will result in impoverishment of ecosystems and human well-being in an age where those who acquire wealth are seemingly impervious to destitution and resistant to any moves to reduce their net wealth (Dorling, 2014). This is, frankly, a tall order, especially as the notion of sustainability is nowadays so universally misinterpreted and miscomprehended that it contains no universal meaning.

What is to be done? One option is for the ecosystem services community to gather together their best evidence and focus on corporate support. In the UK, an example is the Prince of Wales' Corporate Leaders' Group of the Cambridge Institute for Sustainability Leadership (www.cisl.cam.ac.uk). This group is active in the calculation of natural capital and could offer an important front for the ecosystem community to gain corporate ground. Neither governments nor the energetic non-governmental organisations are able to pursue this task. They do not have the punch-power of business, nor do they have the freedom (in the case of governments) or the clout (in the case of the NGOs) to make a sufficiently credible case.

Sustainability nowadays is shifting ground. There is a resigned recognition that a society which is treated unfairly and unjustly cannot embrace sustainability. It is too distracted, too fragmented, too angry and in turmoil, and too immediate in its gratification to address either planetary boundaries (even with social justice thrown in (Rockström and Krum, 2012)), or the mesmerising displays of wide-ranging possibilities of social betterment. So the focus of attention in the sustainability stakes rests on inequality, on enforced vulnerability, on the scope for ecosystem viable enterprise, and on a new moral framework for living and consuming based on pride and community approval (O'Riordan, 2014).

Another target for the ecosystems services community would be to concentrate firepower on the evolution of global Sustainable Development Goals (SDG) being put before the UN General Assembly for adoption by every nation from 2015. All being well, there will be a stocktake of national commitments to meeting these SDGs every five years thereafter. Here, surely, is a golden opportunity for the ecosystem services community to leverage its influence.

There is as yet no full summary of what these SDGs might look like. But the sustainability community is looking for five outcomes.

1. *Coping with poverty*. SDGs should not just seek to reduce overall poverty in a broad measure. They should also seek to connect component parts of poverty as a set of linked themes. These relate to improved measures of inequality, especially indicators of increasing inequality; to the role of women in entering into enterprise, to ways of relating cultural influences on poverty promotion (especially on factors controlling marriage, procreation, and sex prioritisation of children); and to measures of health, particularly for the mother and child. There is also huge scope for including degradation (and enhancement) of ecosystems services in these measures of poverty alleviation: indeed, this is a vital component for women's well-being.
2. *Patterns and rates of urbanisation*. There is an explosion of cities, especially in emerging and developing economies. This is giving rise to significant problems of air, water, waste, and toxic substances pollution for public ill-health; consumption and resources demands in extensive infrastructure, and well-being generally (weak and sometimes antagonistic social cohesion, and strengthening patterns of alienation and communal violence). Again, widening the scope for interpreting natural and social capital (see Parkin, 2013) is an essential element.

3 *Vulnerability to risk and danger.* Vulnerability applies to conditions where those exposed to general threat are not enabled to cope, not being able to avoid "price spikes" (in food and energy), not being offered ways to establish resilience in the face of ecosystem degradation, and are being forced into a myriad of informal social and economic measures to maintain survival usually at the expense of ecosystem integrity. Vulnerability is a feature of economic, cultural, social and political powerlessness. It is a symptom of non-sustainability of political systems and of the geography of hazard, as vulnerable people are forced to locate to, or to remain in, hazardous zones, often where natural processes of safeguard are destroyed.

4 *Capacities and capabilities.* Adaptation to any stress (health, hazard, climate change, social turmoil, cultural violence, personal harm) depends on the capacities and capabilities of the governing system and of communities in given localities to be able to provide the depth and resilience of provision of care. If climate change alters the distribution of disease of fresh water supply, what matters for SDGs is not the crisis of the resulting hazard, but the capability of good governance to foresee and to establish the capacity to adopt the necessary resilience measures. So the SDG process should address the manner in which societies and governments are able to manage and be proactive in the face of avoidable adverse change. This in turn applies to adaptive provision of infrastructure, transport facilities, utilities, flexible digital systems, and the integrated planning process. Right now there are no proper measures of effectiveness in the planning and design staging of interconnected infrastructure capacity. Here is a golden gap for the ecosystem services community to fill.

5 *Learning for sustainability leadership.* There is a breeze of change wafting through education. This is energised through the commitment of many young people to improved learning capabilities for leadership for sustainability. An important SDG is the measure of the capacities of learning for monitoring social and environmental change using the smart phone in particular: as a basis for organised and competent observation; for being much more self-aware of the various forces which impede the transition to sustainability; and of the scope for providing learning ambassadorships for the next generation of citizens to be at the heart of transformative change for sustainability. Incorporating the triumph of full ecological and social valuation of planetary services is a vital element of a successful learning.

Here is the agenda for the ecosystem services community to the end of the decade. It is clear from this exposition that the social dimensions of such services are an intrinsic component of their manifestation and acceptance. Here is where sustainability science must surely come of age.

References

Dorling, D. (2014). *Inequality and the Top 1%.* Verso Press, London and New York.

O'Riordan, T. (2014). Sustainability beyond austerity: possibilities for a successful transition to a wellbeing society. *Analise Social*, vol 49, no 11, pp 417–520.

Parkin, S. (2013). Leadership for sustainability: the search for tipping points. In: O'Riordan, T. and Lenton, T. M. (eds) *Addressing Tipping Points for a Precarious Future.* Oxford University Press, Oxford, pp. 194–212.

Rockström, J., and Klum, M. (2012). *The Human Quest: Prospering within Human Boundaries.* Langenskoilds, Stockholm.

51

ECOSYSTEM SERVICES

Where is the discipline heading?

Georgina Mace

The Millennium Ecosystem Assessment (MA) first stimulated a new and much deeper level of interest in ecosystem services, and this has continued to develop in both research and policy environments. Prior to this there had been a growing understanding of the importance of nature's goods and services for people, and much work had gone into the Ecosystem Approach as a focus for policy, especially through the Convention on Biological Diversity. The MA, however, placed ecosystems firmly at the core of environmental science-policy discussions, and while it drew on earlier work, it provided a more explicit conceptual framing from which later analytical methods could develop, and it fostered more interdisciplinary and transdisciplinary approaches (see also Hauck, 2016). Developments since the publication of the MA in 2005 have been especially rapid (Abson et al., 2014), and the framing for ecosystem services has matured, developed and spread into a range of other related areas, although not always with coherence and clarity. Here I revisit a few of the areas of developing interest and continuing confusion. Many of these are resolvable if addressed in a consistent way; others represent areas of future research need, while yet others draw on much broader societal concerns to do with equity, fairness and sustainability, emphasising the need to place science within a broader political context.

Ecosystem services and natural capital are sometimes treated as if they mean the same thing. In fact, because some commentators find the terminology of ecosystem services difficult, there is increasing use of the term 'natural capital' as an equivalent alternative term. But there are important differences between ecosystem services and natural capital if, as seems logical given the terminology, natural capital is regarded as the stock from which ecosystem services flow. This is an important and useful distinction that resonates with other kinds of stocks and flows, such as between the capital held in the bank and the flow of interest payments, or between the size of a fishery stock and the flow of sustainable catch. The natural capital stock can become depleted if the flow of ecosystem services is too high, or if the stock is not well managed. Examining only the flows of services may not be a timely or efficient way of discovering that there is a problem. Making this distinction then confirms the need to consider the stock-flow relationships explicitly, and not just the flow of ecosystem services. This can be very important where the stock is close to some kind of threshold or in some kind of a weak state, because it may jeopardise the flow of services (Mace et al., 2015). Management should be able to restore services given a clear enough understanding of the stock-flow relationship, and of the necessary level of services required. While there has been a history of stock-flow relationships in natural resource management (e.g. fisheries and forestry), the science is relatively poorly developed for other kinds of ecosystem services, and especially so where there are multiple ecosystem services, representing an important research gap.

Stock-flow relationships are important because without understanding them, the only way to identify systems that are close to their limits will come as they start to fail. Research in this area is therefore likely to increasingly reveal cases where the system is unable to do more, or where demands for multiple ecosystem services from the same area place a limit that is impossible to avoid. Ecosystem services therefore cannot be understood without a broader ecosystem understanding that draws on what is being used and what is needed.

A need to focus on use then raises a question about the *demand* versus the *supply* of ecosystem services. The majority of ecosystem service analyses consider the supply of services with little attention to the demand either now or in the future, and whether that demand might be either moderated or met in some other way. In past times, resources appeared to be abundant or effectively infinite, and directed management could usually increase supply. Hence, hundreds of years ago, for example, the supply of timber could be increased by harvesting new woodlands, and then by planting trees. Decades ago, the world's demand for food was met by planting more productive crops, by adding nutrients and treating crops with pesticides. But over time the limits to some of these actions have become more apparent, as have the costs for other equally significant ecosystem services that have been taken for granted. The MA first demonstrated that regulating services (e.g. water quality, climate regulation, hazard regulation) were overused and depleted to a greater extent than many provisioning services (e.g. food production, aquaculture) and this pattern has been documented at regional and local scales in many other areas since then. In the future, land and sea areas will need to supply an even wider range of provisioning and regulating services, and may require attention to the demand more than the supply side (Yahdjian et al., 2015). For example, if demand for some ecosystem service is being met, or if it seems to be over-supplied, or if the demand can be met in another way, then the ecosystem could be managed to better meet the demands for services that are under-supplied or difficult to replace. Conceivably, areas of agricultural land used for food crops that are surplus to requirements could be better used for regulating services, such as carbon sequestration, flood or other hazard control; all of which are costly to society and often difficult to provide.

Consideration of the demand side quickly leads to another important area for future research. This concerns who the beneficiaries from ecosystem services are, and who decides who receives what. These questions are complicated by the wide range of ecosystem services, some of which are always local (e.g. fresh air) while others are global (climate regulation); some have monetary value and are material goods that can be bought and sold in markets (e.g. food and timber), while others are globally mixed, but may be traded (e.g. carbon fluxes). Perhaps more difficult is the fact that there are very significant distributional issues for certain ecosystems upon which some of the world's poorest people depend, while those same areas represent significant intact ecosystems upon which certain earth systems (e.g. climate regulation, ocean fisheries) also rely. Daw et al. (2011) identified four different ways that different groups of people derive well-being benefits from ecosystem services. First, different groups derive well-being benefits from different ecosystem services, so that winners and losers are created as ecosystems change. Second, local and regional access mechanisms determine who can benefit even when there is ample supply of ecosystem services. Third, individuals' contexts and needs determine the extent to which any ecosystem service actually benefits them; the same provision will mean more to a needy than to a less needy individual. Fourth, aggregated estimates of ecosystem service benefits may neglect crucial poverty alleviation mechanisms, such as cash-based livelihoods, so that a disaggregated analysis is needed when interventions are intended to contribute to poverty alleviation (see also Schaasfsma and Fisher, 2016; Jax, 2016; Sikor et al. 2016).

Securing ecosystem service delivery can therefore still fail the poor, or the system as a whole, unless the full socio-ecological system is carefully considered. Trade-offs among ecosystem services are a recurring research theme (Bennett et al., 2009). In a recent literature review of cases where trade-offs or synergies (win-wins) between ecosystem services had reported, Howe et al. (2014) found trade-offs to be recorded almost three times as often as synergies. Examination of the case studies where trade-offs or synergies were recorded revealed several clear patterns. In cases where there were trade-offs among beneficiaries, there was a significant effect that at least one beneficiary had a private interest in the natural resources available. In other analyses comparing trade-offs and synergies in comparable ecosystems, it was clear that the involvement of provisioning ecosystem services and at least one of the stakeholders acting at the local scale also increased the likelihood that there were trade-offs involved, rather than the synergies that the ecosystem service literature so often promises. More research on how ecosystems can continue to deliver multiple services to different beneficiaries is needed to avoid the over-use of some services and, at worse, more frequent collapses of significant services that are hard to restore or recover.

Ecosystem services has its origins in the natural sciences, its adoption in social sciences is far from complete, and the struggle to maintain the necessary interdisciplinarity for the social and ecological systems referred to above remains a challenge (Abson et al., 2014). But there are also still major gaps in scientific understanding within many core disciplinary areas, ranging from basic ecological and ecosystem science, the links and synergies with biodiversity conservation through to the effective integration into the economy. Even now, as conceptualisations of the nature of the relationships between people and nature continue to evolve, so too does the underpinning science and the core objectives involved (Figure 51.1).

Following Mace (2014), Figure 51.1 summarises evolving ideas for relationships between people and nature that have unfolded over the past 50 years or so. Early on, 'Nature for itself' largely promoted pristine areas, free of people, and led to the development of many protected areas with a high value for wilderness. This ideal continues today, but several other views have been added as the pressure on natural areas has increased, as along with the understanding of people's dependence upon natural systems. As Figure 51.1 illustrates, there were major phases where the focus was on threats to nature ('Nature despite people'), soon to be followed by a more utilitarian view of nature that was in part stimulated by ecosystem services thinking ('Nature for people'). The most recent phase ('People and nature') is the one that makes it clear that people are part of ecosystems, not apart from them, and that there is great benefit to be had from a full understanding of the dynamic links between human societies and the ecosystems they depend upon, sometimes referred to as 'socio-ecological systems'. This framing seems a mature and coherent basis from which the discipline can grow, although many gaps and challenges in the science remain (Carpenter et al., 2009).

Despite the many uncertainties referred to above, ecosystem services have become established and deployed in national and international policies. With the recent establishment of the Intergovernmental Science Policy Platform on Biodiversity and Ecosystem Services (IPBES), they are now being used at inter-governmental level, too. Given this, it behooves all serious students of ecosystem service science and practice to continue to consider the fundamental ideas and broader objectives of sustainability in socio-ecological systems that were the original intent, and not simply to take simple prescriptions or policies for granted. There are many important steps, both big and small, to be taken before ecosystem services start to fulfil their great promise, and much work still to be done.

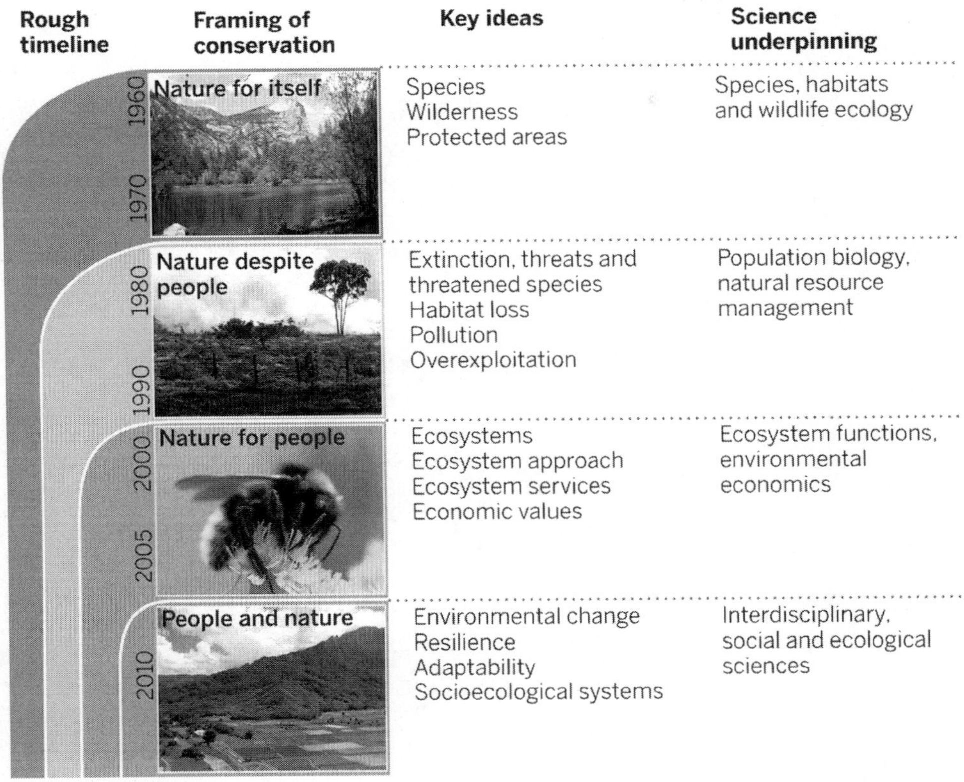

Figure 51.1 Changing views of nature and conservation. Over the past 50 years, the prevailing view of conservation has changed several times, resulting, for example, in a shift in emphasis from species to ecosystems. None of the framings has been eclipsed as new ones have emerged, resulting in multiple framings in use today.

Source: adapted from Mace (2014)

References

Abson, D. J., von Wehrden, H., Baumgärtner, S., Fischer, J., Hanspach, J., Härdtle, W., Heinrichs, H., Klein, A. M., Lang, D. J., Martens, P., and Walmsley, D. (2014). Ecosystem services as a boundary object for sustainability. *Ecological Economics*, vol 103, pp 29–37.

Bennett, E. M., Peterson, G. D., and Gordon, L. J. (2009). Understanding relationships among multiple ecosystem services. *Ecology Letters*, vol 12, pp 1394–1404.

Carpenter, S. R., Mooney, H. A., Agard, J., Capistrano, D., DeFries, R. S., Diaz, S., Dietz, T., Duraiappah, A. K., Oteng-Yeboah, A., Pereira, H. M., Perrings, C., Reid, W. V., Sarukhan, J., Scholes, R. J., and Whyte, A. (2009). Science for managing ecosystem services: beyond the Millennium Ecosystem Assessment. *Proceedings of the National Academy of Sciences of the United States of America*, vol 106, pp 1305–1312.

Daw, T., Brown, K., Rosendo, S., and Pomeroy, R. (2011). Applying the ecosystem services concept to poverty alleviation: the need to disaggregate human well-being. *Environmental Conservation*, vol 38, pp 370–379.

Hauck, J. (2016). Transdisciplinarity. In: Potschin, M., Haines-Young, R., Fish, R. and Turner, R. K. (eds) *Routledge Handbook of Ecosystem Services*. Routledge, London and New York, pp 301–303.

Howe, C., Suich, H., Vira, B., and Mace, G. M. (2014). Creating win-wins from trade-offs? Ecosystem services for human well-being: a meta-analysis of ecosystem service trade-offs and synergies in the real world. *Global Environmental Change*, vol 28, pp 263–275.

Jax, K. (2016). Ecosystem services and ethics. In: Potschin, M., Haines-Young, R., Fish, R. and Turner, R. K. (eds) *Routledge Handbook of Ecosystem Services*. Routledge, London and New York, pp 301–303.

Mace, G. M. (2014). Whose conservation? *Science*, vol 345, pp 1558–1560.

Mace, G. M., Hails, R. S., Cryle, P., Harlow, J., and Clarke, S. J. (2015). Towards a risk register for natural capital. *Journal of Applied Ecology*, vol 52, no 3, pp 641–653.

Sikor, T., Martin, A., Fisher, J. and He, J. (2016). Ecosystem services and justice. In: Potschin, M., Haines-Young, R., Fish, R. and Turner, R. K. (eds) *Routledge Handbook of Ecosystem Services*. Routledge, London and New York, pp 299–301.

Yahdjian, L., Sala, O. E., and Havstad, K. M. (2015). Rangeland ecosystem services: shifting focus from supply to reconciling supply and demand. *Frontiers in Ecology and the Environment*, vol 13, pp 44–51.

52

ECOSYSTEM SERVICES

Never waste the opportunity offered by a good crisis

Robert Fish, Marion Potschin, R. Kerry Turner and Roy Haines-Young

In elaborating the tenets of an ecosystem services perspective, a *Handbook* of this kind naturally straddles the concern to not only integrate and synthesise, but also to make distinctions between preoccupying themes and concerns. The idea of ecosystem services has been presented as a unifying, harmonising concept; a 'statement of the art' crafted around an orderly structure of theories, methods, and applications to make sense of its work. In our scene-setting chapter we noted too an accumulating concern to recognise, understand and better manage human dependencies on nature, and the many and diverse antecedents for the perspective in the wider canon of environmental thought. Even so, the mutually configuring worlds of research and policy that seek to explicate and apply the concept of ecosystem services are also mutable, unruly and highly fluid; readers who have scratched beneath the surface will have found a highly differentiated and contested field. Thus, while it is possible to write of a maturing field in terms of willingness to engage, it remains premature, and would be misleading, to write of a field having reached consensus about the precise character of its operating concepts and models of research practice.

This volume has sliced through a vast body of interdisciplinary scholarship at a particularly dynamic moment. All of the chapters have précised a labyrinth of interlinked studies and provide touchstones for those wishing to probe further. Insofar as a community is emerging with interests in critical and constructive engagement, we are still in the early stages of paradigm change, some distance from the point where the focus of scholarship is merely one of refining and translating thinking. As many of the contributions that make up this volume show, there is a steady stream of new theoretical and empirical spaces to open up to take thinking forward. In putting together this volume, it has been interesting to observe how often statements of current understanding offer a glimpse of hitherto uncharted issues for ecosystem services, and how previously established problems can be re-thought anew when tackled from an ecosystem services perspective. Opportunities to have a positive but disruptive influence on the 'state-of-the-art' feel like a real possibility.

What, though, are the ties that bind all this diversity and innovation together? Reflecting on the recent growth of the field, we have reached a stage in the field where the 'noise' created by a small research community has created capacities and resources that command the attention of wider groups: the curious, the perplexed, the angry and the entrepreneurial. The ecosystem services agenda is functioning as a gathering and passing through-point, and this has its own momentum effects, not least in exposing to scrutiny basic claims, arguments and ambitions. More substantively, there is an expansive and subversive element to the ecosystem services agenda that makes it an important attractor for debate. We might say that one of the general virtues of the perspective is that it has steadily challenged prevailing wisdoms about where environmental concerns begin and end. While some encountering the discourse will quibble about precise ways of classifying ecosystem services, or question more fundamentally the utilitarian starting point of the perspective, what is being advanced here is a provocation to place the natural world unequivocally at the centre of decision-making; that is, at the centre of decisions about the institutions we build within our communities and across our political systems; about the businesses we grow; and about the decisions we make as citizens and economic agents.

If this is a bracing new world for some, we would argue that it is also an invitation to affiliate with environmental concerns in different and exciting new ways. Importantly, it is an invitation to help challenge the discredited, though still persistently asserted, idea that the environmental agenda is somehow a constraint on, and counterpoint to, the real stuff of human progress: environment traded off against economy, against development, against food, against culture, and so forth. From an ecosystem services starting point, all of this is turned on its head. Though the idea of counterpoints and trade-offs remains a common and vexed preoccupation within this emerging field, what is brought forth here is how to make sense of different combinations and patterns of ecosystem service provision. Thus, fundamentally, the choice is not between the natural world or not, but between different models of the natural environment that social systems wish to propagate in the context of different models of human well-being. This is more than simply a conceptual re-badging of societal choice. The important point is that natural capital is explicitly recognised as the resource around which every other choice must turn, and around which the limits of our apparently unfettered freedom to procure and arrange flows of benefits from ecosystems must be negotiated. These limits are social as well as environmental, and about reflecting back on norms as much as they are about understanding biophysical thresholds.

A world recast as one of relationships between ecosystem service providers and beneficiaries is a very different world indeed, for, as the contributors to this volume have shown, we are being encouraged to test our assumptions about what we think our natural world is for, and what roles, actions and responsibilities we are presumed to identify with, and take on, with respect to the many and diverse benefits of ecosystems. Conceptually at least, there is an indifference to where priorities for management may lie with respect to human well-being, for at its starkest, the framework is an exercise in translating the natural world into a set of standardised units around which comparisons and choices can then be explored. From the perspective of decision-making, the difficulty here is how to address the framework's in-built relativism. The alignment of the perspective to the practice of valuation has provided one major and prominent way into this problem, though the centrality of this practice to the development of the agenda signals for some an unhelpful concession of the environment to an economic mind-set, where money and markets become the context in which norms and limits are defined. For others, it is to challenge the conventional purview of the economic; a positively disruptive force in a world where an economic starting point holds sway. What we can say is that ecosystem services agenda will be one of the venues in which debates about the economic and more-than-economic basis of environmental policy- and decision-making are played out. And in these debates the need for a strong sensitivity to understanding who stands to gains or lose from decisions, and building institutions that can account and mitigate for the distributional consequences of decisions, will be key.

In drawing to a close a volume on ecosystem services, it is useful to reflect once again on the challenge of mainstreaming the perspective. There have been many claims and asides made in this vein throughout the volume, including some in our own scene-setting introduction. Let us pursue some further thoughts here, but perform a different final manoeuvre. In the puzzle-solving, puzzle-framing world of ecosystem services, some of the basic parameters and approaches of the field remain open to debate at the same time as ambitions for practical uptake are being apparently realised with alacrity. This situation has a long pedigree. Some three decades ago, Soulé (1985) remarked that: 'In crisis disciplines one must act before knowing all the facts; crisis disciplines are thus a mixture of science and art, and their pursuit requires intuition as well as information'. His comments were written as a manifesto for the new 'synthetic discipline' of 'conservation biology', an important forbearer of an ecosystem services perspective in its own right. It was also one designed to align the production and utilisation of ecological knowledge to explicitly normative and political ends: namely 'providing principles and tools for preserving biological diversity' (*ibid.* 727).

The sentiments of Soulé capture something of the transdisciplinary dynamic at the heart of innovation in the field of ecosystem services – another crisis discipline – where practical questions and problems of resource management have often run ahead of a settled evidence base. This type of reasoning has, of course, a wider and more pervasive academic pedigree, not least foreshadowing Funtowicz and Ravetz's (1991) idea of a post-normal science and their advocacy of developing procedures for scientific inquiry and intervention for situations where 'facts are uncertain, values in dispute, stakes high and decisions urgent'. The principles and tools to have emerged under the rubric of an ecosystem services perspective provide one version of this post-normal world of environmental governance. In thinking about the issue of mainstreaming, we need, however, to ensure that ecosystem services research is not simply about a process of rolling out tools and techniques to handle analytical and procedural uncertainties in the production, procurement, application and integration of ecological knowledges. Amongst the many challenges of the 21st century is the need to construct a sustainable partnership between people and nature. Ecosystem services are, we conclude, fundamental to thinking deeply and imaginatively about what it is 'we' want ecosystems 'to do', and what it is we are expecting decision-makers to create, combine and integrate knowledge about. In short, the appropriate response is the Machiavellian one, that we should *never waste the opportunity offered by a good crisis*.

As the idea of ecosystem services tightens its grip on policy and practice discourse, we see that it is researchers, *as much as* practitioners, who will find themselves inheriting and confronting questions about our relationships to nature in a modern world. Mainstreaming is not only a process of moving outwards towards questions of utility and application, but also one of moving inwards towards active processes of critical reflection and re-invention in research. This is only partly about researchers developing a durable set of concepts, a credible set of data infrastructures or a robust set of decision support tools and instruments to inform choices. In general, we should be wary of viewing ecosystem services research as an exercise in steadily working out the details of a single all-encompassing master plan. The extant world of resource management is messier than any such an aspiration would allow. The signal of a field in good health is one encouraging scrutiny and alternatives within the practice of scholarship if the intention is to cross and explore divides and to create conversations that are true to the perspective's holistic ambitions. Drawing lines too prescriptively around ideas, methods and application leads to stifling constructions of the world that have little currency in practice. The field will grow where creativity and openness are starting points, and where the paradigm of ecosystem services remains 'a work in progress'. An adaptive view of the world requires an adaptive research agenda, critical in its mission, and mindful that solutions are never total and beyond revision.

References

Funtowicz, S. O., and Ravetz, J. R. (1991). A new scientific methodology for global environmental issues. In: Costanza, R. (ed.) *Ecological Economics: The Science and Management of Sustainability*. Columbia University Press, New York.

Soulé, M. E. (1985). What is conservation biology? A new synthetic discipline addresses the dynamics and problems of perturbed species, communities, and ecosystems. *BioScience*, vol 35, no 11, pp 727–734.

INDEX